ALGEBRA

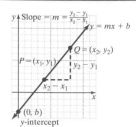

Let $P(x_1, y_1)$ and $Q(x_2, y_2)$.

Distance between P and Q is $\sqrt{(x_2 - x_1)^2 + (y_2 - y_1)^2}$.

Pythagorean theorem: If $\triangle ABC$ is a right triangle with sides a and b, and hypotenuse c (the side opposite the right angle) then $a^2 + b^2 = c^2$.

Quadratic formula: If $ax^2 + bx + c = 0$, $a \neq 0$, then $x = \dfrac{-b \pm \sqrt{b^2 - 4ac}}{2a}$.

Factoring rules:
$$x^2 - y^2 = (x - y)(x + y)$$
$$x^3 - y^3 = (x - y)(x^2 + xy + y^2)$$
$$x^3 + y^3 = (x + y)(x^2 - xy + y^2)$$

Factorial: $n! = n(n-1)(n-2)\cdots 3 \cdot 2 \cdot 1;\quad 0! = 1;\quad n! = n(n-1)!$

Binomial expansion:
$$(a + b)^2 = a^2 + 2ab + b^2$$
$$(a + b)^3 = a^3 + 3a^2 b + 3ab^2 + b^3$$
$$(a + b)^n = \sum_{k=0}^{n} \binom{n}{k} a^{n-k} b^k \text{ where } \binom{n}{k} = \frac{n!}{k!(n-k)!}$$

GEOMETRY

(*A* represents area, *P* represents perimeter, *C* represents circumference, *V* represents volume, *S* represents total surface area)

Triangle

$$A = \frac{1}{2}bh$$
$$P = a + b + c$$

Rectangle

$$A = \ell w$$
$$P = 2\ell + 2w$$

Trapezoid

$$A = \frac{a + b}{2} h$$
$$P = a + b + h(\csc \theta + \csc \phi)$$

Circle

$$A = \pi r^2$$
$$C = 2\pi r$$

Sector

$$A = \frac{\theta r^2}{2}$$
$$s = r\theta$$

Rectangular box

$$V = abc$$
$$S = 2(ab + bc + ac)$$

Cone

$$V = \frac{1}{3}\pi r^2 h$$
$$S = 2r\ell + \pi r^2$$

Cylinder

$$V = \pi r^2 h$$
$$S = 2\pi r h + 2\pi r^2$$

Sphere

$$V = \frac{4}{3}\pi r^3$$
$$S = 4\pi r^2$$
$$x^2 + y^2 + z^2 = r^2$$

TRIGONOMETRY

Trigonometric functions Let θ be an angle in standard position with $A = (a, b)$ the point of intersection of θ and the unit circle. Then the trigonometric functions are defined by:

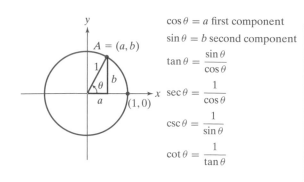

$\cos \theta = a$ first component

$\sin \theta = b$ second component

$\tan \theta = \dfrac{\sin \theta}{\cos \theta}$

$\sec \theta = \dfrac{1}{\cos \theta}$

$\csc \theta = \dfrac{1}{\sin \theta}$

$\cot \theta = \dfrac{1}{\tan \theta}$

Important trigonometric evaluations:

Angle θ	0	$\frac{\pi}{6}$	$\frac{\pi}{4}$	$\frac{\pi}{3}$	$\frac{\pi}{2}$	π	$\frac{3\pi}{2}$
$\cos \theta$	1	$\frac{\sqrt{3}}{2}$	$\frac{\sqrt{2}}{2}$	$\frac{1}{2}$	0	-1	0
$\sin \theta$	0	$\frac{1}{2}$	$\frac{\sqrt{2}}{2}$	$\frac{\sqrt{3}}{2}$	1	0	-1
$\tan \theta$	0	$\frac{\sqrt{3}}{3}$	1	$\sqrt{3}$	undefined	0	undefined
$\sec \theta$	1	$\frac{2}{\sqrt{3}}$	$\sqrt{2}$	2	undefined	-1	undefined
$\csc \theta$	undefined	2	$\sqrt{2}$	$\frac{2}{\sqrt{3}}$	1	undefined	-1
$\cot \theta$	undefined	$\sqrt{3}$	1	$\frac{1}{\sqrt{3}}$	0	undefined	0

Basic trigonometric identities:
$$\sin^2 \theta + \cos^2 \theta = 1 \qquad \sin(-\theta) = -\sin \theta \qquad \sin(\alpha \pm \beta) = \sin \alpha \cos \beta \pm \cos \alpha \sin \beta$$
$$\sin\left(\theta + \frac{\pi}{2}\right) = \cos \theta \qquad \cos(-\theta) = \cos \theta \qquad \cos(\alpha \pm \beta) = \cos \alpha \cos \beta \mp \sin \alpha \sin \beta$$

PROBABILITY

The **probability** of an event E occurring on any trial of an experiment (e.g. flip of a coin or roll of a die) is equal to the proportion of times the event actually occurs in n independent trials of the experiment as $n \to \infty$.

If a trial of an experiment has a discrete set of outcomes (e.g. being dealt a particular one of 52 cards, or the sum of two rolled dice) then the "value" of the outcome is said to be a **discrete random variable** X.

If a trial of an experiment has a continuum of outcomes (e.g. the height or weight of a randomly chosen person) then the "value" of the outcome is said to be a **continuous random variable** X.

A **probability density function (PDF)** of a continuous random variable X is a nonnegative function such that the probability of X lying in the interval $[a, b] = \int_a^b f(x)\, dx$.

The **cumulative distribution function (CDF)** of a random variable X is the function F defined by $F(x) = P(X \le x)$, i.e. the probability X is less than or equal to x. If X describes a data set, then $F(x)$ equals the fraction of data in the interval $(-\infty, x]$.

For a continuous random variable X with PDF $f(x)$, the **mean** of X is given by $\int_{-\infty}^{\infty} x f(x)\, dx$ provided that the improper integral is convergent.

For a continuous random variable X with PDF $f(x)$ and mean μ, the **variance** of X is given by $\sigma^2 = \int_{-\infty}^{\infty} (x - \mu)^2 f(x) dx$ provided the improper integral converges. The **standard deviation** of X is given by σ, the square root of the variance.

Distribution:	Uniform	Exponential	Pareto	Laplace
Parameter(s):	$a < b$	$c > 0$	$p > 1$	$b > 0$
Probability density function (PDF):	$\begin{cases} \dfrac{1}{b-a} & \text{for } a \le x < b \\ 0 & \text{otherwise} \end{cases}$	$\begin{cases} ce^{-cx} & \text{for } x \ge 0 \\ 0 & \text{otherwise} \end{cases}$	$\begin{cases} \dfrac{p-1}{x^p} & \text{for } x \ge 1 \\ 0 & \text{otherwise} \end{cases}$	$\dfrac{be^{-b|x|}}{2}$
Cumulative density function (CDF):	$\begin{cases} 0 & \text{for } x < a \\ \dfrac{x-a}{b-a} & \text{for } a \le x < b \\ 1 & \text{for } x \ge b \end{cases}$	$\begin{cases} 1 - e^{-cx} & \text{for } x \ge 0 \\ 0 & \text{otherwise} \end{cases}$	$\begin{cases} 1 - \dfrac{1}{x^{p-1}} & \text{for } x \ge 1 \\ 0 & \text{otherwise} \end{cases}$	$\begin{cases} \dfrac{1}{2}e^{bx} & \text{for } x \le 0 \\ \dfrac{1}{2} + \dfrac{1}{2}e^{-bx} & \text{otherwise} \end{cases}$
Mean:	$\mu = \dfrac{1}{2}(a+b)$	$\mu = c^{-1}$	$\mu = \dfrac{p-1}{p-2},\ p > 2$	$\mu = 0$
Variance:	$\sigma^2 = \dfrac{1}{12}(b-a)^2$	$\sigma^2 = c^{-2}$	$\sigma^2 = \dfrac{p-1}{p-3} - \left(\dfrac{p-1}{p-2}\right)^2,\ p > 3$	$\sigma^2 = \dfrac{2}{b^2}$

Distribution:	Logistic	Normal	Lognormal
Parameter(s):	$r > 0,\ a$	$\mu,\ \sigma > 0$	$\mu,\ \sigma > 0$
Probability density function (PDF):	$f(x) = \dfrac{re^{a-rt}}{(1 + e^{a-rt})^2}$	$\dfrac{1}{\sqrt{2\pi}\sigma}\exp\left[-\dfrac{(x-\mu)^2}{2\sigma^2}\right]$	$\dfrac{1}{x\sqrt{2\pi}\sigma}\exp\left[-\dfrac{(\ln x - \mu)^2}{2\sigma^2}\right]$
Cumulative density function (CDF):	$y(x) = \dfrac{1}{1 + e^{a-rx}}$	$\dfrac{1}{2}\left[1 + \operatorname{erf}\left(\dfrac{x-\mu}{\sqrt{2\sigma^2}}\right)\right]$	$\dfrac{1}{2} + \dfrac{1}{2}\operatorname{erf}\left(\dfrac{\ln x - \mu}{\sqrt{2}\sigma}\right)$
Mean:	$\mu = \dfrac{a}{r}$	μ	$e^{\mu + \sigma^2/2}$
Variance:	$\sigma^2 = \dfrac{1}{3}\left(\dfrac{\pi}{r}\right)^2$	σ^2	$(e^{\sigma^2} - 1)e^{2\mu + \sigma^2}$

The **Gauss error function** (or simply **error function**) is a function defined as $\operatorname{erf}(x) = \dfrac{1}{\sqrt{\pi}}\int_0^x e^{-t^2}\, dt$ for all $x \ge 0$.

Standard Normal Distribution Values (z-scores). See Table 7.3 on page 566.

WileyPLUS

WileyPLUS is a research-based online environment for effective teaching and learning.

WileyPLUS builds students' confidence because it takes the guesswork out of studying by providing students with a clear roadmap:

- what to do
- how to do it
- if they did it right

It offers interactive resources along with a complete digital textbook that help students learn more. With *WileyPLUS*, students take more initiative so you'll have greater impact on their achievement in the classroom and beyond.

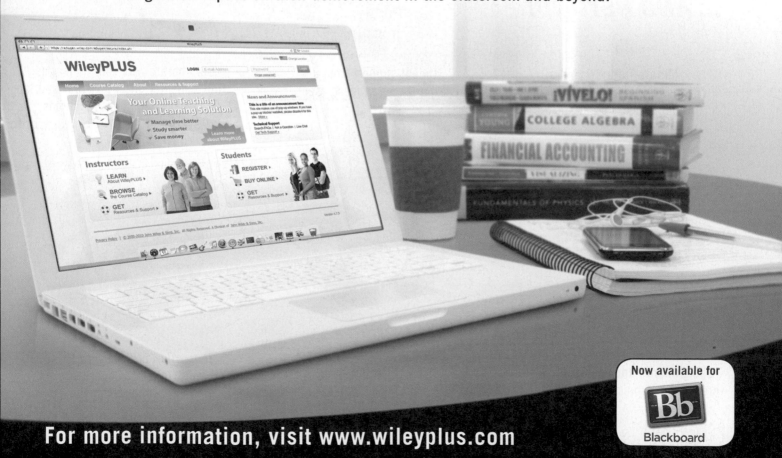

For more information, visit www.wileyplus.com

WileyPLUS

ALL THE HELP, RESOURCES, AND PERSONAL SUPPORT YOU AND YOUR STUDENTS NEED!

www.wileyplus.com/resources

1st DAY OF CLASS ...AND BEYOND!

2-Minute Tutorials and all of the resources you and your students need to get started

WileyPLUS

Student Partner Program

Student support from an experienced student user

Wiley Faculty Networ

Collaborate with your colleagues, find a mentor, attend virtual and live events, and view resources
www.WhereFacultyConnect.com

WileyPLUS

Quick Start

Pre-loaded, ready-to-use assignments and presentations created by subject matter experts

Technical Support 24/7
FAQs, online chat, and phone support
www.wileyplus.com/support

© Courtney Keating/iStockphoto

Your *WileyPLUS* Account Manager, providing personal training and support

Calculus for the Life Sciences

Sebastian J. Schreiber
University of California, Davis

Karl J. Smith
Professor Emeritus, Santa Rosa Junior College

Wayne M. Getz
University of California, Berkeley

WILEY

PUBLISHER   Laurie Rosatone
ACQUISITIONS EDITOR   Shannon Corliss
FREELANCE PROJECT EDITOR   Anne Scanlan-Rohrer
ASSISTANT CONTENT EDITOR   Jacqueline Sinacori
EDITORIAL ASSISTANT   Michael O'Neal
SENIOR CONTENT MANAGER   Karoline Luciano
SENIOR PRODUCTION EDITOR   Kerry Weinstein
PRODUCTION MANAGEMENT SERVICES   Sherrill Redd, Aptara Inc.
MARKETING MANAGER   Melanie Kurkjian
SENIOR PRODUCT DESIGNER   Tom Kulesa
EDITORIAL OPERATIONS MANAGER   Melissa Edwards
SENIOR PHOTO EDITOR   Mary Ann Price
SENIOR DESIGNER   Madelyn Lesure
COVER PHOTO   Franklin Kappa/Caiaimage/Getty Images Inc.

This book was set in 10/12 TimesTen by Aptara and printed and bound by
Courier/Kendallville. The cover was printed by Courier Kendallville.

Library of Congress Cataloging in Publication Data:

Schreiber, Sebastian J., author.
  Calculus for the life sciences / Sebastian J. Schreiber, Karl J. Smith, and Wayne M. Getz.
    pages cm
  Includes bibliographical references and index.
    ISBN 978-1-118-16982-7 (cloth : alk. paper)   1. Calculus–Textbooks.   I. Smith, Karl J., author.
II. Getz, Wayne Marcus, author.   III. Title.
    QA303.2.S38   2014
    515–dc23
                                                                                    2013034154

ISBN 978-1-118-16982-7 (cloth)
ISBN 978-1-118-18066-2 (Binder Ready Version)

Printed in the United States of America

10 9 8 7 6 5 4 3 2 1

Sebastian J. Schreiber received his B.A. in mathematics from Boston University in 1989 and his Ph.D. in mathematics from the University of California, Berkeley in 1995. He is currently Professor of Ecology and Evolution at the University of California, Davis. Previously, he was an associate professor of mathematics at the College of William and Mary, where he was the 2005 recipient of the Simon Prize for Excellence in the Teaching of Mathematics, and Western Washington University. Professor Schreiber's research on stochastic processes, nonlinear dynamics, and applications to ecology, evolution, and epidemiology has been supported by grants from the U.S. National Science Foundation, the U.S. National Oceanic and Atmospheric Administration, the Bureau for Land Management, and the U.S. Fisheries and Wildlife Service. He is the author or co-author of over seventy scientific papers in peer-reviewed mathematics and biology journals, including papers co-authored with undergraduate students. Professor Schreiber is currently on the editorial boards of the research journals *Discrete and Continuous Dynamical Systems B*, *Ecology*, *Journal of Biological Dynamics*, *Journal of Mathematical Biology*, *Mathematical Medicine and Biology*, and *Theoretical Ecology*.

Karl J. Smith received his B.A. and M.A. (in 1967) degrees in mathematics from UCLA. He moved to northern California in 1968 to teach at Santa Rosa Junior College, where he taught until his retirement in 1993. Along the way, he served as department chair, and he received a Ph.D. in 1979 in mathematics education at Southeastern University. A past president of the American Mathematical Association of Two-Year Colleges, Professor Smith is active nationally in mathematics education. He was founding editor of Western AMATYC News, a chairperson of the Committee on Mathematics Excellence, and an NSF grant reviewer. He was a recipient in 1979 of an Outstanding Young Men of America Award, in 1980 of an Outstanding Educator Award, and in 1989 of an Outstanding Teacher Award. Professor Smith is the author of over 60 successful textbooks. Over two million students have learned mathematics from his textbooks.

Wayne M. Getz received his undergraduate and Ph.D. (1976) degrees in applied mathematics from the University of the Witwatersrand, South Africa. In 1979 he immigrated to the United States to take a faculty position at the University of California, Berkeley, where he is currently the A. Starker Leopold Professor of Wildlife Ecology and Biomathematician in Agricultural Experiments Station. Professor Getz has a D.Sc. from the University of Cape Town, South Africa and has honorary appointments in the Mammal Research Institute at the University of Pretoria and in the School of Mathematical Sciences at the University of KwaZulu-Natal, both in South Africa, and he is a founder and trustee of the South African Centre for Epidemiological Modeling and Analysis. Recognition for his research in biomathematics and its application to various areas of physiology, behavior, ecology, and evolution includes an Alexander von Humboldt US Senior Scientist Research Award in 1992, and election to the American Association for the Advancement of Science (1995), the California Academy of Sciences (2000), and the Royal Society of South Africa (2003). He was appointed as a Chancellor's Professor at UC Berkeley from 1998 to 2001. Professor Getz has served as a consultant to both government and industry, and his research over the past thirty years has been funded by various government institutions and private foundations. Professor Getz has published a book titled *Population Harvesting* in the Princeton Monographs in Population Biology series, edited other books and volumes, and is an author or coauthor on more than 250 scientific papers in over fifty different peer-reviewed applied mathematics and biology journals.

I dedicate this book to my son, Dimitri, for his irrepressible curiosity about all things great and small, even calculus.

SJS

I dedicate this book to Ben Becker, my son-in-law. Without him, this book would not exist.

KS

I dedicate this book to my grandchildren Talia, Kaela, Ariana, and Benjamin in the hope that at least one of them will one day use it.

WMG

"If the 20th century belonged to physics, the 21st century may well belong to biology. Just 50 years after the discovery of DNA's chemical structure and the invention of the computer experiment, a revolution is occurring in biology, driven by mathematical and computational science."

Jim Austin, US editor of *Science*, and Carlos Castillo-Chavez, professor of biomathematics, *Science*, February 6, 2004

Calculus was invented in the second half of the seventeenth century by Isaac Newton and Gottfried Leibniz to solve problems in physics and geometry. Calculus heralded in the "age of physics" with many of the advances in mathematics over the past 300 years going hand-in-hand with the development of various fields of physics, such as mechanics, thermodynamics, fluid dynamics, electromagnetism, and quantum mechanics. Today, physics and some branches of mathematics are obligate mutualists: unable to exist without one another. The history of the growth of this obligate association is evident in the types of problems that pervade modern calculus textbooks and contribute to the canonical lower division mathematics curricula offered at educational institutions around the world.

The "age of biology" is most readily identified with two seminal events: the publication of Charles Darwin's *On The Origin of Species*, in 1859; and, almost 100 years later, Francis Crick and James Watson's discovery in 1953 of the genetic code. About mathematics, Darwin stated

"I have deeply regretted that I did not proceed far enough at least to understand something of the great leading principles of mathematics; for men thus endowed seem to have an extra sense."

Despite Darwin's assertion, mathematics was not as important in the initial growth of biology as it was in physics. Over the past three decades, however, dramatic advances in biological understanding and experimental techniques have unveiled complex networks of interacting components and have yielded vast sets of data about the structure of genomes and the variation and distribution of organisms in space and time. To extract meaningful patterns from these complexities, mathematical methods applied to the study of such patterns is crucial to the maturation of many fields of biology. Mathematics will function as a tool to dissect out the complexities inherent in biological systems rather than be used to encapsulate physical theories through elegant mathematical equations.

Mathematics will ultimately play a different type of role in biology than in physics because the units of analysis in biology are extraordinarily more complex than those of physics. The difference between an ideal billiard ball and a real billiard ball or an ideal beam and a real beam dramatically pales in comparison with the difference between an ideal and a real *Salmonella* bacterium, let alone an ideal and a real elephant. Biology, unlike physics, has no axiomatic laws that provide a precise and coherent theory upon which to build powerful predictive models. The closest biology comes to this ideal is in the theory of enzyme kinetics associated with the simplest cellular processes and the theory of population genetics that only works for a small handful of discrete, environmentally insensitive, individual traits determined by the particular alleles occupying discrete identifiable genetic loci. Eye color in humans provides one such example.

This complexity in biology means that accurate theories are much more detailed than those in physics, and precise predictions, if possible at all, are much more computationally demanding than comparable precision in physics. Only with the advent of extremely powerful computers can we aspire to use mathematical models to solve the problems of how a string of peptides folds into an enzyme with predicted catalytic properties, how a neuropil structure recognizes and categorizes an object, or how the species composition of a lake changes with an influx of heat, pesticides, or fertilizer.

It is critical that all biologists involved in modeling are properly trained to understand the meaning of output from mathematical models and to have a proper perspective on the limitations of the models themselves to address real problems. Just as we would not allow a butcher with a fine set of scalpels to perform exploratory surgery for cancer in a human being, so we should be wary of allowing biologists poorly trained in the mathematical sciences to use powerful simulation software to analyze the behavior of biological systems. Consequently, the time has come for all biologists who are interested in more than just the natural history of their subject to obtain a sufficiently rigorous grounding in mathematics and modeling, so that they can appropriately interpret models with an awareness of their meaning and limitations.

About This Book

It is no longer adequate for biologists to study either an engineering calculus or a watered-down version of the calculus. The application of mathematics to biology has progressed sufficiently far in the past three decades and mathematical modeling is sufficiently ubiquitous in biology to justify an overhaul of how mathematics is taught to students in the life sciences. In a recent article titled *Math and Biology: Careers at the Interface,** the authors state,

"Today a biology department or research medical school without 'theoreticians' is almost unthinkable. Biology departments at research universities and medical schools routinely carry out interdisciplinary projects that involve computer scientists, mathematicians, physicists, statisticians, and computational scientists. And mathematics departments frequently engage professors whose main expertise is in the analysis of biological problems."

In other words, mathematics and biology departments at universities and colleges around the world can no longer afford to build separate educational empires; instead, they need to provide coordinated training for students wishing to experience and ultimately contribute to the explosion of quantitatively rigorous research in ecology, epidemiology, genetics, immunology, physiology, and molecular and cellular biology. To meet this need, interdisciplinary courses are becoming more common at both large and small universities and colleges.

In this text, we present the basic canons of first-year calculus—but motivated through real biological problems. When combined with a course in statistics, students will have the quantitative foundation needed for research in the biological sciences and the background to take further courses in mathematics. In particular, this book can be viewed as a gateway to the exciting interface of mathematics and biology. As a calculus-based introduction to this interface, the main goals of this book are twofold:

- To provide students with a thorough grounding in concepts and applications, analytical techniques, and numerical methods of calculus

- To have students understand how, when, and why calculus can be used to model biological phenomena

To achieve these goals, the book has several important features.

Features

Concepts Motivated through Applications

First, and foremost, topics are motivated where possible by significant biological applications, several of which appear in no other introductory calculus texts. Significant applications include CO_2 buildup at the Mauna Loa observatory in Hawaii, scaling of metabolic rates with body size, optimal exploitation of resources in patchy environments, insect developmental rates and degree days, rapid decline of populations, velocities of stooping peregrine falcons, drug infusion and accumulation rates, measurement of cardiac output, *in vivo* HIV dynamics, mechanisms of memory formation, and spread of disease in human populations. Many of these examples involve real-world data and whenever possible, we use these examples to motivate and develop formal definitions, procedures, and theorems. Since we learn by doing, every section ends with a set of applied problems that expose students to additional applications, as well as recurring applications that are further developed as more knowledge is gained. These applied problems are always preceded by a set of drill problems designed to provide students with the practice they need to master the methods and concepts that underlie many of the applied problems.

Chapter Projects

Second, for more in-depth applications, each chapter includes one or more projects that can be used for individual or group work. These projects are diverse in scope, ranging from a study of enzyme kinetics to heart rates in mammals to disease outbreaks.

Early Use of Sequences and Difference Equations

Third, sequences, difference equations, and their applications are interwoven at the sectional level in the first four chapters. We include sequences in the first half of the book for three reasons. The first reason is that difference equations are a fundamental tool in modeling and give rise to a variety of exciting applications (e.g. population genetics), mathematical phenomena (e.g. chaos) and numerical methods (i.e. Newton's method and Euler's method). Hence, students are exposed to discrete dynamical models in the first half of the book and continuous dynamical models in the second half. The second reason for including sequences is that two of the most important concepts, limits and derivatives, provide fundamental ways to explore the behavior of difference equations (e.g., using limits to explore asymptotic

---

* Jim Austin and Carlos Castillo-Chavez, "Math and Biology: Careers at the Interface," *Science*, February 6, 2004.

behavior and derivatives to linearize equilibria). The third reason is that integrals are defined as limits of sequences. Consequently, it only makes sense to present sequences before discussing integrals. The material on sequences is placed in clearly marked sections so that instructors wishing to teach this topic during the second semester can do so easily.

Inclusion of Bifurcation Diagrams and Life History Tables

Fourth, we introduce two topics, bifurcation diagrams and life history tables that are often not covered in other calculus books. Bifurcation diagrams for univariate differential equations are a conceptually rich yet accessible topic. They provide an opportunity to illustrate that small parameter changes can have large dynamical effects. Life history tables provide students with an introduction to age structured populations and the net reproductive number R_0 of a population or a disease.

Historical Quests

Fifth, throughout the text there are problems labeled as 𝕳istorical 𝕼uests. These problems are not just historical notes to help students see mathematics and biology as living and breathing disciplines; rather, they are designed to involve students in the quest of pursuing great ideas in the history of science. Yes, they provide some interesting history, but they also lead students on a quest that should be rewarding for those willing to pursue the challenges they offer.

Multiple Representations of Topics

Sixth, throughout the book, concepts are presented visually, numerically, algebraically, and verbally. By using these different perspectives, we hope to enhance as well as reinforce understanding of and appreciation for the main ideas.

Review Sections

Seventh, we include review questions at the end of each chapter that cover concepts from each section in that chapter.

Content

Chapter 1: This chapter begins with a brief overview of the role of modeling in the life sciences. It then focuses on reviewing fundamental concepts from precalculus, including power functions, the exponential function, inverse functions and logarithms. While most of these concepts are familiar, the emphasis on modeling and verbal, numerical and visual representations of concepts will be new to many students. The chapter includes a strong emphasis on working with real data including fitting linear and periodic functions to data. This chapter also includes an introduction to sequences through an emphasis on elementary difference equations.

Chapter 2: In this chapter, the concepts of limits, continuity, and asymptotic behavior at infinity are first discussed. The notion of a derivative at a point is defined and its interpretation as a tangent line to a function is discussed. The idea of differentiability of functions and the realization of the derivative as a function itself are then explored. Examples and problems focus on investigating the meaning of a derivative in a variety of contexts.

Chapter 3: In this chapter, the basic rules of differentiation are first developed for polynomials and exponentials. The product and quotient rules are then covered, followed by the chain rule and the concept of implicit differentiation. Derivatives for the trigonometric functions are explored and biological examples are developed throughout. The chapter concludes with sections on linear approximation (including sensitivity analysis), higher order derivatives and l'Hôpital's rule.

Chapter 4: In this chapter, we complete our introduction to differential calculus by demonstrating its application to curve sketching, optimization, and analysis of the stability of dynamic processes described through the use of derivatives. Applications include canonical problems in physiology, behavior, ecology, and resource economics.

Chapter 5: This chapter begins by motivating integration as the inverse of differentiation and in the process introduces the concept of differential equations and their solution through the construction of slope fields. The concept of the integral as an "area under a curve" and net change is then discussed and motivates the definition of an integral as the limit of Riemann sums. The concept of the definite integral is developed as a precursor to presenting the fundamental theorem of calculus. Integration by substitution, by parts, and through the use of partial fractions is discussed with a particular focus on biological applications. The chapter concludes with sections on numerical integration and additional applications, including estimation of cardiac output, survival-renewal processes, and work as measured by energy output.

Chapter 6: In this chapter, we provide a comprehensive introduction to univariate differential equations. Qualitative, numerical, and analytical approaches are covered, and a modeling theme unites all sections. Students are exposed via phase line diagrams, classification of equilibria, and bifurcation diagrams to the modern approach of studying differential equations. Applications to *in vivo* HIV dynamics, population collapse, evolutionary games, continuous drug infusion, and memory formation are presented.

Chapter 7: In this chapter, we introduce applications of integration to probability. Probability density functions are motivated by approximating histograms of real-world data sets. Improper integration is presented and used as a tool to compute expectations and variances. Distributions covered in the context of describing real-world data include the uniform, Pareto, exponential, logistic, normal, and lognormal distributions. The chapter concludes with a section on life history tables and the net reproductive number of an age-structured population.

Chapter 8: In this chapter, we introduce functions of two variables, particularly in the context of the representation of surfaces in three-dimensional space, which has general relevance as well as biological modeling relevance. We then provide an introduction to 2-by-2 matrices, 2-D vectors, and related eigenvalues and eigenvectors—purely in terms of their relevance to modeling 2-D systems using linear differential equations and finding their equilibria. The chapter concludes with a section on phase-plane methods used to explore the behavior of nonlinear 2-D differential equation models, with examples drawn from pharmacology, cell biology, ecology, and epidemiology.

Supplementary Materials for Students and Instructors

Instructor's Solutions Manual This supplement, written by Tamas Wiandt, provides worked-out solutions to most exercises in the text (ISBN 9781118645567).

Student Solutions Manual This supplement, written by Tamas Wiandt, provides detailed solutions to most odd-numbered exercises (ISBN 9781118645598).

Instructor's Manual This supplement, written by Eli Goldwyn, contains teaching tips, additional examples, and sample assignments (ISBN 9781118676981).

WileyPLUS

WileyPLUS is a research-based online environment for effective teaching and learning.

WileyPLUS builds students' confidence because it takes the guesswork out of studying by providing students with a clear roadmap: what to do, how to do it, and whether or not they did it right. Students will take more initiative so you'll have greater impact on their achievement in the classroom and beyond. Please ask your Wiley sales representative for details.

Acknowledgments

Two of us (Getz and Smith) owe a debt of gratitude to Ben Becker, Michael Westphal, and George Lobell: without their efforts more than fifteen years ago, the seeds of the project that culminated in the production of this book would never have been sown. The three of us are even more deeply indebted to our families for the tolerance shown while we scrambled to meet deadlines; they patiently waited for a seemingly endless project to reach completion. We thank the editors at Wiley for their belief in our book, particularly our acquisitions editors David Dietz and Shannon Corliss, and project editor, Ellen Keohane, who picked up the text at a time when the publishing industry as a whole was going through a recession. Through Ellen's tireless efforts and her orchestration of a highly professional group of designers, illustrators, and typesetters, we were able to turn our LaTeX manuscript into a beautifully designed and produced book. We reserve special thanks to Celeste Hernandez and Marie Vanisko for scrutinizing our text and checking for errors in our formulations and solutions. We also thank Tamas Wiandt, Erin Roberts, Ben Weingartner, Sarah Day, David Brown, and Omar Shairzay for their corrections and suggestions. All remaining errors are our responsibility, and we apologize to readers for any of these that may have given them cause for pause. Toward the end of the project, Ellen moved on and was replaced by Anne Scanlan-Rohrer. Anne has our admiration for the professional way she kept our project on track and saw it through to completion. Finally, we thank all our colleagues who patiently reviewed different parts of the text at various stages of writing and provided invaluable feedback that has greatly improved the final form of this book.

Reviewers and Class-Testers

Olcay Akman, *Illinois State University*
Linda J. S. Allen, *Texas Tech University*
Martin Bonsangue, *California State University, Fullerton*
Eduardo Cattani, *University of Massachusetts, Amherst*
Lester Caudill, *University of Richmond*
Natalia Cheredeko, *University of Toronto*
Casey T. Cremins, *University of Maryland*
Sarah Day, *College of William and Mary*
Alice Deanin, *Villanova University*
Anthony DeLegge, *Benedictine University*
Dan Flath, *Macalaster College*
William Fleischman, *Villanova University*
Guillermo Goldsztein, *Georgia Institute of Technology*
Edward Grossman, *City College of New York*
Hongyu He, *Louisiana State University*
Shandelle M. Henson, *Andrews University*
Yvette Hester, *Texas A & M University*
Alberto Jimenez, *California Polytechnic State University, San Luis Obispo*
Timothy Killingback, *University of Massachusetts Boston*

M. Drew LaMar, *College of William and Mary*
Glenn Ledder, *University of Nebraska*
Alun L. Lloyd, *North Carolina State University*
J. David Logan, *University of Nebraska*
Yuan Lou, *The Ohio State University*
Joseph Mahaffy, *San Diego State University*
Edward Migliore, *University of California, Santa Cruz*
Laura Miller, *University of North Carolina*
Florence Newberger, *California State University,*
   Long Beach
Timothy Pilachowski, *University of Maryland,*
   College Park
Victoria Powers, *Emory University*

Michael Price, *University of Oregon*
Karen Ricciardi, *University of Massachusetts, Boston*
Yevgenya Shevtsov, *University of California,*
   Los Angeles
Patrick Shipman, *Colorado State University*
Nicoleta E. Tarfulea, *Purdue University, Calumet*
Ramin Vakilian, *California State University,*
   Northridge
Rebecca Vandiver, *St. Olaf College*
David Brian Walton, *James Madison University*
James Wright, *Green Mountain College*
Justin Wyss-Gallifent, *University of Maryland*
Mary Lou Zeeman, *Bowdoin College*

CONTENTS

APPLICATIONS AND MODELS INDEXS

Figure 1 Mathematics has been used to model many biological systems on Earth, ranging from global climatic processes to viral dynamics.

NOAA/NASA GOES Project

Preview of Modeling and Calculus

Is calculus relevant?

"The interface between mathematics and biology presents challenges and opportunities for both mathematicians and biologists. Unique opportunities for research have surfaced within the past ten to twenty years because of the explosion of biological data with the advent of new technologies and because of the availability of advanced and powerful computers that can organize the data. For biology, the possibilities range from the level of the cell and molecule to the biosphere. For mathematics, the potential is great in traditional applied areas such as statistics and differential equations, as well as in such nontraditional areas as knot theory."

"These challenges: aggregation of components to elucidate the behavior of ensembles, integration across scales, and inverse problems, are basic to all sciences, and a variety of techniques exist to deal with them and to begin to solve the biological problems that generate them. However, the uniqueness of biological systems, shaped by evolutionary forces, will pose new difficulties, mandate new perspectives, and lead to the development of new mathematics. The excitement of this area of science is already evident, and is sure to grow in the years to come."

(Executive Summary, National Science Foundation-sponsored workshop led by Simon Levin, 1990).

The above quotation, which is as true today as when it was written, hints at the exciting opportunities that exist at the interface of mathematics and biology. The goal of this course is to provide you with a strong grounding in calculus, while at the same time introducing you to various research areas of mathematical biology and inspiring you to take more courses at this interdisciplinary interface. In Chapter 1, we will set the tone for the entire book and review some of the skills needed to work at this interface. But first, in this preview, the idea of mathematical modeling is introduced to give you an underlying understanding of this important concept. Throughout the book you will find real-life problems that can be solved using mathematics. For example, in Olympic weightlifting, medals are awarded to individuals in different weight classes, as heavier individuals tend to lift more weight. In Chapter 1, we use mathematics and a basic physiological principle to predict how much an individual can lift scales with their body weight. Using the resulting mathematical model, we identify one of the greatest weightlifters of all time, Pocket Hercules (see Figure 1.1).

Models come in many guises: Architects make buildings models that are either small-scale replicas or, more recently, visual images created using computer-aided design packages. Political scientists, through debate and discussion, create verbal and written models that simulate the potential outcomes of a proposed policy. Artists make sketches and small-scale sculptures before starting a large-scale project. Flight simulators allow people to gain skills in piloting without the dangers associated with flying. Scientists in many disciplines (e.g., physics, biology, economics, chemistry, sociology, and even psychology) use mathematical models to investigate important phenomena.

Real-world problems inspired the creation of quantitative tools to grapple with their complexity. The counting and division of flocks of birds influenced the early development of number theory. The measurement and division of land led to the development of geometry. Understanding the motion of the planets and the forces of electricity, magnetism, and gravity resulted in the development of calculus. More recently, the study of the dynamics of population growth and population genetics led to many of the basic topics in stochastic processes. The immense success of mathematical models in understanding physical processes was recognized by E. P. Wigner in "The Unreasonable Effectiveness of Mathematics in the Natural Sciences" (*Communications in Pure and Applied Mathematics* 13 (1960): 1–14)—his now famous essay—in which he states:

"The miracle of the appropriateness of the language of mathematics for the formulation of the laws of physics is a wonderful gift which we neither understand nor deserve. We should be grateful for it, and hope that it will remain valid in future research and that it will extend, for better or for worse, to our pleasure even though perhaps also to our bafflement, to wide branches of learning."

As highlighted in the quotation from the NSF-sponsored workshop, one of the areas to which mathematics has extended most dramatically over the past half century is the biological sciences. The importance of this mathematics-biology interface is threefold. First, field and laboratory experiments are generating an explosion of data at both the cellular and environmental levels of study. To make these data meaningful requires extracting patterns within the data (e.g., correlations among variables, clustering of points in time and space). Mathematics, which from one viewpoint is the study of patterns (e.g., numerical, geometrical), provides a powerful methodology to identify and extract these patterns. This power of mathematics is reflected in the following statement of one of the founders of calculus, Sir Isaac Newton (1642–1727), in his book.

"The latest authors, like the most ancient, strove to subordinate the phenomena of nature to the laws of mathematics."

Second, mathematics is a language that permits the precise formulations of assumptions and hypotheses. Consider the words of another founding father of calculus, Gottfried Wilhelm Leibniz (1646–1716).

"In symbols one observes an advantage in discovery which is greatest when they express the exact nature of a thing briefly and, as it were, picture it; then indeed the labor of thought is wonderfully diminished."

Third, mathematics provides a logical, coherent framework to deduce the implications of one's assumptions.

One of the goals of this book is to help you understand how, when, and why calculus can be used to model biological phenomena. To achieve this understanding, you will be expected to develop simple models, to understand more complicated models sufficiently well to slightly modify them, to determine the appropriate techniques to analyze the models (e.g., numerical vs. analytical, stability vs. bifurcation analysis), and to interpret the results of your analysis. Examples of biological phenomena that we will encounter include epidemic outbreaks, blood flow, population extinctions, tumor regrowth after chemotherapy, population genetics, regulatory genetic networks, mechanisms for memory formation, enzyme kinetics, and evolutionary games.

A second goal of this book is provide you with a thorough grounding in calculus, one of the greatest intellectual achievements of humankind. Calculus provides an analytic framework for studying the rates at which things change and the accumulation of change over time. This book provides you with a detailed tour through the key concepts and applications of calculus, its analytical techniques, and its numerical methods. In the remainder of this introduction, we briefly address two basic questions: What is mathematical modeling? What is calculus?

What is mathematical modeling?

A real-life situation is usually far too complicated to be precisely or mathematically defined. When confronted with a problem in the real world, therefore, it is usually necessary to develop a mathematical framework based on certain assumptions about the real world. This framework can then be used to find a solution that will tell us something about the real world. The process of developing this mathematical framework is referred to as mathematical modeling.

What, precisely, is a mathematical model? It is an abstract description of a real-life problem that does not have an obvious solution. The first step involves abstraction in which certain assumptions about the real world are made, variables are defined, and appropriate mathematical expressions are developed.

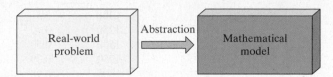

In this text, we discuss modeling biological systems. Consequently, as we progress through the book, we spend some time identifying the features associated with molecular, physiological, behavioral, life history, and population-level processes of many species and biological processes. After abstraction, the next step in modeling is to simplify the mathematics or derive related mathematical facts from the mathematical model.

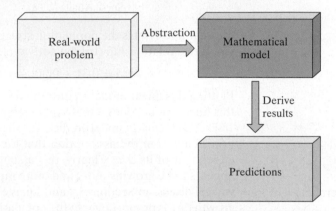

The results derived from the mathematical model should lead us to some predictions about the real world. The next step is to gather data from the situation being modeled and to compare those data with the predictions. If the two do not agree, then the gathered data are used to modify the assumptions underlying the model, and the process repeats.

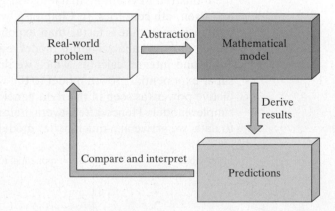

One use of the modeling process is predicting world population size, that is, the total number of humans living on Earth. Predicting world population size is of critical importance in determining future impacts of humans on their environment and future needs for energy, food, shelter, and other resources. Table 1 reports world population sizes over the past millennium.

It is clear that the world population is growing. But how fast is it growing? As a first guess, we might try modeling with a linear function $N(t) = a + bt$, where t is

Table 1 Human world population sizes over the past 1000 years

Year AD	Population size (in billions)	Year AD	Population size (in billions)
1000	0.31	1930	2.07
1250	0.40	1940	2.30
1500	0.50	1950	2.52
1750	0.79	1960	3.02
1800	0.98	1970	3.70
1850	1.26	1980	4.44
1900	1.65	1990	5.27
1910	1.75	1999	5.98
1920	1.86	2010	6.86

Source: http://en.wikipedia.or/wiki/World_population_estimates#cite_note_6. Data taken from the column headed: United Nations Department of Economic and Social Affairs (2008).

the year and $N(t)$ is the population size. Using the first two data points from Table 1 yields this relationship (try this for yourself!):

$$N(t) = 0.31 + 0.00036\,t \text{ billion individuals in year } t$$

Plotting this linear model against the data yields the left panel of Figure 2. Although this linear model does a reasonable job of predicting population sizes from 1000 to 1500 AD, it clearly underpredicts for the remainder of the millennium. This limitation stems from the assumption that the population size is increasing at a rate independent of its size. Clearly, this assumption is not reasonable as our population appears to be growing more and more rapidly over the centuries. A better model, as we will discuss in Sections 1.4 and 1.6, is exponential growth in which the population growth rate is proportional to the population size. The exercises of Section 1.4 will ask you to fit an exponential model to the data. This exponential model fitted to the first few data points slightly improves the predictive power of the model, as seen in the middle panel of Figure 2. However, the new model also substantially underpredicts world population sizes after 1750 AD. One possible explanation for this underprediction is that the model does not account for successive cultural revolutions, such as the Industrial Revolution, that led to surges in population size during the nineteenth and twentieth centuries. In Chapter 6, we examine a model that accounts for population growth that is faster than exponential: so-called super-exponential model. Developing and analyzing this model requires simultaneously the tools from differential and integral calculus, which we discuss next. In Section 6.2, you will fit this super-exponential growth model to the world population data and find that its predictive power (as seen in the right panel of Figure 2) is significantly better than the simpler models. Hence, after several iterations of model formulation and comparison to data, we arrive at a much better model for predicting future population sizes.

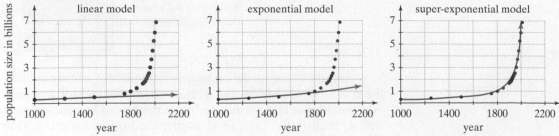

Figure 2 Three models of human population growth of increasing complexity: a simple linear model, an exponential growth model discussed in Sections 1.4 and 1.6, and a more sophisticated super-exponential growth model discussed in Section 6.2.

What is calculus?

Very likely, you have enrolled in a course that requires that you use this book. If you read the preface, you know that the intended audience is students who wish to learn about calculus but are majoring in an area related to biology. You might think of calculus as the culmination of all of your mathematical studies. To a certain extent, that is true, but it is also the beginning of your study of mathematics as it applies to how the real world changes in time and across space. All your prior work in mathematics is elementary. With calculus, you cross the dividing line between using elementary and advanced mathematical tools for studying a variety of applied topics. Calculus is the mathematics of motion and change over time and space.

What distinguishes calculus from your previous mathematics courses of algebra, geometry, and trigonometry is the transition from discrete static applications to those that are dynamic and often continuous. For example, in elementary mathematics you considered the slope of a line, but in calculus we define the (nonconstant) slope of a nonlinear curve. In elementary mathematics you found average changes in quantities such as the position and velocity of a moving object, but in calculus we can find instantaneous changes in the same quantities. In elementary mathematics you found the average of a finite collection of numbers, but in calculus we can find the average value of a function with infinitely many values over an interval.

The development of calculus in the seventeenth century, independently by Newton and Leibniz, was the result of their attempt to answer some fundamental questions about the world and the way things work. These investigations led to two fundamental concepts of calculus, namely, the idea of a *derivative*, which deals with rates of change, and that of an *integral*, which deals with accumulated change. The breakthrough in the development of these concepts was the formulation of a mathematical tool called a *limit*.

1. **Limit.** The limit is a mathematical tool for studying the *tendency* of a function as its variable *approaches* some value.

2. **Derivative.** The derivative is defined in the context of a limit. One of its uses is to compute rates of change and slopes of tangent lines to curves. The study of derivatives is called **differential calculus**. Derivatives can be used in sketching graphs and in finding the extreme (largest and smallest) values of functions. Biologists use derivatives to calculate, for example, the rates of change to the biochemical states of cells within individuals, rates of growth of populations, and rates of the spread of disease within populations.

3. **Integral.** The integral is found by taking a special limit of a sum of terms, and it is used initially to compute the accumulation of change. The study of this process is called **integral calculus**. Area, volume, work, and degree-days (the latter used to monitor the development of plants and "cold-blooded" animals) are a few of the many quantities that can be expressed as integrals. Biologists can use integrals to calculate, for example, the amount of fat bears store before going into hibernation, the time it takes an insect to develop from an egg into an adult as a function of temperature, the probability that an individual will die before a certain age, or the number of infected people as a disease spreads through a population.

Let us begin our study by taking an intuitive look at each of these three essential ideas of calculus.

The Limit

Zeno (ca. 500 BC) was a Greek philosopher who is known primarily for his famous paradoxes. One of these concerns a race between Achilles, a legendary Greek hero, and a tortoise. When the race begins, the (slower) tortoise is given a head start, as shown in Figure 3.

Is it possible for Achilles to overtake the tortoise? Zeno pointed out that by the time Achilles reaches the tortoise's starting point, $a_1 = t_0$, the tortoise will have

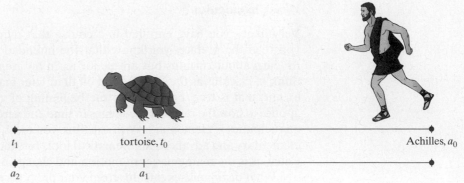

t_1 tortoise, t_0 Achilles, a_0

a_2 a_1

Figure 3 Achilles and the tortoise.

moved ahead to a new point, t_1. When Achilles gets to this next point, a_2, the tortoise will be at a new point, t_2. The tortoise, even though much slower than Achilles, keeps moving forward. Although the distance between Achilles and the tortoise is getting smaller and smaller, the tortoise will apparently always be ahead.

Of course, common sense tells us that Achilles will overtake the slow tortoise, but where is the error in this reasoning? The error is in the assumption that an infinite amount of time is required to cover a distance divided into an infinite number of segments. This discussion gets at an essential idea in calculus, the notion of a limit.

Consider the successive positions for both Achilles and the tortoise:

Achilles: $a_0, a_1, a_2, a_3, a_4, \cdots$

Tortoise: $t_0, t_1, t_2, t_3, t_4, \cdots$

After the start, the positions for Achilles, as well as those for the tortoise, form sets of positions that are ordered by the counting numbers. Such ordered listings are called **sequences**, which we introduce in Section 1.7. As we discuss in our first example below, and explore further in Chapter 2, the limit of a sequence of values $t_0, \cdots, t_n$ can be bounded above by some value T, say, so that for all values of n, no matter how large, we have $t_n < T$. In the context of the Achilles paradox, this means that Achilles will pass the tortoise at time T, if T is the smallest of all values that satisfies this inequality.

Example 1   Sequences: an intuitive preview

The sequence

$$\frac{1}{2}, \frac{2}{3}, \frac{3}{4}, \frac{4}{5}, \cdots$$

can be described by writing the general form of the n-th term: $a_n = \dfrac{n}{1+n}$ where $n = 1, 2, 3, 4, \cdots$. Can you guess the value L that a_n approaches as n gets large? This value is called the *limit of the sequence*.

Solution   We say that L is the number that $\dfrac{n}{n+1}$ tends toward as n becomes large without bound. We define a notation to summarize this idea:

$$L = \lim_{n \to \infty} \frac{n}{n+1}$$

As we consider larger and larger values for n, we find a sequence of fractions:

$$\frac{1}{2}, \frac{2}{3}, \frac{3}{4}, \cdots, \frac{1,000}{1,001}, \frac{1,001}{1,002}, \cdots, \frac{9,999,999}{10,000,000}, \cdots$$

It is reasonable to guess that the sequence of fractions is approaching the number 1.

Hence, for Zeno's paradox with

$$t_0 = a_1 = 1/2, t_1 = a_2 = 2/3, \ldots, t_{n-1} = \frac{n}{n+1}, \cdots$$

Achilles would pass the tortoise at position $L = 1$.

The Derivative: Rates of Change

The derivative provides information about the rate of change over small intervals (in fact, infinitesimally small!) of time or space. For instance, in trying to understand the role of humans in global climate change, we may be interested in the rate at which carbon dioxide levels are changing. In Section 1.2, we show that it is possible to come up with a function that describes how carbon dioxide levels (in parts per million) vary as a function of time. The relationship between this function and the data is illustrated in Figure 4.

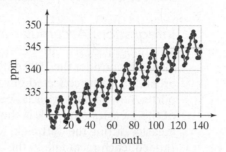

Figure 4 Carbon dioxide levels (in parts per million) as a function of months after April 1974.

In a scientific discussion about carbon dioxide levels, we might be interested in the rate of change of carbon dioxide levels at a particular time, say the second month (June 1974) of this data set. To find the rate of change from the second to tenth month, we could find the change in carbon dioxide levels, $331.8 - 331.0 = 0.8$ parts per million, and divide it by the change in time, $10 - 2 = 8$ months, to get the rate of change

$$\frac{331.8 - 331.0}{10 - 2} = 0.1 \text{ ppm per month}$$

over this eight-month period. Note that this rate of change corresponds to the slope of the *secant line* passing through the points $P = (2, 331.0)$ and $Q = (10, 331.8)$ as illustrated in Figure 5a. Although this rate of change describes what happens over the eight-month period, it clearly does not describe what is happening around the second month. Indeed, during the second month, the carbon dioxide levels are decreasing not increasing. Consequently, we expect the rate of change to be negative.

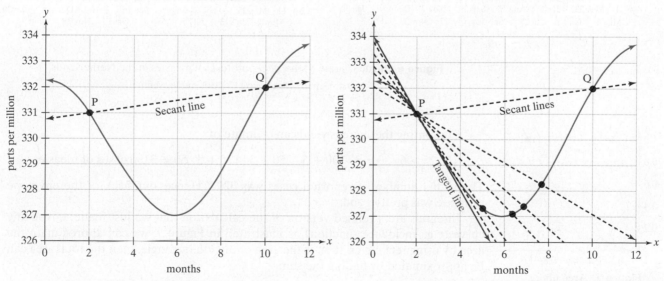

a. Secant line whose slope is a rate of change

b. Limit of secant lines is the tangent line.

Figure 5 The tangent line.

To get the *instantaneous rate of change* at the beginning of the second month, we can consider moving the point Q along the curve to the point P. As we do so, the points P and Q define secant lines that appear to approach a limiting line. This limiting line, as illustrated in Figure 5b, is called the **tangent line**.

The slope of this line corresponds to the instantaneous rate of change for carbon dioxide levels at the beginning of the second month of the data set. Later you will be able to find this instantaneous rate of change, which is approximately -1.24 ppm per month. The slope of this limiting line is also known as the derivative at P. The study of the derivative is called **differential calculus**.

Integration: Accumulated Change

The integral deals with accumulated change over intervals of time or space. For instance, consider the 1999 outbreak of measles in the Netherlands. During this outbreak, scientists collected information about the **incidence rate**: the number of reported new cases of measles per day. How this incidence rate varied over the course of the measles outbreak is shown in Figure 6. To find the total number of cases of measles during the outbreak, we want to find the area under this incidence "curve." Indeed, each rectangle in the left-hand side of this figure has a base of width "one day" and a height with units of measles per day. Hence, the area of each of these rectangles corresponds to the number of measles cases in one day. Summing up the area of these rectangles gives us the total number of measles cases during the outbreak. To get a rough estimate of this accumulated change, we can approximate the area under the incidence curve using the six larger rectangles imposed on the data, as illustrated in the right-hand side of Figure 6.

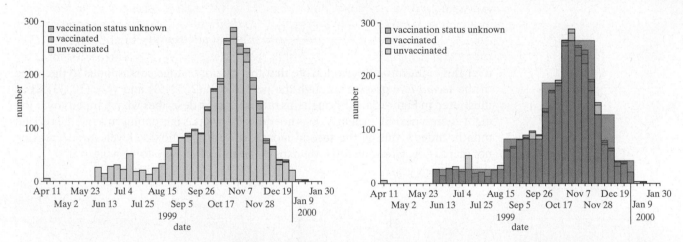

Figure 6 Incidence rate of the 1999 outbreak of measles in the Netherlands.
Source: Centers for Disease Control and Prevention.

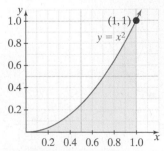

Figure 7 Area under a curve.

Computing these areas yields an estimate of

$$11 \cdot 25 + 6 \cdot 80 + 3 \cdot 200 + 5 \cdot 250 + 3 \cdot 125 + 3 \cdot 50 = 3130 \text{ cases of measles}$$

The actual number of reported cases was 3292. Hence, our back of the envelope estimate was pretty good.

Integrals are a refined version of the calculation that we just made. Given any curve (e.g., incidence function) as illustrated in Figure 7, we can approximate the area by using rectangles. If A_n is the area of the nth rectangle, then the total area can be approximated by finding the sum

$$A_1 + A_2 + A_3 + \cdots + A_{n-1} + A_n$$

This process is shown in Figure 8. To get better estimates of the area, we use more rectangles with smaller bases. The limit of this process leads to the *definite integral*, the key concept for **integral calculus**.

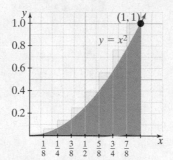 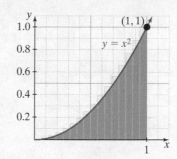

a. Eight approximating rectangles   **b.** Sixteen approximating rectangles

Figure 8 Approximating the area using circumscribed rectangles

In Chapter 1, we introduce some modeling concepts while reviewing basic mathematical concepts such as real numbers and functions—including linear, periodic, power, exponential, and logarithmic functions. Using these functions, we model the cyclic rise of carbon dioxide concentrations in the atmosphere, dangers facing large versus small organisms, population growth, and the binding of receptor molecules. We also introduce the basic notions of sequences and difference equations. Using these constructs, we encounter the dynamics of oscillatory populations, drug delivery, and gene frequencies. In the ensuing chapters, we develop the ideas of differential and integral calculus, and along the way, build necessary skills in biological modeling. Using these ideas and skills, you will model a diversity of biological topics such as disease outbreaks, blood flow, population extinctions, tumor regrowth after chemotherapy, genetics of populations, genetic networks, mechanisms for memory formation, enzyme kinetics, and evolutionary games.

Figure 1.1 Mathematical models are used in Section 1.3 to identify Pocket Hercules as one of the all-time greatest weightlifters.

CHAPTER 1

Modeling with Functions

Preview

"Mathematicians do not study objects, but relations between objects."

Henri Poincare, 1854–1912.

Although all readers taking a first course in calculus have a background in algebra, geometry, and trigonometry, the depth of exposure and choice of material covered can be quite variable. The material in this chapter is designed to provide a common framework upon which to build an introductory course in calculus for students who have a strong interest in the life sciences. In reviewing real numbers and functions, our intention is also to develop the notation we will use throughout the book. As students, you must become familiar with this notation if you want to be fluent in reading the mathematical text in this book. We also introduce data—and concepts around working with data—early on, because this component of the mathematical modeling process is critical to testing model predictions in the context of real-world problems (as discussed in the introduction to this book). We pay particular attention to power, exponential, and logarithmic functions since these all play a critical role in the development of differential and integral calculus. Trigonometric functions are important but less fundamental, and they have been dealt with extensively in precalculus mathematics courses. Thus, we provide only a brief review; we expect students who are rusty on this topic to go back and review trigonometry functions themselves. The topics dealt with in the function building section and inverse function subsection provide the kinds of skills that are needed in model building. Finally, we introduce the notion of sequences. Sequences are important both for introducing the concept of limits and in the context of dynamics, where consecutive terms in sequences can be used to represent the changes in the state of some object over time—an idea that is central to the application of calculus to all branches of science.

Real Numbers and Functions

You may have had a medical test in which an electrocardiograph, as shown in Figure 1.2, was used to check whether your heart was beating normally. In order to analyze graphs such as this, we need to seek unifying ideas relating graphs, data, tables, and equations. The mathematical concept that unifies these elements is the notion of a *real-valued* function, which is at the core of the development of both differential and integral calculus.

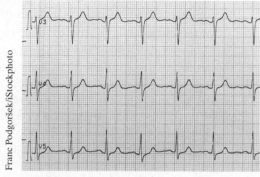

Figure 1.2 Portion of an electrocardiograph.

In this section, we discuss real numbers, functions, and basic properties of functions. To do this, we use the set notation {x : statement}, which means the set of all values of x or points x on the number line that satisfy or are defined by the statement following the colon.

Real numbers

Historically, the concept of numbers arose to address a need to count and keep exact records of land and property and to facilitate commerce. This process began with the counting numbers, now referred as the **natural numbers**, as depicted in Table 1.1.

Table 1.1 Sets of numbers

Name	Symbol	Set	Examples
Natural numbers	$\mathbb{N}$	$\{1, 2, 3, 4, 5, \cdots\}$	$6793, \sqrt{4}$
Integers	$\mathbb{Z}$	$\{\cdots -2, -1, 0, 1, 2, \cdots\}$	$-\sqrt{25}, \dfrac{18}{6}$
Rational numbers	$\mathbb{Q}$	$\left\{\dfrac{p}{q} : p, q \text{ are integers}, q \neq 0\right\}$	$\dfrac{5}{3} = 1.\bar{6}, -7.46\overline{31}$, where the overbar indicates a repeating decimal
Irrational numbers	$\mathbb{Q}'$	Numbers whose decimal representation does not terminate or repeat	$\sqrt{2}, -\sqrt{3}, \pi, e$
Real numbers	$\mathbb{R}$	All rational and irrational numbers	

It took a surprisingly long time for human civilization to add zero to this group to obtain the whole numbers. Negative numbers, which by some historical accounts first appeared in India and China around the seventh century, were then added to obtain the **integers**. The integers, however, are not closed under the operation of division: for example, $-4/2 = -2$ is an integer, but $4/3$ is not. Ancient Egyptian surveyors were well aware of fractional numbers, but only after negative numbers were widely accepted could the set of all positive and negative fractions, called the **rational numbers**, be defined.

Rational numbers are extremely useful for the measurements of "continuous" traits such as weight, height, humidity, and temperature, which are often measured by counting. For instance, we measure lengths by counting the number of marked intervals (e.g., inches, centimeters) on a tape measure. By subdividing these intervals into smaller and smaller fractions, we obtain more and more accurate measurements. We might expect that if we allow for all possible fractional divisions, then we can measure the precise length of anything. It came as a shock to the Greeks that this expectation is wrong! For instance, the Greeks proved that the length of the diagonal of a unit square (i.e., sides of length one) cannot be expressed as a rational number (see the 𝔥𝔦𝔰𝔱𝔬𝔯𝔦𝔠𝔞𝔩 𝔔𝔲𝔢𝔰𝔱 in Problem Set 1.1). Because this length corresponds to a number that cannot be found in the set of rational numbers, it is called *irrational* (not rational). It is denoted by the symbol $\sqrt{2}$ and its value can be approximated as precisely as we want by bounding it above and below by sequences of rational numbers that approach it in the limit! Intuitively, if we have a ruler with all fractional divisions, we can measure arbitrarily close approximations of this length.

To deal with irrational numbers, mathematicians extended the rational numbers to a larger set of numbers that we call the **real numbers**, $\mathbb{R}$. One can think of the real numbers as living on the edge of an infinitely long ruler with demarcations at all powers of ten. A real number is a point on this line and can be represented in a decimal form with its integer part before the decimal and tenths, hundredths, thousandths, ten thousandths, and so on, after the decimal. Rational numbers on this line have decimal representations that terminate or repeat, while the irrational numbers have decimal representations that do not terminate (see Table 1.1). For example, $\pi = 3.141592\ldots$ has a decimal representation that does not terminate or repeat and, consequently, is an irrational number—as is the Euler number e that we will encounter later in this chapter. Since $\sqrt{2} = 1.4142135\ldots$ is irrational, its decimal representation also does not terminate or repeat itself.

Intervals of real numbers arise so frequently in calculus that it is worthwhile giving them special names and notations. An **open interval** from a to b is denoted

$$(a, b) = \{x : a < x < b\}$$

Notice that this interval includes all the real numbers between a and b but does not include a and b themselves. A **closed interval** from a to b is denoted

$$[a, b] = \{x : a \leq x \leq b\}$$

Unlike an open interval, a closed interval includes the end points. In addition to these finite intervals, we are often interested in **infinite intervals**. These are intervals where either the right side of the interval extends infinitely far in the positive direction or the left side extends infinitely far in the negative direction, or both. In the first case, to denote this situation, we use the symbol ∞ on the right side of the interval, and in the second case we use the symbol $-\infty$ on the left side of the interval, as follows:

$$(a, \infty) = \{x : x > a\}, [a, \infty) = \{x : x \geq a\}, (-\infty, b) = \{x : x < b\}, \quad \text{and} \quad (-\infty, b] = \{x : x \leq b\}$$

The typical graphical depictions of these intervals on the real line is shown in Figure 1.3. For infinite intervals, it is important to realize there is no number "∞"

or "$-\infty$." These symbols are only used to indicate numbers in the interval whose magnitudes are arbitrarily large and positive or large and negative, respectively.

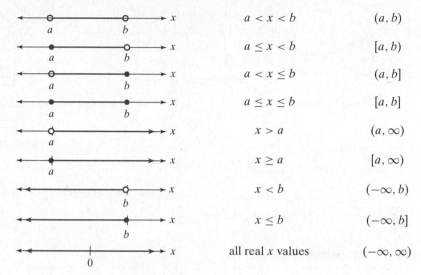

Figure 1.3 Graphical representations of intervals.

Often domains of functions are not a single interval but consist of the union of two or more intervals. We denote the **union of two (or more) intervals** with the notation ∪. For instance,

$$(a, b) \cup (c, d) = \{x : a < x < b \quad \text{or} \quad c < x < d\} \quad \text{and}$$

$$[a, b) \cup (c, d] = \{x : a \leq x < b \quad \text{or} \quad c < x \leq d\}$$

Functions

Biologists, mathematicians, and other researchers often study relationships between two quantities. The mathematical study of such relationships involves the concept of a function.

Function

A **function** $f : X \to Y$ is a rule that assigns to each element x of a set X (called the **domain** D) a unique element y of a set Y. The element y is called the **image** of x under f and is denoted by $f(x)$, read as "f of x." The set of all images $f(x)$ for x in X is called the **range** R of f.

A function can also be regarded as follows:

- A rule that assigns a unique "output" in the set Y to each "input" from the set X (Figure 1.4a)

- A graph that corresponds to the set of ordered pairs $\{(x, f(x)) : x \in D\}$ in the xy plane (Figure 1.4b)

- A machine that into which values of x are inserted and, after some internal operations are performed, a unique value $f(x)$ is produced (Figure 1.4c)

- An algebraic equation (Figure 1.4d)

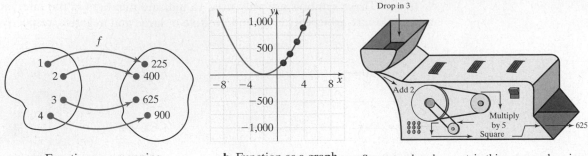

a. Function as a mapping

b. Function as a graph

Some number drop out; in this case, we drop in a 3 and 625 drops out; this number is called *f* of 3 and is written $f(3) = 625$. The output for the other members of the domain is shown:

$$f(x) = [(x + 2)5]^2$$

d. Function as an equation

c. Function as a machine

Figure 1.4 Different representations of a function.

Example 1   Identifying functions

Determine whether the following rules are functions. If one is a function, identify its domain and (if possible) its range.

a. To the real number r assign the area of a circle with radius r.

b. To each person in Atlanta assign his or her telephone number.

c. To the irrational reals, assign the value 1; to the rational reals, assign the value 0.

d. To each month from May 1974 to December 1985 assign the average carbon dioxide (CO_2) concentration measured at the Mauna Loa Observatory of Hawaii. The data are graphed in Figure 1.5.

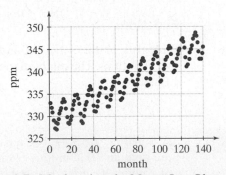

Figure 1.5 CO_2 (ppm) at the Mauna Loa Observatory.

Data Source: Komhyr, W. D., Harris, T. B., Waterman, L. S., Chin, J. F. S. and Thoning, K. W. (1989), Atmospheric carbon dioxide at Mauna Loa Observatory 1. NOAA Global monitoring for climatic change measurements with a nondispersive infrared analyzer, 1974–1985; J. Geop. Res., v. 94, no. D6, pp. 8533–8547. Data can be downloaded at http://www.seattlecentral.org/qelp/sets/016/016.html

e. To each (adjusted) income of a married individual filing jointly assign the federal tax rate for 2009.

Solution

a. This function can be expressed algebraically as

$$\text{area} = \pi r^2$$

Since a radius of a circle can only be nonnegative, the domain of this function is the nonnegative reals, $[0, \infty)$. The range of this function is also $[0, \infty)$.

b. Assigning telephone numbers to individuals in Atlanta is not a function for two reasons. First, not everyone has a phone number. For these individuals no assignment can be made. Second, many people may have more than one phone number, in which case the rule does not specify which of these phone numbers to associate with such an individual. By appropriately shrinking the domain, this rule becomes a function. For instance, if the domain is restricted to individuals in Atlanta with a single home phone number, assigning the home phone numbers to these individuals is a function.

c. Assigning 1 to irrationals and 0 to rationals defines a function whose domain is the reals and whose range is the set $\{0, 1\}$. This function is known as the *Dirichlet function* and is effectively impossible to graph for reasons that will become clearer in Chapter 2.

d. Assigning average monthly CO_2 concentrations from May 1974 to December 1985 is a function whose domain is the set

$$\{\text{May } 1974, \text{June } 1974, \text{July } 1974, \ldots, \text{March } 1986, \text{April } 1985\}$$

Alternatively, if we identify any natural number n with n months after April 1974 until December 1985, then the domain of this function is

$$\{1, 2, 3, \ldots, 140\}$$

as there are eleven years and eight months of monthly data recordings. To determine the range, we would have to find the values of the collected data. These data are illustrated in Figure 1.5 and suggest the range is contained in the interval $[327, 350]$. While these data, in themselves, cannot be precisely described by a simple algebraic formula, we shall see in Section 1.3 that this function is well approximated by a simple algebraic formula.

e. Assign each adjusted income for a married individual filing jointly in 2009 the federal tax rate. Since each adjusted income for a married individual filing jointly has one and only one tax rate, this rule, which is described in the tax tables, is a function. For instance, an adjusted income of greater than $372,950 is assigned a tax rate of 35%. We will reconsider this in Example 6.

As the preceding example and figure illustrate, functions can be represented in a variety of ways: verbally, algebraically, numerically, or graphically. Being able to move freely between these representations of a function is a skill that this book tries to cultivate.

Example 2   From words to algebraic representations

For regular strength Tylenol, each tablet contains 325 mg of acetaminophen. According to the *Handbook of Basic Pharmacokinetics*,[*] approximately 67% of the drug is removed from the body every four hours. Suppose Professor Schreiber had x mg of acetaminophen in his body four hours ago and just swallowed two more tablets of regular strength Tylenol. Write a formula in terms of x for the amount A of acetaminophen in Schreiber's body.

[*]W. A. Ritschel, *Handbook of Basic Pharmacokinetics*, 2nd ed. (Hamilton, IL: Drug Intelligence Publications, 1980): 413–426.

Solution   Since $100\% - 67\% = 33\%$ of the acetaminophen from four hours ago remains in Schreiber's body now, $0.33x$ is the amount of acetaminophen that is still in his body from four hours ago. Since Schreiber just swallowed two tablets of 325 mg per tablet, the total amount of acetaminophen in his body is

$$A = 0.33x + 325 \cdot 2 = 0.33x + 650 \text{ mg}$$

We will use this function in Section 1.7 to examine how the amount of acetaminophen in Schreiber's body varies in time whenever he takes two tablets every four hours. ◼

In this book, unless otherwise specified, the domain of a function is the set of real numbers for which the function is a well-defined real number determined by the context of the problem. We call this the **implied domain** convention. For example, if $f(x) = \dfrac{1}{x-2}$ and $g(y) = \sqrt{y}$, we need $x \neq 2$ and $y \geq 0$, respectively. Alternatively, if n is the number of people in an elevator, the context requires that n is a whole number.

Example 3   From algebraic expressions to graphs

Find the domain D of the following functions.

a. $y = \sqrt{1-x}$

b. $y = \dfrac{1}{(1-x)(x-2)}$

Solution

a. Because the argument of square roots must be nonnegative whenever we are dealing with real numbers, the domain in this case is $D = \{x : 1 - x \geq 0\}$. Equivalently $D = (-\infty, 1]$.

b. Because we cannot divide by zero, the domain consists of x such that $x \neq 1, 2$. Equivalently, the domain is $D = (-\infty, 1) \cup (1, 2) \cup (2, \infty)$. ◼

Plants use light energy, in the form of photons, to synthesize glucose from carbon dioxide and water while excreting oxygen as a byproduct of this process called *photosynthesis*. Plants then use the sugars to fuel other processes associated with their maintenance and growth while the oxygen is used by animals and other creatures for respiration. Thus, photosynthesis is a key process not only for plants but also for animal life on Earth!

Example 4   From verbal descriptions to graphs

Let $P(t)$ denote the photosynthetic activity of a leaf as function of t, where t is the number of hours after midnight. Sketch a rough graph of this function. Assume the sunrise is at 6 A.M. and the sunset is at 8 P.M.

Solution   Noting that there is no photosynthetic activity prior to the sunrise, we have $P(t) = 0$ for $0 \leq t \leq 6$. At sunrise, the photosynthetic activity slowly increases with the availability of light and reaches some maximum during midday. As the sun begins to set, the photosynthetic activity of the plant declines to zero and remains zero for the rest of the day. The graph of this function is shown in Figure 1.6.

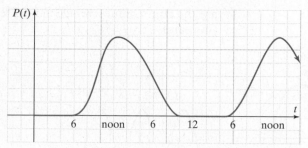

Figure 1.6 Sample graph of photosynthetic activity.

In Example 1, you were asked to identify functions. We extend this question to deciding if a given graph is the graph of a function. By looking at the definition of a function, we see that its graph has one point for a given element of the domain. Graphically, this idea can be stated in terms of the following **vertical line test**.

Vertical Line Test

A set of points in the xy-plane is the graph of a real valued function if and only if every vertical line intersects the graph at, at most, one point.

Example 5   Vertical line test in action

Determine which of the given graphs is the graph of a function.

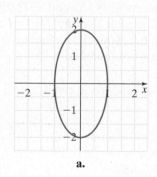

a.

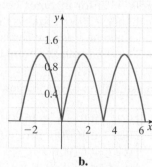

b.

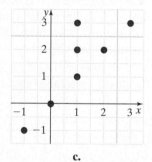

c.

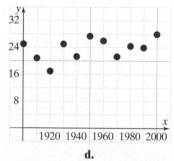

d.

Solution   In panel **a** (next page), a vertical line intersects the curve at two points for $x = -0.5$. Hence, this curve fails the vertical line test and is not the graph of a function. In fact, this curve is an ellipse given by the set of points that satisfy

$$x^2 + \frac{y^2}{4} = 1$$

The upper and lower halves of this ellipse can be described by the *pair* of functions

$$y = 2\sqrt{1 - x^2} \quad \text{and} \quad y = -2\sqrt{1 - x^2}$$

In panel **b** (next page), this curve does satisfy the vertical line test for all points x, as shown below for $x = 1$. In fact, recalling your trigonometric functions (see the next section), it is the graph of the function $y = |\sin x|$.

In panel **c** (next page), this set of points is not the graph of a function as the vertical line at $x = 1$ intersects three points.

In panel **d** (next page), this set of points is the graph of a function as it passes the vertical line test for all x, as shown below for $x = 1970$. In fact, these points are the graph of the average annual temperature in New York as a function of time (in years).

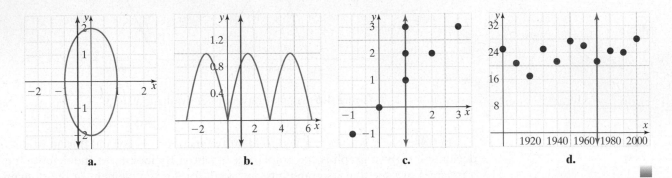

a.   b.   c.   d.

Piecewise-defined functions

In the real world sometimes functions must be defined with more than one formula; therefore, these are called **piecewise-defined functions** or just piecewise functions for short.

Example 6   Income tax rates

The federal income tax rates for married filing jointly in 2009 can be described as 10% for (adjusted) incomes up to $16,700, 15% for incomes between $16,701(rounding up) and $67,900, 25% for incomes between $67,901 and $137,050, 28% for incomes between $137,051 and $208,850, 33% for incomes between $208,851 and $372,950, and 35% for incomes greater than $372,950. Express the income tax rate $f(x)$ for an individual in 2009 with adjusted income x as a piecewise function. Graph the income tax rates over the interval $(0, 500000]$ (note the point 0 is not included).

Solution   An algebraic representation of this piecewise function is given by

$$f(x) = \begin{cases} 0.1 & \text{if } 0 < x \le 16{,}700 \\ 0.15 & \text{if } 16{,}700 < x \le 67{,}900 \\ 0.25 & \text{if } 67{,}900 < x \le 137{,}050 \\ 0.28 & \text{if } 137{,}050 < x \le 208{,}850 \\ 0.3 & \text{if } 208{,}850 < x \le 372{,}950 \\ 0.35 & \text{if } x > 372{,}950 \end{cases}$$

The graph of this piecewise function over the interval $(0,500,000]$ is shown in Figure 1.7. This graph consists of linear pieces with jumps between income brackets.

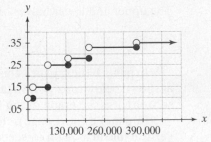

Figure 1.7 Graph of 2009 income tax rates for married filing jointly.

A particularly important piecewise-defined function is the *absolute value function*.

Absolute Value Function

The **absolute value function** $y = |x|$ is defined by

$$|x| = \begin{cases} x & \text{if } x \geq 0 \\ -x & \text{if } x < 0 \end{cases}$$

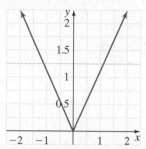

Figure 1.8 Graph of $y = |x|$.

When x is nonnegative, the absolute value of x is itself. When x is negative, the absolute value of x is the negative of itself. Hence, the graph of the absolute value function is shown in Figure 1.8.

Increasing and decreasing functions

Functions may exhibit different properties on intervals within their domain, as we now describe.

Increasing and Decreasing Functions

Let I be an interval in the domain of a function. Then

f is **increasing** on I if $f(x) < f(y)$ for all $x < y$ in I; that is, its graph rises from left to right on I

f is **decreasing** on I if $f(x) > f(y)$ for all $x < y$ in I; that is, its graph falls from left to right on I

f is **constant** on I if $f(x) = f(y)$ for every x and y in I; that is, the graph is flat on I

These classifications are shown graphically in Figure 1.9.

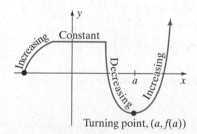

Figure 1.9 Classifications of functions.

Example 7 Classifying a function

Consider the function f defined by the following graph on the interval $I = [-2, 3]$.

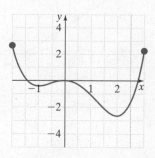

Find the intervals on which f is increasing and the intervals on which f is decreasing.

Solution   The function f is decreasing on $[-2, -1)$, increasing on $(-1, 0)$, decreasing on $(0, 2)$, and increasing on $(2, 3]$. Note that the interval is open at points where the function switches from increasing to decreasing or vice versa.   ■

PROBLEM SET 1.1

Level 1 DRILL PROBLEMS

Determine whether the descriptions in Problems 1 to 6 represent functions. If a description is a function, find the domain and (if possible) the range.

1. **a.** $\{(4, 7), (3, 4), (5, 4), (6, 9)\}$

   b. $\{6, 9, 12, 15\}$

2. **a.** $\{(5, 2), (7, 3), (1, 6), (7, 4)\}$

   b. $\{(x, y) : y = 4x + 3\}$

3. **a.** $\{(x, y) : y \leq 4x + 3\}$

   b. $\{(x, y) : y = 1$ if x is positive and $y = -1$ if x is negative$\}$

4. **a.** $\{(x, y) : y$ is the closing price of IBM stock on July 1 of year $x\}$

   b. $\{(x, y) : x$ is the closing price of Apple stock on July 1 of year $y\}$

5. **a.** $\{(x, y) : (x, y)$ is a point on a circle of radius 4 passing through $(2, 3)\}$

   b. $\{(x, y) : (x, y)$ is a point on an upward-opening parabola with vertex $(-3, -4)\}$

6. **a.** $\{(x, y) : (x, y)$ is a point on a line passing through $(2, 3)$ and $(4, 5)\}$

   b. $\{(x, y) : (x, y)$ is a point on a line passing through $(4, 5)$ and $(-3, 5)\}$

Use the vertical line test in Problems 7 to 9 to determine whether the curve is a function. Also state the probable domain and range.

7. **a.**

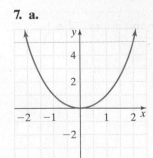

b.

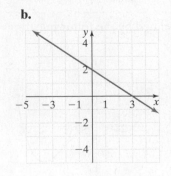

8. **a.**

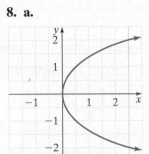

b.

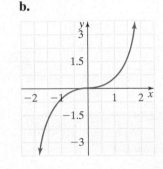

9. **a.**

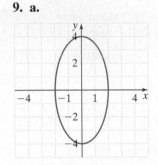

b.

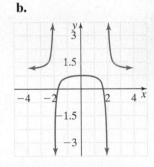

Problems 10 to 12 show the output you might find for a calculator graph. The Xmin and Xmax values show the input x values and the Ymin and Ymax values show the input y values. The scale for each tick mark on the x and y axis is also shown. These values determine the extent of the box that is shown. Using these calculator images use the vertical line test to determine whether the curve is a function. Also state the probable domain and range if you assume that a curve reaching a boundary of the frame continues in the same fashion as it continues beyond the shown screen.

10. **a.**

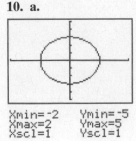

```
Xmin=-2    Ymin=-5
Xmax=2     Ymax=5
Xscl=1     Yscl=1
```

b.

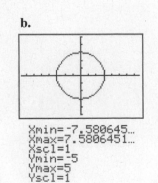

```
Xmin=-7.580645...
Xmax=7.5806451...
Xscl=1
Ymin=-5
Ymax=5
Yscl=1
```

11. a.

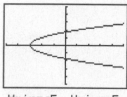

Xmin=-5 Ymin=-5
Xmax=5   Ymax=5
Xscl=1   Yscl=1

b.

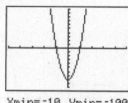

Xmin=-10 Ymin=-100
Xmax=10   Ymax=100
Xscl=2   Yscl=10

12. a.

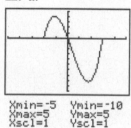

Xmin=-5 Ymin=-10
Xmax=5   Ymax=5
Xscl=1   Yscl=1

b.

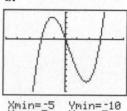

Xmin=-5 Ymin=-10
Xmax=5   Ymax=5
Xscl=1   Yscl=1

In Problems 13 to 18 find the domain of f and compute the indicated values or state that the corresponding x-value is not in the domain.

13. $f(x) = -x^2 + 2x + 3$; $f(0)$, $f(1)$, $f(-2)$

14. $f(x) = 3x^2 + 5x - 2$; $f(1)$, $f(0)$, $f(-2)$

15. $f(x) = \dfrac{(x+3)(x-2)}{x+3}$; $f(2)$, $f(0)$, $f(-3)$

16. $f(x) = (2x - 1)^{-3/2}$; $f(1)$, $f\left(\dfrac{1}{2}\right)$, $f(0)$

17. $f(x) = \begin{cases} -2x + 4 & \text{if } x \le 0 \\ x + 1 & \text{if } x > 0 \end{cases}$; $f(3)$, $f(1)$, $f(0)$

18. $f(x) = \begin{cases} 3 & \text{if } x < -1 \\ x + 1 & \text{if } -1 \le x \le 5 \\ \sqrt{x} & \text{if } x > 5 \end{cases}$; $f(-6)$, $f(5)$, $f(16)$

19. Consider a function machine

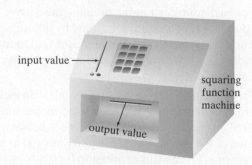

input value ⟶

squaring function machine

output value

that yields the table of values

Input values	Output values
1	1
2	4
3	9
-5	25

Algebraically, define the simplest function F you can think of for input values x from the domain $D = \mathbb{R}$.

20. Consider a function machine

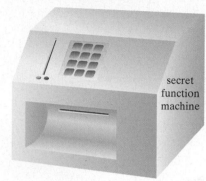

secret function machine

that yields the table of values

Input values	Output values
0	3
1	5
2	7
3	9
4	11

Algebraically, define the simplest function S you can think of for input values t from the domain $D = [0, \infty)$.

21. Suppose you are given a machine that multiplies the input value by 3 and then subtracts 7. Complete the table of values given below

Input values	Output values
3	2
5	
0	
-3	

and algebraically, define a function M for input values x from the domain $D = \mathbb{R}$.

22. Suppose there is super-secret machine the produces the table

Input values	Output values
0	5
1	6
2	9
3	14
4	21

Algebraically define the simplest function, T, for input values t from the domain $D = \mathbb{R}$.

Find the domain and range for the graphs indicated in Problems 23 to 28. Also tell where the function is increasing, decreasing, and constant.

23.

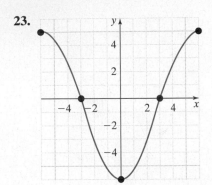

24.

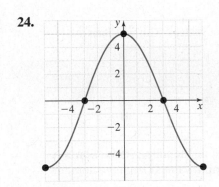

25.

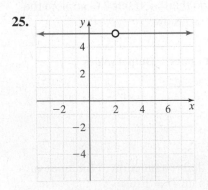

26.

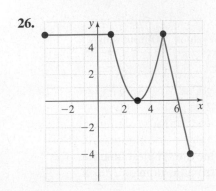

27.

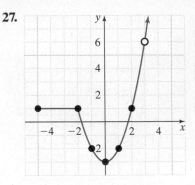

28.

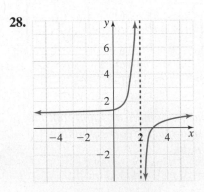

For each verbal description in Problems 29 to 36, write a rule in the form of an equation, state the domain, and then graph the function.

29. For each number x in the domain, the corresponding range value, y, is found by multiplying by 3 and then subtracting 5.

30. For each number x in the domain, the corresponding range value, y, is found by squaring and then subtracting 5 times the domain value.

31. For each number x in the domain, the corresponding range value, y, is found by subtracting the domain value from 5 and then taking the square root.

32. For each number x in the domain, the corresponding range value, y, is found by taking 5 added to 5 times the domain value and then dividing this by the result of adding 1 to the domain value.

33. From a square whose side has length x (in inches), create a new square whose side is 5 inches longer. Find an expression for the difference between the area of the two squares (in square inches) as a function of x. Graph this expression for $0 \leq x \leq 10$.

34. From a square whose side has length x (in meters), create a new square whose side is 10 meters longer. Find an expression for the sum of the areas of the two squares (in square meters) as a function of x. Graph this expression for $0 \leq x \leq 10$.

35. Find the area of a square as a function of its perimeter P.

36. Find the area of a circle as a function of its circumference C.

Level 2 APPLIED AND THEORY PROBLEMS

37. Recall from Example 2, a tablet of regular strength Tylenol contains 325 mg of acetaminophen and approximately 67% of the drug is removed from the body every four hours. Suppose Professor Schreiber had x mg of acetaminophen in his body four hours ago and just swallowed two more tablets of regular strength Tylenol. Write a formula in terms of x for the amount A of acetaminophen in Schreiber's body four hours from now.

38. Diazepam is a medication used for the management of anxiety disorders. Approximately 68% of the drug is removed from the body every 24 hours. Suppose a patient had x mg of diazepam in his body 24 hours ago and just took an oral dose of 30 mg. Write a formula in terms of x for the amount A of diazepam in the patient's body 24 hours from now.

39. Professor Getz mows his backyard lawn every Saturday. Draw a graph of the height of the grass in the lawn over a two-week period, beginning just after Getz mowed his lawn one Saturday.

40. Continuous morphine infusion (CMI) is a means of providing a continuous dosage of medication (morphine) for acute pain. Morphine is administered continuously by a computerized pump connected to the patient by an intravenous tube (IV). Graph the concentration of morphine in a patient's blood (in milligrams/liter) over a three-day period. Assume the patient initially had no morphine in the blood stream and the concentration of morphine was relatively constant at 1 mg/liter on the last day.

41. Biologists have found that the speed of blood in an artery is a function of the distance of the blood from the artery's central axis (Figure 1.10). According to *Poiseuille's law*, the speed (centimeters/second) of blood that is r centimeters from the central axis of an artery is given by the function

$$S(r) = C(R^2 - r^2)$$

where R is the radius of the artery and C is a constant that depends on the viscosity of the blood and the pressure between the two ends of the blood vessel. (The law and the unit *poise*, a unit of viscosity, are named for the French physician Jean Louis Poiseuille, 1799–1869.) Suppose that for a certain artery

$$C = 1.76 \times 10^5 \text{ cm/s}$$

and

$$R = 1.2 \times 10^{-2} \text{ cm}$$

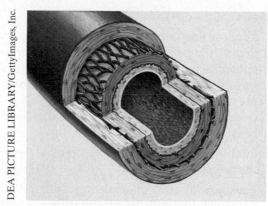

DEA PICTURE LIBRARY/GettyImages, Inc.

Figure 1.10 Cutaway view of an artery.

a. Compute the speed of the blood at the central axis of this artery.

b. Compute the speed of the blood midway between the artery's wall and central axis.

c. What is the domain for the function defined by the ordered pairs (r, S)?

d. Graph this function for $S \geq 0$.

42. The reaction rate of an autocatalytic reaction is given by the formula

$$R(x) = kx(a - x)$$

for $0 \leq x \leq a$, where a is the initial concentration of substance A and x is the concentration of X.

a. What is the domain?

b. Graph this function for $k = 3$ and $a = 8$.

43. Consider the function defined to study the rate at which animals learn when a psychology student performed an experiment in which a rat was sent repeatedly through a laboratory maze. Suppose that the time (in minutes) required for the rat to traverse the maze on the nth trial is modeled by the function

$$f(n) = 3 + \frac{12}{n}$$

a. What is the domain of the function f if n is a continuous variable?

b. What is the domain of the function f in the context of the psychology experiment?

c. Graph the function f defined in part **b** for n on interval $[1, 20]$.

d. What will happen to the time required for the rat to traverse the maze as the number of trials increases? Will the rat ever be able to traverse the maze in less than three minutes?

44. Consider the function defined by

$$f(x) = \frac{150x}{200 - x}$$

a. What is the domain of the function f?

b. Suppose that during a nationwide program to immunize the population against a certain form of influenza, public health officials found the cost (in millions of dollars) of inoculating $x\%$ of the population is modeled by f. For what values of x does $f(x)$ have a practical interpretation in this context?

c. Graph the function for its interpretable range for the problem at hand.

d. Compare the cost of inoculating the first 50% of the population with the cost for the second 50%.

45. *Friend's rule* is a method for calculating pediatric drug dosages in terms of a child's age (up to $12\frac{1}{2}$ years). If A is the adult dose (in milligrams) and n is the age of the child (in years), then the child's dose is given by

$$D(n) = \frac{2}{25}nA$$

a. What is the domain for the function defined by (n, D)?

b. Graph this function for continuous $0 \le n \le 12.5$ and $A = 100$.

c. If a 3-year-old child receives 100 mg of a certain drug, what is the corresponding dose for a 5-year-old child?

46. *Young's rule* is another method for calculating pediatric drug dosages in terms of a child's age. If A is the adult dose (in milligrams) and n is the age of the child (in years), then the child's dose is given by

$$D(n) = \frac{n}{n + 12}A$$

a. What is the domain for the function defined by (n, D) if the formula is applied to individuals up to age 16?

b. Graph this function for continuous $0 \le n \le 16$ and $A = 100$.

c. If a 6-year-old child receives 120 mg of a certain drug, what is the corresponding dose for an 8-year-old child?

47. *Clark's rule* is a method for calculating pediatric drug dosages based on a child's weight (w) in pounds (lb). If A denotes the adult dose (in milligrams) then the corresponding child's dose is given by

$$D(w) = \left(\frac{w}{150}\right) A$$

a. What is the domain for the function defined by (w, D)?

b. Graph this function for $A = 200$ mg.

c. If a 70-lb child receives 90 mg of a certain drug, what is the corresponding dose for an adult?

48. Table 1.2 tabulates the estimated number of HIV/AIDS cases diagnosed each year in the United States from 1999 to 2002.

Table 1.2 Number of diagnosed cases of HIV/AIDS by year

Age at diagnosis (years)	1999	2000	2001	2002
< 13	187	163	206	162
13–14	28	31	33	30
15–24	2,646	2,803	2,926	2,926
25–34	7,817	7,386	7,221	7,338
35–44	9,115	9,289	9,119	9,450
45–54	3,887	4,212	4,408	4,675
55–64	1,112	1,250	1,303	1,450
> 64	382	386	427	432

Source: Survey Report Volume 14 from the Center of Disease control, Division of HIV/AIDS Prevention.

a. Use these data to draw a graph of the number of cases being diagnosed each day during the period starting at the beginning of 1999 and ending at the end of 2002 for the age group 25–34. This should be done by assuming that the average daily rate each year holds at the beginning of the year and then joining these points by a "continuous" curve (i.e., a curve with no jumps or breaks). The concept of continuity will be made more precise in the Chapter 2.

b. Use these data to draw a graph of the number of new cases diagnosed each day for all age groups.

49. 𝔥𝔦𝔰𝔱𝔬𝔯𝔦𝔠𝔞𝔩 𝔔𝔲𝔢𝔰𝔱

Courtesy of Karl Smith

Pythagoras
(ca. 569–475 BC)

Throughout the text, you will find problems called Historical Quest. These problems are not just historical notes to help you see mathematics and biology as living disciplines; rather, these problems are designed to involve you in the quest of pursuing great ideas in the history of science. Yes, they relate some interesting history, but they will also lead you on a quest that you may find interesting.

Even though we know little about the man himself, we do know that Pythagoras was a Greek philosopher who is sometimes described as the first true mathematician in the history of mathematics. He founded a philosophical and religious school in Croton and attracted many followers, known today as the Pythagoreans. The Pythagoreans were a secret society who had their own philosophy, religion, and way of life. This group investigated music, astronomy, geometry, and number properties. Because of their strict secrecy, much of what we know about them is legend, and it is difficult to tell what work can be attributed to Pythagoras himself. We also know that it was considered impious for a member of the Pythagoreans to claim any discovery for himself. Instead, each new idea was attributed to their founder Pythagoras. No doubt you know the Pythagorean theorem, but do you know that the Pythagoreans believed that all things are numbers and that by a *number* they meant the ratio of two whole numbers? For this $\mathfrak{Historical\ Quest}$ you are to use these two ideas to prove that $\sqrt{2}$ is an irrational number.

There is a legend (not historical fact) that one day a group of Pythagoreans were out in a boat seeking truth. One person on board came up with the following argument: Construct a right triangle with legs of length 1 unit. By the Pythagorean theorem, the length of the hypotenuse is (using modern notation) exactly $\sqrt{2}$ units long. Is the length of this side a rational number or an irrational number? Let $\sqrt{2} = \frac{p}{q}$. (Remember, they believed that all numbers could be expressed as the ratio of two whole numbers; thus, assume that $\sqrt{2}$ is a rational number.) Assume that $\frac{p}{q}$ is a reduced fraction (if it is not reduced, simply reduce it and work with the reduced form). See if you can reproduce the work done in the boat; that is, show the details outlined here. Square both sides of the equation and prove that p is an even number. If p is even, then it can be written as $p = 2k$. Use this fact to show that q is even. Thus, the fraction $\frac{p}{q}$ is not reduced. Now, if you understand logic as did the Pythagoreans, you can see the contradiction. What is it? How can you use this information to prove that $\sqrt{2}$ is irrational? Legend has it that this contradiction bothered those on the boat so much that they tossed the person who came up with this argument overboard—and pledged themselves to secrecy!

1.2   Data Fitting with Linear and Periodic Functions

In the previous section we presented data about carbon dioxide (CO_2) collected at the top of the Mauna Loa volcano since 1958 by the U.S. government's Climate Monitoring Diagnostics Laboratory. These data are plotted in Figure 1.5. Scientists routinely collect data involving two variables x and y and refer to such data as *bivariate*. In many cases, a list of bivariate points, such as the Mauna Loa CO_2 data, can be modeled by a relatively simple functional relationship of the form $y = f(x)$ that passes, if not through all points, then close by all points. The advantages of the model are that it describes the data more concisely than a list, it can make predictions for uncollected data values, and it can generate hypotheses. For instance, if we had a function that did a good job of describing how carbon dioxide concentrations fluctuate in time, then we could make predictions about future levels of carbon dioxide concentrations. The importance of these predictions stems from the fact that carbon dioxide is a greenhouse gas. It prevents the escape of heat radiating from the Earth. Consequently, carbon dioxide in the atmosphere influences the Earth's temperature, and many people would like to know what the temperature might be twenty or fifty years from now so that they can plan accordingly.

The most commonly fitted function is a *linear function*: a function that depicts a constant rate of change with respect to unit changes in the argument of the function. In this section, we review basic facts about linear functions and briefly discuss how to fit linear functions to data sets, a process referred to in statistics as *linear regression*. For instance, the data in Figure 1.5 suggest that carbon dioxide concentrations in the atmosphere are tending to increase across years. Using linear regression, we can determine at what rate this increase across years is occurring. In addition to exhibiting a linear trend, the carbon dioxide data clearly exhibit seasonal fluctuations. These seasonal fluctuations can be modeled by periodic functions. Consequently, the section continues by reviewing basic properties of periodic and trigonometric functions and fitting trigonometric functions to data sets. Using a combination of linear and trigonometric functions, we arrive at surprisingly good model of CO_2 fluctuations.

Linear functions

Linear functions play a fundamental role in differential calculus as they can be used to approximate functions locally (i.e., over a relatively small interval of the domain of the variable x). A **linear function** is a function of the form

$$y = f(x) = mx + b$$

where m is the **slope** and b is the **vertical** or **y-intercept** of the linear function. The vertical intercept b is the value of y when x equals zero. Equivalently, it is the y-value at which the graph of $y = f(x)$ intercepts the y-axis; that is, $b = f(0)$. In contrast, the slope m of the line tells us that if we increase the x-value by an increment, say 0.2, then the corresponding y-value increases by m times that increment, $0.2m$. Equivalently, the change in y divided by the corresponding change in x is always the constant m. This leads us to a **slope formula**.

Slope of a Line	A nonvertical line that contains the points $P_1 = (x_1, y_1)$ and $P_2 = (x_2, y_2)$ has **slope** $$m = \frac{y_2 - y_1}{x_2 - x_1}$$

When the function $y = mx + b$ is regarded as a relationship between the paired variables (x, y), x is called the **independent variable** and y the **dependent variable**, because the relationship is designed to answer this question: What value of y corresponds to a given value for x?

Example 1   From graphs to equations

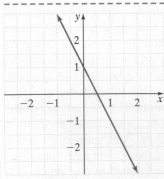

Figure 1.11 The equation $y = -2x + 1$.

Let $y = f(x)$ be the linear function whose graph is shown in Figure 1.11. Find the equation for $f(x)$.

Solution   Looking at the graph, we see that the y intercept is given by $b = 1$. Since $y = 1$ when $x = 0$ and $y = 0$ when $x = 0.5$, we see that y decreases by 1 when x increases by 0.5. Thus,

$$m = \frac{1 - 0}{0 - 0.5} = -2$$

and the equation of this line is

$$y = -2x + 1$$

Example 2   From equations to graphs

Let $y = f(x)$ be a linear function such that $f(2) = 3$ and $f(-2) = -1$. Write a formula for $f(x)$ and sketch the graph.

Solution   Since $f(x)$ is linear, we write $f(x) = mx + b$ where we need to determine the constants m and b. The slope is given by

$$m = \frac{f(2) - f(-2)}{2 - (-2)} = \frac{3 - (-1)}{4} = 1$$

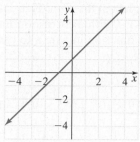

Figure 1.12 Graph of $y = x + 1$.

Therefore, $f(x) = x + b$. To find b, we solve

$$f(2) = 3$$
$$b + 2 = 3$$
$$b = 1$$

Hence, $y = f(x) = x + 1$. To graph this function, it suffices to draw a line that passes through the points $(-2, -1)$ and $(2, 3)$ as shown in Figure 1.12. ∎

Fitting linear functions to data

Many data sets exhibit trends that can be reasonably described by linear functions. We can fit linear functions to data using either formal or informal approaches. Informal approaches include eyeballing how well a selected line passes through a given set of data or fitting a line to two suitably chosen points in the data set. Formal statistical methods provide ways for finding the *best-fitting line* in some well-defined mathematical sense, which we describe after the next example.

Example 3   Carbon dioxide output from electric power plants

In Figure 1.13 the carbon dioxide emissions of most of the electricity generation plants in California are plotted as a function of the heat input for the year 1997. The heat input units are a million British thermal units (i.e., 10^6 BTU or 1 MMBTU) and CO_2 emissions are measured in metric tons.

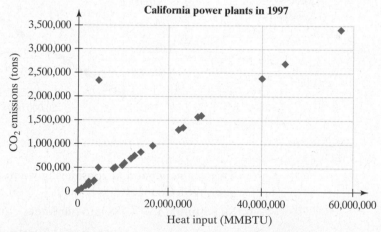

Figure 1.13 Data from the Emissions and Generation Resource Integrated Database.

In Table 1.3, six points that appear in Figure 1.13 are listed.

a. Since the data in Figure 1.13 look linear, use the first two data points in Table 1.3 to find a line that passes through the data. Graph this line.

Table 1.3 California power plants in 1997

Heat input (MMBTU)	CO_2 output (tons)
45.179×10^6	2.685×10^6
1.00×10^6	0.058×10^6
1.902×10^6	0.113×10^6
3.334×10^6	0.197×10^6
0.086×10^6	0.005×10^6
13.897×10^6	0.826×10^6
$\vdots$	$\vdots$

b. One data point in Figure 1.13 looks like it does not fit the rest of the data. This data point corresponds to a heat input of 4.488×10^6 MMBTU with a corresponding output of around 2.3×10^6 metric tons. Use the linear function in part **a** to estimate the CO_2 output for this plant.

Solution

a. To find the line $y = mx + b$ that passes through $(45.179 \times 10^6, 2.685 \times 10^6)$ and $(1 \times 10^6, 0.058 \times 10^6)$, we first solve for the slope:

$$m = \frac{2.685 - 0.058}{45.179 - 1} \approx 0.059$$

Using the point-slope formula (see Problem 17 in Problem Set 1.2) for a line yields

$$y - 2.685 \approx 0.059(x - 45.179)$$
$$y \approx 0.059x - 0.019 \times 10^6 \text{ tons of } CO_2$$

Sketching this line over the data graph shown in Figure 1.13 yields the following graph.

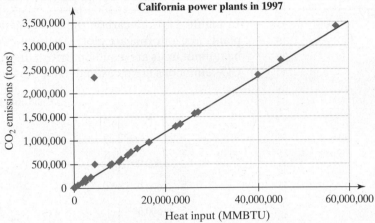

Line fitted to the California power plant data.

This is a very good fit considering we just used the first two data points. Such a good fit does not always happen.

b. Substituting $x = 4.488$ into our linear equation yields

$$y = 0.059(4.488) - 0.019 \approx 0.25 \times 10^6 \text{ tons of } CO_2$$

This is significantly smaller than the value of 2.3×10^6 tons of CO_2 given in the data. Thus, the power plant represented by this point on the graph pollutes almost ten times as much as it should compared with other power plants of similar energy output.

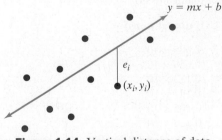

Figure 1.14 Vertical distance of data from a line.

Sometimes we can get a good fit to data by appropriately choosing two data points and finding the line that passes through these points. However, this method is quite ad hoc, because it depends on the two points selected and thus yields many different possible lines. Statisticians have solved this problem by inventing a method called **linear regression**. It is used to find a line that best fits that data in the following sense: The slope parameter m and y-intercept parameter b are chosen to minimize the sum of the squared vertical distances e_i of the data from the line (see Figure 1.14). The values e_i are called the *residuals* because they represent "what is left over once the linear fit has been taken into account."

Why squared distances? To find the answer to this question and to learn the statistical underpinnings of linear regression, you need to take an introductory statistics course! However, we note without further details (see any elementary statistics text for details) that a *sum-of-squares* measure of the fit leads to relatively simple formulas for the slope and y-intercept of the best-fitting line (which can be easily computed with calculators, computer software, and on line web applications). Part of this simple formula is derived in Chapter 4 as an application of differentiation.

Example 4 CO$_2$ concentrations in Hawaii

Table 1.4 describes how CO_2 concentrations (in ppm) have varied from May 1974 to December 1985 at the Mauna Loa Observatory in Hawaii. A plot of these data (where time is measured in months) was given by Example 1 of Section 1.1.

Table 1.4 CO_2 concentrations at the Mauna Loa Observatory of Hawaii

Month	CO$_2$	Month	CO$_2$	Month	CO$_2$	Month	CO$_2$	Month	CO$_2$	Month	CO$_2$
1	333.2	25	334.8	49	338.0	73	341.5	97	344.3	121	347.5
2	332.1	26	334.1	50	338.0	74	341.3	98	343.4	122	346.8
3	331.0	27	332.9	51	336.4	75	339.4	99	342.0	123	345.4
4	329.2	28	330.6	52	334.3	76	337.8	100	339.8	124	343.2
5	327.4	29	329.0	53	332.4	77	336.0	101	337.9	125	341.3
6	327.3	30	328.6	54	332.3	78	336.1	102	338.1	126	341.5
7	328.5	31	330.1	55	333.8	79	337.2	103	339.3	127	342.8
8	329.5	32	331.6	56	334.8	80	338.3	104	340.7	128	344.4
9	330.7	33	332.7	57	336.2	81	339.4	105	341.5	129	345.0
10	331.4	34	333.2	58	336.7	82	340.5	106	342.7	130	345.9
11	331.8	35	334.9	59	337.8	83	341.7	107	343.2	131	347.5
12	333.3	36	336.0	60	339.0	84	342.5	108	345.2	132	348.0
13	333.9	37	336.8	61	339.0	85	343.0	109	345.8	133	348.7
14	333.4	38	336.1	62	339.2	86	342.5	110	345.4	134	348.1
15	331.8	39	334.8	63	337.6	87	340.8	111	344.0	135	346.6
16	329.9	40	332.5	64	335.5	88	338.6	112	342.0	136	344.6
17	328.6	41	331.3	65	333.8	89	337.0	113	340.0	137	343.0
18	328.5	42	331.2	66	334.1	90	337.1	114	340.2	138	342.9
19	329.3	43	332.4	67	335.3	91	338.5	115	341.4	139	344.2
20	*	44	333.5	68	336.7	92	339.9	116	343.0	140	345.6
21	331.7	45	334.7	69	337.8	93	340.9	117	343.9		
22	332.7	46	335.2	70	338.3	94	341.7	118	344.6		
23	333.5	47	336.5	71	340.1	95	342.8	119	345.2		
24	334.8	48	337.8	72	340.9	96	343.7	120	347.1		

Data source: Komhyr, W. D., Harris, T. B., Waterman, L. S., Chin, J. F. S. and Thoning, K. W. (1989), Atmospheric carbon dioxide at Mauna Loa Observatory 1. NOAA Global monitoring for climatic change measurements with a nondispersive infrared analyzer, 1974–1985; *J. Geop. Res.*, v. 94, no. D6, pp. 8533–8547.

a. Find the best-fitting line to the CO_2 data. Plot this line against the data.

b. Determine at what rate (in ppm/year) the concentration of CO_2 has been increasing.

c. Estimate the CO_2 concentration for December 2004 using your best-fitting line. How does this compare with the average level of 338 ppm over the period May 1974 to December 1985?

d. For the CO_2 concentration in each data point, subtract the CO_2 concentration predicted by the best-fitting line. Plot the resulting residuals. What do you notice?

Solution

a. Entering the data into a graphing calculator or to a computer spreadsheet, and then running a linear regression routine, yields the best-fitting curve

$$y = 0.1225x + 329.3$$

STOP: Do not just read this—do it! Plotting this line against the data results in Figure 1.15.

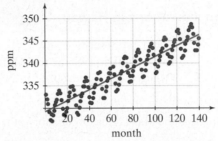

Figure 1.15 Best-fitting line for CO_2 at the Mauna Loa Observatory in Hawaii.

b. Since the slope of the line is 0.1225, the rate that CO_2 concentration has been increasing on average is 0.1225 ppm/month. Multiplying by 12 yields an annual rate of 1.47 ppm/year.

c. The number of months between December 2004 and May 1974 is $12 \cdot 30 + 8 = 368$. Substituting $x = 368$ into the best-fitting line yields a prediction of

$$y = 0.1225 \cdot 368 + 329.3 \approx 374.4$$

The estimated CO_2 concentration for December 2004 is 374.4 ppm. This is $374.4 - 338 = 36.4$ ppm higher than the average level from May 1974 to December 1985.

d. Subtracting the best-fitting line from the data and plotting the first five years yields Figure 1.16. This figure illustrates that with the removal of the linearly increasing trend, the residual CO_2 concentrations exhibit reasonably well-defined oscillations that seem to roughly repeat themselves over a twelve-month cycle.

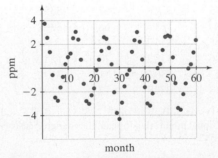

Figure 1.16 Residuals for the CO_2 at the Mauna Loa Observatory in Hawaii once the values predicted by the best-fitting line have been subtracted from the data.

Periodic and trigonometric functions

Many biological and physical time series exhibit oscillatory behavior, as just shown by Example 4. These types of data sets can be described by *periodic functions* that repeat their values at evenly spaced intervals. More formally, we make the following definition.

Periodic Function

A real-valued function f is **periodic** if there is a real number $T > 0$ such that

$$f(x) = f(x + T)$$

for all x. The smallest possible value of T is called the **period** of f. The **amplitude** (if it exists) of a periodic function is half of the difference between its largest and smallest values.

Example 5   Estimating periods and amplitudes

Estimate the period and amplitude for the CO_2 data in Figure 1.16.

Solution   A quick examination of the CO_2 data reveals that the time between peaks is approximately twelve months, so the period is a year. From the plot of the residuals in Figure 1.16, we see that the largest values of the data seem to be around 3 ppm, while the smallest values are typically around -3 ppm. Hence, the amplitude is approximately $(3 - (-3))/2 = 3$ ppm. ∎

Two important periodic functions that you have encountered previously in pre-calculus mathematics studies are the **cosine** and **sine** functions. The cosine function, $y = \cos x$, is defined for all reals and has a range of $[-1, 1]$. Hence, the amplitude of cosine is 1. As with all trigonometric functions used here, we assume x is measured in radians. You may recall the an angle of $90°$ is a right angle, which is equal to $\pi/2$ radians. Consequently, the full period of the sine and cosine function is 2π. In other words, $\sin(x) = \sin(x + 2\pi)$ and $\cos(x) = \cos(x + 2\pi)$ for any value of x measured in radians. The graphs of both functions are shown in Figure 1.17. Since the graph of sine is the graph of cosine shifted to the right by $\pi/2$, it follows that

$$\sin x = \cos\left(x - \frac{\pi}{2}\right)$$

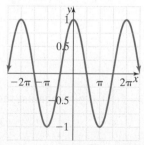
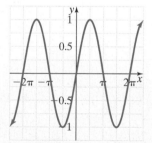

a. cosine curve period: 2π; amplitude: 1   **b. sine curve** period: 2π; amplitude: 1

Figure 1.17 Graphs of cosine and sine.

Curves with the shape of sine or cosine functions are called **sinusoidal**. An important two-parameter family of sinusoidal functions are functions of the form

$$y = f(x) = a \cos(bx)$$

where a is a real number and b is a nonzero real number. Since the range of $f(x)$ is $[-|a|, |a|]$, the amplitude of $f(x)$ is $|a|$. To find the period $T > 0$ of $f(x)$, we need to find the smallest $T > 0$ such that

$$a \cos(bx) = a \cos(b(x + T)) = a \cos(bx + bT)$$

This occurs when $bT = 2\pi$. Therefore, the period of f is $2\pi/b$. In the following example, we put this information to use.

Example 6   Fitting the CO_2 data

Consider

$$y = f(x) = a\cos(bx)$$

where a and b are positive constants.

a. Write an equation $f(x)$ that provides a good fit to the CO_2 residual data shown in Figure 1.16 from the Mauna Loa Observatory and plot this equation against the given data.

b. Let $g(x) = 0.1225x + 329.3$ be the best-fitting line shown in Figure 1.15 and $f(x)$ the equation you have just obtained. Plot $h(x) = f(x) + g(x)$ against the data shown in Table 1.4. Use h to predict the carbon dioxide level in March 2006, and compare to what you find online.

Solution

a. We found in Example 5 that the amplitude for the data in Figure 1.16 is approximately 3 ppm and the period is 12 months. Therefore, we need to choose $a = 3$ and find b. To get a period of 12 months, we need $2\pi/b = 12$. Therefore, $b = \pi/6$ and we get the equation

$$f(x) = 3\cos\left(\frac{\pi}{6}x\right)$$

The graph of this equation against the data is shown below.

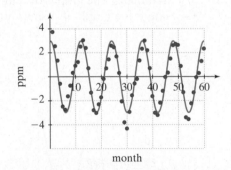

b. Plotting $h(x) = f(x) + g(x)$ against the data yields the following graph.

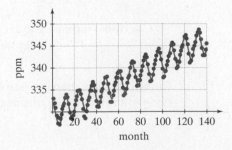

A truly remarkable fit! Next, calculate $h(12 \cdot 31 + 11) = h(383) = f(383) + g(383) = 2.60 + 376.2 = 378.8$. According to one website, the March measurement was 381 ppm. Hence, CO_2 may well be increasing slightly faster than predicted by the model, possibly due to an accelerating rate of CO_2 emissions.

Other important trigonometric functions that you will encounter in this book are given by taking either reciprocals or ratios of the sine and cosine functions. For example, the tangent function is

$$y = \tan x = \frac{\sin x}{\cos x}$$

Because $\cos x = 0$ for odd integer multiples of $\pi/2$, the domain of tangent is all real numbers except these odd integer multiples of $\pi/2$. Furthermore, as we discuss further in Chapter 2, $\tan x$ approaches $+\infty$ as x approaches $\pi/2$ from the left and approaches $-\infty$ as x approaches $\pi/2$ from the right. Consequently, the range of tangent is the entire reals. Also it does not have a well-defined amplitude but is periodic, with period π, as shown in Figure 1.18. Like tangent, the other trigonometric functions—namely, cotangent $\cot x = \dfrac{\cos x}{\sin x}$, secant $\sec x = \dfrac{1}{\cos x}$, and cosecant $\csc x = \dfrac{1}{\sin x}$—are not defined for all reals, have no well-defined amplitude, but also have a well-defined period as shown in Figure 1.18. Since these functions are all expressed in terms of the sine and cosine functions, and cosine is the sine function with a $\pi/2$ shift in its argument, the properties of all the trigonometric functions can be directly deduced from the properties of the sine function.

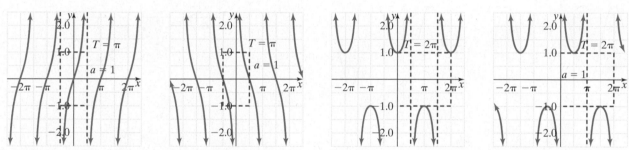

Figure 1.18 Graphs of the trigonometric functions $y = \tan(x)$, $y = \cot(x)$, $y = \sec(x)$, and $y = \csc(x)$.

PROBLEM SET 1.2

Level 1 DRILL PROBLEMS

Solve for y as a function of x and graph the resulting function for Problems 1 to 10.

1. $5x - 4y - 8 = 0$

2. $x - 3y + 2 = 0$

3. $100x - 250y + 500 = 0$

4. $2x - 5y - 200 = 0$

5. $3x + y - 2 = 0, \quad -7 \le x \le 1$

6. $2x - 2y + 6 = 0, \quad 1 \le x \le 5$

7. $y = 4\cos x$

8. $y = \cos(4x)$

9. $y = (\sin x)/2, \quad -8 \le x \le 8$

10. $y = \sin(x/2), \quad -8 \le x \le 8$

Using the information in Problems 11 to 16, find the formula for the line $y = mx + b$.

11. Slope 3, passing through $(1, 3)$

12. Slope $\dfrac{2}{5}$; passing through $(5, -2)$

13. Passing through $(-1, 2)$ and $(0, 1)$

14. Passing through $(5, 6)$ and $(7, 6)$

15. y-intercept 4 passing through $(3, 4)$

16. horizontal line through $(-2, 5)$

17. Show that
$$y - k = m(x - h)$$
is the equation of the line passing through the point (h, k) with slope m.

18. Derive the equation of vertical line passing through (h, k). Does this set of points represent a function?

Classify each graph in Problems 19 to 24 as a linear function or a periodic function. If it is linear, estimate the slope and write an equation of the form $y = mx + b$. If it is periodic, estimate the period and the amplitude and write an equation of the form $y = a\cos(bx)$, $a > 0$.

A

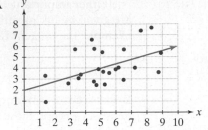

19.

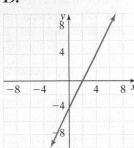

20.

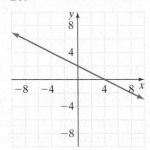

B

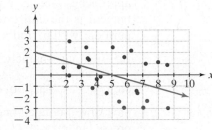

21.

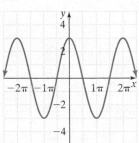

22.

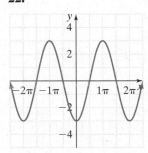

C

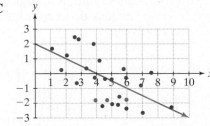

23.

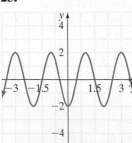

24.

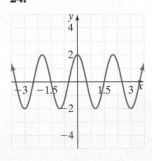

D

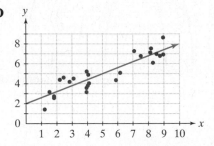

E

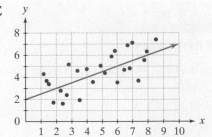

Match the equations in Problems 25 to 30 along with the scatter diagrams and best-fitting lines in figure panels A–F below.

25. $y = 0.6x + 2$

26. $y = 0.5x + 2$

27. $y = 0.4x + 2$

28. $y = -0.4x + 2$

29. $y = -0.5x + 2$

30. $y = -0.7x + 2$

F

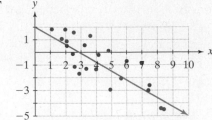

Consider some standard trigonometric curves shown in Figure 1.19. Specify the period and amplitude for each graph (if exists) in Problems 31 to 36, and graph each curve.

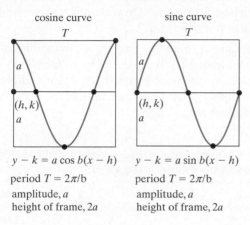

cosine curve

sine curve

$y - k = a \cos b(x - h)$

period $T = 2\pi/b$

amplitude, a

height of frame, $2a$

$y - k = a \sin b(x - h)$

period $T = 2\pi/b$

amplitude, a

height of frame, $2a$

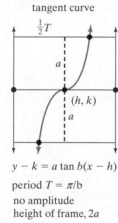

tangent curve

$y - k = a \tan b(x - h)$

period $T = \pi/b$

no amplitude

height of frame, $2a$

Figure 1.19 Standard cosine, sine, and tangent curves.

31. $y = \dfrac{1}{2} \cos(x + \pi/6)$

32. $y = 2 \sin(x - \pi/4)$

33. $y = 2 \sin 2\pi x$

34. $y = 3 \cos 3\pi x$

35. $y = \tan(2x - \pi/2)$

36. $y = \tan(x/2 + \pi/3)$

Level 2 APPLIED AND THEORY PROBLEMS

37. A life insurance table indicates that a woman who is now A years old can expect to live E years longer. Suppose that A and E are linearly related and that $E = 50$ when $A = 24$ and $E = 20$ when $A = 60$.

   a. At what age may a woman expect to live 30 years longer?

   b. What is the life expectancy of a newborn female child?

   c. At what age is the life expectancy zero?

38. In certain parts of the world, the number of deaths N per week has been observed to be linearly related to the average concentration x of sulfur dioxide in the air. Suppose there are 97 deaths when $x = 100$ mg/m^3 and 110 deaths when $x = 500$ mg/m^3.

   a. What is the functional relationship between N and x?

   b. Use the function in part **a** to find the number of deaths per week when $x = 300$ mg/m^3. What concentration of sulfur dioxide corresponds to 100 deaths per week?

   c. Research data on how air pollution affects the death rate in a population. You may find the following articles helpful: D. W. Dockery, J. Schwartz, and J. D. Spengler, "Air Pollution and Daily Mortality: Associations with Particulates and Acid Aerosols," *Environmental Research* 59 (1992): 362–373; Y. S. Kim, "Air Pollution, Climate, Socioeconomics Status and Total Mortality in the United States," *Science of the Total Environment* (1985): 245–256. Summarize your results in a one-paragraph essay.

39. The chart in Figure 1.20, based on data reported in the November 1987 issue of *Scientific American*, shows the fat intake compared with death rates in various regions around the world.

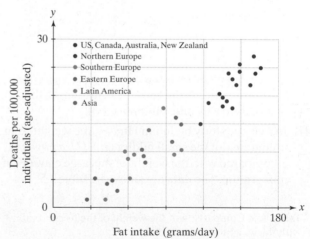

Figure 1.20 Fat intake and age adjusted death rates due to cancer for countries in various regions of the world.

Source: Data are selected from the following article: Leonard Cohen, "Diet and Cancer," *Scientific American*, November 1987, p44.

Which of the following lines best fits the data?

   A. $y = 0.139x$

   B. $y = 0.231x - 3$

   C. $y = 0.981x + 1$

Use your choice to estimate the number of deaths per 100,000 population to be expected from an average fat intake of 150 g/day (roughly the fat intake in the United States during the period the data was collected).

40. The chart in Figure 1.21 shows a comparison of Foude number with stride length for humans, kangaroos, and others.

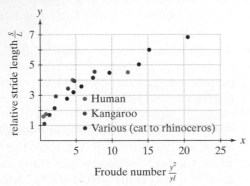

Figure 1.21 Comparison of Froude number with stride length.

Data Source: Graph by Patricia J. Wynne, from "How Dinosaurs Ran," by R. McNeill Alexander, *Scientific American*, April 1991, p. 132 ©1991 by Scientific American, Inc. All rights reserved.

It can be shown that the best-fitting line is one of the following:

A. $y = 0.31x$

B. $y = 0.221x + 2$

C. $y = 0.29x + 1$

Which do you think is the correct one? Use your choice to estimate the relative stride length that corresponds to a Froude number $x = 4$.

41. In a classic study by Julian Huxley, the weight X, in milligrams, of the small fiddler crab (*Uca pugnax*) is compared with the weight of the large claw (Y, in milligrams). The data are shown in Table 1.5.

Table 1.5 Comparison of the weight of the fiddler crab with the weight of its large claw

X	Y	X	Y
57.6	5.3	355.2	104.5
80.3	9.0	420.1	135.0
109.2	13.7	470.1	164.9
156.1	25.1	535.7	195.6
199.7	38.3	617.9	243.0
238.3	52.5	680.6	271.6
270.0	59.0	743.3	319.2
300.2	78.1		

Data source: Julian S. Huxley, 1932. Problems of Relative Growth. Reprinted by The Johns Hopkins University Press, 1993.

a. Plot the points in the table. Does this look like a linear model to you?

b. Plot the line $y = 0.47x - 49$ on the axis for the points you plotted in part **a**. Does this look like a best-fitting line? Do you think you can find a better fitting line?

42. The data in Table 1.6 compare the mandibles of the male stag-beetle (*Cyclommatus tarandus*) where X is the total length (body and mandibles) in millimeters and Y is the length of the mandibles in millimeters.

Table 1.6 Comparison of body weight with the length of the mandibles of the male stag-beetle

X	Y	X	Y
20.38	3.88	36.13	12.08
24.01	5.31	37.32	12.73
26.38	6.33	38.44	14.11
27.76	7.32	39.26	14.70
29.65	8.17	41.34	15.84
32.20	9.73	43.22	17.39
33.11	10.71	45.51	18.83
35.01	11.49	46.32	19.19

a. Plot these points. Does this look like a linear model to you?

b. Plot the line $y = 0.62x - 9.7$ on the axis for the points you plotted in part **a**. Does this look like a best-fitting line? Do you think you can find a better-fitting line?

43. Table 1.7 shows the census figures (in millions) for the U.S. population since the first census.

Table 1.7 U.S. population

Year	Population	Year	Population
1780	2.8	1900	76.0
1790	3.9	1910	92.0
1800	5.3	1920	105.7
1810	7.2	1930	122.8
1820	9.6	1940	131.7
1830	12.9	1950	150.7
1840	17.1	1960	179.3
1850	23.2	1970	203.3
1860	31.4	1980	226.5
1870	39.8	1990	248.7
1880	50.2	2000	281.4
1890	62.9	2010	310.5

a. Plot these points where 1780 represents $x = 0$. Does this look like a linear model can provide a good fit?

b. Plot the line $y = 1.3x - 50$ on the axis for the points you plotted in part **a**. Does this look like a best-fitting line? Do you think you can find a better-fitting line?

44. Ethyl alcohol is metabolized by the human body at a constant rate (independent of concentration). Suppose the rate is 10 milliliters per hour.

    a. Express the time t (in hours) required to metabolize the effects of drinking ethyl alcohol in terms of the amount A of ethyl alcohol consumed (in milliliters).

    b. How much time is required to eliminate the effects of a liter of beer containing 3% ethyl alcohol?

    c. Discuss how the function in part **a**. can be used to determine a reasonable "cutoff" value for the amount of ethyl alcohol A that each individual may be served at a party.

45. In a 1971 study by Savini and Bodhaine, data for velocity of water versus depth were collected for the Columbia River below Grand Coulee Dam. The data are reported in Table 1.8 and were measured 13 feet from the shoreline.

    Table 1.8 Depth and flow of Grand Coulee Dam

Depth (feet)	Velocity (feet/second)
0.7	1.55
2.0	1.11
2.6	1.42
3.3	1.39
4.6	1.39
5.9	1.14
7.3	0.91
8.6	0.59
9.9	0.59
10.6	0.41
11.2	0.22

    Data source: Savini, J. and Bodhaine, G. L. (1971), Analysis of current meter data at Columbia River gaging stations, Washington and Oregon; USGS Water Supply Paper 1869-F.

    a. Plot these points.

    b. Find the lines defined by the first two data points and the first and last data points. Plot these lines against the data and decide which fits the data better.

    c. Use technology to find the best-fitting data line.

    d. Estimate the velocity of the river at a depth of 12 feet and 20 feet. Discuss the answers you obtain.

46. Eighty-eight samples of shells of the native butter clam (*Saxidomus giganteus*) were collected.

These clams grow to lengths of 12–13 cm and live for more than 20 years. A scatter plot of their data is given in Figure 1.22.

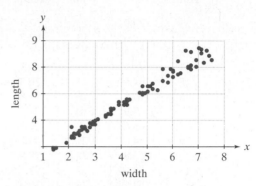

Figure 1.22 Plot of length and width of clam samples.
Source: **Quantitative Environmental Learning Project (QELP)** Web Site at http://seattlecentral.edu/qelp/index.html.

    a. A pair of points on this data set are given by (1.3, 1.7) and (7.3, 8.9). These two points are drawn in black in the Figure 1.22. Sketch the line passing through these points and find the formula for this line.

    b. Use your line to estimate the width of a butter clam whose length is 12 cm.

47. Temperature fluctuations in many parts of the world exhibit sinusoidal patterns. Consider, for example, the average monthly temperature in Chappaqua, New York, reported in Table 1.9.

    Table 1.9

Month	Temp.	Month	Temp.
August	77	February	35.5
September	70	March	42.5
October	59	April	53
November	48.5	May	63
December	39	June	72.5
January	35	July	77.5

    a. Plot these data.

    b. Approximate the amplitude of the data and the period.

    c. Find a and b such that $56.25 + a \cos(bx)$ has the same amplitude and period as your data, and plot this function against your data. Let x represent months after July.

48. Problem 47 illustrates temperature oscillations on a yearly time scale corresponding to the seasons. Temperatures also vary daily: cooler at night and warmer at noon. Consider, for example, the average

July hourly temperature in Abeerdeen, New York, shown in the following table.

Hour	Temp.	Hour	Temp.
1	79.9	13	62.3
2	78.6	14	62.7
3	76.4	15	65.1
4	73	16	67.7
5	69.9	17	70.3
6	67.8	18	72.8
7	65.9	19	76
8	66.6	20	76.8
9	64.6	21	77.7
10	63.8	22	78.9
11	63	23	79.7
12	62.1	24	79.8

a. Plot these data.

b. Approximate the amplitude of the data and the period.

c. Find a and b such that $70.8 + a \cos(bx)$ has the same amplitude and period as your data, and plot this function against your data. Let x represent months after July.

1.3   Power Functions and Scaling Laws

Why can an ant lift a hundred times its weight whereas a typical man can only lift about six-tenths of his weight? Why is getting wet life-threatening for a fly but not for a human? Why can a mouse fall from a skyscraper and still scurry home, while a human who falls is likely to be killed? Why are elephants' legs so much thicker relative to their length than are gazelles' legs? A class of functions called *power functions* provides a means to answer these questions.

Power functions and their properties

Power Functions

A function $f(x)$ is a **power function** if it is of the form

$$y = f(x) = ax^b$$

where a and b are real numbers. The variable x is called the **base**, the parameter b is called the **exponent**, and the parameter a is called the **constant of proportionality**.

Note that $\frac{5}{7}x^{-1}$ and x^3 are power functions, while 3^x is not because, in this latter case, the exponent rather than the base is the variable.

Example 1   Graphing power functions

Graph each of the following sets of functions and discuss how they differ from one another and what properties they have in common.

a. $y = x^2$, $y = x^4$, and $y = x^6$

b. $y = x^3$, $y = x^5$, and $y = x^7$

c. $y = x^{1/2}$, $y = x$, and $y = x^{3/2}$

d. $y = \dfrac{1}{x}$ and $y = \dfrac{1}{x^2}$

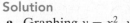

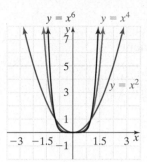

Solution

a. Graphing $y = x^2$, $y = x^4$, and $y = x^6$ gives the figure on the left.

All of these graphs tend to "bend" upward and are "U-shaped." All three of these graphs intersect at the points $(0, 0)$, $(-1, 1)$, and $(1, 1)$. On the interval $[-1, 1]$ the function with the smallest exponent grows most rapidly as you move away from $x = 0$, and on the intervals $(-\infty, 1)$ and $(1, \infty)$ the function with the largest exponent increases most rapidly.

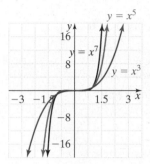

b. Graphing $y = x^3$, $y = x^5$, and $y = x^7$ gives the figure on the left.

All of these graphs are "seat shaped," bending downward for negative x and bending upward for positive x. All three of these graphs intersect at the points $(0, 0)$, $(-1, -1)$ and $(1, 1)$. On the interval $[-1, 1]$ the function with the smallest exponent grows most rapidly as you move away from $x = 0$, and on the intervals $(-\infty, 1)$ and $(1, \infty)$ the function with the largest exponent grows most rapidly.

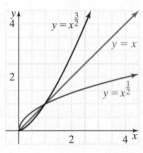

c. Graphing $y = x^{1/2}$, $y = x$, and $y = x^{3/2}$ gives the figure on the left.

We graphed over the domain $[0, \infty)$ of $y = x^{1/2}$ and $y = x^{3/2}$. All of these graphs increase as x increases and pass through the points $(0, 0)$ and $(1, 1)$. The graph of $x^{1/2}$ becomes steeper and steeper at 0, while the graph of $x^{3/2}$ becomes flatter and flatter. Moreover, the graph of $x^{1/2}$ bends downward, while the graph of $x^{3/2}$ bends upward.

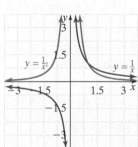

d. Graphing $y = \dfrac{1}{x}$ and $y = \dfrac{1}{x^2}$ gives the figure on the left.

Both of these functions pass through the point $(1, 1)$ and approach positive or negative infinity (i.e., have a vertical asymptotes) as x approaches 0, although the function $\dfrac{1}{x}$ does so by approaching $-\infty$ in the third quadrant while the "other branch" of $\dfrac{1}{x^2}$ approaches $+\infty$ in the second quadrant. Branches of the graphs lying above the x axis, bend upward, while parts lying below bend downward. ∎

To algebraically manipulate power functions, we review some properties of exponents.

Laws of Exponents

Let x, y, a, and b be any real numbers. Then provided that both sides of the equality are well defined, the following five rules govern the use of exponents:

1. Addition law: $x^a x^b = x^{a+b}$

2. Subtraction law: $\dfrac{x^a}{x^b} = x^{a-b}$

3. Multiplication law: $(x^a)^b = x^{ab}$

4. Distributive law over multiplication: $(xy)^a = x^a y^a$

5. Distributive law over division: $\left(\dfrac{x}{y}\right)^a = \dfrac{x^a}{y^a}$

Example 2   Using Laws of Exponents

Simplify the following expressions using the laws of exponents.

a. $\dfrac{x^2}{x}$ **b.** $(x^3)^{1/3}x$ **c.** $\left(\dfrac{1}{\sqrt{x}}\right)^4$

Solution

a. Since $\frac{1}{x} = x^{-1}$, we obtain

$$\frac{x^2}{x} = x^2 x^{-1}$$
$$= x^{2-1} = x \qquad \text{by the addition law for } x \neq 0$$

b. We have

$$(x^3)^{1/3}x = x^{3/3}x \qquad \text{by the multiplication law}$$
$$= x^1 x^1 = x^2 \qquad \text{by the addition law}$$

c. Since $\sqrt{x} = x^{1/2}$, we have

$$\left(\frac{1}{\sqrt{x}}\right)^4 = (x^{-1/2})^4 \qquad \text{by the subtraction law for } x > 0$$
$$= x^{-2} \qquad \text{by the multiplication law for } x > 0$$

Proportionality and geometric similarity

In his essay, "On Being the Right Size,"* John B. S. Haldane (1892–1964), noted biologist and one of the founders of the field of population genetics, wrote:

"A man coming out of a bath carries with him a film of water of about one-fiftieth of an inch in thickness. This weighs roughly a pound. A wet mouse has to carry about its own weight in water. A wet fly has to lift many times its own weight and, as everybody knows, a fly once wetted by water or any other liquid is in a very serious position indeed."

You might wonder how Haldane came up with these conclusions. To see why, consider power laws in the context of proportionality, which is defined as follows.

Proportionality
We say that y is **proportional** to x if there exists some constant $a > 0$ such that $y = a x$ for all $x > 0$. When y is proportional to x, we write $$y \propto x$$

Example 3   Geometric similarity

Figure 1.23 A cubical critter (a C^3).

Imagine a world in which all individuals are cubical critters of different types: one such critter is drawn in Figure 1.23. The size of each critter can be characterized using one measurement, L meters, which denotes the length of the critter in any of its three dimensions.

a. Argue that the surface area, S, and volume, V, of the cubical critter are proportional to L^b for appropriate choices of b.

b. If we assume that these cubical critters are essentially "ugly bags of mostly water" (*Star Trek* fans may remember this line as an alien's description of humans who are mostly water encased in a bag of skin; the "ugly" part is a matter of extraterrestrial taste), argue that body

---

*John B. S. Haldane, "On Being the Right Size," *The Harper's Monthly*, March 1926, pp. 424–427.

mass, M, is also proportional to L^b for an appropriate choice of b. In your argument, you may use the fact that 1 cubic meter of water has a mass of 1000 kilograms.

Solution

a. Since the surface area of a cube is $6L^2$ and the volume of a cube is L^3,

$$S \propto L^2 \qquad V \propto L^3$$

In other words, surface area is proportional to length squared and volume is proportional to length cubed.

b. Since we are assuming the cubical critters are made of water and the density of water is 1000 kg/m^3, the mass is $M = 1000 \cdot V = 1000 \cdot L^3$. Hence,

$$M \propto L^3$$

Notice that this proportionality would not change even if we used a different density constant. ■

Geometrical similarity is not confined to cubical critters. So long as all dimensions of an organism scale in the same way, the organisms are geometrically similar. Moreover, for any measurement of length L (e.g., height, arm length, chest circumference), surface area S (e.g., palm surface area, cross-sectional area of a muscle), and mass M (e.g., mass of a hair or the entire body), the relationships $S \propto L^2$ and $M \propto L^3$ continue to hold.

To work with proportionality relationships, we need to remember a few basic rules. Essentially these rules have this effect: we can treat a proportionality symbol for manipulative purposes like an equality sign.

Rules of Proportionality.

- **Transitive property:** If $x \propto y$ and $y \propto z$, then $x \propto z$.
- **Power-to-root property:** If $y \propto x^b$ with $b \neq 0$, then $x \propto y^{1/b}$.
- **General transitive property:** If $x \propto y^b$ and $y \propto z^c$, then $x \propto z^{bc}$.

Example 4   Rules of proportionality
- -

Demonstrate that proportionality satisfies the properties listed in the box above.

Solution

Transitive property: Since $x \propto y$, then there exists a constant $a > 0$ such that $x = ay$. Since $y \propto z$, then there exists a constant $b > 0$ such that $y = bz$. Therefore,

$$x = ay = a(bz) = (ab)z$$

This equality implies that $x \propto z$ with proportionality constant ab.

Power-to-root property: If $y \propto x^b$, then there exists a constant $a > 0$ such that $y = ax^b$. Solving for x in terms of y yields

$$x = \left(\frac{y}{a}\right)^{1/b} = a^{-1/b}y^{1/b}$$

Hence, $x \propto y^{1/b}$ with proportionality constant $a^{-1/b} > 0$.

General transitive property: This property is really just a simple extension of the transitive property, but it is easily demonstrated directly. If $x \propto y^b$ and $y \propto z^c$, then there exist $a_1 > 0$ and $a_2 > 0$ such that $x = a_1 y^b$ and $y = a_2 z^c$. Therefore, $x = a_1 (a_2 z^c)^b = a_1 a_2^b z^{bc}$. Hence, $x \propto z^{bc}$ with proportionality constant $a_1 a_2^b$. ■

Example 5   Dangers of getting wet

To understand the dangers of getting wet, it is reasonable to assume that the mass, W, of the water on your body of mass M after getting wet is proportional to the surface area, S, of your body.

a. For cubical critters find the value of b such that $W \propto M^b$.

b. Suppose you had two cubical critters: a man-sized cubical critter with mass 60 kg, and a mouse-sized cubical critter with mass 0.01 kg. Moreover, assume when the man gets wet, the mass of water clinging to his skin is 0.6 kg. Using proportionality, find the mass of water on the mouse. Compare the ratios $\dfrac{W}{M}$ for the two critters.

c. Graph the ratio $\dfrac{W}{M}$ as a function of M and discuss its implications for the danger of getting wet.

Solution

a. To solve this problem, we use the rules of proportionality found in Example 4. Since we have assumed that $W \propto S$ and $S \propto L^2$, the transitive property implies $W \propto L^2$. Since $M \propto L^3$, the power-to-root property implies $M^{1/3} \propto L$. The general transitive property implies that

$$S \propto L^2 \propto (M^{1/3})^2 = M^{2/3}$$

In other words, $W \propto M^b$ for $b = 2/3$.

b. Since W is proportional to $M^{2/3}$, there exists some number $a > 0$ such that

$$W = a M^{2/3}$$

The man-sized cubical critter has mass $M = 60$ with $W = 0.6$. Substituting these values into $W = a M^{2/3}$ allows us to solve for a:

$$0.6 = a60^{2/3}$$
$$a \approx 0.04$$

The ratio of water mass to body mass for the man is

$$\frac{W}{M} = \frac{0.6}{60} = 1\%$$

The mouse-sized critter has mass $M = 0.01$ kg with $W = 0.04(0.01)^{2/3} \approx 0.00186$ kg. The ratio of water mass to body mass for the mouse is

$$\frac{W}{M} \approx \frac{0.00186}{0.01} \approx 18.6\%$$

We see that the wet cubical man has to lift only 1% of his body mass while the wet cubical mouse has to lift approximately 19% of its body mass.

Note that this calculation does not take into account that a mouse is hairier than a man and therefore likely to retain more water per surface area of body than a man. This calculation assumes that the retention properties for each unit of surface area are the same.

c. We have

$$\frac{W}{M} = \frac{a M^{2/3}}{M} = a M^{-1/3} \approx 0.04 M^{-1/3}$$

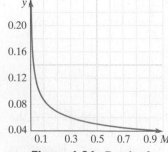

Figure 1.24 Graph of $y = 0.04\,M^{-1/3}$.

The graph of the ratio of water mass to body mass, $y = 0.04 M^{-1/3}$, is shown in Figure 1.24.

This graph illustrates that the bigger creature (i.e., M becomes larger), the amount of water one has to carry relative to one's body mass decreases. Hence, getting wet is much worse for a fly than a human. ◼

The previous example shows how we can use *geometric similarity* to understand the implications of getting wet for critters of vastly different sizes—from humans to flies. In Problem 43 of Problem Set 1.3, we pose a counterpoint analysis of how smaller animals are favored when it comes to the dangers of falling from high places. Although it is true that organisms are often geometrically quite dissimilar, it turns out that in many cases analyses using the approximation of geometric similarity are quite good.

Example 6   Olympic weightlifting

The heaviest weight classes are excluded, as individuals in this class have no weight restriction and therefore are often not geometrically similar to their lighter counterparts.

Table 1.10 reports the body mass and the winning lifts (in kilograms) for the male gold medalists in the 1988, 1992, and 1996 Olympic Games. In this example, we develop a simple model relating body mass to mass lifted.

Table 1.10 Body mass versus winning lift (Olympic gold medalists)

| | 1988 | | | 1992 | | | 1996 | |
Class	Mass	Lift	Class	Mass	Lift	Class	Mass	Lift
≤ 52	51.85	270.0	≤ 52	51.8	265	≤ 54	53.91	287.5
≤ 56	55.75	292.5	≤ 56	55.9	287.5	≤ 59	58.61	307.5
≤ 60	59.7	342.5	≤ 60	59.9	320	≤ 64	63.9	335
≤ 67.5	67.2	340.0	≤ 67.5	67.25	337.5	≤ 70	69.98	375.5
≤ 75	74.8	375.0	≤ 75	74.5	357.5	≤ 76	75.91	367.5
≤ 82.5	82.15	377.5	≤ 82.5	81.8	370	≤ 83	82.06	392.5
≤ 90	89.45	412.5	≤ 90	89.25	412.5	≤ 91	90.89	402.5
≤ 100	99.7	425.0	≤ 100	97.25	410	≤ 99	96.78	420
≤ 110	109.55	455.0	≤ 110	109.4	432.5	≤ 108	107.32	430

a. A basic physiological principle is that the strength of a muscle is proportional to the cross-sectional area of that muscle. Assuming that Olympic male weightlifters are geometrically similar, as illustrated in Figure 1.25, argue that for a lifter of mass M the amount ℓ he can lift is proportional to M^b, and find the value of b for which this should be true.

Heavy weight   Middle weight   Light weight

Lars Baron/Getty Images, Inc.

Figure 1.25 Geometrically similar weightlifters.

b. The relationship $\ell \propto M^b$ from part **a** implies $\ell = a\,M^b$ for some $a > 0$. Find a using the data point $(\ell, M) = (287.5, 53.91)$ (table entry for the category ≤ 54 in year 1996 that leads to a good fit). Plot $\ell = a\,M^b$ for the values of a and b that you obtain.

c. Since the power law you find in part **a.** does a relatively good job of predicting lift as a function of body weight, you can use it to determine an overall winner among the weight classes. Namely, associate a score

$$y = \text{lift}/(\text{body mass})^b$$

with each weightlifter and declare the individual with the largest score to be the overall winner. Use this approach to find the overall winner in the 1988 Olympics.

Solution

a. Let L be a measurement of length (e.g., height), M the mass, and S the cross-sectional area of the weightlifter. Since we assume that weightlifters are geometrically similar, we have $M \propto L^3$ and $S \propto L^2$. Thus,

$$S \propto L^2 \propto (M^{1/3})^2 = M^{2/3}$$

Since we have assumed that $\ell \propto S$, we can conclude that

$$\ell \propto M^{2/3}$$

b. We substitute the data point $(\ell, M) = (287.5, 53.91)$ into the equation $\ell = a\,M^{2/3}$ to find

$$287.5 = a(53.91)^{2/3}$$
$$a \approx 20.15$$

The plot $\ell = 20.15\,M^{2/3}$ against the data as shown in Figure 1.26 illustrates a remarkable fit of the model to the data.

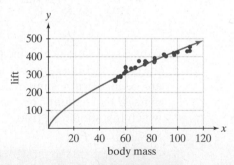

Figure 1.26 Graph showing data points and graph of $\ell = 20.15\,M^{2/3}$.

c. Calculating the individuals scores y for each of the 1988 Olympic lifters in Table 1.10, we get the values 19.42 $(= 270.0/51.85^{2/3})$, 20.04, 22.42, 20.57, 21.12, 19.98, 20.62, 19.77, 19.87. The overall winner here is the Gold Medal winner in the third lightest weight class with a score of 22.42. A quick search shows that this medal winner is Naim Süleymanoğlu, pictured in Figure 1.1, who has been nicknamed Pocket Hercules because of his feats of strength for his small size.

Allometric scaling

Although geometric similarity works wonders, it is not universal. When the shape of an animal (or organ or bone) deviates from geometrical similarity with size, then we say that it scales **allometrically** (*allo* = different, *metric* = measure) if we can find a power law that relates one particular measure (e.g., length of the mammalian femur) denoted by, say, x to another (e.g., cross-sectional area of the mammalian femur) denoted by, say, y as the size of an individual increases. In this case, the fundamental **allometric formula** posits the relationship

$$y = ax^b \qquad \text{or} \qquad y \propto x^b$$

for some constants $a > 0$ and b.

Haldane made the key observation in "On Being the Right Size" that a structure breaks when a load that is proportional to the volume of an organism (cubic dimension) acts on the cross-sectional area (square dimension) of the structure supporting this organism. The essence of this issue is presented in the next example.

Example 7   Breaking bones

a. From physics we know that the force per unit area at the base of a cube, which we denote here using the symbol K, is given by:

$$K = \text{gravitational acceleration} \times \frac{\text{density} \times \text{volume}}{\text{area}}$$

Calculate the dimensions of a sugar cube that would crush under its own weight at the surface of the Earth where the gravitational acceleration is 9.81 m/s^2, given that the sugar cube's density is 1040 kg/m^3 and its crushing strength (the maximum value of K that it can resist) is 5.17×10^6 newtons/m^2.

b. Thomas McMahon collected data on lengths L and diameters D of bones for various cloven-hoofed animals. If these animals were geometrically similar, we would expect $L \propto D$. However, the data suggest that $L \propto D^{2/3}$, as illustrated in Figure 1.27, where lengths are measures in millimeters (mm).

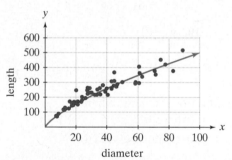

Figure 1.27 Lengths L and diameters D in millimeters of bones for various cloven-hoofed animals and fitted curve.

Data Source: T. A. McMahon (1975). "Allometry and Biomechanics: Limb Bones in Adult Ungulates." *The American Naturalist*, Vol. 109, No. 969, pp. 547–563.

In this data set, the humerus bone of an African impala has a length of 173 mm and a diameter of 22.5 mm. Use this information to estimate the length of a wildebeest humerus whose diameter is 42.6 mm.

Solution

a. From the formula, the force per unit area at the base of the sugar cube is $K = 9.81 \times 1040 L^3 / L^2 = 10202.4L$ newtons/m^2 where L is the length of one side of the base of the cube. Since the crushing strength of sugar is 5,170,000 newtons/m^2, the cube gets crushed under its own weight if

$$10202L \geq 5{,}170{,}000$$

$$L \geq 506.74 \text{ meters}$$

b. Assume that $L = aD^{2/3}$ where L is length and D is diameter measured in millimeters. For an African impala, we are given that $L = 173$ and $D = 22.5$. Solving for the proportionality constant a yields

$$173 = a(22.5)^{2/3}$$

$$a = 173/(22.5)^{2/3} \approx 21.7$$

Using the relationship $L = 21.7D^{2/3}$ with $D = 42.6$ yields $L \approx 264.7$ mm for the length of the wildebeest humerus. The actual value from the data set is 256. Hence, our estimate from the scaling law is not too bad. ∎

PROBLEM SET 1.3

Level 1 DRILL PROBLEMS

Simplify the functions in Problems 1 to 9, and determine whether the functions are power functions. If a function is a power function, write it in the form $y = ax^b$.

1. a. $y = \dfrac{x}{3}$ **b.** $y = \dfrac{1}{3x}$ **c.** $y = 3^x$

2. a. $y = 10$ **b.** $y = x^{10}$ **c.** $y = 10^x$

3. $y = \dfrac{1}{3} + \dfrac{1}{x}$ **4.** $y = \dfrac{2x + 15}{5x}$

5. $y = \dfrac{1}{\sqrt{16x^3}}$ **6.** $y = \dfrac{5\sqrt{x}}{7x^2}$

7. $y = 2^x 3^{2x} 5^x$ **8.** $y = \dfrac{\sqrt{36x}}{6x^5}$

9. $y = \dfrac{\sqrt{144x^3}}{2x^2}$ **10.** $y = (2x^3)^2$

11. If $y \propto x^2$ and y increases from 10^3 to 10^{15}, what happens to x?

12. If $y \propto 6x$ and $x \propto t$, how does t change when y increases from 2×10^2 to 6×10^4?

13. If $y \propto 10x^3$ how is y proportionally related to x?

14. If $x \propto 100y$ and $y \propto 45z$, then how does z change as x decreases from 95 to 12?

15. If $x \propto y^2$ and $y \propto z^3$, then how is x proportionally related to z?

16. If $x \propto \sqrt{y}$ and $y \propto z^2$, then how is x proportionally related to z?

Graph the functions in Problems 17 to 22. By inspection, state the intervals where the function is increasing and the intervals where it is decreasing.

17. $y = 2x^2$ **18.** $y = \dfrac{1}{8}x^4$

19. $y = -x^3$ **20.** $y = 0.1x^5$

21. $y = 12x^{1/2}$ **22.** $y = \dfrac{2}{x}$

23. The linear function

$$y = 3x + b$$

represents a **family of functions** whose graphs all look the same except for the relative placement with respect to the y-axis. On the same coordinate axis, graph the members of this family for the values $b = 0, 4, -3,$ and $\sqrt{2}$ and state which of these are power functions.

24. The quadratic function $y = ax^2$ represents a *family of functions* whose graphs all look the same except for the relative placement with respect to the y-axis. On the same coordinate axis, graph the members of this family for the given parameter, $a = 0, 4 - 3, \sqrt{2},$ and state which of these are power functions.

25. A spherical cell of radius r has volume $V = \frac{4}{3}\pi r^3$ and surface area $S = 4\pi r^2$. Express V as a function of S. If S is quadrupled, what happens to r?

26. Consider a cylinder of radius r and height $5r$. Express the volume and surface area of this cylinder as a function r. If r is doubled, what happens to the volume? If S is quadrupled, what happens to r?

27. Consider a cone of height h and radius $h/2$ at the top. Express the volume and surface area of this cone as a function of h. If h is doubled, what happens to S?

Drug doses for dogs and cats are known to scale with their surface area S. When body mass W is measured in kilograms, then surface area S in square meters is given by

$$S = \frac{K \times W^{2/3}}{100},$$

where for dogs $K = 10.1$ and for cats $K = 10.4$. Further, when converting human drug doses of an average adult to pet drug doses, this formula is used:

$$\text{pet's drug dose} = \frac{\text{pet's } S}{1.73} \times \text{human adult drug dose}$$

In Problems 28 to 33, the human adult dose of a drug is given. Calculate the drug dose (rounded to the nearest milligram) that you would give your dog or cat of the indicated weight.

28. 100 mg of aspirin and your dog weighs 7 kg

29. 200 mg of aspirin and your cat weighs 4.6 kg

30. 250 mg of an antibiotic and your dog weighs 16 kg

31. 500 mg of a renal drug and your cat weighs 5.3 kg

32. 50 mg of an anticoagulant and your dog weighs 31 kg

33. 50 mg of an anticoagulant and your cat weighs 4.8 kg

Level 2 APPLIED AND THEORY PROBLEMS

34. An ant weighs approximately 1/500 ounce and can lift 1/5 ounce, which is approximately 100 times its weight. Assume that strength is proportional to the cross-section of a muscle and that all organisms on Earth (ants and humans) are geometrically similar. Using these assumptions, determine how much a 150-pound person on Earth can lift.

35. A comic book explained Superman's strength by stating that on Krypton an organism's strength is directly proportional to its body mass. Based on this assumption and assuming that Krypton ants are like Earth ants (see Problem 34), how much can a 150-pound person on Krypton lift?

36. In a sample of twenty-six trees of a particular species, wood density, D (kg/m^3), is related to breaking strength, S (MPa), according to the relationship $D \propto S^{0.91}$. If one of the points that this relationship passed through was $(D, S) = (300, 10)$, find the equation and sketch its graph.

37. A sample based on nineteen mountain ash trees of different sizes yielded a relationship between the leaf area, A (m^2), of the tree and the stem diameter at breast height (DBH), d (cm). The relationship obtained was $A \propto d^{2.99}$. If one of the points that this relationship passed through was $(d, A) = (30, 78)$, find the equation and sketch its graph.

38. In Julian Huxley's classic book *Problems of Relative Growth*, there are data showing an allometric relationship between the mass (C milligrams) of the large claw (chela) and that of the rest of the body (B milligrams) in the male fiddler crab (*Uca pugnax*). The exponent of this relationship is approximately 1.6, that is $C = aB^{1.6}$, and it passes through the point $(B, C) = (1000, 500)$. Calculate the parameter a and then graph the relationship for the growth of a large claw mass as a function of an individual's body mass (excluding claw) over the range $50 \leq x \leq 2200$ mg.

39. In 1936, Sinnott showed that there is an allometric relationship between the length (L) and width (W) of gourds, when observed from ovary to maturity. (See Roger V. Jean, *Differential Growth, Huxley's Allometric Formula and Sigmoid Growth* (COMAP, Incorporated, Lexington MA, 1984) UMAP Module 635, p. 421.) He obtained the exponents of $m = 0.95$ for pumpkins (*Cucurbita pepo*) to $m = 2.2$ for the snake gourd (*Trichosanthes*). Plot these relationships on the same graph for both types of gourds over the interval $[1, 50]$

centimeters, if they both pass through the point $(L, W) = (10, 10)$.

40. Professor Smith's house (10 m wide, 20 m long, 4 m high—just a hovel, really) has a 30,000 watt furnace that just barely keeps him warm on cold winter nights. He's thinking of building a larger house to accommodate his growing insect collection and needs advice on the output of the new furnace. The new house will be three times as high, three times as wide, and three times as long.

    a. If he assumes that the furnace size should be proportional to the volume of the house, then what size furnace should he install?

    b. If heat loss depends on the surface area of exterior walls, roof, and floor exposed to the winter cold rather than on the volume of the house, then what size furnace would you recommend?

41. Consider the following quote from Jonathan Swift's *Gulliver's Travels*:

    "The reader may be pleased to observe, that, in the last article of the recovery of my liberty, the emperor stipulates to allow me a quantity of meat and drink sufficient for the support of 1724 Lilliputians. Some time after, asking a friend at court how they came to fix on that determinate number, he told me that his majesty's mathematicians, having taken the height of my body by the help of a quadrant, and finding it to exceed theirs in the proportion of twelve to one, they concluded from the similarity of their bodies, that mine must contain at least 1724 of theirs, and consequently would require as much food as was necessary to support that number of Lilliputians. By which the reader may conceive an idea of the ingenuity of that people, as well as the prudent and exact economy of so great a prince."

    Let F denote the amount of food an individual eats and L the height of an individual. This quotation implicitly assumes that $F \propto L^b$ for an appropriate choice of b. Find this b value and provide a biological explanation for this choice of b.

42. Suppose the main loss of energy is heat loss through the surface. For the quotation in Problem 41, determine the appropriate choice of b so that $F \propto L^b$. Under the assumption, how much should the Lilliputians feed Gulliver?

43. The following quote from Haldane (*On Being the Right Size*, p. 424) illustrates the dangers of being large:

    To the mouse and any smaller animal, [gravity] presents practically no dangers. You can drop a mouse in a thousand-yard mine shaft; and, on arriving at the bottom, it gets a slight shock and walks away. A rat would be probably killed, though it can fall safely from the eleventh story of a building; a man is killed, a horse splashes. For the resistance presented to movement by air is proportional to the surface of a moving object. Divide an animal's length, breadth, and height each by ten; its weight is reduced to a thousandth, but its surface only to a hundredth. So the resistance to falling in the case of the small animal is relatively ten times greater than the driving force.

Consider a cubical critter being dropped down a mine shaft. Let A denote the force due to air resistance that the cubical critter experiences and let M denote the critter's weight. Assume that A is proportional to surface area and M is proportional to volume.

a. Determine the value of b for which $\dfrac{M}{A} \propto M^b$.

b. Graph $y = M^b$ and discuss the implications for a falling cubical critter.

1.4

Exponential Growth

Without doubt, the linear function $y = ax + b$ is the most important elementary function in mathematics. In the context of calculus, its importance is equaled only by the function we introduce in this section, the exponential function. Just why this function is so critical in calculus will become apparent once we introduce the concept of a derivative. In this section, we show that the exponential function is suitable for describing how populations, income, beer froth, and the radioactivity of unstable isotopes change over time.

Exponential growth and exponential functions

The following table provides data on the growth of the United States from 1815 until 1895.

Year	Population (in millions)
1815	8.3
1825	11.0
1835	14.7
1845	19.7
1855	26.7
1865	35.2
1875	44.4
1885	55.9
1895	68.9

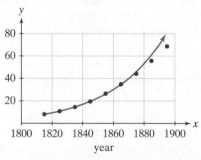

Figure 1.28 Population of the United States.

These data, which are plotted in Figure 1.28, indicate around a tenfold (also referred to as an *order of magnitude*) increase in the U.S. population size during the nineteenth century. To get a finer understanding of the actual rate of growth, we can divide the size of the population in any given year by its size one decade earlier. For example,

$$\frac{\text{population in 1825}}{\text{population in 1815}} = \frac{11}{8.3} \approx 1.3253$$

and

$$\frac{\text{population in 1835}}{\text{population in 1825}} = \frac{14.7}{11.0} \approx 1.3363$$

These calculations tell us that population increased by a factor of approximately 33% over both decades. Let us assume that the population increases by 33% every decade. If t corresponds to the number of decades that have elapsed since 1815 and t is a positive integer, then we might estimate the population size by

$$(8.3) \underbrace{1.33 \times 1.33 \times \cdots \times 1.33}_{t \text{ terms}} = 8.3(1.33)^t$$

More generally, for any real t, we can model the population size $N(t)$ at t decades after 1815 by the exponential function

$$N(t) = 8.3(1.33)^t \text{ million people}$$

The graph of $N(t)$ is plotted in Figure 1.28 against the data, and reasonably approximates the data until 1880, after which it begins to overestimate the population size.

In the previous section, we introduced power functions $y = x^a$ for which the independent variable x is raised to some fixed power. In contrast, the function $N(t)$ has its exponent as the independent variable t and its base is a fixed constant. Such functions are termed *exponential functions*.

Exponential Function

An **exponential function** is a function of the form

$$y = f(x) = a^x$$

where the parameter $a \neq 1$ (the base) is a positive real number and the variable x (the exponent) is a real number.

The graphs of exponential functions have three different shapes, depending on the value of the base, as shown in the following example.

Example 1 Sketching exponential functions

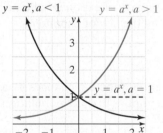

$y = a^x, a < 1$ $y = a^x, a > 1$

$y = a^x, a = 1$

Figure 1.29 Graph for $y = a^x$.

Sketch the exponential function

$$y = a^x$$

where $a > 1$, $a = 1$, and $0 < a < 1$ on the same coordinate axes, and comment on each of the graphs.

Solution The graphs are shown in Figure 1.29.
The graph of $y = a^x$ passes through $(0, 1)$ for all values of a. We also notice:

- If $a < 1$, the graph is increasing for all x.
- If $a = 1$, the graph is a horizontal line (a constant function).
- If $a < 1$, the graph is decreasing for all x.

Example 1 illustrates that if the base of the exponential function is greater than one, then the exponential function is an increasing function. In the context of population change, this increase corresponds to population growth. A fundamental quantity associated with population growth is the *doubling time*; how long before the population size doubles. The following example illustrates this concept.

Example 2 Malthus's estimate for doubling time

In *An Essay on the Principal of Population* (http://www.gutenberg.org/files/4239/4239-h/4239-h.htm), Thomas Malthus wrote:

In the United States of America, where the means of subsistence have been more ample, the manners of the people more pure, and consequently the checks to early marriages fewer, than in any of the modern states of Europe, the population has been found to double itself in twenty-five years.

Let $N(t) = 8.3(1.33)^t$ be our model of population growth in the United States from 1815 onward.

a. Determine whether the population size doubles from 1815 until 1840. Recall that the units of t are decades.

b. Determine whether the population size doubles over every twenty-five year period for our model $N(t)$.

Solution

a. Since t is decades after 1815, 25 years after 1815 corresponds to $t = 2.5$. To determine whether the population doubles between 1815 and 1840, we compute the ratio of the population sizes in those years

$$\frac{N(2.5)}{N(0)} = \frac{8.3(1.33)^{2.5}}{8.3} = 1.33^{2.5} \approx 2.04$$

Hence the population increases by just over a factor of 2.

b. Consider any time t. To determine whether the population doubles between t and $t + 2.5$, we compute the ratio of the population sizes in those years

$$\frac{N(t + 2.5)}{N(t)} = \frac{8.3(1.33)^{t+2.5}}{8.3(1.33)^t} = 1.33^{2.5} \approx 2.04 \qquad \textit{using laws of exponents}$$

We see that Malthus's prediction conforms reasonably well with our model. Notice that we could not test the prediction directly with data, as data are reported only in ten-year intervals. ∎

In Example 2, we used a law of exponents (discussed earlier in Section 1.3). These laws are extremely useful for manipulating exponential functions, so you may review them. For example, using these laws, we uncover a key property of exponential functions. Namely, if $f(x) = a^x$ and h is a real number, then by the law of exponents

$$\frac{f(x + h)}{f(x)} = \frac{a^{x+h}}{a^x} = a^{x+h-x} = a^h$$

In other words, over any interval of length h, the exponential function changes by a fixed factor a^h. In the case of Example 2, this observation implies that the population approximately doubles over any twenty-five-year period.

Exponential growth is much faster than polynomial growth, as we explore in the next example.

Example 3   Exponential growth versus polynomial growth

Use technology to graph the functions $y = 2^x$ and $y = x^4$. Which function takes on larger values when x is large?

Solution   Figure 1.30 shows both functions plotted on a the intervals $[0, 1.8]$, $[0, 10]$, and $[0, 18]$. Over the $[0, 1.8]$ interval $y = 2^x$ initially takes on larger values than $y = x^4$, but at the end of this interval $y = x^4$ takes on the larger values. The graphs on the intervals $[0, 10]$ suggest that $y = x^4$ continues to be larger. On the interval $[0, 18]$, we see the curves again cross at $x = 16$, and for $x > 16$, we see that $y = 2^x$ is larger.

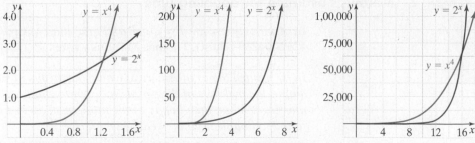

Figure 1.30 Comparing the exponential and quartic growth over different intervals. ∎

The fact that exponential functions with base greater than one grow faster than any polynomial led the economist Thomas Malthus to make a dire prediction about the future of humankind.

Example 4 Malthus's law of misery

In *An Essay on the Principle of Population*,* Thomas Malthus wrote:

"Let us then take this for our rule, though certainly far beyond the truth, and allow that, by great exertion, the whole produce of the Island might be increased every twenty-five years, by a quantity of subsistence equal to what it at present produces. The most enthusiastic speculator cannot suppose a greater increase than this. In a few centuries it would make every acre of land in the Island like a garden."

To illustrate the meaning of this quote, consider the United States to be the "island" and farms to be the "garden." Let $N(t) = 8.3(1.33)^t$ (in millions) be the population size t decades after 1815. Assume that in 1815, the amount of food produced in this year is equivalent to 10 million yearly rations. Further, assume, as suggested by Malthus, that the production of food in the United States will increase every twenty-five years by 10 million yearly rations.

a. Write a formula for the number $R(t)$ of yearly rations (in millions) produced over time.

b. Graph and compare the functions $R(t)$ and $N(t)$.

c. Determine the first year in which there is just enough food to provide everyone with one ration per year.

Solution

a. As the amount of yearly rations increases by 10 million every twenty-five years, $R(t)$ is a linear function with slope $\frac{10}{2.5} = 4$. Since $R(0) = 10$, the intercept of this linear function is 10 and we have

$$R(t) = 10 + 4t$$

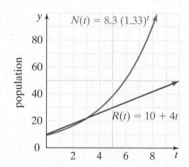

b. Using technology to plot $R(t)$ and $N(t)$ gives the graph shown on the left. In the long term, the predicted population size is much greater than the availability of food.

c. By inspection, it looks like the graphs of $N(t)$ and $R(t)$ intersect at $t = 4$. Hence in forty years, every individual in the population will get one yearly ration every year.

Table 1.11 Froth height decay

Time t (seconds)	Froth height H (centimeters)
0	17.0
15	16.1
30	14.9
45	14.0
60	13.2
75	12.5
90	11.9
105	11.2
120	10.7

Exponential decay

When the base of an exponential function is less than one, the exponential function is decreasing and exhibits so-called *exponential decay*. An amusing example of exponential decay resulted in Arnd Leike, a professor of physics at University of Munich, winning the 2002 Ig Nobel Prize. The Ig Nobel Prize is annually awarded to scientists who firstly make people laugh, and secondly make them think. Leike received his award for his paper, "Demonstration of the Exponential Decay Law Using Beer Froth" (*European Journal Physics* 23 (2002): 21–26.) This paper reports an experiment that Leike performed with a mug of the German beer Erdinger Weissbier. After pouring the beer, Leike measured the height of the beer froth at regular time intervals. The measured values are shown in Table 1.11.

*http://www.gutenberg.org/files/4239/4239-h/4239-h.htm.

If we consider the ratios of heights at subsequent time intervals, we find

$$\frac{\text{height at 45 seconds}}{\text{height at 30 seconds}} = \frac{14}{14.9} \approx 0.94$$

and

$$\frac{\text{height at 60 seconds}}{\text{height at 45 seconds}} = \frac{13.2}{14} \approx 0.94$$

Note that 0.94 represents 6% decay. If we assume, as the data suggest, every 15 seconds the height of the froth decays by a factor of 6%, then we can write an expression (formula) for the froth height and see how well it fits the data.

Example 5   Modeling the decay of beer froth

Find values for the parameters a and b of the function

$$H(t) = ab^t$$

that ensure the function passes through the first data point in Table 1.11 and that the height of the froth declines 6% every 15 seconds. Use technology to graph $H(t)$ alongside the data. How well does the function fit the data? Assume that t is measured in seconds.

Solution   Since the initial height of the froth is 17 cm and $H(0) = ab^0 = a$, we set $a = 17$. On the other hand, assuming that the froth decays by a factor of 6% every 15 seconds means that

$$0.94 = \frac{H(15)}{H(0)} = \frac{ab^{15}}{a} = b^{15}$$

Hence, $b = 0.94^{1/15} \approx 0.99588$. Therefore, we have (in centimeters)

$$H(t) = 17(0.99588)^t$$

The graph is shown in Figure 1.31 and appears to fit the data very well.

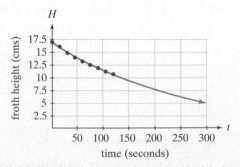

Figure 1.31 Froth height equation plotted with data points.

One way of understanding this exponential decay is to think of the froth as a large collection of bubbles. According to our calculations, approximately every 15 seconds, 6% of the bubbles will pop, leaving only 94% of the original head of froth. As the bubbles continue to pop, there are fewer and fewer that can pop. Consequently, as shown in Figure 1.31, the number of bubbles left to pop declines to zero over time in a way that seems to be modeled rather well by a function that has a variable appearing as the exponent of some base value. For this reason, the decline is called *exponential decay*. In the problem set and in the next section, you will see that exponential decay arises in many biological contexts: exponential decay of a tumor following radiation therapy, exponential decay of a drug in the body, and exponential decay of endangered populations.

The number e

There is one choice of a base a for exponential functions a^x that plays a particularly important role in calculus. It is Euler's number e named after the Swiss mathematician Leonhard Euler (1707–1783). Its importance will become apparent only when we get into the machinery of calculus. The following example, however, introduces you to this constant, and in providing an economic interpretation of its value.

Example 6   Continuously compounded interest

Jacob Bernoulli (1655–1705), another Swiss mathematician, discovered the irrational number e while exploring the compound interest on loans. Consider a bank account starting with one dollar. If the bank (never to be seen in the real world!) gives you 100% interest on this dollar after a year, you will have $2.00 after one year. What happens to your initial dollar if the bank compounds the interest more frequently?

a. How much money will you have in the account one year from now if the bank gives you 50% interest six months from now and another 50% interest on the total amount in your account a year from now? This corresponds to compounding a 100% interest rate twice a year.

b. How much will you have in the account if the bank compounds the interest quarterly? In other words, it gives you 25% interest on the total amount in your account four times a year.

c. Create a table corresponding to compounding your dollar monthly, daily, hourly, and every minute. Does the amount in your bank account after one year appear to approach a limiting value?

Solution

a. After the first six months, you will have $1.00(1.5) = \$1.50$. After a year, the total amount of money $1.00(1.5) in your account gets 50% interest and you will have $1.00(1.5)^2 = \$2.25$.

b. After the first three months, you will have $1.00(1.25) = \$1.25$. After the first six months, you have $1.00(1.25)^2 = \$1.5625$. After the first nine months, $1.00(1.25)^3 = \$1.953125$. After one year, $1.00(1.25)^4 \approx \$2.44$.

c. Following the approach from parts **a** and **b**, we get the following table of values.

Number of times compounded	Dollar amount at end of year
1 (annually)	$2.00
2 (half-yearly)	$1.00(1 + 1/2)^2 = \$2.25$
4 (quarterly)	$1.00(1 + 1/4)^4 \approx \$2.4414$
12 (monthly)	$1.00(1 + 1/12)^{12} \approx \$2.61304$
365 (daily)	$1.00(1 + 1/365)^{365} \approx \$2.71457$
8760 (hourly)	$1.00(1 + 1/8760)^{8760} \approx \$2.71813$
525600 (minutely)	$1.00(1 + 1/525600)^{525600} \approx \$2.71828$

The table suggests that the amount in the account is approaching some value near $2.71828 as you compound more and more frequently.

This example suggests that the quantity $(1 + 1/n)^n$ approaches a limiting value near 2.71828 as n gets very large. The actual limiting value is Euler's number e. We state this definition formally using the concept of a limit discussed in the introduction to this book. Limits are discussed extensively in Chapter 2.

Euler's Number e

$$e = \lim_{n \to \infty} \left(1 + \frac{1}{n}\right)^n \approx 2.7182818\ldots$$

Based on Example 6, we can interpret e as the value of one dollar a year from now in a continuously compounded bank account at a rate of 100%. The exponential function $y = e^x$ has many properties that are particularly convenient for calculus. We will see some of these properties in Chapter 2 when we discuss derivatives.

With continuous compounding, we compound interest not quarterly, or monthly, or daily, or even every second, but *instantaneously*, so that the future amount of money A in the account grows continuously. If the account initially has P dollars and an annual interest rate of r, then we define the future money A in the account t years from now as the limiting value of

$$P\left(1 + \frac{r}{n}\right)^{nt}$$

as the number of compounding periods n grows without bound. We denote by $A(t)$ the future value after t years. As with the definition of e, we denote this by the limit notation $n \to \infty$:

$$A(t) = \lim_{n \to \infty} P\left(1 + \frac{r}{n}\right)^{nt} \qquad \text{Let } k = \frac{n}{r}, \text{ so that } krt = nt \text{ and } \frac{r}{n} = \frac{1}{k};$$
$$\text{also } k \to \infty \text{ as } n \to \infty$$

$$= \lim_{k \to \infty} P\left(1 + \frac{1}{k}\right)^{krt}$$

$$= \lim_{k \to \infty} P\left[\left(1 + \frac{1}{k}\right)^k\right]^{rt}$$

$$= P\left[\lim_{k \to \infty} \left(1 + \frac{1}{k}\right)^k\right]^{rt} \qquad \textit{Scalar rule for limits (see Chapter 2)}$$

$$= Pe^{rt} \qquad \textit{Definition of } e$$

These observations are now summarized.

Future Value

If P dollars are compounded n times per year at an annual rate r, then the **future value** after t years is given by

$$A(t) = P\left(1 + \frac{r}{n}\right)^{nt}$$

and if the compounding is continuous, the future value is

$$A(t) = Pe^{rt}$$

Since $e > 1$, the exponential function $y = e^x$ increases rapidly. To get a sense of how quickly this exponential function increases, you will use technology in the next example to determine how long it would take the continuously compounded bank account with one dollar to yield $10,000. An analytical approach to solving this problem is given in Section 1.6, after we introduce logarithm functions.

Example 7   Naïve approach to solving an exponential equation

Graph $y = e^x$ and $y = 10,000$ to solve the equation $e^x = 10,000$.

Solution   Graphing $y = e^x$ and $y = 10,000$ yields the following graph (with units on the y-axis in units of 10,000).

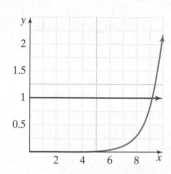

We estimate the x value at which the intersection occurs to be $x \approx 9$. Hence, it would take only 9 years for a dollar to become \$10,000 dollars in an account compounding continuously at a rate of 100% per year.

PROBLEM SET 1.4

Level 1 DRILL PROBLEMS

Graph the exponential functions in Problems 1.4 to 6.

1. $y = 2^x$

2. $y = \left(\dfrac{1}{2}\right)^x$

3. $y = 3^{-x}$

4. $y = e^{2x}$

5. $y = 0.1^{-x}$

6. $y = \pi^{ex}$

Graph the exponential function $y = f(x)$ and the polynomial function $y = g(x)$ using the same domain (x-values) and range (y-values) in Problems 7 to 10. Estimate for what value of x that $f(x) \geq g(x)$.

7. $f(x) = 2^x$ and $g(x) = 2x$.

8. $f(x) = e^x$ and $g(x) = 2x$.

9. $f(x) = \pi^x$ and $g(x) = x^4 - 4$.

10. $f(x) = 1.1^x$ and $g(x) = 5x^5 + x + 1$.

Use a graphical approach to estimate the solution to the equations specified in Problems 11 to 18. Note that the solution may be negative.

11. $e^{2x} = 10$

12. $e^{x/2} = 1/2$

13. $e^{-x} = x$

14. $e^{-x} = -3x$

15. $e^{x/3} = x$

16. $e^{-x} = -x^2$

17. $e^x = -x^2 + 2$

18. $e^x = x^2 + 2x$

Assume that \$1,000 is invested for t years at r percent interest compounded in the frequencies given in Problems 19 to 24. Calculate the future value.

19. $t = 25, r = 0.07$; annual compounding

20. $t = 10, r = 0.12$; semiannual compounding

21. $t = 1, r = 0.01$; continuous compounding

22. $t = 5, r = 0.02$; continuous compounding

23. $t = \frac{1}{3}$ (4 months), $r = 0.16$; monthly compounding

24. $t = \frac{1}{2}$ (6 months), $r = 0.08$; monthly compounding

Find the equation for the exponential function $f(x) = b\,a^x$ that passes through the two indicated points in Problems 25 to 28.

25. $(0, 2)$ and $(1, 5)$

26. $(-2, 32)$ and $(2, 8)$

27. $(1/2, 3)$ and $(1, 1)$

28. $(-2, 1/2)$ and $(2, 2)$

Level 2 APPLIED AND THEORY PROBLEMS

29. In Example 5, we modeled the height of the beer froth (in centimeters) as $H(t) = 17(0.99588)^t$ where t is measured in seconds. Plot this function over a four-minute time interval and estimate at what time the froth is one half of its original height.

30. In Example 2, we modeled the population size (in millions) of the United States as $N(t) = 8.3(1.33)^t$ where t is measured in decades after 1815. Plot this function over a fifty-year period and estimate the time at which the population would triple in size. How does this compare to the actual time that the population tripled in size?

31. Consider $100 in a bank account that has annual interest rate of 20%.

   a. Compute the amount in the bank account one year later if the money is compounded once a year, twice a year, and four times a year.

   b. Find an expression for the amount of money in the bank account if the money is compounded n times a year.

   c. Evaluate this expression for larger and larger n and estimate the value it appears to be approaching.

32. Consider $1,000 in a bank account that has annual interest rate of 5%.

   a. Compute the amount in the bank account one year later if the money is compounded once a year, twice a year, and four times a year.

   b. Find an expression for the amount of money in the bank account if the money is compounded n times a year.

   c. Evaluate this expression for larger and larger n and estimate the value it appears to be approaching.

33. Consider a bacterial species that produces ten offspring per day. Assume that you start with one bacterial cell, no cells die, and all cells reproduce at the same rate.

   a. Compute the bacterial population size in one year if the cells reproduce only once per day, twice per day, and four times per day.

   b. Find an expression for the population size if the population reproduces n times per day.

   c. Evaluate this expression for larger and larger n and estimate the value it appears to be approaching.

34. Consider a bacterial species that produces five offspring per day. Assume that you start with twenty bacterial cells, no cells die, and all cells reproduce at the same rate.

   a. Compute the bacterial population size in one year if the cells reproduce only once per day, twice per day, and four times per day.

   b. Find an expression for the population size if the population reproduces n times per day.

   c. Evaluate this expression for larger and larger n and estimate the value it appears to be approaching.

35. The following functions give the population size $P(t)$ in millions for four fictional countries where t is the number of decades since 1900.

   Country 1: $P_1(t) = 3(1.5)^t$

   Country 2: $P_2(t) = 10(1.1)^t$

   Country 3: $P_3(t) = 20(0.95)^t$

   Country 4: $P_4(t) = 2(1.4)^t$

   a. Which country had the largest population size in 1900?

   b. Which country has the fastest population growth rate? By what percentage does this population grow every decade?

   c. Is any of these populations decreasing in size? If so, which one and by what fraction does the population size decrease every decade?

36. The following functions give the froth height (in centimeters) of three fictional beers where t represents time (in seconds).

   Beer 1: $H_1(t) = 20(0.99)^t$

   Beer 2: $H_2(t) = 40(0.9)^t$

   Beer 3: $H_3(t) = 15(0.98)^t$

   a. Which beer has the highest froth initially? What is the height?

   b. Which beer has the slowest decay of froth? For this beer, what percentage of the height is lost in ten seconds? Twenty seconds?

   c. Which beer has the highest froth height after ten seconds?

37. Hyperthyroidism is a condition in which the thyroid gland makes too much thyroid hormone. The condition can lead to difficulty concentrating, fatigue, and weight loss. One treatment for hypothyroidism is the administration of replacement thyroid hormone such as thyroxine (T4). The concentration of this hormone in a patient's body exhibits exponential decay with a half-life of about seven days. Consider an individual that has taken 100 mcg of T4.

   a. How much of the T4 is in the body after fourteen days?

   b. Write an expression for the amount of T4 in the body after t days.

   c. Use a graph to estimate the time required for the amount of T4 to be reduced to 10 mcg.

38. In Problem 37, we discussed hyperthyroidism. For individuals with this conditions, the body converts thyroxine (T4) into triiodothyronine (T3), which is the hormone that the body uses at the cellular level. The bodies of some individuals are unable to do this conversion; consequently these people are given injections of T3. The half-life of T3 is about ten hours. Consider an individual who has taken 100 mcg of T3.

a. How much of the T3 is in the body after thirty hours?

b. Write an expression for the amount of T3 in the body after t hours.

c. Use a graph to estimate the time required for the amount of T3 to be reduced to 10 mcg.

39. Carbon-14 has a half-life of 5,730 years. How much is left of 500 g of C-14 after t years?

40. If a bacterial population initially has twenty individuals and doubles every 9.3 hours, then how many individuals will it have after three days?

41. "Whale Numbers up 12% a Year" was a headline in a 1993 Australian newspaper. In a thirteen-year study, beginning in 1981, scientists estimated that the humpback whale (*Megaptera novaeangliae*) population off the coast of Australia was increasing by 12% a year from a level of 350 individuals.

a. Write an expression for $P(t)$, the populations size at t years after 1981, assuming $P(0) = 350$.

b. Estimate the size of the population in 2010.

42. The population size (in millions) of Mexico in the early 1980s is reported in Table 1.12.

Table 1.12 Population in Mexico

Year	Population (in millions)
1980	67.38
1981	69.13
1982	70.93
1983	72.77
1984	74.66
1985	76.60

a. Assume the population growth in Mexico is exponential. Use the first two data points to find a formula for $P(t)$, the population size (in millions) t years after 1980.

b. Plot $P(t)$ against the data. Discuss the quality of the fit.

c. Estimate the size of the population in 2004.

d. Look up Mexico's actual population size in 2004. Does your model overpredict or underpredict the population size? Discuss your answer.

43. Consider an instant lottery game in which you buy a scratch-off card and there is a certain chance p of winning a prize. If you buy N scratch-off cards, the chance of *not* winning a prize is $(1 - p)^N$, that is, the chance you didn't win the prize on the first card times the chance you didn't win the prize on the second card, and so on. Therefore, the chance of winning a prize with N cards is $1 - (1 - p)^N$.

a. If there is a 1/10 chance of winning a prize and you buy ten cards, what is the chance you win a prize?

b. If there is a chance 1/100 of winning a prize and you buy 100 cards, what is the chance you win a prize?

c. If there is chance $1/N$ of winning a prize and you buy N cards, what is the chance you win a prize? What value does this approach for large N?

44. In an experimental study performed at Dartmouth College, two groups of mice with tumors were treated with the chemotherapeutic drug cisplatin. Prior to the therapy, the tumor consisted of proliferating cells (also known as clonogenic cells) that grew exponentially with a doubling time of approximately 2.9 days. Assume the initial tumor size was 0.1 cm³.

a. Write an expression for the tumor size after t days.

b. Use graphing to estimate at what time the tumor is 0.5 cm³ in size.

45. In the experimental study described in Problem 44, each of the mice was given a dosage of 10 mg/kg of cisplatin. At the time of the therapy, the average tumor size was approximately 0.5 cm³. Assume all the cells became quiescent (i.e., no longer dividing) and decay with a half-life of approximately 5.7 days.

a. Write an expression for the tumor size after t days of treatment.

b. Use graphing to estimate at what time the tumor is 0.05 cm³ in size.

46. By comparing the graphs of $\sqrt{x}^{\pi}$ and $\pi^{\sqrt{x}}$ determine which is larger:

a. $(\sqrt{3})^{\pi}$ or $\pi^{\sqrt{3}}$

b. $(\sqrt{5})^{\pi}$ or $\pi^{\sqrt{5}}$

c. $(\sqrt{6})^{\pi}$ or $\pi^{\sqrt{6}}$

d. From parts **a–c**, notice that $(\sqrt{x})^{\pi}$ is larger for some values of x, while $\pi^{\sqrt{x}}$ is larger for others. For $x = \pi^2$

$$(\sqrt{x})^{\pi} = \pi^{\sqrt{x}}$$

is obviously true (since $\pi^{\pi} = \pi^{\pi}$). Using a graphical method, find another value (approximately) for which the given statement is true.

In calculus the number e is sometimes introduced using slopes. In Problems 47 to 49 explore this idea.

47. a. Draw the graph of $y = 2^x$ and plot the points $(0, 1)$ and $(2, 4)$, which are on the graph.

b. Consider the secant line passing through these points. Now, consider the slope of the secant line as the point $(2, 4)$ slides along the curve toward the point $(0, 1)$. Draw the line that you think will result when the point $(2, 4)$ reaches the point $(0, 1)$. This is the *tangent line* to the curve $y = 2^x$ at $(0, 1)$.

c. Using the tangent line, and the fact that the slope of a line is RISE/RUN, estimate (to the nearest tenth) the slope of the tangent line.

48. a. Draw the graph of $y = 3^x$ and plot the points $(0, 1)$ and $(2, 9)$, which are on the graph.

b. Consider the secant line passing through these points. Now, consider the slope of the secant line as the point $(2, 9)$ slides along the curve toward the point $(0, 1)$. Draw the line that you think will result when the point $(2, 9)$ reaches the point $(0, 1)$. This is the *tangent line* to the curve $y = 2^x$ at $(0, 1)$.

c. Using the tangent line, and the fact that the slope of a line is RISE/RUN, estimate (to the nearest tenth) the slope of the tangent line.

49. a. Draw a line passing through $(0, 1)$ with slope 1.

b. Compare the graphs of $y = 2^x$ (Problem 47) and $y = 3^x$ (Problem 48). Now, it seems reasonable that there exists a number between 2 and 3 with the property that the slope of the tangent through $(0, 1)$ is 1. Draw such a curve.

c. On the same coordinate axes, draw the graph of $y = e^x$. How does this curve compare with the curve you drew in part **b**? In calculus, the number between 2 and 3 with the property that the slope of the tangent through $(0, 1)$ is 1 is used as the number e.

1.5   Function Building

We have reviewed basic properties of linear, periodic, exponential, and power functions. By combining these functions, we can greatly enlarge our "toolbox" of functions. With this larger toolbox, we can describe more data sets and model more biological processes. For instance, in this section we develop models of the waxing and waning of tides and the rates at which microbes consume nutrients.

Shifting, reflecting, and stretching

The simplest way to create the graph of a new function from the graph of another function is to shift the graph vertically or horizontally.

Horizontal and Vertical Shifts	Let $y = f(x)$ be a given function with $a > 0$.
	Horizontal shifts:
	$y = f(x - a)$ shifts the graph of $y = f(x)$ to the right a units
	$y = f(x + a)$ shifts the graph of $y = f(x)$ to the left a units
	Vertical shifts:
	$y = f(x) + a$ shifts the graph of $y = f(x)$ upward a units
	$y = f(x) - a$ shifts the graph of $y = f(x)$ downward a units

To understand why these shifts occur, consider $y = f(x - a)$. Substituting $x + a$ for x yields $y = f(x + a - a) = f(x)$. Hence, the function $y = f(x - a)$ has the same value as the function $y = f(x)$ when you "shift x" to the right by a.

Example 1   Shifting graphs

Consider the function $y = f(x)$ whose graph is given by Figure 1.32.

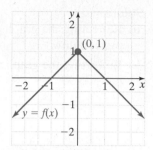

Figure 1.32 Graph of $y = f(x)$.

Sketch the graphs of $y = f(x - 0.5)$, $y = f(x) - 0.5$, and $y = f(x + 1) + 1$.

Solution $y = f(x - 0.5)$ shifts the graph right 0.5 units. $y = f(x) - 0.5$ shifts the graph down 0.5 units. $y = f(x + 1) + 1$ shifts the graph left 1 unit and up 1 unit. These graphs are shown in Figure 1.33

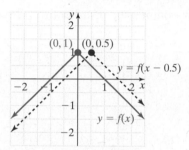
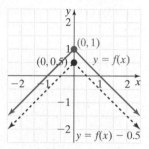
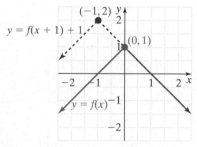

a. Shift to the right 0.5 units   b. Shift down 0.5 units   c. Shift left 1 unit and up 1 unit

Figure 1.33 Shifting a graph.

In addition to shifting graphs, we can reflect graphs across axes.

Reflections

Let $y = f(x)$ be a given function.

The graph of $y = -f(x)$ is the **reflection across the x axis**. It is found by replacing each point (x, y) on the graph with $(x, -y)$.

The graph of $y = f(-x)$ is the **reflection across the y axis**. It is found by replacing each point (x, y) on the graph with $(-x, y)$.

Example 2   Reflecting a function

Consider the function $y = f(x)$ whose graph is given by

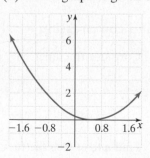

a. Sketch $y = f(-x)$.

b. Sketch $y = -f(x)$.

Solution

a. Reflecting the graph about the y axis yields the desired graph in red in the left panel below.

b. Reflecting the graph about the x axis yields the desired graph in red in the right panel below.

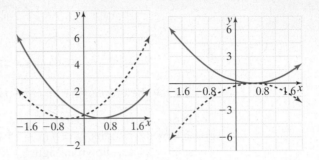

A curve can be stretched or compressed in either the x-direction, the y-direction, or both, as shown in Figure 1.34.

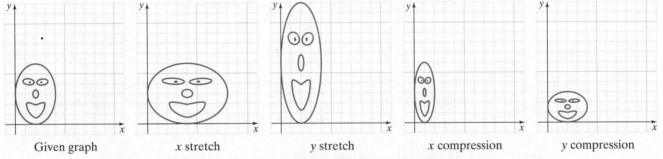

| Given graph | x stretch | y stretch | x compression | y compression |

Figure 1.34 Stretching and compressing a graph.

Stretching and Compressing

Let $y = f(x)$ be a given function.

To sketch the graph of $y = f(bx)$, replace each point (x, y) with $(\frac{1}{b}x, y)$. If $0 < b < 1$, then we call the transformation an **x-dilation** (or x stretch). If $b > 1$, then we call the transformation an **x-compression**.

To sketch the graph of $y = c\,f(x)$, replace each point (x, y) with (x, cy). If $c > 1$, then we call the transformation a **y-dilation** (or y stretch). If $0 < c < 1$, then we call the transformation a **y-compression**.

Example 3 Stretching and compressing

Consider the function $y = f(x)$ defined by

$$f(x) = \begin{cases} 0 & \text{if } x < 0 \\ x & \text{if } 0 \le x < 1 \\ 1 & \text{if } x \ge 1 \end{cases}$$

Find and sketch the functions $y = f(2x)$ and $y = 3f(2x)$.

Solution   We begin by sketching the function $y = f(x)$ which is 0 for negative x, linear with slope 1 from $x = 0$ to $x = 1$, and equal to 1 for $x \geq 1$ to obtain the left panel below.

Since $2x < 0$ if and only if $x < 0$, we get the function $y = f(2x)$ equals 0 for $x < 0$. Since $0 \leq 2x < 1$ if and only if $0 \leq x < 1/2$, $y = f(2x) = 2x$ for $0 \leq x < 1/2$. Finally, $y = f(2x) = 1$ for $x \geq \dfrac{1}{2}$. Therefore, we have shown

$$f(2x) = \begin{cases} 0 & \text{if } x < 0 \\ 2x & \text{if } 0 \leq x < 1/2 \\ 1 & \text{if } x \geq 1/2 \end{cases}$$

Plotting this function, we get the center panel below. Hence, $y = f(2x)$ compresses the function $y = f(x)$ by a factor of 2 in the horizontal direction.

The function $y = 3f(2x)$ stretches the function $y = f(2x)$ by a factor of 3 in the vertical direction. Hence, we get

$$3f(2x) = \begin{cases} 0 & \text{if } x < 0 \\ 6x & \text{if } 0 \leq x < 1/2 \\ 3 & \text{if } x \geq 1/2 \end{cases}$$

Plotting this function, we get the right panel below.

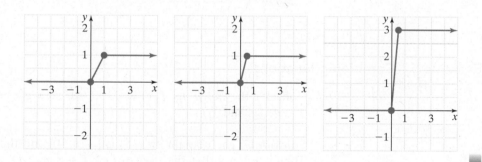

By compressing and stretching sinusoidal functions, we can model periodic phenomena like tidal movements.

Example 4   Modeling tidal movements

The tides for Toms Cove in Assateague Beach, Virginia, on August 19, 2004, are listed in the table on the left.

Assume that this can be modeled by

$$T(t) = A\cos[B(t + C)] + D$$

where T denotes the height (in feet) of the tide t hours after midnight. Find values of A, B, C and D such that the function fits the Assateague tide data.

Time	Height) (ft)	Tide
5:07 A.M.	0.4	Low
10:57 A.M.	4.0	High
5:23 P.M.	0.4	Low

Solution   The data suggest that the period of T is approximately twelve hours, in which case $B = \dfrac{2\pi}{12} = \dfrac{\pi}{6}$. The amplitude of the tide is given by $A = \dfrac{4 - 0.4}{2} = 1.8$. Since the graph of cosine is always centered around the horizontal axis, we need to vertically shift the graph up by the midtide height, $D = (4.0 + 0.4)/2 = 2.2$. Finally, since the high tide occurs approximately at $t = 11$, we can choose $C = -11$ to shift

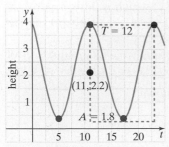

Figure 1.35 Graph fitting the Assateague tide data.

the graph left by 11. Putting this all together yields

$$T(t) = 1.8 \cos\left[\frac{\pi}{6}(t - 11)\right] + 2.2$$

To graph this function, we note that $T(t)$ has a vertical shift of 2.2 and a horizontal shift of 11 of the cosine function. Additionally, the amplitude of the cosine function is 1.8 and the period is 12, as shown in Figure 1.35.

Adding, subtracting, multiplying, and dividing

The easiest way to create new functions is to perform arithmetic operations on old functions. The first three of these operations result in a function whose domain is the intersection of the domains of the original functions, where the symbol ∩ is used to denote *set intersection*. Since division by zero is not permitted, division can further reduce the domain of the new function.

Functional Arithmetic

Let f and g be functions with domains A and B, respectively. Then,

Addition: $f + g$ is defined by $(f + g)(x) = f(x) + g(x)$ with domain $A \cap B$

Subtraction: $f - g$ is defined by $(f - g)(x) = f(x) - g(x)$ with domain $A \cap B$

Multiplication: fg is defined by $(fg)(x) = f(x)g(x)$ with domain $A \cap B$

Division: f/g is defined by $(f/g)(x) = f(x)/g(x)$ with domain consisting of points x in $A \cap B$ such that $g(x) \neq 0$

Example 5   Combining functions

Consider the functions $f(x) = \sqrt{100 - x^2}$ and $g(x) = \sin x$. Find the domains for $f + g$, fg, and f/g.

Solution   The domain of $f(x)$ is $[-10, 10]$ and the domain of $g(x)$ is $(-\infty, \infty)$. Thus it follows that the domains of $f + g$ and fg are $[-10, 10]$. Multiplying the graphs pointwise yields a graph that does not alter the domain, but taking the quotient f/g requires that we remove points on $[-10, 10]$ where $g(x) = 0$; that is the values $0, \pm\pi, \pm 2\pi, \pm 3\pi$.

Two important classes of functions that we get by adding, multiplying, and dividing power functions are polynomials and rational functions.

Polynomials and Rational Functions

Let n be a whole number. A **polynomial function of degree n** is a function of the form

$$y = a_0 + a_1 x + a_2 x^2 + \cdots + a_n x^n$$

where $a_0, a_1, \ldots, a_n$ are constants.

A **rational function** is a function of the form

$$y = \frac{a_0 + a_1 x + a_2 x^2 + \cdots + a_n x^n}{b_0 + b_1 x + \cdots + b_m x^m}$$

where division by zero is excluded, n, m, are natural numbers, $a_0, a_1, a_2, \ldots, a_n$, $b_0, b_1, b_2, \ldots, b_m$ are constants.

Rational functions arise in biology in many ways, particularly in the context of the rate at which organisms extract resources from their environments. Bacteria, for example, have special molecular receptors embedded in their cell membrane to ingest nutrients, such as glucose, into their cell bodies. These receptors "capture" nutrient molecules outside of the cell and transport them into the cell body. This process is illustrated in Figure 1.36.

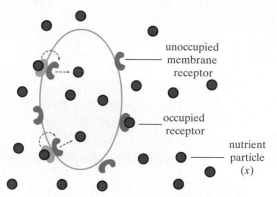

Figure 1.36 Cell body and receptors.

The rate at which nutrients can be brought into the cell body is called the *uptake rate*. The uptake rate is limited by the number of receptors and the time it takes a receptor to bring a nutrient particle into the cell body. The next example derives the Michaelis-Menton uptake function named after biochemists Leonor Michaelis (1875–1947) and Maud Menten (1879–1960). In addition to describing nutrient uptake, this function is used to describe enzyme kinetics and the consumption rates of organisms such as foxes eating rabbits, or ladybugs eating aphids.

Example 6   Michaelis-Menten glucose uptake rate

This problem is divided into two parts: the derivation of the Michaelis-Menten uptake function in part **a** and application of this function in part **b**.

a. To find the amount of glucose brought into the cell per hour, let T be the amount of time a receptor is unbound and can bind to a glucose molecule. The total number of receptors on the cell is R and x is the concentration of nutrients. Assume the number N of glucose molecules brought into the cell per hour by one receptor is proportional to T and x (i.e., more time for binding or higher glucose concentrations lead to more glucose uptake):

$$N = aTx$$

where $a > 0$ is the proportionality constant. If a bound receptor requires b hours (b is much less than 1) to bring a glucose molecule into the cell, then T satisfies

$$T = \text{one hour} - \text{time spent bound to glucose} = 1 - bN = 1 - baTx$$

Solve for T in terms of x and derive an expression for the total uptake rate $f(x)$ in terms of x.

b. In the 1960s, scientists at Woods Hole Oceanographic Institution measured the uptake rate of glucose by bacterial populations from the coast of Peru.* In one

*R. F. Vaccaro and H. W. Jannasch (1967). "Variations in uptake kinetics for glucose by natural populations in seawater." *Limnology and Oceanography*. 12:540–542.

field experiment, they collected the following data:

Glucose concentration (micrograms per liter)	Uptake rate for 1 liter of bacteria (micrograms per hour)
0	0
20	12
40	16
60	18
80	19
100	20

By an appropriate change of variables (see Problem set 1.5), one can use linear regression to estimate the parameters for the uptake function. Doing so yields $f(x) = \dfrac{1.2078x}{1 + 0.0506x}$. How does this relate to the expression for $f(x)$ derived in part **a**? Use technology to plot this function against the data. How good is the fit?

Solution

a. Solving for T, we get

$$T = 1 - baTx$$
$$T + baTx = 1$$
$$T(1 + bax) = 1$$
$$T = \frac{1}{1 + bax}$$

Substituting this expression into the expression $N = aTx$, we get

$$N(x) = \frac{ax}{1 + bax}$$

Since N is the number of molecules handled by each receptor and the cell has R receptors, the total uptake rate of the cell is

$$f(x) = \frac{NRax}{1 + bax}$$

b. The function $f(x)$ has an identical form to the function derived in part **a**, with $ba = 0.0506$ and $aNR = 1.2078$. Using technology to plot the function against the data yields the following graph.

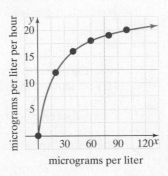

The fact that the function fits the data so well gives us confidence that the arguments used to construct the function are sound! One interesting question to ask is what happens to the uptake rate $f(x)$ as x gets very large (i.e., approaches $+\infty$)? In the next chapter, we will develop ideas to tackle this question.

Composing

Situations often arise in biology where the relationship between two variables x and z is mediated by a third variable y. For example, the rate z at which a population of mice or shrews grows is related to the number y of insects the animals consume per unit time, and this rate y is related to the density x of insects in the area where these animals feed. Let f be the function that relates consumption rate y to resource density x; that is $y = f(x)$. Let g be the function that relates the per capita population growth rate z of the population to the consumption rate y; that is, $z = g(y)$. Then by substitution, we obtain $z = g(f(x))$. We have expressed the growth rate as a function of resource density through the process of taking a function of a function. This process is known as **composition** and is shown in Figure 1.37.

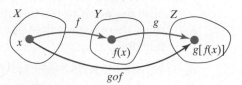

Figure 1.37 Composition of two functions.

Composite Functions

Let f and g be functions with domains A and B, respectively. The **composite function** $g \circ f$ is defined by

$$(g \circ f)(x) = g[f(x)]$$

The domain of $g \circ f$ is the subset of A for which $g \circ f$ is defined.

To visualize how functional composition works, think of $f \circ g$ in terms of an assembly line in which f and g are arranged in series, with output f becoming the input of g.

Example 7   Composing functions

Let $f(x) = 2x + 1$ and $g(x) = \sqrt{x}$. Find the composite functions $g \circ f$ and $f \circ g$ and their domains.

Solution   The function $g \circ f$ is defined by

$$g[f(x)] = g(2x + 1) = \sqrt{2x + 1}$$

Notice that $g \circ f$ means that f is applied first, then g is applied. Since $g \circ f$ is defined only for $2x + 1 \geq 0$ or $x \geq -\frac{1}{2}$, the domain of $g \circ f$ is $\left[-\frac{1}{2}, \infty\right)$.

The function $f \circ g$ is defined by

$$f(g(x)) = f(\sqrt{x}) = 2\sqrt{x} + 1$$

In this part, first apply g then apply f. Since $f \circ g$ is defined only for $x \geq 0$, we see the domain of $f \circ g$ is $[0, \infty)$.

Example 7 illustrates that functional composition is not, in general, commutative. That is, in general,

$$f \circ g \neq g \circ f$$

Sometimes it can be useful to express a function as the composite of two simpler functions.

Example 8   Decomposing functions

Express each of the following functions as the composite $f \circ g$ of two functions f and g.

a. $\sin^2 x$ **b.** $(2 + \cos x)^3$

Solution

a. A good way of thinking about this is to think about how you would use a calculator to evaluate this expression. You would first find sine of x, and then square the result. Hence, let $g(x) = \sin x$ and $f(x) = x^2$ so that

$$f[g(x)] = f(\sin x) = (\sin x)^2 = \sin^2 x$$

b. To evaluate the function $y = (2 + \cos x)^3$, first take cosine of x, add two, and raise the result to the power x. Since the evaluation of this function takes three steps, there is more than one way that we can represent it as a composition of two functions.

Let $g(x) = \cos x + 2$ and $f(x) = x^3$. Then,

$$f[g(x)] = f(\cos x + 2) = (\cos x + 2)^3$$

Alternatively, let $g(x) = \cos x$ and $f(x) = (2 + x)^3$. Then,

$$f[g(x)] = f(\cos x) = (2 + \cos x)^3$$

The next example involves the composition of two well-known functions in ecology: (1) the consumption function $y = f(x)$ that relates the rate at which an organism is able to consume a resource of density x in the environment (also known as the *functional response*) and (2) the per capita growth rate of an organism $g(y)$ that is a function of the consumption rate y.

The example is based on data pertaining to the daily rates at which individual short-tailed shrews (*Blarina brevicauda*), as shown in Figure 1.38, gather cocoons of the European pine sawfly (*Neodiprion sertifer*) buried in forest-floor litter. These data, as a function of cocoon density x per thousandth acre (i.e., acres $\times 10^{-3}$), can be fitted reasonably well by the function

$$y = f(x) = \frac{320\,x}{110 + x} \qquad \text{cocoons per day.}$$

This function is equivalent to the Michaelis-Menten uptake function derived in Example 6.

© FLPA/S & D & K Maslow/AgeFotostock America, Inc.

Figure 1.38 Short-tailed shrew (*Blarina brevicauda*).

Example 9   Short-tailed shrews exploiting cocoons

Consider the shrew population studied by the Canadian ecologist C. S. Holling, "Some Characteristics of Simple Types of Predation and Parasitism," *The Canadian Entomologist* 91(1959): 385–398. Suppose we are given the information that under ideal conditions (i.e., when the number of sawfly cocoons per shrew is essentially unlimited) each pair of shrews produces an average of around twenty female and twenty male progeny per year.

a. Use these data to estimate the maximum per capita growth rate r per day in the growth rate function* $g(y) = r\left(1 - \dfrac{b}{y}\right)$, where y is the number of cocoons

*W. M. Getz, "Metaphysiological and Evolutionary Dynamics of Populations Exploiting Constant and Interactive Resources: *r-K* Selection Revisited," *Evolutionary Ecology* 7(1993): 287–305.

consumed per day per shrew, r is the *maximal per capita growth rate of the population*, and $b = 100$ is the *growth rate break even point*, that is, $g(b) = 0$.

b. Use functional composition on this growth rate function and Holling's response function $f(x) = \dfrac{320x}{110 + x}$ on the daily rate at which shrews collect cocoons to find the daily per capita growth rate $G = g \circ f$ as a function of cocoon density x.

Solution

a. Under ideal conditions, the per capita growth rate is given by 20 pairs/year. To convert this yearly rate to a daily rate, we divide by 365 days/year to get an estimate for the maximal per capita growth rate

$$r = \frac{20}{365} \approx 0.055 \text{ pairs/day}$$

Setting $b = 100$, we get the per capita growth rate as a function of cocoons consumed per day y:

$$g(y) = 0.055 \left(1 - \frac{100}{y} \right)$$

b. Taking functional composition of g and f now yields the per capita growth rate as a function of the cocoon density:

$$G(x) = g[f(x)] = 0.055 \left(1 - \frac{100}{f(x)} \right)$$
$$= 0.055 \left(1 - \frac{100(110 + x)}{320x} \right)$$
$$= 0.055 \left(\frac{320x - 100x - 11{,}000}{320x} \right)$$
$$= \frac{0.055(220x - 11{,}000)}{320x} \text{ pairs/day}$$

PROBLEM SET 1.5

Level 1 DRILL PROBLEMS

Let $y = f(x)$ be the function whose graph is given by Figure 1.39. Sketch the graph of the functions in Problems 1 to 6.

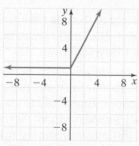

Figure 1.39 Graph of f.

1. $y = f(x) + 2$

2. $y = f(x + 1)$

3. $y = f(x - 2) + 1$

4. $y = 2f(x + 2)$

5. $y = -f(x)$

6. $y = f(-x + 2)$

Sketch the graph of the functions in Problems 7 to 10 by appropriately shifting, stretching, or translating the graph of $y = \cos x$.

7. $y = \cos \left(x - \dfrac{\pi}{2} \right)$

8. $y = 3 \cos(2x)$

9. $y = 3 \cos \dfrac{x}{2}$

10. $y - 2 = 2 \cos \left(x + \dfrac{2\pi}{3} \right)$

In Problems 11 to 19 sketch the graph of each function without the aid of technology.

11. $y = (x - 2)^2$

12. $y = (x - 2)^2 + 1$

13. $y = -1.25(x - 2)^2$

14. $y = (x + 1)^3 + 5$

15. $y = (2x + 1)^3 + 5$

16. $y = (2x + 1)^3/2$

17. $y = 1 - e^{-x}$ **18.** $y = -|x|/2$

19. $y = 1 - 2|x|$

20. Find the indicated values given the functions
$f = \{(0, 1), (1, 4), (2, 7), (3, 10)\}$ and
$g = \{(0, 3), (1, -1), (2, 1), (3, 3)\}$

  a. $(f + g)(1)$

  b. $(f - g)(2)$

  c. $(fg)(2)$

  d. $(f/g)(0)$

  e. $(f \circ g)(2)$

21. Find the indicated values given the functions

$$f(x) = \frac{2x^2 - 5x + 2}{x - 2}$$

and

$$g(x) = x^2 - x - 2$$

  a. $(f + g)(-1)$

  b. $(f - g)(2)$

  c. $(fg)(9)$

  d. $(f/g)(99)$

  e. $(f \circ g)(0)$

22. Let $f(t)$ be a periodic function with period $p = 2\pi$ and amplitude $a = 1$. Show that the given functions are periodic and find their period and amplitude.

  a. $f(t - 1) + 2$

  b. $5f(t)$

  c. $f\left(\dfrac{t}{\pi}\right)$

  d. $2f\left(t + \dfrac{\pi}{2}\right) - 3$

23. Let $f(t)$ be a periodic function with period $p = T$ and amplitude $a = A$. Show that the following functions are periodic and find their period and amplitude.

  a. $f(t + 1) - 2$

  b. $4f(t)$

  c. $-2f(3t)$

  d. $2f(t - 4) + 1$

Express each of the functions in Problems 24 to 29 as the composition $f \circ g$ of two functions f and g. (Answers are not unique.)

24. $y = (2x^2 - 1)^4$

25. $y = \sqrt{1 - \sin x}$

26. $y = e^{-x^2}$

27. $y = e^{1-x^2}$

28. $y = |x + 1|^2 + 6$

29. $y = (x^2 - 1)^3 + \sqrt{x^2 - 1} + 5$

For each of the functions in Problems 30 to 33, find $f + g$, fg, and, f/g. Also give the domain of each of these functions.

30. $f(x) = \dfrac{x - 2}{x + 1}$ and $g(x) = x^2 - x - 2$

31. $f(x) = \dfrac{2x^2 - x - 3}{x + 1}$ and $g(x) = x^2 - x - 2$

32. $f(x) = \sqrt{1 - x}$ and $g(x) = \sqrt{4 - x^2}$

33. $f(x) = \sqrt{4 - x^2}$ and $g(x) = \sin(\pi x)$

Level 2 APPLIED AND THEORY PROBLEMS

34. The tides for Hell Gate, Wards Island, New York, on September 6, 2004, are given by the following table:

Time	Height (ft)	Tide
12:08 A.M.	2.1	Low
5:19 A.M.	5.8	High
12:00 noon	2.1	Low

Let

$$T(t) = A\cos[B(t + C)] + D \text{ feet}$$

denote the height of the tide t hours after midnight. Find values of A, B, C, and D such that the function fits the Hell Gate tide data.

35. The tides for Bodega Bay, California, on March 10, 2005, are given by the following table:

Time	Height (ft)	Tide
4:36 A.M.	1.1	Low
10:43 A.M.	5.8	High
5:02 P.M.	-0.4	Low

Let

$$T(t) = A\cos[B(t + C)] + D \text{ feet}$$

denote the height of the tide t hours after midnight. Find values of A, B, C, and D such that the function fits the Bodega Bay tide data.

36. Enzymes are nature's catalysts because they are compounds that enhance the rate (speed) of biochemical reactions. Enzymes are used according to the body's need for them. Some enzymes aid in blood clotting and some aid in digestion. Even enzymes within the cell are needed for specific reactions. In this problem, you will derive a model of a biochemical reaction where there is a substance (e.g., glucose) that is converted to a new substance (e.g., fructose) by an enzyme (e.g., isomerase). Let $f(x)$ be the amount of substance produced per minute as a function of the substrate concentration x. To model this reaction rate, assume that enzymes are either "occupied" (i.e., processing a substrate particle) or are "unoccupied" (i.e., waiting to bind to another substrate particle).

a. Let t be the fraction of time that an enzyme is unoccupied. Assuming that $1 - t$ is proportional to x, find t as a function of x.

b. Assuming that $f(x)$ is proportional to $1 - t$, find an expression for $f(x)$.

c. Below are data for glucose-6-phosphate converted to fructose-6-phospate by the enzyme phosphoglucose isomerase.

Substrate concentration (micromolar)	Reaction rate (micromolar/minute)
0.08	0.15
0.12	0.21
0.54	0.70
1.23	1.1
1.82	1.3
2.72	1.5
4.94	1.7
10.00	1.8

Using linear regression on the transformed data, the uptake rate can be approximated by $f(x) = \dfrac{1.95x}{1 + 0.95x}$. Graph this function against the data.

37. In several applications it has been useful to fit a function of the form $y = f(x) = \dfrac{bx}{1 + ax}$ to a data set. Consider the change of variables given by $t = 1/x$ and $z = 1/y$.

a. Write an expression for z in terms of t.

b. Consider the following data set

Substrate concentration y (micromolar)	Reaction rate x (micromolar/minute)
0.08	0.15
0.12	0.21
0.54	0.70
1.23	1.1
1.82	1.3
2.72	1.5
4.94	1.7
10.00	1.8

Take the reciprocals of the (x, y) data values to get the corresponding (t, z) values. Use technology to fit a line to the (t, z) data. If this line is given by $z = c + dt$, use your work in **a** to find the parameters a and b in $y = \dfrac{bx}{1 + ax}$.

38. Environmental studies are often concerned with the relationship between the population of an urban area and the level of pollution. Suppose it is estimated that when p hundred thousand people live in a certain city, the average daily level of carbon monoxide in the air is

$$L(p) = 0.07\sqrt{p^2 + 3}$$

ppm. Further, assume that in t years there will be

$$p(t) = 1 + 0.02t^3$$

hundred thousand people in the city. Based on these assumptions, what level of air pollution should be expected in four years?

39. The volume, V, of a certain cone is given by

$$V(h) = \frac{\pi h^3}{12}$$

Suppose the height is expressed as a function of time, t, by $h(t) = 2t$.

a. Find the volume when $t = 2$.

b. Express the volume as a function of elapsed time by finding $V \circ h$.

c. If the domain of V is $[0, 6]$, find the domain of h; that is, what are the permissible values for t?

40. The surface area, S, of a spherical balloon with radius r is given by

$$S(r) = 4\pi r^2$$

Suppose the radius is expressed as a function of time t by $r(t) = 3t$.

a. Find the surface area when $t = 2$.

b. Express the surface area as a function of elapsed time by finding $S \circ r$.

c. If the domain of S is $(0, 8)$, find the domain of r; that is, what are the permissible values for t?

41. The ecologist C. S. Holling mentioned in Example 9 also collected data on the daily rates at which individual masked shrews (*Sorex cinereus*), gathered European pine sawfly cocoons in

forest-floor litter. His data for this species are fitted by the functional response

$$f(x) = 110\frac{x^4}{300^4 + x^4} \qquad \text{cocoons per day}$$

where x is the density of cocoons on the forest floor. If breeding pairs for this species produce approximately four female and four male progeny per year under favorable conditions and the growth breakeven point is $b = 40$ cocoons per day, then find the specific form of the per capita hyperbolic growth rate r per day: $g(y) = r(1 - b/y)$ for this species. Use this to derive the composite per capita growth rate function $G = (g \circ f)(x)$. Plot a graph of this composite function plotted over the interval $[200, 400]$.

42. Suppose the number of hours between sunrise and sunset in Los Angeles, is modeled by

$$H = 12.17 + 1.5\sin\left(\frac{2\pi n}{365} - 1.5\right)$$

where n is the number of the day in the year ($n = 1$ on January 1 and $n = 365$ on December 31, except in leap years when $n = 366$). On what days of the year in 2011 were there approximately twelve hours of daylight in Los Angeles?

43. According to the model in Problem 42, when will the length of the day in Los Angeles be about thirteen hours?

44. In an experimental study performed at Dartmouth College, two groups of mice with tumors were treated with the chemotherapeutic drug cisplatin. In Problems 44 and 45 in Section 1.4, we modeled the growth of the volume of these tumors before and after chemotherapy. A key feature missing in the model was that tumors often consist of a mixture of quiescent (i.e., nondividing) cells and proliferating cells. To add this important component to the model, assume (as was observed in the experiments), that 99% of cells are quiescent and 1% are proliferating following chemotherapy. Furthermore, assume that the volume of tumors at the time of therapy is 0.5 cm^3, the quiescent cells have a half-life of 5.7 days, and the proliferating cells have a doubling time of 2.9 days.

   a. Write a function of the tumor growth as a function of t in days.

   b. Graph this function over a twenty-day period and describe what you see.

45. The following graph posted at the Global Warming Art website indicates regular oscillations to observed sunspot activity over the past 170 years.

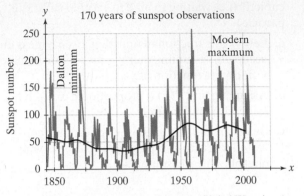

Source: Image created by Robert A. Rohde/Global Warming Art. Reprinted with permission.

Find a trigonometric function of the form $y = a + b\cos(cx + d)$, where x is in years, that provides a good fit to the data from the minimum that occured around 1844 to the minimum that occurred in 1997.

46. The following is a graph of a famous data set on the number of lynx pelts handled by the Hudson Bay Company from 1821 to 1910.

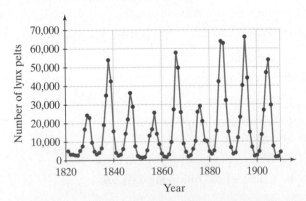

Data Source: Taken from Krebs, C J., R. Boonstra, S. Boutin and A. R. E. Sinclair, "What Drives the 10-year Cycle of Snowshoe Hares?" *BioScience* 51(1)(2001): 25–35. First published by Elton, C., and M. Nicholson, "The Ten-Year Cycle in Numbers of the Lynx in Canada," *Journal of Animal Ecology* 11 (1942): 215–244.

Write a trigonometric function of the form $y = a + b\cos(cx + d)$, where x is in years, that provides a good fit to these data.

1.6 Inverse Functions and Logarithms

Table 1.13 U.S. population size N as a function of time t

Year t	$N = f(t)$ Population size (in millions)
1815	8.3
1825	11.0
1835	14.7
1845	19.7
1855	26.7
1865	35.2
1875	44.4
1885	55.9
1895	68.9

Table 1.14 Year t as a function of U.S. population size N

N Population size (in millions)	Year $t = f^{-1}(N)$
8.3	1815
11.0	1825
14.7	1835
19.7	1845
26.7	1855
35.2	1865
44.4	1875
55.9	1885
68.9	1895

Sometimes when we are given the output of a function, we want to know what inputs could generate the observed output. For instance, consider the function that assigns to each gene the protein that it encodes. If in an experimental study we observe certain proteins at high abundance, we might want to know what genes might have been expressed. At another time we might be interested in how long before a population doubles in size or reaches a specific size. Here, we introduce the concepts of one-to-one functions and inverse functions to tackle such questions. A particularly important family of inverse functions consists of logarithm functions. As you will discover, logarithmic functions are extremely convenient for examining questions about scaling laws, population growth, and radioactive decay.

Inverse functions

Let's reexamine the U.S. population growth data introduced at the beginning of Section 1.4 and shown again in Table 1.13.

We can view this table as defining a function $f(t)$ that associates each year t with the U.S. population size $N = f(t)$ in that year. The domain of this function is {1815, 1825, 1835, 1845, 1855, 1865, 1875, 1885, 1895} and its range is {8.3, 11, 14.7, 19.7, 26.7, 35.2, 44.4, 55.9, 68.9}. An important feature of this function is for each value N in its range, there is only a single year t in the domain such that $f(t) = N$. For example, there are 19.7 million people in the United States only in 1845 and 55.9 million people in the United States only in 1885. Functions that have this feature are called *one-to-one*. This one-to-one feature of $f(t)$—that is, two different inputs lead to two different outputs (some people refer to this as *two-to-two*)—allows us to define a new function $t = f^{-1}(N)$ that sends each population size N (in millions) to the year t corresponding to that population size N. This function is the *inverse function of N* and is shown in Table 1.14.

While the U.S. population data function $N = f(t)$ is one-to-one and that allows us to define an inverse function, not all functions are one-to-one. For example, consider the quadratic function $y = f(x) = x^2$. Since there are two values of $x = 2, -2$ such that $f(x) = 4$, this function is not one-to-one.

One-to-One (Two-to-Two)

A function $f : X \to Y$ is **one-to-one** if $f(a) = f(b)$ for some a and b in X implies that $a = b$ (this can also be thought of as two different x in X to two different y in Y).

Example 1 Checking for one-to-one

Determine whether each of the following functions is one-to-one.

a. $y = f(x) = x^3$

b. $y = f(x) = \sin x$

Solution

a. Suppose that $a^3 = b^3$. Taking the cube root of both sides of this equation, the laws of exponents imply $a = (a^3)^{1/3} = (b^3)^{1/3} = b$. Therefore, $y = x^3$ is one-to-one as we have shown that $f(a) = f(b)$ implies $a = b$.

b. We know that sine takes on the value 0 infinitely often. In particular, $\sin 0 = \sin \pi = 0$. Since we found $a = 0$ and $b = \pi$ such that $a \neq b$ but $f(a) = f(b)$, the function $f(x) = \sin x$ is not one-to-one.

In Section 1.1, we used the vertical line test to determine if a given relation is a function. We have a similar test, called the **horizontal line test**, to determine if a given function is one-to-one.

| Horizontal Line Test | A function f is one-to-one if and only if every horizontal line over the function's domain intersects the graph of $y = f(x)$ in at most one point. |

Example 2   Using the horizontal line test

Determine which of the following functions are one-to-one.

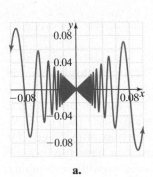

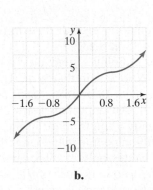

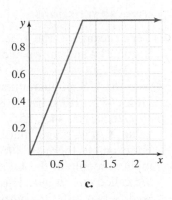

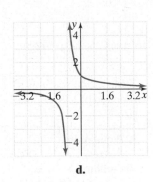

a.

b.

c.

d.

Solution

a. Since any horizontal line would intersect multiple points on the graph of $y = f(x)$, this function is not one-to-one.

b. Since any horizontal line only intersects the graph of $y = f(x)$ in one point, this function is one-to-one.

c. Since the horizontal line $y = 1$ passes through an infinite number of points of the graph, this function is not one-to-one.

d. Since any horizontal line intersects the graph in at most one point, this function is one-to-one.

When a function $y = f(x)$ is one-to-one, we can associate a single value in its domain with a single value in its range. This association defines the **inverse function** of $f(x)$.

| Inverse Function | Let f be a one-to-one function with domain D and range R. The **inverse** f^{-1} of f is the function with domain R and range D such that $$f^{-1}(y) = x \text{ if and only if } y = f(x)$$ Equivalently, $$f^{-1}(f(x)) = x \text{ for all } x \text{ in } D$$ |

The second characterization of an inverse function states that if we take the output from the function f as the input to the function f^{-1}, then we get back the input into the function f. Hence, the inverse function f^{-1} can be viewed as a "machine that undoes" the work of the function f.

Example 3 Using the definitions of the inverse function

a. Find $f^{-1}(2)$ and $f^{-1}(-3)$ for a one-to-one function $y = f(x)$ satisfying $f(-3) = 4$, $f(1) = 2$, $f(2) = 0$, and $f(4) = -3$.

b. Show that $g(x) = (x - 3)^{1/3}$ is the inverse function of $f(x) = x^3 + 3$.

Solution

a. To find $f^{-1}(2)$, we need to find x such that $f(x) = 2$. This is $x = 1$. Therefore, $f^{-1}(2) = 1$. To find $f^{-1}(-3)$, we need to find x such that $f(x) = -3$. This is $x = 4$. Therefore, $f^{-1}(-3) = 4$.

b. Since the range of f is all of the real numbers, the domain of g should be all of the reals. For any real number x, we have that

$$(g \circ f)(x) = (f(x) - 3)^{1/3}$$
$$= (x^3 + 3 - 3)^{1/3}$$
$$= (x^3)^{1/3} = x$$

Thus, g is f^{-1}.

In part **b** of the previous example, we expressed the inverse function g as a function of x instead of as a function of y. This choice corresponds to the convention that x is considered the independent variable and y is the dependent variable. For the U.S. population growth example at the beginning of the section, we did not interchange the names of N and t, as N and t had different units, that is, population size (in millions) and years. In general, it is not necessary to interchange the names of x and y if we are comfortable expressing the inverse function as $x = f^{-1}(y)$.

To find the inverse of a function algebraically, there is a simple procedure to follow.

Finding the Inverse

Let $y = f(x)$ be a one-to-one function with domain D and range R. To find the inverse function, follow these steps:

Step 1. Solve the equation $y = f(x)$ for x in terms of y (provided this is possible).

Step 2. (Optional) To express the inverse function as a function of x, interchange the xs and ys in the solution from Step 1. This results in the equation $y = f^{-1}(x)$.

Example 4 Finding inverses

Find the inverses for the following functions.

a. The function defined by the equation

$$y = f(x) = \frac{1}{1 + x}$$

b. The function defined by this verbal description: to every $r \geq 0$ associate the area of a circle of radius r

Solution

a. We begin by solving for x in terms of y for the equation $y = f(x)$:

$$y = \frac{1}{1+x} \qquad assuming\ x \neq -1$$

$$1 + x = \frac{1}{y}$$

$$x = \frac{1}{y} - 1$$

Interchanging the roles of x and y, we get $y = f^{-1}(x) = \frac{1}{x} - 1$ for all $x \neq 0$. Notice that the range of $f(x)$ is all the real numbers but zero, and this range corresponds to the domain of the function $f^{-1}(x)$ that we found.

b. The area A of a circle of radius $r \geq 0$ is given by $A = \pi r^2$. The range of A is $[0, \infty)$. To find the inverse, for every $A \geq 0$, solve for r in terms of A

$$A = \pi r^2$$

$$r = \sqrt{\frac{A}{\pi}}$$

In words, the radius of a circle is the square root of its area divided by π. Since there are no conventions associated with the variable names A and r, we do not bother to interchange the names, especially because A stands for *area* and r stands for *radius* and we do not want to mix these up.   ▪

The idea of interchanging the roles of x and y to find the equation of an inverse function also can be used to graph an inverse function. Indeed, if $(x, y) = (a, b)$ is a point on the graph of $y = f(x)$, then $f(a) = b$. Therefore, $a = f^{-1}(b)$ and $(x, y) = (b, a)$ is a point on the graph of $y = f^{-1}(x)$.

Graphing Inverses

If f is one-to-one, then the graph of its inverse $y = f^{-1}(x)$ is given by reflecting the graph of $y = f(x)$ about the line $y = x$.

Example 5   Graphing inverses

Consider the function $y = f(x)$ whose graph is shown below,

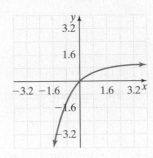

Sketch the graph of $y = f^{-1}(x)$ by hand.

Solution   If we sketch the line $y = x$ in black, then reflecting the graph about the line $y = x$ yields the graph of $y = f^{-1}(x)$, shown in red.

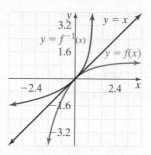

Logarithms

Consider the exponential function $y = a^x$ where $a > 0$. Suppose that $x_1 \neq x_2$ are such that $a^{x_1} = a^{x_2}$. Then by the laws of exponents, $1 = a^{x_1}a^{-x_2} = a^{x_1-x_2}$. Since $x_1 - x_2 \neq 0$, a must equal 1. Hence, we may conclude that $y = a^x$ is a one-to-one function provided that $a \neq 1$. Therefore, $y = a^x$ has an inverse function provided that $a \neq 1$. This inverse is the *logarithm function with base a*.

Logarithm

Let $a > 0$ and $a \neq 1$. Then

$$y = \log_a x \text{ if and only if } a^y = x$$

$y = \log_a x$ is the **logarithm of x with base a**.

The statement "$y = \log_a x$" should be read as "y is the exponent on a base a that gives the value x." *Do not forget that a logarithm is an exponent.*

Example 6   Using the definition of logarithm

Find x such that

a. $x = \log_2 16$ **b.** $x = \log_4 16$ **c.** $\log_{10} x = 3$ **d.** $\log_e x = 2$

Solution
a. "x is the exponent on a base 2 that gives 16"; Since $2^4 = 16$, $x = 4$.
b. "x is the exponent on a base 4 that gives 16"; Since $4^2 = 16$, $x = 2$.
c. "3 is the exponent on a base 10 that gives x"; $x = 10^3 = 1{,}000$.
d. "2 is the exponent on a base e that gives x"; $x = e^2$.

In elementary work, the most commonly used base is 10, so we call a logarithm to the base 10 a **common logarithm** and agree to write it without using a subscript 10. Thus, part **c** of the previous example is usually written $\log x = 3$. In biological applications dealing with natural growth or decay, the base e is more common. A logarithm to the base e is called a **natural logarithm** and is denoted by $\ln x$. The expression $\ln x$ is often pronounced "ell en ex" or "lawn ex." In some texts, especially those pertaining to information theory in computer science, the function $\log_2 x$ is of theoretical importance and it is written simply as $\lg x$. Its use, however, is not common in differential or integral calculus.

Logarithmic Notations

- **Common logarithm**: $\log x$ means $\log_{10} x$.
- **Natural logarithm**: $\ln x$ means $\log_e x$.

To **evaluate** a logarithm means to find an exact answer if possible (e.g., $\log_9 81 = 2$), or to calculate a decimal approximation to a required number of decimal places. Find the keys labeled $\boxed{\text{LOG}}$ and $\boxed{\text{LN}}$ on your calculator. Verify the following calculator evaluations using your own calculator:

$$\log 5.03 \approx 0.7015679851 \qquad \ln 3.49 \approx 1.249901736 \qquad \log 0.00728 \approx -2.137868621$$

Since logarithms are exponents, the following properties of logarithms follow immediately from the properties of exponents and the definition of logarithms.

Laws of Logarithms	

Additive law: $\log_a x + \log_a y = \log_a xy$

Subtractive law: $\log_a x - \log_a y = \log_a \dfrac{x}{y}$

Multiplicative law: $y \log_a x = \log_a (x^y)$

Change of base: $\log_b x = \dfrac{\log_a x}{\log_a b}$

Cancellation properties: $\log_a a^x = x$
$a^{\log_a x} = x, \; x > 0$

Example 7   Graphing logarithmic functions

Use technology to graph the logarithmic functions $y = \log x$, $y = \ln x$, and $y = \log_2 x$ on the same coordinate axes. Discuss the common properties of these graphs.

Solution   The graphs (using technology) are shown in Figure 1.40.

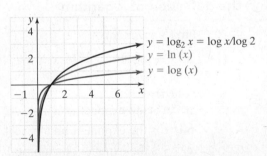

Figure 1.40 Graphs of logarithmic functions.

In all cases, the function has a domain of $(0, \infty)$ and range of $(-\infty, \infty)$—that is, the range is the real number line **R**. The x-intercept is $(1, 0)$. Additionally, the graph appears to be increasing from $-\infty$ as x increases from 0, and the steepness of the graph is decreasing with increasing x.

Example 8   Solving exponential equations

Approximate the solutions to two decimal places.

   a. $10^x = 0.5$ **b.** $e^x = 10,000$ **c.** $1.33^t = 2$ **d.** $\ln(2x) = 1$ **e.** $\log_2 4^x = 3$

Solution   Be sure to duplicate the results below using your calculator.

   a. This means x is the exponent on a base 10 which gives 0.5: in symbols, $x = \log 0.5$. Then evaluate with your calculator to find $x = \log 0.5 \approx -0.30$.

b. Using the definition, $x = \ln 10{,}000 \approx 9.21$.

c. We note t is the exponent on a base 1.33 which is 2. That is,

$$t = \log_{1.33} 2 = \frac{\log 2}{\log 1.33} \approx 2.43$$

d. We have

$$\ln(2x) = 1$$
$$e^1 = 2x \qquad \textit{definition of logarithm}$$
$$x = \frac{e}{2} \approx 1.36$$

e. We have

$$\log_2 4^x = 3$$
$$x \log_2 4 = 3 \qquad \textit{by cancellation proprieties}$$
$$x \cdot 2 = 3$$
$$x = \frac{3}{2} = 1.5$$

Logarithmic functions are key to finding half-lives of exponentially decaying quantities and to finding doubling times for exponentially growing quantities:

Exponential half-life. In a process $x(t)$ decaying exponentially, its half-life is the time $t = T$ that it takes to change from its current size $x(0)$ to $x(T) = x(0)/2$.

Exponential doubling time. In a process $x(t)$ growing exponentially, its doubling time is the time $t = T$ that it takes to change from its current size $x(0)$ to $x(T) = 2x(0)$.

In the next example, we solve for these quantities for processes introduced in Section 1.4.

Example 9 Half-life and doubling time

a. In Example 5 of Section 1.4, the function $H(t) = 17(0.99588)^t$ denoted the height of the beer froth at time t seconds. Find the time T at which half of the froth has been lost.

b. In Section 1.4, we modeled population growth in the United States with the function $N(t) = 8.3(1.33)^t$ where N is measured in millions of individuals and t is decades after 1815. In Example 2b of Section 1.4, we estimated the doubling time for the population as twenty-five years. Find a more precise estimate for the doubling time T.

Solution
a. The half-life is the solution T to the equation

$$\frac{H(T)}{H(0)} = 0.5 \qquad \textit{given equation}$$

$$(0.99588)^T = 0.5 \qquad \textit{evaluate functions}$$

$$T = \log_{0.99588} 0.5 \qquad \textit{definition of logs}$$

$$= \frac{\log 0.5}{\log 0.99588} \qquad \textit{change of base}$$

$$= 167.89 \qquad \textit{evaluate}$$

Thus, it takes the froth around 168 seconds, which is almost three minutes, to decay to half its height!

b. The doubling time T is the solution to the equation

$$\frac{N(T)}{N(0)} = 2 \qquad \textit{given equation}$$

$$1.33^T = 2 \qquad \textit{evaluate functions}$$

$$\ln 1.33^T = \ln 2 \qquad \textit{take logs}$$

$$T \ln 1.33 = \ln 2 \qquad \textit{multiplicative law}$$

$$T = \frac{\ln 2}{\ln 1.33} \approx 2.431 \qquad \textit{evaluate}$$

Since T is in decades, the doubling of the population occurs approximately in twenty-four years and four months.

Table 1.15 Body sizes of different organisms

Organism	Approximate body size
Mycoplasma	< 0.1 picogram
Average bacterium	0.1 nanogram
Large amoeba	0.1 milligram
Bee	100 milligrams
Hamster	100 grams
Human	100 kilograms
Blue whale	100 metric tons
Sequoia	5,000 metric tons

The logarithmic scale

Organisms vary greatly in their body mass. On the one hand, mycoplasma, which are bacteria without cell walls, weigh less than a tenth of a picogram (i.e., less than 10^{-13} grams). On the other hand, an ancient sequoia weighs as much as 5,000 tons (i.e., 10^{10} grams). Body sizes of other species are shown in Table 1.15. Marking all of these body sizes on an axis would be exceedingly difficult (try it!). However, if we take the log of these body sizes in grams, we can mark these sizes easily on a single axis as shown in Figure 1.41. An equivalent means of representing the data is to mark the body sizes on a *logarithmic scale* where powers of 10 are equally spaced. Doing so yields the lower panel in Figure 1.41, which simply corresponds to replacing the log body sizes in the upper panel of Figure 1.41 with the actual body sizes. Each unit increase on the logarithmic scale thus represents a tenfold increase in the underlying quantity.

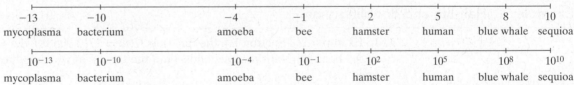

Figure 1.41 Log body sizes (upper panel) and body sizes on a logarithmic scale (lower panel) of the organisms listed in Table 1.15.

Example 10   Using the log scale

Mark the numbers 0.00005, 0.1, 20, and 60,000 on a logarithmic scale.

Solution   Applying log to the numbers 0.00005, 0.1, 20, and 60,000, we get -4.3, -1, 1.3, and 4.8. Marking these log values on an x-axis yields

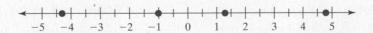

To turn this figure into a logarithmic scale, we replace the exponents x with 10^x.

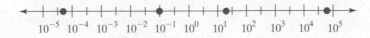

When plotting functions in the xy plane, it can be useful to plot one or both axes on the logarithmic scale.

Semi-log plots correspond to using the logarithmic scale on one axis and arithmetic scale on the other axis.

Log-log plots correspond to using the logarithmic scale on both axes.

The next example illustrates how plotting on semi-log or log-log plots can transform the graphs of nonlinear functions into the graphs of linear functions.

Example 11   Exponentials and power functions on logarithmic scales

a. Plot the function $y = 8.3(1.33)^x$ with the y axis on a logarithmic scale and the x axis on an arithmetic scale.

b. Plot the function $y = 5/x$ with the y and x axes on logarithmic scales.

Solution

a. To plot the y values on a logarithmic scale, we take the log of y

$$\log y = \log(8.3(1.33)^x)$$
$$= \log(8.3) + \log(1.33)x \quad \textit{using laws of logarithms}$$

Plotting $\log y$ against x yields the left panel below, while replacing the values on the $\log y$ axis with ten raised to the power of these values produces the semi-log plot in the right panel below.

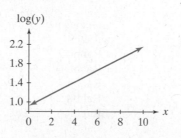 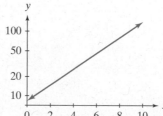

b. Taking the log of y yields

$$\log y = \log(5/x) = \log 5 - \log x$$

Plotting $\log y$ as a function of $\log x$ yields the left panel below, while replacing the values on the $\log y$ axis and $\log x$ axis with ten raised to the power of these values produces the log-log plot in the right panel below.

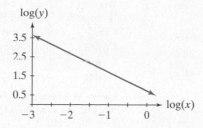 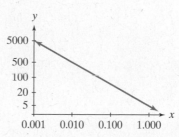

The previous example illustrates how the graphs of exponential functions and power functions are linear when using logarithmic scales appropriately.

Exponentials, Power Functions, and the Logarithmic Scale	If $y = ba^x$ is an exponential function with $b > 0$, then $\log y = \log b + x \log a$ is a linear function of x. Therefore, *exponential functions appear linear on a semi-log plot.* If $y = ax^b$ is a power function with $a > 0$, then $\log y = \log a + b \log x$ is a linear function of $\log x$. Therefore, *power functions appear linear on log-log plots.*

These observations allow us to apply data fitting techniques discuss in Section 1.2 to exponential and power functions.

Example 12   Linear regression on a logarithmic scale

The metabolic rate of an organism is the rate at which it builds up (anabolism) and breaks down (catabolism) the organic materials that constitute its body. A famous data set exhibiting an allometric scaling law for relating metabolic rate y to body mass x was first published by Max Kleiber and is reproduced here in Table 1.16.

Table 1.16 Metabolic data

Animal	Weight	kCal/day
Mouse	0.021	3.6
Rat	0.282	28.1
Guinea pig	0.410	35.1
Rabbit	2.980	167
Rabbit 2	1.520	83
Rabbit 3	2.460	119
Rabbit 4	3.570	164
Rabbit 5	4.330	191
Rabbit 6	5.330	233
Cat	3.0	152
Monkey	4.200	207
Dog	6.6	288
Dog 2	14.1	534
Dog 3	24.8	875
Dog 4	23.6	872
Goat	36	800
Chimpanzee	38	1,090
Sheep	46.4	1,254
Sheep 2	46.8	1,330
Woman	57.2	1,368
Woman 2	54.8	1,224
Woman 3	57.9	1,320
Cow	300	4,221
Cow 2	435	8,166
Cow 3	600	7,877
Heifer	482	7,754

Source: M. Kleiber, *The Fire of Life* (Krieger Pub. Co., Huntington, NY, Revised edition (June 1975), 205.

Since the data should exhibit allometry, we would expect that there exist real numbers $a > 0$ and b such that

$$y = ax^b$$

a. Apply the natural log to all of the data in Table 1.16 and use technology to find the best-fitting line for the converted data.

b. 1 data point missing from Table 1.16 is for the elephant. Use the best-fitting line to estimate the metabolic rate of an African elephant with mass of 6,800 kilograms.

Solution

a. Taking logarithms of the masses and metabolic rates in Table 1.16 and plotting the new data yields the red dots in Figure 1.42. This figure illustrates that the data on a logarithmic scale appear linear.

As before, we use technology to find the best-fitting line:

$$\ln y = 0.755917 \ln x + 4.20577$$

The $\ln y$-intercept is $(0, 4.20577)$ and the slope is $0.755917 \approx \dfrac{3}{4}$. There have been many theoretical attempts to explain this scaling exponent.

b. To predict the metabolic rate, y, for an elephant of mass $x = 6,800$, we substitute this x value into the equation for the best-fitting line and solve for y.

$$\ln y = 0.755917 \ln x + 4.20577$$
$$\ln y = 0.755917 \ln 6,800 + 4.20577$$
$$\ln y = 10.8765$$
$$y = e^{10.8765} \approx 52,918$$

The elephant will burn off approximately 53,000 kilocalories per day.

In the problem set that follows you will see that a similar approach can be used for fitting exponential functions to data.

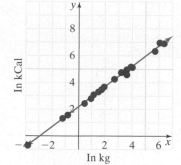

Figure 1.42 Metabolic rates on a logarithmic scale with the best-fitting line.

PROBLEM SET 1.6

Level 1 DRILL PROBLEMS

Determine whether the functions defined by the tables in Problems 1 to 4 are one-to-one. For the functions that are one-to-one, write the inverse function.

1.

x	$f(x)$
1	0
2	2
3	4
4	2
5	6

2.

x	$f(x)$
2	0
−3	14
44	22
5	6

3.

x	$f(x)$
11.9	0
17	1
−2	4
4	2
5	6

4.

x	$f(x)$
1	2
3	4
5	6
4	4
−1	0

Use the horizontal line test to determine which of the functions in Problems 5 to 8 is one-to-one. For the functions that are one-to-one, sketch the inverse.

5.

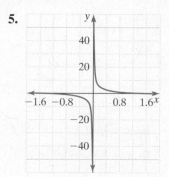

6.

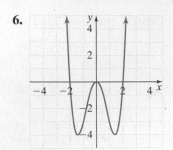

7.

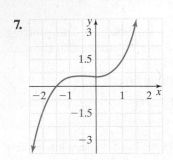

8.

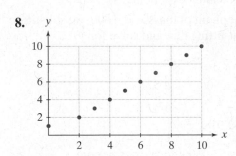

Find the inverse of the functions in Problems 9 to 14. State the domain and range of the inverse.

9. $y = \dfrac{x}{1+x}$

10. $y = e^{2x+1}$

11. $y = (x+1)^3 - 2$

12. $y = \exp(x^2)$ on $[0, \infty)$

13. $y = \exp(x^2)$ on $(-\infty, 0]$

14. $y = \sqrt{\ln x}$

Find x in Problems 15 to 19 using the definition of logarithm (do not use a calculator).

15. a. $x = \log 10$ **b.** $x = \log 0.001$

16. a. $x = \ln e^2$ **b.** $x = \ln e^{-4}$

17. a. $x = \log_5 125$ **b.** $x = \log_8 64$

18. a. $5 = \log x$ **b.** $18 = \ln x$

19. a. $\ln x = 3$ **b.** $\log x = 4.5$

Simplify the expressions given in Problems 20 to 22.

20. a. $2^{8 \log_2 x}$ **b.** $3^{3 \log_3 x}$ **c.** $5^{-2 \log_5 x}$

   d. $2^{3 \log_{1/2} x}$ **e.** $3^{-\log_{1/3} x}$

21. a. $\log_2 8^x$ **b.** $\log_3 81^x$ **c.** $\log_4 64^x$

   d. $\log_{1/2} 32^x$ **e.** $\log_3 9^{-x}$

22. a. $e^{4 \ln x}$ **b.** $e^{3 \ln(x^2+1)}$ **c.** $e^{-2 \ln(x^2-1)}$

   d. $e^{-3 \ln(1/x)}$ **e.** $e^{-\ln(1/(x^2+1))}$

In Problems 23 to 25 write the expressions in terms of base e and simplify where possible.

23. a. 5^x **b.** $\dfrac{1}{2^x}$ **c.** $5^{1/x}$

   d. 4^{x^2} **e.** 3^{x^e}

24. a. 3^{1-x} **b.** 3^{x+2} **c.** $2^{1/x+e}$

   d. 4^{x^2} **e.** 3^{-3x-2}

25. a. $\log(x+1)$ **b.** $\log(ex+e)$

   c. $\log_2(x^2-2)$ **d.** $\log_7(2x-3)$

Simplify the expressions in Problems 26 to 30 using the definition of logarithm (do not use a calculator).

26. $\log 100 + \log \sqrt{10}$

27. $\ln e + \ln 1 + \ln e^{542}$

28. $\log_8 4 + \log_8 16 + \log_8 8^{2.3}$

29. $10^{\log 0.5}$

30. $\ln e^{\log 1,000}$

Sketch the indicated points on a logarithmic scale in Problems 31 to 34.

31. $0.002, 0.5, 10, 25000$

32. $0.0003, 0.01, 0.1, 1, 200$

33. $0.00004, 0.2, 10, 200000$

34. $7, 10, 2000, 10000000$

Level 2 APPLIED AND THEORY PROBLEMS

35. In Example 6 of Section 1.5, you modeled the uptake rate of glucose by bacterial populations with the Michaelis-Menten function $U = \dfrac{1.2078\,C}{1 + 0.0506\,C}$ micrograms/hour where C is the concentration of glucose (mg/l) (micrograms/liter).

   a. If the observed uptake rate of glucose was 1 mcg/hour, find the concentration of glucose per liter.

b. Find the inverse function of the uptake rate function and be sure to identify the appropriate units for this function.

36. In Example 9 of Section 1.5, you modeled the per capita growth rate for a shrew population with the function $F = \dfrac{0.055(220\,C - 11{,}000)}{320\,C}$ pairs/day where C is the cocoon density (cocoons per thousandth acre).

   a. If the observed shrew per capita growth rate was 0.02 pairs/day, find the density of cocoons.

   b. Find the inverse function of the per capita growth rate function and be sure to identify the appropriate units for this function.

37. In Problem 41 of Section 1.4, you modeled the population size of humpback whales off the coast of Australia with the exponential function $N(t) = 350(1.12)^t$ where t is measured in years since 1981. Estimate the doubling time for this population of whales.

38. In Problem 42 of Section 1.4, you modeled the population size (in millions) of Mexico with the function $P(t) = 67.38(1.026)^t$, where t is years after 1980. Find the doubling time for the population.

39. In an experimental study performed at Dartmouth College, two groups of mice with tumors were treated with the chemotherapeutic drug cisplatin. Prior to the therapy, the tumor consisted of proliferating cells (also known as *clonogenic cells*) that grew exponentially with a doubling time of approximately 2.9 days. In Problem 44 from Section 1.4, you modeled the volume of the tumor with the function $V(t) = 0.1(2)^{t/2.9}$ cm^3, where t is measured in days. Find an exact expression for the time at which the tumor size is 0.5 cm^3.

40. In the experimental study described in Problem 39, each of these mice was given a dose of 10 mg/kg of cisplatin. At the time of the therapy, the average tumor size was approximately 0.5 cm^3. Assume all the cells became quiescent (i.e., no longer dividing) and assume decay with a half-life of approximately 5.7 days. In Problem 45 from Section 1.4, you modeled the volume of this tumor with the function $V(t) = 0.5(1/2)^{t/5.7}$. Find an exact expression for when the tumor size is 0.1 cm^3.

41. Figure 1.43 shows a plot of the weight W (in grams) versus length L (in meters) for a sample of 158 male and 167 female western hognose snakes (*Heterodon nasicus*) from Harvey County, Kansas. The females are represented by open circles, and the males by closed circles. The scale is log-log.

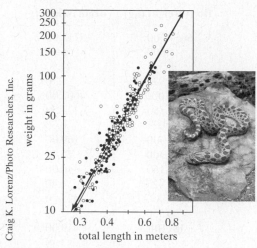

Figure 1.43 Regression line of weight versus length.
Data Source: D. R. Platt, *Natural History of the Hognose Snakes Heterodon platyrhinos and Heterodon nasicus* (Natural History Museum of the University of Kansas. Reprinted with permission.)

It appears that when $L = 0.4$ cm, the corresponding weight on the best-fitting line is $W = 28$ g; likewise, $L = 0.6$ m appears to correspond to $W = 100$ g. Assuming an allometric relationship $W = cL^m$, we have

$$28 = c(0.4)^m \quad \text{and} \quad 100 = c(0.6)^m$$

Find the allometric relationship between weight and length (round c to the nearest integer).

42. It is known that fluorocarbons have the effect of depleting ozone in the upper atmosphere. Suppose the amount Q of ozone in the atmosphere is depleted by 15% per year, so that after t years, the amount of original ozone Q_0 that remains may be modeled by

$$Q = Q_0(0.85)^t$$

   a. How long (to the nearest year) will it take before half the original ozone is depleted?

   b. Suppose through the efforts of careful environmental management, the ozone depletion rate is decreased so that it takes 100 years for half the original ozone to be depleted, what is the new rate (to the nearest hundredth of a percent)?

43. Allison and Cicchetti reported data on body weight (in kilograms) and corresponding brain weight (in grams) for sixty-two different terrestrial mammals (no whales). A partial list of the data is given below.

Body weight (kg)	Brain weight (g)
6654.000	5712.00
1.000	·6.60
3.385	44.50
0.920	5.70
2547.000	4603.00
10.550	179.50
0.023	0.30
160.000	169.00
3.300	25.60
52.160	440.00
0.425	6.40
465.000	423.00
0.550	2.40
187.100	419.00
0.075	1.20
3.000	25.00
0.785	3.50
0.200	5.00
1.410	17.50
60.000	81.00

Data source: Allison, T., and D. V. Cicchetti. (1976). "Sleep in Mammals: Ecological and Constitutional Correlates." *Science* 194, pp. 732–734.

a. Plot the log of brain size against the log body weight.

b. Find the best-fitting line using least squares regression on the log-transformed data.

44. Rivers and streams carry small solid particles of rock and mineral downhill, either suspended in the water column ("suspended load") or bounced, rolled, or slid along the river bed ("bed load"). Solid particles are classified according to their mean diameter from smallest to largest as clay, silt, sand, pebble, cobble, and boulder. During low velocity flow, only very small particles (clay and silt) can be transported by the river, whereas during high velocity flow, much larger particles may be transported, as documented in the table below.

Diameter of objects moved (mm)	Speed of current (m/sec)	Classification of objects
0.2	0.10	Mud
1.3	0.25	Sand
5	0.50	Gravel
11	0.75	Coarse gravel
20	1.00	Pebbles
45	1.50	Small stones
80	2.50	Large stones (fist sized)
180	3.50	Boulders

Data soure: Nielsen A. (1950). "The Torrential Invertebrate Fauna," *Oikos* 2: 176–196.

a. Plot the log of diameter against the log of the speed.

b. Find the best-fitting line using least squares regression on the log-transformed data.

45. Rainbow trout taken from four different localities along the Spokane River (eastern Washington) during July, August, and October 1999 were analyzed for heavy metals for the Washington State Department of Ecology. As part of this study, the length (in millimeters) and weight (in grams) of each trout were measured, as documented in the Table below; age determinations using scales are currently underway.

Length (mm)	Weight (g)
457	855
405	715
455	975
460	895
335	472
365	540
390	660
368	581
385	609
360	557
346	433
438	840
392	623
324	387
360	479
413	754
276	235
387	538
345	438
395	584

Data source: Johnson, A. *Results from Analyzing Metals in 1999 Spokane River Fish and Crayfish Samples* (Washington State Dept. of Ecology report, 2000).

a. Plot the log of weight against length.

b. Find the best-fitting line using least squares regression on the transformed data.

46. Consider the first four entries presented in the table below, which represent one estimate of the world's population levels over the second millennium AD:

Year AD (t)	Population size x (in billions)
1000	0.31
1250	0.40
1500	0.50
1750	0.79

Provide a semi-log plot of the points $(t, \ln x)$ and find and graph the best-fitting line through these points on the same plot. From this line, provide an estimate of the average growth rate exponent r for the population size function $x(t) = ce^{rt}$ over this period of 750 years.

47. Consider the entries presented in the table below, which represent one estimate of the world's population levels over the period 1750 to 1920:

Year AD (t)	Population size x (in billions)
1750	0.79
1800	0.98
1850	1.26
1900	1.65
1910	1.75
1920	1.86

Provide a semi-log plot of the points $(t, \ln x)$ and find and graph the best-fitting line through these points on the same plot. From this line, provide an estimate of the average growth rate exponent r for the population size function $x(t) = ce^{rt}$ over this 170-year period.

48. Consider the entries presented in the table below, which represent one estimate of the world's population levels over the period 1920 to 2010:

Year AD (t)	Population size x (in billions)
1920	1.86
1930	2.07
1940	2.30
1950	2.52
1960	3.02
1970	3.70
1980	4.44
1990	5.27
1999	5.98
2010	6.86

Provide a semi-log plot of the points $(t, \ln x)$ and find and graph the best-fitting line through these points on the same plot. From this line, provide an estimate of the average growth rate exponent r for the population size function $x(t) = ce^{rt}$ over this 90-year period.

1.7 Sequences and Difference Equations

Often, experimental measurements are collected at discrete intervals of time. For example, the number of elephants in wildlife park in Africa may be counted every year to ensure that poachers are not exterminating the population. Blood may be drawn on a weekly basis from a patient infected with HIV and the number of CD4+ cells produced by the patient's immune system counted to monitor patient response to treatment. Data obtained in this regular fashion can be represented by a sequence of numbers over time. In this section, we describe the basic properties of such *sequences* and demonstrate that some sequences can be generated recursively using a relationship called a *difference equation*. These equations are formulated using a function from the natural numbers to the real numbers.

Sequences

We begin with the idea of a sequence, which is simply a succession of numbers that are listed according to a given prescription or rule. Specifically, if n is a natural number, the sequence whose nth term is the number a_n can be written as

$$a_1, a_2, a_3, \ldots, a_n, \ldots$$

The number a_1 is called the first term, a_2 the second term, $\ldots$, and a_n the nth term.

Sequence	A **sequence** is a real-valued function whose domain is the set of natural numbers.

When working with sequences, we alter the usual functional notation. For a function a from the natural to the real numbers we should write $a(1)$, $a(2)$, $a(3)$, $\ldots$, but for convenience we write a_1, a_2, a_3, $\ldots$. The function $a(n)$ is written a_n and is called the *general term*.

Example 1   Finding the sequence, given the general term

Find the first five terms of the sequences whose general term is given.

a. $a_n = n$ **b.** $a_n = \sin \dfrac{\pi n}{2}$ **c.** $a_n = \dfrac{n}{1+n}$

d. a_n is the digit in the nth decimal place of the number π

Solution

a. Since n is the general term, we have 1, 2, 3, 4, and 5 for the first five terms.

b. For $n = 1$, $\sin \dfrac{\pi}{2} = 1$; for $n = 2$, $\sin \dfrac{2\pi}{2} = 0$; for $n = 3$, $\sin \dfrac{3\pi}{2} = -1$; for $n = 4$,

$\sin \dfrac{4\pi}{2} = 0$; and for $n = 5$, $\sin \dfrac{5\pi}{2} = 1$. Thus the first five terms are 1, 0, 1, 0, 1.

c. Take the first five natural numbers (in order) to find: $\dfrac{1}{1+1} = \dfrac{1}{2}$, $\dfrac{2}{1+2} = \dfrac{2}{3}$,

$\dfrac{3}{1+3} = \dfrac{3}{4}$, $\dfrac{4}{1+4} = \dfrac{4}{5}$, and $\dfrac{5}{1+5} = \dfrac{5}{6}$

d. Since $\pi \approx 3.141592 \cdots$, we see the first five terms of this sequence are: 1, 4, 1, 5, and 9. ∎

To visualize a sequence, one can graph the sequence of points

$$(1, a_1), (2, a_2), (3, a_3), \ldots$$

in the coordinate plane. The first several terms of the first four sequences from Example 1 are graphed in Figure 1.44. Since the domain consists of the natural numbers, the graph consists of discrete points.

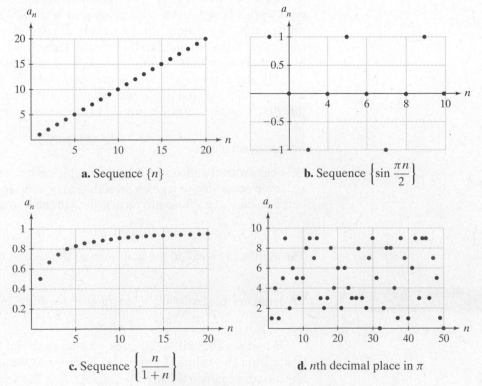

a. Sequence $\{n\}$

b. Sequence $\left\{\sin \dfrac{\pi n}{2}\right\}$

c. Sequence $\left\{\dfrac{n}{1+n}\right\}$

d. nth decimal place in π

Figure 1.44 Graphs of sequences.

Difference equations

Beyond specifying a sequence by its general term, sequences can also be generated term by term using a rule called a *difference equation*, which specifies how to calculate each term in the sequence from the values of preceding terms. For example, the difference equation

$$a_{n+1} = ra_n \quad \text{and} \quad a_1 = r$$

generates the *geometric sequence*

$$a_1 = r$$
$$a_2 = ra_1 = r^2$$
$$a_3 = ra_2 = r^3$$
$$a_4 = ra_3 = r^4$$
$$\vdots$$
$$a_n = ra_{n-1} = r^n$$
$$\vdots$$

Similarly, the difference equation

$$a_{n+1} = a_n + d \quad \text{and} \quad a_1 = d$$

generates the *arithmetic sequence*

$$a_1 = d$$
$$a_2 = a_1 + d = 2d$$
$$a_3 = a_2 + d = 3d$$
$$a_4 = a_3 + d = 4d$$
$$\vdots$$
$$a_n = nd$$
$$\vdots$$

Example 2   Geometric decay of acetaminophen

Example 2 of Section 1.1 stated that a tablet of regular strength Tylenol contains 325 mg of acetaminophen and that approximately 67% of the drug in the body is removed from the body every four hours. Assume Professor Schreiber just swallowed two tablets of Tylenol. Let a_n be the amount in milligrams of acetaminophen in his body $4n$ hours after taking the two tablets.

a. Find a_1 and find r such that $a_{n+1} = ra_n$.

b. Find and plot the amount of acetaminophen in Professor's Schreiber's body over the next twenty-four hours, that is, $n = 1, 2, 3, 4, 5, 6$.

Solution

a. Initially, there is 650 mg in his body. Hence, after four hours, there is only $(1 - 0.67)650 = 214.5$ mg. Since the fraction of acetaminophen remaining in the body after each four hour time interval is 0.33, we have $a_{n+1} = 0.33a_n$ and $r = 0.33$.

b. Using the difference equation from part **a**, we get

$$a_1 = 214.5 \qquad\qquad a_2 = 0.33a_1 = 70.785$$
$$a_3 = 0.33a_2 \approx 23.359 \qquad a_4 = 0.33a_3 \approx 7.708$$
$$a_5 = 0.33a_4 \approx 2.544 \qquad a_6 = 0.33a_5 \approx 0.839$$

Plotting these values produces the figure on the next page.

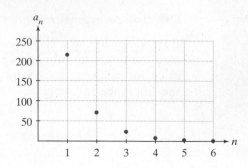

More generally, for any real-valued function f, we can make the following definition.

Difference Equation	Let f be a real-valued function. Then

$$a_{n+1} = f(a_n)$$

is a **difference equation**.

A sequence $a_1, a_2, a_3, \ldots$ satisfying this equation for all n is a **solution to the difference equation**. By specifying a particular value of a_1 and applying the $a_{n+1} = f(a_n)$ inductively, one can generate solutions to the difference equation, provided f is well-defined at every step.

Difference equations allow us to describe how quantities evolve over discrete intervals of time. For example, the difference equation $a_{n+1} = 0.33a_n$ in Example 2 described the decay of a drug in the body. A similar equation could describe the weekly growth of a bacterial culture in a laboratory, or even a population of California condors that were reintroduced to a wild area where they had previously become extinct because of use of the pesticide DDT.

From a modeling perspective, discrete intervals of time implied by the iteration of the difference equation (e.g., daily, weekly, or annual growth rules) correspond to one of two factors: synchronized events of the system (e.g., daily injections of a drug, annual reproductive cycles in a population) or intervals separating experimental measurements of the system (e.g., daily blood cell counts, annual population counts).

Example 3   The difference equation implicit in taking repeated square roots

Enter any nonzero number into your calculator. Press the square root key $\boxed{\sqrt{}}$ and record your answer. Press again and record repeatedly. Let a_n denote the nth number displayed on the screen.

a. Find a difference equation for a_n.

b. Graph the first twenty terms of the sequence when $a_1 = 4$ and then when $a_1 = 0.1$. Discuss what happens to a_n in each case as n gets very large.

c. What happens when $a_1 = 1$?

Solution

a. For any selected value a_1, after pressing the square root key, the calculator generates the number $a_2 = \sqrt{a_1}$. Similarly, after the second iteration the number $a_3 = \sqrt{a_2} = \sqrt{\sqrt{a_1}}$ is obtained. Proceeding inductively yields

$$a_{n+1} = \sqrt{a_n}.$$

Thus, the difference equation in this case is $a_{n+1} = f(a_n)$ with $f(x) = \sqrt{x}$.

b. Plotting the first twenty terms of the sequence with $a_1 = 4$ and $a_1 = 0.1$ respectively yields Figures 1.45a and 1.45b. Both plots suggest that as n gets larger, a_n approaches the value 1 (but does not become 1).

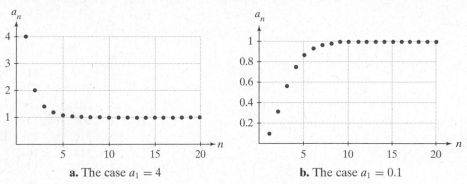

a. The case $a_1 = 4$ **b.** The case $a_1 = 0.1$

Figure 1.45 Graph of the sequence $a(n+1) = \sqrt{(a_n)}$.

c. If $a_1 = 1$, then $a_2 = \sqrt{a_1} = \sqrt{1} = 1$. Proceeding inductively, $a_n = 1$ for all integers $n \geq 1$. ∎

Difference equations can be used to model a variety of biological phenomena. The next two examples illustrate their usage in modeling repeated drug dosages and the purging of a lethal recessive gene from a population.

Example 4   Drug delivery

In Example 2, Professor Schreiber took only two tablets of Tylenol. However, the directions recommend taking two tablets every four to six hours and not taking more than ten tablets in twenty-four hours. Suppose Schreiber takes two tablets every four hours. To model how the amount of drug in Schreiber's body changes in time, let a_n be the amount of drug in his body right before taking the nth dose.

a. Write a difference equation for a_n.

b. Find a_1, a_2, a_3.

c. What is the maximum amount of acetaminophen in Schreiber's body during the first twelve hours of taking Tylenol?

d. Suppose contrary to the directions, Schreiber kept on taking doses every four hours for several day (which you should not do unless directed to do so by your physician). What value does a_n seem to approach?

Solution

a. If a_n is the amount of drug in the body just before taking the nth dose, then the amount of drug in the body after taking the nth dose is $a_n + 650$ mg. Since we know from Example 2 that 67% of the drug leaves the body in four hours, the amount of drug left in the body before taking the next dose is $(1 - 0.67)$ $(a_n + 650) = 0.33\,a_n + 214.5$. Therefore,

$$a_{n+1} = 0.33\,a_n + 214.5$$

b. Without being told, there is no way for us to know what the value of a_1 is. It is reasonable to assume that before taking the first dose, Schreiber has no acetaminophen in his body, in which case $a_1 = 0$. In this case, for $n = 2$ and $n = 3$, we obtain $a_2 = 0.33 \cdot 0 + 214.5 = 214.5$ mg and $a_3 = 0.33(214.5) + 214.5 = 285.285$ mg.

c. The maximum amount of acetaminophen in the body occurs right after taking a dose. The amounts of acetaminophen in the body after taking the first, second, and third dose are $650, 214.5 + 650 = 864.5$ mg, and $285.285 + 650 = 935.285$ mg. Hence, the maximum is given by 935.2853 mg.

d. Computing a_n for $n = 1, 2, \ldots, 20$, yields this table of values:

n	a_n	n	a_n
1	0	11	320.14
2	214.50	12	320.15
3	285.29	13	320.15
4	308.64	14	320.15
5	316.35	15	320.15
6	318.9	16	320.15
7	319.74	17	320.15
8	320.01	18	320.15
9	320.10	19	320.15
10	320.14	20	320.15

This table suggests that a_n is approaching a value that rounded to two decimal places is 320.15 mg. ∎

The difference equation $a_{n+1} = 0.33a_n + 214.5$ in Example 4 is an example of a *linear difference equation*: the right-hand side of the difference equation depends linearly on a_n. In Problem Set 1.7, you are asked to write explicit solutions for linear difference equations.

Difference equations arise in biology whenever we consider how certain quantities change over regular, discrete intervals of time:

- From one fifteen-second period to the next, as in the beer froth problem considered in Section 1.4

- From one four-hour period to the next, as in this example

- From one day to the next, as in the tumor growth problems presented in Problem Sets 1.4–1.6

- From one month to the next, as in the carbon dioxide concentration problem considered in Section 1.2

- From one year to the next, or even one decade to the next, as discussed in the U.S. population growth problem in Section 1.4

Now we consider a model of how a particular quantity changes from *one generation to the next*.

The quantity to model is the proportion of a particular *allele* (i.e., a variant of a particular gene) responsible for a genetic disease, such as Tay-Sachs or cystic fibrosis, that has a lethal effect when untreated. The model we present is the simplest example of a class of models that traces the proportion of a particular allele a in a *diploid* organism that has two possible alleles a and A associated with the gene in question and thus has *genotypes aa, aA = Aa* and *AA*. These models are *only valid for large populations*, where the assumption that one can replace the concept of probabilities with proportions holds.

Specifically, if one flips a coin four times and represents the proportion of heads using the variable x, then it is unreasonable to assume that half of the flips were heads and half were tails (i.e., $x = 0.5$), since quite often one might land up with three heads and five tails ($x = 3/8$), five heads and three tails ($x = 5/8$), or values of x even closer to 0 or 1. On the other hand, if one flipped the coin a million times, then one can safely assume, to a very good approximation, that half of the flips were heads and half were tails, that is, $x = 0.50$.

In such models, we apply the following principle of Gregor Mendel (1822–1884), derived from his work on plant hybridization

Random mating and Mendelian inheritance principle: Under the assumptions of individuals choosing mates at random, and alleles segregating randomly and independently, it follows that if x and $(1 - x)$ are the proportion of alleles A and a in a population, then the proportion of *genotypes* among the progeny, before evaluating their ability to survive, are as follows:

- x^2 for type aa
- $2x(1 - x)$ for type Aa (the 2 arises because $aA = Aa$)
- $(1 - x)^2$ for type AA

This accounts for all possible genotypes, which we check by adding these three genotype frequencies to obtain the value 1:

$$x^2 + 2x(1 - x) + (1 - x)^2 = x^2 + 2x - 2x^2 + 1 - 2x + x^2 = 1$$

Example 5   Lethal recessive genes

Suppose a disease in humans is primarily due to the existence of a *lethal recessive allele a*. By **lethal recessive**, we mean that individuals of type aa die from the disease, whereas individuals of type AA and Aa are not affected by the disease.

a. If x_n denotes the proportion of alleles a in generation n in a population of size N (i.e., for a total of $2N$ alleles), where N is very large, write an expression for x_{n+1} in terms of x_n under a random mating and Mendelian inheritance assumption. Also, assume all aa genotypes die while genotypes Aa and AA successfully reproduce one copy of themselves each.

b. Use technology to plot x_n for $n = 1, 2, 3, \ldots, 100$.

Solution

a. If x_n is the proportion of alleles in generation n, then under random mating and Mendelian inheritance principles we expect the proportion of genotypes aa, Aa, and AA in the progeny to be x_n^2, $2x_n(1 - x_n)$ and $(1 - x_n)^2$. However, since all the aa genotypes die so that a occurs only in the heterozygote genotypes Aa, in generation $n + 1$ the proportion of a alleles in a population of size N is

$$x_{n+1} = \frac{\text{number of } a \text{ alleles}}{\text{total number of alleles}}$$

$$= \frac{x_n(1 - x_n)N}{2x_n(1 - x_n)N + (1 - x_n)^2 N}$$

$$= \frac{x_n}{2x_n + (1 - x_n)} \qquad \textit{dividing top and bottom by } (1 - x_n)N$$

$$= \frac{x_n}{1 + x_n}$$

b. Using technology to compute $x_2, \ldots, x_{100}$, we get the plot illustrated in Figure 1.46. This plot shows two things. First, when the initial proportion of the lethal allele is high, the proportion of this lethal allele initially decreases very rapidly. However, as the proportion of the allele gets low, it decreases much less rapidly (e.g., in Problem Set 1.7, you will be asked to show that it takes approximately 1000 generations for the alleles to reach a proportion of 0.1%).

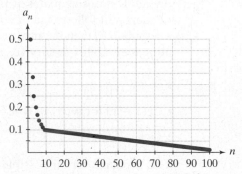

Figure 1.46 Rate of decline of a recessive lethal gene over n generations when initially at a proportion of 0.5 in the population.

In Figure 1.47, experiments on the fruit fly show that the difference equation $x_{n+1} = \dfrac{x_n}{1 + x_n}$ does a reasonable job of describing observed frequencies of the lethal allele, *Glued*, in fruit flies. The observed trajectories illustrate that even if you start with the same initial conditions (i.e., 50% with Glued), random birth and death events can result in different experimental trajectories. Hence, the model can only be expected to describe what happens for the "average" experiment.

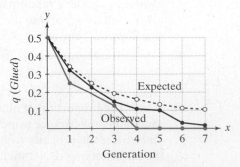

Figure 1.47 Data on the decline of the Glued gene in fruit flies compared with the expected rate predicted by the model in Example 5.

Source: Genetics 83: 793–810 August, 19iG Dynamics of Correlated Genetic Systems. I. Selection in the Region of the Glued Locus of Drosophila Melanogaster, M. T. Clegg, J. F. Kidwell, M. G. Kidwell and N. J. Daniel.

In Examples 3 and 5 we saw that for certain initial values the difference equations generating the sequences in question produced a string of constant values. Specifically, in Example 3 the difference equation $a_{n+1} = \sqrt{a_n}$ produced the sequence $1, 1, 1, \ldots$ for $a_1 = 1$ (i.e., the square root of 1 is 1) and in Example 5 the difference equation $x_{n+1} = \dfrac{x_n}{1 + x_n}$ produced the sequence $0, 0, 0, \ldots$, when $x_1 = 0$ (i.e., if the lethal allele is not present initially, it never appears). Such starting values are called **equilibria** for the equations in question.

Equilibrium

An **equilibrium** of the difference equation

$$a_{n+1} = f(a_n)$$

is an initial value a_1 such that $f(a_1) = a_1$. From this it easily follows that $a_1 = a_2 = a_3 = \cdots$.

Example 6   Finding equilibria

Find the equilibria for the following three difference equations. Discuss how the answers you find relate to what was observed in Examples 3, 4, and 5.

a. $a_{n+1} = \sqrt{a_n}$ **b.** $a_{n+1} = 0.33a_n + 214.5$ **c.** $x_{n+1} = \dfrac{x_n}{1 + x_n}$

Solution

a. To find the equilibria, we need to solve $a = \sqrt{a}$. Since the only numbers whose square roots are themselves are 0, 1, the equilibria for this difference equation are given by 0 and 1. In Example 3, we saw that for various positive initial conditions, the sequence a_n approaches the equilibrium 1 as n gets large.

b. To find the equilibria, we need to solve

$$a = 0.33a + 214.5$$
$$0.67a = 214.5$$
$$a \approx 320.15$$

In Example 4, we observed that for the initial condition $a_1 = 0$, the sequence a_n would approach this equilibrium value.

c. To find the equilibria, we need to solve, $x = \dfrac{x}{1 + x}$. One solution to this equation is $x = 0$. Any other solution must satisfy $1 = \dfrac{1}{1 + x}$. Cross-multiplying yields $1 + x = 1$. Hence, $x = 0$ is the only equilibrium. In Example 5 it appeared the sequence corresponding to $x_1 = 0.5$ might be approaching this equilibrium. However, since the approach seems quite slow, it is not obvious that x_n becomes arbitrarily close to zero. ∎

In Chapter 2, we explore more carefully the sequence approach that identified equilibria in Example 6. The next example illustrates that an equilibrium is not always approached. It is based on a model for population biology.

In 1981, Thomas Bellows investigated how the survivorship of different species of stored grain beetles depended on the population abundance x. Some of the data from this experiment are illustrated in Figure 1.48. Bellows showed that the function $s(x) = \dfrac{1}{1 + (ax)^b}$ with x corresponding to population density, $a > 0$ and $b > 0$, could describe all of these data sets. The function $s(x)$ describes the fraction of grain beetles surviving as a function of population abundance. If $r > 0$ is the average number of progeny produced by an individual, then the population model arising from this is

$$x_{n+1} = r x_n s(x_n)$$

with the specific form considered by Bellows given in the next example.

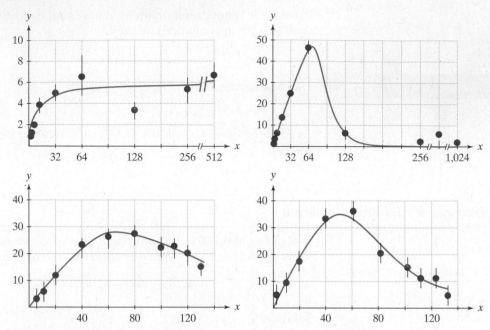

Figure 1.48 Relationship between number of survivors and initial egg density for four species of stored product beetles.

Source: After T. S. Bellows, "The Descriptive Properties of Some Models for Density Dependence," *Journal of Animal Ecology* 50(1)(1981): 139–156. Reprinted with permission.

Example 7 Generalized Beverton-Holt dynamics

If x_n is the population density in generation n, then the population model

$$x_{n+1} = \frac{r\,x_n}{1 + (ax_n)^b}$$

(sometimes referred to as the generalized Beverton-Holt model) produces solutions with behavior that depends on the three parameters r, a, and b.

a. For $r = 2$ and $a = 0.01$, find the equilibria of the model and show that they do not depend on the specific value of b.

b. For $b = 3$ and $b = 6$, with r and a as above, compute and graph for the first fifty terms of the sequence starting with the initial condition $x_1 = 99$. Compare the sequences obtained for $b = 3$ and $b = 6$.

Solution

a. To find the equilibria, we need to solve

$$x = \frac{2x}{1 + (x/100)^b}$$

for x. Clearly, $x = 0$ is a solution. For $x \neq 0$, we obtain

$$1 = \frac{2}{1 + (x/100)^b}$$
$$1 + (x/100)^b = 2$$
$$(x/100)^b = 1$$
$$x/100 = 1$$
$$x = 100$$

Thus, $x = 100$ is an equilibrium value regardless of the value of $b > 0$.

b. Using technology for $b = 3$:

$$x_{n+1} = \frac{2x_n}{1 + (x_n/100)^3}$$

and $x_1 = 99$, yields Figure 1.49a. It appears that the sequence is approaching the equilibrium value of $x = 100$.

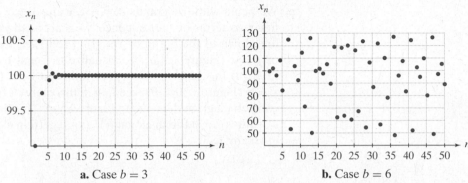

a. Case $b = 3$ **b.** Case $b = 6$

Figure 1.49 Sequences generated from the Beverton-Holt model for different values of the parameter b.

Using technology for $b = 6$:

$$x_{n+1} = \frac{2x_n}{1 + (x_n/100)^6}$$

and $x_1 = 99$, yields Figure 1.49b. Despite starting near the equilibrium abundance of $x = 100$, this sequence exhibits oscillatory bursts of population growth and decline without any other characterizable pattern of behavior. In Chapter 4, we will discuss methods to distinguish among these different outcomes. ∎

Cobwebbing

Another way to visualize sequences determined by a difference equation

$$a_{n+1} = f(a_n)$$

is via a graphical technique known as **cobwebbing**.

Before describing this technique, we provide a graphical characterization of equilibria.

| **Finding Equilibria Graphically** | To find equilibria of $a_{n+1} = f(a_n)$, it suffices to look for intersection points of the graphs of $y = x$ and $y = f(x)$ |

| **Cobwebbing** | To create a **cobweb** for the difference equation $a_{n+1} = f(a_n)$ with initial condition a_1, follow these steps:

Step 1. Graph the functions $y = f(x)$ and $y = x$ in the xy plane.

Step 2. Draw a vertical line segment from (a_1, a_1) to $(a_1, f(a_1))$ and draw a horizontal line segment from $(a_1, f(a_1))$ to $(f(a_1), f(a_1))$. Since $a_2 = f(a_1)$, you will have ended at the point (a_2, a_2).

Step 3. Repeat this procedure as desired. More specifically, if you are the point (a_n, a_n), then draw a vertical line segment from (a_n, a_n) to $(a_n, f(a_n))$ and draw a horizontal line segment from $(a_n, f(a_n))$ to $(f(a_n), f(a_n)) = (a_{n+1}, a_{n+1})$. |

Example 8   Cobwebbing square roots

Consider the difference equation $a_{n+1} = f(a_n)$ where $f(x) = \sqrt{x}$. Use cobwebbing to visualize the first ten terms of the sequence determined by

a. $a_1 = 4$ **b.** $a_1 = 0.1$

Solution

a. We begin with the graphs of $y = \sqrt{x}$ and $y = x$ (Figure 1.50**a**). To visualize the first two terms of the sequence, we start at the point $(4, 4)$ and draw a vertical line down to the graph of $y = f(x)$ followed by a horizontal line to the graph of $y = x$ (Figure 1.50**b**). To visualize the next term, draw a vertical down from $(2, 2)$ to the graph of $y = f(x)$ followed by a horizontal line to the graph of $y = x$ (Figure 1.50**c**). Proceeding in this manner for seven more iterates gives the cobweb diagram depicted in Figure 1.50**d**. This figure shows that the sequence of a_n values down the diagonal $y = x$ are getting closer to the value 1, as we found in Example 3.

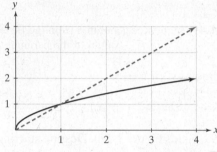
a. Draw the functions $y = x$ and $y = \sqrt{x}$

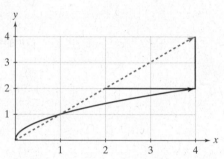
b. Start cobweb above the equilibrium

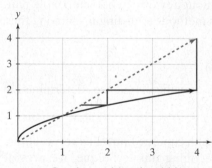
c. Continue with cobwebbing

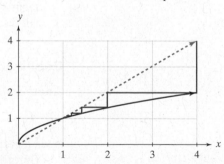
d. Solution approaches the equilibrium

Figure 1.50

b. To visualize the first ten terms of the sequence with $a_1 = 0.1$, start at $(0.1, 0.1)$, draw a vertical line to the graph of $y = f(x)$ and then a horizontal line to the graph of $y = x$ (Figure 1.51**a**). As you continue, the cobwebbing shows that the sequence of a_n values are getting closer to the value 1 (Figure 1.51**b**).

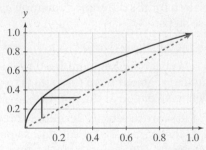
a. Start cobweb below the equilibrium

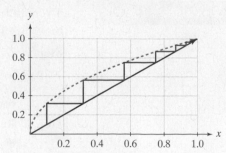
b. Solution approaches the equilibrium

Figure 1.51 Cobwebbing for the difference equation $a_{n+1} = \sqrt{a_n}$.

Cobwebbing an increasing function, such as the square root function, is relatively simple. The cobweb diagram gets more complicated when the function is both increasing and decreasing over its relevant domain (see next example).

Example 9   Cobwebbing a hump-shaped function

Use cobwebbing to visualize the first forty terms of the sequence determined by the equation

$$a_{n+1} = \frac{3a_n}{1 + (a_n/100)^6}$$

from starting value $a_1 = 50$. Discuss the primary difference between this example and Example 8.

Solution   We begin by drawing the graphs of $y = f(x) = \dfrac{3x}{1 + (x/100)^6}$ and $y = x$ (Figure 1.52**a**). To visualize the first two terms of the sequence, start at $(50, 50)$ and draw a vertical line up to the graph of $y = f(x)$ followed by a horizontal line to the graph of $y = x$ (Figure 1.52**b**). To visualize the next term, draw a vertical down from $(150, 150)$ to the graph of $y = f(x)$ followed by a horizontal line to the graph of $y = x$ (Figure 1.52**c**). Unlike our previous cobwebbing, we see that the sequence is already exhibiting some oscillation. In fact, continuing for the remaining thirty-seven terms yields the wild web depicted in Figure 1.52**d**.

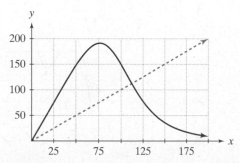

a. Draw functions $y = x$ and $y = f(x) = \dfrac{3x}{1 + (x/100)^6}$

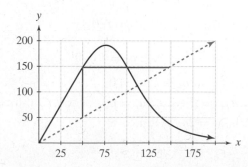

b. Start cobweb at a convenient initial value

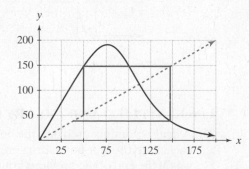

c. Continue with cobwebbing

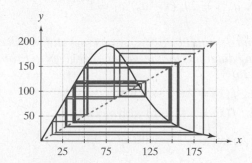

d. Solution oscillates wildly around the equilibrium

Figure 1.52 Cobwebbing for the difference equation $a_{n+1} = \dfrac{3a_n}{1 + (a_n/100)^6}$.

PROBLEM SET 1.7

Level 1 DRILL PROBLEMS

Find and graph the first five terms for the sequences in Problems 1 to 10.

1. $a_n = 1 - \dfrac{1}{n}$

2. $a_n = (-1)^{n+1}$

3. $a_n = \cos\left(\dfrac{\pi n}{2}\right)$

4. $a_n = \dfrac{\cos(2n\pi)}{n}$

5. a_n is the nth digit of the decimal representation of the number $\dfrac{1}{7}$

6. a_n is the nth digit of e

7. $a_1 = 256, a_{n+1} = \sqrt{a_n}$

8. $a_1 = 2, a_{n+1} = a_n^2, n \geq 2$

9. $a_1 = -4, a_2 = 6, a_n = a_{n-1} + a_{n-2}, n \geq 3$

10. $a_1 = 1$ and $a_2 = 2, a_{n+1} = a_n a_{n-1}, n \geq 3$

Find a_5 for each difference equation in Problems 11 to 20.

11. $a_{n+1} = a_n + 8; a_1 = 0$

12. $a_{n+1} = 3a_n; a_1 = 1$

13. $a_n = \dfrac{1}{2} a_{n-1} + 2; a_1 = 100$

14. $a_n = \dfrac{1}{10} a_{n-1} + 2; a_1 = 1,000$

15. $a_{n+1} = 5a_n + 2; a_1 = 0$

16. $a_{n+1} = 1 - 2a_n; a_1 = 0$

17. $a_{n+1} = 2a_n + 1; a_1 = 8$

18. $a_{n+1} = 1 - \dfrac{1}{2} a_n; a_1 = 0$

19. $a_{n+1} = \dfrac{2a_n}{1 + a_n}; a_1 = 1$

20. $a_{n+1} = 2a_n(1 - a_n); a_1 = 1$

Find the equilibria of $a_{n+1} = f(a_n)$ and sketch cobwebbing diagrams for the values of a_1 given in Problems 21 to 26.

21. $f(x) = 2x(1 - x)$ with $a_1 = 0.1$

22. $f(x) = x(2 - x)$ with $a_1 = 0.4$

23. $f(x) = \dfrac{3x}{1 + x}$ with $a_1 = 0.1$

24. $f(x) = \dfrac{3x}{1 + x}$ with $a_1 = 3$

25. $f(x) = 1 + x/2$ with $a_1 = 0$

26. $f(x) = \dfrac{1}{1 + x}$ with $a_1 = 3$

Find the equilibria of $a_{n+1} = f(a_n)$ where the graph of $y = f(x)$ is shown in Problems 27 to 30, and sketch the cobwebbing diagrams starting with the given a_1 value.

27. $a_1 = 1$

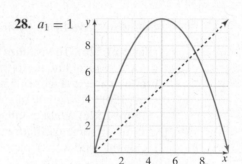

28. $a_1 = 1$

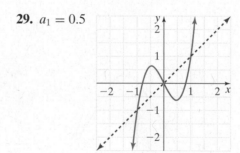

29. $a_1 = 0.5$

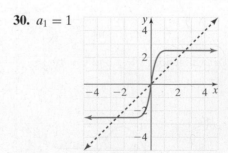

30. $a_1 = 1$

Level 2 APPLIED AND THEORY PROBLEMS

31. A drug is administered into the body. At the end of each hour, the amount of drug present is half what it was at the end of the previous hour. What percentage of the drug is present at the end of four hours? At the end of n hours?

32. A friend has a really bad headache. He decides to take 500 mg of aspirin every four hours. At the end

of each four-hour period, the body clears out 80% of the aspirin in his body. Let a_n denote the amount of aspirin in your friend's body at the time he takes the nth aspirin.

a. Write a difference equation for a_n and identify the value of a_1.

b. Write the first five terms of a_n assuming initially there is no aspirin in your friend's body.

c. Find the equilibrium of this difference equation.

33. Consider the general case of a patient who is taking a drug for a health issue. Let A be the amount the patient takes each time and c be the fraction of the drug cleared by the body between doses. Define a_n to be the amount of drug in the body *immediately prior* to taking the nth dose.

a. Write a difference equation for a_n.

b. Find the equilibrium of the difference equation in **a**.

c. Determine under what conditions the equilibrium amount of drug in the body (prior to taking the next dose) is greater than the dose A being taken by the patient.

34. A doctor has prescribed a drug for a patient. Let A be the amount the patient takes each time and c be the fraction of drug cleared by the body between doses. Define a_n to be the amount of drug in the body *immediately after* taking the nth dose.

a. Write a difference equation for a_n.

b. Find the equilibrium of the difference equation in part **a**.

c. Determine under what conditions the equilibrium amount of drug in the body (right after taking a dose) is greater than twice the dose A being taken by the patient.

35. The wildebeest (or gnu) is a ubiquitous species in the Serengeti of Africa. The following data about wildebeest abundance were collected by the Serengeti Research Institute.

year	1961	1963	1965	1967	1971	1972	1977	1978
population size in thousands	263	357	439	483	693	773	1444	1249

a. Assuming $x_{n+1} = a\,x_n$ can be used to model the data, find the constant value of a that would cause the population to grow from size 263 in 1961 to size 1,249 seventeen years later.

b. Generate the sequence $x_{n+1} = a\,x_n$ for $n = 1$ corresponding to year 1961 to $n = 18$ corresponding to the year 1978. Plot this sequence together with the data on the same diagram. How well does the model fit the data?

c. Suppose poachers kill 150 wildebeest each year. Modify the difference equation found in part **a** to account for this poaching and calculate the sequence of values it predicts from 1978 to 1997 (twenty years), if poaching started in 1978. From your answer, what can you deduce?

36. The Ricker model of a dynamic salmon population is given by
$$a_{n+1} = b\,a_n\,e^{-c\,a_n}$$
where b is the total number of progeny produced per individual per generation and $e^{-c\,a_n}$ represents the fraction of progeny that survives after accounting for the effects of adult cannibalism of very young fish. Find all the equilibria for this model and determine under what conditions they are positive. Sketch cobwebbing diagrams for $b = 0.9$, $b = 2.0$, $b = 8.0$, and $b = 20.0$. In these diagrams, let $c = 1.0$ and $a_1 = 2$.

37. A simple **continued fraction** is an expression of the form
$$b_0 + \cfrac{1}{b_1 + \cfrac{1}{b_2 + \cfrac{1}{b_3 + \dots}}}$$
where $b_0, b_1, \dots$ are real numbers. The simplest continued fraction occurs when $1 = b_0 = b_1 = b_2 = \dots$. This continued fraction is generated by the sequence
$$a_{n+1} = 1 + \frac{1}{a_n} \qquad a_1 = 1$$

a. Find the first five terms of this sequence in "expanded form" (i.e., no algebraic reductions) and in simplified form.

b. Find the equilibrium of the difference equation.

c. Use cobwebbing to determine the asymptotic behavior of a_n for the case $a_1 = 1$.

38. Historical Quest

Courtesy of Karl Smith

Fibonacci
1170–1250

Leonardo of Pisa, also known as Fibonacci, was one of the best mathematicians of the Middle Ages. He played an important role in reviving ancient mathematics and introduced the Hindu-Arabic place-value decimal system to Europe. His book, *Liber abaci*, published in 1202, introduced Arabic numerals, as well as the famous *rabbit problem*, for which he is best remembered today. To describe Fibonacci's rabbit problem, we consider a sequence whose *n*th term is defined by a difference equation. Suppose rabbits breed in such a way that each pair of adult rabbits produces a pair of baby rabbits each month.

The first month after birth, the rabbits are adolescents and produce no offspring. However, beginning the second month, the rabbits are adults, and each pair produces a pair of offspring every month. The sequence of numbers describing the number of rabbits is called the *Fibonacci sequence*, and it has applications in many areas, including biology and botany.

Number of months	Number of pairs	Pairs of Rabbits (pairs shown in color are ready to reproduce in the next month)
Start	1	
1	1	
2	2	
3	3	
4	5	
5	8	

Same pair
(rabbits never die)

In this 𝔥𝔦𝔰𝔱𝔬𝔯𝔦𝔠𝔞𝔩 𝔔𝔲𝔢𝔰𝔱 you are to examine some properties of the Fibonacci sequence. Let a_n denote the number of pairs of rabbits in the "colony" at the end of n months.

a. Explain why $a_1, a_2 = 1, a_3 = 2, a_4 = 3$, and, in general,

$$a_{n+1} = a_{n-1} + a_n$$

for $n = 2, 3, 4, \ldots$

b. The *growth rate* of the colony during the $(n+1)^{th}$ month is

$$r_n = \frac{a_{n+1}}{a_n}$$

Compute r_n for $n = 1, 2, 3, \ldots, 10$.

c. Show that r_n satisfies the difference equation
$$r_{n+1} = 1 + \frac{1}{r_n}$$ (Hint: combine the difference equations in parts **a** and **b**) and solve for the equilibrium of this difference equation.

39. Consider the difference equation $x_{n+1} = \dfrac{x_n}{1 + x_n}$ introduced in Example 5. Let x_1 be given.

a. Write explicit expressions for x_2, x_3, x_4, and x_5 in terms of x_1.

b. Use part **a** to find a reasonable guess for an explicit expression of x_n in terms of x_1.

c. Verify your guess by making sure it satisfies the difference equation.

40. A biologist discovers that a gene has a lethal allele *a* that is not purely recessive: genotypes of the form *aa* all die before reproducing and half the genotypes of the form *Aa* also die before reproducing.

a. Show, in contrast to Example 5, that the difference equation describing the proportion x_n of the lethal gene from one generation to the next is now given by the difference equation

$$x_{n+1} = \frac{x_n}{2}$$

b. Calculate the first ten terms of the resulting sequence starting from $x_1 = 0.5$.

c. Find all equilibrium solutions.

d. Compare the sequence you obtain in part **b** with the first ten terms of the sequence obtained in Example 5 (you have to calculate these). What do you notice about how rapidly the allele disappears?

41. A biologist discovers that a gene has a lethal allele *a* that is not purely recessive: genotypes of the form *aa* all die before reproducing and two thirds of the genotypes of the form *Aa* also die before reproducing.

a. Show, in contrast to Example 5, that the difference equation describing the proportion x_n of the lethal gene from one generation to the next is now given by the difference equation

$$x_{n+1} = \frac{x_n}{3 - x_n}$$

b. Calculate the first ten terms of the resulting sequence starting from $x_1 = 0.5$.

c. Find all equilibrium solutions.

d. Compare the sequence you obtain in part **b** with the first ten terms of the sequence obtained in Example 5 (you have to calculate these). What do you notice about how rapidly the allele disappears?

42. A biologist discovers that a gene has a lethal allele *a* that is not purely recessive: genotypes of the form *aa* all die before reproducing and one third genotypes of the form *Aa* also die before reproducing.

   a. Show, in contrast to Example 5, that the difference equation describing the proportion x_n of the lethal gene from one generation to the next is now given by the difference equation

$$x_{n+1} = \frac{2x_n}{3 + x_n}$$

b. Calculate the first ten terms of the resulting sequence starting from $x_1 = 0.5$.

c. Find all equilibrium solutions.

d. Compare the sequence you obtain in part **b** with the first ten terms of the sequence obtained in Example 5 (you have to calculate these). What do you notice about how rapidly the allele disappears?

43. Compare the first ten terms of the sequences obtained from the difference equations derived in Example 5 and in Problems 40, 41, and 42. What do you conclude about the effect of a lethal allele in the population when it has a partial effect on the genotypes *Aa*? What happens when the lethal allele kills all *Aa* genotypes before they have a chance to reproduce?

CHAPTER 1 REVIEW QUESTIONS

1. Let $y = \sqrt{\log x + 1}$.

   a. State the domain and range of this function.

   b. Find the inverse of this function. State its domain and range as an inverse function.

2. The maximum temperature on a fall day in Woodland, California, was 92°F at 5 PM. The minimum temperature was 52°F at 5 AM. Let $T(t)$ be the temperature in degrees at the tth hour after midnight. Write a sinusoidal function of the form $T(t) = A + B\cos(C(t - D))$ that could be used to describe the temperatures on this warm fall day.

3. Find the equilibrium of $a_{n+1} = f(a_n)$ where the graph of $y = f(x)$ is shown. Sketch the cobwebbing diagram for $a_1 = -0.5$.

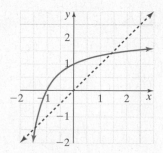

4. Sketch the x values $0.0001, 0.1, 0.3$, and $3,000,000$ on a logarithmic axis.

5. Calculate the residuals of the graphs $y = x$ and $y = \frac{x}{2} + 1$ with respect to the data set

$\{(0, 0.7), (1, 1.1), (2, 2.2)\}$. Use the sum-of-squares criterion to decide which of these two curves fits the data best.

6. The time for cooking a roast is approximately proportional to its surface area.

   a. How does the cooking time scale with the weight of the roast?

   b. How much longer should it take to cook a 20-lb roast compared to a 10-lb roast?

7. John Damuth hypothesized a power function relationship $y = cx^m$ between the body mass x of individuals and the density y of populations belonging to related species.* The data he presented for five species of apes, where the units of x are kilograms and y are numbers/square kilometer: the western lowland gorilla, (127, 1.8); the common chimpanzee, (45.0, 2.5); the bonobo chimpanzee, (22.7, 4.0); the Bornean orangutan, (53.0, 2.0); and the agile gibbon, (5.9, 5.1). By plotting these data on a semi-log scale, what would you estimate to the nearest integer that the power of the allometric relationship is between abundance and body mass?

*John Damuth, "Interspecific Allometry of Population Density in Mammals and Other Animals: The Independence of Body Mass and Population Energy-use," *Biological Journal of the Linnean Society*, 31(2008): 193–246.

8. Consider a bank account with initially $1,000. The annual interest rate for this account is 10%.

   a. Find the amount in the account one year from now if the interest is compounded once a year, twice a year, three times a year.

   b. What value is the amount in the account approaching as the number n of times the money is compounded in the year approaches ∞?

9. An exponential function $y = ab^x$ is plotted below on a semi-log plot. Find a and b.

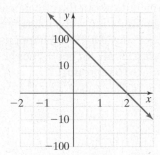

10. Hypothyroidism is a condition in which the thyroid gland makes too little thyroid hormone. One treatment for hypothyroidism is administration of replacement thyroid hormone such as thyroxine (T4). The concentration of this hormone in a patient's body exhibits exponential decay with a half-life of about seven days. Consider an individual who takes 100 mcg of T4 once a week. Let a_n denote the amount of T4 in the individual's body right before the nth dose.

    a. Write a difference equation for a_n.

    b. Find the equilibrium for the difference equation in part a and describe in words what this equilibrium means.

    c. In the long term, what is the greatest amount of T4 in the individual's body?

11. Consider a population with two alleles, A and a, at a single locus such that all individuals born with the genotype aa die immediately, 10% of individuals with genotype Aa die immediately, and individuals with the genotype AA always survive to reproduce. Let x_n denote the fraction a alleles in the population in generation n.

    a. Write a difference equation for x_n.

    b. Find x_2, x_2, x_4 given that the initial proportion of a is $x_1 = 0.9$.

12. Data points with a curve fitted to those points are shown. Decide whether the data in parts a–d are better modeled by an exponential or a logarithmic function.

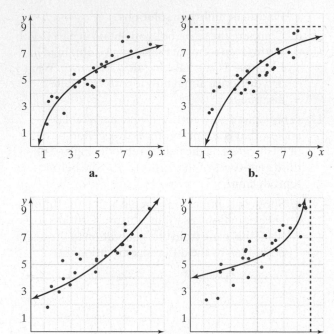

a.   b.

c.   d.

13. Let $y = f(x)$ be the function whose graph is given in Figure 1.53.

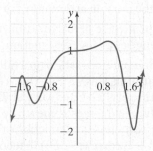

Figure 1.53 Graph of f.

Mix and match the following functions with their corresponding graphs.

a. $y = f\left(\dfrac{x}{2}\right)$

b. $y = 2f(x)$

c. $y = f(-x)$

d. $y = -f(x)$

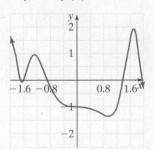
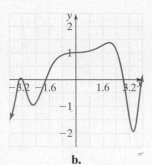

a.   b.

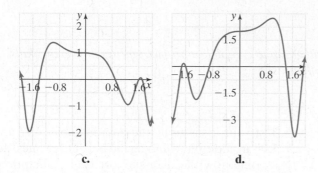

c.   **d.**

14. As a result of recovery efforts, the size of the whooping crane population in Wood Buffalo Park grew from 20 individuals in 1941 to 518 individuals in 2007. Assume that you can model the population growth of the whooping cranes as $N(t) = ae^{rt}$ individuals where t is years after 1941.

  a. Find a and r.

  b. Use the model to estimate the doubling time for the whooping crane population.

15. Find the first five terms for the given sequences.

  a. $a_n = 2 - \dfrac{n}{n+1}$

  b. $a_n = \left(\dfrac{1}{2}\right)^{n-1}$

  c. a_n is the nth prime number.

16. The pollution level, P, in Lake Bowegon varies during a typical year according to the formula

$$P(t) = 50 - 30\cos\left(\frac{2\pi t}{365}\right)$$

where t is the number of days from the beginning of the year. A treatment program initiated by the Department of Wildlife is 50% effective against this pollution. When does the model predict, for the first time, that the pollution will be at a level of 40?

17. The number of apples, n, in a tree is a function of the population density, d, of bees pollinating the apple orchard. This function can be modeled by the formula

$$n(d) = \frac{500d}{6+d}$$

The average weight, w, in grams, of an apple at time of harvest is the following decreasing function of the number apples:

$$w(n) = 70 - \frac{n}{10}$$

Are either of these linear functions?

  a. Graph the weight of the apple as a function of the density of bees. What is the domain of w?

  b. As the number of bees increases, what can you say about the average weight of an apple?

18. The amount of solids discharged from the MWRA (Massachusetts Water Resources Authority) sewage treatment plant on Deer Island (near Boston Harbor) is given by the function

$$f(t) = \begin{cases} 130 & \text{if } 0 \le t \le 1 \\ -30t + 130 & \text{if } 1 < t \le 2 \\ 100 & \text{if } 2 < t \le 4 \\ -5t^2 + 25t + 80 & \text{if } 4 < t \le 6 \\ 1.25t^2 - 26t + 161 & \text{if } 6 < t \le 10 \end{cases}$$

where $f(t)$ is measured in tons per day, and t is measured in years, from 1992.

  a. How many more tons per day were discharged in 2002 than in 1996?

  b. Sketch the graph of f.

19. A female moth (*Tinea pellionella*) lays nearly 150 eggs and then dies. In one year, up to five generations may be born, and each female larva eats about 20 mg of wool. Assume that two thirds of the eggs die, and 50% of the remaining moths are females. Use an exponential population growth model to estimate the largest amount of wool that may be destroyed by the female descendants of one female over a period of one year.

20. The level of a certain pollutant in the Los Angeles area has been decreasing linearly since 1990 when a new pollution control program began. The level of pollution was 0.17 ppm in 2000 and had fallen to 0.11 ppm in 2010.

  a. Let P be the level of pollutant (in parts per million) at time t (years after 2000). Express P as a function of t.

  b. Air with a pollutant level of 0.05 ppm is considered clean. If the present trend continues, when will this clean level be achieved?

GROUP PROJECTS

Seeing a project through on your own, or working in a small group to complete a project, teaches important skills. The following projects provide opportunities to develop such skills.

Project 1A   Heart rates in mammals

Smaller mammals and birds have faster heart rates than larger ones. If we assume that evolution has determined the best rate for each, why isn't there a single best rate? Is there a model that leads to a correct rule relating heart rates? A warm-blooded animal uses large quantities of energy to maintain body temperature because of heat loss through its body surfaces. Cold-blooded animals require very little energy when they are resting. The major energy drain on a resting warm-blooded animal seems to be maintenance of body temperature.

The amount of energy available is roughly proportional to blood flow through the lungs—the source of oxygen. Assuming the least amount of blood needed is circulated, the amount of available energy will equal the amount used. In this project, you are to develop a model of blood flow and heart pulse rates as a function of body size and validate the model using the data in Tables 1.17 and 1.18. Be sure to address the following points:

- Set up a model based on geometric similarity relating body weight to basal (resting) blood flow through the heart. State your assumptions.

- There are many animals for which pulse rate data are available but not blood flow data. Set up a model based on geometric similarity that relates body weight to basal pulse rate.

- Test your ideas using the data in Tables 1.17 and 1.18. In addition to finding the best-fitting lines, determine how the data support your assumptions about geometric similarity.

Table 1.17 Weight and blood flow data on humans and some mammals

Animal	Weight (kg)	Blood flow (deciliters/min)
Human (age 5)	18	23
Human (age 10)	31	33
Human (age 16)	66	52
Human (age 25)	68	51
Human (age 33)	70	43
Human (age 47)	72	40
Human (age 60)	70	46
Rabbit	4.1	5.3
Goat	24	31
Dog 1	16	22
Dog 2	12	12
Dog 3	6.4	11

Data source: W. S. Spector, *Handbook of Biological Data* (1956).

Table 1.18 An across species comparison of the mammalian heart

Animal	Weight (kg)	Pulse (1/min)	Heart weight (g)	Ventricle length(cm)
Shrew	0.004	660		
Mouse	0.025	670	0.13	0.55
Rat	0.2	420	0.64	1.0
Guinea pig	0.3	300		32.00
Rabbit	2	205	5.8	2.2
Dog	5	120		
Dog 2	30	85	102	4.0
Sheep	50	70	210	6.5
Human	70	72		
Horse	450	38	3900	16
Ox	500	40	2030	12
Elephant	3000	48		

Data source: A. J. Clark, *Comparative Physiology of the Heart*, Macmillan (1972).

Project 1B: The mouse to elephant curve

The most universal feature of living organisms is their turnover of energy. Animals, with few exceptions, obtain energy by the oxidation of organic compounds, and the rate of energy turnover (the metabolic rate) is often measured by the rate of oxygen consumption. The fact that there is a regular relationship between the metabolic rate, or rate of oxygen consumption, and the body size of animals is thoroughly familiar to biologists. In the early part of the twentieth century, French scientists realized that the heat dissipation from warm-blooded animals must be roughly proportionate to their free surface. Since smaller animals have a larger relative surface, they must also have a higher relative rate of heat production than larger animals. In this project, you are to develop a model to explore this relation. Use the data in Table 1.19 to assess the accuracy of your model and its assumptions. The project needs to address the following point:

- Develop a model based on geometric similarity to describe surface area of a mammal as a function of body size. Be sure to state all of your assumptions.

- Develop a model based on geometric similarity to describe the metabolic rate of an organism as a function of body size.

- Test your models using the data in Table 1.19.

- The curve described by the data in Table 1.19 is called the *mouse-to-elephant curve*. However, the original data set from which the curve was derived did not include the relevant numbers for the elephant. Find the weight of an elephant and determine what the models predict for its metabolic rate. If possible, compare this prediction with the actual value.

- In 1847, Carl Bergmann, a German biologist, formulated what later became known as *Bergmann's*

rule, which states that animals that live in colder climates are of larger body size than their relatives from warmer climates. Based on your analysis, does this rule make sense?

Table 1.19 Data on the relationship between the weight of individuals and their basal metabolic rate

Animal	Body mass (kg)	kCal/day
Mouse	0.021	3.6
Rat	0.282	28.1
Guinea pig	0.410	35.1
Rabbit	2.980	167
Rabbit 2	1.520	83
Rabbit 3	2.460	119
Rabbit 4	3.570	164
Rabbit 5	4.330	191
Rabbit 6	5.330	233
Cat	3.0	152
Monkey	4.200	207
Dog	6.6	288
Dog 2	14.1	534
Dog 3	24.8	875
Dog 4	23.6	872
Goat	36	800
Chimpanzee	38	1,090
Sheep	46.4	1,254
Sheep 2	46.8	1,330
Woman	57.2	1,368
Woman 2	54.8	1,224
Woman 3	57.9	1,320
Cow	300	4,221
Cow 2	435	8,166
Cow 3	600	7,877
Heifer	482	7,754

Project 1C: Golden Ratio

Around 300 BC, the greatest of the ancient Greek geometers, Euclid of Alexandria, defined what he called the "extreme and mean ratio," now better known as the *golden ratio*. (See Mario Livio, *The Golden Ratio: the Story of Phi*, the Worlds Most Astonishing Number, New York: Broadway Books, 2002, p. 3.) Euclid's ratio states:

A straight line is said to have been cut in extreme and mean ratio when, as the whole line is to the greater segment, so is the greater segment to the lesser segment.

Specifically, if we look at the line illustrated in Figure 1.54, this statement can be expressed mathematically as

$$(a + b)/a = a/b$$

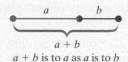

a + b is to *a* as *a* is to *b*

Figure 1.54 Line segments used by Euclid to define "extreme and mean ratio."

The *golden rectangle* is the rectangle whose sides conform to the golden ratio. The beauty, and astonishing perfection, of the golden rectangle arises from this fact: If you add one additional edge parallel to a short side of the rectangle to form a square within the rectangle, the smaller rectangle so formed (now oriented at 90 degrees to the original rectangle) is also a golden rectangle, as illustrated in Figure 1.55.

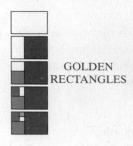

GOLDEN RECTANGLES

Figure 1.55 This panel is a golden rectangle. If we cut off the red square, as in the second panel, the vertical residual rectangle is also golden. Continuing to cut off squares as in the third to fifth panels leaves ever smaller residual golden rectangles. This leaves a smaller rectangle which again is cut and so on infinitum.

A spiral can be constructed passing through the corners of all embedded squares of the preceding construction in a such way that the spiral is *equiangular*—also known as the *logarithmic spiral*, which has the form shown in Figure 1.56.

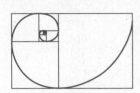

Figure 1.56 If we inset quarter circles into the squares obtained by repeatedly reducing golden rectangles as illustrated in this construction, we obtain a very good approximation of a true logarithmic or equiangular spiral, which is much more difficult to construct accurately.

Leonardo da Vinci (1452–1519), one of the greatest painters of all time, so valued the aesthetic proportions of the golden rectangle that aspects of figures and forms in many of his paintings conform to golden rectangle proportions. In the twentieth century, the noted American architect Frank Lloyd Wright (1867–1959) used the logarithmic spiral in designing the Guggenheim Museum in New York City.

1. List five paintings that art historians regard as compositions containing golden rectangles.

2. The shell of the nautilus mollusk, *Nautilus pompilius*, has the shape of a logarithmic spiral. Find a list of at least five other natural objects that contain shapes conforming to logarithmic spirals.

3. Define, using the concept of a tangent to a curve, what is meant by an *equiangular spiral*.

4. From Euclid's statement regarding the extreme and mean ratio, commonly denoted by the Greek letter phi (ϕ), show that $\phi = \dfrac{1+\sqrt{5}}{2}$.

5. If

$$\phi_1 = \sqrt{1} = 1,$$

$$\phi_2 = \sqrt{1+\sqrt{1}} = \sqrt{2} = 1.141421\ldots.$$

$$\phi_3 = \sqrt{1+\sqrt{1+\sqrt{1}}} = \sqrt{1+\sqrt{2}} = 1.553773\ldots.$$

$$\phi_4 = \sqrt{1+\sqrt{1+\sqrt{1+\sqrt{1}}}} = 1.598053\ldots.$$

$$\vdots$$

$$\phi_n = \sqrt{1+\sqrt{1+\ldots 1+\sqrt{1}}} \quad n \text{ square roots deep}$$

then use technology to calculate ϕ_i, $i = 5, \ldots, 10$. Use the definition of ϕ_n to generate a relationship of the form $\phi_{n+1} = f(\phi_n)$ and demonstrate that an equilibrium solution $\phi = f(\phi)$ is the golden ratio. To how many decimal places do the numerical values of ϕ and ϕ_{10} coincide?

6. If

$$\phi'_0 = 1,$$

$$\phi'_1 = 1 + \frac{1}{1} = \frac{2}{1} = 2,$$

$$\phi'_2 = 1 + \frac{1}{1+\frac{1}{1}} = \frac{3}{2} = 1.5$$

$$\phi'_3 = 1 + \frac{1}{1+\frac{1}{1+\frac{1}{1}}} = \frac{5}{3} = 1.666\ldots$$

$$\phi'_4 = 1 + \frac{1}{1+\frac{1}{1+\frac{1}{1+\frac{1}{1}}}} = \frac{8}{5} = 1.666\ldots$$

$$\phi'_5 = 1 + \frac{1}{1+\frac{1}{1+\frac{1}{1+\frac{1}{1+\frac{1}{1}}}}} = \frac{13}{8} = 1.625$$

$$\vdots$$

$$\phi'_n = 1 + \frac{1}{1+\frac{1}{1+\frac{1}{1+\cdots 1+\frac{1}{1}}}} \quad n \text{ denominators deep}$$

can you find a relationship of the form $\phi'_{n+1} = f(\phi'_n)$? Demonstrate that an equilibrium solution $\phi' = f(\phi')$ is the golden ratio. Notice that the denominators and numerators of the consecutive fractions $\phi'_1 = \frac{2}{1}$, $\phi'_2 = \frac{3}{2}$, $\phi'_3 = \frac{5}{3}, \ldots$ are the Fibonacci sequence discussed in the 𝕳istorical 𝕼uest , Problem 38 of Section 1.7. Use this fact to write the expression for ϕ'_{10}. Draw a conclusion.

7. Compare the value of ϕ'_{10} obtained in the preceding question with the golden ratio ϕ and ϕ_{10} obtained from the question before that. Which of ϕ_{10} and ϕ'_{10} provides the better approximation to ϕ? Can you generalize this statement to ϕ_n and ϕ'_n as an approximation to ϕ for any n?

CHAPTER 2

Limits and Derivatives

Figure 2.1 Biologists count spawning adult sockeye salmon to obtain a stock-recruitment relationship (see Figure 2.43).

© Darryl Leniuk/Age Fotostock America, Inc.

Preview

"Mathematics, in one view, is the science of infinity."

Philip Davis and Reuben Hersh,
The Mathematical Experience, Birkhäuser
Boston, 1981, p. 152.

Calculus, one of the great intellectual achievements of humankind, came to fruition through the work of Sir Isaac Newton (1642–1727) and Gottfried Wilhelm Leibniz (1646–1716). It consists of two parts, *differential calculus* and *integral calculus*, both of which hinge on the concept of a *limit* in which the behavior of a function is described as its argument approaches a selected *limiting value*. In this chapter, we first discuss some of the basic properties of a limit and the associated concept of *continuity*. Using limits, we then introduce the main concept in differential calculus: the *derivative*. This concept is one of the fundamental ideas in mathematics and a cornerstone of modern scientific thought. It allows us to come to grips with the phenomenon of change in the value of variables over increasingly small intervals of time or space (or any other independent variable—variables representing such things as the velocity of a stooping falcon or the density of a population of fish). By finding the derivative of functions, we calculate the slopes of tangent lines to these functions and examine rates of change defined by these functions. Applications considered in this chapter include enzyme kinetics, biodiversity, foraging of hummingbirds, wolf predation, and population dynamics of the sockeye salmon depicted in Figure 2.1.

2.1 Rates of Change and Tangent Lines

One of the fundamental concepts in calculus is that of the limit. Intuitively, "taking a limit" corresponds to investigating the value of a function as you get closer and closer to a specified point without actually reaching that point. To "motivate limits," we begin by showing how they arise when we consider rates of change and the tangent lines to graphs of functions.

Rates of change

Rates of change describe how quantities change with respect to a variable such as time or body mass. For instance, in countries where overpopulation is an issue, projections are constantly made about the population growth rate. For a patient receiving drug treatment, physicians perform experiments to estimate the clearance rate of the drug (i.e., the rate at which the drug leaves the blood stream). To calculate a rate of change, we first select an interval of interest and then compute the average rate of change over that interval using the equation in the box below. To calculate the rate of change at a point of interest requires taking limits, as discussed after the next example.

Average Rate of Change	The **average rate of change** of f over the interval $[a, b]$ is $$\frac{f(b) - f(a)}{b - a}$$

Example 1   Mexico population growth

The (estimated) population size of Mexico (in millions) in the early 1980s is reported in the following table.

Year	Time lapsed t	Population $N(t)$
1980	0	67.38
1981	1	69.13
1982	2	70.93
1983	3	72.77
1984	4	74.66
1985	5	76.60

From this table, we see that $N(t)$ denotes the population size in millions t years after 1980.

a. Compute the population's average rate of change (i.e., the average population growth rate) for the interval of time $[3, 5]$. Identify the units for this average rate of change and interpret this rate of change.

b. Compute the average population growth rate for the interval $[1, 3]$. How does this compare to your answer in **a**? What does this imply about the population growth?

c. Compute the average population growth rates for the intervals $[0, 5]$, $[0, 4]$, $[0, 3]$, $[0, 2]$, and $[0, 1]$. Discuss the trend of these growth rates.

Solution
a. The average growth rate for $N(t)$ over $[3, 5]$ is given by

$$\frac{N(5) - N(3)}{5 - 3} = \frac{76.6 - 72.77}{2} = 1.915$$

The units of this growth rate are millions per year. Hence, between the years 1983 and 1985, the population increases on average by 1.915 million individuals per year.

b. The average growth rate for $N(t)$ over $[1, 3]$ is given by

$$\frac{N(3) - N(1)}{3 - 1} = \frac{72.77 - 69.13}{2} = 1.82$$

The population is growing on average of approximately 1.8 million per year. Thus, the average growth rate from 1981 to 1983 is less than the average growth rate from 1983 to 1985. The population growth rate in Mexico appears to be increasing over the time period 1983 to 1985.

c. Computing the average growth rates over the requested time intervals yields

Time interval	Average growth rate
[0, 5]	1.84
[0, 4]	1.82
[0, 3]	1.80
[0, 2]	1.78
[0, 1]	1.75

The average population growth rate is decreasing over the smaller time intervals and appears to converge to a value close to, but a little below, 1.75 million per year. ∎

The *instantaneous rate of change* of $f(x)$ at $x = a$ (if it exists) is defined to be the limiting value of the average rate of change of f on smaller and smaller intervals starting at $x = a$. For instance, in Example 1, we would estimate the instantaneous rate of change of the population in Mexico in 1980 to be 1.75 million per year. In other words, the population is growing at a rate of 1.75 million per year in 1980. More precisely, we have the following definition.

Instantaneous Rate of Change

The **instantaneous rate of change** of f at $x = a$ is given by

$$\lim_{b \to a} \frac{f(b) - f(a)}{b - a}$$

where the symbol "$\lim_{b \to a}$" is interpreted as taking b arbitrarily close but not equal to a.

How to calculate such limits is one of the challenges of differential calculus, a problem that we tackle further in Sections 2.2 and 2.3.

Example 2 Rate of change of carbon dioxide

In Example 4 in Section 1.1, we initially approximated the concentration of CO_2 in parts per million at the Mauna Loa Observatory of Hawaii with the linear function

$$L(t) = 329.3 + 0.1225\,t$$

where t is measured in months after April 1974. We then refined our approximation with the function

$$F(t) = 329.3 + 0.1225\,t + 3\cos\left(\frac{\pi t}{6}\right)$$

Using the definition introduced in this section, estimate the instantaneous rate of change of the functions $L(t)$ and $F(t)$ at $t = 3$.

Solution To estimate the instantaneous rate of change of L at $t = 3$, we can look at the average rate of change over the interval $[3, b]$ for values of b progressively closer to 3. From our definition for the average rate of change, we have

$$\frac{L(b) - L(3)}{b - 3} = \frac{329.3 + 0.1225b - (329.3 + 0.1225 \cdot 3)}{b - 3}$$

$$= \frac{0.1225(b - 3)}{b - 3}$$

$$= 0.1225$$

Since L is a linear function of b, the average rate of change of L is independent of b, as the above algebra reveals, and equals the slope of $L(t)$ for all $b \neq 3$. Hence, we find that the instantaneous rate of change of $L(t)$ at $t = 3$ is 0.1225 ppm/month.

On the other hand, the function $F(b)$ is nonlinear in the variable t. To estimate the instantaneous rate of change of $F(t)$ at $t = 3$, we look at the average rate of change

$$\frac{F(b) - F(3)}{b - 3}$$

over intervals $[3, b]$ for values of b values progressively closer to 3, but satisfying $b > 3$. Carrying out a sequence of such calculations yields the following table:

Interval	$b - 3$	Average rate of change of $F(t)$
$[3, 4]$	1	-1.3775
$[3, 3.5]$	0.5	-1.43041
$[3, 3.1]$	0.1	-1.44758
$[3, 3.01]$	0.01	-1.44829
$[3, 3.001]$	0.001	-1.44830

This table suggests the instantaneous rate of change of $F(t)$ is a little less than -1.45 ppm/month.

You might ask why there is a difference in the signs between the answers for $L(t)$ and $F(t)$. Recall that $L(t)$ only described the linear trend of the CO_2 data that was increasing. On the other hand, $F(t)$ captured the seasonal fluctuations of the CO_2 levels, which are sometimes increasing and sometimes decreasing. Turning back to Figure 1.5 in Section 1.1, we see that, indeed, in the third month of the data, the level of CO_2 was decreasing.   ■

This example indicates that a linear fit to an oscillating function may provide a reasonable estimate of rates of change averaged over several oscillations; however, it is a poor estimate of the instantaneous rate of change because that depends on the particular stage of each oscillation.

Velocity

Now we consider concepts central to the history of calculus, concepts that motivated much of the work of Newton: the *velocity* (considered here) and the *acceleration* (considered in Sections 3.6 and 5.1) of an object moving along a line. For example, the object may be an athlete running along a racing track, or a coffee mug falling to the ground.

Average and Instantaneous Velocity

Let $f(t)$ be the position of an object at time t. The **average velocity** of an object from time t to time $t + h$ is given by the formula

$$\text{AVERAGE VELOCITY} = \frac{\text{displacement}}{\text{time elapsed}} = \frac{f(t + h) - f(t)}{h}$$

while the **instantaneous velocity** of an object at time t is given by the formula

$$\text{INSTANTANEOUS VELOCITY} = \lim_{h \to 0} \frac{\text{displacement}}{\text{time elapsed}} = \lim_{h \to 0} \frac{f(t + h) - f(t)}{h}$$

If we define $b = t + h$ and $a = t$, then our definitions of average velocity and instantaneous velocity are equivalent to the average rate of change of the object's position and instantaneous rate of change of the object's position.

If you have ever watched a track meet, you may have wondered how fast the winners ran during their races. In the next example, we find out exactly how fast the world's best 100-meter sprinters are and how their velocities change during the race.

Julia Vynokurova/Getty Images, Inc.

Figure 2.2 Usian Bolt is widely regarded as the fastest person ever, earning him the nickname "Lightning Bolt."

Example 3   The fastest humans

Among the fastest 100-meter races ever run are the performances of Ben Johnson at the 1988 Seoul Olympics and Usain Bolt at the 2008 Beijing Olympics. The reaction times at the start of the race and the times taken to run each 10-meter split (the first split is the time to reach 10 meters minus the reaction time) are given in Table 2.1. Ben Johnson was disqualified three days after the event when the steroid Stanozolol was found in his urine.

a. From these data, estimate Usain Bolt's velocities at the end of each of the 10-meter splits during the course of the race. (For calculation of Ben Johnson's velocities, see Problem 51 in Problem Set 2.1.)

Table 2.1 Reaction times to start running (RT) and times (in seconds) take to run each of the 10-meter splits in the 1998 and 2008 Olympic 100-meter races

Split time	Ben Johnson 1988 (s)	Usain Bolt 2008 (s)
RT	0.13	0.17
0–10 m	1.70	1.68
10–20 m	1.04	1.02
20–30 m	0.93	0.91
30–40 m	0.86	0.87
40–50 m	0.84	0.85
50–60 m	0.83	0.82
60–70 m	0.84	0.82
70–80 m	0.85	0.82
80–90 m	0.87	0.83
90–100 m	0.90	0.90
Total time	9.79	9.69

b. From the data calculated in part **a**, discuss how Usain Bolt's velocities changed through the course of the race.

Solution

a. From Table 2.1 and the definition of average velocity given above, Usain Bolt's average velocities in meters per second (m/s) at the end of each split can be shown to be

Velocity (m/s) at	0 m	10 m	20 m	30 m	40 m	50 m	60 m	70 m	80 m	90 m	100 m
Usain Bolt	0	5.9	9.8	11.0	11.5	11.8	12.2	12.2	12.2	12.0	11.1

b. From Figure 2.3, we see that Bolt's velocity keeps increasing until the end of the sixth 10-meter split. His velocity then remains constant for the next two 10-meter splits but declines noticeably over the final 10-meter split.

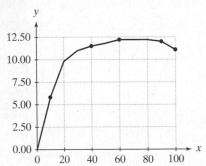

Figure 2.3 Usain Bolt's average velocities at the end of all the 10-meter splits are plotted here with values at each split joined by the solid blue line.

We have all seen objects fall for different periods of time, whether it be a coffee mug falling off a table or a peregrine falcon stooping from high altitudes. The next example uses a basic physical principle (developed in greater detail in Section 5.1) to determine how the distance traveled by a falling object changes in time.

Example 4   Instantaneous velocity of a falling object

Consider a coffee mug falling off a 32-foot ladder. Ignoring air resistance, the height of this coffee mug from the ground at time t seconds can be shown to be

$$f(t) = 32 - 16t^2 \text{ feet}$$

Determine what time the mug is halfway to the ground and its instantaneous velocity at this time.

Solution   The coffee mug is halfway to the ground when the height above the ground is 16 feet. Therefore, we need to solve $16 = f(t) = 32 - 16t^2$. Equivalently, $t^2 = 1$, which has solutions $t = -1, +1$. The only solution that is relevant physically is $t = 1$ seconds.

To find the instantaneous velocity of $f(t)$ at $t = 1$, we need the displacement over the time interval $t = 1$ to $t = 1 + h$. This displacement is given by

$$f(1+h) - f(1) = 32 - 16(1+h)^2 - (32 - 16)$$
$$= -16(1+h)^2 + 16$$
$$= -16 - 32h - 16h^2 + 16 = -32h - 16h^2$$

Therefore,

$$\frac{\text{displacement}}{\text{elapsed time}} = \frac{-32h - 16h^2}{h} = -32 - 16h$$

Letting the elapsed time h go to zero yields an instantaneous velocity of $-32 - 16(0) = -32$ feet/second. This velocity is approximately -22 miles per hour! The negative sign on the velocity corresponds to the fact that the mug is falling downward.

Tangent lines

A linear function is a function with constant slope, which raises this question: What is the slope of a nonlinear function at a point? To answer this question, we need to solve the *tangent problem*, whose solution resides in the following words of Leibniz quoted by David Berlinski (*Infinite Ascent: A Short History of Mathematics* (2008) Random House, NY):

We have only to keep in mind that to find a tangent means to draw the line that connects two points of the curve at an infinitely small distance

As illustrated in Figure 2.4 and stated by Leibniz, the tangent line of $y = f(x)$ at a point $(a, f(a))$ (in blue) can be approximated by secant lines (in red) passing through the points $(a, f(a))$ and $(a + h, f(a + h))$ as $h \neq 0$ gets closer and closer to 0. The slope of this secant line is given by

$$\text{SLOPE OF SECANT LINE} = \frac{f(a + h) - f(a)}{h}$$

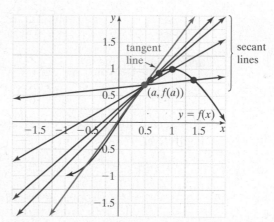

Figure 2.4 Approximating the tangent line with a secant line.

By letting h get arbitrarily close to 0, we get the

$$\text{SLOPE OF TANGENT LINE} = \lim_{h \to 0} \frac{f(a + h) - f(a)}{h}$$

where the symbol "$\lim_{h \to 0}$" can be interpreted as taking $h > 0$ arbitrarily close to, but not equal to 0. This limit may or may not exist, so not every curve will have a tangent line at every point.

Example 5   Approximating a tangent line

Approximate the tangent line of $y = \ln x$ at the point $x = 1$ using the method of secants with decreasing values of $h = 1, 0.5, 0.1, 0.01,$ and 0.001.

Solution   As a first approximation to the tangent line, for the case $h = 1$, we consider the secant line passing through the point $(1, \ln 1) = (1, 0)$ and $(2, \ln 2)$. The slope of this secant line is given by

$$\text{SLOPE OF SECANT LINE} = \frac{\ln 2 - 0}{2 - 1} = \ln 2 \approx 0.693$$

Using the point slope formula for a line, the equation of the secant line is

$$y = (\ln 2)(x - 1)$$

Graphing $y = \ln x$ and $(\ln 2)(x - 1)$ yields the plots depicted in Figure 2.5.

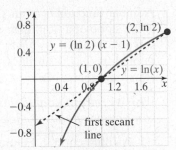

Figure 2.5 Plots of the functions $y = \ln x$ and $y = (\ln 2)(x - 1)$.

To obtain a better approximation, we now move closer to $(1, 0)$ by letting $h = 0.5$. This secant line passes through $(1, 0)$ and $(1.5, \ln 1.5)$. The slope of this secant line is

$$\frac{\ln 1.5 - 0}{1.5 - 1} = 2\ln 1.5 \approx 0.811$$

and the secant line is

$$y = 2(\ln 1.5)(x - 1)$$

This secant looks like a better approximation to a tangent line, as illustrated in Figure 2.6.

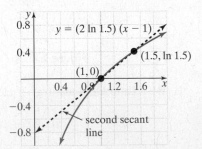

Figure 2.6 Plots of the functions $y = \ln x$, $y = (\ln 2)(x - 1)$, and $y = (2\ln 1.5)(x - 1)$.

To obtain better approximations, repeat the exercise of finding the slope of the secant line passing through $(1, 0)$ and $(1 + h, \ln(1 + h))$ for $h = 0.1$, 0.01, and 0.001 to obtain the values reported to three decimal places (3 dp) in the following table:

h	Slope of secant line (to 3 dp)
1	0.693
0.5	0.811
0.1	0.953
0.01	0.995
0.001	1.000

This table suggests that as h gets closer to 0, the slope of the corresponding secant line approaches 1. Hence, it seems reasonable to approximate the tangent line by $y = x - 1$, as shown in Figure 2.7.

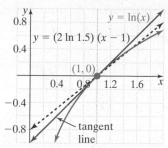

Figure 2.7 Plots of the functions $y = \ln x$, $y = 2(\ln 1.5)(x - 1)$ and $y = x - 1$.

We will see later that this is actually the tangent line.

Sometimes it is possible to algebraically determine the slope of the tangent line.

Example 6 Tangent line for a parabola

Find the equation of the tangent line passing through the point $(1, 1)$ on the curve $y = x^2$.

Solution   To find the tangent line, we first need its slope. The slope of the secant line passing through the point $(1, 1)$ and $(1 + h, (1 + h)^2)$ is given by

$$\frac{(1+h)^2 - 1}{1 + h - 1} = \frac{1 + 2h + h^2 - 1}{h}$$
$$= \frac{2h + h^2}{h}$$
$$= 2 + h \quad for\ h \neq 0$$

The slope of the tangent line is $\lim_{h \to 0}(2 + h)$. Since $2 + h$ gets arbitrarily close to 2 as h gets close to 0, the slope of the tangent line is 2. Using the point-slope formula, we find the equation of the tangent line:

$$y = 2(x - 1) + 1 = 2x - 1$$

Plotting this line against $y = x^2$ in Figure 2.8 shows that the tangent line just "kisses" the parabola at $(1, 1)$, touching it in a single point in this case.

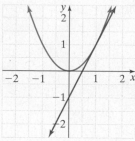

Figure 2.8 Graph of the tangent to $y = x^2$ at $(1, 1)$.

PROBLEM SET 2.1

Level 1 DRILL PROBLEMS

Find the average rate of change for the functions in Problems 1 to 6 on the specified intervals.

1. $f(x) = 4 - 3x$ on $[-3, 2]$
2. $f(x) = 5$ on $[-3, 3]$
3. $f(x) = 3x^2$ on $[1, 3]$
4. $f(x) = -2x^2 + x + 4$ on $[1, 4]$
5. $f(x) = \sqrt{x}$ on $[4, 9]$
6. $f(x) = \dfrac{-2}{x+1}$ on $[1, 5]$

Approximate the instantaneous rate of change for the functions in Problems 7 to 12 at the indicated point. These are the same functions as those given in Problems 1 to 6.

7. $f(x) = 4 - 3x$ at $x = -3$
8. $f(x) = 5$ at $x = 3$
9. $f(x) = 3x^2$ at $x = 1$
10. $f(x) = -2x^2 + x + 4$ at $x = 4$
11. $f(x) = \sqrt{x}$ at $x = 4$
12. $f(x) = \dfrac{-2}{x+1}$ at $x = 1$

Find the instantaneous velocity of a cup falling from the specified height h feet and time t second in Problems 13 to 16. Use the fact that the height of the cup as a function of time t is given by $f(t) = h - 16t^2$ feet.

13. $h = 64$ feet; $t = 1$ seconds
14. $h = 64$ feet; $t = 2$ seconds
15. $h = 4$ feet; $t = 1/2$ seconds
16. $h = 1,600$ feet; $t = 10$ seconds

Trace the curves in Problems 17 to 22 onto your own paper and draw the secant line passing through P and Q. Next, imagine $h \to 0$ and draw the tangent line at P, assuming that Q moves along the curve to the point P. Finally, estimate the slope of the curve at P using the slope of the tangent line you have drawn.

17.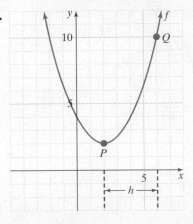

18.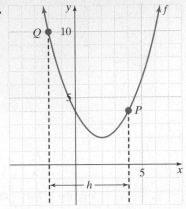

19.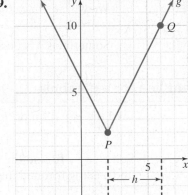

20.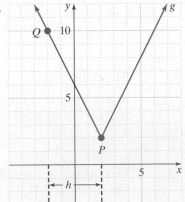

21.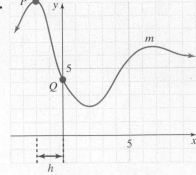

22.

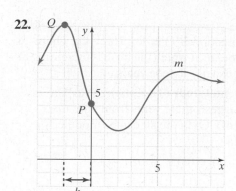

Estimate the tangent line for $y = f(x)$ in Problems 23 to 36 by approximating secant slopes to estimate the limiting value. Graph both the function and the tangent line on the same plot.

23. $f(x) = x^2 - x + 1$ at $x = -1$

24. $f(x) = 4 - x^2$ at $x = 0$

25. $f(x) = \sin \dfrac{\pi x}{2}$ at $x = 0.5$

26. $f(x) = \cos \dfrac{\pi x}{2}$ at $x = 0.5$

27. $f(x) = \tan \dfrac{\pi x}{2}$ at $x = 0.5$

28. $f(x) = \sin \dfrac{\pi x}{2}$ at $x = 1$

29. $f(x) = \dfrac{1}{x + 3}$ at $x = 3$

30. $f(x) = e^x$ at $x = 0$

31. $f(x) = \ln x$ at $x = 9$

32. $f(x) = \tan x$ at $x = 0$

33. $f(x) = \tan \dfrac{\pi x}{2}$ at $x = 0.95$

34. $f(x) = \ln \dfrac{1}{x}$ at $x = 1$

35. $f(x) = \ln \dfrac{1}{x}$ at $x = 0.1$

36. $f(x) = e^{-x}$ at $x = 0$

Algebraically determine the tangent line for $y = f(x)$ at the point specified in Problems 37 to 42. Graph both $y = f(x)$ and the tangent line on the same plot.

37. $f(x) = 3x - 7$ at $x = 3$

38. $f(x) = x^2$ at $x = -1$

39. $f(x) = 3x^2$ at $x = -2$

40. $f(x) = x^3$ at $x = 1$

41. $\sqrt{x}$ at $x = 9$. Hint: Multiply by $\dfrac{\sqrt{x+h} + \sqrt{x}}{\sqrt{x+h} + \sqrt{x}}$

42. $\sqrt{5x}$ at $x = 5$

Level 2 APPLIED AND THEORY PROBLEMS

Find the average rate of change of the given functions over the specified intervals in Problems 43 to 46. Be sure to specify units and briefly state the meaning of the average rate of change.

43. $P(t) = 8.3 \times 1.33^t$ is the number (in millions) of people living in the United States t decades after 1815 over intervals $[0, 2]$ and $[2, 4]$.

44. $L(x) = 20.15\,x^{2/3}$ is the number of kilograms lifted by Olympic Gold Medal weightlifters weighing x kilograms over intervals $[56, 75]$ and $[100, 110]$.

45. The height $H(t)$ of beer froth after t seconds over intervals $[0, 30]$ and $[60, 90]$. Use the data in Table 2.2.

Table 2.2 Height of beer froth as a function of time after pouring

Time t (seconds)	Froth height H (centimeters)
0	17.0
15	16.1
30	14.9
45	14.0
60	13.2
75	12.5
90	11.9
105	11.2
120	10.7

Data Source: Arnd Leike, "Demonstration of the Exponential Decay Law Using Beer Froth," *European Journal Physics* 23 (2002): 21–26.

46. The height $f(x)$, in feet, of the tide at time x hours, where the graph of $y = f(x)$ is provided in Figure 2.9, over intervals $[0.25, 0.50]$ and $[1, 2]$

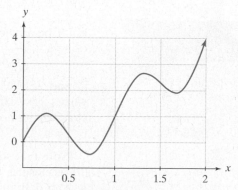

Figure 2.9 Tidal height.

Approximate the instantaneous rates of change of the given functions at the points specified in Problems 47 to 50. These are the same functions as those in Problems 43 to 46.

47. $P(t) = 8.3(1.33)^t$ is the number (in millions) of people living in the United States t decades after 1815 at the points $t = 0$ and $t = 2$.

48. $L(x) = 20.15\, x^{2/3}$ is the number of kilograms lifted by Olympic Gold Medal weightlifters weighing x kilograms at the points $x = 56$ and $x = 100$.

49. Use the data in Table 2.2 to estimate the instantaneous rate of change of height $H(t)$ of beer froth after 0 and 60 seconds respectively.

50. The height $f(x)$, in feet, of the tide at time x hours where the graph of $y = f(x)$ is in given in Figure 2.9 at the points $x = 0.25$ and $x = 1$.

51. From the data presented in Table 2.1 in Example 3, calculate Ben Johnson's velocities at the end of each of the 10-meter splits during the course of his race and discuss how his velocity changes over the course of the race.

52. The population of a particular bacterial colony was determined to be given by the function

$$P(t) = 84 + 61t + 3t^2$$

thousand individuals t hours after observation began. Find the rate at which the colony was growing after exactly five hours.

53. The biomass of a particular bush growing in a field is given by the function $B(t) = 1 + 10t^{1/2}$ kg, where t is years. Find the rate at which the bush is increasing in biomass after exactly sixteen years.

54. An environmental study of a certain suburban community suggests that t years from now, the average level of CO_2 in the air can be modeled by the formula

$$q(t) = 0.05t^2 + 0.1t + 3.4$$

parts per million.

a. By how much will the CO_2 level change in the first year?

b. By how much will the CO_2 level change over the next (second) year?

c. At what rate will the CO_2 level be changing with respect to time exactly one year from now?

2.2   Limits

In defining the instantaneous rate of change and the slope of a tangent line, we use the notation for a limit in Section 2.1. The concept of a *limit* is one of the foundations of both differential and integral calculus. Thus, we devote this section to exploring the concept of limits in the context of functions before proceeding to the calculus itself.

Introduction to Limits

We begin our study of limits with the following mathematically "informal" definition.

Limit (Informal)	Let f be a function. The notation $$\lim_{x \to a} f(x) = L$$ is read as "the **limit** of $f(x)$ as x approaches a is L" and means that the functional values $f(x)$ can be made arbitrarily close to L by requiring that x be sufficiently close to, but not equal to, a.

Note that sometimes $f(x)$ may not be defined at $x = a$ but may be defined for all x as close as you like to the value of a.

This definition is made mathematically precise at the end of this section. At several other places in this book we favor informal over formal definitions, using the following statement of the historian E. T. Bell as our justification (see *Men of Mathematics*, New York: Simon & Schuster, 1937, p. 98):

To the early developers of the calculus the notions of variables and limits were intuitive; to us they are extremely subtle concepts hedged about with thickets of semi-metaphysical mysteries concerning the nature of numbers . . . .

Our goal is to provide definitions that make the concepts usable without getting caught up in the technicalities of formal definitions.

Example 1   Finding limits

Find the following limits using the informal definition provided above.

a. Graph the function $y = x^2$ and find $\lim_{x \to 2} x^2$ for $x > 2$.

b. Numerically evaluate the function $y = \dfrac{x - 2}{x^2 - 4}$ to find $\lim_{x \to 2} \dfrac{x - 2}{x^2 - 4}$.

c. Use the informal definition of a limit to find $\lim_{x \to 0} e^{-1/x^2}$.

Solution

a. Graph $y = x^2$. Choose several values of x (getting closer to 2, but not equal to 2) and then corresponding y values. As illustrated in Figure 2.10, we can see that as x gets closer to 2 from above, the value of the function gets closer to 4.

In fact, if we zoom in around the point $x = 2$, we obtain the plot illustrated in Figure 2.11.

These graphs (correctly) suggest that $\lim_{x \to 2} x^2 = 4$. This limit corresponds to simply evaluating x^2 at $x = 2$. This is the idea of *continuity*, which is described in the next section.

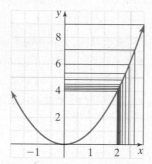

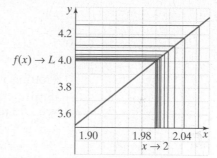

Figure 2.10 Illustration of the process of finding $\lim_{x \to 2} x^2$.

Figure 2.11 Illustration of the process of finding $\lim_{x \to 2} x^2$ when zooming in around the point $x = 2$.

b. The function $f(x) = \dfrac{x - 2}{x^2 - 4}$ is not defined at $x = 2$ (division by zero), so we cannot simply evaluate this function at $x = 2$ as suggested at the end of part **a**. Instead, we can only consider values near (but not equal to) 2. Since $x^2 - 4 = (x - 2)(x + 2)$, we have

$$f(x) = \frac{x - 2}{(x - 2)(x + 2)} = \frac{1}{x + 2} \quad \text{for } x \neq 2$$

Evaluating f at x values near 2 yields the following table:

x	$f(x)$	x	$f(x)$
1.0	0.333333	3.0	0.200000
1.5	0.285714	2.5	0.222222
1.9	0.256410	2.1	0.243902
1.99	0.250627	2.01	0.249377
1.999	0.250063	2.001	0.249938
1.9999	0.250006	2.0001	0.249994

This table suggests that $\lim_{x \to 2} \dfrac{x - 2}{x^2 - 4} = 0.25$, which corresponds to evaluating $\dfrac{1}{x + 2}$ at $x = 2$.

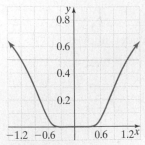

Figure 2.12 Graph of the function e^{-1/x^2}.

c. The function $f(x) = e^{-1/x^2}$ is not defined at $x = 0$. However, if x is close to zero, then $-\dfrac{1}{x^2}$ is very large and negative. Consequently, if x is close to zero, e^{-1/x^2} is close to zero. This suggests that $\lim\limits_{x \to 0} e^{-1/x^2} = 0$. We can reinforce this conclusion by looking at the graph of f, as shown in Figure 2.12. ∎

The existence of a limit $\lim\limits_{x \to a} f(x) = L$ can be interpreted in terms of choosing the appropriate window for viewing a function.

Example 2   Choosing the correct viewing window

For the following limits $\lim\limits_{x \to a} f(x) = L$, determine how close x needs to be to a to ensure that $f(x)$ is within 0.01 and 0.00001 of L. In each case plot the function in the appropriate window to illustrate your findings.

a. $\lim\limits_{x \to 4}(2x - 1) = 7$ **b.** $\lim\limits_{x \to 0} e^{-1/x^2}$

Solution

a. To have $2x - 1$ within 0.01 of 7, we need

$$6.99 \le 2x - 1 \le 7.01$$
$$7.99 \le 2x \le 8.01 \quad \textit{adding 1 to all sides of the inequality.}$$
$$3.995 \le x \le 4.005 \quad \textit{dividing all sides of the inequality by 2.}$$

Plotting $y = 2x - 1$ in the window $[3.995, 4.005] \times [6.99, 7.01]$, yields the graph illustrated in Figure 2.13. Notice that the graph of the function plotted in Figure 2.13 just fits in this window!

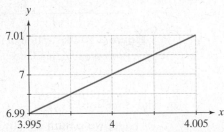

Figure 2.13 Plot of the function $y = 2x - 1$ in the window $[3.995, 4.005] \times [6.99, 7.01]$.

To have $2x - 1$ within 0.00001 of 7, we need

$$6.99999 \le 2x - 1 \le 7.00001$$
$$7.99999 \le 2x \le 8.00001$$
$$3.999995 \le x \le 4.000005$$

Plotting $y = 2x - 1$ in the window $[3.999995, 4.000005] \times [6.99999, 7.00001]$ yields Figure 2.14. Again, notice that the graph of the function plotted in Figure 2.14 just fits in this window.

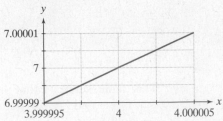

Figure 2.14 Plot of the function $y = 2x - 1$ in the window $[3.999995, 4.000005] \times [6.99999, 7.00001]$.

b. To ensure that e^{-1/x^2} is within 0.01 of 0, we need

$$-0.01 \le e^{-1/x^2} \le 0.01$$

Since the left-hand inequality is always true, we can ignore it. Also, since the natural logarithm is an increasing function, we have

$$\ln e^{-1/x^2} \le \ln 0.01$$

$$\frac{-1}{x^2} \le -\ln 100$$

$$\frac{1}{x^2} \ge \ln 100$$

$$\frac{1}{\ln 100} \ge x^2 \qquad since\ x^2 > 0\ for\ x \neq 0$$

$$0.46599 \approx \sqrt{\frac{1}{\ln 100}} \ge |x| \qquad since\ \sqrt{x}\ is\ increasing$$

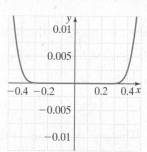

Figure 2.15 Plot of the function $y = e^{-1/x^2}$ in the window $[-0.46599, 0.46599] \times [-0.01, 0.01]$.

Thus, if we plot $y = e^{-1/x^2}$ in the window $[-0.46599, 0.46599] \times [-0.01, 0.01]$, we obtain the graph illustrated in Figure 2.15. In this case, the upper half of the function just fits the window.

To ensure that e^{-1/x^2} is within 0.00001 of 0, we need

$$-0.00001 \le e^{-1/x^2} \le 0.00001$$

Since the left-hand inequality is always true, we can ignore it. Also, since the natural logarithm is an increasing function, we have

$$\ln e^{-1/x^2} \le \ln 0.00001$$

$$\frac{-1}{x^2} \le -\ln 100{,}000$$

$$\frac{1}{\ln 100{,}000} \ge x^2 \qquad since\ x^2 > 0\ for\ x \neq 0$$

$$0.294718 \approx \sqrt{\frac{1}{\ln 100{,}000}} \ge |x| \qquad since\ \sqrt{x}\ is\ increasing$$

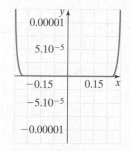

Figure 2.16 Plot of the function $y = e^{-1/x^2}$ in the window $[-0.294718, 0.294718] \times [-0.00001, 0.00001]$.

Thus, if we plot e^{-1/x^2} in the window $[-0.294718, 0.294718] \times [-0.00001, 0.00001]$ we obtain the graph illustrated in Figure 2.16, whose upper half also just fits the window.

The statement $\lim_{x \to a} f(x) = L$ can fail in two ways. First, $\lim_{x \to a} f(x)$ exists but does not equal L. Second, the $\lim_{x \to a} f(x)$ does not exist. In other words, no matter what L we choose, the statement $\lim_{x \to a} f(x) = L$ is false. You typically encounter the first failure when someone is testing you on this material or when someone has defined a function such that $f(a) \neq \lim_{x \to a} f(x)$. The second failure is more interesting.

Example 3 Limit failures

Determine whether the following limits exist.

a. $\displaystyle \lim_{x \to 0} \cos \frac{2\pi}{x}$

b. $\displaystyle \lim_{x \to 0} x \cos \frac{2\pi}{x}$

Solution

a. To understand $\displaystyle \lim_{x \to 0} \cos \frac{2\pi}{x}$, we might begin by evaluating $\cos \dfrac{2\pi}{x}$ at smaller and smaller x values. But we need to be careful!

- If we evaluate at $x = 1, x = 0.1, x = 0.01, x = 0.001, \ldots$, we obtain $\cos 2\pi = 1$, $\cos 20\pi = 1$, $\cos 200\pi = 1, \ldots$. This suggests $\lim\limits_{x \to 0} \cos \dfrac{2\pi}{x} = 1$.

- If we evaluate at $x = 2$, $x = 2/3$, $x = 2/5 \ldots$, we obtain $\cos \pi = -1$, $\cos 3\pi = -1$, $\cos 5\pi = -1, \ldots$. This suggests that $\lim\limits_{x \to 0} \cos \dfrac{2\pi}{x} = -1$.

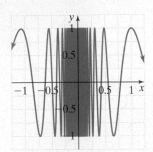

Figure 2.17 Graph of $f(x) = \cos \dfrac{2\pi}{x}$.

Both of these statements cannot be true simultaneously. Lets consider the statement $\lim\limits_{x \to 0} \cos \dfrac{2\pi}{x} = 1$; this requires that $\cos \dfrac{2\pi}{x}$ can be made *arbitrarily close to* 1 *for all x sufficiently close (but not equal to)* 0. However, there are x's arbitrarily close but not equal to 0 (namely, $x = 2/3, 2/5, 2/7, \cdots$) such that $\cos \dfrac{2\pi}{x} = -1$, which is 2 units away from 1. Hence, $\lim\limits_{x \to a} \cos \dfrac{2\pi}{x} \neq 1$. This argument can be refined to show that $\lim\limits_{x \to 0} \cos \dfrac{2\pi}{x} \neq L$ for any choice of L. Therefore, the limit does not exist.

Graphing this function in Figure 2.17 illustrates the dramatic nature of this nonexisting limit.

b. To understand $\lim\limits_{x \to 0} x \cos \dfrac{2\pi}{x}$, we begin by noticing that cosine takes on values between -1 and 1. Hence, for $x \neq 0$, $-1 \leq \cos \left(\dfrac{2\pi}{x} \right) \leq 1$ and thus

$$-|x| \leq x \cos \frac{2\pi}{x} \leq |x|$$

for all $x \neq 0$. Therefore, by choosing x sufficiently close to 0 but not equal to 0, we can make $|x|$ as close to 0 as we want, so that $x \cos \dfrac{2\pi}{x}$ becomes arbitrarily close to 0. Therefore,

$$\lim_{x \to 0} x \cos \frac{2\pi}{x} = 0$$

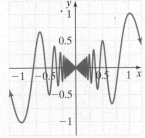

Figure 2.18 Graph of $f(x) = \cos \dfrac{2\pi}{x}$.

as the graph of $y = x \cos \dfrac{2\pi}{x}$ in Figure 2.18 illustrates. In the next section, we will see that this example is an application of the squeeze theorem.

We occasionally rely on technology to compute limits. Although technology almost always steers us in the right direction, cases exist where it drives us to incorrect conclusions.

Example 4   A computational dilemma

Consider the function

$$f(x) = \frac{\sqrt{1 + x^2} - 1}{x^2}$$

a. Use technology to evaluate $f(x)$ at $x = \pm 0.1, \pm 0.01, \pm 0.001, \pm 0.0001$. Based on these evaluations, formulate a conclusion about $\lim\limits_{x \to 0} f(x)$.

b. Use technology to evaluate $f(x)$ at $x = \pm 10^{-5}, \pm 10^{-6}, \pm 10^{-7}, \pm 10^{-8} \pm 10^{-9}$. Based on these evaluations, formulate a conclusion about $\lim\limits_{x \to 0} f(x)$. Compare your results to those of part **a**.

Solution

a. We begin with a table of values.

x	$f(x)$
± 0.1	0.498756
± 0.01	0.499988
± 0.001	0.500000
± 0.0001	0.500000

This table suggests that the limit is 0.5. Moreover, plotting this function in Figure 2.19 over the interval $-1 \leq x \leq 1$ reaffirms this conclusion.

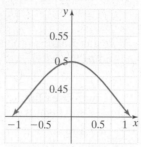

Figure 2.19 Graph of numerical solution to $f(x) = \dfrac{\sqrt{1+x^2}-1}{x^2}$ over the range $[-1, 1]$.

b. Next, we evaluate f for even smaller values of x.

x	$f(x)$
$\pm 10^{-5}$	0.500000
$\pm 10^{-6}$	0.500044
$\pm 10^{-7}$	0.488498
$\pm 10^{-8}$	0.000000
$\pm 10^{-9}$	0.000000

It appears that f is approaching the value 0, not 0.5. Plotting the graph of $y = f(x)$ over this smaller range of x values yields Figure 2.20, which suggests that the limiting value is 0—very strange indeed!

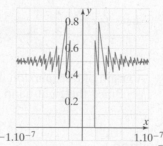

Figure 2.20 Graph of numerical solution to $f(x) = \dfrac{\sqrt{1+x^2}-1}{x^2}$ over the range $[-10^{-7}, 10^{-7}]$.

So we have a dilemma. Should the answer be 0 or 0.5? Later, we will develop more reliable methods that will show this limit is 0.5. Hence, when you use technology, always be aware that technology may mislead you.

One-sided limits

The definition of the limit of $f(x)$ as x approaches a requires that $f(x)$ approach the same value independent of whether x approaches a from the right or the left. In this sense, $\lim_{x \to a} f(x)$ is a "two-sided" limit. One-sided limits, on the other hand, only require that $f(x)$ approach a value as x approaches a from the left or the right.

One-sided Limits (Informal)

Right-hand limit: We write

$$\lim_{x \to a^+} f(x) = L$$

if we can make $f(x)$ as close to L as we please by choosing x sufficiently close to a and *to the right of a*.

Left-hand limit: We write

$$\lim_{x \to a^-} f(x) = L$$

if we can make $f(x)$ as close to L as we please by choosing x sufficiently close to a and *to the left of a*.

A two-sided limit cannot exist if the corresponding one-sided limits are different. Conversely, it can be shown that if the two one-sided limits of a given function f as $x \to a^-$ and $x \to a^+$ both exist and are equal, then the two-sided limit, $\lim_{x \to a} f(x)$, exists. These observations are so important that we restate them as follows:

Matching Limits

Let f be a function. Then

$$\lim_{x \to a} f(x) = L \text{ if and only if } \lim_{x \to a^+} f(x) = \lim_{x \to a^-} f(x) = L$$

Example 5   Finding one-sided limits

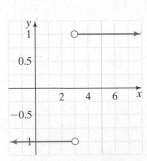

Figure 2.21 Graph of
$f(x) = \dfrac{x - 3}{|x - 3|}$.

Consider the function $f(x) = \dfrac{x - 3}{|x - 3|}$. Find the right-hand limit as $x \to 3^+$ and the left-hand limit as $x \to 3^-$; discuss whether $\lim_{x \to 3} f(x)$ exists.

Solution   Since $f(x) = \dfrac{x - 3}{|x - 3|} = 1$ whenever $x > 3$, we have

$$\lim_{x \to 3^+} f(x) = 1$$

Since $f(x) = \dfrac{x - 3}{|x - 3|} = -1$ whenever $x < 3$, we have

$$\lim_{x \to 3^-} f(x) = -1$$

A graph of this function is given in Figure 2.21.

Since the right-hand and left-hand limits are not the same, $\lim_{x \to 3} f(x)$ does not exist.

Example 6   The floor function

The **floor function**, sometimes called the **step-function**, is the function that returns the largest integer less than or equal to x. The function is typically denoted by $\lfloor x \rfloor$. For instance, $\lfloor 3 \rfloor = 3$, $\lfloor \pi \rfloor = 3$, $\lfloor \frac{1}{3} \rfloor = 0$, and $\lfloor -1.1 \rfloor = -2$.

a. Graph the $y = \lfloor x \rfloor$ over the interval $[-\pi, \pi]$.

b. Determine at what values of a, $\lim_{x \to a} \lfloor x \rfloor$ does not exist.

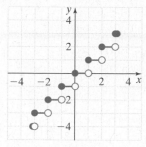

Figure 2.22 Graph of $\lfloor x \rfloor$ on $\lfloor -\pi, \pi \rfloor$.

Solution

a. The graph of $y = \lfloor x \rfloor$ is shown in Figure 2.22. Notice that the closed circles include the end point and the open circles exclude the end point.

b. From Figure 2.22, we see that the limit will not exist at integer values. That is, at the points $a = -3, -2, -1, 0, 1, 2, 3$, we have

$$\lim_{x \to a^+} \lfloor x \rfloor = \lim_{x \to a^-} \lfloor x \rfloor + 1$$

Therefore, by the matching limits property, the limit does not exist at $a = -3, -2, -1, 0, 1, 2, 3$. ∎

One-sided limits are particularly useful when considering piece-wise defined functions (see Section 1.1 for a definition).

In the next example, the piece-wise defined function describes the feeding behavior of the copepod *Calanus pacificus*, illustrated in Figure 2.23. Planktonic copepods are small crustaceans found in the sea. These organisms play an important role in global ecology as they are a major food source for small fish, whales, and sea birds. It is believed that they form the largest animal biomass on Earth. Given their importance, scientists are interested in understanding how their feeding rate depends on availability of resources. In a classic ecology paper, C. S. Holling classified feeding rates into three types. (See his article, "The Functional Response of Invertebrate Predators to Prey Density," *Memoirs of the Entomological Society of Canada* 48 (1966): 1–86.) The first type, so-called type I, assumes that organisms consume at a rate proportional to the amount of food available until they achieve a maximal feeding rate. The type II feeding rate was studied in Example 6 of Section 1.6.

Courtesy Dr. David Pond, Scottish Marine Institute

Figure 2.23 The planktonic copepod *Calanus pacificus* as seen under an electron microscope. Copepods such as this are believed by some scientists to form the largest animal biomass on Earth.

Example 7 Type I functional response

In the 1970s, B. W. Frost, a scientist at the Department of Oceanography at the University of Washington, measured feeding rates of the planktonic copepod *Calanus pacificus* in the lab. In one of his experiments, *C. pacificus* were offered different concentrations of the diatom species *Coscinodiscus anstii*. Frost found that *C. pacificus* reached its maximal feeding rate of 1,250 cells/hour when the concentration of *C. anstii* was approximately 200 cells/milliliter (see Figure 2.24). If you assume that the feeding rate is proportional to the concentration x of *C. anstii* until they achieve their maximal feeding rate, then the feeding rate as a function of x is of the form

$$f(x) = \begin{cases} ax \text{ cells/hour} & \text{if } x \leq 200 \\ 1{,}250 \text{ cells/hour} & \text{if } x > 200 \end{cases}$$

where $a > 0$ is a proportionality constant.

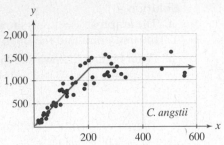

Figure 2.24 Feeding rate I (cells/hour) of a copepod as a function of the density of the diatoms (cells/milliliter) upon which it feeds.

Data Source: B. W. Frost, 1972. "Effects of Size and Concentration of Food Particles on the Feeding Behavior of the Marine Planktonic Copepod *Calanus pacificus.*" *Limnology and Oceanography* 17(6): 805–815.

a. Find $\lim\limits_{x \to 200^+} f(x)$ and $\lim\limits_{x \to 200^-} f(x)$.

b. Determine for what choice of a, $\lim\limits_{x \to 200} f(x)$ exists.

Solution

a. Since $f(x) = 1{,}250$ for all $x > 200$, we find

$$\lim_{x \to 200^+} f(x) = 1{,}250$$

On the other hand, $f(x) = ax$ for all $x \leq 200$. Hence, as x increases to 200, $f(x)$ approaches $200a$ and

$$\lim_{x \to 200^-} f(x) = 200a,$$

b. By the matching limit property, $\lim\limits_{x \to 200} f(x)$ exists if and only if the left- and right-hand limits are equal. Therefore, we need to ensure that $1{,}250 = 200a$ or $a = 6.25$. In which case, $\lim\limits_{x \to 200} f(x) = 1{,}250$. The graph of this function, along with the data as plotted in Figure 2.24, illustrates that by choosing $a = 6.25$, the linear function and constant function are pasted together in such a way that their values agree at $x = 200$.

Limits: A formal perspective (optional)

This section can be omitted by those not going on to major in mathematics at the undergraduate level. Our informal definition of the limit provides valuable intuition that allows you to develop a working knowledge of this fundamental concept. For theoretical work, however, the intuitive definition will not suffice, because it gives no precise, quantifiable meaning to the terms "arbitrarily close to L" and "sufficiently close to a." In the nineteenth century, leading mathematicians, including Augustin-Louis Cauchy (1789–1857) and Karl Weierstrass (1815–1897), sought to put calculus on a sound logical foundation by giving precise definitions for the foundational ideas of calculus. The following definition, derived from the work of Cauchy and Weierstrass, gives precision to the **definition of a limit**.

Limit (Formal Definition)

Let f be a real-valued function.

$$\lim_{x \to a} f(x) = L$$

if for every $\epsilon > 0$ there is some $\delta > 0$ such that $|f(x) - L| \leq \epsilon$ whenever $0 < |x - a| < \delta$.

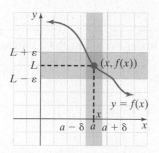

Figure 2.25 The epsilon-delta definition of limit.

Behind the formal language is a fairly straightforward idea. Given any $\epsilon > 0$ specifying a desired degree of proximity to L, a number $\delta > 0$ is found that determines how close x must be to a to ensure that $f(x)$ is within ϵ of L. This is shown in Figure 2.25.

Because the Greek letters ϵ (epsilon) and δ (delta) are traditionally used in this context, the formal definition of limit is sometimes called the epsilon-delta definition of the limit. The goal of this subsection is to show how it can be used rigorously to establish a variety of results.

One can view this definition as setting up an adversarial relationship between two individuals. One person shouts out a value of $\epsilon > 0$. The opponent has to come up with a $\delta > 0$ such that $f(x)$ is within ϵ of L whenever x is within δ of a. This relationship is illustrated in Figure 2.26.

For each $\epsilon > 0$ This forms an interval around L on the y axis.	**there is a $\delta > 0$** This forms an interval around a on the x axis.	**if $0 < |x - c| < \delta$.** This says that if x is in the δ interval on the x axis...	**then $|f(x) - L| < \epsilon$.** ...then $f(x)$ is in the ϵ interval on the y axis

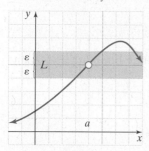 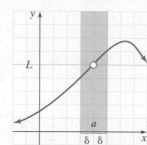 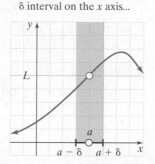 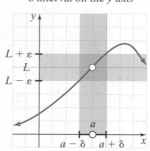

Figure 2.26 Formal definition of limit: $\lim_{x \to a} f(x) = L$.

Note that $f(a)$ does not need to equal the value of L in order for $\lim_{x \to a} f(x) = f(a)$. If $\lim_{x \to a} f(x) = f(a)$, then the epsilon-delta argument confirms that $f(x)$ is continuous at $x = a$. On the other hand, if $\lim_{x \to a} f(x) \neq f(a)$, then the epsilon-delta argument confirms that $f(x)$ is not continuous at $x = a$.

Notice that whenever x is within δ units of a (but not equal to a), the point $(x, f(x))$ on the graph of f must lie in the rectangle (shaded region) formed by the intersection of the horizontal band of width 2ϵ centered at L and the vertical band of width 2δ centered at a. The smaller the ϵ interval around the proposed limit L, generally the smaller the δ interval will need to be for $f(x)$ to lie in the ϵ interval. If such a δ can be found no matter how small ϵ is, then L must be the limit. The following examples illustrate epsilon-delta proofs, one in which the limit exists and one in which it does not.

Example 8   An epsilon-delta proof of a limit statement

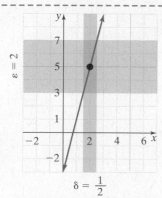

Figure 2.27 Illustration of the epsilon-delta proof of a limit statement.

Show that $\lim_{x \to 2}(4x - 3) = 5$.

Solution   The object is to prove that the limit is 5. We have

$$|f(x) - L| = |4x - 3 - 5|$$
$$= |4x - 8|$$
$$= 4|x - 2| \quad \textit{this must be less than } \epsilon \textit{ whenever } |x - 2| < \delta$$

For a given $\epsilon > 0$ choose $\delta = \dfrac{\epsilon}{4}$. Then

$$|f(x) - L| = 4|x - 2| < 4\delta = 4\left(\frac{\epsilon}{4}\right) = \epsilon$$

This process is illustrated for $\epsilon = 2$ in Figure 2.27.

Example 9   Limit of a constant
- -

Use an epsilon-delta proof to show that the limit of a constant is a constant; that is, show $\lim\limits_{x \to a} c = c$.

Solution   Let $f(x) = c$

$$|f(x) - c| = |c - c|$$
$$= 0$$

Thus, for any given $\epsilon > 0$, pick any $\delta > 0$. Then, if $|x - a| < \delta$, we have

$$|f(x) - c| = 0 < \epsilon$$

Example 10   An epsilon-delta proof to show that a limit does not exist
- -

Show that $\lim\limits_{x \to 0} \dfrac{1}{x}$ does not exist.

Solution   Let $f(x) = \dfrac{1}{x}$ and L be any number. Suppose that $\lim\limits_{x \to 0} f(x) = L$. Consider the graph of f plotted in Figure 2.28.

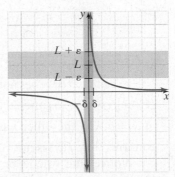

Figure 2.28 Illustration of the epsilon-delta proof that the limit $\lim\limits_{x \to 0} f(x) = L$ exists.

It would seem that no matter what value $\epsilon > 0$ is chosen, it would be impossible to find a corresponding $\delta > 0$. Indeed, suppose that

$$\left| \frac{1}{x} - L \right| < \epsilon$$

Then

$$-\epsilon < \frac{1}{x} - L < \epsilon$$

and

$$L - \epsilon < \frac{1}{x} < L + \epsilon$$

This inequality holds whenever

$$\frac{1}{|x|} < |L| + \epsilon$$

and hence will be violated whenever

$$|x| < \frac{1}{|L| + \epsilon}$$

Thus, no matter what value of $\epsilon > 0$ we choose, we cannot find a $\delta > 0$ such that $|1/x - L| < \epsilon$ for all $0 < |x - 0| < \delta$. Since this statement is true for any L we chose, it follows that the limit does not exist.

PROBLEM SET 2.2

Level 1 DRILL PROBLEMS

Given the functions f and g defined by the graphs in Figure 2.29, find the limits in Problems 1 to 4.

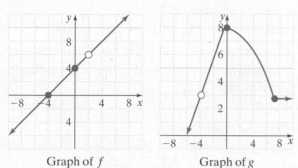

Graph of f Graph of g

Figure 2.29 Graphs of the functions f and g.

1. **a.** $\lim\limits_{x \to -4} f(x)$ **b.** $\lim\limits_{x \to 0} f(x)$

2. **a.** $\lim\limits_{x \to 7} g(x)$ **b.** $\lim\limits_{x \to 0} g(x)$

3. **a.** $\lim\limits_{x \to 2} f(x)$ **b.** $\lim\limits_{x \to -4} g(x)$

4. **a.** $\lim\limits_{x \to 2^-} f(x)$ **b.** $\lim\limits_{x \to -4^+} g(x)$

Given the functions defined by the graphs in Figure 2.30, find the limits, if they exist, in Problems 5 to 8. If the limits do not exist, discuss why.

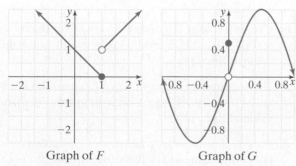

Graph of F Graph of G

Figure 2.30 Graphs of the functions F and G.

5. **a.** $\lim\limits_{x \to 1^-} F(x)$ **b.** $\lim\limits_{x \to 1^+} F(x)$ **c.** $\lim\limits_{x \to 1} F(x)$

6. **a.** $\lim\limits_{x \to -1^-} F(x)$ **b.** $\lim\limits_{x \to -1^+} F(x)$ **c.** $\lim\limits_{x \to -1} F(x)$

7. **a.** $\lim\limits_{x \to 0^-} G(x)$ **b.** $\lim\limits_{x \to 0^+} G(x)$ **c.** $\lim\limits_{x \to 0} G(x)$

8. **a.** $\lim\limits_{x \to 0.5^-} G(x)$ **b.** $\lim\limits_{x \to 0.5^+} G(x)$ **c.** $\lim\limits_{x \to 0.5} G(x)$

Describe each figure in Problems 9 to 12 with a one-sided limit statement. For example, for Problem 9, the answer is $\lim\limits_{x \to 1^+} f(x) = 2$.

9.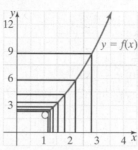

10.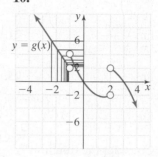

11.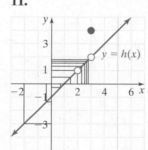

12.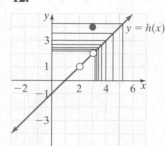

Approximate the limits by filling in the appropriate values in the tables in Problems 13 to 15 using a one-sided statement.

13. $\lim\limits_{x \to 5^-} f(x)$ where $f(x) = (3x - 2)$

x	2	3	4	4.5	4.9	4.99
$f(x)$	4					

14. $\lim\limits_{x \to 2^-} g(x)$ where $g(x) = \dfrac{x^3 - 8}{x^2 + 2x + 4}$

x	1	1.5	1.9	1.99	1.999	1.9999
$g(x)$	-1					

15. $\lim\limits_{x \to 2} h(x)$ where $h(x) = \dfrac{3x^2 - 2x - 8}{x - 2}$

x	1	1.5	1.9	1.99	1.999	1.9999
$h(x)$	7					

x	3	2.5	2.1	2.01	2.001	2.0001
$h(x)$	13					

Determine the limits in Problems 16 to 24. If the limit exists, explain how you found the limit. If the limit does not exist, explain why.

16. $\lim\limits_{x\to 5}\dfrac{1}{x}$

17. $\lim\limits_{x\to -3^+}\dfrac{|x+3|}{x+3}$

18. $\lim\limits_{x\to -1}\cos x$

19. $\lim\limits_{x\to -1}\cos(\pi x)$

20. $\lim\limits_{x\to 1}\dfrac{\ln x}{x-1}$

21. $\lim\limits_{x\to 2}\dfrac{\sqrt{x+2}-2}{x-2}$

22. $\lim\limits_{x\to 0}\dfrac{x}{|x|}$

23. $\lim\limits_{x\to 0}\dfrac{x^2}{|x|}$

24. $\lim\limits_{x\to 4}\dfrac{(x-4)^2}{|x-4|}$

25. Consider the function

$$f(x)=\sqrt{x}\cos\left(\frac{1}{x}\right)$$

whose graph is given Figure 2.31. Does $\lim\limits_{x\to 0^+}f(x)$ exist? If so, what is it? If not, why not?

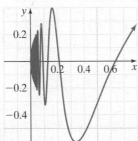

Figure 2.31 Graph of the function $f(x)=\sqrt{x}\cos\left(\frac{1}{x}\right)$.

26. Consider the function

$$f(x)=|x|\sin\left(\frac{1}{x}\right)$$

whose graph is given Figure 2.32. Does $\lim\limits_{x\to 0}f(x)$ exist? If so, what is it? If not, why not?

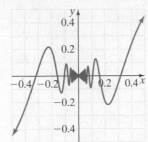

Figure 2.32 Graph of the function $f(x)=|x|\sin\left(\frac{1}{x}\right)$.

27. Consider the statement $\lim\limits_{x\to 1}(4+x)=5$. How close does x need to be to 1 to ensure that $4+x$ is within the given distance of 5?

   a. 0.1   **b.** 0.01   **c.** 0.001

28. Consider the statement $\lim\limits_{x\to 2}x^2=4$. How close does x need to be to 2 to ensure that x^2 is within the given distance of 4?

   a. 0.1   **b.** 0.01   **c.** 0.001

29. Consider the statement $\lim\limits_{x\to 0^+}\sqrt{x}=0$. How close does x need to be to 0 to ensure that $\sqrt{x}$ is within the given distance of 0?

   a. 0.1   **b.** 0.01   **c.** 0.001

30. Consider the statement $\lim\limits_{x\to e}\ln x=1$. How close does x need to be to e to ensure that $\ln x=1$ is within the given distance of 1?

   a. 0.1   **b.** 0.01   **c.** 0.001

31. Consider the function

$$f(x)=\frac{\sqrt{4-x^2}-2}{x^2}$$

   a. Use technology to graph $y=f(x)$ over the interval $[-2,2]$.

   b. Use technology to graph $y=f(x)$ over the interval $[-0.1,0.1]$. Based on your graph, guess the value of $\lim\limits_{x\to 0}f(x)$.

   c. Use technology to graph $y=f(x)$ over the interval $[-10^{-7},10^{-7}]$. Based on your graph, guess the value of $\lim\limits_{x\to 0}f(x)$.

   d. Most technologies can keep track of sixteen or fewer digits. In light of this observation, discuss what might be happening in parts **b** and **c**.

32. Consider the function $f(x)=\dfrac{\ln(1+x^2)}{x^2}$.

   a. Use technology to graph $y=f(x)$ over the interval $[-2,2]$.

   b. Use technology to graph $y=f(x)$ over the interval $[-0.1,0.1]$. Based on your graph, guess the value of $\lim\limits_{x\to 0}f(x)$.

   c. Use technology to graph $y=f(x)$ over the interval $[-10^{-7},10^{-7}]$. Based on your graph, guess the value of $\lim\limits_{x\to 0}f(x)$.

   d. Most technologies can keep track of sixteen or fewer digits. In light of this observation, discuss what might be happening in parts **b** and **c**.

In Problems 33 to 38, prove the limit exists using the formal definition of the limit.

33. $\lim\limits_{x\to 5}(x+1)=6$

34. $\lim\limits_{x\to 5}(1-3x)=-14$

35. $\lim\limits_{x\to 2}\dfrac{1}{x}=\dfrac{1}{2}$

36. $\lim\limits_{x\to 1}(x^2-3x+2)=0$

37. $\lim_{x \to 2}(x^2 + 2) = 6$

38. $\lim_{x \to 1}(x^2 + 1) = 2$

Level 2 APPLIED AND THEORY PROBLEMS

39. The federal income tax rates for singles in 2010 is shown in Table 2.3.

Table 2.3 Schedule X – Single

If taxable income is over	But not over	The tax is
$ 0	$ 8,350	10% of the amount over $0
$ 8,350	$ 33,950	$835 plus 15% of the amount over $8,350
$ 33,950	$ 82,250	$4,675 plus 25% of the amount over $33,950
$ 82,250	$171,550	$16,750 plus 28% of the amount over $82,250
$171,550	$372,950	$41,754 plus 33% of the amount over $171,550
$372,950	no limit	$108,216 plus 35% of the amount over $372,950

Express the income tax $f(x)$ for an individual in 2010 with adjusted income x dollars as a piecewise defined function.

a. Graph $y = f(x)$ over the interval $[0, 500,000]$.

b. Determine at what values of a, $\lim_{x \to a} f(x)$ does not exist.

40. In 2011, the U. S. postal rates were 44 cents for the first ounce or fraction of an ounce, and 17 cents for each additional ounce or fraction of an ounce up to 3.5 ounces. Let p represent the total amount of postage (in cents) for a letter weighing x ounces.

a. Graph $y = p(x)$ over the interval $[0, 3.5]$ ounces.

b. Determine at what values of a, $\lim_{x \to a} f(x)$ does not exist.

41. A wildlife ecologist who studied the rate at which wolves kill moose in Yellowstone National Park found that when moose were plentiful, wolves killed moose at the rate of one moose per wolf every twenty-five days. (Note this doesn't mean that wolves only eat every twenty-five days, because they hunt in packs and share kills.) However, when the density of moose drops below $x = 3$ per km^2, then the rate at which wolves kill moose is proportional to the density. Construct a type I functional response $f(x)$ (see Example 7) such that $f(x)$ has a limit at $x = 3$.

42. A student looking at Figure 2.24 decided that the following function might provide a better fit to the data:

$$f(x) = \begin{cases} 6.25x \text{ cells/hour} & \text{if } x \le 150 \\ ax + b \text{ cells/hour} & \text{if } 150 < x < 300 \\ 1,300 \text{ cells/hour} & \text{if } x \ge 300 \end{cases}$$

Find values for the parameters a and b that ensure $f(x)$ has limits at $x = 150$ and $x = 300$.

2.3   Limit Laws and Continuity

Having defined limits, we are ready to develop some tools to verify their existence and to compute them more readily. In some cases, taking the limit of a function reduces to evaluating the function at the limit point, and in other cases we cannot find the limit by evaluation. In this section, we find when evaluation is acceptable and when it is not.

Properties of limits

With a definition of a limit in hand, it is important to understand how the definition acts under functional arithmetic. For instance, if $\lim_{x \to a} f(x) = L$ and $\lim_{x \to a} g(x) = M$, then $f(x)$ and $g(x)$ can be made arbitrarily close to L and M, respectively, for all x sufficiently close but not equal to a. Hence, $f(x)g(x)$ must be arbitrarily close to LM for all x sufficiently close but not equal to a. Therefore, it is reasonable to conjecture that the limit of the product $f \cdot g$ is the product LM of the limits. Indeed, this is true and can be proved using the formal definition of the limit. In fact, limits satisfy all the arithmetic properties that you would think they should, as summarized in the following limit laws.

Limit Laws	Let f and g be functions such that $\lim\limits_{x \to a} f(x) = L$ and $\lim\limits_{x \to a} g(x) = M$. Then

Sums $\lim\limits_{x \to a} (f(x) + g(x)) = L + M$

Differences $\lim\limits_{x \to a} (f(x) - g(x)) = L - M$

Products $\lim\limits_{x \to a} f(x)g(x) = LM$

Quotients $\lim\limits_{x \to a} \dfrac{f(x)}{g(x)} = \dfrac{L}{M}$ provided that $M \neq 0$

Example 1   Using limit laws

Using the limit laws, find the following limits. You may assume that $\lim\limits_{x \to 4} x = 4$ and $\lim\limits_{x \to 4} 1 = 1$.

a. $\lim\limits_{x \to 4} x^2$　　　　　　**b.** $\lim\limits_{x \to 4} (x^2 + x)$　　　　　　**c.** $\lim\limits_{x \to 4} \left(\dfrac{1}{x} - \dfrac{1}{x^2} \right)$

Solution

a.
$$\lim_{x \to 4} x^2 = (\lim_{x \to 4} x)(\lim_{x \to 4} x) \quad \textit{product law}$$
$$= 4 \times 4 \quad \textit{given value}$$
$$= 16$$

b.
$$\lim_{x \to 4} (x^2 + x) = \lim_{x \to 4} x^2 + \lim_{x \to 4} x \quad \textit{sum law}$$
$$= \left[\lim_{x \to 4} x \right]^2 + \lim_{x \to 4} x \quad \textit{product law}$$
$$= [4]^2 + 4 \quad \textit{given value}$$
$$= 20$$

c.
$$\lim_{x \to 4} \left(\frac{1}{x} - \frac{1}{x^2} \right) = \lim_{x \to 4} \frac{1}{x} - \lim_{x \to 4} \frac{1}{x^2} \quad \textit{difference law}$$
$$= \lim_{x \to 4} \frac{1}{x} - \left[\lim_{x \to 4} \frac{1}{x} \right]^2 \quad \textit{product law}$$
$$= \frac{1}{4} + \left[\frac{1}{4} \right]^2 \quad \textit{quotient law}$$
$$= \frac{3}{16}$$

The preceding example illustrates that applying the product and sum limit laws repeatedly allows us to quickly compute limits of polynomials and rational functions as x approaches a by evaluating them at the value a, provided a is in the domain.

Limits of Polynomials and Rational Functions	Let f be either a polynomial or a rational function. If a is in the domain of f, then $$\lim_{x \to a} f(x) = f(a)$$

Proof. We have previously shown that $\lim\limits_{x \to a} c = c$ (the limit of a constant is a constant) and $\lim\limits_{x \to a} x = a$. By applying the limit law for products repeatedly, we have $\lim\limits_{x \to a} x^n = a^n$ for $n = 1, 2, 3, \ldots$. Let $p(x) = b_0 + b_1 x + b_2 x^2 + \ldots b_n x^n$ be a polynomial. Then

$$\lim_{x \to a} p(x) = \lim_{x \to a} b_0 + \lim_{x \to a} b_1 x + \cdots + \lim_{x \to a} b_n x^n \qquad \textit{limit law for sums}$$

$$= b_0 \lim_{x \to a} 1 + b_1 \lim_{x \to a} x + \cdots + b_n \lim_{x \to a} x^n \qquad \textit{limit law for products}$$

$$= b_0 + b_1 a + \ldots b_n a^n$$

$$= p(a)$$

Thus, we have shown $\lim\limits_{x \to a} p(x) = p(a)$ for any polynomial. You will be asked to prove the result for a rational function in problem 34 in Problem Set 2.3. ∎

Example 2 Finding limits algebraically

Find the limits and show each step of your derivation.

a. $\lim\limits_{x \to 2}(2x^4 - 5x^3 + 2x^2 - 5)$

b. $\lim\limits_{x \to 2} \dfrac{x^2 - 4}{x + 2}$

c. $\lim\limits_{x \to -2} \dfrac{x^2 - 4}{x + 2}$

Solution

a. Since $2x^4 - 5x^3 + 2x^2 - 5$ is a polynomial, it is sufficient to evaluate the polynomial at $x = 2$:

$$\lim_{x \to 2}(2x^4 - 5x^3 + 2x^2 + 5) = 2(2)^4 - 5(2)^3 + 2(2)^2 - 5$$

$$= 32 - 40 + 8 - 5$$

$$= -5$$

b. Since $\dfrac{x^2 - 4}{x + 2}$ is a rational function and $x = 2$ is in the domain, it is sufficient to evaluate the rational function at $x = 2$:

$$\lim_{x \to 2} \frac{x^2 - 4}{x + 2} = \frac{(2)^2 - 4}{2 + 2}$$

$$= \frac{0}{4}$$

$$= 0$$

c. Since $x = -2$ is not in the domain, we cannot simply evaluate the function at $x = -2$ to determine the limit. However, we can factor and then evaluate at $x = -2$:

$$\lim_{x \to -2} \frac{x^2 - 4}{x + 2} = \lim_{x \to -2} \frac{(x - 2)(x + 2)}{x + 2}$$

$$= \lim_{x \to -2} (x - 2) \qquad \textit{now it is a polynomial}$$

$$= -4$$

In Example 3b of Section 2.2, we computed a limit by "squeezing" a function for an unknown limit between functions whose limits we understand. The squeeze theorem provides a general approach for computing limits in this manner. This theorem, stated below, is sometimes called the *sandwich theorem* or *pinching theorem*.

Theorem 2.1 Squeeze theorem

Let f, g, and h be functions such that

$$f(x) \le g(x) \le h(x) \quad \text{for all } x \text{ near but not equal to } a$$

If $\lim\limits_{x \to a} f(x) = L$ and $\lim\limits_{x \to a} h(x) = L$, then $\lim\limits_{x \to a} g(x) = L$.

Example 3 Using the squeeze theorem

Find $\lim\limits_{x \to 5} \dfrac{x^2 + (x-5)^2 \sin\left(\dfrac{1}{x-5}\right)}{x+1}$

Solution Since $-1 \le \sin\left(\dfrac{1}{x-5}\right) \le 1$ for $x \ne 5$, we have

$$-(x-5)^2 \le (x-5)^2 \sin\left(\frac{1}{x-5}\right) \le (x-5)^2$$

for $x \ne 5$. Therefore,

$$\frac{x^2 - (x-5)^2}{x+1} \le \frac{x^2 + (x-5)^2 \sin\left(\dfrac{1}{x-5}\right)}{x+1} \le \frac{x^2 + (x-5)^2}{x+1}$$

for all $x > -1$ and $x \ne 5$. Define

$$f(x) = \frac{x^2 - (x-5)^2}{x+1}$$

$$g(x) = \frac{x^2 + (x-5)^2 \sin\left(\dfrac{1}{x-5}\right)}{x+1}$$

$$h(x) = \frac{x^2 + (x-5)^2}{x+1}$$

Since f and h are rational functions with $x = 5$ in their domain,

$$\lim_{x \to 5} f(x) = f(5) = 25/6 \quad \text{and} \quad \lim_{x \to 5} h(x) = h(5) = 25/6$$

Since $f(x) \le g(x) \le h(x)$ for x near 5, the squeeze theorem implies

$$\lim_{x \to 5} g(x) = 25/6$$

Continuity at a point

Example 2 illustrated how easy it is to find limits $\lim\limits_{x \to a} f(x)$ when f is a polynomial or rational function and $x = a$ is in the domain of f. One simply evaluates f at $x = a$. When one is able to evaluate a limit so easily, a function is *continuous* at $x = a$. The idea of *continuity* corresponds to the intuitive notion of a curve "without breaks or jumps."

Continuity at a Point	A function f is **continuous at the point** $x = a$ if f is defined at $x = a$ and $\lim\limits_{x \to a} f(x) = f(a)$. A function f is **discontinuous at the point** $x = a$ if it is not continuous at the point $x = a$.

Example 4 Checking continuity

Test the continuity of each of the following functions at $x = 0$ and $x = 1$. If the function is not continuous at the point, explain. Discuss whether the function can be redefined at points of discontinuity to make it continuous.

a. The function f is defined by the graph $y = f(x)$:

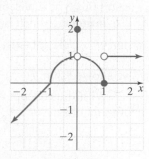

b. $g(x) = \dfrac{x^2 + 2x - 3}{x - 1}$ if $x \neq 1$, $g(x) = 6$ if $x = 1$

Solution

a. Since $f(x)$ approaches 1 from both sides of $x = 0$, we see $\lim\limits_{x \to 0} f(x) = 1$. However, as $f(0) = 2$, we see that

$$\lim_{x \to 0} f(x) \neq f(0)$$

and f is not continuous at $x = 0$. However, this discontinuity can be fixed by redefining $f(0)$ to be 1.

At $x = 1$, we see

$$\lim_{x \to 1^-} f(x) = 0 \quad \text{and} \quad \lim_{x \to 1^+} f(x) = 1$$

Thus the limit does not exist, so f is not continuous at $x = 1$. Since the left- and right-hand side limits do not agree, there is no way redefine f at $x = 1$ to repair this discontinuity.

b. At $x = 0$ we use the limit law for quotients

$$\lim_{x \to 0} \frac{x^2 + 2x - 3}{x - 1} = \frac{0^2 + 2(0) - 3}{0 - 1} = 3$$

and $g(0) = 3$, so g is continuous at $x = 0$. At $x = 1$ we see $g(1) = 6$. We cannot use the limit of a quotient law because of division by zero at $x = 1$. However, if we factor the numerator and then take the limit, we get

$$\lim_{x \to 1} \frac{x^2 + 2x - 3}{x - 1} = \lim_{x \to 1} \frac{(x - 1)(x + 3)}{x - 1}$$
$$= \lim_{x \to 1}(x + 3)$$
$$= 4$$

Since $\lim\limits_{x \to 1} g(x) \neq g(1)$, g is not continuous at $x = 1$. However, this discontinuity is reparable by redefining $g(1) = 4$.

In general, continuity of a function $f(x)$ at $x = a$ can fail in three ways. First, the limit $\lim_{x \to a} f(x) = L$ may be well defined, but either $f(a)$ is not defined or $f(a) \neq L$. When this occurs, f has a **point discontinuity at** $x = a$. These discontinuities can be repaired by redefining f at $x = a$ by $f(a) = L$. For example, in Example 4a, we repaired a point discontinuity at $x = 0$. The second type of discontinuity is a **jump discontinuity** where $\lim_{x \to a^+} f(x) = L$ and $\lim_{x \to a^-} f(x) = M$ with $L \neq M$. The left and right limits don't agree, there is no way to repair the discontinuity. Example 4a illustrates this type of discontinuity at $x = 1$. The final type of discontinuity is an **essential discontinuity**, which occurs when one or both of the one-sided limits, $\lim_{x \to a^+} f(x)$ or $\lim_{x \to a^-} f(x)$, do not exist or are infinite. In Example 3a and Figure 2.17 in Section 2.2, we showed that $f(x) = x \cos \dfrac{2\pi}{x}$ exhibits an essential discontinuity at $x = 0$.

With the concept of continuity, we are able to get a limit law for a composition of functions.

Composition Limit Law

Let f and g be functions such that $\lim_{x \to a} f(x) = L$ and g is continuous at $x = L$. Then

$$\lim_{x \to a} g[f(x)] = g[f(a)]$$

The next examples illustrate why $\lim_{x \to L} g(x)$ existing does not suffice for $\lim_{x \to a} g[f(x)]$ to exist.

Example 5   Compositional limits

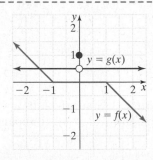

Consider the functions f and g, whose graphs are shown on the left in blue and red, respectively.

a. Find $\lim_{x \to 1} f(x)$ and $\lim_{x \to 0} g(x)$.

b. Does $\lim_{x \to 1} g[f(x)]$ exist? If it doesn't exist, determine whether you can redefine one of the functions at a point to ensure the limit exists.

Solution

a. In the graph $f(x)$ approaches 0 as x approaches 1. Hence, $\lim_{x \to 1} f(x) = 0$. Since $g(x) = 0.5$ for all $x \neq 0$, $\lim_{x \to 0} g(x) = 0.5$.

b. Consider the right-sided limit, $\lim_{x \to 1^+} g[f(x)]$. Since $f(x) < 0$ for $x > 1$ and $g(x) = 0.5$ for $x \neq 0$, we have $g[f(x)] = 0.5$ for all $x > 1$. Therefore, $\lim_{x \to 1^+} g[f(x)] = 0.5$.

Now, consider the left-sided limit, $\lim_{x \to 1^-} g[f(x)]$. Since $f(x) = 0$ for x in $[-1, 1]$, $g[f(x)] = 1$ for x in $[-1, 1]$. Therefore, $\lim_{x \to 1^-} g[f(x)] = 1$.

Since the right-sided and left-sided limits are not equal, the limit $\lim_{x \to 1} g[f(x)]$ does not exist. However, if we redefine $g(0) = 0.5$. Then,

$$\lim_{x \to 1^-} g[f(x)] = 1/2 = \lim_{x \to 1^+} g[f(x)]$$

and the limit does exist. Notice that by redefining g to be 0.5 at $x = 0$, we made g continuous at $x = 0$ and, consequently, resulted in the limit of the composition becoming well defined.

Using the limit laws, we can derive some laws of continuity.

Continuity Laws

Let f and g be functions that are continuous at $x = a$. Then

Sums $f + g$ is continuous at $x = a$

Differences $f - g$ is continuous at $x = a$

Products $f \cdot g$ is continuous at $x = a$

Quotients f/g is continuous at a provided that $g(a) \neq 0$

Composition $g \circ f$ is continuous at $x = a$, provided g is continuous at $x = f(a)$

Proof. We will illustrate the proof of the continuity property for products. All other parts follow in a similar manner. Assume that f and g are continuous at a. Then $\lim_{x \to a} f(x) = f(a)$ and $\lim_{x \to a} g(x) = g(a)$. Hence

$$\lim_{x \to a}(fg)(x) = \lim_{x \to a} f(x)\, g(x)$$

$$= \lim_{x \to a} f(x) \lim_{x \to a} g(x) \quad \text{limit law for products}$$

$$= f(a)g(a) \quad \text{continuity of } f \text{ and } g \text{ at } x = a$$

$$= (fg)(a)$$

Therefore, fg is continuous at $x = a$. ■

Since we have shown that $\lim_{x \to a} f(x) = f(a)$ for polynomial and rational functions at points in their domain, these functions are continuous at all points on their domain. As it turns out, this statement holds for all elementary functions.

Theorem 2.2   Continuity of elementary functions theorem

Let f be a polynomial, a rational function, a trigonometric function, a power function, an exponential function, or a logarithmic function. Then f is continuous at all points in its domain.

Armed with the tools of continuity, we can readily calculate many limits.

Example 6   Quick limits

Use the results of this section to find the given limits, and justify each step of your derivation.

a. $\lim_{x \to 1}(\ln x - \sin(\pi x) + x^3)$ **b.** $\lim_{x \to 4} \dfrac{\ln \sqrt{x}}{1 + x}$

Solution

a. $\lim_{x \to 1}(\ln x - \sin(\pi x) + x^3) = \lim_{x \to 1} \ln x - \lim_{x \to 1} \sin(\pi x) + \lim_{x \to 1} x^3$ *sum and difference limit laws*

$$= \ln 1 - \sin \pi + 1^3 \quad \text{continuity of elementary functions}$$

$$= 0 - 0 + 1$$

$$= 1$$

b. $\displaystyle\lim_{x\to 4}\frac{\ln\sqrt{x}}{1+x}=\frac{\displaystyle\lim_{x\to 4}\ln\sqrt{x}}{\displaystyle\lim_{x\to 4}(1+x)}$ *quotient limit law*

$\displaystyle\qquad\qquad=\frac{\ln\sqrt{4}}{1+4}$ *composition limit law and continuity*
of elementary functions

$\displaystyle\qquad\qquad=\frac{1}{5}\ln 2$

Combining continuity theorems with the limit laws, we can compute limits that we could not otherwise find.

Example 7   Technology vanquished

Recall in Example 4 of Section 2.2, we used technology to study the limit

$$\lim_{x\to 0}\frac{\sqrt{1+x^2}-1}{x^2}$$

and this study was inconclusive. Now find this limit using algebra and the results of this section.

Solution   To work with the expression $f(x)=\dfrac{\sqrt{1+x^2}-1}{x^2}$, we need to simplify it. One way to simplify is to multiply the numerator and denominator by $\sqrt{1+x^2}+1$.

$$\frac{\sqrt{1+x^2}-1}{x^2}=\frac{\sqrt{1+x^2}-1}{x^2}\cdot\frac{\sqrt{1+x^2}+1}{\sqrt{1+x^2}+1}$$

$$=\frac{1+x^2-1}{x^2(\sqrt{1+x^2}+1)}$$

$$=\frac{1}{\sqrt{1+x^2}+1}$$

We now turn to evaluating the limit.

$$\lim_{x\to 0}\frac{\sqrt{1+x^2}-1}{x^2}=\lim_{x\to 0}\frac{1}{\sqrt{1+x^2}+1}\qquad\text{\textit{from the above simplification}}$$

$$=\frac{\displaystyle\lim_{x\to 0}1}{\displaystyle\lim_{x\to 0}(\sqrt{1+x^2}+1)}\qquad\text{\textit{limit law for quotients}}$$

$$=\frac{1}{\sqrt{1+0^2}+1}\qquad\text{\textit{$\sqrt{x}$ is continuous}}$$

$$=\frac{1}{2}$$

Notice that this value of $\dfrac{1}{2}$ corresponds to our initial guess of 0.5 in Example 4 of Section 2.2, based on using technology with $x=10^{-5}$ but not when using technology with $x\le 10^{-8}$.

Intermediate Value Theorem

The function f is said to be **continuous on the open interval** (a, b) if it is continuous for each number (i.e., at each point) in this interval. Note that the end points are not part of open intervals. If f is also continuous from the right at a, we say it is *continuous on the half-open interval* $[a, b)$. Similarly, f is continuous on the half-open interval $(a, b]$ if it is continuous at each number between a and b and is continuous from the left at the end point b. Finally, f is **continuous on the closed interval** $[a, b]$ if it is continuous at each number between a and b and is both continuous from the right at a and continuous from the left at b.

Example 8   Intervals of continuity

For the following functions, determine on which intervals the function is continuous.

a. $\dfrac{1}{1 - x^2}$ **b.** $\dfrac{x + 3}{|x + 3|}$ **c.** $\tan x$

Solution

a. Since $\dfrac{1}{1 - x^2}$ is a rational function, it is continuous on its domain, that is, whenever its denominator is nonzero. Since $1 - x^2 = 0$ if and only if $x = \pm 1$, $\dfrac{1}{1 - x^2}$ is continuous on the open intervals $(-\infty, -1)$, $(-1, 1)$, and $(1, \infty)$.

b. Since $\dfrac{x + 3}{|x + 3|}$ equals 1 for all $x > -3$ and equals -1 for all $x < -3$, $\dfrac{x + 3}{|x + 3|}$ is continuous on the open intervals $(-\infty, -3)$ and $(3, \infty)$.

c. Since $\tan x = \dfrac{\sin x}{\cos x}$ is a quotient of the elementary functions $\sin x$ and $\cos x$, it is continuous at all points where $\cos x \neq 0$. Therefore, $\tan x$ is continuous on all intervals of the form $(\pi/2 + k\pi, 3\pi/2 + k\pi)$ where k is an integer. ◼

The graphs of functions that are continuous on an interval cannot have any breaks or gaps as shown by the following theorem.

Theorem 2.3   Intermediate value theorem

Let f be continuous on the closed interval $[a, b]$. If L lies strictly between $f(a)$ and $f(b)$, then there exists at least one number c in the open interval (a, b) such that $f(c) = L$.

This theorem says that if f is a continuous function on some *closed* interval $[a, b]$, then $f(x)$ must take on all values between $f(a)$ and $f(b)$. The intermediate value theorem is extremely useful in ensuring that we can solve certain nonlinear equations. Consider the following example.

Example 9   Proving the existence of roots

Use the intermediate value theorem to prove that there exists a solution to

$$x^5 - x^2 + 1 = 0$$

Use technology to estimate one of the solutions.

Solution   Let $f(x) = x^5 - x^2 + 1$. Since f is a polynomial, it is continuous at all points on the real number line. To use the intermediate value theorem, we need to find an interval $[a, b]$ such that $f(a)$ and $f(b)$ have opposite signs (since 0 is a value between a positive number and a negative one). A little experimentation reveals that $f(-1) = -1 < 0$ and $f(1) = 1 > 0$. Hence, there must be a c in $(-1, 1)$ such that $f(x) = 0$. Using technology, we see that there is a solution around $x = -0.8$. ◼

While the intermediate value theorem allows us to prove when a function has a root, sometimes we need to solve for these roots numerically. One numerical approach is the bisection method.

The Bisection Method

Let $f(x)$ be a continuous function on $[a, b]$ where $f(a)$ and $f(b)$ have opposite signs. By the intermediate value theorem, there is a root in $[a, b]$. To find the root,

Step 1. Calculate $f\left(\dfrac{a+b}{2}\right)$. If $f\left(\dfrac{a+b}{2}\right) = 0$, then you have found a root and you are done.

Step 2. If $f\left(\dfrac{a+b}{2}\right)$ has the same sign as $f(a)$, then by the intermediate value theorem, the root lies between $f\left(\dfrac{a+b}{2}\right)$ and $f(b)$. In this case, rename the interval $\left[\dfrac{a+b}{2}, b\right]$ as $[a, b]$ and repeat the process.

Step 3. If $f\left(\dfrac{a+b}{2}\right)$ has the same sign as $f(b)$, then by the intermediate value theorem, the root lies between $f(a)$ and $f\left(\dfrac{a+b}{2}\right)$. In this case, rename the interval $\left[a, \dfrac{a+b}{2}\right]$ as $[a, b]$ and repeat the process.

Step 4. Keep repeating until the width $b - a$ of the interval is smaller than the desired accuracy for the root.

Example 10   Bisection method for solving equations

Given a plot of the polynomial

$$x^5 - 10x^3 + 21x + 4 = 0$$

and a calculator use the bisection method to find the largest root of this equation correct to two decimal places (two decimal places).

Solution   We see from a plot of this polynomial in Figure 2.33 that it has five roots, one each, respectively, on the integer intervals $[-3, -2]$, $[-2, -1]$, $[-1, 0]$, $[1, 2]$, and $[2, 3]$. Table 2.4 presents the calculations for the bisection method for the interval $[2, 3]$, which contains the largest root. After ten iterations of the method, Table 2.4 implies that the root lies in the interval $[2.5615, 2.5621]$. Thus, to two decimal places, the root is $x = 2.56$. If we wanted to find the root to more than two decimal places, we could keep going, as illustrated in Table 2.4, until the desired accuracy is obtained.

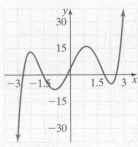

Figure 2.33 Graph of $f(x) = x^5 - 10x^3 + 21x + 4$.

Table 2.4 Bisection method for finding roots of a nonlinear function

Iteration	a	$\dfrac{a+b}{2}$	b	$f(a)$	$f\left(\dfrac{a+b}{2}\right)$	$f(b)$
0	2	2.5	3	−2	−2.09375	40
1	2.5	2.75	3	−2.09375	11.05761	40
2	2.5	2.625	2.75	−2.09375	2.882965	11.05761
3	2.5	2.5625	2.625	−2.09375	0.037423	2.882965
4	2.5	2.53125	2.5625	−2.09375	−1.112400	0.037423
5	2.53125	2.546875	2.5625	−1.112400	−0.559168	0.037423
6	2.546875	2.5546875	2.5625	−0.559168	−0.266371	0.037423
7	2.5546875	2.55859375	2.5625	−0.266371	−0.115859	0.037423
8	2.55859375	2.560546875	2.5625	−0.115859	−0.039565	0.037423
9	2.560546875	2.5615234375	2.5625	−0.039565	−0.001158	0.037423
10	2.5615234375	2.56201171875	2.5625	−0.001158	0.018111	0.037423

Equations that take the form of finding the roots of polynomials, such as the problem in the previous example of finding the largest root of a fifth-order polynomial, are known as *algebraic equations*. Equations that involve exponential or trigonometric functions, such as $x \sin x - 1/2 = 0$ or $e^{\sqrt{x}} - \pi = 0$, are known as *transcendental equations* and generally have no analytical solutions but need to be solved numerically.

Example 11   Limiting global warming

According to an article in the *New Scientist*,* recent research suggests that stabilizing carbon dioxide concentrations in the atmosphere at 450 parts per million (ppm) could limit global warming to 2°C. In Section 1.2, we modeled carbon dioxide concentrations in the atmosphere with the function (which we now present to higher precision to make more transparent the numerical details of the convergence process)

$$f(x) = 0.122463x + 329.253 + 3\cos\frac{\pi x}{6} \text{ ppm}$$

where x is months after April 1974. Use the bisection method to find the first time that the model predicts carbon dioxide levels of 450 ppm. Get a prediction that is accurate to 2 decimal places.

Solution   Solving $f(x) = 450$ is equivalent to solving $g(x) = 0$ where $g(x) = f(x) - 450$. Using technology to plot $g(x)$, we see from the left panel in Figure 2.34 that $g(x)$ first equals zero around $t = 970$ months.

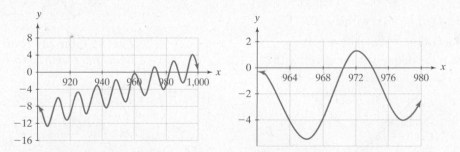

Figure 2.34 Plots of the function $g(x) = f(x) - 450$ on relatively large and small intervals containing $x = 970$ ppm, where $f(x)$ is defined in Example 11.

---

*Catherine Brahic, "Carbon Emissions Rising Faster than Ever," New Scientist.com news service 17 (November 10, 2006): 29.

There appear to be several zeros in the interval [900, 1000]. Zooming into the interval [960, 980], we get the right panel in Figure 2.34. Since the first zero appears to be between 965 and 972, we can set $a = 965$ and $b = 972$ and apply the bisection method, which yields the following table of values (where all the values have all been rounded to 4 decimal places throughout the calculations):

Iteration	a	$\dfrac{a+b}{2}$	b	$f(a)$	$f\left(\dfrac{a+b}{2}\right)$	$f(b)$
0	965	968.5	972	−5.1683	−2.9180	1.2870
1	968.5	970.25	972	−2.9180	−0.1010	1.2870
2	970.25	971.125	972	−0.1010	0.8705	1.2870
3	970.25	970.6875	971.125	−0.1010	0.4453	0.8705
4	970.25	970.4688	970.6875	−0.1010	0.1859	0.4453
5	970.25	970.3594	970.4688	−0.1010	0.0457	0.1859
6	970.25	970.3047	970.3594	−0.1010	−0.0269	0.0457
7	970.3047	970.3320	970.3594	−0.0269	0.0095	0.0457
8	970.3047	970.3184	970.3320	−0.0269	−0.0086	0.0095
9	970.3184	970.3252	970.3320	−0.0086	0.0005	0.0095
10	970.3184	970.3218	970.3252	−0.0086	−0.0041	0.0005
11	970.3218	970.3235	970.3252	−0.0041	−0.0018	0.0005

Hence, the model predicts that carbon dioxide concentrations will reach 450 ppm in 970.32 months, which is 80 years and 10 months, after April 1974. In other words, in February 2055. ■

PROBLEM SET 2.3

Level 1 DRILL PROBLEMS

Determine the limits $\lim\limits_{x \to a^-} f(x)$, $\lim\limits_{x \to a^+} f(x)$ *and* $\lim\limits_{x \to a} f(x)$ *in Problems 1 to 6. If they do not exist, discuss why.*

1. $f(x) = \begin{cases} x^2 - 2 & \text{if } x > 1 \\ 2x - 3 & \text{if } x \le 1 \end{cases}$ with $a = 1$

2. $f(x) = \begin{cases} 3x + 2 & \text{if } x \le 1 \\ 5 & \text{if } 1 < x \le 3 \\ 3x^2 - 1 & \text{if } x > 3 \end{cases}$ with $a = 1$

3. $f(x) = x/|x|$ with $a = 0$.

4. $f(x) = x^2/|x|$ with $a = 0$.

5. $f(x)$ defined by the graph with $a = 1$.

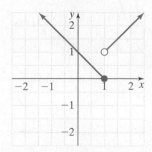

6. $f(x)$ defined by the graph with $a = 0$.

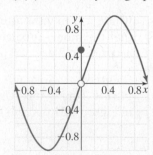

Find the limits in Problems 7 to 14. Justify each step with limit laws and the appropriate results from this section.

7. $\lim\limits_{x \to 3} \dfrac{x^2 + 3x - 10}{3x^2 + 5x - 7}$

8. $\lim\limits_{x \to 0} \dfrac{(x + 1)^2 - 1}{x}$

9. $\lim\limits_{t \to -3} \dfrac{t^2 + 5t + 6}{t + 3}$

10. $\lim\limits_{t \to 0} \dfrac{\sqrt{4 - t^2} - 2}{t^2}$

11. $\lim\limits_{s \to 3} \dfrac{s - 2}{s + 2} + \sin s$

12. $\lim\limits_{s \to 1} s + \sin(\ln s)$

13. $\lim\limits_{x \to \pi} \dfrac{1 + \tan x}{2 - \cos x}$

14. $\lim\limits_{x \to 1/3} \dfrac{x \sin(\pi x)}{1 + \cos(\pi x)}$

The graph of a function f is shown in Problems 15 to 18. Determine at what points f is not continuous and whether f can be redefined at these points to make it continuous. Explain briefly.

15.

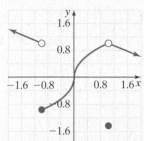

16.

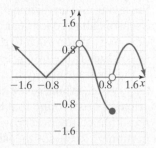

17.

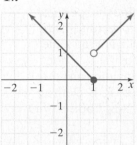

18.

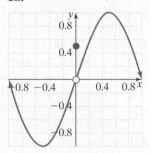

Each function in Problems 19 to 22 is defined for all x > 0, except at x = 2. In each case, find the value that should be assigned to f(2), if any, to guarantee that f will be continuous at 2. Explain briefly.

19. $f(x) = \dfrac{x^2 - x - 2}{x - 2}$

20. $f(x) = \sqrt{\dfrac{x^2 - 4}{x - 2}}$

21. $f(x) = \begin{cases} 2x + 5 & \text{if } x > 2 \\ 15 - x^2 & \text{if } x < 2 \end{cases}$

22. $f(x) = \dfrac{\frac{1}{x} - 1}{x - 2}$

In Problems 23 to 28, use the intermediate value theorem to prove the following equations have at least one solution.

23. $-x^7 + x^2 + 4 = 0$

24. $\dfrac{1}{x + 1} = x^2 - x - 1$

25. $\sqrt[3]{x} = x^2 + 2x - 1$

26. $\sqrt[3]{x - 8} + 9x^{2/3} = 29$

27. $x2^x = \pi$

28. $1 + \sin x + x^3 = 0$

29. Use the bisection method explicated in Example 10 to find the following roots of the polynomial
$$x^5 - 10x^3 + 21x + 4 = 0$$
to an accuracy of 3 decimal places.

  a. smallest root

  b. second smallest root

  c. largest negative root

  d. smallest positive root

In Problems 30 to 33, use the squeeze theorem, limit laws, and continuity of elementary functions to find the limit.

30. $\lim\limits_{x \to 1}(x - 1) \sin \dfrac{1}{x - 1}$

31. $\lim\limits_{x \to 0} \dfrac{5 + x^2 - x \cos(1/x)}{2 + x}$

32. $\lim\limits_{x \to 1} \ln x \sin(\ln x) + 3$

33. $\lim\limits_{x \to \pi/2} \dfrac{x}{8 + \cos x \sin(\cos x)}$

Level 2 APPLIED AND THEORY PROBLEMS

34. Prove that if p and q are polynomial functions with $q(a) \neq 0$, then
$$\lim_{x \to a} \frac{p(x)}{q(x)} = \frac{p(a)}{q(a)}$$

35. Use limit laws to prove if f and g are continuous at $x = a$, then $f + g$ is continuous at $x = a$.

36. Use limit laws to prove if f and g are continuous at $x = a$, then $f - g$ is continuous at $x = a$.

37. Use limit laws to prove if f and g are continuous at $x = a$ and $g(a) \neq 0$, then $\dfrac{f}{g}$ is continuous at $x = a$.

38. Use limit laws to prove if f is continuous at $x = a$ and g is continuous at $x = f(a)$, then $g \circ f$ is continuous at $x = a$.

39. Why does the cubic equation $x^3 + ax^2 + bx + c = 0$ have at least one root for any values of $a, b,$ and c?

40. For any constants a, b, why does the equation $x^n = a + b\cos(x)$ always have at least one real root when n is an odd integer but not necessarily when n is an even integer?

41. Consider an organism that can move freely between two spatial locations. In one location, call it patch 1, the number of progeny produced per individual is
$$f(N) = \frac{100}{1 + N}$$
where N is the population size. In the other location, call it patch 2, the number of progeny produced per

individual is always 5. Assume that all individuals in the population move to the patch that allows them to produce the greatest number of progeny. Let $g(N)$ represent the number of progeny produced per individual for such a population.

a. Write an explicit expression for $g(N)$ such that $g(N) = f(N)$ whenever $N < c$ for some constant $c > 0$ and $g(N) = 5$ whenever $N \geq c$.

b. Determine the value of c that ensures $g(N)$ is continuous for $N \geq 0$.

42. As discussed in Example 11, recent research suggests that stabilizing carbon dioxide concentrations in the atmosphere at 450 parts per million (ppm) could limit global warming to $2°C$. Use the bisection method and the carbon dioxide concentration model

$$f(x) = 0.122463x + 329.253 + 3\cos\frac{\pi x}{6} \text{ ppm}$$

where x is months after December 1973, to find the *second* time that this model predicts carbon dioxide levels of 450 ppm. Get a prediction that is accurate to 2 decimal places.

43. Use the bisection method to find the *last* time that the model in Problem 42 predicts carbon dioxide levels of 450 ppm. Get a prediction that is accurate to 2 decimal places.

44. Scientists believe that it will be extremely difficult to rein in carbon emissions enough to stabilize the atmospheric CO_2 concentration at 450 parts per million, as discussed in Example 11, and think that even 550 ppm will be a challenge. Use the bisection method to find the *first* time that the model in Problem 42 predicts carbon dioxide levels of 550 ppm. Get a prediction that is accurate to 2 decimal places.

45. Fisheries scientists often use data to establish a *stock-recruitment* relationship of the general form $y = f(x)$, where x is the number of adult fish participating in the *spawning* process (i.e., the laying and fertilizing of eggs) that occurs on a seasonal basis each year and y is the number of young fish *recruited* to the fishery as a result of hatching from the eggs and surviving through to the life stage at which they become part of the fishery (i.e., available for harvesting). Two fisheries scientists found that the following stock-recruitment function provides a good fit to data pertaining to the southeast Alaska pink salmon fishery:

$$y = 0.12x^{1.5}e^{-0.00014x}$$

Use the bisection method to find the spawning stock level x that is expected to recruit 10,000 individuals to the fishery. (Hint: For the value of y in question, find the root of the equation $y - f(x) = 0$.) For more information on the research study, see T. J. Quinn and R. B. Deriso, *Quantitative Fish Dynamics* (New York: Oxford University Press, 1977).

46. Use the bisection method to find the spawning stock level x that is expected to recruit 20,000 individuals to the fishery modeled by the stock-recruitment function given in Problem 45.

47. Use the bisection method to find the spawning stock level x that is expected to recruit 5,000 individuals to the fishery modeled by the stock-recruitment function given in Problem 45.

2.4   Asymptotes and Infinity

In Chapter 1, we introduced the notion of *infinity* and represented it with the symbols ∞ and $-\infty$. This symbol was used by the Romans to represent the number 1,000 (a *big* number to them). It was not until 1650, however, that it was first used by John Wallis (1616–1703) to represent an uncountably large number. From childhood many of us come to think of infinity as endlessness, which in a sense it is since infinity is a not a number. To mathematicians, however, infinity is a much more complicated idea than simply endlessness. As the famous mathematician David Hilbert (1862–1943) said, "The infinite! No other question has ever moved so profoundly the spirit of man" (in J. R. Newman, ed., *The World of Mathematics*, New York: Simon & Schuster, 1956, 1593).

In this section, we tackle limits involving the infinite in two ways. First, we determine under what conditions functions approach a limiting value as their argument becomes arbitrarily positive or arbitrarily negative. Second, we study functions whose value becomes arbitrarily large as their arguments approach a finite value where the function is not well defined.

Horizontal asymptotes

To understand the behavior of functions as their argument becomes very positive or negative (i.e., further from the origin in either direction), we introduce horizontal asymptotes.

Horizontal Asymptotes (Informal Definition)

Let f be a function. We write

$$\lim_{x \to \infty} f(x) = L$$

if $f(x)$ can be made arbitrarily close to L for all x sufficiently large. We write

$$\lim_{x \to -\infty} f(x) = L$$

if $f(x)$ can be made arbitrarily close to L for all x sufficiently negative. Whenever one of these limits occurs, we say that $f(x)$ has a **horizontal asymptote** at $y = L$.

Example 1   Finding horizontal asymptotes

Find the following limits involving a given function f. In each, indicate how positive or negative x needs to be to ensure that $f(x)$ is within one ten-millionth of the limiting value L.

a. $\displaystyle\lim_{x \to \infty} \left(2 + \frac{1}{x}\right)$ 　　　**b.** $\displaystyle\lim_{x \to -\infty} e^x$ 　　　**c.** $\displaystyle\lim_{x \to \infty} \frac{10x}{1 + 5x}$

Solution   To help visualize the solutions, plots of these three functions are given in Figure 2.35 over domains of x that illustrate the asymptotic behavior of these functions.

a. For x sufficiently large, $\dfrac{1}{x}$ is arbitrarily close to 0. Hence

$$\lim_{x \to \infty} 2 + \frac{1}{x} = 2$$

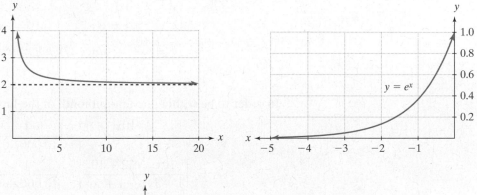

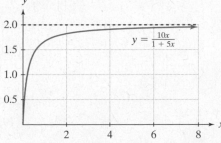

Figure 2.35 Plot of functions. Left panel: $f(x) = \left(2 + \dfrac{1}{x}\right)$; right panel: $f(x) = e^x$; bottom panel: $f(x) = \dfrac{10x}{1 + 5x}$ for $0 \le x \le 10$.

To say that $f(x)$ is within one ten-millionth of the limiting value, L, is to say that

$$|f(x) - L| < \frac{1}{10,000,000}$$

For positive x, we need

$$\left|\left(2 + \frac{1}{x}\right) - 2\right| < \frac{1}{10,000,000}$$

$$\frac{1}{x} < \frac{1}{10,000,000}$$

$$x > 10,000,000$$

Thus, if x is greater than 10,000,000, then $f(x)$ will be within one ten-millionth of $x = 2$.

b. For x sufficiently negative, e^x is arbitrarily small. Hence, we would expect that

$$\lim_{x \to -\infty} e^x = 0$$

To see how negative x needs to be to ensure that e^x is less than one ten-millionth, we have

$$|e^x - 0| < \frac{1}{10,000,000}$$

$$e^x < \frac{1}{10,000,000}$$

$$x < \ln \frac{1}{10,000,000}$$

$$< -16.2$$

Thus, if x is less than -16.2, then $f(x)$ will be within one ten-millionth of zero.

c. To find the limiting value of this function, we can divide the numerator and denominator of $f(x)$ by x:

$$\lim_{x \to \infty} \frac{10x}{1 + 5x} = \lim_{x \to \infty} \frac{10}{\frac{1}{x} + 5}$$

$$= \frac{10}{5} \qquad \textit{since } 1/x \textit{ approaches } 0 \textit{ as } x \textit{ approaches } \infty$$

$$= 2$$

In order to be within one ten-millionth of the limiting value of 2, we need

$$\left|\frac{10x}{1 + 5x} - 2\right| < \frac{1}{10,000,000}$$

$$\left|\frac{10x}{1 + 5x} - \frac{2 + 10x}{1 + 5x}\right| < \frac{1}{10,000,000}$$

$$\left|\frac{-2}{1 + 5x}\right| < \frac{1}{10,000,000}$$

$$\frac{2}{1 + 5x} < \frac{1}{10,000,000} \qquad \textit{for } x > 0$$

$$20,000,000 < 1 + 5x$$

$$19,999,999 < 5x$$

$$x > \frac{19,999,999}{5}$$

Thus, if x is greater than about 4,000,000, then $f(x)$ will be within one ten-millionth of two.

Understanding the asymptotic behavior of a function can help us graph and interpret it. The next example involves dose–response curves that arise in many kinds of biological experiments. For example, the x axis of a dose–response curve may represent concentration of a drug or hormone delivered at various "doses." The y axis represents the response, which could be many things, depending on the experiment. For example, the response might be the activity of an enzyme, accumulation of an intracellular messenger, voltage drop across a cell membrane, secretion of a hormone, increase in heart rate, or contraction of a muscle. In the next example, dose is the amount of a histamine (measured in millimoles) administered to a patient, and response is the percentage of patients exhibiting above-normal temperatures.

Example 2   Dose–response curves

The percentage of patients exhibiting an above-normal temperature response to a specified dose of a histamine is given by the function

$$R(x) = \frac{100\,e^x}{e^x + e^{-5}}$$

where x is the natural logarithm of the dose in millimoles (mmol or mM).

a. Find the horizontal asymptotes of $R(x)$.

b. Show that $R(x)$ is increasing and sketch $y = R(x)$.

c. Calculate how large x needs to be to ensure that it is within 1% of its asymptotic value.

Solution

a. To find the horizontal asymptotes, we find

$$\lim_{x \to \infty} \frac{100\,e^x}{e^x + e^{-5}} = \lim_{x \to \infty} \frac{100\,e^x}{e^x + e^{-5}} \cdot \frac{e^{-x}}{e^{-x}} \qquad \textit{multiply by one in the form } \frac{e^{-x}}{e^{-x}}$$

$$= \lim_{x \to \infty} \frac{100}{1 + e^{-x-5}}$$

$$= \frac{100}{1 + 0} \qquad \textit{since the value } e^{-x} \textit{ approaches zero as x becomes very positive}$$

$$= 100$$

and

$$\lim_{x \to -\infty} \frac{100\,e^x}{e^x + e^{-5}} = \frac{0}{e^{-5} + 0} \qquad \textit{since the value } e^x \textit{ approaches zero as x becomes very negative}$$

$$= 0$$

Thus, the horizontal asymptotes are $y = 100$ and $y = 0$.

b. To show $R(x)$ is increasing, notice that

$$R(x) = \frac{100\,e^x}{e^x + e^{-5}} \frac{e^{-x}}{e^{-x}}$$

$$= \frac{100}{1 + e^{-5-x}}$$

Since $1 + e^{-5-x}$ is a decreasing function, $R(x)$ is an increasing function of x. The y-intercept of $R(x)$ is $R(0) = \dfrac{100\,e^0}{e^{-5} + e^0} \approx 100$. Thus, the graph of the function $R(x)$ has the form illustrated in Figure 2.36. Notice that this curve fits the data fairly well.

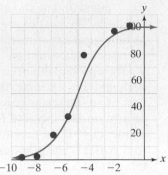

Figure 2.36 Percentage of patients responding to a dose of histamine. The x axis corresponds to the natural logarithm of the doses in millimoles.

Data Source: K. A. Skau, "Teaching Pharmacodynamics: An Introductory Module on Learning Dose–Response Relationships," *American Journal of Pharmaceutical Education* 68 (2004), article 73.

c. To find when $R(x)$ is within 1% of 100, that is when it reaches the value 99, we first note that $R(x) < 100$ for all x. Hence, we only need to solve

$$99 < \frac{100\,e^x}{e^x + e^{-5}}$$

$$99(e^x + e^{-5}) < 100\,e^x \quad \textit{after multiplying both sides by } e^x + e^{-5}$$

$$99\,e^{-5} < e^x \quad \textit{after subtracting } 99\,e^x \textit{ from both sides}$$

$$\ln 99 - 5 < x$$

$$-0.40 < x \quad \textit{evaluating on a calculator}$$

Limits at infinity can be very nonintuitive, as the following example shows.

Example 3   A difference of infinities

Find

$$\lim_{x \to \infty} \sqrt{x^2 + x + 1} - x$$

Solution   To deal with this limit, it is useful to multiply and divide by the conjugate $\sqrt{x^2 + 1} + x$ of $\sqrt{x^2 + 1} - x$.

$$\left(\sqrt{x^2 + 1} - x\right)\frac{\sqrt{x^2 + 1} + x}{\sqrt{x^2 + 1} + x} = \frac{x^2 + 1 - x^2}{\sqrt{x^2 + 1} + x}$$

$$= \frac{1}{\sqrt{x^2 + 1} + x}$$

Since $\sqrt{x^2 + 1} + x$ gets arbitrarily large as x gets arbitrarily large,

$$\lim_{x \to \infty} \sqrt{x^2 + 1} - x = \lim_{x \to \infty} \frac{1}{\sqrt{x^2 + 1} + x} = 0$$

This asymptote is somewhat surprising, as one might first guess that $\sqrt{x^2 + x}$ is much bigger than x for large x and therefore the limit is arbitrarily large. However, our calculations show that this initial guess is wrong.

Vertical asymptotes

Many functions, such as rational functions, logarithms, and certain power functions, are not defined at isolated values. As the argument of the function gets close to these isolated values, the function may become arbitrarily large or arbitrarily negative and exhibit a vertical asymptote.

Vertical Asymptotes (Informal Definition)

Let f be a function. We write

$$\lim_{x \to a^-} f(x) = \infty$$

if $f(x)$ can be made arbitrarily large for all x sufficiently close to a and to the left of a. We write

$$\lim_{x \to a^+} f(x) = \infty$$

if $f(x)$ can be made arbitrarily large for all x sufficiently close to a and to the right of a. We write

$$\lim_{x \to a^-} f(x) = -\infty$$

if $f(x)$ can be made arbitrarily negative for all x sufficiently close to a and to the left of a. We write

$$\lim_{x \to a^+} f(x) = -\infty$$

if $f(x)$ can be made arbitrarily negative for all x sufficiently close to a and to the right of a.

Whenever any one of these limits occurs, we say that $f(x)$ has a **vertical asymptote** at $x = a$.

Example 4 Finding vertical asymptotes

Find $\lim_{x \to a^-} f(x)$ and $\lim_{x \to a^+} f(x)$ for the given functions, then sketch the graph of $y = f(x)$ near $x = a$.

a. $f(x) = \dfrac{1}{x}$ with $a = 0$

b. $f(x) = \dfrac{1}{(x-2)^2}$ with $a = 2$

c. $f(x) = \tan x$ with $a = \dfrac{\pi}{2}$

Solution

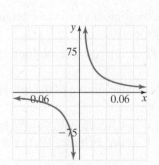

Figure 2.37
Graph of the function $\dfrac{1}{x}$ on $[-1, 1]$.

a. $\lim_{x \to 0^-} f(x) = -\infty$ since for $x < 0$ sufficiently close to 0, $\dfrac{1}{x}$ is arbitrarily negative.

$\lim_{x \to 0^+} f(x) = \infty$ since for $x > 0$ sufficiently near 0, $\dfrac{1}{x}$ is arbitrarily large.

Since $y = \dfrac{1}{x}$ is decreasing for all $x \neq 0$, the graph of $y = \dfrac{1}{x}$ near $x = 0$ is as shown in Figure 2.37.

b. $\lim_{x \to 2^-} \dfrac{1}{(x-2)^2} = \infty$ since for $x < 2$ and sufficiently close to 2, $\dfrac{1}{(x-2)^2}$ is arbitrarily large.

$\lim_{x \to 2^+} \dfrac{1}{(x-2)^2} = \infty$ since for $x > 2$ and sufficiently close to 2, $\dfrac{1}{(x-2)^2}$ is arbitrarily large.* The graph of $y = \dfrac{1}{(x-2)^2}$ close to the vertical asymptote $y = 2$ is illustrated in Figure 2.38.

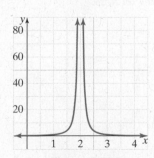

Figure 2.38
Graph of the function $\dfrac{1}{(x-2)^2}$ on $[0, 4]$.

*If you want to support this statement with a technical argument, suppose you want $\dfrac{1}{(x-2)^2} \geq 1,000,000$. Taking the square root of both sides and cross-multiplying yields $\dfrac{1}{1,000} \geq |x - 2|$. Hence, $f(x) \geq 1,000,000$ whenever $0 < |x - 2| < \dfrac{1}{1,000}$.

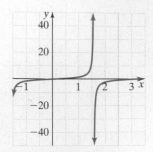

Figure 2.39
Graph of the function
$\tan x$ on $[-1, 4]$.

c. $\lim\limits_{x \to \frac{\pi}{2}^-} \tan x = \lim\limits_{x \to \frac{\pi}{2}^-} \dfrac{\sin x}{\cos x} = \infty$ since for $x < \dfrac{\pi}{2}$ and arbitrarily close to $\dfrac{\pi}{2}$, $\sin x$ is arbitrarily close to 1 and $\cos x$ is positive and arbitrarily close to 0, so the quotient of sine and cosine is arbitrarily large.

$\lim\limits_{x \to \frac{\pi}{2}^+} \dfrac{\sin x}{\cos x} = -\infty$ since for $x > \dfrac{\pi}{2}$ and arbitrarily close to $\dfrac{\pi}{2}$, $\sin x$ is arbitrarily close to 1 and $\cos x$ is negative and arbitrarily close to 0, so the quotient of sine and cosine is arbitrarily negative. The graph of $y = \tan x$ close to the vertical asymptote $x = \pi/2$ is illustrated in Figure 2.39. ∎

Combining the information about horizontal and vertical asymptotes can provide a relatively complete sense of the graph of a function, as illustrated in the next example. This example, described by Francois Messier ("Ungulate Population Models with Predation: A Case Study with the North American Moose." *Ecology* 75 (1994): 478–488), examines wolf-moose interactions over a broad spectrum of moose densities throughout North America. One of his primary objectives was to determine how the predation rate of moose by wolves depends on moose density. Messier found that a Michaelis-Menton function provides a good fit to moose-killing rates by wolves as a function of moose density. Recall that we encountered the Michaelis-Menton function in Example 6 of Section 1.5 for modeling bacterial nutrient-uptake rates.

Example 5   Wolves eating moose

The rate $f(x)$ at which wolves kill moose, as a function of moose density x (numbers per square kilometer), can be described by the function

$$f(x) = \frac{3.36x}{0.46 + x} \text{ moose killed per wolf per hundred days}$$

Here, we examine the shape of the function $f(x)$ over the biologically relevant range of values $x \geq 0$, as well as the biologically irrelevant range $x < 0$.

a. Find all horizontal and vertical asymptotes for $y = f(x)$. Discuss the biological meaning of one of the horizontal asymptotes.

b. Sketch its graph $y = f(x)$ for all x and discuss the biological meaning of the graph for nonnegative x.

c. Relate the graph to the following statement of Sir Winston Churchill (1874–1965) (quoted in H. Eves, *Return to Mathematical Circles*, Boston: Prindle, Weber & Schmidt, 1988):

"I had a feeling once about Mathematics—that I saw it all. Depth beyond depth was revealed to me—the Byss and Abyss. I saw—as one might see the transit of Venus or even the Lord Mayor's Show—a quantity passing through infinity and changing its sign from plus to minus. I saw exactly why it happened and why the tergiversation was inevitable but it was after dinner and I let it go."

Solution

a. First, let us find the horizontal asymptotes.

$$\lim_{x \to \infty} \frac{3.36x}{0.46 + x} = \lim_{x \to \infty} \frac{3.36x}{0.46 + x} \cdot \frac{\frac{1}{x}}{\frac{1}{x}}$$

$$= \lim_{x \to \infty} \frac{3.36}{\frac{0.46}{x} + 1}$$

$$= \frac{3.36}{0 + 1}$$

$$= 3.36$$

Thus, $f(x)$ has a horizontal asymptote $y \approx 3.36$, which $f(x)$ approaches as x approaches ∞; this means that when the moose density is very large, the wolf killing rate stabilizes around 3.36 moose per wolf per 100 days. Similarly (without the corresponding biological meaning) for $x < 0$, we obtain

$$\lim_{x \to -\infty} \frac{3.36x}{0.46 + x} = 3.36$$

Next, let us find the vertical asymptotes. Since $f(x)$ is not defined at $x = -0.46$, there is a possible vertical asymptote at $x = -0.46$. We have $\lim\limits_{x \to -0.46^-} f(x) = \infty$ because when $x < -0.46$, but close to -0.46, x is negative and $0.46 + x$ is arbitrarily small and also negative. Alternatively, we have $\lim\limits_{x \to -0.46^+} f(x) = -\infty$ because when $x > -0.46$, but close to -0.46, x is negative and $0.46 + x$ is arbitrarily small and positive.

b. We begin by drawing the asymptotes: $y = 3.36$ and $x = -0.46$. The y-intercept is found at $x = 0$ as $f(0) = 0$. Our observation that $\dfrac{3.36x}{0.46 + x} = \dfrac{3.36}{0.46/x + 1}$ for $x \neq 0$ implies that this function is increasing for all $x \neq 0$. Using this information, we draw the graph shown in Figure 2.40.

Looking at the nonnegative portion of this graph, we see that the killing rate is zero at $x = 0$ (i.e., moose cannot be killed if they are not around) and that this rate increases with increasing moose density. The rate, however, saturates at approximately $x = 3.36$ moose per wolf per 100 days. Biologists refer to this saturation rate as the killing rate when moose are "not limiting."

c. As viewed from left to right, the function passes from positive infinity to negative infinity as it passes through the value $x = -0.46$, which is Churchill's "quantity passing through infinity and changing its sign from plus to minus." Perhaps Churchill saw a wolf after his dinner and that's why he let it go.

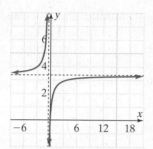

Figure 2.40
Graph of the function $\dfrac{3.36x}{0.46}$.

Infinite limits at infinity

As x gets larger and larger without bound, the value of f might also get larger and larger without bound. In such a case, it is natural to say that $f(x)$ approaches infinity as x approaches infinity.

Infinity at Infinity (Informal Definition)

Let f be a function. We write

$$\lim_{x \to \infty} f(x) = \infty$$

if $f(x)$ can be made arbitrarily large for all x sufficiently large. We write

$$\lim_{x \to -\infty} f(x) = \infty$$

if $f(x)$ can be made arbitrarily large for all x sufficiently negative. We write

$$\lim_{x \to \infty} f(x) = -\infty$$

if $f(x)$ can be made arbitrarily negative for all x sufficiently large. We write

$$\lim_{x \to -\infty} f(x) = -\infty$$

if $f(x)$ can be made arbitrarily negative for all x sufficiently negative.

Example 6   Limits to infinity

Find the following limits.

a. $\lim\limits_{x \to \infty} x^2$ **b.** $\lim\limits_{x \to \infty} (x - x^2)$ **c.** $\lim\limits_{x \to \infty} \dfrac{x^2}{1,000,000 + 10x}$

Solution

a. For large x the number x^2 can be made arbitrarily large for all sufficiently large x, so we say $\lim\limits_{x \to \infty} x^2 = \infty$.

b. It is tempting to use a limit law here and write

$$\lim_{x \to \infty} (x - x^2) = \lim_{x \to \infty} x - \lim_{x \to \infty} x^2$$
$$= \infty - \infty$$
$$= 0$$

However, this is incorrect! Limit laws do not apply to infinite limits. Indeed, $\infty - \infty$ is not a meaningful statement as ∞ is not a real number. Luckily, we can deal with this by noticing that for large x, $x - x^2 = x(1 - x)$ is the product of two numbers such that for large x, one of these numbers is large and positive and the other can be made arbitrarily negative. Thus, for sufficiently large x, $x(1 - x)$ is arbitrarily negative. Hence, $\lim\limits_{x \to \infty} (x - x^2) = -\infty$.

c. Again, it is tempting to use a limit law to conclude the limit is $\dfrac{\infty}{\infty}$. This is meaningless. However, if we divide the numerator and denominator by x, we find (for $x \neq 0$)

$$\frac{x^2}{1,000,000 + 10x} = \frac{x}{\dfrac{1,000,000}{x} + 10}$$

Since $\dfrac{1,000,000}{x} + 10$ approaches $0 + 10 = 10$ as x approaches ∞, we find

$$\frac{x^2}{1,000,000 + 10x} \approx \frac{x}{10} \text{ for } x \text{ sufficiently large. Therefore}$$

$$\lim_{x \to \infty} \frac{x^2}{1,000,000 + 10x} = \infty$$

Part **b** of the previous example and Example 3 illustrate the subtlety of taking the limits of the form $\lim\limits_{x \to \infty} f(x) - g(x)$ when $\lim\limits_{x \to \infty} f(x) = \infty$ and $\lim\limits_{x \to \infty} g(x) = \infty$. In the previous example with $f(x) = x$ and $g(x) = x^2$, the limit of the difference was $-\infty$. In Example 3 with $f(x) = \sqrt{x^2 + 1}$ and $g(x) = x$, the limit of the difference was 0. The following example illustrates that any limiting value is possible.

Example 7   Another limit of an infinite difference

Find $\lim\limits_{x \to \infty} \sqrt{x^2 + 2ax + 1} - x$ where is a constant.

Solution   Multiplying and dividing by the conjugate $\sqrt{x^2 + 2ax + 1} + x$ yields

$$(\sqrt{x^2 + 2ax + 1} - x)\frac{\sqrt{x^2 + 2ax + 1} + x}{\sqrt{x^2 + 2ax + 1} + x} = \frac{x^2 + 2ax + 1 - x^2}{\sqrt{x^2 + 2ax + 1} + x}$$

$$= \frac{2ax + 1}{\sqrt{x^2 + 2ax + 1} + x}\frac{1/x}{1/x}$$

multiplying and dividing by $\dfrac{1}{x}$

$$= \frac{2a + 1/x}{\sqrt{1 + 2a/x + 1/x^2} + 1}$$

Since $\dfrac{1}{x}$ and $\dfrac{1}{x^2}$ go to zero as x gets very positive, we have

$$\lim_{x\to\infty} \sqrt{x^2 + 2ax + 1} - x = \lim_{x\to\infty} \frac{2a + 1/x}{\sqrt{1 + 2a/x + 1/x^2} + 1}$$

$$= \frac{2a + 0}{\sqrt{1 + 0 + 0} + 1} = a$$

We have shown that the limit of a difference of terms can take on any value! ∎

Example 8   Unabated population growth

At the beginning of Section 1.4, we modeled population growth in the United States with the function

$$N(t) = 8.3(1.33)^t \text{ millions}$$

where t represents the number of decades after 1815.

a. Find $\lim_{t\to\infty} N(t)$.

b. Determine how large t has to be to ensure that $N(t)$ is greater than 300,000,000. Discuss how your answer relates to the current U.S. population size.

Solution

a. Since $8.3(1.33)^t$ gets arbitrarily large for large t, we have that $\lim_{t\to\infty} 8.33(1.33)^t = \infty$.

b. We want $N(t) \geq 300,000,000$. Solving for t in this inequality yields

$$8.3(1.33)^t \geq 300,000,000$$

$$1.33^t \geq \frac{300,000,000}{8.3} \approx 36,145,000$$

$$t \ln 1.33 \geq \ln\left(\frac{300,000,000}{8.3}\right)$$

$$t \geq \frac{17.4}{\ln 1.33} \approx 61$$

Therefore, the model predicts that $t = 61$ decades after 1815, in other words in the year 2425, there will be approximately $N(t) = 300$ million people in the United States. Given that the population size in January 2007 was over 300 million, we can see that the model from the 1800s considerably underestimated the future growth of the U.S. population. ∎

PROBLEM SET 2.4

Level 1 DRILL PROBLEMS

In Problems 1 to 24, find the specified limits.

1. $\lim_{x\to-\infty} e^x$

2. $\lim_{x\to 0^+} \ln x$

3. $\lim_{x\to 2^+} \dfrac{1}{x-2}$

4. $\lim_{x\to 2^-} \dfrac{1}{x-2}$

5. $\lim_{x\to 3^-} \left(3 + \dfrac{2x}{x-3}\right)$

6. $\lim_{x\to\infty} \sin x$

7. $\lim_{x\to 3^+} \left(3 + \dfrac{2x}{x-3}\right)$

8. $\lim_{x\to 1^-} \dfrac{x-1}{|x^2-1|}$

9. $\lim_{x\to 3^+} \dfrac{x^2-4x+3}{x^2-6x+9}$

10. $\lim\limits_{x \to \infty} \dfrac{x^3}{1+x^3}$

11. $\lim\limits_{x \to \infty} \cos x^2$

12. $\lim\limits_{x \to -\infty} \dfrac{x^3}{1+x^3}$

13. $\lim\limits_{x \to \infty} \dfrac{(2x+5)(x-2)}{(7x-2)(3x+1)}$

14. $\lim\limits_{x \to \infty} \dfrac{(2x^2-5x+7)}{x^2-9}$

15. $\lim\limits_{x \to \infty} \dfrac{\sin x}{1+x}$

16. $\lim\limits_{Q \to \infty} \dfrac{aQ^2+Q}{1-Q^2}$ where a is a constant.

17. $\lim\limits_{x \to \infty} \dfrac{Ae^x+3}{Be^{2x}+4}$ where $A > 0$ and $B > 0$ are constants.

18. $\lim\limits_{x \to -\infty} \dfrac{1+ax+3x^3}{1+5x-5x^3}$ where a is a constant.

19. $\lim\limits_{x \to \infty} \dfrac{1+5e^{ax}}{7+2e^{ax}}$ where $a > 0$ is a constant.

20. $\lim\limits_{x \to \infty} \dfrac{1+5e^{ax}}{7+2e^{ax}}$ where $a < 0$ is a constant.

21. $\lim\limits_{x \to \infty} \sqrt{x+1} - \sqrt{x}$

22. $\lim\limits_{x \to \infty} x - \sqrt{x}$

23. $\lim\limits_{x \to \infty} \sqrt{x^4+ax^2+x+1} - x^2$

24. $\lim\limits_{x \to -\infty} \sqrt{x^2+1} + x$

For $\lim\limits_{x \to a^+} f(x)$ *in Problems 25 to 30, determine how close* $x > a$ *needs to be to a to ensure that* $f(x) \geq 1{,}000{,}000$.

25. $\lim\limits_{x \to 2^+} \dfrac{1}{x-2}$

26. $\lim\limits_{x \to 0^+} \ln \dfrac{1}{x}$

27. $\lim\limits_{x \to 1^-} \dfrac{1}{1-x}$

28. $\lim\limits_{x \to 3^+} \dfrac{1}{(x-3)^2}$

29. $\lim\limits_{x \to 0^-} \dfrac{-1}{\sin x}$

30. $\lim\limits_{x \to 1^-} \ln \dfrac{1}{x-1}$

For $\lim\limits_{x \to -\infty} f(x) = L$ *in Problems 31 to 34, determine how negative x needs to be to ensure that* $|f(x) - L| \leq 0.05$.

31. $\lim\limits_{x \to -\infty} \dfrac{1}{x^2} = 0$

32. $\lim\limits_{x \to -\infty} (e^x+5) = 5$

33. $\lim\limits_{x \to -\infty} \dfrac{x}{1+x} = 1$

34. $\lim\limits_{x \to -\infty} \dfrac{1}{\ln x^2} = 0$

For the limit $\lim\limits_{x \to \infty} f(x) = \infty$ *in Problems 35 to 38, determine how large x needs to be to ensure that* $f(x) > 1{,}000{,}000$.

35. $\lim\limits_{x \to \infty} x^2$

36. $\lim\limits_{x \to \infty} (e^x+5)$

37. $\lim\limits_{x \to \infty} \dfrac{x^2}{1+x}$

38. $\lim\limits_{x \to \infty} \ln x$

Level 2 APPLIED AND THEORY PROBLEMS

39. In Example 8, we showed that $\lim\limits_{t \to \infty} N(t) = \infty$ where $N(t) = 8.3(1.33)^t$ represents U.S. population size in millions t decades after 1815. To see that $N(t)$ can get arbitrarily large for x sufficiently large, do the following:

   a. Determine how large t needs to be to ensure that $N(t) \geq 500{,}000{,}000$.

   b. Determine how large t needs to be to ensure that $N(t) \geq 1{,}000{,}000{,}000$.

40. In Example 5 of Section 1.4, we modeled the height of beer froth with the function $H(t) = 17(0.99588)^t$ cm where t is measured in seconds.

   a. Determine L such that $\lim\limits_{t \to \infty} H(t) = L$.

   b. Determine how large t needs to be to ensure that $H(t)$ is within 0.1 of L.

   c. Determine how large t needs to be to ensure that $H(t)$ is within 0.01 of L.

41. In Example 6 of Section 1.5, we modeled the uptake rate of glucose by bacterial populations with the function $f(x) = \dfrac{1.2708x}{1+0.0506x}$ mg per hour where x is measured in milligrams per liter.

   a. Find the horizontal and vertical asymptotes of $f(x)$. Interpret the horizontal asymptote(s).

   b. Graph $f(x)$ for all values of x.

42. In Example 5, we examined how the predation rate of wolves depended on moose density. Messier also studied how wolf densities in North America depend on moose densities. He found that the following function provides a good fit to the data:

$$f(x) = \dfrac{58.7(x-0.03)}{0.76+x} \text{ wolves per } 1{,}000 \text{ km}^2$$

where x is number of moose per square kilometer.

   a. Find the horizontal and vertical asymptotes of $f(x)$. Interpret the horizontal asymptotes.

   b. Graph $f(x)$ for all values of $x > 0$.

43. In Problem 42, you were asked to find L such that $\lim\limits_{x \to \infty} f(x) = L$.

   a. Determine how large x needs to be to ensure that $f(x)$ is within 0.1 of L.

b. Determine how large x needs to be to ensure that $f(x)$ is within 0.01 of L.

44. The von Bertalanffy growth curve is used to describe how the size L (usually in terms of length) of an animal changes with time. The curve is given by

$$L(t) = a(1 - e^{-b(t-t_0)})$$

where t measures time after birth and a, b, and t_0 are positive parameters. We will derive this curve in Chapter 6. To better understand the meaning of the parameters t_0 and b, carry out these steps.

a. Evaluate $L(t_0)$. What does this imply about the meaning of t_0?

b. Find $\lim_{t \to \infty} L(t)$. What does this limit say about the biological meaning of a?

c. Graph $L(t)$ and discuss how an organism grows according to this curve.

45. At the beginning of the twentieth century, several notable biologists, including G. F. Gause and T. Carlson, studied the population dynamics of yeast. For example, Carlson grew yeast under constant environmental conditions in a flask; he regularly monitored their population densities.* In Chapter 6, we will show that the following function describes the growth of the population:

$$N(t) = \frac{9.7417e^{0.53t}}{1 + 0.01476e^{0.53t}}$$

where $N(t)$ is the population density and t is time in hours. Find $\lim_{t \to \infty} N(t)$ and discuss the meaning of this limit.

46. The following equation is used to calculate the average firing rate f of a neuron (in spikes per second) as a function of the concentration x of neurotransmitters perfusing its synapses.

$$f(x) = \frac{20\,e^{3x}}{2.1 + e^{3x}}$$

Find the horizontal asymptote; then find the values of x such that $f(x)$ is within 0.5% of its asymptotic values.

47. The following equation is used to calculate the average firing rate f of a neuron (in spikes per second) as a function of the concentration x of neurotransmitters perfusing its synapses.

$$f(x) = \frac{16\,e^{5x}}{3.2 + e^{5x}}$$

Find the values of x such that $f(x)$ is within 0.5% of its asymptotic values.

48. Compare the solutions obtained to Problems 46 and 47 and decide which of these represents a tighter on-off switch of the neuron, that is, from firing at its maximum rate to being inactive. What do you conclude in terms of which of the parameters a, b, and c in the function

$$f(x) = \frac{ae^{cx}}{b + e^{cx}}$$

controls the narrowness of the range of x over which on-off switching occurs? Note that this function is called the *logistic function* and will be encountered in many different examples in upcoming chapters.

* T. Carlson, "Uber Geschwindigkeit und Größe der Hefevermehrung in Würze," *Biochem* Z57 (1913): 313–334.

2.5   Sequential Limits

In Section 1.7, we considered sequences $a_1, a_2, \ldots$ of real numbers, which can be used to model drug concentrations, population dynamics, and population genetics. In some cases, these sequences converged to a limiting value as n got very large. In this section, we study the limits of sequences, their relationship to continuity, and a convergence theorem, as well as how these concepts can be used to understand the asymptotic behavior of difference equations. While limits of functions form the basis of differentiation as we shall soon see, limits of sequences form the basis of integration (as we will discuss in Chapter 5).

Sequential limits and continuity

For sequences, there is only one type of limit to consider: the **sequential limit**, defined as the limiting value of a_n as $n \to \infty$.

Sequential Limits (Informal Definition)	Let $a_1, a_2, a_3, \ldots$ be a sequence. We write

$$\lim_{n \to \infty} a_n = L$$

provided that we can make a_n arbitrarily close to L for all n sufficiently large. In this case, we say the sequence converges to L.

We write

$$\lim_{n \to \infty} a_n = \infty$$

provided that we can make a_n arbitrarily large for all n sufficiently large.

We write

$$\lim_{n \to \infty} a_n = -\infty$$

provided that we can make a_n arbitrarily negative for all n sufficiently large.

Example 1   Finding sequential limits

In each of the following, if it exists, calculate $\lim_{n \to \infty} a_n$ where

a. $a_n = 2^n$ **b.** $a_n = 1 + \dfrac{1}{n}$ **c.** $a_n = \dfrac{2n^2 + 3n - 1}{5n^2 - n + 8}$

d. $a_n = \cos \dfrac{n\pi}{2}$ **e.** $a_n = \dfrac{1}{n} \cos \dfrac{n\pi}{2}$

Solution

a. Since 2^n gets arbitrarily large as n gets arbitrarily large, $\lim_{n \to \infty} 2^n = \infty$.

b. Since $\dfrac{1}{n}$ approaches zero as n gets arbitrarily large, $\lim_{n \to \infty} 1 + \dfrac{1}{n} = 1$.

c. Since the numerator and denominator are polynomials in n, we divide the numerator and denominator by the term with largest exponent, namely, n^2.

$$\lim_{n \to \infty} \frac{2n^2 + 3n - 1}{5n^2 - n + 8} = \lim_{n \to \infty} \frac{2n^2 + 3n - 1}{5n^2 - n + 8} \times \frac{\dfrac{1}{n^2}}{\dfrac{1}{n^2}}$$

$$= \lim_{n \to \infty} \frac{2 + \dfrac{3}{n} - \dfrac{1}{n^2}}{5 - \dfrac{1}{n} + \dfrac{8}{n^2}}$$

$$= \frac{2}{5}$$

d. Since $\cos \dfrac{n\pi}{2}$ alternates between the values 0, 1, and -1, this sequence does not have a limit. There is no unique value that the sequence approaches.

e. Since $\left| \dfrac{1}{n} \cos \dfrac{n\pi}{2} \right| \leq \dfrac{1}{n}$ and we can make $\dfrac{1}{n}$ arbitrarily close to 0 for n sufficiently large,

$$\lim_{n \to \infty} \frac{1}{n} \cos \frac{n\pi}{2} = 0$$

Graphing this sequence, illustrated in the figure below, confirms this convergence to zero.

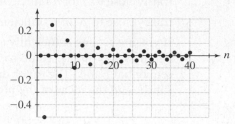

As in our previous limit definitions, the existence of a sequential limit implies that we can make a_n as close to L as we like, provided that n is sufficiently large. But how do we verify this statement? What is meant by *sufficiently large*? The following example illustrates the answer to this question.

Example 2 Finding sufficiently large *n*

Consider $a_n = \dfrac{n}{2+n}$.

a. Find $\lim\limits_{n\to\infty} a_n = L$.

b. Determine how large n needs to be to ensure that $|a_n - L| < 0.002$.

Solution

a. Dividing the numerator and denominator by n yields

$$\lim_{n\to\infty} \frac{n}{2+n} = \lim_{n\to\infty} \frac{1}{2/n + 1}$$
$$= 1$$

Hence, $L = 1$.

b. We want that

$$\left| \frac{n}{2+n} - 1 \right| < 0.002 \qquad \textit{this is } |a_n - L| < 0.002$$

$$\left| \frac{n}{2+n} - \frac{2+n}{2+n} \right| < 0.002$$

$$\left| \frac{-2}{2+n} \right| < 0.002$$

$$\frac{2}{2+n} < 0.002 \qquad \textit{absolute value of a negative divided by a positive number}$$

$$\frac{2}{0.002} < 2+n \qquad \textit{multiply both sides by } 2+n, \textit{ and divide both sides by 0.002}$$

$$998 < n \qquad \textit{simplify and subtract 2 from both sides}$$

The number n must be greater than 998.

There is an important relationship between limits of sequences and limits of functions. This relationship is most useful for proving discontinuity of a function.

Theorem 2.4 Sequential continuity theorem

Let f be a function. Then $\lim\limits_{x\to a} f(x) = L$ if and only if $\lim\limits_{n\to\infty} f(a_n) = L$ for any sequence satisfying $\lim\limits_{n\to\infty} a_n = a$.

One direction of this theorem is clear. If $\lim_{n\to\infty} a_n = a$, then a_n can be made arbitrarily close to a for n sufficiently large. Therefore, if $\lim_{x\to a} f(x) = L$, then $f(a_n)$ is arbitrarily close to L for n sufficiently large. Hence, if $\lim_{x\to a} f(x) = L$, then $\lim_{n\to\infty} f(a_n) = L$. If you are feeling sufficiently adventuresome, try proving this direction using formal definitions of limits. To do so, you will have to come up with a formal definition of sequential limits. The other direction of the sequential continuity theorem is more subtle, and the ideas of the proof are beyond the scope of this text.

Example 3   Proving nonexistence of limits

Show that $\lim_{x\to 0} \sin \dfrac{1}{x}$ does not exist.

Solution   Let $f(x) = \sin \dfrac{1}{x}$. Our goal is to find two sequences a_n and b_n satisfying $\lim_{n\to\infty} a_n = 0$ and $\lim_{n\to\infty} b_n = 0$, but $\lim_{n\to\infty} f(a_n) = L$ and $\lim_{n\to\infty} f(b_n) = M$ with $L \neq M$. Then, we can apply the sequential continuity theorem to conclude that the limit $\lim_{x\to a} f(x)$ does not exist. For this example, we let $a_n = \dfrac{1}{\pi n}$ and $b_n = \dfrac{2}{\pi(4n+1)}$. Then,

$$\lim_{n\to\infty} \frac{1}{\pi n} = 0 \quad \text{and} \quad \lim_{n\to\infty} \frac{2}{\pi(4n+1)} = 0$$

We now find the limits of $f(a_n)$ and $f(b_n)$:

$$\lim_{n\to\infty} f(a_n) = \lim_{n\to\infty} \sin \frac{1}{a_n} = \lim_{n\to\infty} \sin(\pi n) = 0$$

and

$$\lim_{n\to\infty} f(b_n) = \lim_{n\to\infty} \sin \frac{1}{b_n} = \lim_{n\to\infty} \sin\left[\frac{\pi(4n+1)}{2}\right] = 1$$

Since $\lim_{n\to\infty} f(a_n) \neq \lim_{n\to\infty} f(b_n)$, it follows from the sequential continuity theorem that $\lim_{x\to 0} \sin \dfrac{1}{x}$ does not exist because it cannot be equal to both 0 and 1 at the same time.

Asymptotic behavior of difference equations

When we introduced sequences in Section 1.7, we considered a special class of sequences that arises through a difference equation

$$a_{n+1} = f(a_n)$$

where a_1 is specified and f is a function. In some instances, we can actually find explicit expressions for the sequence defined by the difference equation and take the limit.

Example 4   Finding the limit of a sequence

Find an explicit expression for the sequences defined by the following difference equations and find the limit as n becomes large.

   a. $a_{n+1} = 0.1 a_n$ with $a_1 = 0.1$ **b.** $a_{n+1} = \sqrt{a_n}$ with $a_1 = 2$

Solution
   a. We have $a_1 = 0.1$, $a_2 = 0.1 a_1 = 0.1^2$, and $a_3 = 0.1 a_2 = 0.1^3$. Hence, we can see inductively that $a_n = 0.1^n$. Since a_n gets arbitrarily small as n gets sufficiently large, we obtain $\lim_{n\to\infty} a_n = 0$.

b. We have,

$$a_1 = \sqrt{2} = 2^{1/2}, a_2 = a_1^{1/2} = 2^{1/4}, a_3 = a_2^{1/2} = 2^{1/8}, \ldots, a_n = a_{n-1}^{1/2} = 2^{1/2^n}$$

To find this limit, consider the logarithm of this sequence, that is,

$$\ln a_n = \frac{\ln 2}{2^n}$$

Clearly, $\lim_{n \to \infty} \ln a_n = 0$. Thus, by the continuity of e^x and the sequential continuity theorem, we get that

$$\lim_{n \to \infty} a_n = \lim_{n \to \infty} e^{\ln a_n} = e^{\lim_{n \to \infty} \ln a_n} = e^0 = 1$$

∎

In part **b** of Example 4, we saw that sometimes it is useful to find $\lim_{n \to \infty} a_n$ by finding $\lim_{n \to \infty} f(a_n)$ for an appropriate choice of a continuous one-to-one function f.

Example 5 Lethal recessives revisited

In Example 5 of Section 1.7, we modeled the proportion x_n of a lethal recessive allele in a population at time n with the difference equation:

$$x_{n+1} = \frac{x_n}{1 + x_n}$$

Keeping with the notation x_n rather than a_n, assume that the initial proportion of the lethal allele is $x_1 = 0.5$.

a. Verify that $x_n = \dfrac{1}{1 + n}$ satisfies the difference equation.

b. Determine $\lim_{n \to \infty} x_n$. Discuss the implication for the proportion of the lethal recessive allele in the long term.

c. Determine how large n needs to be to ensure that $x_n \leq 0.1$.

d. Determine how large n needs to be to ensure that $x_n \leq 0.01$. Discuss the implications.

Solution

a. First we verify that the formula holds for $n = 1$, namely, $x_1 = \dfrac{1}{1 + 1} = 0.5$.

To verify that $x_n = \dfrac{1}{1 + n}$ satisfies the difference equation for any $n > 1$, we proceed inductively and substitute our expression for x_n into both sides of the difference equation and show that they are equal. Evaluating the proposed solution on the left-hand side of the difference equation $x_{n+1} = \dfrac{x_n}{1 + x_n}$ gives us

$$x_{n+1} = \frac{1}{1 + (n + 1)} = \frac{1}{n + 2}.$$

Evaluating the proposed solution on the left-hand side of the difference equation equation give us

$$\frac{x_n}{1 + x_n} = \frac{\dfrac{1}{1 + n}}{1 + \dfrac{1}{1 + n}}$$

$$= \frac{1}{(n + 1) + 1} \quad \textit{after multiplying by } 1 = \frac{\frac{1}{n+1}}{\frac{1}{n+1}}$$

$$= \frac{1}{n + 2}$$

Hence, we have shown that the formula $x_n = \dfrac{1}{1+n}$ satisfies the difference equation $x_{n+1} = \dfrac{1}{1+x_n}$ for all n.

b. Since $\dfrac{1}{1+n}$ gets arbitrarily small as n gets arbitrarily large, $\lim\limits_{n\to\infty} x_n = 0$. Hence, in the long term, we expect the lethal recessive genes to vanish from the population.

c. We want

$$\frac{1}{1+n} \le \frac{1}{10}$$

$$10 \le 1+n$$

$$9 \le n$$

Hence, after nine generations the proportion of lethal recessives is less than 0.1.

d. We want

$$\frac{1}{1+n} \le \frac{1}{100}$$

$$100 \le 1+n$$

$$99 \le n$$

Hence, after ninety-nine generations the proportion of lethal recessives is less than 0.01. These calculations suggest that initially the proportion of lethal recessives decreases rapidly, but further decreases in the proportion occur more and more slowly.

■

Returning to our a_n notation, recall from Section 1.7, that a point a is an equilibrium of a difference equation $a_{n+1} = f(a_n)$ if $f(a) = a$. In Example 4**a** and Example 5, the only solution to the equation $f(a) = a$ is $a = 0$, and the sequences generated by these difference equations converge to this equilibrium. In Example 4**b**, the equation $\sqrt{a} = a$ has two solutions: $a = 0$ and $a = 1$, and the sequence we examined converged to the latter rather than former equilibrium.

To see why this convergence to equilibria occurs, consider a sequence a_n that satisfies the difference equation

$$a_{n+1} = f(a_n)$$

where f is a continuous function. Assume that $\lim\limits_{n\to\infty} a_n = a$. By the sequential continuity theorem, we have

$$a = \lim_{n\to\infty} a_n$$

$$= \lim_{n\to\infty} a_{n+1} \qquad n+1 \to \infty \text{ if and only if } n \to \infty$$

$$= \lim_{n\to\infty} f(a_n)$$

$$= f\Big(\lim_{n\to\infty} a_n\Big) \qquad \text{by sequential continuity}$$

$$= f(a)$$

Hence, the limiting value a is an equilibrium for this difference equation.

Limits of Difference Equations	Let f be a continuous function and a_n be a sequence that satisfies

$$a_{n+1} = f(a_n)$$

If $\lim\limits_{n\to\infty} a_n = a$, then $f(a) = a$. In other words, a is an equilibrium.

Example 6   To converge or not to converge

Find the equilibria of the following difference equations and use technology to determine whether the specified sequence converges to one of the equilibria.

a. $a_{n+1} = \dfrac{1}{1 + a_n}$ with $a_1 = 1$

b. $a_{n+1} = 2a_n(1 - a_n)$ with $a_1 = 0.1$

c. $a_{n+1} = 3.5a_n(1 - a_n)$ with $a_1 = 0.1$

Solution

a. To find the equilibria, we solve

$$a = \frac{1}{1 + a}$$

$$a(1 + a) = 1 \qquad \textit{after multiplying by } 1 + a$$

$$a^2 + a - 1 = 0$$

$$a = -\frac{1}{2} \pm \frac{\sqrt{5}}{2} \qquad \textit{by the quadratic formula}$$

Hence, if the sequences determined by this difference equation have well-defined limits, then these limits are either $-\dfrac{1}{2} + \dfrac{\sqrt{5}}{2} \approx 0.6180$ or $-\dfrac{1}{2} - \dfrac{\sqrt{5}}{2} \approx -1.6180$. Computing the first twenty terms of the difference equation with $a_1 = 1$ and plotting yields the following graph.

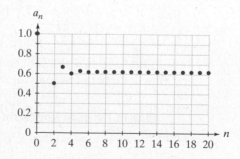

It appears that the sequence is converging to the positive equilibrium.

b. To find the equilibrium, we solve

$$a = 2a(1 - a)$$

$$2a^2 - a = 0$$

$$a(2a - 1) = 0 \qquad \textit{by factoring the common term } a$$

$$a = 0 \quad \text{and} \quad a = \frac{1}{2}$$

Computing and plotting the first twenty terms of the difference equation with $a_1 = 0.1$ and plotting yields the following graph.

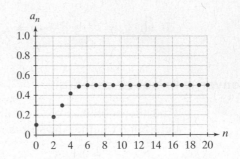

It appears that the sequence is converging to $\dfrac{1}{2}$.

c. To find the equilibrium, we solve

$$a = 3.5a(1 - a)$$
$$3.5a^2 - 2.5a = 0$$
$$a(3.5a - 2.5) = 0 \qquad \textit{by factoring the common term a}$$
$$a = 0 \quad \text{or} \quad a = \frac{5}{7}$$

Computing and plotting the first one hundred terms of the difference equation with $a_1 = 0.1$ and plotting yields the following graph.

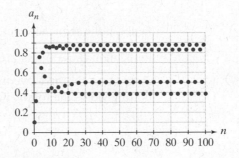

It appears that the sequence does not converge; rather, it seems to eventually oscillate between four values.

One of the most important models in population biology is the discrete logistic model:

$$a_{n+1} = a_n + r a_n \left(1 - \frac{a_n}{K}\right)$$

where the parameter $r > 0$ is called the *intrinsic rate of growth* and $K > 0$ is called the *environmental carrying capacity*.

Example 7   Dynamics of the discrete logistic model

a. Find the equilibrium solutions associated with the discrete logistic equation. What do you observe about the roles of the parameters r and K in determining this equilibrium?

b. Calculate the first twenty points of the sequence $a_{n+1} = a_n + 0.3 a_n(1 - a_n)$ with $a_1 = 0.1$.

c. Repeat part **b** with $a_1 = 1.5$.

d. Calculate the first twenty points of the sequence $a_{n+1} = a_n + 1.9a_n(1 - a_n)$ with $a_1 = 0.6$.

e. Calculate the first twenty points of the sequence $a_{n+1} = a_n + 2.2a_n(1 - a_n)$ with $a_1 = 0.6$.

Solution

a. The equilibria are solutions to the equation

$$a = a + ra\left(1 - \frac{a}{K}\right)$$

$$ra\left(1 - \frac{a}{K}\right) = 0$$

$$a = 0 \quad \text{and} \quad a = K.$$

From this it is clear that the value of r does not influence the value of the equilibria, one of which is equal to K.

b. The values in this sequence are plotted in the left-hand side of Figure 2.41. The sequence appears to be converging to the positive equilibrium $K = 1$.

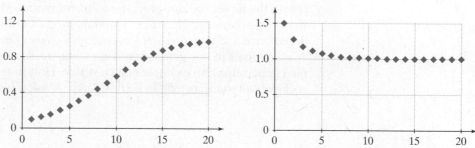

Figure 2.41 Plots of the sequences $a_{n+1} = a_n + 0.3a_n(1 - a_n)$ from starting values $a_1 = 0.1$ (left panel) and $a_1 = 1.5$ (right panel).

c. The values in this sequence are plotted in the right-hand side of Figure 2.41. The sequence appears to be converging to the equilibrium $K = 1$.

d. The values in this sequence are plotted in the left-hand side of Figure 2.42. The sequence exhibits dampened oscillations around the equilibrium $K = 1$. It still appears to be converging to $K = 1$.

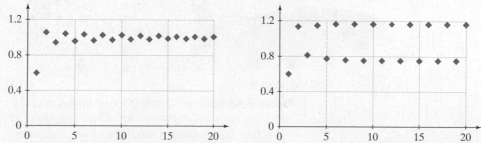

Figure 2.42 Plots of the sequences $a_{n+1} = a_n + ra_n(1 - a_n)$ from the starting value $a_1 = 0.6$ for the cases $r = 1.9$ (left panel) and $r = 2.2$ (right panel).

e. The values in this sequence are plotted in the right-hand side of Figure 2.42. The sequence oscillates between a lower and higher population density. The oscillations are not dampened and the sequence does not appear to be converging to the equilibrium $K = 1$.

Examples 6 and 7 illustrate that the existence of equilibria for a difference equation does not ensure the convergence of the sequences generated by it. This raises the question, when do the sequences generated by a difference equation converge to an equilibrium? In general, this is a hard question. The following theorem, stated without proof, provides a criterion that ensures convergence of solutions of a difference equation. Later, when we have covered the basics of derivatives of functions, we will present another criterion that ensures convergence of a sequence to an equilibrium. We make two definitions before stating this theorem. We say that a sequence is **increasing** (respectively, **decreasing**) if $a_1 \leq a_2 \leq a_3 \leq \ldots$ (respectively, $a_1 \geq a_2 \geq a_3 \ldots$).

Theorem 2.5   Monotone convergence theorem

Let f be a continuous, increasing function on an interval $I = [a, b]$ such that the image of f lies in I (i.e., if x is in I, then $f(x)$ is a value that also lies within I). If $a_1, a_2, a_3, \ldots,$ is a sequence that satisfies $a_{n+1} = f(a_n)$, then the sequence is either increasing or decreasing. Moreover, $\lim\limits_{n \to \infty} a_n = a$ in I exists and satisfies $f(a) = a$.

The next example illustrates the application of the monotone convergence theorem to the Beverton-Holt stock-recruitment model. This model has been used extensively by fisheries scientists to formulate models of fish, such as Pacific salmon, that breed once and then die. *Stock-recruitment curves* describe how a spawning stock of N individuals in one generation contributes recruits R (i.e., new individuals) to the next generation. An example of a Beverton-Holt function fitted to data obtained for sockeye salmon spawning in Karluk Lake, Alaska, is shown in Figure 2.43.

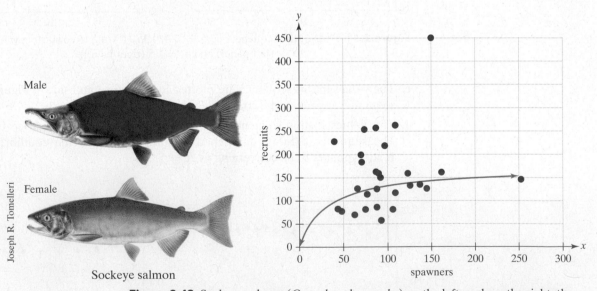

Figure 2.43 Sockeye salmon (*Oncorhynchus nerka*) on the left; and, on the right, the relationship between recruits and stock for sockeye salmon in Karluk Lake, Alaska.

Data Source: John A. Gulland, *Fish Stock Assessment: A Manual of Basic Methods* (New York: Wiley, 1983).

Example 8   Beverton-Holt sockeye salmon dynamics

The function plotted in the right panel of Figure 2.43 is given by

$$R(N) = \frac{N}{0.006\,N + 0.2}$$

Note that it is more natural for us to continue using N_n rather then the sequence notation a_n, since N is the symbol most commonly used to denote population size in the ecology literature.

where N is the stock size (spawners) in a particular generation and $R(N)$ is the number of recruits produced for the next generation. Since the number of recruits determines the size of the stock that will be spawners in the next generation, it follows that the salmon dynamics can be modeled by the difference equation

$$N_{n+1} = R(N_n)$$

where N_n is the stock size of the nth generation.

a. Find the equilibria of this difference equation.

b. Graph $R(N)$ and $y = N$.

c. Apply the monotone convergence theorem to determine what happens to N_n when $N_1 = 10$ and when $N_1 = 200$. Use cobwebbing to illustrate your results.

(Note that it is possible to find an explicit solution of this difference equation. This is explored in Problem Set 2.5.)

Solution

a. To find the equilibria, we solve

$$N = \frac{N}{0.006\,N + 0.2}$$

$$N(0.006\,N + 0.2) = N$$

$$N(0.006\,N - 0.8) = 0$$

It may seem strange for population size to have a fractional value, but not if size is measured as a density—that is, numbers per unit area—or is in units of millions of individuals.

$$N = 0, \frac{0.8}{0.006}$$

Hence, the equilibria are given by 0 and $\dfrac{0.8}{0.006} = 133\dfrac{1}{3}$.

b. Plotting the two functions yields the following graph.

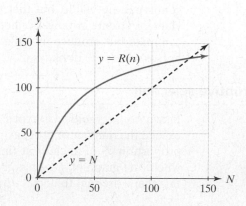

The equilibria correspond to the points where the functions intersect.

c. Since $R(N)$ is increasing on $I = [0, \infty)$ and the image of I under f is I, we can apply the monotone convergence theorem.

Assume $N_1 = 10$. Since $N_2 = \dfrac{10}{0.06 + 0.2} \approx 38.46 \geq N_1$, the monotone convergence theorem implies that N_n is increasing. On the other hand, since the graph of $R(N)$ is saturating at $166\dfrac{2}{3}$, we have $10 \leq N_{n+1} = R(N_n) \leq 166\dfrac{2}{3}$ for all $n \geq 1$. Therefore, by the monotone convergence theorem $\lim\limits_{n \to \infty} N_n$ must equal the equilibrium $133\dfrac{1}{3}$. Cobwebbing with $N_1 = 10$ illustrates this convergence in Figure 2.44 (left).

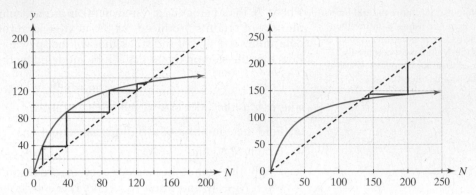

Figure 2.44 Cobwebbing solutions to salmon model for initial values $N_1 = 10$ (left) and $N_1 = 200$ (right).

Assume $N_1 = 200$. Since $N_2 = \dfrac{200}{0.006 \cdot 200 + 0.2} \approx 142.85 \le N_1$, the monotone convergence theorem implies that N_n is a decreasing sequence. On the other hand, $200 \ge N_n \ge 133\dfrac{1}{3}$ for all $n \ge 1$. Therefore, by the monotone convergence theorem, $\lim\limits_{n \to \infty} N_n$ must equal the equilibrium $133\dfrac{1}{3}$ (from part **a**). Cobwebbing with $N_1 = 200$ illustrates this convergence in Figure 2.44 (right). ■

In Example 5 of Section 1.7, we developed a population genetics model under the assumption that individuals could carry and pass on a recessive lethal allele. In formulating this model, we assumed that two alleles, A and a, exist and determine three possible genotypes: AA, Aa, and aa. If the allele a is the recessive lethal, then this implies that genotype aa is the least viable since, by definition, aa individuals all die before they are able to reproduce. In the next example, we assume that all three genotypes are viable, but that the so-called *heterozygous* genotype Aa is the least viable. Extreme instances of heterozygous inviability arise when different genotypes can mate but produce infertile offspring, such as mules, which are produced when horses mate with donkeys.

Example 9   Disruptive selection

If two *homozygous* genotypes AA and aa produce equal numbers of progeny that exceed the number of progeny produced by the *heterozygous* Aa genotype, then the proportion x_n of allele a at time n can be shown to be modeled by $x_{n+1} = f(x_n)$, where the graph of $f(x)$ has the S-shaped curve depicted in Figure 2.45. Curves of this shape are said to depict *disruptive selection*.

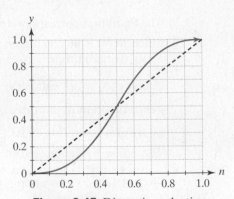

Figure 2.45 Disruptive selection.

Using cobwebbing methods, determine from this graph what happens to x_n in the long term for the two cases given in parts **a** and **b** below.

a. $x_1 = 0.6$ **b.** $x_1 = 0.4$

c. As reported in a 1972 *Science* article, Foster and others experimentally examined changes in two chromosomal proportions in *Drosophila melanogaster*. Data from a set of experiments is graphed in Figure 2.46.

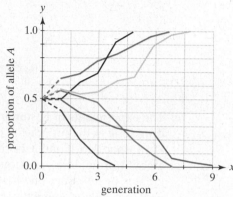

Figure 2.46 Time series for disruptive selection in chromosomal proportions for a species of fruit fly.

Source: G. G. Foster, M. J. Whitten, T. Prout, and R. Gill, "Chromosome Rearrangements for the Control of Insect Pests," *Science* 176: 875–880. Reprinted with permission.

These experimentally determined graphs show how the proportion of an allele changes over generations for initial conditions that through a small amount of random variation lead to different population levels at time 1 on the *x*-axis. Discuss whether these experiments are consistent with the model predictions.

Solution

a. Since the graph of f is increasing, we can apply the monotone convergence theorem. Since $f(0.6) > 0.6$, the sequence x_n is increasing if $x_1 = 0.6$. Since $x_n \leq 1$ for all n and $f(1) = 1$ is the only equilibrium greater than 0.6, x_n converges to 1 as n increases. In other words, the proportion of a alleles approaches one. Cobwebbing reaffirms this prediction, as shown in the following graph.

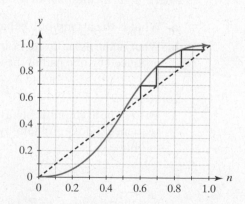

b. Since the graph of f is increasing, we can apply the monotone convergence theorem. Since $f(0.4) < 0.4$, the sequence x_n is decreasing if $x_1 = 0.4$. Since $x_n \geq 0$ for all n and $f(0) = 0$ is the only equilibrium less than 0.4, x_n converges to 0 as *n* increases. In other words, the proportion of *a* alleles approaches zero. Cobwebbing reaffirms this prediction, as shown in this graph.

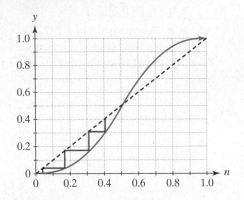

c. The experiments of Foster and others are consistent with the model predictions, if we look at the proportion of allele *a* at the start of the second generation (i.e., at 1 rather than 0 on the horizontal axis). Specifically, within the limits of a small amount of random variation, the data show that if the proportion of allele *a* is greater or less than one-half, then the proportion of allele *a*, respectively, approaches one or goes extinct.

Fibonacci famously posed the following problem (reworded here) over 800 years ago. Suppose a newly born pair of rabbits, one male and one female, are put in an enclosed, but very large, field. Further, suppose all rabbits are able to mate at the age of one month and that the impregnated female gives birth to a male-female pair one month later. Ignoring questions relating to inbreeding, assuming that rabbits never die and there is always enough food for them to eat, what is the asymptotic annual rate of increase of the rabbits in the enclosed field? This is the problem we address in the next example.

Example 10   Fibonacci and the growth rate of rabbits

Fibonacci's rabbit model (see 𝔥𝔦𝔰𝔱𝔬𝔯𝔦𝔠𝔞𝔩 𝔔𝔲𝔢𝔰𝔱, Problem 38 of Section 1.7) has the form

$$R_n = R_{n-1} + R_{n-2}.$$

where R_n is the number of rabbit pairs in month *n*.

a. What is the asymptotic behavior of R_n as $n \to \infty$?

b. What is the annual rate of increase in the population? Calculate the population produced by one pair at the end of the first year (i.e., calculate R_{12} when $R_0 = 1$).

Solution

a. To find the asymptotic behavior of R_n, we can transform the model into a familiar sequence model problem by dividing both sides of the above equation by R_{n-1} to obtain

$$\frac{R_n}{R_{n-1}} = 1 + \frac{R_{n-2}}{R_{n-1}}$$

and then define $a_n = R_n/R_{n-1}$. In this case a_n, the ratio of the number of rabbits in month *n* to those in month $n-1$, satisfies the equation

$$a_n = 1 + \frac{1}{a_{n-1}}$$

The equilibrium solution to this equation is

$$a = 1 + 1/a$$

$$a^2 = a + 1 \quad \text{multiplying both sides by } a$$

$$a^2 - a - 1 = 0$$

$$a = \frac{1}{2} \pm \frac{\sqrt{5}}{2}. \quad \text{by the quadratic formula}$$

Only the positive solution $a \approx 1.6180$ applies here, and it can be shown that the sequence converges to this value as the number of months increases (see Problem 44 in Problem Set 2.5). Thus, although R_n increases without bound, the ratio $a_n = R_n/R_{n-1}$ approaches a constant.

b. To get the annual rate of increase we need to calculate $(1.618)^{12} \approx 322$, a really stunning rate of growth. In Table 2.5, we list the first 13 terms and note that the rate of increase over the first 12 iterations is 233 rather than 322 because the equilibrium value for the rate of increase applies asymptotically rather than initially.

Table 2.5 Fibonacci rabbit growth

Month	Number of pairs
0	1
1	1
2	2
3	3
4	5
5	8
6	13
7	21
8	34
9	55
10	89
11	144
12	233

PROBLEM SET 2.5

Level 1 DRILL PROBLEMS

Determine whether the sequential limits in Problems 1 to 8 exist. If they exist, find the limit. If they do not exist, explain briefly why.

1. $\lim\limits_{n\to\infty} a_n$ where $a_n = \dfrac{n^2 - n}{1 + 3n^2}$

2. $\lim\limits_{n\to\infty} a_n$ where $a_n = \dfrac{5 - 2n}{6 + 3n}$

3. $\lim\limits_{n\to\infty} a_n$ where $a_n = \dfrac{e^n}{1 + e^n}$

4. $\lim\limits_{n\to\infty} a_n$ where $a_n = 2^{3/n}$

5. $\lim\limits_{n\to\infty} a_n$ where $a_{n+1} = -a_n$ and $a_1 = 2$

6. $\lim\limits_{n\to\infty} a_n$ where $a_{n+1} = -a_n^{-1}$ and $a_1 = 3$

7. $\lim\limits_{n\to\infty} a_n$ where $a_n = \cos n$

8. $\lim\limits_{n\to\infty} a_n$ where $a_n = [1 + (-1)^n]$

Consider the sequences defined in Problems 9 to 14.

a. Find $\lim\limits_{n\to\infty} a_n$.

b. Determine how large n needs to be to ensure that $|a_n - L| < 0.001$.

9. $\lim\limits_{n\to\infty} a_n$ where $a_n = \dfrac{n}{3 + n}$

10. $\lim\limits_{n\to\infty} a_n$ where $a_n = \dfrac{2n}{n - 1}$

11. $\lim\limits_{n\to\infty} a_n$ where $a_n = \dfrac{1,000}{n}$

12. $\lim\limits_{n\to\infty} a_n$ where $a_n = \dfrac{n + 1}{1,000n}$

13. $\lim\limits_{n\to\infty} a_n$ where $a_n = \dfrac{n^2 + 1}{n^3}$

14. $\lim\limits_{n\to\infty} a_n$ where $a_n = e^{-n}$

All the sequences in Problems 15 to 18 satisfy $\lim\limits_{n\to\infty} a_n = \infty$. Determine how large n has to be to ensure that $a_n \geq 1,000,000$.

15. $a_n = 2n$

16. $a_n = n^2$

17. $a_n = 2^n - 10,000$

18. $a_n = \dfrac{n^2}{1 + n}$

Find the sequences determined by the difference equation $x_{n+1} = f(x_n)$ with the initial condition x_1 specified in Problems 19 to 24. Determine $\lim\limits_{n\to\infty} a_n$. Justify your answer.

19. $f(x) = x + 2$ with $x_1 = 0$

20. $f(x) = \dfrac{x}{3}$ with $x_1 = 27$

21. $f(x) = \sqrt{x}$ with $x_1 = 100$

22. $f(x) = x^2$ with $x_1 = 1.00001$

23. $f(x) = x^2$ with $x_1 = 0.99999$

24. $f(x) = 4x^2$ with $x_1 = 1$

Find the equilibrium of the difference equations in Problems 25 to 28 and use technology to determine which of the specified sequences converge to one of the equilibria.

25. $a_{n+1} = \dfrac{3}{2 + a_n}$ with $a_1 = 1$

26. $a_{n+1} = \dfrac{1}{5 - a_n}$ with $a_1 = 1$

27. $a_{n+1} = 3a_n(1 - a_n)$ with $a_1 = 0.1$

28. $a_{n+1} = 5.5(1 - a_n)$ with $a_1 = 0.2$

Use the monotone convergence theorem in Problems 29 to 34 to determine the limits of the following specified sequences.

29. $a_{n+1} = \dfrac{2a_n}{1 + a_n}$ with $a_1 = 0.5$

30. $a_{n+1} = \dfrac{2a_n}{1 + a_n}$ with $a_1 = 2$

31. $a_{n+1} = 2 \ln a_n$ with $a_1 = 1$

32. $a_{n+1} = 2 \ln a_n$ with $a_1 = 100$

33. $a_{n+1} = \sqrt{5 + a_n}$ with $a_1 = 0$

34. $a_{n+1} = \sqrt{5 + a_n}$ with $a_1 = 20$

Level 2 APPLIED AND THEORY PROBLEMS

35. In Example 5 of Section 2.5 we introduced a model for the frequency of lethal recessive alleles in a population. The model is given by

$$x_{n+1} = \frac{x_n}{1 + x_n}$$

where x_n is the frequency of the recessive allele in the population.

a. If $x_1 = 0.25$, then verify that $x_n = \dfrac{1}{n + 3}$ satisfies the difference equation.

b. Find $\lim\limits_{n \to \infty} x_n$.

c. Determine how large n needs to be to ensure that $x_n \leq 0.1$.

d. Determine how large n needs to be to ensure that $x_n \leq 0.001$.

36. Let us consider the lethal recessive allele model in greater generality.

a. Verify that $x_n = \dfrac{x_1}{1 + (n - 1)x_1}$ satisfies the difference equation $x_{n+1} = \dfrac{x_n}{1 + x_n}$ for any choice of x_1.

b. For any x_1, find $\lim\limits_{n \to \infty} x_n$.

c. Assuming that x_1 lies in $(0, 1)$, determine how large n needs to be to ensure that $x_n \leq 0.01$.

37. In Example 5 of Section 1.7, we discussed a population genetics model under the assumption that there were two alleles, A and a, that determined three possible genetic types, AA, Aa, and aa. We assume that genetic type Aa (so called heterozygote) is the least viable. If the genotype aa produces nine times more progeny than genotype AA progeny, then the frequency x_n of allele a at time n can be modeled by $x_{n+1} = f(x_n)$ where the graph of f is given by

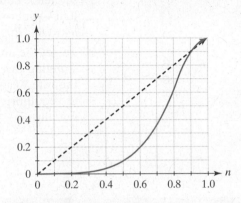

a. Determine what happens to x_n in the long term if $x_1 = 0.91$.

b. Determine what happens to x_n in the long term if $x_1 = 0.89$.

c. As reported in a 1972 *Science* article, Foster and others experimentally examined changes in two chromosomal frequencies in *Drosophila melanogaster*. Data from a set of experiments are graphed next:

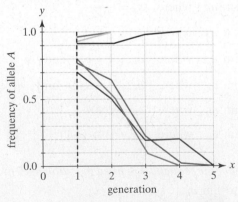

Data Source: Foster and others, "Chromosome Rearrangements," 875–880.

These experimentally determined graphs show how the frequency of an allele changes over generations for different initial conditions. Discuss whether these experiments are consistent with the model predictions.

38. The Beverton-Holt model has been used extensively by fisheries. This model assumes that populations are competing for a single limiting resource and reproduce at discrete moments in time. If we let N_n denote the population abundance in the nth year (or generation), $r > 0$ the maximal per capita growth rate, and $a > 0$ as a competition coefficient, then the model is given by

$$N_{n+1} = \frac{r N_n}{1 + a N_n}$$

with $r > 0$ and $a > 0$.

a. Assume $N_1 = 1$ and find the first four terms of the sequence.

b. Guess the explicit expression for the sequence and verify that your guess is correct.

c. Find the equilibria of this difference equation.

d. By considering the $\lim_{n \to \infty} N_n$, determine under what conditions the population goes extinct (i.e., converges to 0) versus under what conditions it is able to persist (i.e., converge to a positive equilibrium $N > 0$) if it is initially at $N_1 = 1$.

In the Fibonacci rabbit problem laid out in Example 10, suppose only a proportion p of the females that could become pregnant actually do become pregnant each month. (We assume the population starts with a large number of pairs so that when we refer to proportions, we are actually thinking of whole numbers of pairs rather than, say, one-third of two pairs, which makes no biological sense.) What is the annual rate of increase in Problems 39 to 43 if p equals the specified value?

39. 3/4   **40.** 2/3   **41.** 1/2   **42.** 1/3   **43.** 1/4

44. Consider the Fibonacci sequence in Example 10. Show that

$$a_{n+2} = 1 + \frac{1}{a_{n+1}} = 1 + \frac{a_n}{1 + a_n}$$

and hence apply the monotone convergence theorem to find out what happens to the sequences of even elements of a_n.

45. Use technology to calculate the first twenty points of the sequence $a_{n+1} = a_n + r a_n(1 - a_n)$ with $a_1 = 0.5$ for the cases $r = 0.9$, $r = 1.5$ and $r = 2.1$. How does this fit in with the discussion in the solution to part **e** of Example 7.

46. Revisiting the sockeye salmon stock-recruitment relationship considered in Example 8, we see from Figure 2.43 that we could just as credibly fit the Ricker function

$$y = f(x) = 3.7xe^{-0.01x}$$

If x is the stock in one generation and y is the stock recruited in the next generation, then this relationship is actually a population model of the form

$$x_{n+1} = 3.7x_n e^{-0.01x_n}$$

If the population is now at the level $x_1 = 100$ individuals, use technology to generate the number of individuals that you expect in the next ten generations. Deduce the equilibrium value and check this value by using technology or the bisection method to solve the equation $x = 3.7xe^{-0.01x}$. In Figure 2.47, the solid line is fit to the data using a Ricker functional form that is similar to the Beverton-Holt form considered in Example 8.

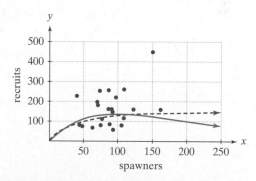

Figure 2.47 Sockeye salmon stock-recruitment.
Data Source: John A. Gulland, 1983. *Fish Stock Assessment: A Manual of Basic Methods*, Wiley.

2.6   Derivative at a Point

In this section, we introduce one of the major concepts in calculus, the idea of a *derivative*. Although functions are fundamental and limits are essential, they simply laid the foundation for the excitement to come. To motivate the idea of the derivative at a point, let us recast Example 1 of Section 2.1 using different notation. If we let $f(x)$ represent the population of Mexico in year x, then the average population between the years 1980 and 1985 can be found by

$$\frac{f(a+h)-f(a)}{h} = \frac{f(1980+5)-f(1980)}{5} = \frac{76.6-67.38}{5} = 1.844 \text{ million per year}$$

where $a = 1980$ (the base year) and $h = 5$ (the duration of the time interval). We also found the average rate of change over smaller and smaller intervals to guess that the instantaneous rate of change of the Mexico population in 1980 would be 1.75 million per year. This idea can be written as

$$\lim_{h \to 0} \frac{f(1980+h)-f(1980)}{h} \approx 1.75 \text{ million per year}$$

In Chapter 1 and Section 2.1, we previewed the notion of the derivative at a point by defining the tangent line for $f(x)$ at a point $x = a$ to be the limit of the slope of the secant lines

$$\lim_{h \to 0} \frac{f(a+h)-f(a)}{h}$$

Slopes of tangent lines and instantaneous rates of changes have the same formula, and it is this limiting process that is the basis for the concept of the derivative.

Derivative at a Point

The **derivative** of function f at a point $x = a$, denoted by $f'(a)$, is

$$f'(a) = \lim_{h \to 0} \frac{f(a+h)-f(a)}{h}$$

provided this limit exists. If the limit exists, we say that f is **differentiable at** $x = a$.

Example 1   Finding derivatives using the definition

Use the definition of a derivative to find the following derivatives.

a. $f'(3)$ where $f(x) = 1$
b. $f'(2)$ where $f(x) = 3x$
c. $f'(1)$ where $f(x) = 1 + 3x^2$

Solution
a. Let $f(x) = 1$ and $a = 3$.

$$f'(3) = \lim_{h \to 0} \frac{f(3+h)-f(3)}{h}$$
$$= \lim_{h \to 0} \frac{1-1}{h}$$
$$= \lim_{h \to 0} \frac{0}{h}$$
$$= 0$$

b. Let $f(x) = 3x$ and $a = 2$.

$$f'(2) = \lim_{h \to 0} \frac{f(2+h)-f(2)}{h}$$
$$= \lim_{h \to 0} \frac{3(2+h)-6}{h}$$
$$= \lim_{h \to 0} \frac{6+3h-6}{h}$$
$$= \lim_{h \to 0} \frac{3h}{h}$$
$$= 3$$

c. Let $f(x) = 1 + 3x^2$ and $a = 1$.

$$f'(1) = \lim_{h \to 0} \frac{f(1+h) - f(1)}{h}$$

$$= \lim_{h \to 0} \frac{[1 + 3(1+h)^2] - [1 + 3(1)^2]}{h}$$

$$= \lim_{h \to 0} \frac{1 + 3 + 6h + 3h^2 - 4}{h}$$

$$= \lim_{h \to 0} \frac{6h + 3h^2}{h}$$

$$= \lim_{h \to 0} (6 + 3h)$$

$$= 6$$

Example 1 illustrates two facts. First, the derivative of a constant function is zero. Intuitively this makes sense, as a constant function by definition does not change and, consequently, its rate of change should be zero. Second, the derivative of a linear function is the slope of the linear function. Intuitively, this makes sense as the rate at which the function is increasing is given by the slope of the function. More interestingly, Example 1 illustrates that we can explicitly compute the slope (equivalently, the instantaneous rate of change) of a quadratic function.

The derivatives in Example 1 were pretty straightforward to compute, but other derivatives require certain algebraic procedures to compute, as illustrated by the next example.

Example 2   Algebraic steps to find a derivative

Find the following derivatives algebraically.

a. $f'(4)$ where $f(x) = \sqrt{x}$ **b.** $f'(5)$ where $f(x) = \dfrac{1}{1+x}$

Solution

a. To find this derivative, we multiply the numerator and denominator of the quotient by the "conjugate" of the original numerator.

$$\frac{f(4+h) - f(4)}{h} = \frac{\sqrt{4+h} - \sqrt{4}}{h}$$

$$= \frac{\sqrt{4+h} - \sqrt{4}}{h} \times \frac{\sqrt{4+h} + \sqrt{4}}{\sqrt{4+h} + \sqrt{4}} \qquad \textit{multiplying by 1}$$

$$= \frac{4+h-4}{h(\sqrt{4+h} + \sqrt{4})} \qquad \textit{multiplying out the numerator}$$

$$= \frac{h}{h(\sqrt{4+h} + 2)} \qquad \textit{simplifying}$$

$$= \frac{1}{\sqrt{4+h} + 2} \qquad \textit{since } h \neq 0 \textit{ in the limit}$$

Hence, taking the limit as h goes to 0 yields $f'(4) = \dfrac{1}{\sqrt{4} + 2} = \dfrac{1}{4}$.

b. To find this derivative, we can multiply by the common denominator in the numerator.

$$\frac{f(5+h)-f(5)}{h} = \frac{1/(1+(5+h))-1/6}{h}$$

$$= \frac{1/(6+h)-1/6}{h} \quad simplifying$$

$$= \frac{1/(6+h)-1/6}{h} \times \frac{(6+h)6}{(6+h)6} \quad multiplying\ by\ 1$$

$$= \frac{6-(6+h)}{h(6+h)6} \quad simplifying$$

$$= \frac{-h}{h(6+h)6} \quad simplifying\ more$$

$$= \frac{-1}{(6+h)6} \quad since\ h \neq 0\ in\ the\ limit$$

Taking the limit as h goes to 0 yields $f'(5) = -\dfrac{1}{36}$.

Slopes of tangent lines

The definition of the derivative was inspired directly by the slope of the tangent line. Using derivatives, we can find the tangent line.

> **Tangent Line**
>
> Let f be a function that is differentiable at the point $x = a$. The **tangent line of f at $x = a$** is the line with slope $f'(a)$ that passes through the point $(a, f(a))$.

Example 3   Tangent line to a parabola

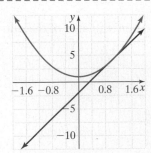

Figure 2.48 Graph of the parabola $y = 1 + 3x^2$ with tangent line at $(1, 4)$.

Find the tangent line to $f(x) = 1 + 3x^2$ at $x = 1$. Sketch the parabola and the tangent line.

Solution   In Example 1 we found that the slope of the tangent line is $f'(1) = 6$. Since the tangent line passes through $(1, f(1)) = (1, 4)$, we can use the point-slope formula to find the equation of the tangent line:

$$y - 4 = 6(x - 1)$$
$$y = 6x - 2$$

The graph of the parabola, along with the tangent line at $(1, 4)$ is shown in Figure 2.48.

In Example 3, the tangent line intersects the graph of the function in exactly one point. This unique intersection is not typical, as illustrated in the next example.

Example 4   Multiple intersections

Find the tangent line to the cubic $f(x) = x^3$ at $x = 1$. Sketch the cubic and the tangent line.

Solution We first find the slope of the tangent line:

$$f'(1) = \lim_{h \to 0} \frac{f(1+h) - f(1)}{h}$$

$$= \lim_{h \to 0} \frac{(1+h)^3 - 1}{h}$$

$$= \lim_{h \to 0} \frac{1 + 3h + 3h^2 + h^3 - 1}{h}$$

$$= \lim_{h \to 0} (3 + 3h + h^2)$$

$$= 3$$

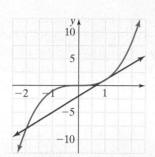

Figure 2.49 Graph of the cubic $y = x^3$ and its tangent line at $(1, 1)$.

The tangent line passes through $(1, f(1)) = (1, 1)$, so the equation of the tangent line is

$$y - 1 = 3(x - 1)$$

Equivalently, $y = 3x - 2$. The graph of $y = x^3$ and the tangent line at $(1, 1)$ is shown in Figure 2.49.

Instantaneous rates of change

In Section 2.1, we defined

$$\frac{f(b) - f(a)}{b - a}$$

to be the average rate of change of f over the interval $[a, b]$. Taking the limit as b approaches a yields the instantaneous rate of change

$$\lim_{b \to a} \frac{f(b) - f(a)}{b - a}$$

In the next example, we relate this definition of the instantaneous rate of change to the derivative.

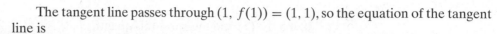

Example 5   Instantaneous rates of change

Show that

$$\lim_{b \to a} \frac{f(b) - f(a)}{b - a} = f'(a)$$

provided that the limits exist.

Solution   Let $h = b - a$. Then $b = a + h$ so that

$$\frac{f(b) - f(a)}{b - a} = \frac{f(a + h) - f(a)}{h}$$

Since b approaching a is equivalent to h approaching 0, we have

$$\lim_{b \to a} \frac{f(b) - f(a)}{b - a} = \lim_{h \to 0} \frac{f(a + h) - f(a)}{h}$$

provided the limits exist. By the definition of a derivative, we get

$$f'(a) = \lim_{b \to a} \frac{f(b) - f(a)}{b - a}$$

The solution to Example 5 allows us to equate the derivative with an instantaneous rate of change.

Instantaneous Rate of Change as a Derivative	Let f be a function that is differentiable at $x = a$. The **instantaneous rate of change of f at $x = a$** is $f'(a)$.

Example 6   Instantaneous velocity

On a calm day, a seed cone of a coastal California redwood tree drops from one of its high branches 305 feet above the ground. From physics, we know that the distance s in feet an object falls after t seconds, when air resistance is negligible, is given by the formula

$$s(t) = 16t^2$$

a. Find $s'(1)$ and interpret this quantity.

b. Find the velocity (instantaneous rate of change) of the cone at the moment it hits the ground.

Solution

a.

$$s'(1) = \lim_{h \to 0} \frac{s(1+h) - s(1)}{h}$$

$$= \lim_{h \to 0} \frac{16(1+h)^2 - 16(1)^2}{h}$$

$$= \lim_{h \to 0} \frac{16 + 32h + 16h^2 - 16}{h}$$

$$= \lim_{h \to 0} \frac{32h + 16h^2}{h}$$

$$= \lim_{h \to 0} (32 + 16h)$$

$$= 32$$

After one second, the cone is falling at a velocity of 32 feet per second (or 32 ft/s).

b. First, we need to find how long it takes the cone to fall to the ground. When the cone hits the ground, it has fallen 305 feet. Hence, we need to solve $305 = s(t) = 16t^2$, which yields $t = \dfrac{\sqrt{305}}{4}$. To find $s'\left(\dfrac{\sqrt{305}}{4}\right)$, we use the definition of a derivative

$$s'\left(\frac{\sqrt{305}}{4}\right) = \lim_{h \to 0} \frac{s\left(\frac{\sqrt{305}}{4} + h\right) - s\left(\frac{\sqrt{305}}{4}\right)}{h}$$

$$= \lim_{h \to 0} \frac{16\left(\frac{\sqrt{305}}{4} + h\right)^2 - 16\left(\frac{\sqrt{305}}{4}\right)^2}{h}$$

$$= \lim_{h \to 0} \frac{8\sqrt{305}\, h + 16h^2}{h}$$

$$= \lim_{h \to 0} (8\sqrt{305} + 16h)$$

$$= 8\sqrt{305}$$

$$\approx 139.7$$

Hence, at the moment the cone hits the ground it is falling at a velocity of 139.7 ft/s. This is equivalent to 95.3 miles per hour (mi/h). Of course, if the effects of air resistance are taken into account, velocity will be less.

Example 7   Melting Arctic sea ice

One of the important consequences of increasing temperature on Earth is that sea ice is melting at both the North and South poles. In 2012, the extent of Arctic sea ice was approximately 3.61 million square kilometers, the lowest in the past thirty years. Figure 2.50 contains the plot of the average Arctic sea ice extent for the past thirty years. These data can be approximated by a quadratic function of the form

$$S(t) = 7.292 + 0.023t - 0.004t^2 \text{ million square kilometers}$$

where t is years after 1980.

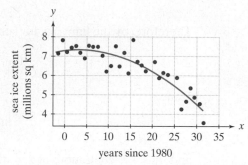

Figure 2.50   Sea ice extent as a function of years since 1980.

a. Find $S'(32)$.

b. Determine the units of this derivative and discuss their meaning.

Solution

a. Using the definition of a derivative, we get

$$S'(32) = \lim_{h \to 0} \frac{S(32 + h) - S(32)}{h}$$

$$= \lim_{h \to 0} \frac{[7.292 + 0.023(32 + h) - 0.004(32 + h)^2] - [7.292 + 0.023 \cdot 32 - 0.004 \cdot 32^2]}{h}$$

$$= \lim_{h \to 0} \frac{0.023h - 0.004(64h + h^2)}{h}$$

$$= \lim_{h \to 0} \frac{-0.233h - 0.004h^2}{h}$$

$$= \lim_{h \to 0} -0.233 - 0.004h = -0.233$$

b. The units of $S'(32)$ are millions of square kilometers per year. Since $S'(32) = -0.233$, we conclude that we are losing 233 thousands of square kilometers of Arctic sea ice per year.

Differentiability and continuity

If a function $f(x)$ is differentiable at a point a, then it is also continuous at a point a, as stated in the following theorem.

Theorem 2.6   Differentiability implies continuity theorem

If f is differentiable at the point $x = a$, then f is continuous at $x = a$.

Proof. To prove this theorem, assume that f is differentiable at $x = a$. Then

$$\lim_{x \to a}[f(x) - f(a)] = \lim_{h \to 0}[f(a + h) - f(a)]$$

$$= \lim_{h \to 0} \frac{h}{h}[f(a + h) - f(a)] \quad \textit{multiplying by one}$$

$$= \lim_{h \to 0} h\left[\frac{f(a + h) - f(a)}{h}\right]$$

$$= \lim_{h \to 0} h \cdot \lim_{h \to 0} \frac{f(a + h) - f(a)}{h} \quad \textit{limit law for product}$$

$$= 0 \cdot f'(a) \quad \textit{definition of derivative}$$

$$= 0$$

Therefore, by the limit law for sums,

$$\lim_{x \to a} f(x) - \lim_{x \to a} f(a) = 0$$

Hence,

$$\lim_{x \to a} f(x) = \lim_{x \to a} f(a)$$

and thus f is continuous at $x = a$. ∎

The reverse of this theorem, however, is not true. To fully appreciate differentiability, it is useful to understand examples of where continuity holds but the function is not differentiable.

Example 8　Absolute value functions have nondifferentiable corners

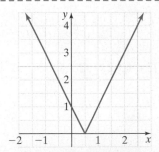

Figure 2.51 Graph of the function $f(x) = |ax - b|$ for the case $a = 2$ and $b = 1$.

The absolute value function $f(x) = |ax - b|$, for positive values of a and b, is defined to be

$$f(x) = \begin{cases} ax - b & \text{if } x \geq b/a \\ b - ax & \text{if } x < b/a \end{cases}$$

Show that $f(x)$, which is illustrated in Figure 2.51 for the case $a = 2$ and $b = 1$, is continuous but not differentiable at $x = b/a$.

Solution Since $\lim\limits_{x \to b/a^+}(ax - b) = \lim\limits_{x \to b/a^-}(b - ax) = 0$, $f(x)$ is continuous at $x = b/a$.

Now consider the limit from the left in the definition of a derivative applied to the given function:

$$\lim_{h \to 0^-} \frac{f(b/a + h) - f(b/a)}{h} = \lim_{h \to 0^-} \frac{b - a(b/a + h) - 0}{h} = \lim_{h \to 0^-} \frac{-ah}{h} = -a$$

Similarly consider the limit from the right in the definition of a derivative applied to the given function:

$$\lim_{h \to 0^+} \frac{f(b/a + h) - f(b/a)}{h} = \lim_{h \to 0^+} \frac{a(b/a + h) - b - 0}{h} = \lim_{h \to 0^+} \frac{ah}{h} = a$$

Since these two limits are not equal, the function is not differentiable at $x = a$. ∎

A biological model where we have continuity but nondifferentiability at a point is illustrated in the following example.

Example 9   Continuous but not differentiable

Examine the continuity and differentiability of $f(x)$ at $x = a$ for the following two functions.

a.
$$f(x) = \begin{cases} 6.25\,x \text{ cells/hour} & \text{if } x \le 200 \\ 1{,}250 \text{ cells/hour} & \text{if } x > 200, \end{cases} \quad a = 200$$

where we recall from Example 7 in Section 2.2 that this function models the feeding rate of planktonic copepods and x is the concentration of planktonic cells per liter.

b. $f(x) = x^{1/3}$ and $a = 0$.

Solution

a. Since

$$\lim_{x \to 200^+} f(x) = \lim_{x \to 200^+} 1250 = 1250$$

and

$$\lim_{x \to 200^-} f(x) = \lim_{x \to 200^-} 6.25x = 1250$$

we see f is continuous at $x = 200$.
On the other hand, for $h < 0$,

$$\lim_{h \to 0^-} \frac{f(200 + h) - f(200)}{h} = \lim_{h \to 0^-} \frac{6.25(200 + h) - 6.25(200)}{h}$$

$$= \lim_{h \to 0^-} \frac{6.25\,h}{h}$$

$$= 6.25$$

and for $h > 0$,

$$\lim_{h \to 0^+} \frac{f(200 + h) - f(200)}{h} = \lim_{h \to 0^+} \frac{1250 - 1250}{h}$$

$$= \lim_{h \to 0^+} \frac{0}{h}$$

$$= 0$$

Since the left- and right-hand limits are not equal, the limit does not exist, so f is not differentiable at $x = 200$. As you can see in Figure 2.52, the function is continuous but is still not differentiable at $x = 200$.

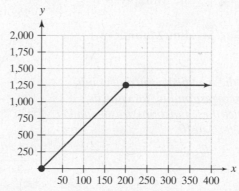

Figure 2.52 Feeding rate of planktonic copepods.

b. Since $f(x) = x^{1/3}$ is arbitrarily close to 0 for all x sufficiently close to 0, $\lim\limits_{x \to 0} f(x) = 0$. Since $f(0) = 0$, f is continuous at $x = 0$. To determine the derivative of $x^{1/3}$ at $x = 0$, we need to consider

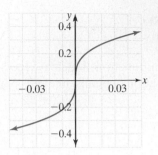

$$\lim_{h \to 0} \frac{f(0 + h) - f(0)}{h} = \frac{(0 + h)^{1/3} - 0^{1/3}}{h}$$

$$= \lim_{h \to 0} h^{-2/3}$$

$$= \infty$$

Hence, the derivative is not defined, as the limit is not finite. Graphing $y = x^{1/3}$ reveals that the slope of the tangent line at $x = 0$ is infinite; that is, it is vertical, as illustrated on the left.

Example 9 illustrates that continuity does not ensure differentiability and that differentiability can fail in at least two ways. The limit of the slopes of the secant lines might not converge or this limit may become infinitely large. While continuity does not imply differentiability, Theorem 2.6 ensures that the opposite is true; that is, differentiability ensures continuity. Hence, differentiability can be viewed as an improvement over continuity.

PROBLEM SET 2.6

Level 1 DRILL PROBLEMS

Using the definition of a derivative, find the derivatives specified in Problems 1 to 10.

1. $f'(-2)$ where $f(x) = 3x - 2$
2. $f'(3)$ where $f(x) = 5 - 2x$
3. $f'(1)$ where $f(x) = -x^2$
4. $f'(0)$ where $f(x) = x + x^2$
5. $f'(-4)$ where $f(x) = \dfrac{1}{2x}$
6. $f'(2)$ where $f(x) = \dfrac{1}{x + 1}$
7. $f'(-1)$ where $f(x) = x^3$
8. $f'(2)$ where $f(x) = x^3 + 1$
9. $f'(9)$ where $f(x) = \sqrt{x}$. Hint: Multiply by 1 (think conjugate).
10. $f'(5)$ where $f(x) = \sqrt{5x}$. Hint: Multiply by 1 (think conjugate).

Find the tangent line at the specified point and graph the tangent line and the corresponding function in Problems 11 to 20. Notice these functions are the same as those given in Problems 1 to 10.

11. $f(x) = 3x - 2$ at $x = -2$
12. $f(x) = 5 - 2x$ at $x = 3$
13. $f(x) = -x^2$ at $x = 1$
14. $f(x) = x + x^2$ at $x = 0$
15. $f(x) = \dfrac{1}{2x}$ at $x = -4$
16. $f(x) = \dfrac{1}{x + 1}$ at $x = 2$
17. $f(x) = x^3$ at $x = -1$
18. $f(x) = x^3 + 1$ at $x = 2$
19. $f(x) = \sqrt{x}$ at $x = 9$
20. $f(x) = \sqrt{5x}$ at $x = 5$

Determine at which values of x in Problems 21 to 26 that f is not differentiable. Explain briefly.

21.

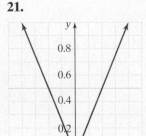

22.

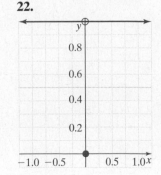

23.

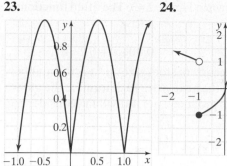

24.

25. $f(x) = |x - 2|$

26. $f(x) = 2|x + 1|$

27. Let $f(x) = \begin{cases} -2x & \text{if } x < 1 \\ \sqrt{x} - 3 & \text{if } x \geq 1 \end{cases}$

    a. Sketch the graph of f.

    b. Show that f is continuous, but not differentiable, at $x = 1$.

28. Give an example of a function that is continuous on $(-\infty, \infty)$ but is not differentiable at $x = 5$.

29. Consider the function defined by

$$f(x) = \begin{cases} \dfrac{1}{-x + 1} & \text{if } x \leq 0 \\ \dfrac{1}{x + 1} & \text{if } x > 0 \end{cases}$$

Sketch this graph and find all points where the graph is continuous but not differentiable.

30. Consider the function defined by

$$f(x) = \begin{cases} 0 & \text{if } x \leq 0 \\ x & \text{if } 0 < x < 1 \\ \dfrac{1}{x} & \text{if } x \geq 1 \end{cases}$$

Sketch this graph and find all points where the graph is continuous but not differentiable.

Level 2 APPLIED AND THEORY PROBLEMS

31. A baseball is thrown upward and its height at time t in seconds is given by

$$H(t) = 64t - 16t^2 \text{ feet}$$

    a. Find the velocity of the baseball after two seconds.

    b. Find the time at which the baseball hits the ground.

    c. Find the velocity of the baseball when it hits the ground.

32. A ball is thrown directly upward from the edge of a cliff and travels in such a way that t seconds later, its height above the ground at the base of the cliff is

$$H(t) = -16t^2 + 40t + 24$$

feet.

    a. Find the velocity of the ball after two seconds.

    b. When does the ball hit the ground, and what is its impact velocity?

    c. When does the ball have a velocity of zero? What physical interpretation should be given to this time?

33. If the data in Figure 2.53 represents a set of measurements relating enzyme activity to temperature in degrees Celsius, and the quadratic equation

$$A(x) = 11.8 + 19.1\,x - 0.2\,x^2$$

provides a good fit to this data, then find $A'(50)$ and discuss its meaning.

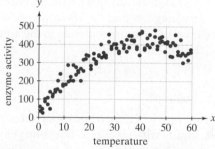

Figure 2.53 Enzyme activity as a function of temperature.

34. In Example 7 we discussed how the extent of Arctic sea ice is modeled by the function

$$S(t) = 7.292 + 0.023t - 0.004t^2 \text{ million square kilometers}$$

where t is years after 1980. Use the definition of the derivative to compute $S'(20)$. How does this value compare to $S'(30)$ found in Example 7? What does this comparison suggest about the rate at which arctic ice is being lost?

35. In 2010, W. B. Grant published an article about the prevalence of multiple sclerosis in three U.S. communities and the role of vitamin D. The best-fitting quadratic relationship to the data published in this article, relating prevalence of multiple sclerosis (MS) to latitude (exposure to sunlight, hence vitamin D synthesis decreases with latitude) is shown in Figure 2.54. This quadratic function is given by

$$P(x) = 499 - 30.8x + 0.5x^2 \text{ cases per 100,000}$$

where x is latitude.

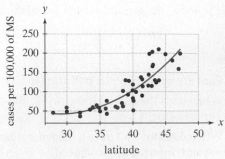

Figure 2.54 Prevalence of MS (in cases per 100,000) as a function of latitude in the United States.

Data Source: W. B. Grant, "The Prevalence of Multiple Sclerosis in 3 US Communities: The Role of Vitamin D" [letter]. *Prevention of Chronic Diseases* 7 (2010): A89.

a. Find $P'(40)$.

b. Find $P'(45)$.

c. Interpret and compare the numbers that you found in **a** and **b**.

36. An environmental study of a certain suburban community suggests that t years from now, the average level of carbon monoxide in the air can be modeled by the formula

$$f(t) = 0.05t^2 + 0.1t + 3.4$$

parts per million.

a. At what rate will the carbon monoxide level be changing with respect to time one year from now?

b. By how much will the carbon monoxide level change during the first year?

37. Perelson and colleagues studied the viral load of HIV patients during antiviral drug treatment.* They estimated the viral load of the typical patient to be

$$V(t) = 216e^{-0.2t}$$

particles per milliliter on day t after the drug treatment.

a. Estimate $V'(2)$.

b. Describe the units of $V'(2)$ and interpret this quantity.

38. Stock-recruitment data and a fitted Beverton-Holt function for sockeye salmon in Karluk Lake, Alaska,

were was shown in Figure 2.43. The fitted function was

$$y = f(x) = \frac{x}{0.006\,x + 0.2}$$

where x is the current stock size and y is the number of recruits for the next year. To determine the number of recruits produced per individual, consider the function

$$y = g(x) = \frac{f(x)}{x} = \frac{1}{0.006\,x + 0.2}$$

a. Algebraically find $g'(10)$.

b. Describe the units of $g'(10)$, and discuss the meaning of this quantity.

39. In Example 6 of Section 1.5, we developed the Michaelis-Menton model for the rate at which an organism consumes its resource. For bacterial populations in the ocean, this model was given by

$$f(x) = \frac{1.2078x}{1 + 0.0506x} \text{ micrograms of glucose per hour}$$

where x is the concentration of glucose (micrograms per liter) in the environment. To determine the rate of glucose consumption per microgram of glucose in the environment, consider the function

$$y = g(x) = \frac{f(x)}{x} = \frac{1.2078}{1 + 0.0506x}$$

a. Algebraically compute $g'(0)$ and $g'(20)$.

b. Describe the meaning of the derivatives that you computed.

40. In Example 5 of Section 2.4, we found that the rate at which wolves kill moose can be modeled by

$$f(x) = \frac{3.36x}{0.42 + x} \quad \begin{array}{l}\text{moose killed per wolf}\\ \text{per hundred days}\end{array}$$

where x is measured in number of moose per square kilometer. To determine the per capita killing rate of moose, consider the function

$$y = g(x) = \frac{f(x)}{x} = \frac{3.36}{0.42 + x}$$

a. Algebraically compute $g'(1)$ and $g'(2)$.

b. Describe the meaning of the derivatives that you computed.

* A. S. Perelson, A. U. Neumann, M. Markowitz, J. M. Leonard, D. D. Ho, "HIV-1 Dynamics in Vivo: Virion Clearance Rate, Infected Cell Lifespan, and Viral Generation Time," *Science* 271 (1996): 1582–1586; and A. S. Perelson and P. W. Nelson "Mathematical Analysis of HIV-1 Dynamics in Vivo," *SIAM Review* 41 (1999): 3–44.

2.7   Derivatives as Functions

Our notion $f'(a)$ for the derivative at the point $x = a$ suggests that we can think of f' as a function. Indeed this is true.

Derivative as a Function

Let f be a function. The **derivative of** f is defined by

$$f'(x) = \lim_{h \to 0} \frac{f(x+h) - f(x)}{h}$$

for all x for which this limit exists.

Example 1 Finding Derivatives

Find the derivatives f' of the following functions f.

a. $f(x) = 1$ **b.** $f(x) = x$ **c.** $f(x) = x^2$ **d.** $f(x) = x^3$

e. Guess the derivative of $f(x) = x^n$ for n a whole number.

Solution

a. If $f(x) = 1$, then $f'(x) = 0$ for every x (see Example 1 of Section 2.6). The derivative of a constant is 0.

b. Use the definition of the derivative of a function.

$$\begin{aligned} f'(x) &= \lim_{h \to 0} \frac{f(x+h) - f(x)}{h} \\ &= \lim_{h \to 0} \frac{x + h - x}{h} \\ &= \lim_{h \to 0} \frac{h}{h} \\ &= 1 \end{aligned}$$

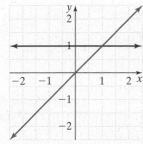

Figure 2.55 The function $f(x) = x$ (blue) and its derivative $f'(x) = 1$ (red) are plotted on the interval $[-2, 2]$.

The function and its derivative are illustrated in Figure 2.55.

c. For $f(x) = x^2$ and a fixed value of x, the definition of a derivative implies

$$\begin{aligned} f'(x) &= \lim_{h \to 0} \frac{f(x+h) - f(x)}{h} \\ &= \lim_{h \to 0} \frac{(x+h)^2 - x^2}{h} \\ &= \lim_{h \to 0} \frac{x^2 + 2hx + h^2 - x^2}{h} \\ &= \lim_{h \to 0} \frac{2hx + h^2}{h} \\ &= \lim_{h \to 0} (2x + h) \\ &= 2x \end{aligned}$$

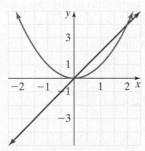

Figure 2.56 The function $f(x) = x^2$ (blue) and its derivative $f'(x) = 2x$ (red).

The function x^2 and its derivative $2x$ are illustrated in Figure 2.56.

d. Again, we use the definition. For a fixed number x

$$\begin{aligned} f'(x) &= \lim_{h \to 0} \frac{f(x+h) - f(x)}{h} \\ &= \lim_{h \to 0} \frac{(x+h)^3 - x^3}{h} \\ &= \lim_{h \to 0} \frac{x^3 + 3hx^2 + 3h^2 x + h^3 - x^3}{h} \\ &= \lim_{h \to 0} \frac{3hx^2 + 3h^2 x + h^3}{h} \\ &= \lim_{h \to 0} (3x^2 + 3hx + h^2) \\ &= 3x^2 \end{aligned}$$

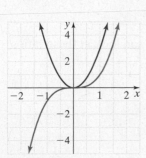

Figure 2.57 The function $f(x) = x^3$ (blue) and its derivative $f'(x) = 3x^2$ (red).

The function x^3 and its derivative $3x^2$ are illustrated in Figure 2.57.

e. The above parts suggest that $f'(x) = nx^{n-1}$ for n a whole number. Indeed, this turns out to be true, as we will show in Chapter 3. ∎

When given numerical data, we can estimate the derivative of the data using the definition of the derivative with the smallest possible h value.

Example 2   Estimating the derivative using a table

In Example 1 of Section 2.1, we considered the population size of Mexico (in millions) in the early 1980s as reported in the following table:

Year	Population (millions)
1980	67.38
1981	69.13
1982	70.93
1983	72.77
1984	74.66
1985	76.60

Let $P(t)$ denote the population size t years after 1980. That is, $t = 0$ corresponds to 1980. For example, we found the average population growth rate in 1980 to be about 1.75 million/year, so we would write this as $P'(0) \approx 1.75$.

a. Estimate $P'(t)$ at $t = 0, 1, 2, 3, 4$ using $h = 1$ in the definition of a derivative.

b. In Problem 28 of Section 1.4, you may have found the population size could be represented by the exponential function $P(t) = 67.37(1.026)^t$. Approximate the derivative with an exponential function and compare it with $P(t)$. What do you notice?

Solution

a. The estimates of $P'(t)$ are calculated from the data as indicated in the third column in Table 2.6.

Table 2.6 Estimates of the rate of population growth in Mexico

Year	t	Estimates of $P'(t)$ from data	$0.026\,P(t)$
1980	0	$\dfrac{P(1) - P(0)}{1} = 69.13 - 67.38 = 1.75$	1.75
1981	1	$\dfrac{P(2) - P(1)}{1} = 70.93 - 69.13 = 1.80$	1.80
1982	2	$\dfrac{P(3) - P(2)}{1} = 72.77 - 70.93 = 1.84$	1.84
1983	3	$\dfrac{P(4) - P(3)}{1} = 74.66 - 72.77 = 1.89$	1.89
1984	4	$\dfrac{P(5) - P(4)}{1} = 76.60 - 74.66 = 1.96$	1.94
1985	5	calculation not possible	–

b. To approximate $P'(t)$ by an exponential function, we can look at these ratios:

$$\frac{P'(1)}{P'(0)} = \frac{1.80}{1.75} \approx 1.029$$

$$\frac{P'(2)}{P'(1)} = \frac{1.84}{1.80} \approx 1.022$$

$$\frac{P'(3)}{P'(2)} = \frac{1.89}{1.84} \approx 1.027$$

$$\frac{P'(4)}{P'(3)} = \frac{1.94}{1.89} \approx 1.026$$

These ratios are all about the same. In fact, the average is 1.026, which is the ratio for the population function itself! We approximate $P'(t)$ by

$$P'(t) \approx 1.75(1.026)^t$$

Comparing this function with the function for the population growth, we see that

$$\frac{P'(t)}{P(t)} \approx \frac{1.75(1.026)^t}{67.37(1.026)^t} \approx 0.026$$

Thus

$$P'(t) \approx 0.026\,P(t)$$

If we use this formula to calculate the derivative at times $t = 0, 1, \ldots, 5$ we see in column 4 of Table 2.6 that values obtained are very close to the estimates obtained directly from the data, with only the $t = 4$ differing by an amount of 0.02. The advantage of having the formula is that we can calculate the derivative at any t value. ■

The solution to part **b** of Example 2 suggests that whenever $P(t)$ has the general exponential form $P(t) = ab^t$, then the derivative has the same form but differing by some constant; that is, $P'(t) = cP(t)$, where for Example 3 we obtained $c = 0.026$. This equation is an example of what is known as a *differential equation*, as it relates a function to its derivative. You will learn more about differential equations in Chapters 6 and 8. In Chapter 3, we will verify that the derivative of an exponential function is a constant multiple of the exponential function.

Notational alternatives

The primed-function notation f' that we have been using to denote derivatives is but one of several used in various texts. Newton used a "dot" notation, which we will not consider here. The notation that mathematicians prefer was developed by Leibniz. This notation is inspired by the following presentation of the derivative.

Let Δx represent a small change in x. The change of $y = f(x)$ over the interval $[x, x + \Delta x]$ is given by

$$\Delta y = f(x + \Delta x) - f(x)$$

The average rate of change of $y = f(x)$ over the interval $[x, x + \Delta x]$ is given by

$$\frac{\Delta y}{\Delta x}$$

Hence, the derivative of f at x is

$$\lim_{\Delta x \to 0} \frac{\Delta y}{\Delta x}$$

Leibniz represented this limit as

$$\frac{dy}{dx} = \lim_{\Delta x \to 0} \frac{\Delta y}{\Delta x}$$

where in some sense dy corresponds to an "infinitesimal" change in y and dx represents an "infinitesimal" change in x. Commonly used variations in notation that you will find in this and other calculus texts include

$$f'(x) = \frac{dy}{dx} = \frac{df}{dx} = \frac{d}{dx}f(x)$$

To indicate the derivative at the point $x = a$ using Leibniz notation, we use the cumbersome expression

$$\frac{dy}{dx}\bigg|_{x=a} = f'(a)$$

Example 3   Using alternative derivative notations

Find the following derivatives.

a. $\left.\dfrac{dy}{dx}\right|_{x=-1}$ where $y = x^3$

b. $\dfrac{df}{dx}$ where $f(x) = x^5$

Solution

a. In Example 1, we found that the derivative of x^3 is $3x^2$. Since $\left.\dfrac{dy}{dx}\right|_{x=-1}$ is the derivative of $y = x^3$ evaluated at $x = -1$, we have

$$\left.\frac{dy}{dx}\right|_{x=-1} = 3x^2\Big|_{x=-1} = 3$$

b. In Example 1, we guessed that the derivative of x^n is nx^{n-1}; in which case, for $n = 5$ we have

$$\frac{df}{dx} = \frac{d}{dx}(x^5) = 5x^4$$

The next example draws upon research undertaken by ecologist Nathan Sanders and colleagues, who assessed the number of local ant species along an elevational gradient, Kyle Canyon, in the Spring Mountains of Nevada to obtain a measure called *species richness*. These data, illustrated in Figure 2.58, are plotted in terms of number of species of ants as a function of elevation (in kilometers).

Example 4   Ant biodiversity

Ecologists, noting that the number of ant species declines at both low (close to sea level) and high (at the tops of mountains) levels, fitted a parabola to the data plotted in Figure 2.58. The fit they obtained is given by

$$S = -10.3 + 24.9\,x - 7.7\,x^2 \text{ species}$$

where x is elevation measures in kilometers.

a. Find $\dfrac{dS}{dx}$.

b. Identify the units of $\dfrac{dS}{dx}$ and interpret $\dfrac{dS}{dx}$.

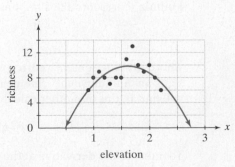

Figure 2.58 Number of species of ants as a function of elevation.

Data Source: N. Sanders, J. Moss, and D. Wagner, "Patterns of Ant Species Richness along Elevational Gradients in an Arid Ecosystem," *Global Ecology and Biogeography*, 12 (2003): 93–102.

Solution

a.
$$\frac{dS}{dx} = \lim_{h \to 0} \frac{S(x+h) - S(x)}{h}$$

$$= \lim_{h \to 0} \frac{[-10.3 + 24.9(x+h) - 7.7(x+h)^2] - [-10.3 + 24.9x - 7.7x^2]}{h}$$

$$= \lim_{h \to 0} \frac{24.9\,h - 15.4\,xh - 7.7h^2}{h}$$

$$= \lim_{h \to 0} (24.9 - 15.4\,x - 7.7\,h)$$

$$= 24.9 - 15.4x.$$

b. The units of $\dfrac{dS}{dx}$ are species per kilometer elevation. $\dfrac{dS}{dx}$ represents the rate of change of species richness with respect to elevation. For elevations less than $24.9/15.4 \approx 1.6$ kilometers, $\dfrac{dS}{dx} > 0$. Consequently, for elevations of less than 1.6 kilometers, an ant-loving entomologist would encounter more species of ants by hiking higher up. However, for elevations greater than 1.6 kilometers, an ant-loving entomologist would encounter more species of ants by walking downward.

Mean Value Theorem

To understand what the derivative tells us about the shape of a function, we need the mean value theorem (MVT). The proof of this theorem is given as a series of challenging exercises in Problem set 2.7.

Theorem 2.7   Mean value theorem

- -

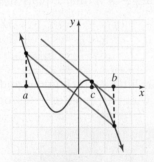

Figure 2.59 Mean value theorem in action. The slope of the tangent line at $x = c$ equals the slope of the secant line from $x = a$ to $x = b$.

Let f be a function that is continuous on the closed interval $[a, b]$ and differentiable on the open interval (a, b). Then there exists c in (a, b) such that

$$f'(c) = \frac{f(b) - f(a)}{b - a}$$

Notice that the right-hand side of this equation is the average rate of change of f over the interval $[a, b]$. Hence, the MVT states that for a differentiable function on an interval $[a, b]$, there is a point in the interval where the instantaneous rate of change equals the average rate of change. Alternatively, we can think of the MVT in geometric terms. Recall that the right-hand side of the MVT equation is the slope of the secant line passing through the points $(a, f(a))$ and $(b, f(b))$. Hence, the MVT asserts that there is a point in the interval such that the slope of the tangent line at this point equals the slope of the secant line. A graphical representation of this interpretation is given in Figure 2.59.

Example 5   Mean value theorem in action

- -

Determine whether the MVT applies for the following functions f on the specified intervals $[a, b]$. If the MVT applies, then find c in (a, b) such that the statement of the MVT holds.

a. $f(x) = x^2$ on the interval $[0, 2]$

b. $f(x) = |x|$ on the interval $[-1, 1]$

Solution

a. Recall that $f'(x) = 2x$ for all x. Hence, f is differentiable on the interval $[0, 2]$. Consequently, the MVT applies and we should be able to find the desired value c. The average rate of change of f on $[0, 2]$ is given by

$$\frac{f(2) - f(0)}{2 - 0} = \frac{2^2 - 0}{2} = 2$$

Solving $f'(x) = 2x = 2$ yields $x = 1$. Hence, the instantaneous rate of change at $x = 1$ equals the average rate of change over the interval $[0, 2]$. The plot on the left with $y = x^2$ in red, the tangent line in blue, and the black line connecting $(0, f(0))$ to $(2, f(2))$ illustrates our calculations.

b. We need to find the derivative of $f(x) = |x|$. Since $f(x) = x$ for $x > 0$, we get

$$\frac{f(x + h) - f(x)}{h} = \frac{x + h - x}{h}$$

$$= \frac{h}{h} = 1$$

whenever every h is sufficiently small but not equal to zero. Hence, $f'(x) = 1$ for $x > 0$. On the other hand, since $f(x) = -x$ for $x < 0$, we get

$$\frac{f(x + h) - f(x)}{h} = \frac{-x - h - (-x)}{h}$$

$$= \frac{-h}{h} = -1$$

whenever h is sufficiently small but not equal to zero. Hence, $f'(x) = -1$ for $x < 0$.

What happens at the point $x = 0$? Our calculations imply that $\lim\limits_{h \to 0^+} \frac{f(h) - f(0)}{h} = 1$ but $\lim\limits_{h \to 0^-} \frac{f(h) - f(0)}{h} = -1$. Since the one-sided limits do not agree, f is not differentiable at $x = 0$ and the MVT need not apply. In fact, since the average rate of change over the interval $[-1, 1]$ equals $\frac{|-1| - |1|}{1 - (-1)} = 0$, there is no instantaneous rate of change of f that equals the average rate of change. ∎

Example 6   Foraging for food

- -

Hummingbirds (Figure 2.60) are small birds that weigh as little as three grams and have an energetically demanding lifestyle. With their wings beating at rates of

Phil Seu Photography/Flickr/Getty Images, Inc.

Figure 2.60 Ruby-throated hummingbird (*Archilochus colubris*).

eighty to a hundred beats per second, the hummingbird can lose 10% to 20% of its body weight in one to two hours. To survive, hummingbirds require relatively large amounts of nectar from flowers. Therefore, they spend much of the day flying between patches of flowers extracting nectar. As a hummingbird extracts nectar in a patch, its energetic gains $E(t)$ in calories increase with time t (in seconds). Figure 2.61 shows a hypothetical graph of energetic gains $E(t)$, in calories, in one patch.

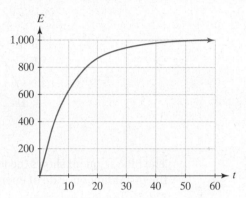

Figure 2.61 Energy gains over time.

a. Approximate the average rate of energy intake over the interval $[0, 60]$.

b. Use the geometric interpretation of the mean value theorem to estimate the time when the instantaneous rate of energy intake equals the average rate of energy intake.

Solution

a. Since $E(0) = 0$ and $E(60) \approx 1000$, we obtain

$$\text{AVERAGE RATE OF ENERGY INTAKE} \approx \frac{1000}{60} \approx 16.7$$

calories per second.

b. Graphing the line connecting the points $(0, 0)$ and $(60, 1000)$ yields the secant line whose slope equals the average rate of change:

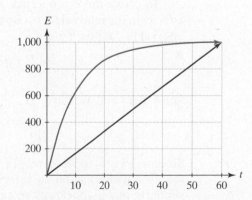

The slope of this line is approximately 16.7. To estimate the time at which $E'(t) = \dfrac{1,000}{60}$, we can place a straightedge on top of the red line segment and slowly slide it upward keeping it parallel to the red segment. If we slide it upward until the straightedge is tangent to the curve $y = E(t)$, then we obtain the following graph:

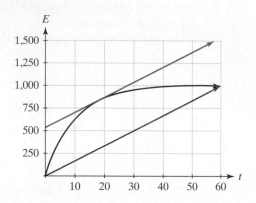

The blue segment shows the location of the tangent line at $t \approx 20$ seconds. Hence, the instantaneous rate of energy intake rate equals the average energy intake rate at $t \approx 20$.

It is worth noting from Figure 2.61 that $E'(t)$ is above the average rate of energy intake for $t < 20$ and below the average for $t > 20$. Hence, as we explore in more detail in Chapter 4, the hummingbird may consider leaving the patch after 20 seconds.

Derivatives and graphs

Using the mean value theorem, we can prove the following two facts about the relationship of the sign of the derivative f' to the graph of $y = f(x)$.

Increasing-Decreasing

Let f be a function that is differentiable on the interval (a, b). If $f'(x) > 0$ for all x in (a, b), then f is increasing on (a, b). If $f'(x) < 0$ for all x in (a, b), then f is decreasing on (a, b).

To prove these properties, assume that $f' > 0$ on (a, b). Take any two points $x_2 > x_1$ in the interval (a, b). By the mean value theorem, there exists a point c in the interval $[x_1, x_2]$ such that

$$f'(c) = \frac{f(x_2) - f(x_1)}{x_2 - x_1}$$

Since $f'(c) > 0$, we have

$$\frac{f(x_2) - f(x_1)}{x_2 - x_1} > 0$$

Since $x_2 - x_1 > 0$, we have $f(x_2) - f(x_1) > 0$. Equivalently, $f(x_2) > f(x_1)$. Therefore, f is increasing on the interval $[a, b]$. The case of $f' < 0$ on $[a, b]$ can be proved similarly, and it appears as an exercise in Problem Set 2.7.

Example 7   Identifying signs of f'
- -
Let the graph of $y = f(x)$ be given by Figure 2.62. For the interval $[-3, 2]$, determine where the derivative of f is positive and where the derivative of f is negative.

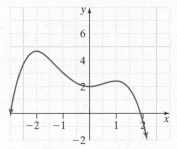

Figure 2.62 Graph of $y = f(x)$.

Solution   Since the graph is increasing on the intervals $(-3, -2)$ and $(0, 1)$, $f' > 0$ on these intervals. Since the graph is decreasing on the intervals $(-2, 0)$ and $(1, 3)$, $f' < 0$ on these intervals.

For a function $y = f(x)$, a **turning point** is an x value where the function switches from increasing to decreasing, or vice versa. More precisely, if f is continuous on (a, b), then c in (a, b) is a turning point provided that either (i) f is increasing on (a, c) and decreasing on (c, b), or (ii) f is decreasing on (a, c) and increasing on (c, b). When f is differentiable on (a, b), turning points correspond to where the derivative f' changes sign. For example, if $f'(x) > 0$ on (a, c) and $f'(x) < 0$ on (c, b), then c is a turning point as the function switches from increasing to decreasing at $x = c$.

Example 8   Mix and match

Match the graphs of $y = f(x)$

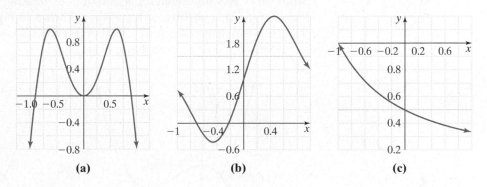

with the graph of their derivatives $y = f'(x)$

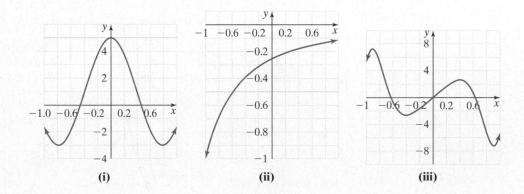

Solution

a. Looking at graph (a), we see three turning points at approximately -0.6, 0, and 0.6. Turning points corresponds to x values where the derivative of function equals zero. Since only graph (iii) of the graphs (i)–(iii) intersects the x axis in three points, the derivative graph for (a) must be (iii).

b. The turning points for graph (b) are at approximately -0.4 and 0.4; the graph of (i) shows the derivative to be 0 at those points. Hence, graph (i) must be the derivative of the graph of (b).

c. There are no turning points on graph (c), so the derivative graph should not cross the x axis. Therefore, the derivative graph is (ii).

Example 9   Reconstructing f from f'

Let the graph of $y = f'(x)$ be given by the graph in Figure 2.63. Sketch a possible graph for $y = f(x)$.

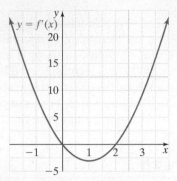

Figure 2.63 Graph of a derivative.

Solution   We can sketch the graph of f by looking at the intervals for which the graph of $f'(x)$ is positive or negative, as shown in Figure 2.64. On intervals where $f'(x)$ is positive, we sketch a curve that is increasing, and on intervals where $f'(x)$ is negative, we sketch a curve that is decreasing.

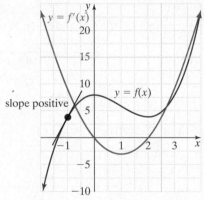

a. On $(-\infty, 0)$ the graph of the derivative is positive, so the graph of f is rising (slope positive).

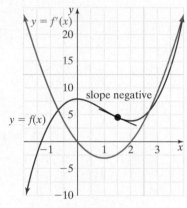

b. On $(0, 2)$ graph of the derivative is negative, so the graph of f is falling.

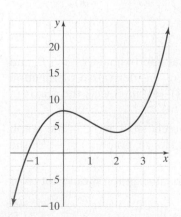

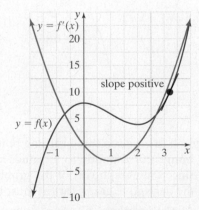

c. On $(2, \infty)$ the graph of the derivative is positive, so the graph of f is rising.

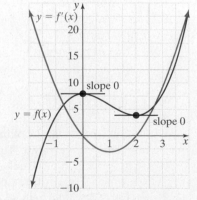

d. The derivative crosses the x-axis at $x = 0$ and $x = 2$, so the graph of f reaches a high point at $x = 0$ and a low point at $x = 2$.

Figure 2.64 Construction of the graph of a function f given its derivative f'.

A possible graph of $y = f(x)$ is shown on the left.

PROBLEM SET 2.7

Level 1 DRILL PROBLEMS

Use the definition of a derivative to find $f'(x)$ for the functions in Problems 1 to 8.

1. $f(x) = 8$

2. $f(x) = 3x - 2$

3. $f(x) = -x^2$

4. $f(x) = x + x^2$

5. $f(x) = x^4$

6. $f(x) = x^3 - x$

7. $f(x) = \dfrac{1}{x}$

8. $f(x) = \dfrac{1}{2x}$

Use the derivatives found in Problems 1 to 8 to find the values requested in Problems 9 to 16.

9. $\left.\dfrac{dy}{dx}\right|_{x=-2}$ where $y = 8$

10. $\left.\dfrac{dy}{dx}\right|_{x=-2}$ where $y = 3x - 2$

11. $\left.\dfrac{dy}{dx}\right|_{x=4}$ where $y = -x^2$

12. $\left.\dfrac{dy}{dx}\right|_{x=4}$ where $y = x + x^2$

13. $\left.\dfrac{dy}{dx}\right|_{x=2}$ where $y = x^4$

14. $\left.\dfrac{dy}{dx}\right|_{x=2}$ where $y = x^3 - x$

15. $\left.\dfrac{dy}{dx}\right|_{x=10}$ where $y = \dfrac{1}{x}$

16. $\left.\dfrac{dy}{dx}\right|_{x=10}$ where $y = \dfrac{1}{2x}$

Find at what point the slope of the instantaneous rate of change equals the average rate of change over the specified intervals in Problems 17 to 22. Also, provide a sketch that illustrates this relationship.

17. $f(x) = 8$ over the interval $[-5, 5]$

18. $f(x) = 3x - 2$ over the interval $[3, 4]$

19. $f(x) = -x^2$ over the interval $[-1, 1]$

20. $f(x) = x + x^2$ over the interval $[0, 1]$

21. $f(x) = \dfrac{1}{x}$ over the interval $[1, 2]$

22. $f(x) = \dfrac{1}{2x}$ over the interval $[1, 4]$.

Mix and match the graphs in Problems 23 to 28 with the graphs labeled (A) to (F), which are the derivative graphs.

23.

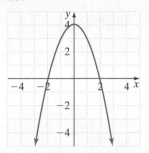

24.

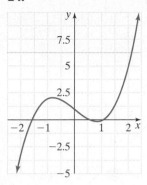

25.

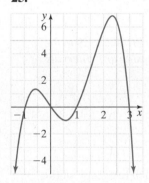

26.

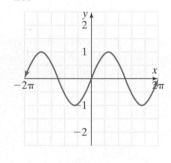

27.

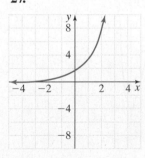

28.

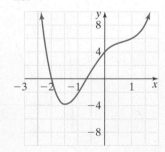

A

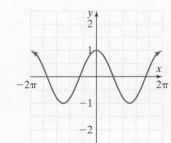

B

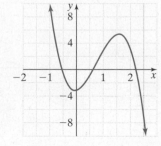

C

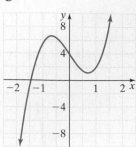

D

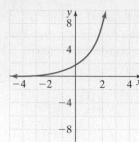

E

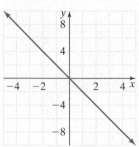

F

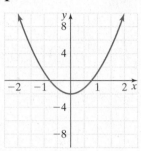

For each of the functions given in Problems 29 to 34, find intervals for which f is increasing and intervals for which f is decreasing.

29. $f(x) = x^2 - x + 1$

30. $f(x) = 5 - x^2$

31. $f(x) = x^3 + x$

32. $f(x) = 8 - x^3$

33. Let f be the function for which the graph of the derivative $y = f'(x) = g(x)$ is given by

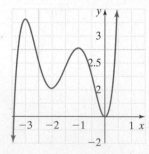

34. Let f be the function for which the graph of the derivative $y = f'(x) = g(x)$ is given by

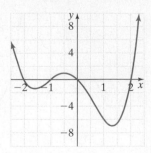

35. For the graph $y = g(x)$ given in Problem 33, estimate all values of c in $[-3, 0]$ such that

$$\frac{g(-3) - g(0)}{3} = g'(c)$$

36. For the graph $y = g(x)$ given in Problem 34, estimate all values of c in $[-2, 2]$ such that

$$\frac{g(-2) - g(2)}{4} = g'(c)$$

In each of Problems 37 to 40, the graph of a function f' is given. Draw a possible graph of f.

37.

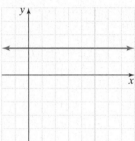

38.

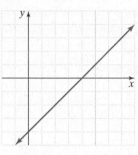

39.

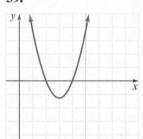

40.

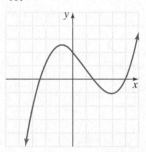

Level 2 APPLIED AND THEORY PROBLEMS

41. Let f be differentiable on the interval (a, b). Use the mean value theorem to prove if $f' < 0$ on $[a, b]$, then f is decreasing on (a, b).

42. **Rolle's theorem**: Let f be differentiable on (a, b) and continuous on $[a, b]$. Assume $f(a) = f(b) = 0$. Without using the mean value theorem, argue that there exists a c in (a, b) such that $f'(c) = 0$.

43. Use Rolle's theorem to prove the mean value theorem.

44. A baseball is hit upward and its height at time t in seconds is given by

$$H(t) = 100t - 16t^2 \text{ feet}$$

a. Find the velocity of the baseball after t seconds.

b. Find the time at which the velocity of the ball is 0.

c. Find the height of the ball at which the velocity is 0.

45. To study the response of nerve fibers to a stimulus, a biologist models the sensitivity, S, of a particular group of fibers by the function

$$f(t) = \begin{cases} t & \text{for } 0 \leq t \leq 3 \\ \dfrac{9}{t} & \text{for } t > 3 \end{cases}$$

where t is the number of days since the excitation began.

a. Over what time period is sensitivity increasing? When is it decreasing?

b. Graph $S'(t)$.

46. During the time period 1905–1940, hunters virtually wiped out all large predators on the Kaibab Plateau near the Grand Canyon in northern Arizona. The data for the deer population, P, over this period of time are as follows:

Year	Deer population	Year	Deer population
1905	4,000	1927	37,000
1910	9,000	1928	35,000
1915	25,000	1929	30,000
1920	65,000	1930	25,000
1924	100,000	1931	20,000
1925	60,000	1935	18,000
1926	40,000	1939	10,000

Source: www.biologycorner.com/worksheets/kaibab.html

a. Estimate $P'(t)$ for $1905 \leq t \leq 1939$.

b. Graph and interpret $P'(t)$.

47. In 1913, Carlson studied a growing culture of yeast (see Problem 45 in Problem Set 2.4 and Section 6.1). The table of population densities $N(t)$ at one-hour intervals is shown here:

Time	Population	Time	Population	Time	Population
0	9.6	6	174.6	12	594.8
1	18.3	7	257.3	13	629.4
2	29.0	8	350.7	14	640.8
3	47.2	9	441.0	15	651.1
4	71.1	10	513.3	16	655.9
5	119.1	11	559.7	17	659.6

a. Estimate $N'(t)$ for $0 \leq t \leq 17$.

b. Graph $N'(t)$ and briefly interpret this graph.

48. Our ruby-throated hummingbird has entered another patch of flowers, and the energy she is getting as a function of time in the patch is plotted below.

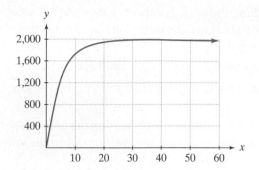

a. Find the average energy intake rate.

b. Estimate at what time the instantaneous energy intake rate equals the average intake rate.

49. Two radar patrol cars are located at fixed positions 6 miles apart on a long, straight road where the speed limit is 65 miles per hour. A sports car passes the first patrol car traveling at 60 miles per hour; then 5 minutes later, it passes the second patrol car going 65 miles per hour. Analyze this situation to show that at some time between the two clockings, the sports car exceeded the speed limit. Hint: Use the MVT.

50. Suppose two race cars begin at the same time and finish at the same time. Analyze this situation to show that at some point in the race they had the same speed.

CHAPTER 2 REVIEW QUESTIONS

1. Find at what point the instantaneous rate of change of the function $f(x) = x^3 - x$ equals the average rate of change over the interval $[-1, 2]$. Provide a sketch that illustrates this relationship.

2. In the 2012 Summer Olympics in London, swimmer Ye Shiwen swam the last 50 meters of her 400-meter individual medley final quicker than the winner of the men's race, Ryan Lochte. The 50-meter split times for both swimmers are shown in the following table:

Distance (meters)	Split time (seconds) for Shiwen	Split time (seconds) for Lochte
50	28.85	25.82
100	33.34	30.06
150	35.34	31.32
200	34.20	31.00
250	38.80	34.68
300	39.22	35.65
350	29.75	29.31
400	28.93	29.22

Data Source: http://www.guardian.co.uk/sport/datablog/2012/aug/02/olympics-2012-ye-shiwen-400-medley-statistics-data#data

a. Find the average velocity of both swimmers in the last split of the race.

b. Find the average velocity of both swimmers over the entire race.

3. Use the definition of derivative to find $f'(x)$ for
$$f(x) = \frac{1}{x^2}.$$

4. Find the average rate of change of $f(x) = x^2 - 2x + 1$ on $[1, 3]$ and the instantaneous rate of change at $x = 1$.

5. Consider $f(x) = 9 - x^2$ and $g(x) = \ln x$.

a. Graph $y = f(x)$ and $y = g(x)$ on the same coordinate axes.

b. Plot the point $P(2, \ln 2)$ on the graph of g. Graphically, estimate the position of the line tangent to g at the point P.

c. Plot the point $Q(2, 5)$ on the graph of f. Algebraically find the line tangent to f at the point Q. Use the equation of this tangent line to show that it "kisses" the graph of f at the point Q.

6. Find $\lim\limits_{x \to 4} \dfrac{16 - x^2}{x - 4}$ with the suggested methods.

a. Graphically

b. Using a table of values

c. By algebraic simplification

d. By using the informal definition of limit

e. Using technology

7. Let $f(x) = \begin{cases} 2 - 2x & \text{if } x < 2, x \neq 1 \\ \dfrac{-1}{x - 4} & \text{if } x > 2 \text{ and } x \neq 7 \end{cases}$

Find the requested limits.

a. $\lim\limits_{x \to 1} f(x)$ b. $\lim\limits_{x \to 2^-} f(x)$

c. $\lim\limits_{x \to 2^+} f(x)$ d. $\lim\limits_{x \to 7} f(x)$

e. Is f continuous at $x = 7$? If not, is this reparable?

8. Evaluate the sequential limits, if they exist.

a. $\lim\limits_{n \to \infty} \dfrac{2n^3 + 4n}{1 - 2n^2 - 5n^3}$

b. $\lim\limits_{n \to \infty} a_n$ where $a_1 = 1$, and $a_{n+1} = \dfrac{1}{a_n}$

c. $\lim\limits_{n \to \infty} a_n$ where $a_1 = 14$, and $a_{n+1} = a_n/2$

9. The graph of a function f' is given. Draw a possible graph of f.

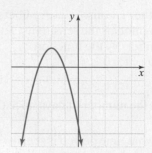

10. An environmental study of a certain suburban community suggests that t years from now, the average level of carbon dioxide in the air can be modeled by the formula
$$q(t) = 0.05t^2 + 0.1t + 3.4$$
parts per million.

a. At what rate will the CO_2 level be changing with respect to time one year from now?

b. By how much will the CO_2 level change in the first year?

c. By how much will the CO_2 level change over the next (second) year?

11. Consider the function $y = \dfrac{2x^2 + 1}{x(x - 2)}$.

a. Find the horizontal asymptote(s) of this function.

b. Find the right-hand side and left-hand side limits of this function at all of its vertical asymptotes.

12. The canopy height (in meters) of a tropical elephant grass (*Pennisetum purpureum*) is modeled by
$$h(t) = -3.14 + 0.142t - 0.0016t^2 + 0.000079t^3 - 0.00000000133t^4$$
where t is the number of days after mowing.

a. Sketch the graph of $h(t)$.

b. Sketch the graph of $h'(t)$.

c. Approximately when was the canopy height growing most rapidly? Least rapidly?

13. The concentration $C(t)$ of a drug in a patient's blood stream is given by

t in minutes	0	0.1	0.2	0.3	0.4	0.5	0.6	0.7	0.8	0.9	1.0
C in milligrams/ milliliter	0	0.2	0.4	0.6	0.8	0.9	1.0	0.9	1.0	0.9	0.7

a. Estimate $C'(t)$ for $t = 0, 0.1, \ldots 0.9$.

b. Sketch $C'(t)$ and interpret it.

14. Suppose that systolic blood pressure of a patient t years old is modeled by
$$P(t) = 39.73 + 23.5 \ln(0.97t + 1)$$
for $0 \leq t \leq 65$, where $P(t)$ is measured in millimeters of mercury.

a. Sketch the graph of $y = P(t)$.

b. Using the graph in part **a**, sketch the graph of $y = P'(t)$.

15. Find the limit $\lim\limits_{x \to \infty} \dfrac{e^x}{2 + 2e^x}$ and determine how positive x needs to be to ensure that $\dfrac{e^x}{2 + 2e^x}$ is within one millionth of the limiting value L.

16. Whales have difficulty finding mates in the vast oceans of the world when their population numbers drop below a critical value. Thus, a model of the growth of whale populations from one whale generation to the next is going to be relatively more robust at intermediate whale densities than at low densities when finding mates is a problem, or at high densities when competition for food is a problem. The form of a hypothetical function f in the difference equation

$$a_{n+1} = f(a_n)$$

that reflects the above properties is illustrated in Figure 2.65, where a_n is the density of the whales in generation n (units are whales per 1000 sq km).

Determine $\lim\limits_{n \to \infty} a_n$ when $a_1 = 55$. Justify your answer.

17. Give a proof that the function $f(x) = xe^{-x} - 0.1$ has at least one positive root.

18. Sketch a graph of a function $y = f(x)$ on the interval $[-2, 2]$ such that $f(-2) = f(0) = f(2) = 0$, $\dfrac{dy}{dx} > 0$ on $[-2, -1]$ and $(1, 2]$, and $\dfrac{dy}{dx} < 0$ on $(-1, 1)$.

19. The average population growth rate of Mexico from 1981 to 1983 was 1.82 million per year and from 1983 to 1985 was 1.915 million per year. Assume the population size $N(t)$ as a function of time and $N'(t)$ are continuous. Prove that at some point in time between 1981 and 1985, the instantaneous rate of population growth was 1.85 million per year.

20. Consider the graph of f shown in Figure 2.66. On the interval $[-3, 4]$ find the points of discontinuity as well as places where the derivative does not exist. Explain your reasoning.

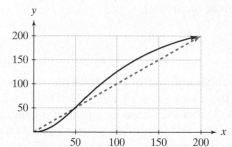

Figure 2.65 A function, f, modeling the growth of a hypothetical whale population.

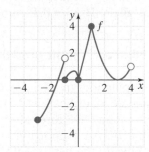

Figure 2.66 A function, f, on the interval $[-3, 4]$.

GROUP PROJECTS

Seeing a project through on your own, or working in a small group to complete a project, teaches important skills. The following projects provide opportunities to develop such skills.

Project 2A   A simple model of gene selection

One of the simplest problems in population genetics is to consider what happens to a particular version of a gene, where each version is referred to as an *allele*, that is being selected for or against because it confers some advantage or disadvantage to carriers of that allele. Examples of disadvantageous or *deleterious alleles* are those associated with genetic diseases such as sickle cell anemia, hemophilia, and Tay-Sachs disease. Most of our genes come in pairs of alleles, and if one allele in the pair is deleterious, then the effect of that allele may often be partially or fully masked by the other allele in the pair.

If a person has a double dose of the deleterious allele, the disease is expressed in its severest form. If a person has a single dose of the deleterious allele (i.e., one normal and one deleterious allele) then, depending on the disease, a milder version of the disease may be expressed (partial masking), or the individual is completely healthy (full masking). In the latter case, the individual is said to be a *carrier* for the disease (e.g., hemophilia).

On the other hand, alleles may confer a strong advantage to an organism that carries them. For example, if an insect carries an allele of a particular gene that allows it to detoxify a pesticide or a virus carries an allele that allows it to neutralize an otherwise effective drug, then we say that these pests and pathogens carry alleles of genes that *confer resistance* to the chemicals that would otherwise control or kill them.

Sometimes individuals who carry two different alleles of a particular gene are better off than individuals who have two copies of the same allele, regardless of which allele it is Biologist call this condition *heterozygous superiority*, and it is associated with the phenomenon called *hybrid vigor*. For example, an individual human is going to be better at fighting disease if he or she has two different alleles at an immune system gene responsible for the production of antibodies that protect against invading pathogens, such as the influenza virus.

Population geneticists have devised a simple model that allows them to assess what happens to such alleles. The form of this model is $p_{n+1} = f(p_n)$, where p_n represents the proportion of the allele in question in the population in the nth generation: If $p_n = 1$, then every individual in the population has a double dose of the allele in question. If $p_n = 0$, then no one has even a single dose of this allele. If $p_n = 0.5$, some individuals do not have the allele, some have a single dose, and some a double dose of the allele, but the proportion in the population of this allele is 1/2.

In this simple allele proportion model, the specific form of $f(p)$ is

$$f(p) = \frac{p\,(ap + (1 - p))}{ap^2 + 2p(1 - p) + b(1 - p)^2}$$

where $a \geq 0$ and $b \geq 0$ are constants that determine whether the allele in question confers an overall advantage or disadvantage or is associated with heterozygote superiority.

In this project, investigate the value of the equilibria that arise for various combinations of a and b, paying particular attention to whether a or b is greater than or less than 1. Interpret the various cases in terms of the limiting values of several sequences of proportions p_n that start at different values p_1 satisfying $0 < p_1 < 1$. Also, describe how these cases correspond to classification of the alleles as advantageous, deleterious, or associated with heterozygote superiority. Find specific cases in the literature, or by searching the Web, to illustrate these three phenomena.

Project 2B   Fibonacci rabbit growth when death is included

In the rabbit population growth process proposed by Fibonacci (Example 10 of Section 2.5), the assumption was that all the rabbits live forever. As an alternative, let's assume that only a proportion s of rabbits alive each month survive to the next month (independent of how old they are or what gender they are) and, of those that survive, only a proportion p of the females from the month before produce a litter that always consists of r male-female pairs. Investigate the growth of this population by carrying out the following tasks.

a. Derive an equation for the rate at which the proportion of pairs increases from month to month as a function of the three population parameters $0 < s \leq 1$, $0 < p \leq 1$, and r a positive integer. Note that getting the correct equation can be a little tricky, so use a diagram similar to the one outlined in Chapter 1: see the Fibonacci 𝕳istorical 𝔔uest, that is, Problem 38 in Section 1.7. In particular, starting out with a suitable number of new-born pairs, draw diagrams for the cases $(s, p, r) = (1/2, 1, 1)$, $(1, 1/2, 1)$, and $(1, 1, 2)$ and use these to construct a general expression for a_n in terms of a_{n-1} and a_{n-2} that contains the three parameters in question.

b. What must hold true for the parameters s, r, and p to ensure the population size remains constant for all time for any initial population size? For the case $r = 1$, express s as a function of p such that the population is neither growing nor declining. Hence, in the square of the positive quadrant of the p-s plane defined by $0 \leq p \leq 1$ and $0 \leq s \leq 1$, shade all points where the rabbit population is growing and all points where it is declining. *Hint:* Write a difference equation for $x_n = a_n/a_{n-1}$.

c. Repeat exercise **b** for the case $r = 2$ and make a general statement about how the two shaded areas change as r increases.

CHAPTER 3

Derivative Rules and Tools

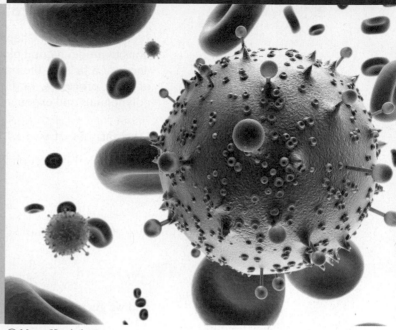

© Muzon/iStockphoto

Preview

"I loved history, but still loved science, and thought maybe you don't need quite as much calculus to be a biology major."

Elizabeth Moon, science fiction writer (b. 1945)
Part 2 of an interview with Jayme Lynn Blaschke
November 1999; http://www.sfsite.com/02b/em75.htm

In the previous chapters we have only been able to compute derivatives by directly appealing to the definition of the derivative. As you may have noticed, computing derivatives in this manner quickly becomes tedious. In this chapter we consider the rules and tools that allow us to quickly compute the derivative of any imaginable function. Learning these rules is critical, as expressed in the following admonition (from Colin Adams, Joel Hass, and Abigail Thompson, *How to Ace Calculus: The Streetwise Guide*, New York: W. H. Freeman, 1998):

"Know these backwards and forwards. They are to calculus what 'Don't go through a red light' and 'Don't run over a pedestrian' are to driving."

The first three sections of this chapter provide essential basic rules for calculating derivatives, and the fourth section focuses on the important trigonometric functions. The last three sections expand these tools so that we can apply them to a variety of applications in biology and the life sciences. Applications in this chapter include predicting the growth of a fetal heart and of the Yellowstone bison population. Also, we use differential calculus to investigate the clearance rate of HIV viral particles (see Figure 3.1) from infected humans, how Northwestern crows break whelk shells, dose-response curves in the context of administering drugs and into rates of mortality due to airborne diseases, and Usain Bolt's record-breaking, 100-meter run in the Olympic Games in Beijing.

3.1   Derivatives of Polynomials and Exponentials

As we have seen in Chapters 1 and 2, we can use polynomial and exponential functions to model natural phenomena ranging from the melting of Arctic sea ice to the decay of a drug in the body. To facilitate finding the instantaneous rates of change of these processes, we now derive general rules for computing the derivatives of polynomials and exponentials.

Derivatives of $y = x^n$

In Example 1 of Section 2.7, we proved that

$$\frac{d}{dx}x = 1$$

$$\frac{d}{dx}x^2 = 2x$$

$$\frac{d}{dx}x^3 = 3x^2$$

and we guessed that $\frac{d}{dx}x^n = nx^{n-1}$. This powerful result is known as the **power rule**.

Power Rule

For any real number $n \neq 0$,

$$\frac{d}{dx}x^n = nx^{n-1}$$

At this point, we are only equipped to prove the power rule for any natural number n. Later, we shall prove the general power rule. The proof when n is a natural number involves the binomial expansion of $(a + b)^n$ (if you don't remember this expansion, look it up on the World Wide Web):

$$(a + b)^n = a^n + na^{n-1}b + \frac{n(n-1)}{2}a^{n-2}n^2 + \cdots + b^n$$

Now to the proof of the power rule for n a natural number:

Proof.   If $f(x) = x^n$, then from the binomial theorem

$$f(x + h) = (x + h)^n$$

$$= x^n + nx^{n-1}h + \frac{n(n-1)}{2}x^{n-2}h^2 + \cdots + h^n$$

From the definition of derivative we have

$$f'(x) = \lim_{h \to 0} \frac{f(x + h) - f(x)}{h}$$

$$= \lim_{h \to 0} \frac{\left[x^n + nx^{n-1}h + \frac{n(n-1)}{2}x^{n-2}h^2 + \cdots + h^n\right] - [x^n]}{h}$$

$$= \lim_{h \to 0} \frac{nx^{n-1}h + \frac{n(n-1)}{2}x^{n-2}h^2 + \cdots + h^n}{h}$$

$$= \lim_{h \to 0} \frac{h\left[nx^{n-1} + \frac{n(n-1)}{2}x^{n-2}h + \cdots + h^{n-1}\right]}{h}$$

$$= \lim_{h \to 0} \left[nx^{n-1} + \frac{n(n-1)}{2}x^{n-2}h + \cdots + h^{n-1}\right]$$

$$= nx^{n-1}$$

Note that if $n = 0$, then $f(x) = x^n = x^0 = 1$, so $f'(x) = 0$ as expected, since 1 is a constant.

Example 1   Using the power rule

Find

a. $\dfrac{d}{dx}x^5\Big|_{x=2}$ **b.** $\dfrac{d}{dQ}Q^{29}$

Solution

a. $\dfrac{d}{dx}x^5\Big|_{x=2} = 5x^4\Big|_{x=2} = 5\cdot 2^4 = 80$ **b.** $\dfrac{d}{dQ}Q^{29} = 29Q^{28}$

Derivatives of sums, differences, and scalar multiples

The limit laws from Chapter 2 allow us to quickly compute the derivatives of a sum, difference, or scalar multiple whenever we know the derivatives for f and g. In stating these laws, it is more succinct to use the "prime" rather than full Leibnitz notation.

Elementary Differentiation Rules

Let f and g be differentiable at x. Let c be a constant. Then

Sum $(f + g)'(x) = f'(x) + g'(x)$

Difference $(f - g)'(x) = f'(x) - g'(x)$

Scalar multiple $(cf)'(x) = cf'(x)$

In other words, the derivative of a sum is the sum of the derivatives, the derivative of a difference is the difference of the derivatives, and the derivative of a scalar multiple is the scalar multiple of the derivative.

Combining these elementary differentiation rules with the power rule allows us to differentiate any polynomial. Note that throughout this and subsequent chapters, we use the verb *differentiate* in a technical sense. To differentiate a function is to "take its derivative" using the methods of differential calculus presented in this book. It does not mean that we are trying to distinguish the function from some other function, unless we specifically say so.

Example 2   Using differentiation rules

Let $f(x) = x^3 + 3x^2 + 10$.

a. Find f'. Justify each step of your differentiation.
b. Determine on what intervals f is increasing and on what intervals f is decreasing.

Solution

a.
$$\frac{d}{dx}(x^3 + 3x^2 + 10) = \frac{d}{dx}x^3 + \frac{d}{dx}3x^2 + \frac{d}{dx}10 \qquad \textit{sum rule}$$

$$= \frac{d}{dx}x^3 + 3\frac{d}{dx}x^2 + \frac{d}{dx}10 \qquad \textit{scalar multiple rule}$$

$$= 3x^2 + 6x + 0 \qquad \textit{power rule}$$

Hence

$$f'(x) = 3x^2 + 6x$$

b. To determine where f is increasing and where f is decreasing, we need to find where $f' > 0$ and $f' < 0$, respectively. Since

$$f'(x) = 3x^2 + 6x = 3x(x + 2)$$

we look at the signs of the factors and the product by looking at a number line.

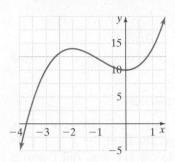

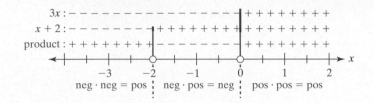

On the interval $(-\infty, -2)$, f is increasing since $f'(x) > 0$; on $(-2, 0)$, f is decreasing since $f'(x) < 0$; on $(0, \infty)$, f is increasing again since $f'(x) > 0$. Graphing $y = f(x)$ confirms these calculations. ∎

Example 3   Growth of a fetal heart

In 1992, a team of cardiologists determined how the left ventricular length L (in centimeters) of the heart in a fetus (Figure 3.2) increases from eighteen weeks until birth. (See J. Tan, N. Silverman, J. Hoffman, M. Villegas, and K. Schmidt, "Cardiac Dimensions Determined by Cross-Sectional Echocardiography in the Normal Human Fetus from 18 Weeks to Term," *American Journal of Cardiology* 70(1992): 1459–1497.) The cardiologists used the following function to model the data

$$L(t) = -2.318 + 0.2356t - 0.002674t^2$$

where t is the age of the fetus (in weeks). Here $t = 18$ means at the end of week 18.

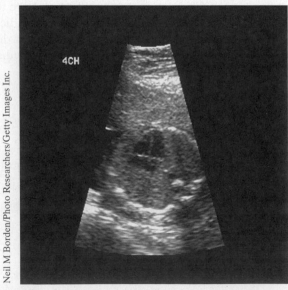

Neil M Borden/Photo Researchers/Getty Images Inc.

Figure 3.2 Fetal echocardiogram reveals a four-chamber heart correctly oriented in the left chest.

a. Find $L'(t)$ for $18 \leq t \leq 38$.

b. Discuss and interpret the units of $L'(t)$.

c. During which week between weeks 18 and 38 is the ventricular length growing most rapidly, and what is the associated rate? When is the ventricular length growing most slowly?

Solution

a.
$$\frac{dL}{dt} = \frac{d}{dt}(-2.318) + 0.2356\frac{d}{dt}(t) - 0.002674\frac{d}{dt}(t^2) \quad \textit{sum and scalar multiple laws}$$

$$= 0 + 0.2356 \cdot 1 - 0.002674 \cdot 2t \quad \textit{power law}$$

$$= 0.2356 - 0.005348\,t$$

b. The units of $L'(t)$ are centimeters per week. $L'(t)$ describes the rate at which the ventricular length is growing.

c. Since $L'(t)$ is a linear function with negative slope, its largest value on the interval $[18, 38]$ is at $t = 18$ and its smallest value on this interval is at $t = 38$. In particular,

$$L'(18) = 0.2356 - 0.005348 \times 18 = 0.139336 \quad \text{cm/week}$$

and

$$L'(38) = 0.2356 - 0.005348 \times 38 = 0.032376 \quad \text{cm/week}$$

Hence, the ventricular length in the last twenty weeks of pregnancy is increasing most rapidly at the beginning of this twenty week period and growing least rapidly at the time of birth.

■

In addition to being used to finding derivatives of all polynomials, the power rule and the scalar multiplication rule can be used to find derivatives of all scaling laws.

Example 4   Back to lifting weights
- -

In Example 6 of Section 1.3, we modeled the amount an Olympic weightlifter could lift as

$$L = 20.15M^{2/3} \text{ kilograms}$$

where M is the body mass in kilograms of the weightlifter. Find and interpret $\dfrac{dL}{dM}$ at $M = 90$ kilograms.

Solution   To compute the derivative, we note that $n = 2/3$, and although n is not an integer, the power rule still applies with $n - 1 = -\dfrac{1}{3}$. Thus we obtain

$$\left.\frac{dL}{dM}\right|_{M=90} = 20.15\frac{d}{dM}M^{2/3}\Big|_{M=90}$$

$$= 20.15 \cdot \frac{2}{3} \cdot M^{-1/3}\Big|_{M=90}$$

$$\approx 2.998 \quad \textit{correct to three decimal places}$$

Hence, for weightlifters weighing close to 90 kilograms, the rate at which the amount lifted increases with mass of the weightlifter is 2.998 kilograms per kilogram of body mass.

■

Derivatives of exponentials

Consider the function $f(x) = a^x$ for some positive constant $a > 0$. To find the derivative, we use the definition of the derivative. Let x be a fixed number.

$$f'(x) = \lim_{h \to 0} \frac{f(x+h) - f(x)}{h} \qquad \textit{definition of derivative, provided the limits exist}$$

$$= \lim_{h \to 0} \frac{a^{x+h} - a^x}{h} \qquad \textit{since } f(x) = a^x$$

$$= \lim_{h \to 0} \frac{(a^h - 1)a^x}{h} \qquad \textit{common factor}$$

$$= \left[\lim_{h \to 0} \frac{a^h - 1}{h} \right] a^x \qquad \textit{property of limits provided } k = \lim_{h \to 0} \frac{a^h - 1}{h} \textit{ exists}$$

$$= ka^x$$

$$= kf(x)$$

Although it is beyond the scope of this book, it can be shown that $k = \lim_{h \to 0} \dfrac{a^h - 1}{h}$ exists whenever $a > 0$. In the following example, we estimate the value of k for the case $a = 2$.

Example 5 Derivative of 2^x

- -

Find $\dfrac{d}{dx} 2^x$ by estimating $\lim\limits_{h \to 0} \dfrac{2^h - 1}{h}$.

Solution   We showed that $\dfrac{d}{dx} 2^x = k2^x$ where $k = \lim\limits_{h \to 0} \dfrac{2^h - 1}{h}$. To estimate k, we can create the following table with a calculator:

h	$\dfrac{2^h - 1}{h}$	h	$\dfrac{2^h - 1}{h}$
0.1	0.717735	−0.1	0.66967
0.01	0.695555	−0.01	0.69075
0.001	0.693387	−0.001	0.692907
0.0001	0.693171	−0.0001	0.693123

Since $k \approx 0.693$, $\dfrac{d}{dx} 2^x \approx (0.693)2^x$. In Example 6, we show that in fact $k = \ln 2 \approx 0.69315$. ∎

Since $f'(x) = kf(x)$ for an appropriate choice of k whenever $f(x) = a^x$, we can ask this question: Is there a value of a such that $k = 1$? It turns out that the number e, which we defined in Section 1.4 as $e = \lim\limits_{n \to \infty} (1 + 1/n)^n$, is the appropriate choice of a. Namely,

$$\frac{d}{dx} e^x = e^x \text{ for all real } x$$

Except for multiplying by a constant, e^x is the only function that remains unchanged under the operation of differentiation; that is, if $f(x) = f'(x)$, then $f(x) = ae^x$ for some real number a. This fact inspired one mathematician to write: "Who has not been amazed to learn that the function $y = e^x$, like a phoenix rising again from its own ashes, is its own derivative?" (Francois l'Lionnais, *Great Currents of Mathematical Thought*, vol. 1, New York: Dover Publications, 1962). Armed with the derivative of e^x, we can use the rules of differentiation to find the derivative of more general exponential functions $f(x) = e^{ax}$.

Derivative of the Natural Exponential	For any real number a, $$\frac{d}{dx}e^{ax} = ae^{ax}$$ Further, for $b > 0$, $$\frac{d}{dx}b^x = (\ln b)b^x$$

Proof. If $a = 0$, then

$$\frac{d}{dx}e^0 = 0$$

so the statement is true. If a is any nonzero real number, then $e^{ax} = (e^a)^x$ and

$$\frac{d}{dx}e^{ax} = \left(\lim_{h \to 0} \frac{e^{a(x+h)} - e^{ax}}{h}\right) \qquad \textit{definition of a derivative}$$

$$= \left(\lim_{h \to 0} \frac{(e^{ax}e^{ah} - e^{ax})}{h}\right) \qquad \textit{addition law of exponents}$$

$$= \left(\lim_{h \to 0} \frac{(e^{ah} - 1)e^{ax}}{h}\right) \qquad \textit{taking out a common factor}$$

$$= \left(\lim_{h \to 0} \frac{e^{ah} - 1}{h}\right)e^{ax} \qquad \textit{Taking constant out of the limit}$$

To find this limit, define $\Delta x = ah$. Since $h = \Delta x/a$ and $h \to 0$ whenever $\Delta x \to 0$,

$$\lim_{h \to 0} \frac{e^{ah} - 1}{h} = \lim_{\Delta x \to 0} \frac{e^{\Delta x} - 1}{\Delta x/a} \qquad \textit{substitute } \Delta x = ah$$

$$= a \lim_{\Delta x \to 0} \frac{e^{\Delta x} - 1}{\Delta x} \qquad \textit{limit law for products}$$

$$= a \cdot 1 \qquad \textit{derivative of } e^x \textit{ evaluated at } x = 0$$

Thus, we have shown that

$$\frac{d}{dx}e^{ax} = ae^{ax}$$

For the last part, find the value of a such that $b = e^a$ and finish the details on your own in Problem 27 in Problem Set 3.1. ∎

Example 6 Déjà Vu

Find the exact value of $\dfrac{d}{dx}2^x$.

Solution $\dfrac{d}{dx}2^x = (\ln 2)2^x$. Since $\ln 2 \approx 0.693$, this agrees with the result obtained in Example 5.

Example 7 Clearance of HIV

Human immunodeficiency virus (shown in Figure 3.1) is a blood-borne pathogen that is typically transmitted through sexual contact or sharing of needles among drug users. HIV attacks the immune system. Understanding how the viral load in the blood of an HIV-infected individual changes with time is critical to treating HIV patients with a "cocktail" of several antiretroviral drugs. Theoretical immunologists have used data from various experiments to model observed changes in the viral load $V(t)$, in particles per milliliter, of an HIV patient undergoing antiretroviral drug therapy for t days.[*] If no new viral particles are generated by the host, they found that the viral load over time can be modeled by the equation

$$V(t) = 216,000\, e^{-0.2t}$$

Find $V'(t)$ and interpret it.

Solution

$$V'(t) = 216,000\frac{d}{dt}e^{-0.2t} \qquad \text{\textit{differentiation for scalar multiples}}$$

$$= 216,000(-0.2)e^{-0.2t} \qquad \text{\textit{derivative of an exponential}}$$

$$= -43,200e^{-0.2t}$$

The units of $V'(t)$ are particles per milliliter per day. $V'(t)$ describes the rate at which the viral load is changing whenever it is not replenished by new viral particles. Since $V'(t) < 0$ for all t, the viral load is decreasing.

Note that in HIV-infected patients, new viral particles are produced in the various cells found in the blood and in lymph tissue. In Chapter 6, we account for this second component of the infection process to obtain a more complete model of viral load dynamics within human hosts. ∎

Example 8   Exponential depletion of resources

In Example 4 of Section 1.4, we projected that the U.S. population would contain $N(t) = 8.3(1.33)^t$ million individuals t decades after 1815. Suppose the amount of food produced each year, measured in terms of rations (i.e. the amount of food needed to sustain one individual for one year), grew linearly during this same period with the amount given by the equation

$$R(t) = 10 + 4t$$

The number of surplus rations $S(t)$ over this period can be found by taking the difference of the above two functions:

$$S(t) = R(t) - N(t)$$
$$= 10 + 4t - 8.3(1.33)^t$$

Determine at what point in time $S(t)$ starts decreasing.

Solution   To find where S changes from increasing to decreasing, we need to determine where $S'(t)$ changes sign from $S'(t) > 0$ to $S'(t) < 0$. In particular, we need to find where $S'(t) = 0$, provided the derivative exists at the point in question. We have

$$S'(t) = \frac{d}{dt}[10 + 4t - 8.3(1.33)^t] \qquad \text{\textit{derivative of both sides of given equation}}$$

$$= \frac{d}{dt}10 + 4\frac{d}{dt}t - 8.3\frac{d}{dt}(1.33)^t \qquad \text{\textit{elementary rules of differentiation}}$$

$$= 0 + 4 - 8.3(\ln 1.33)\,(1.33)^t \qquad \text{\textit{power and exponential rules of differentiation}}$$

$$= 4 - 8.3(\ln 1.33)(1.33)^t$$

[*]A. S. Perelson, A. U. Neumann, M. Markowitz, J. M. Leonard, D. D. Ho, "HIV-1 Dynamics In Vivo: Virion Clearance Rate, Infected Cell Lifespan, and Viral Generation Time," *Science* 271 (1996): 1582–1586. Also, A. S. Perelson and P. W. Nelson, "Mathematical Analysis of HIV-1 Dynamics In Vivo," *SIAM Review* 41(1999):3–44.

If we now solve for the values of t satisfying $S'(t) = 0$, we obtain

$$4 - 8.3(\ln 1.33)(1.33)^t = 0$$

$$(1.33)^t = \frac{4}{8.3 \ln 1.33} \qquad \textit{rearranging terms}$$

$$t \ln 1.33 = \ln\left(\frac{4}{8.3 \ln 1.33}\right) \qquad \textit{taking logarithms}$$

$$t = \frac{\ln\left(\dfrac{4}{8.3 \ln 1.33}\right)}{\ln 1.33} \qquad \textit{dividing by } \ln 1.33$$

$$\approx 1.84$$

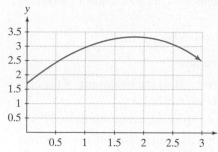

Figure 3.3 Graph of the number of surplus rations.

Evaluating $S'(t)$ at values of t greater than and less than 1.84, we find that $S'(t) > 0$ for $t < 1.84$ and $S'(t) < 0$ for $t > 1.84$. Since the units of time are in decades, we see that in the year $1815 + 18.4 \approx 1833$ the surplus of resources will begin to decline. Plotting $y = S(t)$ reveals that at $t \approx 1.84$, $S(t)$ takes on its largest value and then begins to decrease, as shown in Figure 3.3.

PROBLEM SET 3.1

Level 1 DRILL PROBLEMS

Differentiate the functions given in Problems 1 to 14. Assume that C is a constant.

1. a. $f(x) = x^7$ **b.** $g(x) = 7^x$

2. a. $f(x) = x^4$ **b.** $g(x) = 4^x$

3. a. $f(x) = 3x^5$ **b.** $g(x) = 3(7)^5$

4. a. $f(x) = x^3 + C$ **b.** $g(x) = C^2 + x$

5. a. $f(x) = x^2 + 3\pi + C$ **b.** $g(x) = \pi^2 - 2x - C$

6. $f(x) = 5x^3 - 5x^2 + 3x - 5$

7. $f(x) = x^5 - 3x^2 - 1$

8. $f(x) = 2x^2 - 5x^8 + 1$

9. $s(t) = 4e^t - 5t + 1$

10. $f(t) = 5 - e^{2t}$

11. $f(t) = 5.9(2.25)^t$

12. $f(t) = 82.1(1.85)^t$

13. $g(x) = Cx^2 + 5x + e^{-2x}$

14. $F(x) = 5e^{Cx} - 4x^2$

Determine on what intervals each function given in Problems 15 to 19 is increasing and on what intervals it is decreasing.

15. $f(x) = x^3 - x^2 + 1$

16. $g(x) = \dfrac{1}{3}x^3 - 9x + 2$

17. $f(x) = x^5 + 5x^4 - 550x^3 - 2{,}000x^2 + 60{,}000x$ (round to the nearest tenth)

18. $g(x) = x^3 + 35x^2 - 125x - 9{,}375$

19. $H(w) = 2w - e^w$

20. Let $f(x) = x^{1/2}$.

   a. Find the derivative using the definition of the derivative.

   b. Apply the power rule with $n = 1/2$.

21. Let $f(x) = x^{3/2}$.

   a. Find the derivative using the definition of derivative. Hint: Write $x^{3/2}$ as $x\sqrt{x}$ and rationalize the numerator.

   b. Apply the power rule with $n = 3/2$.

Simplify the functions in Problems 22 to 26 and find their derivative whenever it is well defined.

22. $g(x) = x^2(x^3 - 3x)$

23. $f(x) = \dfrac{x^{1/3}}{x^2}$

24. $f(x) = (e^x - 1)(e^x + 1)$

25. $h(t) = \dfrac{3^t + 3^{-t}}{2^t}$

26. $q(x) = \dfrac{x^2 - 4}{x + 2}$

Level 2 APPLIED AND THEORY PROBLEMS

27. Prove that for any real number $b > 0$

$$\frac{d}{dx}b^x = (\ln b)b^x$$

28. Use the limit laws to prove the sum rule for differentiation:

$$(f + g)' = f' + g'$$

29. Use the limit laws to prove the scalar multiple rule for differentiation:

$$(cf)' = cf'$$

for a constant c.

30. After pouring a mug full of the German beer Erdinger Weissbier (see Section 1.4), German physicist Leike ("Demonstration of Exponential Decay," pp. 21–26) measured the height of the beer froth at regular time intervals. He estimated the height (in centimeters) of the beer froth as

$$H(t) = 1.7(0.94)^t$$

where t is measured in seconds. Find

$$\frac{dH}{dt}\bigg|_{t=25}$$

and interpret this quantity.

31. A drug that influences weight gain was tested on eight animals of the same size, age, and sex. Each animal was randomly assigned to a dose level. After two weeks, the difference W in the end and start weight (measured in decagrams) was calculated. The best-fitting quadratic equation to the data was found to be

$$W = 1.13 - 0.41\,D + 0.17\,D^2$$

where D is the dose level that ranges from 1 to 8.

a. Find $\dfrac{dW}{dD}$ and identify its units.

b. When does weight gain increase with dose level D?

32. Using data from 158 marine species, John Hoenig of the Virginia Institute of Marine Sciences studied how the natural mortality rate M of a species depends on the maximum observed age T ("Empirical Use of Longevity Data to Estimate Mortality Rates," *Fisheries Bulletin* 82 (1983): 898–902). Using linear regression, he found

$$M(T) = e^{1.44 - 9.82\,T}$$

where T is measured in years. Find and interpret

$$\frac{dM}{dT}\bigg|_{T=10}$$

33. During an outbreak of influenza at a school the number of students who became ill after t days is given by

$$N(t) = 50(1 - Ce^{-0.1t})$$

where C is a constant.

a. If ten people were ill at the beginning of the epidemic (when $t = 0$), what is C?

b. At what rate is $N(t)$ increasing when $t = 5$?

34. A glucose solution is administered intravenously into the blood stream of a patient at a constant rate of r milligrams/hour. As the glucose is being administered, it is converted into other substances and removed from the blood stream. Suppose the concentration of the glucose solution after t hours is given by

$$C(t) = r - (r - k)e^{-t}$$

where k is a constant.

a. If C_0 is the initial concentration of glucose (when $t = 0$), what is C_0 in terms of r and k?

b. What is the rate at which the concentration of glucose is changing at time t?

35. In Problem 39 in Problem Set 1.3, we found that the length L (cm) of a pumpkin is related to the width W (cm) of a pumpkin by the allometric equation

$$L = 1.12W^{0.95}$$

How rapidly is length changing with regard to width for pumpkins of size 5 cm compared with those of 50 cm?

36. At the beginning of Section 1.4, we saw that the population model $N(t) = 8.3(1.33)^t$, where N is in millions and t is in decades starting at $t = 0$ representing the year 1815, can be used to describe the size of the U.S. population during most of the nineteenth century. Use this relationship to compare the total rate of growth of the U.S. population in 1815 versus the growth rate in 1865 (i.e., five decades later).

37. In Section 1.3, we found that the amount lifted (in kilograms) by an Olympic weightlifter can be predicted by the scaling law

$$L = 20.15\,M^{2/3}$$

where M is the mass of the lifter in kilograms. Find and interpret $\dfrac{dL}{dM}\bigg|_{M=100}$.

38. In Example 12 of Section 1.6 (changing the names of the variables from x and y to M and R), we found that the metabolic rate (in kilocalories/day) for animals ranging in size from mice to elephants is given by the function $\ln R = 0.75 \ln M + 4.2$ which yields the equation

$$R = e^{4.2} M^{3/4},$$

where M is the body mass of the animal in kilograms.

a. The average California Condor weighs about 10 kg. Find and interpret $\left.\dfrac{dR}{dM}\right|_{M=10}$.

b. The average football player weighs about 100 kg. Find and interpret $\left.\dfrac{dR}{dM}\right|_{M=100}$.

c. Compare and discuss the quantities that you found in parts **a** and **b**.

39. The number of children newly infected with a particular pathogen that is transmitted through contact with their mothers has been modeled by the function

$$N(t) = -0.21t^3 + 3.04t^2 + 44.05t + 200.29$$

where $N(t)$ is measured in thousands of individuals per year, and t is the number of years since 2000. In epidemiology, $N(t)$ is known as an *incidence function*.

a. At what rate is the incidence function N changing with respect to time in the year 2010?

b. When will the incidence start to decline?

3.2   Product and Quotient Rules

Previously, we saw that the derivative of a sum equals the sum of the derivatives, and the derivative of a difference equals the difference of the derivatives. Armed with these elementary differentiation rules, we might guess that the derivative of a product is the product of the derivatives. The following simple example, however, shows this not the case. Let $f(x) = x$ and $g(x) = x^2$, and consider their product

$$p(x) = f(x)g(x) = x^3$$

Because $f'(x) = 1$ and $g'(x) = 2x$, the product of the derivatives is

$$f'(x)g'(x) = (1)(2x) = 2x$$

whereas the actual derivative of $p(x) = x^3$ is $p'(x) = 3x^2$. Hence, our naïve guess is wrong! It is also easy to show that the derivative of a quotient is not the quotient of the derivatives. The goal of this section is to uncover the correct rules for differentiation for products and quotients of functions.

Product rule

To derive a rule for products, we appeal to our geometric intuition by considering areas where $\Delta x > 0$ and $f(x)$ and $g(x)$ are assumed to be increasing, positive differentiable functions of x. Note that the algebraic steps stand alone—without considering area or making the assumptions we made in the previous sentence.

Let

$$\underbrace{p(x)}_{\text{area of rectangle}} = \underbrace{f(x)}_{\text{length}}\underbrace{g(x)}_{\text{width}}$$

This product of p can be represented as the area of a rectangle:

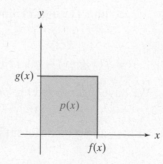

Next, we find

$$\underbrace{p(x + \Delta x)}_{\text{area of larger rectangle}} = \underbrace{f(x + \Delta x)}_{\text{length}} \underbrace{g(x + \Delta x)}_{\text{width}}$$

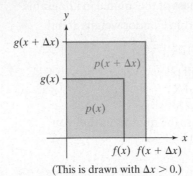

(This is drawn with $\Delta x > 0$.)

The next step gives us the area of the "inverted L-shaped" region:

$$p(x + \Delta x) - p(x) = f(x + \Delta x)g(x + \Delta x) - f(x)g(x)$$

The key to the proof of the product rule is to rewrite this difference. We can see how to do this by looking at the area of the inverted L-shaped region in another way:

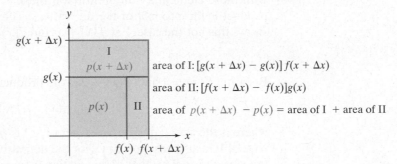

area of I: $[g(x + \Delta x) - g(x)]\, f(x + \Delta x)$

area of II: $[f(x + \Delta x) - f(x)]g(x)$

area of $p(x + \Delta x) - p(x) = $ area of I $+$ area of II

$$\underbrace{p(x + \Delta x) - p(x)}_{\text{area of inverted L-shaped region}} = \underbrace{[g(x + \Delta x) - g(x)]\,f(x + \Delta x)}_{\text{area of region I}} + \underbrace{[f(x + \Delta x) - f(x)]g(x)}_{\text{area of region II}}$$

Divide both sides by Δx (where $\Delta x \neq 0$):

$$\frac{p(x + \Delta x) - p(x)}{\Delta x} = \frac{[g(x + \Delta x) - g(x)]}{\Delta x}f(x + \Delta x) + \frac{[f(x + \Delta x) - f(x)]}{\Delta x}g(x)$$

The last step in deriving the product rule is to take the limit as $\Delta x \to 0$:

$$p'(x) = \lim_{\Delta x \to 0} \frac{p(x + \Delta x) - p(x)}{\Delta x}$$

$$= \lim_{\Delta x \to 0} \left\{ f(x + \Delta x) \left[\frac{g(x + \Delta x) - g(x)}{\Delta x} \right] + g(x) \left[\frac{f(x + \Delta x) - f(x)}{\Delta x} \right] \right\}$$

$$= \lim_{\Delta x \to 0} f(x + \Delta x) \underbrace{\lim_{\Delta x \to 0} \left[\frac{g(x + \Delta x) - g(x)}{\Delta x} \right]}_{\text{This is the derivative of } g.} + g(x) \underbrace{\lim_{\Delta x \to 0} \left[\frac{f(x + \Delta x) - f(x)}{\Delta x} \right]}_{\text{This is the derivative of } f.}$$

$$= f(x)g'(x) + g(x)f'(x) \qquad \lim_{\Delta x \to 0} f(x + \Delta x) = f(x) \textit{ because } f \textit{ is continuous}$$

We have just proved the **product rule**.

Product Rule

Let f and g be differentiable at x. Then

$$(fg)'(x) = f'(x)g(x) + f(x)g'(x)$$

A simple way to remember the product rule is with this mnemonic: "The derivative of the product is the derivative of the first times the second plus the derivative of the second times the first." Or sing the words of the following ditty aloud or in your mind.

Sing the product rule in time,
One prime two plus one two prime.
Isn't mathematics fun,
One prime two plus two prime one.

Example 1   Computing with the product rule

Find $p'(x)$ and determine on what intervals p is increasing.

a. $p(x) = xe^x$ **b.** $p(x) = x^2 2^x$

Solution

a. Let $f(x) = x$ and $g(x) = e^x$. Then $p(x) = f(x)g(x)$. By the product rule,

$$p'(x) = f'(x)g(x) + f(x)g'(x)$$
$$= 1 \cdot e^x + x \cdot e^x$$
$$= (1 + x)e^x$$

Since $e^x > 0$ for all x, we have $p'(x) > 0$ if and only if $1 + x > 0$. Hence, p is increasing on the interval $(-1, \infty)$. Indeed, plotting $y = p(x)$ supports this conclusion:

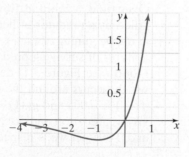

b. Let $f(x) = x^2$ and $g(x) = 2^x$. Then $p(x) = f(x)g(x)$. Recall that $f'(x) = 2x$ and $g'(x) = (\ln 2)2^x$. Hence, by the product rule,

$$p'(x) = f'(x)g(x) + f(x)g'(x)$$
$$= 2x2^x + x^2(\ln 2)2^x$$
$$= x2^x(2 + x \ln 2)$$

Since $p' > 0$ whenever $x > 0$ or $x < -\dfrac{2}{\ln 2}$, p is increasing on these intervals. Indeed, plotting $y = p(x)$ supports this conclusion:

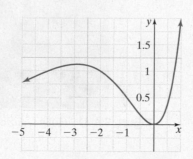

In the next problem, we encounter the notion that many events in biology are not certain but occur with a particular probability. Examples of this is an individual dying over a specified interval of time, a female giving birth to a particular number of individuals (e.g., the litter size of a mouse), or a molecule binding to a receptor site on a cell. You may have already learned, or will learn if you take a course in basic probability theory, that probabilities associated with independent events combine through multiplication. For example, if one flips a coin and the probability of getting heads is $p = 1/2$, then the probability of getting heads twice in a row is $p \times p = 1/2 \times 1/2 = 1/4$. Similarly, if one rolls a fair (honest or unbiased) six-sided die, then the probability of rolling a 3 is $p = 1/6$ and the probability of rolling an even number is $p = 3/6 = 1/2$. Thus the probability of getting a 3 on the first roll and an even number on the second roll is $1/6 \times 3/6 = 3/36 = 1/12$.

Although calculus and probability theory are generally taught as separate courses in college, probability is such an integral part of biological systems that it is useful for us to state and use the following fact for the examples we develop in this text.

Probability of two independent events. If p and q are the probabilities of the events 1 and 2 respectively occurring independent of one another, then pq is the probability that event 1 and event 2 both occur.

In Chapter 7, we revisit probability theory and explore how the integral calculus has become a central tool in the development of this theory.

Example 2   Survival rates

Over time, a zebra in Etosha National Park, Namibia, can die either because it is killed by a predator or because it succumbs to disease (primarily anthrax). Starting at the beginning of each year, suppose $f(t)$ and $g(t)$ respectively represent the probabilities of surviving predation and disease up to week t. Professor Getz's research group (see Figure 3.4) determined that at the height of a typical anthrax season, which occurs around the end of the eleventh week of a typical year (i.e., $t = 11$), $f(11) = 0.965$ and $g(11) = 0.981$. They also determined that the probability of surviving predation and disease at the end of the eleventh week is decreasing by 0.004 and 0.005 per week respectively.

Courtesy of Dr. Wayne Getz

Figure 3.4 Dr. Holly Ganz collecting blood from a zebra in Etosha National Park, Namibia, to test if it died of anthrax.

Assume that the events of dying from predation or disease are independent.

a. Find the probability of surviving to the end of week 11.

b. Find the rate at which this probability is changing at end of week 11.

c. Use your answers from parts **a** and **b** to estimate the probability of surviving to the end of week 12.

Solution

a. Since the events of being killed by a predator and dying from disease are assumed to be independent, then the probability $p(t)$ of surviving until week t is given, using the multiplication rule for independent probabilities, by the equation

$$p(t) = f(t)g(t)$$

Thus, the probability of surviving up to the end of week 11 is given by $p(11) = f(11)g(11) = 0.965 \times 0.981 = 0.9467$ (to 4 decimal places).

b. The problem statement implies that $f'(11) = -0.004$ and $g'(11) = -0.005$ (the negative values arise because the probabilities are decreasing). Using the product rule, we get

$$p'(11) = f'(11)g(11) + f(11)g'(11) = -0.965(0.004) - 0.981(0.005) = -0.0088$$

The probability of surviving is decreasing at rate of $0.0088 = 0.88\%$ per week at the end of week 11.

c. Since the probability of surviving is decreasing at a rate of 0.0088 per week at the end of week 11, we can estimate the probability surviving to the end of week 12 by

$$0.9467 - 0.0088 = 0.9379 = 93.79\%$$

Example 3   Per capita or intrinsic rate of growth

As we have seen in Section 3.1, single species population models can be of the form

$$N_{n+1} = N_n f(N_n) = g(N_n)$$

where N_n is the population abundance in the nth generation, $f(N)$ is the per capita growth rate of the population density as a function of population N, and $g(N)$ is the growth rate of the whole population as a function of N.

Find an expression in terms of f and N for $g'(0)$. Briefly explain what this expression represents.

Solution   Applying the product rule to the relationship $g(N) = Nf(N)$, we have

$$g'(N) = \left(\frac{d}{dN}N\right) f(N) + Nf'(N)$$
$$= f(N) + Nf'(N)$$

Evaluating at $N = 0$,

$$g'(0) = f(0) + 0 \times f'(0) = f(0)$$

Hence, the rate $g'(0)$ at which growth changes at low densities equals the per capita growth rate of the population at low densities.

Quotient rule

Before we derive a quotient rule, we begin with an example for finding the derivative of a reciprocal, which is a special case of a quotient that has 1 in the numerator.

Example 4　Reciprocal rule

Find the derivative of the reciprocal $\dfrac{1}{f(x)}$ of a differentiable function f by using the definition of derivative.

Solution　Let $r(x) = \dfrac{1}{f(x)}$. Then $r(x + h) = \dfrac{1}{f(x + h)}$, so using the definition of derivative we find:

$$r'(x) = \lim_{h \to 0} \frac{r(x + h) - r(x)}{h} \qquad \textit{definition of derivative}$$

$$= \lim_{h \to 0} \frac{\frac{1}{f(x+h)} - \frac{1}{f(x)}}{h}$$

$$= \lim_{h \to 0} \frac{\frac{f(x) - f(x+h)}{f(x) f(x+h)}}{h} \qquad \textit{common denominator for the numerator}$$

$$= \lim_{h \to 0} \frac{f(x) - f(x + h)}{h f(x) f(x + h)} \qquad \begin{array}{l} \textit{multiplying numerator and denominator by} \\ f(x) f(x + h) \end{array}$$

$$= \lim_{h \to 0} \frac{1}{f(x) f(x + h)} \lim_{h \to 0} \frac{f(x) - f(x + h)}{h} \qquad \textit{limit of a product}$$

$$= \frac{1}{[f(x)]^2} \lim_{h \to 0} \left[-\frac{f(x + h) - f(x)}{h} \right] \qquad \begin{array}{l} \textit{since } f \textit{ is continuous and factoring} \\ \textit{out } -1 \end{array}$$

$$= \frac{1}{[f(x)]^2} [-f'(x)] \qquad \textit{definition of derivative}$$

We restate the result of this example for easy reference.

Reciprocal Rule

Let f be differentiable at x. Then

$$\frac{d}{dx} \left[\frac{1}{f(x)} \right] = -\frac{f'(x)}{[f(x)]^2}$$

provided that $f(x) \neq 0$.

Example 5　Using the reciprocal rule

Find the derivative of $g(x) = \dfrac{1}{x^2 + x + 1}$.

Solution　Let $f(x) = x^2 + x + 1$. Then $g(x) = \dfrac{1}{f(x)}$ and $f'(x) = 2x + 1$. By the reciprocal rule,

$$g'(x) = -\frac{f'(x)}{f(x)^2}$$

$$= -\frac{2x + 1}{(x^2 + x + 1)^2}$$

Example 6 Breaking Whelks

Crows feed on whelks by flying up and dropping the whelks (Figure 3.5) on a hard surface to break them.

Figure 3.5 Whelks (*Nucella lamellosa*, pictured here among other species of marine molluscs) are frequently consumed by Northwestern crows.

Biologists have noticed that Northwestern crows consistently drop whelks from about five meters. As a first step to understanding why this might be the case, we consider data collected by the Canadian scientist Reto Zach in which he repeatedly dropped whelks from various heights to determine how many drops were required to break the whelk. The data are shown in Figure 3.6.

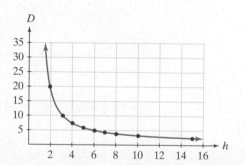

Figure 3.6 Data showing how the number of drops to break a whelk depends on the height of the drops.

Data Source: Reto Zach, 1979. Shell Dropping: Decision-Making and Optimal Foraging in Northwestern Crows. *Behaviour*, 68, pp. 106–117.

A best-fitting curve relating the number of drops, *D*, to the height, *h* (in meters), for these data is given by

$$D(h) = 1 + \frac{20.4}{h - 0.84}$$

a. Find $\dfrac{dD}{dh}$

b. Find $\dfrac{dD}{dh}\Big|_{h=4}$ and interpret this quantity.

In Chapter 4, we shall use this function to determine the optimal height from which to drop whelks.

Solution

a. $\dfrac{dD}{dh} = \dfrac{d}{dh}(1) + 20.4\dfrac{d}{dh}\left[\dfrac{1}{h - 0.84}\right]$ *elementary rules of differentiation*

$= 0 + 20.4\left[\dfrac{-1}{(h - 0.84)^2}\right]$ *reciprocal rule*

$= \dfrac{-20.4}{(h - 0.84)^2}$

b. $\dfrac{dD}{dh}\Big|_{h=4} = \dfrac{-20.4}{(4 - 0.84)^2} \approx -2.04$

At $h = 4$ meters, the required number of drops decreases at a rate of -2.04 per meter. For instance, if we increased the height by approximately 1 meter, we should expect the number of drops to decrease by approximately 2. This can also be seen on the graph in Figure 3.6 from the fact that at $h = 4$, $D \approx 8$, while at $h = 5$, $D \approx 8 - 2 = 6$.   ■

Combining the reciprocal and product rule, we can find the derivative of a quotient of functions. Let f and g be differentiable functions, and assume that $g(x) \neq 0$.

$$\frac{d}{dx}\left[\frac{f(x)}{g(x)}\right] = \frac{d}{dx}\left[f(x) \cdot \frac{1}{g(x)}\right]$$

$$= f'(x) \cdot \frac{1}{g(x)} + f(x)\frac{d}{dx}\left[\frac{1}{g(x)}\right]   \textit{product rule}$$

$$= \frac{f'(x)}{g(x)} + f(x)\left[\frac{-g'(x)}{g(x)^2}\right]   \textit{reciprocal rule}$$

$$= \frac{f'(x)g(x) - f(x)g'(x)}{g(x)^2}   \textit{common denominator}$$

Thus, we have derived what is known as the **quotient rule**.

Quotient Rule

Let f and g be differentiable at x. Then

$$(f/g)'(x) = \frac{f'(x)g(x) - f(x)g'(x)}{g(x)^2}$$

provided $g(x) \neq 0$.

There exist a variety of playful mnemonics that can be used to remember the quotient rule. For example, if we replace f by *hi* and g by *lo*, then we get the limmrick "lo-dee-hi less hi-dee-lo, draw the line and square below."

Example 7   Computing with the quotient rule

Find the following derivatives.

a. $\dfrac{d}{dt}\left[\dfrac{1 + 2t}{3 + 4t}\right]$ **b.** $\dfrac{d}{dx}\left[\dfrac{e^x}{1 + x^2}\right]$

Solution

a. Let $f(t) = 1 + 2t$ and $g(t) = 3 + 4t$. By the quotient rule

$$\frac{d}{dt}\left[\frac{1+2t}{3+4t}\right] = \frac{d}{dt}\left[\frac{f(t)}{g(t)}\right]$$

$$= \frac{f'(t)g(t) - f(t)g'(t)}{g(t)^2}$$

$$= \frac{2(3+4t) - (1+2t)4}{(3+4t)^2}$$

$$= \frac{2}{(3+4t)^2}$$

b. Let $f(x) = e^x$ and $g(x) = 1 + x^2$. By the quotient rule

$$\frac{d}{dx}\left[\frac{e^x}{1+x^2}\right] = \frac{d}{dx}\left[\frac{f(x)}{g(x)}\right]$$

$$= \frac{f'(x)g(x) - f(x)g'(x)}{g(x)^2}$$

$$= \frac{e^x(1+x^2) - e^x 2x}{(1+x^2)^2}$$

$$= \frac{e^x(x^2 - 2x + 1)}{(1+x^2)^2}$$

$$= \frac{e^x(x-1)^2}{(1+x^2)^2}$$

Example 8　Dose–response curves revisited

In Example 2 of Section 2.4, a dose–response curve for patients responding to a dose of histamine was given by

$$R = \frac{100e^x}{e^x + e^{-5}}$$

where x is the natural logarithm of the dosage in millimoles (mmol).

a. Find $\dfrac{dR}{dx}$.

b. Graph $\dfrac{dR}{dx}$ to determine at what logarithmic dosage the response is increasing most rapidly.

Solution

a.
$$\frac{d}{dx}\left[\frac{100e^x}{e^x + e^{-5}}\right] = \frac{100e^x(e^x + e^{-5}) - e^x 100e^x}{(e^x + e^{-5})^2}$$

$$= \frac{100e^{x-5}}{(e^x + e^{-5})^2}$$

b. Graphing $\dfrac{dR}{dx}$ yields

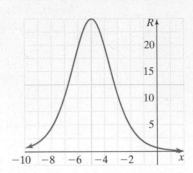

Hence, $\dfrac{dR}{dx}$ takes on its largest value at approximately $x = -5$, and the response increases most rapidly at this logarithmic dose. That is, it increases most rapidly at the dose $e^{-5} \approx 0.0067$ mmol. ∎

PROBLEM SET 3.2

Level 1 DRILL PROBLEMS

Find the derivatives in Problems 1 to 18.

1. $p(x) = (3x^2 - 1)(7 + 2x^3)$

2. $p(x) = (x^2 + 4)(1 - 3x^3)$

3. $q(x) = \dfrac{4x - 7}{3 - x^2}$

4. $q(x) = \dfrac{x + 1}{1 + x^2}$

5. $f(x) = x2^x$

6. $f(x) = x^3 3^x$

7. $f(x) = (1 + x + x^2)e^x$

8. $f(x) = (e^3 + e^2 + e)x^2$

9. $F(L) = (1 + L + L^3 + L^4)(L - L^2)$

10. $G(M) = (M - M^3)(1 - 4M)$

11. $f(x) = (4x + 3)^2$ Hint: Think $(4x + 3)(4x + 3)$

12. $g(x) = (5 - 2x)^2$

13. $f(x) = \dfrac{e^x}{1 + e^x}$

14. $g(t) = \dfrac{1 + te^t}{1 + t}$

15. $f(p) = \dfrac{ap}{1 + 2^p}$ where a is a constant

16. $g(m) = \dfrac{bm}{1 - 3^m}$ where b is a constant

17. $F(x) = \dfrac{2}{3x^2} - \dfrac{x}{3} + \dfrac{4}{5} + \dfrac{x + 1}{x}$

18. $G(x) = x^2 - \dfrac{1}{x^2} + \dfrac{5}{x^4}$

Find the equation for the tangent line at the prescribed point for each function in Problems 19 to 24.

19. $f(x) = (x^3 - 2x^2)(x + 2)$ where $x = 1$

20. $G(x) = (x - 5)(x^3 - x)$ where $x = -1$

21. $F(x) = \dfrac{x + 1}{x - 1}$ where $x = 0$

22. $f(x) = e^x + e^{-x}$ where $x = 0$

23. $F(x) = \dfrac{3x^2 + 5}{2x^2 + x - 3}$ where $x = -1$

24. $g(x) = x \ln x$ where $x = 1$.

25. a. Differentiate the function
$$f(x) = 2x^2 - 5x - 3$$

b. Factor the function in part **a** and differentiate by using the product rule. Show that the two answers are the same.

26. a. Use the quotient rule to differentiate
$$f(x) = \dfrac{2x - 3}{x^3}$$

b. Rewrite the function in part **a** as $f(x) = x^{-3}(2x - 3)$ and differentiate by using the product rule.

c. Rewrite the function in part **a** as $f(x) = 2x^{-2} - 3x^{-3}$ and differentiate using the power rule.

d. Show that the answers to parts **a**, **b**, and **c** are all the same.

Level 2 APPLIED AND THEORY PROBLEMS

27. The body mass index (BMI) for an individual who weighs w pounds and is h inches tall is given by

$$B = \frac{703w}{h^2}$$

$$\text{BMI} = \left\{ \frac{\text{WEIGHT (pounds)}}{\text{HEIGHT (inches)}^2} \right\} \times 703$$

weight in pounds

	120	130	140	150	160	170	180	190	200	210	220	230	240	250
4'6	29	31	34	36	39	41	43	46	48	51	53	56	58	60
4'8	27	29	31	34	36	38	40	43	45	47	49	52	54	56
4'10	25	27	29	31	34	36	38	40	42	44	46	48	50	52
5'0	23	25	27	29	31	33	35	37	39	41	43	45	47	49
5'2	22	24	26	27	29	31	33	35	37	38	40	42	44	46
5'4	21	22	24	26	28	29	31	33	34	36	38	40	41	43
5'6	19	21	23	24	26	27	29	31	32	34	36	37	39	40
5'8	18	20	21	23	24	26	27	29	30	32	34	35	37	38
5'10	17	19	20	22	23	24	26	27	29	30	32	33	35	36
6'0	16	18	19	20	22	23	24	26	27	28	30	31	33	34
6'2	15	17	18	19	21	22	23	24	26	27	28	30	31	32
6'4	15	16	17	18	20	21	22	23	24	26	27	28	29	30
6'6	14	15	16	17	19	20	21	22	23	24	25	27	28	29
6'8	13	14	15	17	18	19	20	21	22	23	24	25	26	28

height in feet and inches

■ healthy weight ■ overweight ■ obese

Source: The U.S. Department of Health and Human Services. See: http://www.surgeongeneral.gov/library/calls/obesity/fact_advice.html

A person with a body mass index greater than 30 is considered obese; see the BMI chart.

a. Consider all adults who are $h = 63$ inches tall. After setting $h = 63$ inches so that B is now a function of w only, find $\dfrac{dB}{dw}$ at $w = 130$ and interpret your results.

b. Consider all children who weigh 60 lb. After setting $w = 60$ lb so that B is now a function of h only, find $\dfrac{dB}{dh}$ at $h = 54$ and interpret your results.

28. In addition to zebra, Professor Getz's group studied springbok (a type of antelope) in Etosha National Park. In a study paralleling Example 2 of this section, they found that the probability of a springbok surviving predation and disease to the end of the eighth week of a typical year is given by, respectively, 94.9% and 98.8%. Suppose they also determined that at the end of this eight weeks the probability of dying from predation and disease is decreasing respectively by 0.6% and 0.25% per week. Assume the events of dying from predation or disease are independent.

a. Find the probability of surviving to the end of week 8.

b. Find the rate of change of probability of surviving at the end of week 8.

c. Estimate the probability of surviving to the end of week 9.

29. Suppose that Professor Getz's group found that the probability of zebra surviving both predation and disease to the end of week 4 was 98.4% and 99.6%, respectively. Suppose they also determined that at the end of week 4 the probability of a zebra dying from predation and disease is decreasing respectively by 0.4% and 0.1% per week (remember when converting percentages to numbers to shift the value by two decimal places). Assume the events of dying from predation or disease are independent.

a. Find the probability of surviving up to the end of week 4.

b. Find the rate of change of probability of surviving at the end of week 4.

c. Estimate the probability of surviving to end of week 5.

30. Consider the generalized Beverton-Holt model of population growth given by

$$N_{n+1} = g(N_n)$$

where

$$g(N) = \frac{N}{1 + (aN)^b}$$

and $a > 0, b > 0$.

a. Find $g'(N)$.

b. Determine what values of $b > 0$ cause g to be increasing for all $N > 0$.

c. When b is outside the range of values found in part **b**, determine on what interval g is increasing and on what interval g is decreasing.

31. A *ligand* is a molecule that binds to another molecule or other chemically active structure (*e.g.*, a receptor on a membrane) to form a larger complex. In a study of two ligands (I and II) competing for the same sites on a substrate, ligand II is added to a substrate solution that already contains ligand I. As the concentration of ligand II is increased, the concentration of ligand I bound to the substrate decreases. This one-site ligand competition process is characterized by this equation:

$$T = a + \frac{b - a}{1 + 10^{x - c}}$$

where T is the concentration of bound ligand I per milligram of tissue and x is the exponent of the concentration of ligand II in the solution. The constants

a and b arise from the relative binding rates of the two ligands and satisfy $a > 0$ and $b > a$.

a. Compute $\dfrac{dT}{dx}$

b. Interpret the quantities a, b, and c and sketch the graph.

32. In the 1960s, scientists at Woods Hole Oceanographic Institution measured the uptake rate of glucose by bacterial populations from the coast of Peru (R. R. Vaccaro and H. W. Jannasch, "Variations in Uptake Kinetics for Glucose by Natural Populations in Sea Water," *Limnol. Oceanogr.* (1967) 12, 540–542.). In one field experiment, they found that the uptake rate can be modeled by $f(x) = \dfrac{1.2078x}{1 + 0.0506x}$ micrograms per hour where x is micrograms of glucose per liter. Compute and interpret $f'(20)$ and $f'(100)$.

33. In Example 5 of Section 2.4, we found that the predation rate of wolves on moose in North America could be modeled by

$$f(x) = \frac{3.36x}{0.46 + x} \text{ moose killed per wolf per 100 days}$$

where x is measured in number of moose per square kilometer. Compute and interpret $f'(0.5)$ and $f'(2.0)$.

34. Cells often use receptors to transport nutrients from outside of the cell membrane to the inner cell. In Example 6 of Section 1.6, we determined that the rate, R, at which nutrients enter the cell depends on the concentration, N, of nutrients outside the cell. The function

$$R = \frac{aN}{b + N}$$

models the amount of nutrients absorbed in one hour where a and b are positive constants.

a. Find R when $N = b$. What does this tell you about b?

b. Compute and interpret $\dfrac{dR}{dN}$. When is R increasing? When is R decreasing?

35. In Problem 42 in Problem Set 2.4, we modeled how wolf densities in North America depend on moose densities with the following function

$$f(x) = \frac{58.7(x - 0.03)}{0.76 + x} \text{ wolves per } 1000 \text{ km}^2$$

where x is number of moose per square kilometer. Determine for what x values $f(x)$ is increasing.

36. In Problem 45 in Problem Set 2.3, two fisheries scientists (T. J. Quinn and R. B. Deriso, "Stock and Recruitment," Chapter 3 (pp. 86–127) in *Quantitative Fish Dynamics*, 1999. Oxford University Press, New York, New York) found

that the following stock-recruitment function provides a good fit to data pertaining to the southeast Alaska pink salmon fishery:

$$y = 0.12x^{1.5}e^{-0.00014x}$$

where y is the number of young fish recruited, and x is the number of adult fish involved in recruitment.

a. Compute $\dfrac{dy}{dx}$.

b. Determine for what x values y is increasing and decreasing. Interpret your results.

37. This problem uses the hyperbolic secant function, denoted $\operatorname{sech} x$, which is an important function in mathematics and is defined by the formula

$$\operatorname{sech} x = \frac{2}{e^x + e^{-x}}$$

We will discuss further in Chapter 4 how two mathematicians, W. O. Kermack and A. G. McKendrick, showed that the weekly mortality rate during the outbreak of the plague in Bombay in 1905–1906 can be reasonably well described by the function

$$f(t) = 890 \operatorname{sech}^2(0.2t - 3.4) \text{ deaths/week}$$

where t is measured in weeks. Determine when the mortality rate is increasing and when the mortality rate is decreasing.

38. In a classic paper, V. A. Tucker and K. Schmidt-Koenig modeled the energy expended by a species of Australian parakeet during flight ("Flight of Birds in Relation to Energetics and Wind Directions," *The Auk* 88 (1971): 97–107). They used the following function:

$$E(v) = \frac{[0.074(v - 35)^2 + 22]}{v} \text{ calories per kilometer}$$

where v is the bird's velocity (km/h).

a. Find a formula for the rate of change of energy with respect to v.

b. At what velocity, v, is the energy expenditure neither increasing nor decreasing? Discuss the importance of this velocity.

Australian budgerigar (*Melopsittacus undulatus*)

Chain Rule and Implicit Differentiation

We now move to the next level in terms of developing tools to differentiate functions that can be regarded as the composite of more elementary functions (discussed in Section 1.5). These tools will give us the power to differentiate functions such as the bell-shaped curve $y = e^{-x^2}$, the polynomial $y = (1 + 2x + x^3)^{101}$, and the logarithm function $y = \ln x$.

Chain rule

Suppose we were asked to find the derivative of the function $y = (1 + 2x + x^3)^{101}$. It is not practical to expand this product in order to take the derivative of a polynomial. Instead, we can use a result known as the *chain rule*. In order to motivate this important rule, let us consider an application.

It is known that the carbon monoxide pollution in the air is changing at the rate of 0.02 ppm (parts per million) for each person in a town whose population is growing at the rate of 1000 people per year. To find the rate at which the level of pollution is increasing with respect to time, we must compute the product

$$(0.02 \text{ ppm/person})(1000 \text{ people/year}) = 20 \text{ppm/year}$$

We can generalize this commonsense calculation by noting that the pollution level, L, is a function of the population size, P, which itself is a function of time, t. Thus, L as a function of time is $(L \circ P)(t)$ or, equivalently, $L[P(t)]$. With this notation, the commonsense calculation becomes:

$$\begin{bmatrix} \text{RATE OF CHANGE OF } L \\ \text{WITH RESPECT TO } t \end{bmatrix} = \begin{bmatrix} \text{RATE OF CHANGE OF } L \\ \text{WITH RESPECT TO } P \end{bmatrix} \begin{bmatrix} \text{RATE OF CHANGE OF } P \\ \text{WITH RESPECT TO } t \end{bmatrix}$$

Expressing each of these rates in terms of an appropriate derivative of $L[P(t)]$ in Leibniz form, we obtain the following equation:

$$\frac{dL}{dt} = \frac{dL}{dP}\frac{dP}{dt}$$

These observations anticipate the following important result known as the **chain rule**.

Chain Rule

If $y = f(u)$ is a differentiable function of u, and u, in turn, is a differentiable function of x, then $y = f[u(x)]$ is a differentiable function of x, and its derivative is given by the product

$$\frac{dy}{dx} = \frac{dy}{du}\frac{du}{dx}$$

Equivalently,

$$(f \circ u)'(x) = f'[u(x)]u'(x)$$

Proof. To prove the chain rule, define

$$G(h) = \begin{cases} \dfrac{f[u(x + h)] - f[u(x)]}{u(x + h) - u(x)} & \text{if } u(x + h) \neq u(x) \\ f'[u(x)] & \text{otherwise} \end{cases}$$

It should be intuitive that $G(h)$ is continuous at $h = 0$. (You will be asked to verify this statement in Problem Set 3.3). With this observation in hand, the proof of the chain rule becomes straightforward. By the definition of the derivative.

$$(f \circ u)'(x) = \lim_{h \to 0} \frac{f[u(x+h)] - f[u(x)]}{h} \qquad \textit{definition of derivative; note } h \neq 0$$

$$= \lim_{h \to 0} \left[G(h) \cdot \frac{u(x+h) - u(x)}{h} \right] \qquad \textit{definition of G}$$

$$= \lim_{h \to 0} G(h) \lim_{h \to 0} \frac{u(x+h) - u(x)}{h} \qquad \textit{limit law for products}$$

$$= G(0)\, u'(x) \qquad \textit{continuity of G at 0 and differentiability of u at a}$$

$$= f'[u(x)]\, u'(x) \qquad \textit{definition of G}$$

Example 1 Life made easier

Let $y = \dfrac{d}{dx}(1 + 2x + x^3)^{101}$. Find $\dfrac{dy}{dx}$.

Solution View this as the composition of two functions: the "inner" function $u(x) = 1 + 2x + x^3$ and the "outer" function $f(u) = u^{101}$. Now use the chain rule:

$$\frac{dy}{dx} = \frac{dy}{du} \cdot \frac{du}{dx}$$

$$= 101u(x)^{100}(0 + 2 + 3x^2)$$

$$= 101(1 + 2x + x^3)^{100}(2 + 3x^2)$$

In practice, we usually do not write down a function u, but carry out the above process mentally and write

$$y = (1 + 2x + x^3)^{101}$$

$$\frac{dy}{dx} = \underbrace{101(1 + 2x + x^3)^{100}}_{\substack{\textit{derivative of} \\ \textit{outer function}}} \underbrace{(2 + 3x^2)}_{\substack{\textit{derivative of} \\ \textit{inner function}}}$$

Example 2 Escaping parasitism

Parasitoids, usually wasps or flies, are insects whose young develop on and eventually kill their host, typically another insect. Parasitoids have been extremely successful in controlling insect pests, especially in agriculture. To better understand this success, theoreticians have extensively modeled host-parasitoid interactions. A key term in these models is the so-called *escape function* $f(x)$, which describes the fraction of hosts that escape parasitism when the parasitoid density is x individuals per unit area. If parasitoid attacks are randomly distributed among the hosts, then the escape function is the form $f(x) = e^{-ax}$ where a is the searching efficiency of the parasitoid.

Suppose that a population of parasitoids attacks alfalfa aphids with a searching efficiency of $a = 0.01$. If the density of parasitoids is currently 100 wasps per acre and is increasing at a rate of 20 wasps per acre per day, find the rate at which the fraction of aphids escaping parasitism is initially changing.

Solution Let time in days be denoted by the independent variable t, starting with $t = 0$ at the current time. Since the density of wasps $x(t)$ is changing with time, the fraction of hosts that escape is a composition of two functions $f[x(t)]$. Hence, by the chain rule

$$\frac{df}{dt} = \frac{df}{dx} \cdot \frac{dx}{dt} \qquad \textit{chain rule}$$

$$= (-0.01)e^{-0.01x} \cdot 20 \qquad \frac{dx}{dt} = 20 \textit{ is given}$$

$$= -0.2e^{-0.01x}$$

Evaluating at the current time $t = 0$ at which $x(0) = 100$, we get

$$\frac{d}{dt}\Big|_{t=0} f[x(t)] = -0.2e^{-0.01(100)} \approx -0.074.$$

Thus the fraction of hosts escaping is decreasing at a rate of 0.074 per day at $t = 0$.

In Example 2, we found the derivative of $f(x) = e^{-0.01x}$ by using the derivative of an exponential (Section 3.1), but we could also use the chain rule. It is worthwhile to restate an extended derivative rule for a natural exponential function:

$$\frac{d}{dx}e^{u(x)} = e^{u(x)}\frac{du}{dx}$$

We illustrate this idea with the following example.

Example 3   The bell-shaped curve

Consider the bell-shaped function

$$f(x) = e^{-x^2}$$

a. Find f' and determine where f is increasing and where it is decreasing.

b. Plot f and f' on the same coordinate axes and graphically verify the results from part **a**.

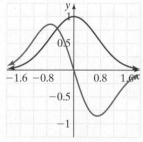

Figure 3.7 Graph of bell-shaped function (red) and its derivative (blue).

Solution

a. Using the extended derivative rule for a natural exponential function with $u = -x^2$, we find

$$f'(x) = e^{-x^2}(-2x) = -2xe^{-x^2}$$

Since $e^{-x^2} > 0$, the derivative is positive when $x < 0$, so the function f is increasing on $(-\infty, 0)$; and it is negative when $x > 0$, so the function is decreasing on $(0, \infty)$.

b. The graph of $y = f(x)$ is shown in red and $y = f'(x)$ in blue in Figure 3.7.

We see that the derivative function (blue) is positive where the bell-shaped curve (red) is rising, and the derivative function is negative where the bell-shaped curve is falling.

Example 4   Chain rule with graphs

Consider the functions $y = f(x)$ and $y = g(x)$ whose graphs in red and blue, respectively, are as shown here:

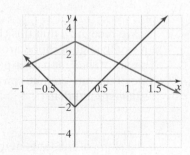

Find

$$(f \circ g)'(-0.5)$$

Solution   By the chain rule,

$$(f \circ g)'(-0.5) = f'[g(-0.5)]g'(-0.5)$$

By inspection, $g(-0.5) = 2$. To find the derivative of g at -0.5, we note that g is linear on the interval $[-1, 0]$, so the derivative is the slope of the line segment which we find by rise/run $= 2/1 = 2$, thus, $g'(-0.5) = 2$.

To find the derivative of f at $g(-0.5) = 2$, we note that f (red curve) is linear on $[0, 2]$ and has slope $m = $ rise/run $8/2 = 4$. Thus, $f'[g(-0.5)] = f'(2) = 4$.

We conclude that

$$\frac{d}{dx}\Big|_{x=-0.5} f(g(x)) = f'(g(-0.5))g'(-0.5) = 4 \times 2 = 8$$

Implicit differentiation

The equation $y = \sqrt{25 - x^2}$ *explicitly* defines $f(x) = \sqrt{25 - x^2}$ as a function of x for $-5 \le x \le 5$. The same function can also be defined *implicitly* by the equation $x^2 + y^2 = 25$, as long as we restrict y by $0 \le y \le 5$ so the vertical line test is satisfied. To find the derivative of the explicit form, we use the chain rule:

$$\frac{d}{dx}\sqrt{25 - x^2} = \frac{d}{dx}(25 - x^2)^{1/2}$$

$$= \frac{1}{2}(25 - x^2)^{-1/2}(-2x) \qquad \begin{array}{l}\textit{chain rule with } f(u) = u^{1/2} \textit{ and} \\ \hspace{3.5em} u(x) = 25 - x^2\end{array}$$

$$= \frac{-x}{\sqrt{25 - x^2}}$$

To obtain the derivative of the same function in its implicit form, we simply differentiate across the equation $x^2 + y^2 = 25$, remembering that y is a function of x and using the chain rule:

$$\frac{d}{dx}(x^2 + y^2) = \frac{d}{dx}(25) \qquad \textit{differentiate both sides}$$

$$2x + 2y\frac{dy}{dx} = 0 \qquad \textit{chain rule for the derivative of } y^2$$

$$\frac{dy}{dx} = -\frac{x}{y} \qquad \textit{solve for } \frac{dy}{dx}$$

$$= -\frac{x}{\sqrt{25 - x^2}} \qquad \textit{write as a function of x, if desired}$$

The procedure we have just illustrated is called **implicit differentiation**.

Example 5   Circular tangents

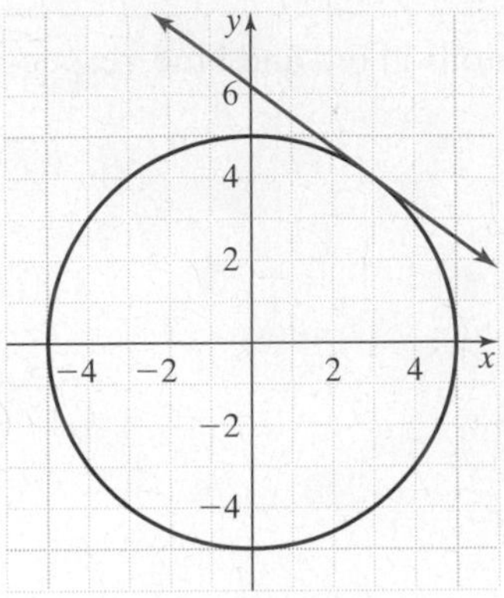

Figure 3.8 Tangent line (red) to a given circle.

Consider a circle of radius 5 centered at the origin. Find the equation of the tangent line of this circle at $(3, 4)$.

Solution   The equation of this circle is

$$x^2 + y^2 = 25$$

We recognize that this circle is not the graph of a function. However, if we look at a small neighborhood around the point $(3, 4)$, as shown in Figure 3.8, we see that this part of the graph does pass the vertical line test for functions. Thus, the required slope of the tangent line can be found by evaluating the derivative of dy/dx at $(3, 4)$. We have found that

$$\frac{dy}{dx} = -\frac{x}{y}$$

so the slope of the tangent at $(3, 4)$ is

$$\frac{dy}{dx} = -\frac{3}{4}$$

Thus, the equation of the tangent line is

$$y - 4 = -\frac{3}{4}(x - 3)$$

$$y = -\frac{3}{4}x + \frac{9}{4} + \frac{16}{4}$$

$$y = -\frac{3}{4}x + \frac{25}{4}$$

More generally, given any equation involving x and y, we can differentiate both sides of the equation, use the chain rule, and solve for $\frac{dy}{dx}$. This becomes particularly important when one cannot (or not easily) express y in terms of x explicitly. The next example is of this type and is included because of its historical importance and aesthetic appeal. You may think that such curves have no application to biology; but, for example, the logarithmic spiral (see Figure 1.56) encountered in Project 1C: Golden Ratio (at the end of Chapter 1) describes the growth of the nautilus mollusk.

Example 6   Limaçon of Pascal

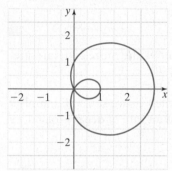

Figure 3.9 Limaçon of Pascal.

The limaçon of Pascal is a famous curve defined by the set of points satisfying

$$(x^2 + y^2 - 2x)^2 = x^2 + y^2$$

The graph is shown in Figure 3.9. This curve was discovered by Étienne Pascal, the father of the more famous Blaise Pascal. The name *limaçon* comes from the Latin *limax*, which means "a snail." Find the equation for the tangent line at the point $(0, 1)$.

Solution   To find the slope of the tangent line, we differentiate both sides implicitly and then evaluate at $(0, 1)$.

$$(x^2 + y^2 - 2x)^2 = x^2 + y^2 \qquad \text{given equation}$$

$$\frac{d}{dx}(x^2 + y^2 - 2x)^2 = \frac{d}{dx}(x^2 + y^2) \qquad \begin{array}{l}\text{differentiate both sides with}\\ \text{respect to } x\end{array}$$

$$2(x^2 + y^2 - 2x)\frac{d}{dx}(x^2 + y^2 - 2x) = 2x + 2y\frac{dy}{dx} \qquad \text{chain rule}$$

$$2(x^2 + y^2 - 2x)(2x + 2y\frac{dy}{dx} - 2) = 2x + 2y\frac{dy}{dx} \qquad \text{chain rule again}$$

$$2(0^2 + 1^2 - 2 \cdot 0)(2 \cdot 0 + 2 \cdot 1\frac{dy}{dx} - 2) = 2(0) + 2(1)\frac{dy}{dx} \qquad \text{evaluate at } (0, 1)$$

$$2(1)(2\frac{dy}{dx} - 2) = 2\frac{dy}{dx} \qquad \text{simplify}$$

$$4\frac{dy}{dx} - 4 = 2\frac{dy}{dx}$$

$$2\frac{dy}{dx} = 4$$

$$\frac{dy}{dx} = 2 \qquad \text{solve for } \frac{dy}{dx}$$

Hence, the slope of the tangent line is 2 and the tangent line is

$$y - 1 = 2(x - 0)$$

$$y = 2x + 1$$

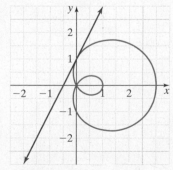

Figure 3.10 Limaçon with tangent line at $(0, 1)$.

The tangent line is shown in Figure 3.10.

Derivatives of logarithms

Implicit differentiation allows us to easily find the derivative of logarithms and power functions.

Derivative of the Natural Logarithm	If $y = \ln x$, then $$\frac{dy}{dx} = \frac{1}{x}$$

Proof.

$$y = \ln x \qquad \textit{given function}$$

$$e^y = x \qquad \textit{exponentiating both sides}$$

$$\frac{d}{dx}e^y = \frac{d}{dx}(x) \qquad \textit{differentiate both sides}$$

$$e^y \frac{dy}{dx} = 1 \qquad \textit{chain rule}$$

$$\frac{dy}{dx} = \frac{1}{e^y} \qquad \textit{solve for } dy/dx$$

$$= \frac{1}{x} \qquad \textit{substitute}$$

In the problem set of this section you are asked to prove the more general statements of this result; namely, if $y = \ln |x|$ then, for $x \neq 0$, $\frac{dy}{dx} = \frac{1}{x}$.

The next example deals with the clearance rate—which, as we saw in Example 7 of Section 3.1 in the context of viral particles, is the constant rate of decay—of a drug from the blood stream of an individual.

Example 7   Clearance of acetaminophen

Scientists have estimated that the clearance rate of the drug acetaminophen in the blood stream of an average adult is 0.28 per hour. This means that after an initial dose of acetaminophen at time $t = 0$, the fraction of acetaminophen in the blood t hours later is $e^{-0.28t}$.

a. Find the time, T, it takes for a fraction x of the drug to clear the body.

b. Find and interpret $\left.\dfrac{dT}{dx}\right|_{x=1/2}$.

Solution

a. Since $e^{-0.28t}$ is the fraction of drug remaining in the body at time t and x is the fraction that has left the body, we need to solve

$$1 - x = e^{-0.28T}$$
$$\ln(1 - x) = -0.28T$$
$$T = -3.57 \ln(1 - x)$$

b. We use the results of part **a** to find

$$\left.\frac{dT}{dx}\right|_{x=1/2} = \left.\frac{(-3.57)(-1)}{1 - x}\right|_{x=1/2}$$

$$= \frac{3.57}{1 - 0.5}$$

$$= 7.14$$

Thus, the time it takes to clear an extra percentage of the drug, given 50% ($x = \frac{1}{2}$ is 50%) has cleared, is approximately $7.14 \times 0.01 = 0.0714$ hours or 4 minutes and 17 seconds. ∎

If the base on the logarithm has a base other than the natural base, e, then we use the following result, which you will be asked to verify in the Problem 41 of Problem Set 3.3:

Derivative of a General Logarithm	If b is a positive number (other than 1) and $x > 0$ then $$\frac{d}{dx} \log_b x = \frac{1}{x \ln b}$$

Example 8   Derivative of a log with base 2

For $x > 0$ differentiate $f(x) = x \log_2 x$.

Solution

$$f'(x) = \left(\frac{d}{dx} x\right) \log_2 x + x \frac{d}{dx} \log_2 x \qquad \text{by the product rule}$$

$$= (1) \log_2 x + x \frac{1}{\ln 2} \frac{1}{x} \qquad \text{by derivative of general logarithm}$$

$$= \log_2 x + \frac{1}{\ln 2}$$

∎

In Section 3.1, we stated the power rule and promised to prove it later in this chapter for all real numbers. We fulfill this promise in the following example.

Example 9   Power law for positive real numbers

Consider $y = x^n$ where $x > 0$ and n is any real number other than 0. Prove that

$$\frac{dy}{dx} = nx^{n-1}$$

Solution   We will prove this for $x > 0$ by taking the natural logarithm of both sides and then differentiating to find the derivative.

$$y = x^n \qquad \text{given equation}$$
$$\ln y = \ln x^n \qquad \text{take the natural logarithm of both sides}$$
$$\ln y = n \ln x \qquad \text{property of logarithms}$$
$$\frac{1}{y} \frac{dy}{dx} = n \frac{1}{x} \qquad \text{by chain rule and derivatives of natural logarithm}$$
$$\frac{dy}{dx} = n \frac{y}{x} \qquad \text{solve for } \frac{dy}{dx}$$
$$= n \frac{x^n}{x} \qquad \text{since } y = x^n$$
$$= nx^{n-1} \qquad \text{property of exponents}$$

You will encounter the proof for $x < 0$ in the problem set at the end of this section.

Example 10   Modeling problem using the chain rule

Table 3.1 Population as a function of time

Time t	Population $p(t)$
0	4.6696
0.25	4.6717
0.5	4.6779
0.75	4.6884
1	4.7032
1.25	4.7225
1.5	4.7463
1.75	4.7751
2	4.8088
2.25	4.8479
2.5	4.8926
2.75	4.9432
3	5.0000

An environmental study of a certain suburban community suggests that when the population is p thousand people, the amount of carbon monoxide in the air can be modeled by the function

$$C(p) = \sqrt{0.5p^2 + 17}$$

where C is measured in parts per million. The population (in millions) at various times (in years) for the last three years is given in Table 3.1.

a. Decide through visual inspection (plotting each of the two graphs together with the data) which of the following functions model the population data most accurately. (Note how subscripts we apply to y allow us to distinguish between the values predicted by the two models, although once the best model is selected, we will rename the left-hand side p.) Compute the derivatives for each of these models.

- Linear: $y_l = 0.109t + 4.618$
- Quadratic: $y_q = 0.039t^2 - 0.009t + 4.672$

b. Using the model you selected in part **a**, find the rate at which the level of pollution is changing at the end of year three.

Solution

a. To determine which of the three proposed models best fits the population data, we plot the graphs shown in Figure 3.11. The rates of change for these models are calculated by finding these derivatives:

- Linear: $y_l' = 0.109$ at $t = 3$, $y_l = 0.109$ ppm
- Quadratic: $y_q' = 0.078t - 0.009$ at $t = 3$, $y_q = 0.225$ ppm

You could just use the residual sum-of-squares as discussed in Section 1.2 to decide which is better. Also, see Project 3A at the end of this chapter.

The quadratic function clearly fits much better than the linear function and predictions differ notably at $t = 3$. Renaming the right-hand-side of the quadratic function $p(t)$, the function we use in our analysis is the quadratic function, which after renaming the left-hand-side is

$$p(t) = 0.039t^2 - 0.009t + 4.672$$

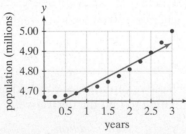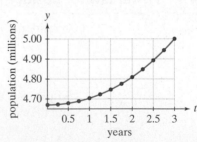

Figure 3.11 Data fitted by the linear (left panel) and quadratic (right panel) functions in part **a** of Example 10.

b. By substituting the quadratic population function $p(t)$ selected in part **a** into the researcher's pollution function $C(p)$, we can represent the level of pollution as $C[p(t)]$, a composite function of time. Applying the chain rule, we find that

$$\frac{dC}{dt} = \frac{dC}{dp}\frac{dp}{dt}$$

$$= \frac{d}{dp}\left[(0.5p^2 + 17)^{1/2}\right]\frac{d}{dt}\left[0.039t^2 - 0.009t + 4.672\right]$$

$$= \left[\frac{1}{2}(0.5p^2 + 17)^{-1/2}(0.5)(2p)\right][0.039(2t) - 0.009]$$

$$= 0.5p(0.5p^2 + 17)^{-1/2}(0.078t - 0.009)$$

When $t = 3$, $p(3) = 0.039(3)^2 - 0.009(3) + 4.672 = 4.996$. Therefore, at $t = 3$ it follows that

$$\left.\frac{dC}{dt}\right|_{t=3} = 0.5(4.996)\left[0.5(4.996)^2 + 17\right]^{-1/2}[0.078(3) - 0.009]$$

$$\approx 0.104$$

Thus, our analysis suggests that after three years, the level of pollution is increasing at the rate of 0.104 parts per million per year.

PROBLEM SET 3.3

Level 1 DRILL PROBLEMS

Use the chain rule to compute the derivative dy/dx for the functions given in Problems 1 to 4.

1. $y = u^2 + 1$; $u = 3x - 2$

2. $y = 2u^2 - u + 5$; $u = 1 - x^2$

3. $y = \dfrac{2}{u^2}$; $u = x^2 - 9$

4. $y = u^2$; $u = \ln x$

Differentiate each function in Problems 5 to 8 with respect to the given variable of the function.

5. **a.** $g(u) = u^5$
   b. $u(x) = 3x - 1$
   c. $f(x) = (3x - 1)^5$

6. **a.** $g(u) = u^3$
   b. $u(x) = x^2 + 1$
   c. $f(x) = (x^2 + 1)^3$

7. **a.** $g(u) = u^{15}$
   b. $u(x) = 3x^2 + 5x - 7$
   c. $f(x) = (3x^2 + 5x - 7)^{15}$

8. **a.** $g(u) = u^7$
   b. $u(x) = 5 - 8x - 12x^2$
   c. $f(x) = (5 - 8x - 12x^2)^7$

Differentiate each function in Problems 9 to 18.

9. $y = (5 - x + x^4)^9$

10. $y = e^{2 + x^2}$

11. $y = \dfrac{1}{(1 + x - x^5)^{12}}$

12. $y = e^{(x+1)^7}$

13. $y = \ln x^2$

14. $y = (2x + 12)^\pi$

15. $y = \ln(2x + 5)$

16. $y = xe^{-x^2}$

17. $y = (x^4 - 1)^{10}(2x^4 + 3)^7$

18. $y = \sqrt{\dfrac{x^3 - x}{4 - x^2}}$

Find dy/dx by implicit differentiation in Problems 19 to 25.

19. $x^2 + y = x^3 + y^3$

20. $xy = 25$

21. $xy(2x + 3y) = 2$

22. $\dfrac{1}{y} + \dfrac{1}{x} = 1$

23. $(2x + 3y)^2 = 10$

24. $\ln(xy) = e^{2x}$

25. $e^{xy} + \ln y^2 = x$

26. Consider the functions $y = f(x)$ and $y = g(x)$ whose graphs in red and blue, respectively, are shown in Figure 3.12.

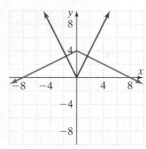

Figure 3.12 Functions $y = f(x)$ (red) and $y = g(x)$ (blue).

   a. Find $\left.\dfrac{d}{dx} f[g(x)]\right|_{x=2}$ **b.** Find $\left.\dfrac{d}{dx} g[f(x)]\right|_{x=2}$

27. The graphs of $u = g(x)$ and $y = f(u)$ are shown in Figure 3.13.

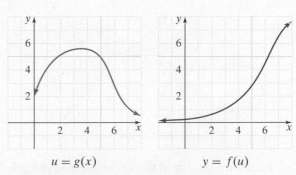

$u = g(x)$ $y = f(u)$

Figure 3.13 Chain rule with graphs.

   a. Find the approximate value of u at $x = 2$. What is the slope of the tangent line at that point?

   b. Find the approximate value of y at $x = 5$. What is the slope of the tangent line at that point?

   c. Find the slope of $y = f[g(x)]$ at $x = 2$.

28. Let $g(x) = f[u(x)]$, where $u(-3) = 5$, $u'(-3) = 2$, $f(5) = 3$, and $f'(5) = -3$. Find an equation for the tangent to the graph of g at the point where $x = -3$.

29. Let f be a function for which

$$f'(x) = \frac{1}{x^2 + 1}$$

a. If $g(x) = f(3x - 1)$, what is $g'(x)$?

b. If $h(x) = f\left(\frac{1}{x}\right)$, what is $h'(x)$?

30. The cissoid of Diocles is a curve of the general form represented by the following particular equation

$$y^2(6 - x) = x^3$$

as illustrated in Figure 3.14.

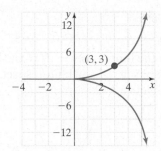

Figure 3.14 Cissoid of Diocles.

Find the equation of the tangent line to this graph at $(3, 3)$.

31. The folium of Descartes is a curve of the general form represented by the following particular equation

$$x^3 + y^3 - \frac{9}{2}xy = 0$$

as illustrated in Figure 3.15.

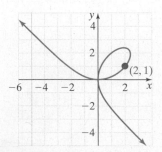

Figure 3.15 Folium of Descartes.

Find the equation of the tangent line to this graph at $(2, 1)$.

32. Another version of the folium of Descartes is given by the equation

$$x^3 + y^3 = 3xy$$

as illustrated in Figure 3.16

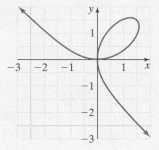

Figure 3.16 Folium of Descartes.

Find at what points the tangent line is horizontal.

Level 2 APPLIED AND THEORY PROBLEMS

33. Arteriosclerosis develops when plaque forms in the arterial walls, blocking the flow of blood; this, in turn, often leads to heart attack or stroke (Figure 3.17). Model the cross-section of an artery as a circle with radius R centimeters, and assume that plaque is deposited in such a way that when the patient is t years old, the plaque is $p(t)$ cm thick, where

$$p(t) = R[1 - 0.009(12{,}350 - t^2)^{1/2}]$$

Find the rate at which the cross-sectional area covered by plaque is changing with respect to time in a sixty-year-old patient.

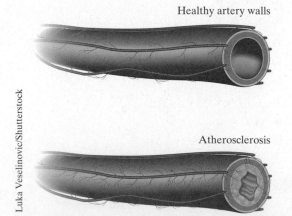

Healthy artery walls

Atherosclerosis

Figure 3.17 Blood flow in a healthy artery (top) and in an artery narrowed by arteriosclerosis (bottom).

34. In Problem 38 in Problem Set 3.2, the energy expended by a species of Australian parakeet during flight was modeled by the function

$$E(v) = \frac{[0.074(v - 35)^2 + 22]}{v} \quad \begin{array}{l}\text{calories per}\\\text{kilometer}\end{array}$$

where v is the bird's velocity (km/h). Assume at one point in time (say $t = 0$) in its flight, a parakeet's velocity is 25 km/hour and its velocity is increasing at a rate of 2 km/hour^2. Find the instantaneous rate of change of this parakeet's energy use at $t = 0$. Be sure to identify the correct units for your answer.

35. In Example 5 of Section 2.4, we found that the predation rate of wolves in North America could be modeled by

$$f(x) = \frac{3.36x}{0.46 + x} \quad \begin{array}{l}\text{moose killed per wolf}\\\text{per hundred days}\end{array}$$

where x is measured in number of moose per square kilometer. If the current moose density is $x = 0.5$ and is increasing at a rate of 0.1 per year, determine the rate at which the predation rate is increasing.

36. In Problem 42 in Problem Set 2.4, we modeled how wolf densities in North America depend on moose densities with the following function

$$f(x) = \frac{58.7(x - 0.03)}{0.76 + x} \text{ wolves per 1000 km}^2$$

where x is the number of moose per square kilometer. If the current moose density is $x = 2.0$ and decreasing at a rate of 0.2 per year, determine the current rate of change of the wolf densities.

37. The proportion of a species of aphid that escapes parasitism is

$$f(x) = e^{-0.02x}$$

where x is the density of parasitoids. If the density of parasitoids is currently 10 wasps per acre and decreasing at a rate 20 wasps per acre per day, find at what rate the likelihood of escaping parasitism is changing.

38. In Example 10 we saw that an environmental study of a certain suburban community suggested the following relationship between the population level p (thousand people) and the amount of carbon monoxide C (ppm) in the air:

$$C(p) = \sqrt{0.5p^2 + 17}$$

Now suppose that the population p (in thousands) at time t is modeled by the function

$$p(t) = 0.10626e^{0.023t}$$

Use the chain rule to find the rate at which the pollution level is changing after three years and compare this with the estimate obtained in Example 10.

39. The bicorn (also called the cocked hat) is a quartic curve studied by mathematician James Sylvester (1814–1897) in 1864. As illustrated in Figure 3.18, it is given by the set of points that satisfy the equation

$$y^2(1 - x^2) = (x^2 + 2y - 1)^2$$

Find the formulas for the two tangent lines at $x = 1/2$.

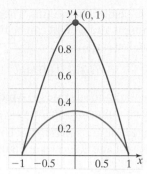

Figure 3.18 Bicorn curve.

40. Prove that if f is differentiable at $u(a)$ and u is continuous at a, then

$$G(h) = \begin{cases} \dfrac{f[u(a + h)] - f(u(a))}{u(a + h) - u(a)} & \text{if } u(a + h) - u(a) \neq 0 \\[2mm] f'[u(a)] & \text{otherwise} \end{cases}$$

is continuous at $h = 0$.

41. Prove that $\dfrac{d}{dx} \log_b x = \dfrac{1}{x \ln b}$ for $b \neq 1, x > 0$.

42. Prove that

$$\frac{dy}{dx} = nx^{n-1}$$

for $y = x^n$ where $x < 0$ and n is any non-zero number for which y is defined.

43. The average height of boys in the United States as a function of age is shown in Figure 3.19. This relationship is well described by the linear function

$$h(t) = 32 + 0.19t \text{ inches}$$

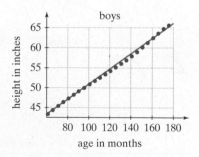

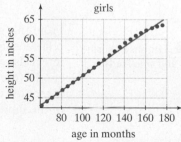

Figure 3.19 Heights in inches of boys and girls in the United States as a function of time in months.

where t is measured in months. The relationship between height and weight can be modeled by the power function

$$W(h) = 0.0024h^{2.6} \text{ lb}$$

Find the instantaneous rates at which height and weight are changing at age ten years.

44. The average height of girls in the United States as a function of age is shown in Figure 3.19. This relationship is well-described by the linear function

$$h(t) = 32 + 0.185t \text{ inches}$$

where t is measured in months. The relationship between height and weight can be modeled by the power function

$$W(h) = 0.0024h^{2.6} \text{ lb}$$

Find the instantaneous rates at which height and weight are changing at age ten years.

3.4   Derivatives of Trigonometric Functions

Many physical and biological processes change periodically over time and consequently are represented by a periodic function. A powerful result in mathematical analysis proves that periodic functions can be represented as a sum of sines and cosines; this is called the *Fourier series representation*. In this section, we find the derivative of these fundamental functions and their functional relatives — tangent, secant, cotangent, and cosecant. Recall from Chapter 1 that we assume the trigonometric functions are functions of real numbers or of angles measured in radians. We make this assumption because the trigonometric differentiation formulas rely on limit formulas that become more complicated if degrees, rather than radians, are used to measure angles.

Derivatives of sine and cosine

Before stating the theorem regarding the derivatives of the sine and cosine functions, we look at the graph of $\dfrac{\sin(x + h) - \sin x}{h}$ for small values of h. In particular, for $h = 0.01$, the expression $\dfrac{\sin(x + 0.01) - \sin x}{0.01}$ has the following graph:

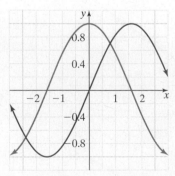

Figure 3.20 Graphs of cosine (blue) and sine (red).

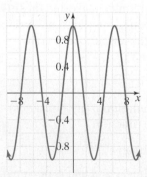

From this graph, it appears that the derivative of $f(x) = \sin x$ is $f'(x) = \cos x$. This relationship appears to match up, as illustrated in Figure 3.20: sine is increasing on intervals where cosine is positive and has turning points where cosine is zero.

Before verifying this assertion, we need two important limits. The first limit is presented numerically in Example 1. You will be asked to give analytical derivations of the limits in Problems 29 and 30 in Problem Set 3.4.

Two Important Trigonometric Limits

$$\lim_{x \to 0} \frac{\sin x}{x} = 1 \quad \text{and} \quad \lim_{x \to 0} \frac{\cos x - 1}{x} = 0$$

Example 1   Numerical approach to one of the trigonometric limits

Find $\lim\limits_{x \to 0} \dfrac{\sin x}{x}$ numerically and graphically using technology.

Solution   Note that

$$\frac{\sin(-x)}{-x} = \frac{-\sin x}{-x} = \frac{\sin x}{x}$$

so the left- and right-hand limits should be the same. Thus, for the numerical approach, we develop a table for x values approaching 0 from the right.

x	$\dfrac{\sin x}{x}$	x	$\dfrac{\sin x}{x}$
$\frac{1}{10}$	0.998334	$\frac{1}{70}$	0.999966
$\frac{1}{20}$	0.999583	$\frac{1}{80}$	0.999974
$\frac{1}{30}$	0.999815	$\frac{1}{90}$	0.999979
$\frac{1}{40}$	0.999896	$\frac{1}{100}$	0.999983
$\frac{1}{50}$	0.999933	$\frac{1}{110}$	0.999986
$\frac{1}{60}$	0.999954	$\frac{1}{120}$	0.999988

The numbers in this table appear to be approaching 1 as x tends toward 0 from the right ($x > 0$). Thus, we might infer from the table that

$$\lim_{x \to 0} \frac{\sin x}{x} = 1$$

Plotting $\sin x / x$ near 0 in Figure 3.21 reaffirms our tabular approach to finding the limit.

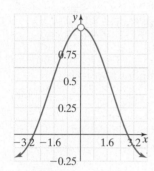

Figure 3.21 Graph of $y = \dfrac{\sin x}{x}$ near the point $x = 0$

We can now state the derivative rule for the sine and cosine functions.

Derivative Rules for Sine and Cosine

The functions $\sin x$ and $\cos x$ are differentiable for all x and

$$\frac{d}{dx} \sin x = \cos x \qquad \text{and} \qquad \frac{d}{dx} \cos x = -\sin x$$

Proof.   We will prove the first derivative formula using the trigonometric identity

$$\sin(x + h) = \sin x \cos h + \cos x \sin h$$

and the definition of derivative. For a fixed x

$$\frac{d}{dx} \sin x = \lim_{h \to 0} \frac{\sin(x + h) - \sin x}{h} \qquad \textit{definition of derivative}$$

$$= \lim_{h \to 0} \frac{\sin x \cos h + \cos x \sin h - \sin x}{h} \qquad \textit{trigonometric identity}$$

$$= \lim_{h \to 0} \frac{\sin x(\cos h - 1) + \cos x \sin h}{h} \qquad \textit{factoring}$$

$$= \lim_{h \to 0} \left[\sin x \left(\frac{\cos h - 1}{h} \right) + \cos x \left(\frac{\sin h}{h} \right) \right]$$

$$= \lim_{h \to 0} \left[\sin x \left(\frac{\cos h - 1}{h} \right) \right] + \lim_{h \to 0} \left[\cos x \left(\frac{\sin h}{h} \right) \right] \qquad \textit{sum limit law}$$

$$= \sin x \lim_{h \to 0} \frac{\cos h - 1}{h} + \cos x \lim_{h \to 0} \frac{\sin h}{h} \qquad \textit{product limit law}$$

$$= (\sin x)(0) + (\cos x)(1) \qquad \textit{important trigonometric limits}$$

$$= \cos x$$

To find the derivative of cosine, we use the trigonometric identities

$$\cos x = \sin\left(x + \frac{\pi}{2}\right) \qquad \text{and} \qquad \cos\left(x + \frac{\pi}{2}\right) = -\sin x$$

and the chain rule

$$\frac{d}{dx}\cos x = \frac{d}{dx}\sin\left(x + \frac{\pi}{2}\right)$$
$$= \cos\left(x + \frac{\pi}{2}\right)$$
$$= -\sin x$$

Example 2   Derivatives involving sine and cosine functions

Differentiate the functions

a. $f(x) = \sin 2x$ **b.** $f(x) = x^2 \sin x$ **c.** $f(x) = \dfrac{\sqrt{x}}{\cos x}$

Solution

a. Setting $u = 2x$ and $y = f(u) = \sin u$, the chain rule implies that

$$\frac{dy}{dx} = f'(u)\frac{du}{dx}$$
$$= (\cos u)2$$
$$= 2\cos 2x$$

b. By the product rule,

$$f'(x) = \sin x \frac{d}{dx}x^2 + x^2\frac{d}{dx}\sin x$$
$$= 2x \sin x + x^2 \cos x$$

c.
$$f'(x) = \frac{d}{dx}\left[\frac{x^{1/2}}{\cos x}\right]$$

$$= \frac{\cos x \frac{d}{dx}(x^{1/2}) - x^{1/2}\frac{d}{dx}\cos x}{\cos^2 x} \qquad \textit{quotient rule}$$

$$= \frac{\frac{1}{2}x^{-1/2}\cos x - x^{1/2}(-\sin x)}{\cos^2 x} \qquad \textit{power rule}$$

$$= \frac{\frac{1}{2}x^{-1/2}(\cos x + 2x \sin x)}{\cos^2 x} \qquad \textit{common factor}$$

$$= \frac{\cos x + 2x \sin x}{2\sqrt{x}\cos^2 x}$$

Example 3   Rate of change of CO_2

In Section 1.2, we initially approximated the concentration of CO_2 (in ppm) at the Mauna Loa Observatory in Hawaii with the function

$$h(t) = 329.3 + 0.1225\,t + 3\cos\left(\frac{\pi t}{6}\right)$$

Find $h'(3)$ and compare it to the approximation found in Example 2 of Section 2.1.

Solution

$$h(t) = 329.3 + 0.1225\,t + 3\cos\left(\frac{\pi t}{6}\right) \qquad \textit{given function}$$

$$h'(t) = 0 + 0.1225 + 3\left[-\sin\left(\frac{\pi t}{6}\right)\left(\frac{\pi}{6}\right)\right] \qquad \textit{elementary derivatives and chain rule}$$

$$= 0.1225 - \frac{\pi}{2}\sin\left(\frac{\pi t}{6}\right)$$

Evaluating at $t = 3$ yields

$$h'(3) = 0.1225 - \frac{\pi}{2}\sin\left(\frac{\pi}{2}\right) = 0.1225 - \frac{\pi}{2} \approx -1.4483$$

This agrees with the numerical solution of Example 2 in Section 2.1.

Periodic fluctuations in biological systems can be internally or externally driven, as the next two examples illustrate.

Example 4   Circadian rhythms

Circadian rhythms are roughly twenty-four-hour cycles in physiological processes of an organism that are primarily driven by the individual's circadian clock. Examples in mammals include heart and breathing patterns and daily fluctuations in body temperature and hormones. In humans, the circadian clock driving these periodic patterns is formed by a cluster of neurons found in the hypothalamus, a part of the brain. However, many brainless organisms, ranging from yeast cells to plants, also exhibit circadian rhythms.

A simple model of circadian rhythm for body temperature in an organism is

$$T(t) = A + B\sin\left(\frac{\pi}{12}(t - C)\right) \text{ degrees Celsius}$$

where t is measured in hours. Assume that $0 \le C \le 6$.

a. Sketch $T(t)$ and discuss the meaning of A, B, and C.

b. Find and sketch $T'(t)$. Discuss what this plot tells you about the circadian rhythm of body temperature.

Solution

a. Using what we studied about trigonometric functions in Chapter 1, we have that $T(t)$ oscillates around the value A with an amplitude of B and a period of 24 hours. The $-C$ term shifts the graph to the right by C hours. Therefore, we get the following plot of $T(t)$:

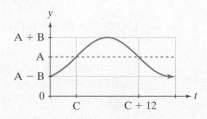

Therefore, the body temperature of $A°$ Celsius occurs at hours C and $C + 12$ each day and the maximum temperature $A + B°$ Celsius occurs at hour $C + 6$ each day.

b. Taking the derivative of T yields

$$T'(t) = \frac{d}{dt}A + B\frac{d}{dt}\sin\left(\frac{\pi}{12}(t - C)\right) \qquad \textit{derivative rules}$$

$$= 0 + B\frac{d}{dt}\sin\left(\frac{\pi}{12}(t - C)\right)$$

$$= 0 + B\cos\left(\frac{\pi}{12}(t - C)\right)\frac{\pi}{12} \qquad \textit{derivative of sine and chain rule}$$

Plotting the derivative using what we learned in Chapter 1 gives

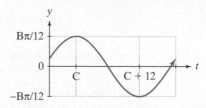

The maximum instantaneous rate of change of body temperature occurs at hour C each day, while the minimum rate of change occurs at hour $C + 12$ each day. Since $T'(t) > 0$ on $[0, C + 6)$ and $T'(t) < 0$ on $(C + 6, C + 18)$, body temperature is increasing between hours 0 and $C + 6$ and decreasing between hours $C + 6$ and $C + 18$.

Example 5   Periodic populations

Many populations live in environments that change in a periodic fashion with time (e.g., diurnal and seasonal cycles). To understand the dynamics of an algal population growing in a climate chamber programmed to have a particular light-dark cycle, a research scientist conducted an experiment in which she found that the algae abundance (in cells per liter) over time t in hours could be modeled by the function

$$N(t) = 10{,}000e^{\sin t}$$

a. Verify that $N(t)$ satisfies the relationship

$$N'(t) = \cos t\, N(t)$$

Explain what this relationship means. Based on the assumption that the growth rate of the population is proportional to the intensity of light, at what times are the light intensity greatest?

b. Determine at what times the population is increasing and at what times the population is decreasing.

Solution

a. Taking the derivative of $N(t)$ with the chain rule, we get

$$N'(t) = 10{,}000\, e^{\sin t}\frac{d}{dt}\sin t = 10{,}000\, e^{\sin t}\cos t$$

Hence, by the definition of $N(t)$, we have

$$N'(t) = \cos t\, N(t)$$

We can interpret $\cos t$ as the per capita growth rate of the population. This per capita growth is greatest (i.e., equals one) at $t = 0, \pm2\pi, \pm4\pi, \ldots$ Hence, at these moments of time the light intensity must be the greatest.

b. The population increases when $N'(t) > 0$. This occurs when $\cos t > 0$, in other words, when t is in the intervals $(0, \pi/2)$, $(3\pi/2, 5\pi/2)$, $\ldots$. On the complementary intervals, the population is decreasing; that is, on these intervals, those algal cells that are dying more quickly than dividing.

Derivatives of other trigonometric functions

If you know the derivatives of sine and cosine and the basic rules of differentiation, then all the other trigonometric derivatives follow.

Derivative Rules for Trigonometric Functions

The six basic trigonometric functions $\sin x$, $\cos x$, $\tan x$, $\csc x$, $\sec x$, and $\cot x$ are all differentiable wherever they are defined and

$$\frac{d}{dx}\sin x = \cos x \qquad\qquad \frac{d}{dx}\cos x = -\sin x$$

$$\frac{d}{dx}\tan x = \sec^2 x \qquad\qquad \frac{d}{dx}\cot x = -\csc^2 x$$

$$\frac{d}{dx}\sec x = \sec x \tan x \qquad \frac{d}{dx}\csc x = -\csc x \cot x$$

All the additional derivative rules are proved by using the quotient rule along with formulas for the derivative of sine and cosine. Here we will obtain the derivative of the tangent function and leave the rest to Problem Set 3.4.

$$\frac{d}{dx}\tan x = \frac{d}{dx}\frac{\sin x}{\cos x} \qquad \textit{trigonometric identity}$$

$$= \frac{\cos x \frac{d}{dx}(\sin x) - \sin x \frac{d}{dx}(\cos x)}{\cos^2 x} \qquad \textit{quotient rule}$$

$$= \frac{\cos x(\cos x) - \sin x(-\sin x)}{\cos^2 x} \qquad \textit{derivatives of sine and cosine}$$

$$= \frac{\cos^2 x + \sin^2 x}{\cos^2 x}$$

$$= \frac{1}{\cos^2 x} \qquad \textit{trigonometric identity}$$

$$= \sec^2 x \qquad \textit{trigonometric identity}$$

Notice that the derivatives of all "co" trig functions except for cosine have the "co-trig" derivative form of their corresponding trigonometric partners, but with a sign change. Thus, for example, because the derivative of tangent is secant squared, this rule implies that the derivative of cotangent is the negative of cosecant squared.

Example 6   Derivative of a product of trigonometric functions

Differentiate $f(x) = \sec x \tan x$.

Solution

$$f'(x) = \frac{d}{dx}(\sec x \tan x)$$

$$= \sec x \frac{d}{dx}(\tan x) + \tan x \frac{d}{dx}(\sec x) \qquad \textit{product rule}$$

$$= \sec x(\sec^2 x) + \tan x(\sec x \tan x)$$

$$= \sec^3 x + \sec x \tan^2 x$$

PROBLEM SET 3.4

Level 1 DRILL PROBLEMS

Differentiate the functions given in Problems 1 to 20.

1. $f(x) = \sin x + \cos x$
2. $g(x) = 2\sin x + \tan x$
3. $y = \sin 2x$
4. $y = \cos 2x$
5. $f(t) = t^2 + \cos t + \cos \dfrac{\pi}{4}$
6. $g(t) = 2\sec t + 3\tan t - \tan \dfrac{\pi}{3}$
7. $y = e^{-x} \sin x$
8. $y = \tan x^2$
9. $f(\theta) = \sin^2 \theta$
10. $g(\theta) = \cos^2 \theta$
11. $y = \cos x^{101}$
12. $y = (\cos x)^{101}$
13. $p(t) = (t^2 + 2)\sin t$
14. $y = x\sec x$
15. $q(t) = \dfrac{\sin t}{t}$
16. $f(x) = \dfrac{\sin x}{1 - \cos x}$
17. $g(x) = \dfrac{x}{1 - \sin x}$
18. $y = \sin(2t^3 + 1)$
19. $y = \ln(\sin x + \cos x)$
20. $y = \ln(\sec x + \tan x)$

Use the given trigonometric identity in parentheses and the basic rules of differentiation to find the derivatives of the functions given in Problems 21 to 24.

21. $f(x) = \sec x \ \left(\sec x = \dfrac{1}{\cos x}\right)$
22. $f(x) = \csc x, \left(\csc x = \dfrac{1}{\sin x}\right)$
23. $f(x) = \cot x, \left(\cot x = \dfrac{1}{\tan x}\right)$
24. $f(x) = \cot x, \left(\cot = \dfrac{\sin x}{\cos x}\right)$
25. Differentiate $y = \dfrac{\sec x + \tan x}{\csc x + \cot x}$
26. Differentiate $y = \dfrac{\sqrt{\sin x}}{\cot x}$
27. Differentiate $y = \ln \sin^2 x$
28. **a.** If $F(x) = \ln|\cos x|$, show that $F'(x) = -\tan x$.
    b. If $f(x) = \ln|\sec x + \tan x|$, show that $f'(x) = \sec x$.

Level 2 APPLIED AND THEORY PROBLEMS

29. Consider three areas as shown in Figure 3.22.

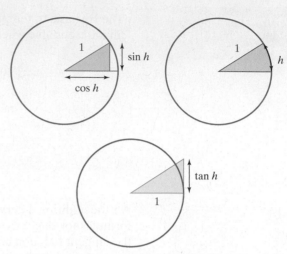

Figure 3.22 Triangles and a unit circle with subtended angle h.

a. What is the area $g(h)$ of the blue-shaded triangle?

b. What is the area $f(h)$ of the pink-shaded sector? Hint: The area of a sector of a circle of radius r and central angle θ measured in radians is
$$A = \frac{1}{2}r^2\theta.$$

c. What is the area $k(h)$ of the green-shaded triangle?

d. Use the fact that $g(h) \le f(h) \le k(h)$ for small $h > 0$ to prove that
$$\lim_{x \to 0} \frac{\sin x}{x} = 1$$
by beginning with the inequality

BLUE AREA $\le$ PINK AREA $\le$ GREEN AREA

30. Prove
$$\lim_{x \to 0} \frac{\cos x - 1}{x} = 0$$
Hint: Multiply by 1 written as $\left(\dfrac{\cos x + 1}{\cos x + 1}\right)$ and use a fundamental trigonometric identity.

31. A researcher studying a certain species of fish in a northern lake models the population after t months of the study by the function
$$P(t) = 100e^{-t}\sin t + 800$$
At what rate is the population changing after two months? Is the population growing or declining at this time?

32. In an experiment with algae, a research scientist manipulated the light in growth chambers so that
$$P(t) = 7{,}000\, e^{\cos(\pi t/12)}$$

described the population density (in cells per liter) as a function of t (in hours).

a. Find a function $r(t)$ such that
$$P'(t) = r(t)\, P(t)$$

b. Under the assumption that the growth rate of the population is proportional to the intensity of light, what is the period of the light fluctuations in the chamber?

c. Determine at what times (if any) the population is decreasing in abundance.

33. In a more ambitious experiment with algae, a research scientist manipulated the light in algal tanks so that
$$P(t) = 5{,}000\, e^{\cos t + t}$$
describes the population density (in cells per liter) as a function of t (in hours).

a. Find a function $r(t)$ such that
$$P'(t) = r(t)\, P(t)$$

b. Under the assumption that the growth rate of the population is proportional to the intensity of light, what is the period of the light fluctuations in the tank?

c. Determine at what times (if any) the population is decreasing in abundance.

34. In Example 4 of Section 1.5, we modeled the tides for Toms Cove in Assateague Beach, Virginia, on August 19, 2004 with the function
$$H(t) = 1.8\cos\left[\frac{\pi}{6}(t - 11)\right] + 2.2$$
where H is the height of the tide (in feet) and t is the time (in hours after midnight). Find and interpret $\left.\dfrac{dH}{dt}\right|_{t=6}$.

35. The temperature in a swimming pool is given by
$$T(t) = 25 - 4\sin(\pi t/12) \text{ degrees Celsius}$$
where t is measured in hours since midnight.

a. Find $T'(12)$ and interpret.

b. During what hours is the temperature of the pool increasing?

36. The human heart goes through cycles of contraction and relaxation (called systoles). During these cycles, blood pressure goes up and down repeatedly; as the heart contracts, pressure rises, and as the heart relaxes (for a split second), the pressure drops. Consider the following approximate function for the blood pressure of a patient:
$$P(t) = 100 + 20\cos\left(\frac{\pi t}{35}\right) \text{ mmHg}$$
where t is measured in minutes. Find and interpret $P'(t)$.

3.5 Linear Approximation

We have seen that the tangent line is the line that just touches a curve at a single point. In this section, we will discover that the tangent line can be used to provide a reasonable approximation to a curve. Using these linear approximations we will be able to make projections about the size of a bison population (Figure 3.23) estimate $\sqrt{10}$, and estimate the effects of measurement error.

Jeff Banke/Shutterstock

Figure 3.23 The North American plain's bison (*Bison bison*) is one of two subspecies of bison that roamed the Great Plains. Their numbers have dropped from tens of millions to thousands over the last 200 years.

Approximating with the tangent line

We begin with an example that illustrates how well a tangent line can approximate a curve.

Example 1   Zooming in at a point

Consider the function $y = \ln x$.

a. Find the tangent line at $x = 1$.

b. Graph $y = \ln x$ and the tangent line over the intervals $[0.1, 2]$, $[0.5, 1.5]$, $[0.9, 1.1]$. Discuss what you find.

Solution

a. Since

$$\frac{d}{dx} \ln x \Big|_{x=1} = \frac{1}{x} \Big|_{x=1} = 1$$

we get that the tangent line is the line of slope 1 through the point $(1, 0)$ — that is, the equation

$$(y - 0) = 1 \cdot (x - 1) \quad \Rightarrow \quad y = x - 1$$

as we claimed in Section 2.1.

b. The graphs $y = \ln x$ (in blue) and $y = x - 1$ (in red) on the intervals $[0.1, 2]$, $[0.5, 1.5]$, and $[0.9, 1.1]$ are shown in Figure 3.24. This figure illustrates that as we zoom into the point $(1, 0)$, the tangent line provides a better and better approximation of our original function.

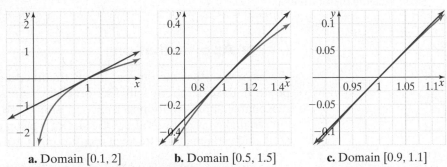

a. Domain $[0.1, 2]$ **b.** Domain $[0.5, 1.5]$ **c.** Domain $[0.9, 1.1]$

Figure 3.24 Zooming in on the graphs of $y = \ln x$ and $y = x - 1$ about the point $(1, 0)$

The difference between a tangent line and the associated curve becomes more and more negligible as you zoom into the point of contact. Thus, it seems quite reasonable to approximate the function with the tangent line in a neighborhood of the point at which this tangent is constructed. This is called the **linearization** of a function.

Linear Approximation

If f is differentiable at $x = a$, then the **linear approximation** of f around a is given by

$$f(x) \approx f(a) + f'(a)(x - a)$$

for x near a.

With linear approximations, we can make predictions about the future, as illustrated in the next example. We use data on the abundance of the North American bison in Yellowstone National Park going back as early as 1902. (Estimates of bison population levels in Yellowstone from 1902–1931 can be found at http://www.seattlecentral.edu/qelp/Data.html.) Annual abundances for bison in Yellowstone for the twenty-nine-year period 1902 to 1931 are shown in Figure 3.25. These data suggest that the bison population was recovering in the first part of the twentieth century after years of intense hunting in the nineteenth century.

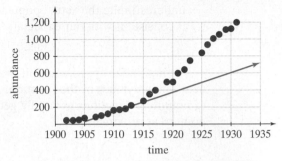

Abundance from 1902 to 1931 and projected
abundance via linear approximation,
extrapolating from years 1908 and 1909
(straight line)

Figure 3.25 Bison abundance in Yellowstone National Park.

In the next example, we use the data only from 1908 to 1915. However, a project involving all the data is outlined at the end of the chapter.

Example 2 Predicting bison abundance

Suppose it is January 1910, and you are the Yellowstone National Park manager.

a. Use a linear function to extrapolate what the abundance of bison might be in 1910 through 1915, given that you know the bison abundance in 1908 and 1909 are 95 and 118, respectively.

b. Compare your estimates to the actual population size in 1910 to 1915 as given in the last column in Table 3.2.

Table 3.2 Estimate and measured abundance of Bison in Yellowstone National Park, 1910–1915

Year	t	Abundance Estimated	Measured
1910	10	141	149
1911	11	164	168
1912	12	187	192
1913	13	210	215
1914	14	233	Unknown
1915	15	256	270

Solution

a. Let $N(t)$ denote the number of bison in year t, and for the sake of simplicity, set $t = 0$ at the year 1900. If we approximate $N(t)$ at $t = 8$ by a linear function we get

$$N(t) \approx N(8) + N'(8)(t - 8)$$

To approximate $N'(8)$, we can use

$$N'(8) \approx \frac{N(9) - N(8)}{9 - 8} = 23$$

Hence,

$$N(t) \approx 95 + 23(t - 8)$$

for t "near" 8. Hence, our approximation yields the predictions shown in Table 3.2, with actual abundance in the last column.

b. As we can see in Figure 3.25, our estimates are pretty good; nevertheless, they underestimate the actual population size more and more as time moves on. This is consistent with our expectation that population growth might be exponential. ∎

Using linear approximation, we can estimate the value of a function at points near known values of the function. As we will see, however, linear approximations of nonlinear functions generally get increasingly worse as we move away from the point of approximation.

Example 3   Approximating $\sqrt{10}$

Consider the function $f(x) = \sqrt{x}$.

a. Find the linear approximation of f at $x = 9$.

b. Use the linear approximation found in part **a** to approximate $\sqrt{10}$. Compare this approximation to a calculator approximation.

c. How well does this same approximation work for $\sqrt{16}$?

Solution

a. $f'(x) = \frac{1}{2}x^{-1/2}$, so $f'(9) = \frac{1}{2}(9)^{-1/2} = \frac{1}{6}$ and the linear approximation for $\sqrt{x}$ at $x = 9$ is

$$\sqrt{x} \approx f(9) + f'(9)(x - 9)$$
$$= 3 + \frac{1}{6}(x - 9)$$

for x near 9.

b. If we now apply the approximation in part **a** to find $\sqrt{10}$, we obtain

$$\sqrt{10} \approx 3 + \frac{1}{6}(10 - 9) = 3\frac{1}{6} \approx 3.16667$$

This is fairly close to the calculator approximation

$$\sqrt{10} \approx 3.16228$$

So the error is 0.004 (to three decimal places)

c. Similarly,

$$\sqrt{16} \approx 4 + \frac{1}{6}(16 - 9) = 4\frac{7}{6} \approx 5.16666$$

Since we know the answer is 4, the approximation now has an error of more than 1.1. ∎

The next example shows that the linear approximation of $\sin x$ in the neighborhood of 0 is very simple, but it fails badly as the approximation is pushed too far beyond 0.

Example 4   Approximating sin x

Consider $y = \sin x$

a. Find the linear approximation of $\sin x$ at $x = 0$.

b. Plot the difference between $y = \sin x$ and its linear approximation on the intervals $[-1, 1]$, $[-0.5, 0.5]$, and $[-0.1, 0.1]$. Discuss the meaning of these plots.

c. Approximate $\sin 2$, $\sin 1$, and $\sin 0.25$ with the linear approximation from part **a**. Compare your approximations to calculator approximations.

Solution

a. Since $\dfrac{d}{dx} \sin x \Big|_{x=0} = \cos 0 = 1$, we get the linear approximation at 0:

$$\begin{aligned} \sin x &\approx f(0) + f'(0)(x - 0) \\ &= 0 + 1(x - 0) \\ &= x \end{aligned}$$

b. The graphs of $\sin x - x$ on the intervals $[-1, 1]$, $[-0.5, 0.5]$ and $[-0.1, 0.1]$ are illustrated in Figure 3.26. These figures illustrate that the difference between $\sin x$ and x gets smaller and smaller as you zoom around the point $x = 0$. Hence, $y = x$ is a better and better approximation for $\sin x$ as x approaches 0.

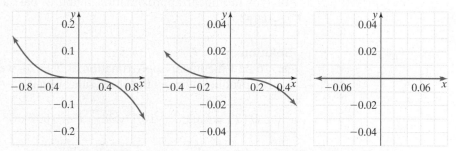

Figure 3.26 Graphs of $\sin x - x$ zooming onto the point $(0, 0)$

c.

linear approximation	calculator approximation	error
$\sin 0.25 \approx 0.25$	$\sin 0.25 \approx 0.247404$	small: $<0.5\%$
$\sin 1 \approx 1$	$\sin 1 \approx 0.841471$	moderate: around 15%
$\sin 2 \approx 2$	$\sin 2 \approx 0.909297$	large: $>100\%$

Development of organisms from immature forms, like eggs or seeds, to adult forms depends on a series of biochemical reactions. These biochemical reactions are often temperature sensitive, breaking down at too low or too high temperatures. For plants and insects whose internal temperatures are largely determined by the temperature of their environment, it follows that developmental rates depend on the temperature of the environment. This temperature dependence can be measured experimentally, as the following example illustrates.

Example 5   Developmental rates of the spider mite destroyer

Roy and colleagues estimated how developmental rates of a beetle (*Stethorus punctillum*) varied with temperature. They modeled the rate at which the fourth stage of development is completed with the following function

$$D(T) = 0.03T(T - 10.7)\sqrt{38 - T} \text{ percent development completed per day}$$

where T is measured in degrees Celsius. This model and the data used to parameterize the model are shown in Figure 3.27. This figure suggests that the developmental rate is approximately linear in T for temperatures near 20°C.

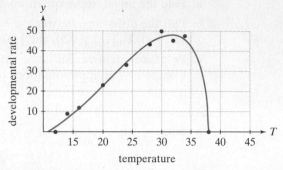

Figure 3.27 Developmental rate (percent completed per day) for a stage of development of the beetle *Stethorus punctillum* (also known as *spider mite destroyer*).

Data Source: Michèle Roy, Jacques Brodeur, and Conrad Cloutier. "Relationship between temperature and developmental rate of *Stethorus punctillum* (Coleoptera: Coccinellidae) and its prey *Tetranychus mcdanieli* (Acari: Tetranychidae)," *Environmental Entomology* 31 (2002), 177–187.

a. Find a linear approximation to $D(T)$ near the value $T = 20$.

b. Plot the linear approximation found in part a. How does this compare to the data?

Solution

a. To find the linear approximation at $T = 20$, we need to compute $D'(20)$. Using the product rule and the chain rule, we get

$$D'(T) = 0.03(T - 10.7)\sqrt{38 - T} + 0.03T\sqrt{38 - T} + 0.03\frac{T(T - 10.7)}{\sqrt{38 - T}}(-1/2)$$

Evaluating this expression at $T = 20$ yields

$$D'(20) \approx 3.07$$

Using the point slope formula for a line, we get the linear approximation

$$D(T) - D(20) \approx D'(20)(T - 20)$$
$$D(T) - 23.67 \approx 3.07(T - 20)$$

Equivalently,

$$D(T) \approx 3.07T - 37.73$$

b. Plotting the linear approximation (red line) from part **a** against $D(T)$ and the data yields

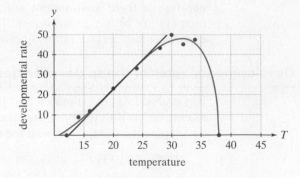

The linear approximation works quite well, and it even appears to provide a better estimate for the lower developmental threshold where the developmental rate equals zero at $T = 12$.

Error analysis

When a scientist makes a measurement, it is always subject to some measurement error. Hence, we have

$$\text{measurement} = \text{actual value} + \text{measurement error}$$

For instance, when the clearance rate of acetaminophen is given as 0.28 per hour, this estimate is the average of a series of measurements, each of which may vary by 0.05 or so. Consequently, when we estimate the half-life of acetaminophen, or the amount in the blood stream several hours after taking the drug, it is important to understand how small variations in the estimate 0.28 influence half-life estimates.

Example 6   Professor Getz's headache

Professor Getz takes 1000 mg of acetaminophen to combat a headache.

a. Solve for the half-life T of the drug as a function of the clearance rate x per hour.

b. Determine the half-life of the acetaminophen, assuming that $x = 0.28$ is a good estimate of the clearance rate.

c. What is the derivative of the half-life T with respect to x and its value at $x = 0.28$?

d. Use linear approximation to estimate the change ΔT in the estimated half-life if the estimate $x = 0.28$ is off by Δx. Interpret this result.

Solution

a. Let $A(t)$ denote the amount of acetaminophen in the body at time t hours. Since the clearance rate is x, we have

$$A(t) = A(0)e^{-xt}$$

The half-life is the time $t = T$ such that

$$A(T) = \frac{A(0)}{2}$$

$$e^{-xT} = \frac{1}{2}$$

$$-xT = \ln \frac{1}{2}$$

$$T = \frac{\ln 2}{x}$$

Hence, the half-life as a function of x is

$$T(x) = \frac{\ln 2}{x}$$

b. Evaluating T at $x = 0.28$ yields $T(0.28) = 2.47553$ hours.

c. Differentiating the half-life function derived in part **a** yields

$$T'(x) = -\frac{\ln 2}{x^2}$$

from which we can calculate $T'(0.28) = -8.84116$.

d. If $x = 0.28 + \Delta x$ where Δx can be viewed as a small measurement error, then by linear approximation we have

$$T(0.28 + \Delta x) \approx T(0.28) + T'(0.28)\Delta x$$

$$= 2.47553 - 8.84116 \, \Delta x$$

Thus,

$$\Delta T = T(0.28 + \Delta x) - T(0.28)$$
$$\approx 2.47553 - 8.84116\,\Delta x - 2.47553$$
$$= -8.84116\,\Delta x$$

Hence, for a measurement error of Δx per hour, half-life changes by approximately $-8.84116\,\Delta x$ hours. For instance, if the measurement error in the clearance rate is $\Delta x = 0.05$, then our estimate of the half-life decreases approximately by $8.84116 \cdot 0.05 = 0.4421$ hours. Thus the estimate of the half-life, T, is quite sensitive to the estimate of the clearance rate.

Example 6 illustrates how an error in the measurement of the independent variable x propagates to an error in the dependent variable y—a process called **error propagation**.

Error Estimates and Sensitivity

Suppose $y = f(x)$ is a quantity of interest and $x = a$ is the true value of x. If there is an error of Δx in measuring $x = a$, then the approximate resulting error in y is given by

$$\Delta y = f(a + \Delta x) - f(a)$$
$$\approx f(a) + f'(a)\Delta x - f(a)$$
$$= f'(a)\Delta x$$

$f'(a)$ is called the **sensitivity** of y to x at $x = a$. The greater the sensitivity the greater the propagation of error.

Consider another example.

Example 7   Estimating metabolic rates

In Example 12 of Section 1.6 (below we changed the names of the variables from x and y to M and R), we discovered that the mouse-to-elephant curve describing how the metabolic rate R (in kilocalories/day) depends on body mass M (in kilograms) is approximately given by

$$\ln R = 0.75 \ln M + 4.2$$

Thus, taking exponentials yields the equation

$$R = e^{4.2} M^{0.75}$$

a. Estimate the metabolic rate of a California condor weighing 10 kg.

b. Determine the sensitivity of your estimate to the measurement of 10 kg. Discuss how a small error ΔM propagates to an error ΔR in your estimate for R.

Solution
a. For the 10 kg condor, we get $R = e^{4.2} \times 10^{0.75} \approx 375$ kilocalories/day.

b. The sensitivity of this estimate to our estimate for the condor weight is

$$R'(10) = 50.01 M^{-0.25}\Big|_{M=10} \approx 28.13$$

Hence, $\Delta R \approx 28.13\Delta M$. For example, an error of $\Delta M = 0.1$ kg yields an error of $\Delta R = 2.81$ kilocalories/day in estimating R.

Elasticity

Often scientists are more interested in the percent error and not the absolute error. For example, a scientist may ask this question: How does a 10% error in the measurement of the clearance rate result in a percentage error in the estimate of the half-life? If $x = a$ is the true value of the independent variable and there is a measurement error of Δx, then the **percent error** in x is

$$\frac{\Delta x}{a} \times 100\%$$

With an error of Δx in the independent variable, we get an error of $\Delta y = f(a + \Delta x) - f(a)$ in y. Hence, the *percent error in y* is

$$\frac{\Delta y}{f(a)} \times 100\%$$

The ratio of the percentage error in y over the percentage error in x is given by

$$\frac{\dfrac{\Delta y}{f(a)} \times 100\%}{\dfrac{\Delta x}{a} \times 100\%} = \frac{\Delta y}{\Delta x} \frac{a}{f(a)}$$

and can be approximated by

$$f'(a) \frac{a}{f(a)}$$

This quantity is used quite commonly in the analysis of biological models and, consequently, has a special name: elasticity.

> **Elasticity**
>
> Let $y = f(x)$ be a function that is differentiable at $x = a$. The **elasticity** of f with respect to x at a is
>
> $$E = f'(a) \frac{a}{f(a)}$$
>
> We can interpret E as stating that for a 1% error in the measurement of $x = a$, there is a $E\%$ error in the measurement of y.

Example 8   Elasticity of metabolic rates

Let us revisit Example 7 where we estimated the metabolic rate of a California condor weighing 10 kg.

a. Find the elasticity of your estimate of the metabolic rate to the estimate of the condor's weight.

b. Interpret your elasticity in terms of 10% error in the estimate of the condor's weight.

Solution

a. To compute the elasticity, recall we found that $R(10) \approx 375$ kilocalories/day and $R'(10) \approx 28.13$. Hence, the elasticity at $M = 10$ is

$$R'(10) \frac{10}{R(10)} \approx 28.13 \frac{10}{375} \approx 0.75$$

b. Since the elasticity is 0.75, a 10% measurement error in the weight of the condor would result in approximately a 7.5% error in the estimate of the metabolic rate.

Using elasticity, we can estimate with what accuracy we need to measure an independent variable to ensure a certain accuracy in the estimate of a dependent variable.

Example 9   Determining measurement accuracy

The body mass index (BMI) for individual weighing w pounds and h inches tall is given by

$$B = \frac{703w}{h^2}$$

a. Determine the elasticity of B with respect to the variable h for a particular weight class w (i.e., we regard w as constant, so that B is a function of the variable h alone).

b. Estimate how accurate your height measurement needs to be to guarantee less than a $\pm 5\%$ error in your BMI measurement.

Solution

a. To compute the elasticity, we first need the derivative

$$\frac{dB}{dh} = -\frac{1406w}{h^3}$$

Hence, the elasticity is

$$\frac{dB}{dh}\frac{h}{B} = -\frac{1406w}{h^3}\frac{h}{703w/h^2}$$
$$= -2$$

Note that this answer does not depend on w or h but is a pure number! Think about why this is the case.

b. Since the elasticity is -2, an $x\%$ error in h results in a $-2x\%$ error in our estimate for BMI. To ensure that our error is no greater than $\pm 5\%$, we need to ensure that the error in the measurement of the height is no greater than $\pm 2.5\%$.

PROBLEM SET 3.5

Level 1 DRILL PROBLEMS

In Problems 1 to 6 find the linear approximation of $y = f(x)$ at the specified point, and use technology to graph the function and its linear approximation to determine whether the linear approximation tends to overestimate or underestimate $y = f(x)$ near the specified point.

1. $y = \cos x$ at $x = \dfrac{\pi}{2}$ **2.** $y = e^x$ at $x = 0$.

3. $y = \sin x$ at $x = \dfrac{\pi}{2}$ **4.** $y = x^2$ at $x = -2$

5. $y = \dfrac{1}{1+x^2}$ at $x = 2$ **6.** $y = xe^{-x}$ at $x = \ln 2$

In Problems 7 to 12, estimate the indicated quantity using a linear approximation around the given point x and compare to the true value obtained using technology.

7. $\sqrt{26}, x = 25$ **8.** $\sqrt{0.99}, x = 1$

9. $\ln 0.9, x = 1$ **10.** $\cos\left(\dfrac{\pi}{2} + 0.01\right), x = \dfrac{\pi}{2}$

11. $\tan 0.2, x = 0$ **12.** $e^{-0.2}, x = 0$

Find the sensitivity of $y = f(x)$ at the point specified in Problems 13 to 18, and use it to estimate Δy for the given measurement error Δx.

13. $y = \sqrt{x}$ at $x = 9$, with $\Delta x = 0.01$

14. $y = \sqrt{2x^2 + 1}$ at $x = -2$, with $\Delta x = 0.01$

15. $y = \ln x$ at $x = 2$, with $\Delta x = -0.2$

16. $y = \cot x$ at $x = \dfrac{\pi}{2}$, with $\Delta x = 0.1$

17. $y = \cos x$ at $x = \dfrac{\pi}{2}$, with $\Delta x = -0.01$

18. $y = \dfrac{1}{x+1}$ at $x = 0$, with $\Delta x = -0.05$

Find the elasticity of $y = f(x)$ at the point specified in Problems 19 to 24 and use it to estimate the percent error in y for the given percent error in x.

19. $y = \sqrt{x}$ at $x = 9$, with 1% error in x

20. $y = \sqrt{2x^2 + 1}$ at $x = -2$, with 8% error in x

21. $y = \ln x$ at $x = 2$, with 5% error in x

22. $y = \cot x$ at $x = \dfrac{\pi}{2}$, with 10% error in x

23. $y = \sin x$ at $x = \dfrac{\pi}{2}$, with 10% error in x

24. $y = \dfrac{1}{x+1}$ at $x = 0$, with 12% error in x

Level 2 APPLIED AND THEORY PROBLEMS

25. In Example 5, we saw that Roy and colleagues estimated how developmental rates of a beetle (*Stethorus punctillum*) varied with temperature. They modeled the rate at which egg development is completed with the following function

$$D(T) = 0.021\,T(T - 11.9)\sqrt{37 - T} \text{ percent}$$
$$\text{development completed per day}$$

where T is measured in degrees Celsius. This model and the data used to parameterize the model are shown in Figure 3.28. This figure suggests that the developmental rate is approximately linear in T for temperatures near 20°C.

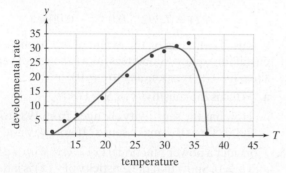

Figure 3.28 Developmental rate (percent development completed per day) for egg development of the beetle *Stethorus punctillum* (spider mite destroyer).

a. Find a linear approximation to $D(T)$ near the value $T = 20$.

b. Sketch the linear approximation on top of the graph in Figure 3.28. How does this compare to the data?

26. In Example 5, we saw that Roy and colleagues estimated how developmental rates of a beetle (*Stethorus punctillum*) varied with temperature. They modeled the rate at which the first stage of larval development is completed with the following function

$$D(T) = 0.07\,T(T - 11.8)\sqrt{37 - T} \text{ percent}$$
$$\text{development completed per day}$$

where T is measured in degrees Celsius. This model and the data used to parameterize the model are shown in Figure 3.29. This figure suggests that the developmental rate is approximately linear in T for temperatures near 20°C.

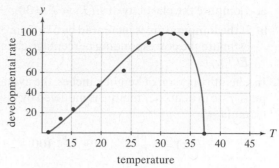

Figure 3.29 Developmental rate (percent completed per day) for the first stage of larval development for the beetle *Stethorus punctillum* (spider mite destroyer).

a. Find a linear approximation to $D(T)$ near the value $T = 20$.

b. Sketch the linear approximation on top of the graph in Figure 3.29. How does this compare to the data?

27. If your measurement of the radius of a circle is accurate to within 3%, approximately how accurate (to the nearest percentage) is your calculation to the area A when the radius is $r = 12$ cm? (Recall the formula $A = \pi r^2$).

28. Suppose a 12-ounce can of Coke® (i.e., Coca-Cola) has a height of 4.75 inches. If your measurement of the radius has an accuracy to within 1%, how accurate is your measurement for the volume? Check your answer by examining a Coke can.

29. An environmental study suggests that t years from now, the average level of carbon monoxide in the air will be

$$Q(t) = 0.05t^2 + 0.1t + 3.4$$

parts per million (ppm). By approximately how much will the carbon monoxide level change during the next six months?

30. A certain cell is modeled as a sphere. If the formulas $S = 4\pi r^2$ and $V = \dfrac{4}{3}\pi r^3$ are used to compute the surface area and volume of the sphere, respectively, estimate the effect of S and V produced by a 1% increase in the radius r.

31. In a model developed by John Helms ("Environmental Control of Net Photosynthesis in Naturally Grown *Pinus Ponderosa* Nets," *Ecology* (Winter 1972, p. 92), the water evaporation $E(T)$ for a ponderosa pine is modeled by

$$E(T) = 4.6e^{17.3T/(T+237)}$$

where T (degrees Celsius) is the surrounding air temperature.

a. Compute the elasticity of $E(T)$ at $T = 30$.

b. If the temperature is increased by 5% from 30°C, estimate the corresponding percentage change in $E(T)$.

32. In a healthy person of height x inches, the average pulse rate in beats per minute is modeled by the formula

$$P(x) = \frac{596}{\sqrt{x}} \qquad 30 \le x \le 100$$

a. Compute the sensitivity of P at $x = 60$.

b. Use your answer to part **a** to estimate the change in pulse rate that corresponds to a height change from 59 to 60 inches.

c. Compute the elasticity of P. Does it depend on x?

d. Determine how accurate the measurement of x needs to be to ensure the estimate for P has an error of less than 10%.

33. In Example 6, we showed that the half-life, T, of a drug with clearance rate x is given by

$$T(x) = \frac{\ln 2}{x}$$

Suppose that the true value of the clearance rate of some drug is given by $x = a$.

a. Find the elasticity of T with respect to x.

b. If you want to estimate the half-life of this drug within an error of 2%, how accurately do you have to measure the clearance rate of the drug?

34. A drug is injected into a patient's blood stream. The concentration of the drug in the blood stream t hours after the drug is injected is modeled by the formula

$$C(t) = \frac{0.12t}{t^2 + t + 1}$$

where C is measured in milligrams per cubic centimeter.

a. Compute the sensitivity of C at $t = 30$.

b. Use your answer to part **a** to estimate the change in concentration over the time period from 30 to 35 minutes after injection.

35. According to Poiseuille's law, the speed of blood flowing along the central axis of an artery of radius R is modeled by the formula

$$S(R) = cR^2$$

where c is a constant. What percentage error (rounded to the nearest percent) will you make in the calculation of $S(R)$ from this formula if you make a 1% error in the measurement of R?

36. The gross U.S. federal debt (in trillions of dollars) from 2000 to 2004 is given in the following table

Year	Gross federal debt
2000	5.629
2001	5.770
2002	6.198
2003	6.760
2004	7.355

Data source: From the historical tables of the Office of Management and Budget, 2006, as downloaded from www.whitehouse. gov/omb/budget/Historicals.

a. Plot the data and the linear approximation of the data at $t = 0$ (2000). Discuss the quality of this approximation.

b. Use a linear approximation to estimate the federal debt in 2010. Look up the actual gross federal debt to see how well the approximation worked.

37. In Example 7 of Section 2.6, the function

$$S(t) = 7.292 + 0.023t - 0.004t^2$$

was used to fit data on the extent of the Arctic sea ice S (million square kilometers) as a function of years t since 1980 (corresponding to $t = 0$).

a. What is the sensitivity of S in 1980?

b. How has the sensitivity changed from 1980 to 2000? Does this make sense? If not, why not?

38. Consider a power function $f(x) = ax^b$ with $a > 0$ and $b \ne 0$. Show that the elasticity of $f(x)$ is independent of the value x. What does it depend on? Use this answer to quickly solve the following problems.

a. If there is a 5% error in estimating the mass M of a weightlifter, what is approximately the percent error in estimating the lift $L = 20.15\,M^{2/3}$ kg of the weightlifter?

b. If there is a 10% error in estimating the mass M of an organism, then what is approximately the percent error in estimating the metabolic rate $R \propto M^{3/4}$ kcal/hour of the organism?

c. If there is a 2% error in measuring the weight W of a person, what is approximately the percent error in estimating the body mass index $B \propto W$ of the person?

39. In Problem 31 in Problem Set 3.2, we used the following function (here we replace c with the constant $k = e^{-c-10}$ and x with $\ln y$)

$$T = a + \frac{b - a}{1 + ky}$$

to model T—the concentration of bound ligand I per milligram of tissue—in terms of y representing the concentration of a second ligand II in the solution.

a. What is the elasticity of T with respect to changes in y?

b. If there is a 10% error in estimating the concentration y of ligand II, then what is the error in calculating the level of ligand I as a function of y for the case $a = 1, b = 2$, and $k = 1$.

40. In Example 2 of Section 2.4, we used the function

$$R(x) = \frac{100e^x}{e^x + e^{-5}}$$

to represent the percent of patients exhibiting an above normal temperature response to dose x millimoles (mM) of a particular histamine.

a. What is the elasticity of R with respect to changes in x?

b. If there is a 5% error in estimating the dose x when $x = -5$ (mM), then what is the percent error in calculating the response R?

3.6 Higher Derivatives and Approximations

The derivative of a function can be interpreted as the instantaneous rate of change, and it yields linear approximations to the function. Since the derivative of a function is also a function, this latter function also has a derivative. What does this derivative of a derivative represent? How useful is it? The goal of this section is to answer these questions—and considerably more.

Second derivatives

The **second derivative** of a function f is the derivative of f' and is denoted f''. In other words,

$$f''(x) = \frac{d}{dx}\left(\frac{d}{dx}f(x)\right)$$

Equivalently, we write

$$f''(x) = \frac{d^2}{dx^2}f(x) = f^{(2)}(x)$$

or if $y = f(x)$,

$$f''(x) = \frac{d}{dx}\left(\frac{dy}{dx}\right) = \frac{d^2y}{dx^2}$$

Note that $\dfrac{d^2}{dx^2}$ is regarded as the symbol for the "operation of taking the second derivative of a function with respect to its argument x."

Example 1   Finding second derivatives

Find $f''(x)$ for the given functions.

a. $f(x) = \sin x$ **b.** $f(x) = x^2$ **c.** $f(x) = x2^x$

Solution

a. Since $f'(x) = \cos x$, we get that $f''(x) = \dfrac{d}{dx}\cos x = -\sin x$.

b. Since $f'(x) = 2x$, we get that $f''(x) = \dfrac{d}{dx}2x = 2$.

c. Since $f'(x) = 2^x + x(\ln 2)2^x = 2^x(1 + x\ln 2)$, we get that

$$f''(x) = 2^x \ln 2 + 2^x \ln 2(1 + x\ln 2)$$
$$= 2^x \ln 2(1 + 1 + x\ln 2)$$
$$= 2^x \ln 2(2 + x\ln 2)$$

What do these second derivatives represent? Consider the following definition.

Concave Up/Concave Down	If the graph of a function f lies above all its tangents on an interval I, then it is said to be **concave up** on I. If the graph of f lies below all of its tangents on I, it is said to be **concave down**.

concave up

concave down

Since f'' is the derivative of f', the mean value theorem implies that if $f'' > 0$ on an interval, then f' is increasing on this interval. What does this mean? In terms of tangent lines, this means that the slope of the tangent line is increasing in the interval. Hence, f is "bending upward" or, equivalently, is concave up on this interval. Alternatively, if $f'' < 0$, then the slope of the tangent line is decreasing and f is "bending downward" or, equivalently, is concave down.

Concavity	Let f be a function whose first and second derivatives are defined at $x = a$. If $f''(a) < 0$, then $y = f(x)$ is concave down near $x = a$. If $f''(a) > 0$, then $y = f(x)$ is concave up near $x = a$.

Example 2   Identifying concavities

Identify the concavities of the function defined by the given graphs. In other words, determine where the graphs are concave up and where they are concave down.

a.
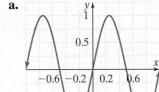

b.
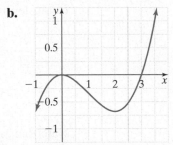

Solution

a. The easiest way to proceed is to place a straight edge (*e.g.,* ruler, pencil) on the graph; keeping it tangent to the curve, move from left to right. Whenever the straight edge is rotating in a counterclockwise fashion, the slope of the tangent line is increasing and $f'' > 0$. Hence, the function is concave up for these values of x. Alternatively, whenever the straight edge is rotating in a clockwise fashion, the slope of the tangent line is decreasing and $f'' < 0$. Hence, the function is concave down for these value of x.

　For the graph in part **a** of the example, there is a clockwise rotation from $x = -1$ to $x = -0.5$ and from $x = 0$ to $x = 0.5$. Hence, the function is concave down on $(-1, -0.5)$ and $(0, 0.5)$. Alternatively, there is a counterclockwise rotation from $x = -0.5$ to $x = 0$ and from $x = 0.5$ to $x = 1$. Hence, the function is concave up on $(-0.5, 0)$ and $(0.5, 1)$.

b. Using the approach described in part **a**, we find that this graph is concave down over $(-1, 1)$ and concave up on $(1, 4)$.

A point on a continuous graph that separates a concave downward portion of a curve from a concave upward portion is called an **inflection point**. The following example illustrates this idea using data from mortality due to airborne diseases.

Example 3   Sigmoidal decay in deaths due to airborne diseases

In a study of deaths in the United States, Ausubel and colleagues found that deaths from aerially transmitted diseases as a fraction of all deaths could be very well described by the sigmoidal function shown in Figure 3.30. Determine where this function is concave up and down. Find the point of inflection. Discuss what these changes in concavity mean.

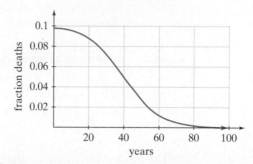

Figure 3.30 Fraction of deaths as a function of time (in years) after 1880.

Data: J. H. Ausubel, P. S. Meyer, and I. K. Wernick, "Death and the Human Environment: The United States in the 20th Century," *Technology in Society* 23(2) (2001): 131–146. Reprinted with permission.

Solution   To estimate the intervals of concavity, we can place a ruler as a tangent to the curve and slowly move it from the left side to the right side. Doing so, we notice that the ruler is rotating clockwise from $x = 0$ to $x \approx 40$ and rotating counterclockwise from $x \approx 40$ to $x = 120$. Hence, the point of inflection appears to be located at $x = 40$, and the fraction of deaths due to airborne diseases is decreasing at a faster and faster rate from 1880 ($x = 0$) to 1920 ($x = 40$); that is, the curve is concave down. The fraction of deaths is decreasing at a slower and slower rate from 1920 ($x = 40$) to 1980 ($x = 100$); that is, the curve is concave up.

The curve in Example 3 is an example of a *sigmoidal curve*, a curve that is monotonic, with horizontal asymptotes at positive and negative infinity and a single point of inflection. Another example of a sigmoidal curve is Figure 1.28 in Section 1.4; this is a model of population growth in the United States where growth is initially exponential and then levels off asymptotically. The following graphs show both forms of sigmoidal curves.

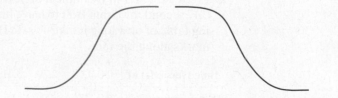

Graphs of increasing (in blue) and decreasing (in red) sigmoidal functions.

Changing rates of change

Recall from Section 2.1 the interpretation of the derivative of a function as the rate of change of that function with respect to increasing values of the function's argument. If we go back to the origins of the calculus, we see that velocity was interpreted by Newton as the instantaneous rate of change over time of the position of a particle in space, and acceleration was interpreted as the instantaneous rate of change over time of velocity. These interpretations led Newton to formulate his famous second law on the relationship between the force acting on an object, the mass of the object, and its resulting acceleration.

Average and Instantaneous Acceleration

Let $v(t)$ be the velocity of an object at time t in a predetermined direction. The **average acceleration** of an object from time t to time $t + h$ is

$$\text{AVERAGE ACCELERATION} = \frac{\text{change in velocity}}{\text{time elapsed}} = \frac{v(t + h) - v(t)}{h}$$

while the **instantaneous acceleration** of an object at time t is

$$\text{INSTANTANEOUS ACCELERATION} = \lim_{h \to 0} \frac{\text{change in velocity}}{\text{time elapsed}} = \lim_{h \to 0} \frac{v(t + h) - v(t)}{h}$$

The field of kinematics—that is, the motion of points or bodies through space—and the application of calculus to physics are developed in physics courses. But no calculus book is complete without at least one example dealing with acceleration, and here we focus on a biological one.

Example 4   When does Usain Bolt slow down?

Example 3 of Section 2.1 provides data on times for the 10-meter splits of Usain Bolt's record-breaking 100-meter win in the 2008 Beijing Olympic Games. Use these data to answer the following questions.

a. Use technology to fit a fifth-order polynomial $s(t)$ to these data that specifies the distance covered by Bolt as a function of time, beginning with the moment he leaves the starting blocks until the end of this race.

b. Calculate and plot the first and second derivatives of $s(t)$ to obtain graphs of Bolt's instantaneous velocity and acceleration during the race.

c. From the plots in part **b**, determine the time t_s during the race that Bolt switches from speeding up to slowing down, and calculate his velocity v_{max}, which is a maximum, at this time.

d. What is Bolt's average deceleration (negative of acceleration) from time t_s until the end of the race?

Solution

a. From Table 2.1 in Example 3 of Section 2.1, and taking into account that it takes 0.17 second for Usain Bolt to leave his starting block, we can construct the following table of how long it takes Usain Bolt to reach each of the successive 10-meter marks along the race.

Time (seconds) at	0	1.68	2.70	3.61	4.48	5.33	6.15	6.97	7.79	8.62	9.52
Distance (meters)	0	10	20	30	40	50	60	70	80	90	100

Source: http://speedendurance.com/2008/08/22/usain-bolt-100m-10-meter-splits-and-speed-endurance/

From these data we can use technology to obtain the following function $s(t)$ that specifies distance covered as time progress from leaving the blocks at $t = 0$ to finishing the race at $t = 9.52$, which with the time it takes for Usain Bolt to leave blocks, gives him a final race time of $9.52 + 0.17 = 9.69$ seconds (each term is specified to five significant figures):

$$s(t) = 2.0288t + 3.0905t^2 - 0.52044t^3 + 0.046520t^4 - 0.0016942t^5$$

This function is illustrated in the right panel of Figure 3.31.

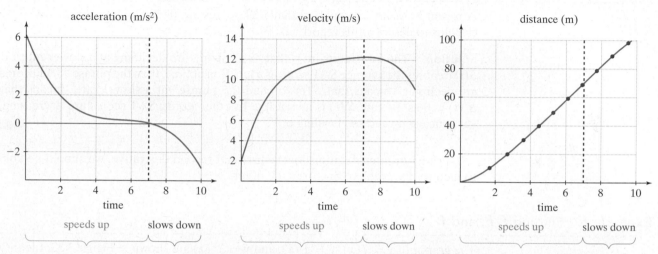

Figure 3.31 Distance covered as a function of time (seconds) by Usain Bolt in his record-breaking 100-meter win in the 2008 Beijing Olympic Games. Right panel: data and five-degree polynomial fit; middle panel: corresponding velocity profile; left panel: acceleration profile. Maximum velocity $v_{max} = 12.28$ is achieved (dotted vertical line) at $t = 7.17$ seconds, when the acceleration curve passes through zero.

b. Taking the first derivative, we obtain the velocity function (middle panel in Figure 3.31)

$$s'(t) = 2.0288 + 6.1811t - 1.5613t^2 + 0.18608t^3 - 0.0084710t^4$$

and taking the second derivative we obtain the acceleration function (left panel in Figure 3.31)

$$s''(t) = 6.1811 - 3.1227t + 0.55824t^2 - 0.033884t^3$$

c. The function $s''(t)$ has a root at $t_s = 7.17$. We see from Figure 3.31 that $s''(t)$ is decreasing and goes from positive on $[0, 7.16]$ to negative on $[7.18, 9.52]$. Thus, the velocity and acceleration plots show that Bolt accelerates, first strongly, and then more slowly until $t_s = 7.17$, at which time he reaches his maximum instantaneous velocity of $v_{max} = 12.28$ meters/second (m/s).

d. The average acceleration from time $t_s = 7.17$ to the end of the race at $t = 9.52$ is, by definition with $v(t) = s'(t)$,

$$\frac{s'(7.17) - s'(9.52)}{7.17 - 9.52} = -0.83$$

so taking the negative sign into account, the average deceleration is 0.83 m/s^2.

The reason for choosing a fifth-order polynomial rather than a fourth or sixth in Example 4 is that the runners accelerate strongly at the beginning and tend to decelerate as they tire at the end. Thus, acceleration is initially positive and eventually negative. Since this type of behavior is best modeled by odd-ordered rather than even-ordered polynomials, we choose an odd-ordered polynomial for distance, as it becomes odd again after taking two derivatives. We choose a fifth-order instead of third-order polynomial as it does a better job modeling the intermediate, close to zero acceleration segment that we see occur approximately around seconds 4 to 7 during the course of the race (see Problem 53 in Problem Set 3.6).

Example 5   Declining rates

A recent news article reported that SAT scores are declining at a slower rate. Use calculus to describe this report.

Solution   The key statement is that "SAT scores are declining at a slower rate." If $S(t)$ denotes the average SAT score as a function of time, then the phrase "SAT scores are declining" means that $S'(t) < 0$ and the phrase "at a slower rate" means that $S''(t) > 0$ as the rate $S'(t)$ is increasing. In other words, SAT scores are decreasing, yet concave up and so "leveling off"! ∎

Using our interpretations of the first and second derivative, we should be able to identify the graph of one from the other.

Example 6   Finding f, f′, and f″

The graphs of $y = f(x)$, $y = f'(x)$, and $y = f''(x)$ are shown in Figure 3.32. Identify f, f', and f''.

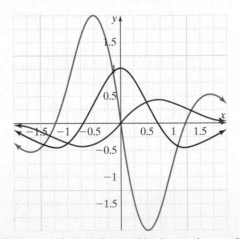

Figure 3.32 Graph of a function and its first and second derivatives.

Solution   To identify f, f', f'', we can start with one graph—say, the black one— and determine when it is rising and falling. We can see that it is falling roughly on the intervals $[-2, -1.25]$ and $[0, 1.25]$ and rising on the complementary intervals. Since the blue curve is negative on $[-2, -1.25]$ and $[0, 1.25]$ and positive on the complementary intervals, the blue curve may be the graph of the derivative of function defined by the black curve. On the other hand, the black curve appears to be negative where the red curve is falling and positive where the red curve is rising. Hence, the black curve appears to be the graph of the derivative of the function defined by the red curve. Therefore, we conclude that $y = f(x)$ is defined by the red curve, $y = f'(x)$ is defined by the black curve, and $y = f''(x)$ is defined by the blue curve. ∎

Second-order approximations

In Section 3.5, we approximated functions with their tangent lines. While a good start, these approximations can be improved upon by using first and second derivatives.

Example 7   Stripping away the tangent line

Consider $y = e^{2x}$.

a. Find the tangent line at $x = 0$.

b. Compute $\dfrac{d^2y}{dx^2}\bigg|_{x=0}$ and determine whether the linear approximation overestimates or underestimates $y = e^{2x}$ near $x = 0$.

c. Plot the difference between $y = e^{2x}$ and its tangent line. Discuss what you notice.

Solution

a. Since $\dfrac{dy}{dx}\bigg|_{x=0} = 2e^{2x}\bigg|_{x=0} = 2e^0 = 2$, the tangent line has slope 2 and passes through the point $(0, 1)$—that is, the line

$$(y - 1) = 2(x - 0) \quad \Rightarrow \quad y = 2x + 1$$

b. $\dfrac{d^2y}{dx^2} = \dfrac{d}{dx} 2e^{2x} = 4e^{2x}$, which is 4 at $x = 0$. Since the second derivative is positive, $\dfrac{dy}{dx}$ is increasing near $x = 0$, and we would expect the tangent line to underestimate (*i.e.*, lie under) $y = e^{2x}$ near $x = 0$. Indeed, graphing the function $y = e^{2x}$ (blue curve) and $y = 2x + 1$ (red curve) confirms this prediction.

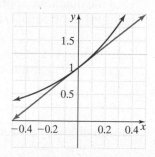

c. Plotting $y = e^{2x} - 2x - 1$ yields a function that looks like a parabola.

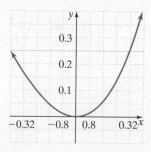

The preceding example suggests that function obtained by taking the difference between the original function and its derivative is approximately parabolic. If we want to approximate this function by a quadratic function, how do we find the best quadratic approximation? To answer this question, we develop the following procedure for finding the best quadratic approximation to a function f around the point $x = 0$. To find

$$f(x) \approx a + bx + cx^2$$

near $x = 0$, we require

$$a = f(0)$$

To have the first derivatives of $f(x)$ and the approximation $a + bx + cx^2$ agree at $x = 0$, we can take derivatives of both sides

$$f'(x) \approx b + 2cx$$

At $x = 0$, we want $f'(0) = b$, so we define $b = f'(x)$. Finally, to have their second derivatives agree at $x = 0$, we differentiate one more time:

$$f''(x) \approx 2c$$

This leads us to define $c = \dfrac{f''(0)}{2}$. This gives a **quadratic** (*second-order* or *parabolic*) **approximation** at $x = 0$:

$$f(x) \approx f(0) + f'(0)x + \frac{1}{2}f''(0)x^2$$

Let us see how well this approximation works.

Example 8   Quadratic approximation

Find the quadratic approximation to $y = e^{2x}$ at $x = 0$. Plot $y = e^{2x}$, its linear approximation, and its quadratic approximation.

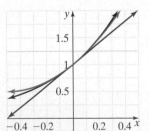

Figure 3.33 Graph of a function $f(x) = e^{2x}$ (in black) along with its linear (in red) and quadratic approximations (in blue).

Solution   Let $f(x) = e^{2x}$, so from the previous example $f'(x) = 2e^{2x}$ and $f''(x) = 4e^{2x}$. The linear approximation is

$$f(x) \approx f(0) + f'(0)(x - 0) = 1 + 2x$$

The quadratic approximation is

$$f(x) = f(0) + f'(0)x + \frac{1}{2}f''(0)x^2$$
$$= 1 + 2x + 2x^2$$

The graphs of $y = e^{2x}$, $y = 2x + 1$, and $y = 1 + 2x + 2x^2$ are shown in Figure 3.33. The quadratic approximation does a significantly better job of approximating the function $y = e^{2x}$.   ▄

In cases where the linear approximation is a horizontal line, the quadratic approximation is the first approximation to give real information about the concavity of the curve in question.

Example 9   Approximating the cosine
---

a. Find the linear and quadratic approximations of $y = \cos x$ at $x = 0$.

b. Use the quadratic approximation to estimate $\cos 1$, $\cos 0.5$, and $\cos 0.1$. Compare your approximations to the answers given by a calculator.

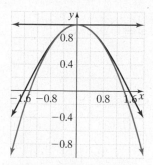

Figure 3.34 Graph of a function $f(x) = \cos x$ (in black) along with its linear (in red) and quadratic approximations (in blue).

Solution

a. Let $f(x) = \cos x$, so $f'(0) = -\sin 0 = 0$ and $f''(0) = -\cos 0 = -1$. The linear approximation is

$$y = 1$$

The quadratic approximation is

$$f(x) = f(0) + f'(0)x + \frac{1}{2}f''(0)x^2$$

$$= 1 - \frac{1}{2}x^2$$

The graph is a downward facing parabola. The graphs of $y = \cos x$, $y = 1$, and $y = 1 - \frac{1}{2}x^2$ are shown in Figure 3.34.

b. We compare the quadratic and calculator approximations.

Quadratic approximation	Calculator approximation	Comment (dp ≡ decimal places)
$\cos 1 \approx 1 - \frac{1}{2} = 0.5$	$\cos 1 \approx 0.540302$	fair: correct to 1 dp
$\cos 0.5 \approx 1 - \frac{1}{2}(0.5)^2 = 0.875$	$\cos 0.5 \approx 0.877583$	better: correct to 2 dp
$\cos 0.1 \approx 1 - \frac{1}{2}(0.1)^2 = 0.995$	$\cos 0.1 \approx 0.995004$	better yet: correct to 5 dp!

The approximations get better and better as you get closer and closer to $x = 0$.

More generally, we may wish to approximate a function near a point $x = a$ with a parabola. As an exercise, you can verify that by forcing the quadratic approximation and the function to agree up to the second derivative at $x = a$, you obtain the following **second-order approximation** of f at $x = a$.

Second-Order Approximation

Let f have a first and second derivative defined at $x = a$. The *second-order approximation* of f around $x = a$ is given by

$$f(x) \approx f(a) + f'(a)(x - a) + \frac{1}{2}f''(a)(x - a)^2$$

Example 10   Professor Getz's headache continues

Recall in Example 6 of Section 3.5 that we found that the half-life for a drug as a function of the clearance rate x per hour is given by

$$T(x) = \frac{\ln 2}{x} \text{ hours}$$

For a dose of 1000 mg of acetaminophen, we estimated a clearance rate of approximately 0.28 per hour.

a. Compute the first- and second-order approximation of $T(x)$ at $x = 0.28$.

b. Plot both approximations together with the function $T(x)$.

c. Discuss whether an error analysis using the sensitivity $T'(0.28)$ overestimates or underestimates the propagation of error from estimating x to estimating $T(x)$.

Solution

a. Computing the first and second derivatives of $T(x)$ at $x = 0.28$ gives

$$T'(0.28) = \frac{d}{dx}\bigg|_{x=0.28} \frac{\ln 2}{x}$$

$$= -\frac{\ln 2}{x^2}\bigg|_{x=0.28} \approx -8.841$$

$$T''(0.28) = \frac{d}{dx}\bigg|_{x=0.28} -\frac{\ln 2}{x^2}$$

$$= \frac{2\ln 2}{x^3}\bigg|_{x=0.28} \approx 63.151$$

Therefore, the first-order approximation is given by

$$T(x) \approx T(0.28) + T'(0.28)(x - 0.28)$$

$$\approx 2.475 - 8.841(x - 0.28)$$

and the second-order approximation is given by

$$T(x) \approx T(0.28) + T'(0.28)(x - 0.28) + \frac{1}{2}T''(0.28)(x - 0.28)^2$$

$$\approx 2.476 - 8.841(x - 0.28) + \frac{1}{2}63.151(x - 0.28)^2$$

b. Plotting $T(x)$ (in black), the first-order approximation (in blue), and the second-order approximation (in red) yields

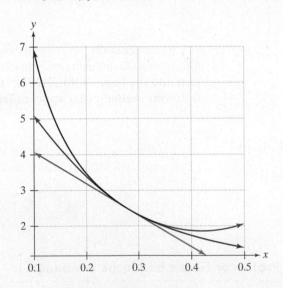

c. Since the sensitivity analysis using $T'(0.28)$ is based on the linear approximation of $T(x)$ near $x = 0.28$ and T is concave up, the sensitivity analysis underestimates the propagation of error from estimating x to estimating $T(x)$. This underestimation corresponds to the tangent line (in black) in part **a** lying under the graph of $T(x)$. ∎

Even higher derivatives

Once we have taken second derivatives, there is no reason to stop. We can attempt to take the derivative of the second derivative, and the derivative of the resulting derivatives until the function obtained is 0; or we can continue indefinitely (e.g., see parts **b** and **c** of the next example). These **higher derivatives** and the associated notations are defined as follows:

- **First derivatives:** $f^{(1)}(x) = f'(x) = \dfrac{d}{dx} f(x) = \dfrac{df}{dx}$

- **Second derivatives:** $f^{(2)}(x) = f''(x) = \dfrac{d}{dx}\left(\dfrac{d}{dx} f(x)\right) = \dfrac{d^2}{dx^2} f(x) = \dfrac{d^2 f}{dx^2}$

- **Third derivatives:** $f^{(3)}(x) = f'''(x) = \dfrac{d}{dx}\left[\dfrac{d}{dx}\left(\dfrac{d}{dx} f(x)\right)\right] = \dfrac{d^3}{dx^3} f(x) = \dfrac{d^3 f}{dx^3}$

- ***n*th derivatives:** $f^{(n)}(x) = \dfrac{d}{dx}\left(f^{(n-1)}(x)\right) = \dfrac{d}{dx}\left(\dfrac{d^{n-1} f}{dx^{n-1}}\right) = \dfrac{d^n f}{dx^n}$

Example 11   Higher derivatives

Find the following higher derivatives.

a. $\dfrac{d^3}{dx^3}(1 + x + x^3)$

b. $\dfrac{d^5}{dx^5} e^{2x}$

c. $f^{(101)}(x)$ where $f(x) = \sin x$

Solution

a. If $y = 1 + x + x^3$, then

$$\frac{dy}{dx} = \frac{d}{dy}(1 + x + x^3) = 1 + 3x^2$$

$$\frac{d^2 y}{dx^2} = \frac{d}{dy}(1 + 3x^2) = 6x$$

$$\frac{d^3 y}{dx^3} = \frac{d}{dy}(6x) = 6$$

b. If $f(x) = e^{2x}$, then

$f'(x) = 2e^{2x}$

$f''(x) = 4e^{2x}$

$f'''(x) = 8e^{2x}$

$f^{(4)}(x) = 16e^{2x}$

$f^{(5)}(x) = 32e^{2x}$

c. At first this problem might seem insane. Take 101 derivatives of a function? However, if we proceed calmly, a pattern will quickly emerge that will dispel this insanity. If $f(x) = \sin x$, then

$f'(x) = \cos x$

$f''(x) = -\sin x$

$f'''(x) = -\cos x$

$f^{(4)}(x) = \sin x$

We are back to where we started, and the derivatives cycle in a fixed pattern of period four; that is, repetition occurs every four derivatives so that
$$f^{(1)}(x) = f^{(5)}(x) = f^{(9)}(x) = \cdots \cos x, \quad f^{(2)}(x) = f^{(6)}(x) = f^{(10)}(x) = \cdots - \sin x,$$
and so on. Hence, $f^{(100)}(x) = \sin x$ and $f^{(101)}(x) = \dfrac{d}{dx}\sin x = \cos x$. ■

We conclude with an application of higher derivatives from politics.

Example 12   Presidential proclamation

In the fall of 1972, President Nixon announced that the rate of increase of inflation was decreasing. This was the first time a sitting president used the third derivative to advance his case for reelection.

So reported Hugo Rossi in "Mathematics Is an Edifice, Not a Toolbox" (*Notices of the AMS*, 43, October 1996). Discuss how a third derivative was being used by President Nixon.

Solution   Let V denote the "value of a dollar" at time t in years. Inflation means that the value of a dollar is decreasing, so $\dfrac{dV}{dt} < 0$. If inflation is increasing, then the value of a dollar is decreasing at a faster rate; that is, $\dfrac{d^2V}{dt^2} < 0$. Finally, if the rate of increase of inflation is decreasing, we get $\dfrac{d^3V}{dt^3} > 0$. Hence, at the level of the third derivative, things were looking less bleak for the value of the dollar! ■

PROBLEM SET 3.6

Level 1 DRILL PROBLEMS

Find the higher derivatives indicated in Problems 1 to 12.

1. $\dfrac{d^2}{dx^2}(xe^{-x})$

2. $\dfrac{d^3}{dx^3}(2^x)$

3. $f^{(4)}(x)$, where $f(x) = 1 + x + x^2 + x^3 + x^4$

4. $f^{(103)}(x)$, where $f(x) = \cos x$

5. $\dfrac{d^{99}}{dx^{99}}(\sin 3x)$

6. $\dfrac{d^3}{dw^3}(1 + w + w^2 + w^3 + w^4)$

7. $\dfrac{d^4}{dt^4}\left(\dfrac{1}{4}t^8 - \dfrac{1}{2}t^6 - t^2 + 2\right)$

8. $\dfrac{d^{n+1}}{dx^{n+1}}x^n$

9. $f^{(10)}(x)$, where $f(x) = (1 + x)^{10}$

10. $f^{(4)}(x)$, where $f(x) = \dfrac{4}{\sqrt{x}}$

11. $\dfrac{d^2}{dw^2}\dfrac{1}{1+w}$

12. $\dfrac{d^2y}{dx^2}$, where $y = (x^2 + 4)(1 - 3x^3)$

In Problems 13 to 18 you are given the function $s(t)$ that specifies the position of an object on a line as a function of time t. Find expressions for the velocity and acceleration as a function of time for $t \geq 0$. Solve for t when the acceleration is 0; use technology where needed to find the values and when multiple values exist, use the closest to $t = 0$.

13. $s(t) = t^2 - 3t$

14. $s(t) = t^3 - 5t^2 - 8t$

15. $s(t) = 1 - \cos t/3$

16. $s(t) = 1 - \cos t^2$

17. $s(t) = te^{-t}$

18. $s(t) = \dfrac{1}{1 + e^t}$

In Problems 19 to 24 find the linear approximations of the given functions $f(x)$ around $x = a$. Using second-order derivatives, determine whether the linear approximation tends to overestimate or underestimate $f(x)$ near $x = a$.

19. $f(x) = e^x$ at $x = 0$

20. $f(x) = \cos x$ at $x = 0$

21. $f(x) = 1 - x^2$ at $x = 2$

22. $f(x) = \tan x$ at $x = \pi$

23. $f(x) = \dfrac{1}{1 + x}$ at $x = 2$

24. $f(x) = xe^{-x}$ at $x = 1$

In Problems 25 to 34 determine on what intervals f is increasing, decreasing, concave up, concave down, and find the points of inflection.

25. $y = 1 - x + x^3$

26. $y = 1 + 2x + 18/x$

27. $y = xe^{-x}$

28. $y = e^{-x^2}$

29. $y = \dfrac{x}{1 + x}$

30. $y = \dfrac{x}{x^2 + 1}$

31. $y = 3x^4 - 2x^3 - 12x^2 + 18x - 5$

32. $y = x^4 + 6x^3 - 12x^2 + 18x - 5$

33. $y = \sec x$

34. $y = x^3 + \sin x$ on $\left[-\dfrac{\pi}{2}, \dfrac{\pi}{2}\right]$

Find the first- and second-order approximations of $y = f(x)$ around $x = a$ in Problems 35 to 40. Use technology to plot the function and its approximations near $x = a$.

35. $y = \sin x$ at $x = 0$

36. $y = 1 + x^2$ at $x = 2$

37. $y = e^x$ at $x = 0$

38. $y = \sec x$ at $x = 0$

39. $y = \sqrt{x}$ at $x = 4$

40. $y = \sqrt[3]{x}$ at $x = 27$

In Problems 41 to 44, identify $y = f(x)$, $y = f'(x)$, and $y = f''(x)$.

41.

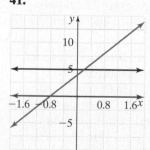

42.

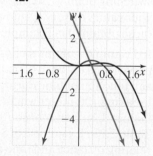

43.

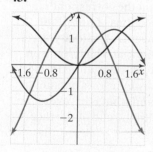

44.

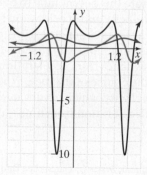

45. Sketch the graph of a function with all of the following properties:

$f'(x) > 0$ when $x < -1$

$f'(x) > 0$ when $x > 3$

$f'(x) < 0$ when $-1 < x < 3$

$f''(x) < 0$ when $x < 2$

$f''(x) > 0$ when $x > 2$

46. Sketch the graph of a function with all of the following properties:

$f'(x) > 0$ when $x < 2$ and when $2 < x < 5$

$f'(x) < 0$ when $x > 5$

$f'(2) = 0$

$f''(x) < 0$ when $x < 2$ and when $4 < x < 7$

$f''(x) > 0$ when $2 < x < 4$ and when $x > 7$

Level 2 APPLIED AND THEORY PROBLEMS

47. The slogan of a particular home improvement company is "Improving Home Improvement." Explain the role of derivatives in this slogan.

48. A politician claims that "Under a new law, prices would rise slower than if the law were not passed." Explain the role of higher derivatives in this statement.

49. At the website **http://www.nlreg.com/aids.htm**, you can find the following figure that graphs the number of new cases of AIDS since 1980.

  a. Estimate where the function is concave up and concave down.

  b. Describe in words what these changes in the concavity mean for the AIDS epidemic.

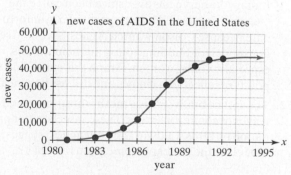

Source: Phil Sherrod at http://www.nlreg.com and http://www.dtreg.com

50. In Example 2 of Section 2.4, a dose-response curve for patients responding to a dose of histamine is given by

$$R = \frac{100e^x}{e^x + e^{-5}}$$

where x is the natural logarithm of the dose in millimoles (mmol).

a. Compute $\dfrac{d^2R}{dx^2}$

b. Determine for what dose ranges R is concave up and concave down. Interpret your results.

51. Historical Quest

Maria Gaetana Agnesi 1718–1799

One of the most famous women in the history of mathematics is Maria Gaetana Agnesi.

She was born in Milan, the first of twenty-one children. Her first publication was at age nine, when she wrote a Latin discourse defending higher education for women. Her most important work was a now classic calculus textbook published in 1748. Agnesi is primarily remembered for a curve defined by the equation

$$y = \frac{a^3}{x^2 + a^2}$$

for a positive constant a. The curve was named *versiera* (from the Italian verb *to turn*) by Agnesi, but John Colson, an Englishman who translated her work, confused the word *versiera* with the word *avversiera*, which means "wife of the devil" in Italian; the curve has ever since been called the "witch of Agnesi." This was particularly unfortunate because Colson wanted Agnesi's work to serve as a model for budding young mathematicians, especially young women. Graph this curve, find the points of inflection (if any), and discuss its concavity.

52. The spruce budworm is a moth whose larvae eat the leaves of coniferous trees. These insects suffer predation by birds. Ludwig and colleagues* suggested a model for the per capita predation rate, $p(x)$:

$$p(x) = \frac{bx^2}{a^2 + x^2}$$

where b is the maximum predation rate and a is the number of budworms at which the predation rate is half its maximum rate. What is the concavity of this curve, and is there a point of inflection?

53. In Example 4 of this section, we fitted a fifth-order polynomial to the data on the distance covered as a function of time during Usain Bolt's record-breaking 100-meter win in the 2008 Beijing Olympic Games. Use technology to fit a third-order polynomial to these data; then differentiate this polynomial to obtain the corresponding velocity and acceleration functions of time. Plot these three functions (displacement, velocity, acceleration) and estimate the time t_s at which Bolt switches from acceleration to deceleration during the course of the race and his velocity v_{max}, which is a maximum, at this switching point. What is the average deceleration from the switching point to the end of the race? Compare your values of t_s, v_{max}, and average deceleration with those obtained in Example 4.

54. In Example 4 of this section, we fitted a fifth-order polynomial to the data on the distance covered as a function of time during Usain Bolt's record-breaking 100-meter win in the 2008 Beijing Olympic Games. Use technology to fit a seventh-order polynomial to these data; then differentiate this polynomial to obtain the corresponding velocity and acceleration functions of time. Plot these three functions (displacement, velocity, acceleration) and estimate the time t_s at which Bolt switches from acceleration to deceleration during the course of the race and his velocity v_{max}, which is a maximum, at this switching point. What is the average deceleration from the switching point to the end of the race? Compare your values of t_s, v_{max}, and average deceleration with those obtained in Example 4.

55. In Example 4 of this section, we fitted a fifth-order polynomial to the data on the distance covered as a function of time during Usain Bolt's record-breaking 100-meter win in the 2008 Beijing Olympic Games. Use Ben Johnson's split times given in Table 2.1 in Example 3 of Section 2.1 to obtain data relating distance and time of Johnson's performance during the 1988 Seoul Olympics 100-meter final. Fit a fifth-order polynomial to the data; then differentiate this polynomial to obtain the corresponding velocity and acceleration functions of time. Plot these three functions (displacement, velocity, acceleration) and estimate the time t_s at which Johnson switches from acceleration to deceleration during the course of the race. What is Johnson's maximum velocity, which occurs at t_s, and what is his average deceleration from time t_s until the end of the race?

---

*D. Ludwig, D. D. Hones, and C. S. Holling, "Qualitative Analysis of Insect Outbreak Systems: The Spruce Budworm and Forest," *Journal of Animal Ecology* 47(1978): 315–332.

Compare your results to Bolt's performance analyzed in Example 4.

56. Let f be a function that is twice differentiable on an interval I containing the point $x = a$. If there exists a $K > 0$ such that $|f''(x)| \leq K$ for all x in I, then show that

$$|f(x) - f(a) - f'(a)(x - a)| \leq \frac{K}{2} |x - a|^2$$

for all x in I. This result gives the error of the first-order approximation. Hint: Pick any point $b \neq a$

in I. Define

$$G(x) = f(x) - f(a) - f'(a)(x - a) - C(x - a)^2$$

where C is chosen such that $G(b) = 0$. Differentiate G and apply the mean value theorem to G and f'.

57. Let f be a function with first- and second-order derivatives at $x = a$. Consider a quadratic of the form $q(x) = b + c(x - a) + d(x - a)^2$. Show that $f(a) = q(a)$, $f'(a) = q'(a)$, and $f''(a) = q''(a)$ if and only if $b = f(a)$, $c = f'(a)$ and $d = f''(a)/2$.

3.7   l'Hôpital's Rule

In models of tumor or population growth, spread of rumors, and risk of being infected, one may encounter limits of the form

$$\lim_{x \to a} \frac{f(x)}{g(x)}$$

where $\lim_{x \to a} f(x)$ and $\lim_{x \to a} g(x)$ are both zero or both infinite. Such limits are called **0/0 indeterminate form** and ∞/∞ **indeterminate form**, respectively, because their value cannot be determined without further analysis. In this section, we study an approach to handling these limits and explore some applications.

The 0/0 and ∞/∞ indeterminate forms

In 1694, French mathematician Guillaume de l'Hôpital (see Historical Quest in Problem Set 3.7) found a useful method for evaluating limits involving indeterminate forms. He considered the special case where $f(a) = g(a) = 0$, f and g are differentiable at $x = a$, and $g'(a) \neq 0$. Under these assumptions, we have

$$\lim_{x \to a} \frac{f(x)}{g(x)} = \lim_{x \to a} \frac{f(x)}{x - a} \frac{x - a}{g(x)} \qquad \textit{multiplying by } \frac{x - a}{x - a}$$

$$= \lim_{x \to a} \frac{f(x)}{x - a} \lim_{x \to a} \frac{x - a}{g(x)} \qquad \textit{product limit law}$$

$$= \lim_{x \to a} \frac{f(x) - f(a)}{x - a} \lim_{x \to a} \frac{x - a}{g(x) - g(a)} \qquad \textit{as } g(a) = f(a) = 0$$

$$= f'(a) \frac{1}{g'(a)} \qquad \textit{by definition of derivative and } g'(a) \neq 0$$

The following example illustrates how replacing the original limit with a limit involving derivatives can be helpful.

Example 1   Using l'Hôpital's argument

Evaluate the following limits.

a. $\displaystyle\lim_{x \to 0} \frac{\sin x}{x}$

b. $\displaystyle\lim_{x \to 2} \frac{x^7 - 128}{x^3 - 8}$

Solution

a. This limit is of indeterminate form because $\sin x$ and x both approach 0 as $x \to 0$. l'Hôpital's rule applies because both $\sin x$ and x are differentiable at $x = 0$. Thus

$$\lim_{x \to 0} \frac{\sin x}{x} = \frac{\cos x}{1}\Big|_{x=0} = 1$$

b. For this example, $f(x) = x^7 - 128$ and $g(x) = x^3 - 8$ and the limit is of the form 0/0. Since $f'(2) = 7 \cdot 2^6$ and $g'(2) = 3 \cdot 2^2 \neq 0$, we can apply l'Hôpital's argument to obtain

$$\lim_{x \to 2} \frac{x^7 - 128}{x^3 - 8} = \frac{f'(2)}{g'(2)} = \frac{7 \cdot 2^6}{3 \cdot 2^2} = \frac{112}{3}$$

This approach to computing limits is known as *l'Hôpital's rule*. The general statement of this rule is given in the following theorem.

Theorem 3.1   l'Hôpital's rule

Let f and g be differentiable functions on an open interval containing a (except possibly at a itself). Suppose $\lim\limits_{x \to a} \dfrac{f(x)}{g(x)}$ produces an indeterminate form $\dfrac{0}{0}$ or $\dfrac{\infty}{\infty}$ and that

$$\lim_{x \to a} \frac{f'(x)}{g'(x)} = L$$

where L is either a finite number, $-\infty$, or ∞. Then

$$\lim_{x \to a} \frac{f(x)}{g(x)} = L$$

The theorem also applies to one-sided limits and to limits at infinity where $x \to \infty$ and $x \to -\infty$.

When we use l'Hôpital's rule, we use the symbol $\overset{H}{=}$ as shown in the following example.

Example 2   Using l'Hôpital's rule

Find the following limits.

a. $\lim\limits_{x \to 3} \dfrac{\sin(x-3)}{x-3}$ **b.** $\lim\limits_{x \to \infty} \dfrac{x + e^{-x}}{2x + 1}$ **c.** $\lim\limits_{x \to \pi/2^+} \dfrac{\cos x}{1 - \sin x}$ **d.** $\lim\limits_{x \to \infty} \dfrac{\sqrt{x}}{\ln x}$

Solution

a. Since $\sin(3-3) = 3 - 3 = 0$, this limit is of an 0/0 indeterminate form. Since both $\sin(x-3)$ and $x - 3$ are differentiable near $x = 3$, we can apply l'Hôpital's rule as follows:

$$\lim_{x \to 3} \frac{\sin(x-3)}{x-3} \overset{H}{=} \lim_{x \to 3} \frac{\cos(x-3)}{1}$$

$$= 1$$

b. Since $\lim\limits_{x \to \infty} x + e^{-x} = \infty$ and $\lim\limits_{x \to \infty} 2x + 1 = \infty$, this limit is an ∞/∞ indeterminate form. Since both $x + e^{-x}$ and $2x + 1$ are differentiable for large x, we can apply l'Hôpital's rule as follows:

$$\lim_{x \to \infty} \frac{x + e^{-x}}{2x + 1} \overset{H}{=} \lim_{x \to \infty} \frac{1 - e^{-x}}{2}$$

$$= \frac{1}{2}$$

c. Since $\cos \pi/2 = 1 - \sin \pi/2 = 0$, this limit is of a $0/0$ indeterminate form. Since both $\cos x$ and $1 - \sin x$ are differentiable near $x = \pi/2$, we can apply l'Hôpital's rule as follows:

$$\lim_{x \to \pi/2^+} \frac{\cos x}{1 - \sin x} \overset{H}{=} \lim_{x \to \pi/2^+} \frac{-\sin x}{-\cos x}$$

$$= -\infty$$

d. Since $\lim_{x \to \infty} \sqrt{x} = \infty$ and $\lim_{x \to \infty} \ln x = \infty$, this limit is an ∞/∞ indeterminate form. Since both $\sqrt{x}$ and $\ln x$ are differentiable for large x, we can apply l'Hôpital's rule as follows:

$$\lim_{x \to \infty} \frac{\sqrt{x}}{\ln x} \overset{H}{=} \lim_{x \to \infty} \frac{\dfrac{1}{2\sqrt{x}}}{\dfrac{1}{x}}$$

$$= \lim_{x \to \infty} \frac{\sqrt{x}}{2} = \infty$$

We consider two applications of l'Hôpital's rule to models of population growth.

Example 3 Exponential versus arithmetic growth

In *An Essay on the Principle of Population*, first published anonymously in 1798, but later attributed to Thomas Malthus, we find the following text:

"Population, when unchecked, increases in a geometrical ratio. Subsistence increases only in an arithmetical ratio. A slight acquaintance with numbers will shew [sic] the immensity of the first power in comparison of the second."

While Example 4 of Section 1.4 explored a special case of this observation, l'Hôpital's rule allows us to fully appreciate the observation of Malthus. Let $P(t) = P_0 c^t$ for some $c > 1$ and $P_0 > 0$ represent the size of a population at time t, and let $F(t) = a + bt$ for some $a > 0$ and $b > 0$ represent the total amount of food available at time t. Find

$$\lim_{t \to \infty} \frac{F(t)}{P(t)}$$

and discuss its implications.

Solution   Since both $P_0 c^t$ and $a + bt$ approach infinity as t approaches infinity, we obtain

$$\lim_{t \to \infty} \frac{a + bt}{P_0 c^t} \overset{H}{=} \lim_{t \to \infty} \frac{b}{P_0 c^t \ln c}$$

$$= 0$$

Hence, as time marches on, the amount of food per individual approaches nothing.

What do tumor growth, sales of mobile phones, spread of rumor or infection, and population growth have in common? They all can be modeled mathematically by specifying how the growth rate of the tumor, rumor, or population depends on its current size, frequency, or abundance. The next example looks an important family of growth functions introduced by F. J. Richards in a 1959 article that appeared in the *Journal of Experimental Botany* [2: 290–301].

Example 4   Generalized logistic growth function and indeterminate form 0/0

Whether it be the spread of a rumor, tumor growth, or population growth, an important class of models describing the rate at which the size N of a population, or a tumor for that matter, changes is

$$G(N) = (r/v)N(1 - (N/K)^v)$$

where $r > 0$, $v > 0$, and $K > 0$ are positive constants that influence the shape of the growth function.

a. For what population sizes N does the population grow?

b. Let $r = 1$ and $K = 100$. Plot $G(N)$ on the interval $[0, 110]$ for $v = 2, 1, 0.1, 0.01$. Discuss what you find.

c. For $K > N > 0$, $r = 1$, and $K = 100$, find

$$\lim_{v \to 0}(r/v)N(1 - N/K)^v)$$

and sketch the resulting curve.

Solution

a. Since r, v, and K are positive, the population growth rate $G(N)$ is positive if

$$1 - (N/K)^v > 0$$
$$1 > (N/K)^v$$
$$1 > N/K$$
$$K > N$$

Therefore, the population grows whenever its population size is positive and less than K.

b. Plotting $G(N)$ for the different v values yields the dashed colored curves in Figure 3.35. Smaller v values correspond to higher curves. Hence, as v gets smaller, the growth rate gets larger. Moreover, the growth function G appears to be approaching a limiting function as v approaches zero.

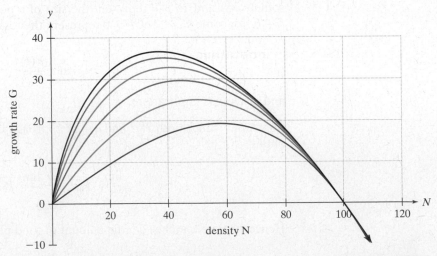

Figure 3.35 Generalized logistic growth function $G(N) = N(1 - (N/100)^v)/v$ for $v = 2$ (in red), 1 (in green), 0.5 (in purple), 0.25 (in cyan), and 0.1 (in magenta). The limiting Gompertz growth function $G(N) = -N\log(N/100)$ is shown as a solid black curve.

c. For $0 < N < K$, $G(N)$ is positive and the limit

$$\lim_{v \to 0} \frac{r N(1 - (N/K)^v)}{v}$$

is of indeterminate form $0/0$ as $r N(1 - (N/K)^v)$ and v equal zero at $v = 0$. Therefore, taking derivatives with respect to v gives

$$\lim_{v \to 0} \frac{r N(1 - (N/K)^v)}{v} \overset{H}{=} \lim_{v \to 0} \frac{-r N \ln(N/K)(N/K)^v}{1} = -r N \ln(N/K)$$

for $N > 0$. This limiting growth function $-r N \ln(N/K)$ is known as the *Gompertz growth function*, which has been used extensively to model tumor growth. Figure 3.35 plots this function (in black) for $K = 100$ and $r = 1$. As anticipated in part **b**, the plots of the generalized logistic growth functions approach the Gompertz growth function as v approaches zero.

Sometimes repeated applications of l'Hôpital's rule are necessary to get anywhere.

Example 5   Applying l'Hôpital's rule twice

Evaluate $\displaystyle \lim_{x \to \infty} \frac{2x^2 - 3x + 1}{3x^2 + 5x - 2}$.

Solution   If we consider the limit of the numerator and denominator separately, we obtain ∞/∞. However, if we apply l'Hôpital's rule twice we obtain

$$\lim_{x \to \infty} \frac{2x^2 - 3x + 1}{3x^2 + 5x - 2} \overset{H}{=} \lim_{x \to \infty} \frac{4x - 3}{6x + 5} \overset{H}{=} \lim_{x \to \infty} \frac{4}{6} = \frac{2}{3}$$

Note that L'Hôpital's rule is not the only way to solve the above example. We could have divided both the numerator and denominator by x^2 to obtain

$$\lim_{x \to \infty} \frac{2x^2 - 3x + 1}{3x^2 + 5x - 2} = \lim_{x \to \infty} \frac{2 - 3/x + 1/x^2}{3 + 5/x - 2/x^2} = \frac{2 - 0 + 0}{3 + 0 - 0} = \frac{2}{3}$$

Most examples in this section, however, do not yield to this simple procedure, so l'Hôpital's rule must be used. Before applying l'Hôpital's rule, however, we must check that the conditions of Theorem 3.1 apply. If they do not hold, then the analysis is not valid, as illustrated by the next two examples.

Example 6   Limit is not an indeterminate form

Evaluate $\displaystyle \lim_{x \to 0} \frac{1 - \cos x}{\sec x}$.

Solution   You must always remember to check that you have an indeterminate form before applying l'Hôpital's rule. The limit is

$$\lim_{x \to 0} \frac{1 - \cos x}{\sec x} = \frac{\displaystyle \lim_{x \to 0} (1 - \cos x)}{\displaystyle \lim_{x \to 0} \sec x} = \frac{0}{1} = 0$$

If you apply l'Hôpital's rule in Example 6, you obtain the WRONG answer:

$$\lim_{x \to 0} \frac{1 - \cos x}{\sec x} \overset{H}{=} \lim_{x \to 0} \frac{\sin x}{\sec x \tan x} \qquad \textit{this is NOT correct}$$

$$= \lim_{x \to 0} \frac{\cos x}{\sec x}$$

$$= \frac{1}{1}$$

$$= 1 \qquad \textit{hence the answer is WRONG}$$

Example 7   Conditions of l'Hôpital's rule are not satisfied

Evaluate $\lim\limits_{x \to \infty} \dfrac{x + \sin x}{x - \cos x}$.

Solution   This limit has the indeterminate form ∞/∞. If you try to apply l'Hôpital's rule, you find

$$\lim_{x \to \infty} \frac{x + \sin x}{x - \cos x} \overset{H}{=} \lim_{x \to \infty} \frac{1 + \cos x}{1 + \sin x}$$

The limit on the right does not exist, because both $\sin x$ and $\cos x$ oscillate between -1 and 1 as $x \to \infty$. Recall that l'Hôpital's rule applies only if $\lim\limits_{x \to c} \dfrac{f'(x)}{g'(x)} = L$ is finite or is $\pm\infty$. This does not mean that the limit of the original expression does not exist or that we cannot find it; it simply means that we cannot apply l'Hôpital's rule. To find this limit, factor out an x from the numerator and denominator and proceed as follows:

$$\lim_{x \to \infty} \frac{x + \sin x}{x - \cos x} = \lim_{x \to \infty} \frac{x\left(1 + \dfrac{\sin x}{x}\right)}{x\left(1 - \dfrac{\cos x}{x}\right)}$$

$$= \lim_{x \to \infty} \frac{1 + \dfrac{\sin x}{x}}{1 - \dfrac{\cos x}{x}}$$

$$= \frac{1 + 0}{1 - 0}$$

$$= 1$$

Other indeterminate forms

Remember that l'Hôpital's rule itself applies only to the indeterminate forms $0/0$ and ∞/∞. Other indeterminate forms, such as 1^∞, 0^0, ∞^0, $\infty - \infty$, and $0 \cdot \infty$, can often be manipulated algebraically, or by taking logarithms, into one of the standard forms $0/0$ or ∞/∞, and then evaluated using l'Hôpital's rule. In a case where we have taken the logarithm to obtain one of the standard forms, we need to remember to transform back by applying exponentiation to our solution.

Example 8   Limit of the form 0^0

Find $\lim\limits_{x \to 0^+} x^{\sin x}$.

Solution   This is a 0^0 indeterminate form. From the graph shown in Figure 3.36, it looks as though the desired limit is 1.

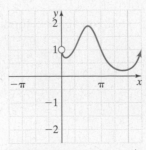

Figure 3.36 Graph of $x^{\sin x}$.

We can verify this conjecture analytically. We proceed as with the previous example by using properties of logarithms.

$$L = \lim_{x \to 0^+} x^{\sin x} \qquad \textit{given equation}$$

$$\textit{taking logarithms} \qquad \ln L = \ln\left[\lim_{x \to 0^+} x^{\sin x}\right]$$

$$= \lim_{x \to 0^+}\left[\ln x^{\sin x}\right] \qquad \textit{the natural logarithm is continuous}$$

$$= \lim_{x \to 0^+}\left[(\sin x)\ln x\right] \qquad \textit{property of logarithms}$$

$$= \lim_{x \to 0^+} \frac{\ln x}{\csc x} \qquad \textit{this is } \frac{\infty}{\infty} \textit{ form}$$

$$\overset{H}{=} \lim_{x \to 0^+} \frac{1/x}{-\csc x \cot x} \qquad \textit{l'Hôpital's rule}$$

$$= \lim_{x \to 0^+} \frac{-\sin^2 x}{x \cos x} \qquad \textit{algebraically simplify}$$

$$= \lim_{x \to 0^+}\left(\frac{\sin x}{x}\right)\left(\frac{-\sin x}{\cos x}\right)$$

$$= (1)(0)$$

$$= 0$$

$$\textit{taking inverse of logarithms} \qquad L = e^0 = 1$$

Example 9   Escaping infection and the indeterminate form $0^{-\infty}$

In models of host-pathogen and host-parasite interactions, the fraction of hosts escaping parasitism is often given by a negative-binomial escape function

$$f(P) = (1 + aP/k)^{-k}$$

where P is the density of the parasites or pathogens, $a > 0$ is the rate at which hosts encounter parasites or pathogens, and $k > 0$ is a clumping parameter. Small values of k correspond to parasites or pathogens being highly aggregated in the environment and large values of k correspond to parasites or pathogens being more evenly distributed across the environment. Assume $a = 0.1$.

a. For $k = 0.1, 1, 5, 10$, plot $f(P)$ over the interval $[0, 10]$. What effect does k have on the risk of being parasitized or infected?

b. For $P > 0$, find $\lim_{k \to \infty} f(P)$.

Solution

a. Plots of $f(P)$ for $k = 0.1, 1, 5, 10$ shown in Figure 3.37 suggest that as k increases the likelihood of escaping parasitism or infection goes down. Hence, as k increases and parasites or pathogens are more evenly distributed across the environment, the risk of parasitism or infection goes up.

b. For $P > 0$, $1 + aP/k$ approaches 0 and $-k$ approaches $-\infty$ as k approaches ∞. Therefore we have an indeterminate form of $0^{-\infty}$. To turn this problem to an indeterminate form of $0/0$, we take the ln of $f(P)$ which yields

$$g(P) = \ln f(P) = -k \ln(1 + aP/k) = \frac{\ln(1 + aP/k)}{-1/k}$$

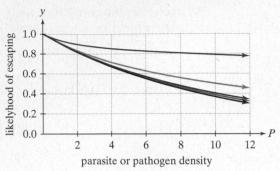

Figure 3.37 Negative binomial escape function $f(P) = (1 + 0.1P/k)^{-k}$ for $k = 0.1$ (in red), 1 (in green), 5 (in purple), and 10 (in cyan). The limiting Poisson escape function $f(P) = \exp(-0.1P)$ is shown as a solid black curve.

Since $\ln(1 + aP/k)$ and $-1/k$ approach zero as k approaches ∞, we can apply l'Hôpital's rule.

$$\lim_{k\to\infty} \frac{\ln(1 + aP/k)}{-1/k} \overset{H}{=} \lim_{k\to\infty} \frac{\dfrac{1}{1 + aP/k}\dfrac{-aP}{k^2}}{\dfrac{1}{k^2}}$$

$$= \lim_{k\to\infty} \frac{-aP}{1 + aP/k} = -aP$$

Therefore, we get $\lim_{k\to\infty} \ln f(P) = -aP$ and by exponentiating

$$\lim_{k\to\infty} f(P) = e^{-aP}$$

for $P > 0$. This limiting escape function is known as the Poisson escape function that corresponds to parasitism or infection events occurring randomly among all hosts. This limiting function is plotted in black in Figure 3.37 and illustrates that the greatest risk of infection or parasitism occurs in this limiting case.

Example 10   Finding a horizontal asymptote and the indeterminate form ∞^0

Find the horizontal asymptote of the graph $f(x) = x^{1/x}$ for $x > 0$.

Solution   To determine if the graph of f has a horizontal asymptote for $x > 0$, we evaluate

$$\lim_{x\to\infty} x^{1/x}.$$

This limit is indeterminate of the form ∞^0. To evaluate it, we take the natural logarithm and proceed as follows:

$$L = \lim_{x\to\infty} x^{1/x}$$

$$\ln L = \ln\left[\lim_{x\to\infty} x^{1/x}\right] \qquad \textit{taking logarithms}$$

$$= \lim_{x\to\infty} \left[\ln x^{1/x}\right]$$

$$= \lim_{x\to\infty} \left[\left(\frac{1}{x}\right)\ln x\right]$$

$$= \lim_{x\to\infty} \frac{\ln x}{x} \qquad \textit{form} \frac{\infty}{\infty}$$

$$\overset{H}{=} \lim_{x\to\infty} \frac{\frac{1}{x}}{1}$$

$$= 0$$

$$L = e^0 = 1 \qquad \textit{taking inverse of logarithms}$$

Thus, $y = 1$ is a horizontal asymptote for the graph of $y = x^{1/x}$, as shown in Figure 3.38.

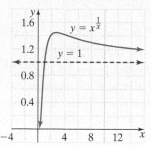

Figure 3.38 Graph of $y = x^{1/x}$ with horizontal asymptote.

We saw in Figure 3.38 that the graph of $f(x) = x^{1/x}$ approaches the line $y = 1$ asymptotically as $x \to \infty$, but how does $f(x)$ behave as $x \to 0^+$? That is, what is

$$\lim_{x \to 0^+} x^{1/x}?$$

It may seem that to answer this question, we need to apply l'Hôpital's rule again, but this limit has the form 0^∞, which is simply 0 and is not indeterminate at all. Other forms that may appear to be indeterminate, but really are not, are $0/\infty, \infty \cdot \infty$, $\infty + \infty$ and $-\infty - \infty$.

PROBLEM SET 3.7

Level 1 DRILL PROBLEMS

1. An incorrect use of l'Hôpital's rule is illustrated in the following limit computations. In each case, explain what is wrong and find the correct value of the limit.

  a. $\displaystyle\lim_{x \to \pi} \frac{1 - \cos x}{x} = \lim_{x \to \pi} \frac{\sin x}{1} = 0$

  b. $\displaystyle\lim_{x \to \pi/2} \frac{\sin x}{x} = \lim_{x \to \pi/2} \frac{\cos x}{1} = 0$

2. Sometimes l'Hôpital's rule leads nowhere. For example, observe what happens when the rule is applied to

$$\lim_{x \to \infty} \frac{x}{\sqrt{x^2 - 1}}$$

Use any method you wish to evaluate this limit.

Find the limits, if possible, in Problems 3 to 18.

3. $\displaystyle\lim_{x \to 1} \frac{x^3 - 1}{x^2 - 1}$

4. $\displaystyle\lim_{x \to 1} \frac{x^{10} - 1}{x - 1}$

5. $\displaystyle\lim_{x \to 0} \frac{1 - \cos^2 x}{\sin^2 x}$

6. $\displaystyle\lim_{x \to 0} \frac{1 - \cos x}{x^2}$

7. $\displaystyle\lim_{x \to \infty} x^{-5} \ln x$

8. $\displaystyle\lim_{x \to 0^+} x^{-5} \ln x$

9. $\displaystyle\lim_{x \to 0^+} \sin x / \ln x$

10. $\displaystyle\lim_{x \to \infty} \frac{\ln(\ln x)}{x}$

11. $\displaystyle\lim_{x \to \infty} \left(1 - \frac{3}{x}\right)^{2x}$

12. $\displaystyle\lim_{x \to \infty} \left(1 + \frac{1}{2x}\right)^{3x}$

13. $\displaystyle\lim_{x \to \infty} (\ln x)^{1/x}$

14. $\displaystyle\lim_{x \to 0^+} (e^x + x)^{1/x}$

15. $\displaystyle\lim_{x \to 0^+} (e^x - 1)^{1/\ln x}$

16. $\displaystyle\lim_{x \to 0} \frac{e^x - 1 - x - x^{3/2}}{x^3}$

17. $\displaystyle\lim_{x \to \infty} (\sqrt{x^2 - x} - x)$

18. $\displaystyle\lim_{x \to 0^+} \left(\frac{1}{x^2} - \ln \sqrt{x}\right)$

In Problems 19 to 22, use l'Hôpital's rule to determine all horizontal asymptotes to the graph of the given function. You are NOT required to sketch the graph.

19. $f(x) = x^{-3} e^{-0.01x}$

20. $f(x) = \dfrac{\ln x^5}{x^{0.02}}$

21. $f(x) = (\ln \sqrt{x})^{2/x}$

22. $f(x) = \left(\dfrac{x + 3}{x + 2}\right)^{2x}$

Verify the statements in Problems 23 to 25.

23. For positive integer n, $\displaystyle\lim_{x \to 0^+} \frac{\ln x}{x^n} = -\infty$

24. For positive integer n, $\displaystyle\lim_{x \to \infty} \frac{\ln x}{x^n} = 0$

25. For positive integer n and any $k > 0$,
$\displaystyle\lim_{x \to \infty} x^n e^{-kx} = 0$

Level 2 APPLIED AND THEORY PROBLEMS

26. Fisheries scientists have found that a Ricker stock-recruitment relationship, which has the form

$$y = axe^{-bx}$$

where y is a measure (also called an *index*) of the number of individuals recruited to the fishery each year (typically one-year-olds), and x is an index of the spawning stock biomass (sometimes measured in terms of eggs produced), provides a reasonable fit to various species. Consider the case where the parameter values are $a = 5.9$ and $b = 0.0018$.

   a. What is the value of the recruitment index as $x \to \infty$?

   b. What is the maximum value of the recruitment index and at what spawning stock index value does it occur?

   c. Over what range of spawning stock index values is the recruitment function concave up and over what values is it concave down?

   d. Use the information obtained in parts **a, b,** and **c** to sketch this function.

27. An agronomist experimenting with a new breed of giant potato has found that individual tubers x months after planting have a biomass in kilograms given by the equation $y(x) = 2e^{-1/(5x)}$ for $x > 0$.

   a. Calculate the rate of growth of the tuber over time and determine what happens to this rate in the limit as $x \to 0$ and $x \to \infty$.

   b. Find the time after planting when the growth rate of the tuber is maximized.

   c. Show that the growth rate is positive for all $x > 0$ and determine the regions over which the growth is accelerating and decelerating.

   d. Sketch the biomass of the potato, as well as its growth rate, indicating the important points and regions calculated in parts **a, b,** and **c.**

28. Determine which function, $f(x) = x^n$ with $n > 0$ or $g(x) = e^{ax}$ with $a > 0$, grows faster at ∞ by computing $\lim_{x \to \infty} \dfrac{f(x)}{g(x)}$.

29. Determine which function, $f(x) = x^n$ with $n > 0$ or $g(x) = \ln x$, grows faster at ∞ by computing $\lim_{x \to \infty} \dfrac{f(x)}{g(x)}$.

30. Consider a drug in the body whose current concentration is 1 mg/liter. In this problem, you investigate the meaning of exponential decay of the drug.

   a. If one-half of the drug particles cleared the body after one hour, determine the concentration of the drug that remains after one hour.

   b. If one-quarter of the drug particles cleared the body every half an hour, determine the concentration of the drug that remains after one hour.

   c. If one-twentieth of the drug particles cleared the body every six minutes, determine the concentration of the drug that remains after one hour.

   d. If $\dfrac{1}{(2n)}$ of the drug particles cleared the body every $1/nth$ of an hour, determine the concentration c_n of the drug that remains after one hour.

   e. Find $\lim_{n \to \infty} c_n$.

31. 𝖧istorical 𝖰uest   The French mathematician Guillaume de l'Hôpital (1661–1704) is best known today for the rule that bears his name, but that rule was discovered by l'Hôpital's teacher, Johann Bernoulli. Not only did l'Hôpital neglect to cite his sources in his book, but there is also evidence that he paid Bernoulli for his results and for keeping their arrangements for payment confidential. In a letter dated March 17, 1694, he asked Bernoulli "to communicate to me your discoveries"—with the request not to mention them to others: "it would not please me if they were made public." (See D. J. Stuik, *A Source Book in Mathematics*, 1200–1800, Cambridge, MA: Harvard University Press, 1969, 313–316.) L'Hôpital's argument, which was originally given without using functional notation, can easily be reproduced:

$$\frac{f(a + dx)}{g(a + dx)} = \frac{f(a) + f'(a)\,dx}{g(a) + g'(a)\,dx}$$

$$= \frac{f'(a)\,dx}{g'(a)\,dx}$$

$$= \frac{f'(a)}{g'(a)}$$

First, place some conditions on the functions f and g that will make this argument true. Next, supply reasons for this argument, and give necessary conditions for the functions f and g.

32. Consider the general logistic growth function

$$G(N) = (r/v)N(1 - (N/K)^v)$$

from Example 4.

   a. For $v > 0$, find the density $N > 0$ that maximizes $G(N)$; that is, solve $G'(N) = 0$. The answer will depend on the parameters r, v, and K.

   b. Take your answer from part **a** and compute its limit as $v \to 0$.

   c. Find density $N > 0$ that maximizes $-rN\log(N/K)$ and compare your answer to what you found in part **b.**

CHAPTER 3 REVIEW QUESTIONS

1. Find $\dfrac{dy}{dt}$ for the following expressions.

a. $x^3 + x\sqrt{x} + \sin 3x$

b. $xy + y^3 = 25$

c. $y = \dfrac{\ln(x^2 - 1)}{\sqrt[3]{x}(2 - x)^3}$

d. $y = x^2 e^{-\sqrt{x}}$

e. $y = x\sqrt{x}\cos 2x$

f. $y = \sin^2\left(\dfrac{\pi x}{4}\right)$

2. Approximate $65^{1/3}$ and determine whether this approximation overestimates or underestimates the true answer. Justify your answers by using derivatives.

3. Find $\dfrac{d^2 y}{dx^2}$ where $y = x^2(2x - 3)^3$.

4. Use the definition of derivative to calculate, showing all details, $\dfrac{d}{dx}(x - 3x^2)$.

5. Find the first- and second-order approximations to $y = e^{x^2}$ at $x = 0$. Graph the function and its approximations.

6. In Figure 3.39, which graph represents the function and which graph the derivative?

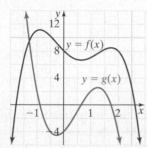

Figure 3.39 A function and its derivative.

7. Sketch the graph of a function with the following properties:

$f'(x) > 0$ when $x < 1$
$f'(x) < 0$ when $x > 1$
$f''(x) > 0$ when $x < 1$
$f''(x) > 0$ when $x > 1$

What can you say about the derivative of f when $x = 1$?

8. The developmental rate of insects and plants as a function of temperature T can be modeled by the Briere model

$$D(T) = a\,T(T - T_L)\sqrt{T_U - T} \quad \text{percent development per day}$$

where T_L is the lower developmental threshold below which an individual does not develop, T_U is the upper developmental threshold above which an individual does not develop, and a is a proportionality constant. Find a linear approximation to $D(T)$ at $T = T_L$, discuss what it means, and discuss where it breaks down.

9. Let f be a function defined by

$$y = x^3 + 35x^2 - 125x - 9{,}375$$

Determine where the function is increasing, where it is decreasing, and where the graph is concave up and where it is concave down.

10. Suppose the proportion of insect hosts escaping parasitism depends on the parasitoid density, d, and is modeled by the function $f(d) = e^{-.05d}$. Does the proportion escaping parasitism increase or decrease with parasitoid density? What is the concavity of this curve, and is there a point of inflection?

11. Determine the concavity and inflection points and use l'Hôpital's rule to find the horizontal asymptotes of the graph of

$$g(t) = \frac{t^2 + t + 1}{t^2 + 1}$$

12. Suppose the concentration in the blood at time t of a drug injected into the body is modeled by

$$C(t) = te^{-2t}$$

Use l'Hôpital's rule to find the horizontal asymptote. Find the time when $C'(t) = 0$. Graph this curve and verify that the largest concentration occurs at the solution to $C'(t) = 0$.

13. Say you want to estimate the height of a tall tree. To do so, you cannot simply drop a tape measure from the top of the tree. However, you can determine the height by using a sextant to determine the angle θ between the ground and the tip of the tree at a distance of 100 feet from the base of the tree.

a. Find the height of the tree, H, as a function of θ.

b. If you measure an angle $\theta = 1.1$ radians, determine the height of the tree.

c. Determine the elasticity of the height in part **b** to θ. Discuss how a 10% error in measuring θ influences the estimate for the height of the tree when $\theta = 1.1$.

14. A bacterial colony is estimated to have a population of P thousand individuals, where

$$P(t) = \frac{24t + 10}{t^2 + 1}$$

and t is the number of hours after a toxin is introduced.

a. At what rate is the population changing when $t = 0$ and $t = 1$?

b. Is the rate increasing or decreasing at $t = 0$ and $t = 1$?

c. At what time does the population begin to decrease?

15. As we saw in Example 6 of Section 1.5, scientists at Woods Hole Oceanographic Institution measured the uptake rate of glucose by bacterial populations from the coast of Peru. In one field experiment, they found that the uptake rate can be modeled by

$$f(x) = \frac{1.2078x}{1 + 0.0506x} \text{ micrograms per hour, where}$$

x is micrograms of glucose per liter. If the current uptake rate is twelve, determine the current level of glucose and, hence, determine the rate at which this uptake rate is itself changing per unit increase in glucose.

16. The gross U.S. federal debt (in trillions of dollars) is plotted below.

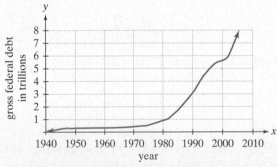

Source: Graph 9, John William's Shadow Government Statistics, No. 455: Special Commentary "Review of Economic, Systematic-Solvency, Inflation, US Dollar and Gold Circumstances," June 12, 2012.

Regarding this debt, President Ronald Reagan stated in 1979 that the United States is "going deeper into debt at a faster rate than we ever have before." Discuss the role of higher-order derivatives in the graph of federal debt from 1950 to the end of the graph in the context of President Reagan's statement.

17. The figure eight curve shown below is defined implicitly by the equation $x^4 = x^2 - y^2$.

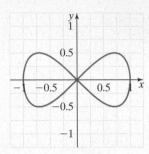

Find all the points on this curve that have horizontal tangents.

18. Consider the functions $y = f(x)$ (in blue) and $y = g(x)$ (in red) whose graphs are shown below.

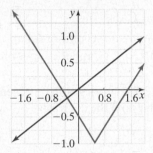

Find $\dfrac{d}{dx} f(g(x))$ at $x = -1$ and $x = 1$.

19. Find the tangent line of $y = (1 + \sin 2x)^{100}$ at $x = \pi/2$.

20. We modeled the concentration of carbon dioxide (in ppm) at the Mauna Loa Observatory of Hawaii with the function

$$h(t) = 329.3 + 0.1225\,t + 3\cos\left(\frac{\pi t}{6}\right)$$

where t is months since May 1974. Find for what months h is increasing and decreasing over the interval $[0, 12]$.

GROUP PROJECTS

Seeing a project through on your own, or working in a small group to complete a project, teaches important skills. The following project provides opportunities to develop such skills.

Project 3 Modeling North American Bison Population

If we look closely at the data plotted in Example 2 of Section 3.5, on the abundance of North American bison in Yellowstone National Park from 1902 to 1931, we get the distinct impression that the data can be represented much better by two linear functions than by one: the first representing data from 1902 to 1915 and the second representing the data from 1915 to 1931. We can fit these two functions by eye, or we can work more precisely using the concept of a *sum-of-squares* measure to gauge how well the functions fit the data. This concept requires that we know the actual values of the data points. Specifically, if we have n data points, indexed by $i = 1, \ldots, n$, then we need to know the values (x_i, y_i) for each data point. For the bison, these data are specified in Table 3.3. (Note that the data for some years are missing. This is not a problem if we just ignore these missing points when indexing the data.)

Table 3.3 Population for the North American bison

Index (i)	Year (x_i)	Abundance (y_i)
1	1902	44
2	1903	47
3	1904	51
4	1905	74
5	1907	84
6	1908	95
7	1909	118
8	1910	149
9	1911	168
10	1912	192
11	1913	215
12	1915	270
13	1916	348
14	1917	397
15	1919	504
16	1920	501
17	1921	602
18	1922	647
19	1923	748
20	1925	830
21	1926	931
22	1927	1008
23	1928	1057
24	1929	1109
25	1930	1124
26	1931	1192

For the bison data, consider piecing together two linear functions so that they both meet at the point $(x_{12}, y_{12}) = (1915, 270)$. Since both lines pass through this point, they must both satisfy the equation

$$\frac{y - 270}{x - 1915} = c$$

for some constant c. If the line fitted to the 1902–1915 date is specified by a constant c_1 and the line fitted to the 1915–1931 data is specified by a constant c_2, then the the actual function fitted to the data is $y = f(x)$ where

$$f(x) = \begin{cases} c_1 x + (270 - 1915 c_1) & 1902 \le x \le 1915 \\ c_2 x + (270 - 1915 c_2) & 1915 \le x \le 1931 \end{cases}$$

The question now is to find the values of c_1 and c_2 that provide the best fit of the function $f(x)$ to the data in the sense of minimizing the sum-of-squares measure, denoted by S, of the fit.

Before we do this, recall, that for any sequence of n points $a_1, a_2, a_3, \cdots, a_n$, the sum of these points can be written as:

$$\sum_{i=1}^{n} a_i = a_1 + a_2 + a_3 + \cdots + a_n$$

(Also note that i does not have to start at $i = 1$, but could start at any integer value less than or equal to n).

Returning to our problem, if we define the value of this measure to be S, where

$$S = \sum_{i=1}^{26} (y_i - f(x_i))^2$$

then we can plot the value of S for different choices of c_1 and c_2. This is best done by considering separately the sums

$$S_1(c_1) = \sum_{i=1}^{11} (y_i - f(x_i))^2$$

and

$$S_2(c_2) = \sum_{i=12}^{26} (y_i - f(x_i))^2$$

1. By calculating S_1 for a range of values of c_1 and S_2 for a range of values of c_2 and then plotting the results, find to two significant figures for the values of c_1 and c_2 that minimize the sum $S = S_1 + S_2$. (Find these by "playing around" with the functions until you find the appropriate intervals over which to plot the two sums.) This is a graphical approach to finding the best-fitting function $f(x)$ defined above.

2. Can you think of a way that you might use differential calculus to solve this problem analytically? Once you find a way to do this, then solve the problem analytically and compare this analytical solution with your graphical solution.

3. What advantages does the analytical solution have over the graphical solution and vice versa?

CHAPTER 4

Applications of Differentiation

Figure 4.1 A great tit is a species of bird whose foraging behavior was studied by biologist Richard Cowie and whose behavior can be predicted by optimal foraging models.

Santiago Bañón / Flickr / Getty Images

Preview

"'If one way be better than another, that you may be sure is nature's way.' Aristotle clearly stated the basic premise of optimization in biology, yet it was almost 2,000 years before the power of this idea was appreciated. The essence of optimization is to calculate the most efficient solution to a given problem, and then to test the prediction. The concept has already revolutionized some aspects of biology, but it has the potential for much wider application."

William Sutherland, on "The best solution" in Nature (2005) 435:569

One of the central ideas in physics, chemistry, and biology is that processes act to optimize some physically or biologically meaningful quantity. For example, from physics we know that light in a vacuum travels along a path that is the shortest distance between two points (taking into account that gravity "bends" space), and from biochemistry we know that proteins fold in a way that minimizes the energy of their constituent amino acid configuration.

Differential calculus is an important tool for analyzing optimization (maximization or minimization). In this chapter we show how optimization applies to various biological problems and processes. Before we do this, however, we study how calculus can be used to construct the graphs of a variety of functions; in particular, we identify where the graph has turning points corresponding to local minimum or maximum values. We then develop procedures for modeling and solving optimization problems, including how to draw the best-fitting line through data plotted on a graph. After considering a number of biological applications, we study how calculus provides insight into dynamic processes such as the growth of populations and the spread of deleterious or mutant genes (e.g., the gene that causes sickle cell anemia) within populations. We end the chapter with an application of difference equations that is at the heart of many numerical methods used by current technologies for finding solutions to nonlinear equations.

4.1   Graphing Using Calculus

In this section, we combine many of the tools that we have studied so far (e.g., limits involving infinity, first and second derivatives) to graph a function. In graphing a function, envision walking along the graph and indicating all the highlights of your walk. For instance, *vertical asymptotes* are places with such a rapid ascent or descent that they make climbing Mount Everest seem like a stroll in a park. *Horizontal asymptotes* are places where the landscape levels out into a never-ending plain. Where the derivative is positive, the graph is ascending; and where the derivative is negative, the graph is descending. Switches in the sign of the slope correspond to either hilltops or valley bottoms along your walk. On ascents where the second derivative is positive, the walk is getting harder. On descents where the second derivative is negative, the descent becomes faster.

Properties of graphs

When graphing the function $y = f(x)$ by hand, follow these procedures to find the highlights of the function shape:

Vertical asymptotes: Determine at what points the function is not well defined (e.g., division by zero). At each of these points, say $x = a$, evaluate the one-sided limits, $\lim_{x \to a^+} f(x)$ and $\lim_{x \to a^-} f(x)$, to determine what the graph looks like near $x = a$. If either of these one-sided limits is $+\infty$ or $-\infty$, then there is a vertical asymptote at this point.

Intervals of increase and decrease: Compute the first derivative $f'(x)$ of $f(x)$ and determine on which intervals $f'(x) > 0$ and on which intervals $f'(x) < 0$. On these intervals, f is increasing and decreasing, respectively.

Intervals of concavity: Compute the second derivative $f''(x)$ and determine on which intervals $f''(x) > 0$ and $f''(x) < 0$. On these intervals, f is concave up and concave down, respectively.

The x and y intercepts Find the x intercepts (i.e., where $f(x) = 0$) and the y intercept (i.e., $y = f(0)$). These points help pin down the placement of the graph.

After identifying these highlights of a function, you are ready to sketch the function.

Example 1   Dropping whelks: a graphical approach

In Example 6 of Section 3.2, we considered how often $D(h)$ a whelk had to be dropped by a crow from a height of h meters before breaking. The function, based on data collected by Reto Zach ("Selection and dropping of whelks, Behaviour," Vol. 67, pp. 134–148, 1978) is given by

$$D(h) = 1 + \frac{20.4}{h - 0.84} \text{ drops}$$

a. Find the horizontal and vertical asymptotes and where D is positive.

b. Find on what intervals D is increasing and on what intervals D is decreasing.

c. Find on what intervals D is concave up and on what intervals D is concave down.

d. Take this information and sketch $D(h)$. Discuss for which h values this function is biologically meaningful.

Solution

a. There is a vertical asymptote at $h = 0.84$. Moreover, $\lim_{h \to 0.84^+} D(h) = +\infty$ and $\lim_{h \to 0.84^-} D(h) = -\infty$. Since $\lim_{h \to \pm\infty} D(h) = 1$, D has a horizontal asymptote of $D = 1$.

$D(h) > 0$ only if $1 + \frac{20.4}{h - 0.84} > 0$, equivalently $\frac{20.4}{h - 0.84} > -1$. If $h > 0.84$, the inequality is always satisfied. If $h < 0.84$, then $h - 0.84$ is negative and multiplying by $h - 0.84$ reverses the direction of the inequality, yielding $20.4 < 0.84 - h$, equivalently $-19.56 > h$. Hence, D is positive on the intervals $(0.84, \infty)$ and $(-\infty, -19.56)$ as indicated by the third ribbon at the bottom of Figure 4.2.

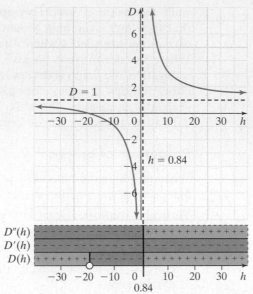

Figure 4.2 Graph of the whelk-dropping function $D(h) = 1 + \dfrac{20.4}{h - 0.84}$. The ribbons below the graph indicate where D, D', and D'' are negative (in red) and positive (in green).

b. Taking the first derivative yields

$$D'(h) = -\frac{20.4}{(h - 0.84)^2}$$

Since $(h - 0.84)^2$ is always positive for $h \neq 0.84$, we get $D'(h) < 0$ for all $h \neq 0.84$, as indicated by the second ribbon at the bottom of Figure 4.2. Therefore, D is decreasing for all $h \neq 0.84$.

c. Taking the second derivative yields

$$D''(h) = \frac{40.8}{(h - 0.84)^3}$$

which is positive for $h > 0.84$ and negative for $h < 0.84$, as indicated by the first ribbon at the bottom of Figure 4.2. Hence, D is concave up for $h > 0.84$ and concave down for $h < 0.84$.

d. Putting all this together yields the graph in Figure 4.2. From the crow's point of view, we require that $h > 0$ (since the crow does not burrow under ground!). Also, this graph is only meaningful for $h > 0.84$, since for $0 < h < 0.84$ the number of drops predicted is negative. For $h > 0.84$, the graph we drew is very similar to the graph shown in Figure 3.6 of Chapter 3. ∎

Many functions have no horizontal asymptotes. Nonetheless, understanding the limits as x approaches $\pm\infty$ may help graph the function.

Example 2   A "W" shaped curve

Consider the function $y = x^4 - 2x^2$.

a. Find the asymptotes, the intervals where the function is increasing or decreasing, the intervals where the function is concave up or down, the roots (x intercepts), and the y intercept.

b. Use all the information found in part **a** to graph the function.

Solution

a. The function is continuous for all real numbers. Hence, there are no vertical asymptotes. We have $\displaystyle\lim_{x \to \pm\infty} x^4 - 2x^2 = \lim_{x \to \pm\infty} x^2(x^2 - 2) = \infty$. Hence, there are no horizontal asymptotes, and y gets arbitrarily large as x approaches $\pm\infty$. Factoring the function $y = x^2(x^2 - 2)$ reveals that it is zero at $x^2 = 0$ and $x^2 = 2$; that is, $x = 0$ and $x = \pm\sqrt{2}$ are roots of the equation. The function is thus negative over the intervals $-\sqrt{2} < x < 0$ and $0 < x < \sqrt{2}$, as depicted in the third ribbon below the plot in Figure 4.3. Further, the y intercept is $y = 0$.

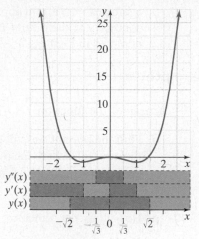

Figure 4.3 Graph of the function $y = x^4 - 2x^2$ with the sign of the function and its first and second derivatives indicated by the green (positive) and red (negative) ribbons below the graph.

To determine where the function is increasing, we first find the roots of the derivative, which solve the equation

$$0 = \frac{dy}{dx} = 4x^3 - 4x$$

$$0 = 4x(x^2 - 1)$$

Hence, the derivative vanishes at $x = 0, \pm 1$. Since $\frac{dy}{dx} = 24$ at $x = 2$, $\frac{dy}{dx} > 0$ on $(1, \infty)$. Since $\frac{dy}{dx} = -\frac{3}{2}$ at $x = \frac{1}{2}$, $\frac{dy}{dx} < 0$ on $(0, 1)$, as depicted in the second ribbon below the plot in Figure 4.3. Since $\frac{dy}{dx} = \frac{3}{2}$ at $x = -\frac{1}{2}$, $\frac{dy}{dx} > 0$ on $(-1, 0)$. Since $\frac{dy}{dx} = -24$ at $x = -2$, $\frac{dy}{dx} < 0$ on $(-\infty, -1)$. Therefore, the function is increasing on the intervals $(-1, 0)$ and $(1, \infty)$ and decreasing on the intervals $(-\infty, -1)$ and $(0, 1)$, as depicted in Figure 4.3.

To determine intervals of concave up and concave down, we determine where the second derivative equals zero.

$$0 = \frac{d^2y}{dx^2} = 12x^2 - 4$$

$$x = \pm\frac{1}{\sqrt{3}}$$

Since $\frac{d^2y}{dx^2} = -4$ at $x = 0$, y is concave down on $\left(-\frac{1}{\sqrt{3}}, \frac{1}{\sqrt{3}}\right)$. Since $\frac{d^2y}{dx^2} = 8$ at $x = \pm 1$, y is concave up on $\left(-\infty, -\frac{1}{\sqrt{3}}\right)$ and $\left(\frac{1}{\sqrt{3}}, \infty\right)$.

b. To sketch the graph using the information from part **a**, we can envision how the graph of the function changes as we move from $-\infty$ to ∞. Since $\lim\limits_{x \to -\infty} y = \infty$, $\frac{dy}{dx} < 0$ on $(-\infty, 1)$, and $y = 0$ at $x = -\sqrt{2}$, the function decreases from $+\infty$, crosses the x axis at $x = -\sqrt{2}$, and continues to decrease to the value $y = -1$ at

$x = -1$. Since $\dfrac{dy}{dx} > 0$ on $(-1, 0)$, the function increases to $y = 0$ at $x = 0$. Since $\dfrac{dy}{dx} < 0$ on $(0, 1)$, the function decreases to $y = -1$ at $x = 1$. Since $\dfrac{dy}{dx} > 0$ on $(1, \infty)$, $y = 0$ at $x = \sqrt{2}$, and $\lim\limits_{x \to \infty} y = \infty$, the function increases, crosses the x axis again at $x = \sqrt{2}$, and approaches $+\infty$ as x approaches $+\infty$. The function changes concavity at the points $\pm 1/\sqrt{3}$. Hence, the graph has the general characteristics depicted in Figure 4.3. ∎

Example 3   Linear asymptotes

Consider the function $y = \dfrac{x^2 + 2}{x}$.

a. Find the asymptotes, the intervals where the function is increasing or decreasing, the intervals where the function is concave up or down, and the x and y intercepts.

b. Use all the information found in part **a** to graph the function.

Solution

a. Rewriting the function as $y = x + \dfrac{2}{x}$ helps us see that a vertical asymptote exists at $x = 0$. In fact, $\lim\limits_{x \to 0^+} y = +\infty$ and $\lim\limits_{x \to 0^-} y = -\infty$. Since $\lim\limits_{x \to \infty} y = \infty$ and $\lim\limits_{x \to -\infty} y = -\infty$, y has no horizontal asymptotes. Indeed since $\dfrac{2}{x}$ goes to zero as x gets large, we expect the graph of $y = x + \dfrac{2}{x}$ to approach the line $y = x$ for large x.

To find the boundaries between the intervals of increase and decrease, we determine where the first derivative equals zero:

$$0 = \frac{dy}{dx} = 1 - \frac{2}{x^2}$$

$$\frac{2}{x^2} = 1$$

$$x^2 = 2$$

$$x = \pm\sqrt{2}$$

Since $\dfrac{dy}{dx} = \dfrac{1}{2}$ at $x = \pm 2$ and $\dfrac{dy}{dx} = -1$ at $x = \pm 1$, we find that y is increasing on the intervals $(-\infty, -\sqrt{2})$ and $(\sqrt{2}, \infty)$ and decreasing on the intervals $(-\sqrt{2}, 0)$ and $(0, \sqrt{2})$.

To determine concavity, we compute the second derivative $\dfrac{d^2y}{dx^2} = \dfrac{4}{x^3}$, which is positive when $x > 0$ and negative when $x < 0$. Hence, y is concave up on $(0, \infty)$ and y is concave down on $(-\infty, 0)$.

There is no y intercept, as the function has a vertical asymptote at $x = 0$. The roots (x intercepts) must satisfy

$$0 = y = x + \frac{2}{x}$$

$$0 = x^2 + 2$$

for which there is no real-valued solution. Hence there are no roots.

b. To graph $y = x + \dfrac{2}{x}$, think about what happens as you move from $-\infty$ to ∞. Since $\lim\limits_{x \to -\infty} y = -\infty$ and $\dfrac{dy}{dx} > 0$ on $(-\infty, -\sqrt{2})$, we find the graph increases from ∞ to $y = -2\sqrt{2}$ at $x = -\sqrt{2}$. Since $\dfrac{dy}{dx} < 0$ on $(-\sqrt{2}, 0)$ and $\lim\limits_{x \to 0^-} y = -\infty$, the graph decreases to infinity as x approaches 0 from the left. Since $\lim\limits_{x\to 0^+} y = \infty$ and $\dfrac{dy}{dx} < 0$ on $(0, \sqrt{2})$, the graph decreases from ∞ to $y = 2\sqrt{2}$ at $x = \sqrt{2}$. Since $\dfrac{dy}{dx} > 0$ on $(\sqrt{2}, \infty)$ and $\lim\limits_{x\to\infty} y = +\infty$, the graph increases toward $+\infty$ as x approaches $+\infty$. Moreover, the concavity only changes at $x = 0$. Finally, since $\lim\limits_{x\to\pm\infty} \dfrac{2}{x} = 0$, it follows that $y = x + \dfrac{2}{x}$ behaves like $y = x$ for sufficiently positive or negative values of x. Using this information, we obtain a sketch that looks something like this, where the dotted red line represents the linear asymptote:

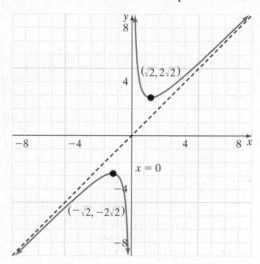

Sometimes just using limits and first derivatives is enough to get a good sense of the graph.

Example 4   Tylenol in the blood stream

As a project for a mathematical biology class, three college students developed a model of how acetaminophen (Tylenol) levels diffuse from the stomach and intestines to the blood stream after taking a dose of 1000 mg. Using data from the Federal Drug Administration (FDA), the students found that the concentration of acetaminophen in the blood was modeled by

$$C(t) = 28.6(e^{-0.3t} - e^{-t}) \text{ micrograms/milliliter}$$

where t is hours after taking the dose. Use information about asymptotes and first derivatives to sketch this function. Discuss the meaning of the graph.

Solution   Since $C(t)$ is continuous everywhere, there are no vertical asymptotes. Since $e^{-0.3t}$ and e^{-t} approach zero as t gets large, $\lim\limits_{t\to+\infty} C(t) = 0$. Therefore, there is a horizontal asymptote at $C = 0$. Alternatively, since $e^{-0.3t} - e^{-t} = e^{-0.7t}(e^{0.4t} - 1)$, $\lim\limits_{t\to-\infty} e^{0.4t} = 0$, and $\lim\limits_{t\to-\infty} e^{-0.7t} = \infty$, we get $\lim\limits_{t\to-\infty} C(t) = -\infty$.

Taking the first derivative yields

$$C'(t) = 28.6\,(e^{-t} - 0.3e^{-0.3t})$$

We have $C'(t) = 0$ if and only if

$$e^{-t} = 0.3e^{-0.3t}$$

$$e^{-0.7t} = 0.3$$

$$-0.7t = \ln 0.3$$

$$t = \frac{\ln 0.3}{-0.7} \approx 1.72 \text{ hours}$$

Since $C'(0) \approx 20$, we have $C'(t) > 0$ on $(-\infty, 1.72)$. Since $C'(t) < 0$ for very large t, we have $C'(t) < 0$ on $(1.72, \infty)$. Hence, as t goes from $-\infty$ to 0, the function increases up from $-\infty$ and passes through 0 at $t = 0$. $C(t)$ increases from $t = 0$ to $t \approx 1.72$, at which point it takes on the value of approximately 12 micrograms/milliliter. For t greater than 1.72, $C(t)$ decreases toward zero as t approaches $+\infty$. Therefore, we can graph the function as follows:

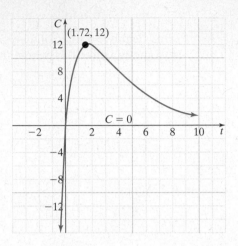

This graph is only biologically meaningful for $t \geq 0$. It shows that initially there is no drug in the blood stream and then that the concentration of drug increases to a maximum concentration of 12 mcg/ml after 1.7 hours. Hence, the maximum effect of Tylenol is felt after approximately two hours. After reaching the maximum value, the concentration decays to zero.

In the previous example, we note that because the function has a maximum above zero and then approaches 0 as $x \to \infty$, the graph necessarily has a point of inflection where $C''(t)$ switches from being positive to negative. In Problem Set 4.1, we ask you to discuss the sign of the second derivative and to solve for the value of x where this point of inflection occurs.

Graphing families of functions

The shape of some functions, such as $f(x) = 3x + a$, does not depend in any critical sense on the value of the parameter a. All a does is move the line of slope 3 up and down the xy plane. As we see in this subsection, however, in more complicated functions the value of a parameter can have a surprising effect, and we can use calculus to discover such effects.

Example 5   To infinity and back

Consider the function $f(x) = \dfrac{1}{x^2 - a}$ with the parameter a. Using first derivatives and asymptotes, determine how the shape of this function depends on the parameter a.

Solution   Since $\lim\limits_{x \to \pm\infty} \dfrac{1}{x^2 - a} = 0$, $f(x)$ has a horizontal asymptote $y = 0$ as $x \to \pm\infty$. If $a < 0$, then $x^2 - a$ is positive for all x and there are no vertical asymptotes. If $a = 0$, then there is a vertical asymptote at $x = 0$. If $a > 0$, then there are vertical asymptotes at $x = \pm\sqrt{a}$. Computing the first derivative yields

$$f'(x) = -\frac{2x}{(x^2 - a)^2}$$

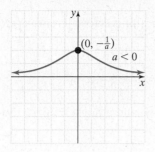

Since the denominator of this expression is positive whenever $x^2 \neq a$, $f'(x) < 0$ for all positive x with $x^2 \neq a$ and $f'(x) > 0$ for all negative x with $x^2 \neq a$.

These computations suggest there are qualitatively three distinctive graphs. First, consider the case where $a < 0$. In this case, there are no vertical asymptotes. There are horizontal asymptotes of 0 as $x \to \pm\infty$. Moreover, $f(x)$ is increasing for negative x and $f(x)$ is decreasing for positive x. Hence, the graph looks something like the figure on the left.

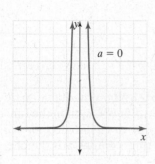

Next, consider the case where $a = 0$. In this case, $f(x) = \dfrac{1}{x^2}$ and there is a vertical asymptote at $x = 0$. In fact, $\lim\limits_{x \to 0} \dfrac{1}{x^2} = \infty$. There are horizontal asymptotes of $y = 0$ as $x \to \pm\infty$. Moreover, $f(x)$ is increasing for negative x and $f(x)$ is decreasing for positive x. Hence the graph, as you by now know, looks something like the figure on the left.

Finally, consider the case where $a > 0$. In this case, there are vertical asymptotes at $x = \pm\sqrt{a}$. In fact, evaluating all the limits as x approaches $\pm\sqrt{a}$ yields

$$\lim_{x \to \sqrt{a}^+} \frac{1}{x^2 - a} = \infty$$

$$\lim_{x \to \sqrt{a}^-} \frac{1}{x^2 - a} = -\infty$$

$$\lim_{x \to -\sqrt{a}^+} \frac{1}{x^2 - a} = -\infty$$

$$\lim_{x \to -\sqrt{a}^-} \frac{1}{x^2 - a} = \infty$$

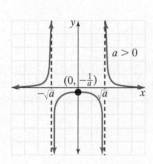

There are horizontal asymptotes of $y = 0$ as $x \to \pm\infty$. Moreover, $f(x)$ is increasing for negative x and $f(x)$ is decreasing for positive x. Hence, if we walk along the graph of f from $+\infty$ to $-\infty$, then we initially ascend from 0. The ascent gets exceptionally steep as x approaches $-\sqrt{a}$. As we cross $x = -\sqrt{a}$, we suddenly fall down to $y = -\infty$. After crossing both infinities, we continue to ascend until reaching a maximum of $y = -\dfrac{1}{a}$ at $x = 0$. From there, we descend through $-\infty$ and skyrocket to $+\infty$ at $x = \sqrt{a}$. After this harrowing jump through infinities, we continue with a descent to zero. In other words, our graph looks something like the figure on the left.

Example 6   Dose-response curves

Dose-response curves can be used to plot the response of an individual to a dose of a drug or hormone. This response can almost be anything. For instance, the response may be heart rate, dilation of an artery, membrane potential, enzyme activity, or the secretion of a hormone. We previously encountered dose-response curves in Example 2 of Section 2.4 and Problem 31 in Problem Set 3.2. A general form of a dose-response curve is

$$y = a + \frac{b - a}{1 + e^{x-c}}$$

where y is the response of the individual and x is the concentration of the dose of drug or hormone. The parameters $a > 0$, $b > 0$, and $c > 0$ affect the shape of the dose-response curve and often can be used to fit the function to particular data sets. Assuming $c = 0$, use limits and first derivatives to determine how the shape of this curve depends on the parameters a and b.

Solution   Consider $f(x) = a + \dfrac{b-a}{1+e^x}$. Since $f(x)$ is continuous for all reals, there are no vertical asymptotes. Since

$$\lim_{x \to \infty} a + \frac{b-a}{1+e^x} = a \quad \text{and} \quad \lim_{x \to -\infty} a + \frac{b-a}{1+e^x} = a + b - a = b,$$

there is a horizontal asymptote $y = a$ as x approaches $+\infty$ and a horizontal asymptote $y = b$ as x approaches $-\infty$.

Taking the first derivative of f, we obtain

$$f'(x) = \frac{(a-b)e^x}{(1+e^x)^2}.$$

This derivative is negative for all x if $b > a$, positive for all x if $b < a$, and zero for all x if $b = a$.

Hence, the graph of $f(x)$ comes in three flavors. If $b > a$, then the function decreases from an asymptote of $y = b$ to an asymptote of $y = a$. If $b < a$, then the function increases from an asymptote of $y = b$ to an asymptote of $y = a$. Finally, if $b = a$, then the function is the constant function $y = a$. These three graphs are sketched below.

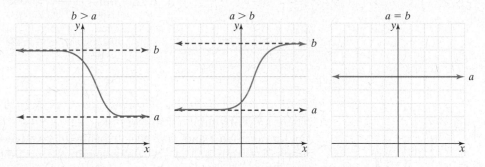

Note that when $b > a$, the graph goes from concave down to concave up and hence passes through an inflection point. Similarly, when $a > b$, the graph goes from concave up to concave down and hence also passes through an inflection point. You are asked to explore this further in Problem Set 4.1.

Example 7   Stock-recruitment curves

In conservation biology and fisheries management, stock-recruitment curves are used to describe the relationship between the current abundance of a population (i.e., the stock) and the number of juveniles entering the system in the next year (i.e., the recruits). A general class of stock-recruitment curves are given by the functions

$$F(N) = \frac{aN^b}{1+N^b}$$

where N is the current population size, $F(N)$ is the number of recruits in the next generation, and a and b are positive parameters. A useful way to categorize these functions is to consider the relationship between the current population abundance and the average of number recruits per individual, that is,

$$f(N) = \frac{F(N)}{N} = \frac{aN^{b-1}}{1+N^b}.$$

Use limits and first derivatives to determine how the parameter b influences the shape of $f(N)$ for $N \geq 0$. Discuss the possible meaning.

Solution   Notice that if $0 < b < 1$, then there is a vertical asymptote at $N = 0$ and $\lim_{N \to 0^+} f(N) = \infty$. If $b \geq 1$, there is no vertical asymptote. To determine the horizontal asymptote as $N \to \infty$, notice that the power of the numerator is less than the power of the higher order term in the denominator. Hence, $\lim_{N \to \infty} f(N) = 0$.

The first derivative of $f(N)$ is given by

$$f'(N) = a\frac{(b-1)N^{b-2}(1+N^b) - bN^{b-1}N^{b-1}}{(1+N^b)^2}$$

$$= a\frac{N^{b-2}(b-1) - N^{2b-2}}{(1+N^b)^2}$$

$$= a\frac{N^{b-2}(b-1-N^b)}{(1+N^b)^2}$$

Hence, if $b \leq 1$, $f'(N) < 0$ for all $N > 0$. However, if $b > 1$, then $f'(N) > 0$ for $0 \leq N \leq (b-1)^{1/b}$ and $f'(N) < 0$ otherwise. Therefore, we get three types of graphs, depending on whether $b < 1$, $b = 1$, or $b > 1$:

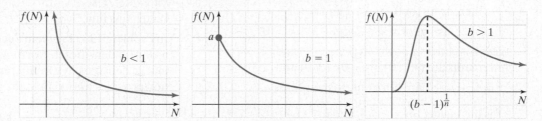

For $b \leq 1$, the number of recruits constantly decreases with stock levels. One interpretation of this fact is at higher population densities, there are fewer resources per individuals and, consequently, fewer recruits produced per individual. For $b > 1$, the number of recruits per individual initially increases and then decreases. One possible explanation is that for $b > 1$, individuals have difficulty finding mates for reproduction. Therefore, at low densities as densities increase, the chance of finding mates increases and the number of recruits produced per individual increases. However, as the population density increases too much (i.e., beyond $(b-1)^{1/b}$), the advantage of finding mates is outweighed by the limited resources available per individual. Consequently, at higher densities, the number of recruits per individual decreases. For $b > 1$, the population exhibits what ecologists call *depensation* or a *strong Allee effect*. Working with data from 128 species, Ran Myers and colleagues used $F(N)$ to evaluate to what extent fish populations exhibit depensation and discussed the implications for populations to recover from environmental disturbances. (See R. A. Myers, N. J. Barrowman, J. A. Hutchings, and A. A. J. Rosenberg, "Population Dynamics of Exploited Fish Stocks at Low Population Levels," *Science* 269 (1995): 1106–1108.)

PROBLEM SET 4.1

Level 1 DRILL PROBLEMS

In Problems 1 to 14, graph the functions by hand by finding asymptotes and using first and second derivatives. Compare your graphs to what you get using technology.

1. $y = x^2 - x$

2. $y = x^2 + 5x - 3$

3. $y = \dfrac{1}{1+x^2}$

4. $y = \dfrac{x}{1+x^2}$

5. $y = x + \dfrac{1}{2+x}$

6. $y = \dfrac{1}{x-1} + x$

7. $y = -12x - \dfrac{9x^2}{2} + x^3$

8. $y = \dfrac{1}{3}x^3 - 9x + 2$

9. $y = e^x + 2e^{-x}$

10. $y = 2e^x + e^{-x}$

11. $y = x - x^3$

12. $y = \dfrac{2+x}{1+x}$

13. $y = \dfrac{x - 3}{x + 1}$ **14.** $y = \dfrac{x^2}{1 + x^4}$

In Problems 15 to 20, graph the families of functions by finding asymptotes and using first and second derivatives. In particular, determine how the graph of the functions depends on the parameter $a > 0$.

15. $y = x^4 - ax^2$ **16.** $y = \dfrac{ax}{x^2 + 1}$

17. $y = ae^x + e^{-x}$ **18.** $y = e^x + ae^{-x}$

19. $y = \dfrac{a + x}{1 + x}$ **20.** $y = ax + \dfrac{1}{x}$

In Problems 21 and 22, sketch the graph of a function with the given properties.

21. $x = 2, x = -2$ are vertical asymptotes

  f is increasing for $0 < x < 2$ and $x > 2$

  f is decreasing for $x < -2$ and $-2 < x < 0$

  graph is concave down on $(-\infty, -2)$ and $(2, \infty)$

  intercepts are $(-1, 0), (-3, 0), (3, 0)$ and $(1, 0)$

22. $y = 1, y = -1$ are horizontal asymptotes

  f is increasing for $x < -\dfrac{3}{2}$ and for $x > \dfrac{3}{2}$

  f is decreasing for $-1 < x < 1$

  graph is concave down for $x < -1$ and for $0 < x < 1$

  graph is concave up for $x > 1$ and for $-1 < x < 0$

Level 2 APPLIED AND THEORY PROBLEMS

23. Consider the graph of $y = ax^2 + bx + c$ for constants $a, b,$ and c. Use second derivatives to determine what happens to the graph as a changes.

24. Consider the graph of $y = \dfrac{e^{ax}}{1 + e^{ax}}$. Use limits and first derivatives to determine how the shape of this curve depends on the parameter a.

25. In Example 6, we saw that the dose-response curve
$$y = a + \frac{b - a}{1 + e^x}$$
has asymptotes $y = a$ and $y = b$, respectively, as x approaches $\pm\infty$. Find the second derivative $y''(x)$ and discuss its properties including an equation that can be used to identify any points of inflection, if they exist.

26. In Example 4 of Section 2.7, we consider patterns of local species richness of ants along an elevational gradient. A function that best fits the data is
$$S = -10.3 + 24.9\,x - 7.7\,x^2$$
where x is elevation measured in kilometers and S is the number of species. Plot this function using information about first derivatives.

27. In Example 6 of Section 1.5, we develop the Michaelis-Menton model for the rate at which an organism consumes its resource. For bacterial populations in the ocean, this model was given by
$$f(x) = \frac{1.2078x}{1 + 0.0506x} \text{ mg of glucose/hr}$$
where x is the concentration of glucose (micrograms per liter) in the environment. Use asymptotes and first derivatives to sketch this function by hand.

28. In Example 5 of Section 2.4 we found that the rate at which wolves kill moose can be modeled by
$$f(x) = \frac{3.36x}{0.42 + x} \text{ moose/wolf/hundred days}$$
where x is measured in number of moose per square kilometer. Use asymptotes and first derivatives to sketch this function.

29. In Problem 42 in Problem Section 2.4, we examined how wolf densities in North America depend on moose densities. We found that the following function provides a good fit to the data:
$$f(x) = \frac{58.7(x - 0.03)}{0.76 + x} \text{ wolves per } 1000\,\text{km}^2$$
where x is number of moose per square kilometer.

  a. Find the horizontal and vertical asymptotes.

  b. Determine on which intervals f is increasing and decreasing.

  c. Determine on which intervals f is concave up and concave down.

  d. Use the information from parts **a–c** to sketch the graph of $f(x)$.

30. Two mathematicians, W. O. Kermack and A. G. McKendrick, showed that the weekly mortality rate during the outbreak of the plague in Bombay (1905–1906) is reasonably well described by the function
$$f(t) = 890\,\text{sech}^2(0.2t - 3.4) \text{ deaths/week}$$
where t is measured in weeks. Sketch this function using information about asymptotes and first derivatives. Recall that
$$\text{sech}\,x = \frac{2}{e^x + e^{-x}}$$

31. In Example 4 we modeled the rate at which acetaminophen diffuses from the stomach and intestines to the blood stream using this equation:
$$C(t) = 28.6(e^{-0.3t} - e^{-t}) \text{ micrograms/milliliter}$$
Calculate the second derivative and discuss its behavior. Identify if the function $C(t)$ has any points of inflection for $t > 0$.

32. In an experiment, a microbiologist introduces a toxin into a bacterial colony growing in an agar dish. The data on the area of the dish covered by living colony members at time t minutes after the introduction of the toxin are given by this equation:

$$A(t) = 5 + e^{-0.04t+1}$$

Sketch the graph of $A(t)$ showing its salient features.

33. Let f be a function that represents the weight of a fish at age t. Write a function that satisfies the following properties.

- The weight of the fish at birth must be positive.
- As the fish ages, the weight increases at decreasing rate.
- No fish can grow bigger than 2 kg.

34. *Aerobic rate* is the rate of a person's oxygen consumption and is sometimes modeled by the function A defined by

$$A(x) = 110 \left(\frac{\ln x - 2}{x} \right)$$

for $x \geq 10$. Graph this function.

35. As a project for their mathematical biology class, three college students developed a model of how acetaminophen levels varied in the blood stream of a child after taking a dose of 325 milligrams. Using FDA data, they found that

$$C(t) = 23.725 \left(-e^{-0.7t} + e^{-0.5t} \right) \text{mg/ml}$$

where t is hours after taking the dose.

a. Use information about asymptotes and first derivatives to sketch this function.

b. Discuss the meaning of your graph. In particular, address when the maximum concentration is achieved and what the maximum concentration is.

36. A naturalist at an animal sanctuary determined that the function

$$f(x) = \frac{4e^{-(\ln x)^2}}{\sqrt{\pi x}}$$

provides a good measure of the number of animals in the sanctuary that are x years old. Sketch the graph of f for $x > 0$.

4.2 Getting Extreme

When viewing the graph of a function as a landscape of hilltops and valley bottoms, each top and each bottom corresponds to a place where the function in question has an extremum. Methods to identify extrema play an important role in applications. For instance, if the function of interest represents how profits due to harvesting a crop depend on the amount of seeds planted, then the farmer would like to know how many seeds per acre yield the greatest profits. In other words, the farmer would like to identify the largest hilltop of the function. Alternatively, if a northwestern crow minimizes the amount of energy required to break whelk shells, then the optimal behavior for a crow corresponds to the deepest valley of a function. In this section, we develop methods to find these hilltops and valleys.

Local extrema

Local Maxima and Minima

Let $f(x)$ be a function of x. We say that f has a **local maximum** at $x = a$ if

$$f(a) \geq f(x)$$

for all x near a. We say that f has a **local minimum** at $x = a$ if

$$f(a) \leq f(x)$$

for all x near a. We say f has a **local extremum** at $x = a$ if there is a local maximum or a local minimum at $x = a$.

Example 1   Finding Extrema

Estimate for what x values, $y = f(x)$, as graphed below, has local maxima and local minima on the domain $D = \{x : -3 \leq x \leq 3\}$.

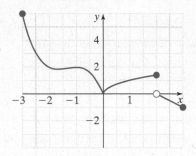

Solution   There are local minima at $x \approx -1.75$, $x = 0$, and $x = 3$. There are local maxima at $x = -3$, $x \approx -1$, and $x = 2$.   ■

The previous example suggests that either extrema occur at end points of the domain, points where f is not differentiable, or points where the derivative of f equals zero. The following theorem verifies these observations.

Theorem 4.1   Fermat's theorem

If f is defined on (a, b) and has a local extremum at $c \in (a, b)$, then either $f'(c) = 0$ or $f'(c)$ is not defined.

Proof.   Suppose that f is defined on (a, b) and has a local extremum at $c \in (a, b)$. This extremum is either a local maximum or a local minimum. Suppose that this extremum is a local minimum. Then we have $f(c) \leq f(x)$ for all x near c. Equivalently, $f(c) \leq f(c + h)$ for all h sufficiently small. Taking a difference quotient yields that

$$\frac{f(c + h) - f(c)}{h} \geq 0$$

for all sufficiently small positive h and

$$\frac{f(c + h) - f(c)}{h} \leq 0$$

for all sufficiently small negative h. Assume $f'(c)$ exists. Then, taking one-sided limits yields

$$f'(c) = \lim_{h \to 0^+} \frac{f(c + h) - f(c)}{h} \geq 0$$

and

$$f'(c) = \lim_{h \to 0^-} \frac{f(c + h) - f(c)}{h} \leq 0$$

Therefore, $f'(c) = 0$. The case of a local maximum can be proved similarly, which you see in Problem 31 in Problem Set 4.2.   ■

Fermat's theorem shows that we can find possible local maxima and local minima by finding points where $f'(x) = 0$ or f' is not defined. Such points have a special name.

Critical Points and Values

If $f'(c) = 0$ or $f'(c)$ is not defined, then c is a **critical point** for f. The value of f at a critical point is called a **critical value**.

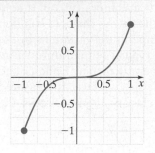

Although all local extrema are critical values, not all critical values are local extrema. Consider, for example, $y = x^3$ plotted on the left for $-1 \leq x \leq 1$.

The derivative is $\dfrac{dy}{dx} = 3x^2$. Hence, $x = 0$ is the only critical point. However as $y = x^3$ increases over all the reals, $x = 0$ is neither a local maximum nor a local minimum.

Example 2   Finding and classifying critical points

Find the critical points of $y = x^3 - 3x^2 - 4$ and determine whether these critical points are local maxima, local minima, or neither.

Solution   We have $\dfrac{dy}{dx} = 3x^2 - 6x = 3x(x - 2)$. Hence, $\dfrac{dy}{dx} = 0$ at $x = 0$ and $x = 2$. These are the critical points of y. To determine whether these critical points correspond to local maxima, local minima, or neither, we can consider how the sign of the derivative varies over the real line. Since $\dfrac{dy}{dx} < 0$ for $0 < x < 2$ and $\dfrac{dy}{dx} > 0$ for $x > 2$, the function decreases over the interval $(0, 2)$ and increases on $(2, \infty)$. Hence, there is a local minimum of $y = -8$ at $x = 2$. Alternatively, since $\dfrac{dy}{dx} > 0$ for $x < 0$ and $\dfrac{dy}{dx} < 0$ for $0 < x < 2$, the function increases until $x = 0$ and then decreases. Thus, there is a local maximum $y = -4$ at $x = 0$. Graphing $y = x^3 - 3x^2 - 4$ with technology in the plot on the left corroborates these statements.

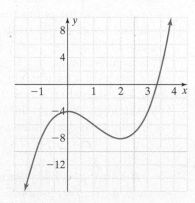

Example 2 illustrates one method for identifying local maxima and local minima; this method is called the *first derivative test*.

First Derivative Test

Assume $f'(c) = 0$ and f is differentiable near $x = c$.

- If the sign of f' changes from positive to negative at $x = c$ (i.e., f changes from increasing to decreasing), then f has a local maximum at $x = c$.

- If the sign of f' changes from negative to positive (i.e., f changes from decreasing to increasing) at $x = c$, then f has a local maximum at $x = c$.

Example 3   Thermodilution

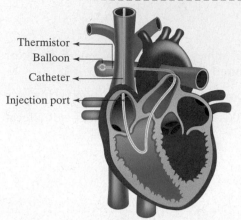

Cardiac output can be determined by thermodilution as illustrated in Figure 4.4. Let's say a doctor injects 10 milliliters (ml) of a cold dextrose solution into a vein entering the heart. As the cold solution mixes with the blood in the heart, the temperature variations in the blood leaving the heart are measured. A typical temperature variation curve (i.e., degrees below normal temperature), may be described by the function $T(t) = 0.2t^2 e^{-t}$ degrees Celsius, where t is measured in seconds. Find the critical points and classify them. Discuss the meaning of your results.

Figure 4.4 Thermodilution involves injecting a cold dextrose solution in the vena cava of the heart and measuring the temperature of the blood leaving the heart (e.g., from the aorta).

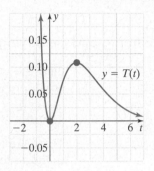

Solution   Taking the derivative yields

$$T'(t) = 0.2(2t)e^{-t} - 0.2t^2 e^{-t}$$

$$= 0.2te^{-t}(2 - t)$$

We have $T'(t) = 0$ at $t = 0$ and $t = 2$. Hence, $t = 0$ and $t = 2$ are the critical points of T. To apply the first derivative test, we need to determine the sign of T' on the intervals $(-\infty, 0)$, $(0, 2)$, and $(2, \infty)$. Since T' is continuous everywhere, it can only have sign changes at the points $t = 0, 2$. Therefore, it suffices to check the sign of T' at one point in each of the intervals. Since $T'(-1) = -0.6e^1 < 0$, T' is negative on $(-\infty, 0)$. Since $T'(1) = 0.2e^{-1} > 0$, T' is positive on $(0, 2)$. Since $T'(3) = -0.6e^{-3} < 0$, T' is negative on $(2, \infty)$. Since at $t = 0$ the sign of T' changes from negative to positive, we have a local minimum at $t = 0$. Since at $t = 2$ the sign of T' changes from positive to negative, at $t = 2$ we have a local maximum. Hence, the temperature of blood leaving the heart drops $T(2) \approx 0.11$ degrees Celsius after two seconds, before returning to its normal temperature.

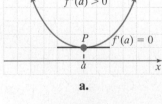

a.

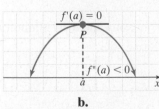

b.

Figure 4.5 The second derivative test determines whether a critical point $x = a$ where $f'(a) = 0$ is a local minimum (a) or a local maximum (b).

Another possibility for identifying local maxima and local minima is using the second derivative. Suppose f has a critical point at $x = a$ and has second-order derivatives at $x = a$. In Section 3.6, we saw that a second-order approximation of $f(x)$ is given by

$$f(x) \approx f(a) + f'(a)(x - a) + \frac{1}{2}f''(a)(x - a)^2$$

Since a is a critical point, $f'(a) = 0$ and the second-order approximation reduces to

$$f(x) \approx f(a) + \frac{1}{2}f''(a)(x - a)^2$$

Provided that $f''(a) \neq 0$, the graph of this approximation is given by a parabola whose vertex is at $x = a$. Furthermore, if $f''(a) > 0$, then this parabola is facing up, which suggests that there is a local minimum at $x = a$ as shown in Figure 4.5**a**.

Alternatively, if $f''(a) < 0$, then this parabola is facing down, which suggests that there is a local maximum at $x = a$ as shown in Figure 4.5**b**. What is suggested by the second approximation can be proven, thereby providing an alternative test for classifying extrema.

Let f have first and second derivatives at $x = a$. Assume that $f'(a) = 0$.

Local maximum If $f''(a) < 0$, then there is a local maximum at $x = a$.

Local minimum If $f''(a) > 0$, then there is a local minimum at $x = a$.

Inconclusive If $f''(a) = 0$, then we can not determine whether the critical point is a maximum or minimum from the second derivative.

Example 4 Using the second derivative test

Find and classify the critical points of $y = -x^3 + 6x^2 + 2$ using the second derivative test.

Solution Computing the first and second derivatives of $y = -x^3 + 6x^2 + 2$ yields

$$\frac{dy}{dx} = -3x^2 + 12x = -3x(x - 4)$$

$$\frac{d^2y}{dx^2} = -6x + 12$$

This derivative always exists, so the critical points correspond to the solutions of $-3x(x - 4) = 0$. Hence, they are given by $x = 0$ and $x = 4$. Evaluating the second derivatives at $x = 0$ and $x = 4$ yields

$$\left.\frac{d^2y}{dx^2}\right|_{x=0} = 12$$

$$\left.\frac{d^2y}{dx^2}\right|_{x=4} = -12$$

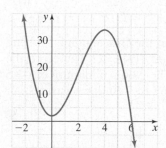

Hence, there is a local minimum at $x = 0$ and a local maximum at $x = 4$. Graphing the function $y = -x^3 + 6x^2 + 2$ demonstrates these conclusions in the graph on the left.

Global extrema

Often in applied problems, we want to find the largest or small value of a function on its domain.

Let f be a function with domain D.

f has a **global minimum** at $x = a$ if

$$f(a) \leq f(x) \text{ for all } x \text{ in } D.$$

f has a **global maximum** at $x = a$ if

$$f(a) \geq f(x) \text{ for all } x \text{ in } D.$$

A global maximum or a global minimum is called a **global extremum.**

Example 5   Finding global extrema

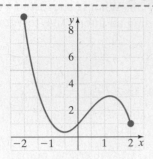

Consider the function whose graph is illustrated on the left. Find the global extrema of this function.

Solution   The global maximum of $y = 9$ occurs at $x = -2$ and the global minimum of approximately 0.1 occurs approximately at $x = -0.6$. ∎

Example 5 illustrates that global extrema may occur at critical points or end points for a continuous function on a closed interval. Thus, we have the following procedure for finding global extrema.

Closed Interval Method

Let f be a continuous function defined on the closed interval $[a, b]$. To find the global extrema of f, do the following:

Find critical points on the interval (a, b).

Evaluate f at all critical points and at end points a and b.

Identify the global extrema The largest value of f at a critical point or end point is the global maximum of f. The smallest value of f at a critical point or end point is the global minimum of f.

Example 6   Using the closed interval method

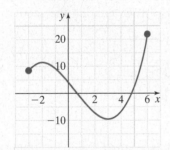

Find the global extrema of $f(x) = \frac{1}{3}x^3 - \frac{1}{2}x^2 - 6x + 4$ on the interval $[-3, 6]$.

Solution   Taking the derivative of f yields

$$f'(x) = x^2 - x - 6 = (x - 3)(x + 2)$$

The critical points are $x = 3$ and $x = -2$. Evaluating f at the critical points and end points yields $f(-3) = 8.5$, $f(-2) = 11\frac{1}{3}$, $f(3) = -9.5$, and $f(6) = 22$. Therefore, the global maximum of 22 occurs at the end point $x = 6$. The global minimum of -9.5 occurs at the critical point $x = 3$. Plotting this function on the left demonstrates our findings. ∎

Example 7   Getting extreme with carbon dioxide

In Example 4 of Section 1.2, we examined how CO_2 concentrations in parts per million (ppm) varied from 1974 to 1985 at the Mauna Loa Observatory in Hawaii. Using linear and periodic functions, we found that the following function gives an excellent fit to the data:

$$f(x) = 0.1225x + 329.3 + 3\cos\left(\frac{\pi x}{6}\right) \text{ ppm}$$

where x is months after April 1974. Using the closed interval method, find the global maximum and minimum CO_2 levels in the one-year interval $[0, 12]$.

Solution   To find the critical points of $f(x)$, we differentiate

$$f'(x) = 0.1225 - \frac{\pi}{2}\sin\left(\frac{\pi x}{6}\right)$$

Although we can solve for the critical points by hand by recalling properties of inverse sine, we circumvent such analysis by using a root finder on a graphing calculator. Finding all the roots of $f'(x)$ on the interval $[0, 12]$ yields $x = 0.149$ and $x = 5.851$. Evaluating f to two decimal points at these critical points and the end points yields

$$f(0) = 332.30$$
$$f(0.149) \approx 332.31$$
$$f(5.851) \approx 327.03$$
$$f(12) \approx 333.77$$

Hence, the global minimum CO_2 level occurred at $x = 5.851$ (sometime in late October 1974). The global maximum occurred at $x = 12$ (in April 1975). Plotting the function over the interval $[0, 12]$ demonstrates these extremes:

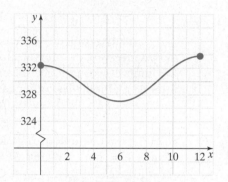

Example 8   Search period of the codling moth

After a codling moth (*Cydia pomonella*, see Figure 4.6) larva hatches from its egg case, it goes looking for an apple in which to burrow. The period between hatching and finding the apple is called the *search period*. For individual larvae, this period will vary, but the average time s that it takes is known to be a function of temperature T in degrees Celsius. A good fit to the available data illustrated in Figure 4.7 is provided by the equation

$$s(T) = \frac{1}{-0.03T^2 + 1.67T - 13.65} \quad \text{for} \quad 20 \leq T \leq 30$$

Use the tools of calculus to find the largest and smallest values of $s(T)$ over the range $20 \leq T \leq 30$.

Figure 4.6 Codling moth adult.

Solution   The polynomial $p(T) = -0.03T^2 + 1.67T - 13.65$ has roots at $T \approx 9.95$ and $T \approx 45.7$. Hence, $p(T) \neq 0$ for $20 \leq T \leq 30$ and $s(T) = \dfrac{1}{p(T)}$ is continuous on this interval. Using the quotient rule, the derivative is

$$s'(T) = \frac{2(0.03T) - 1.67}{(-0.03T^2 + 1.67T - 13.65)^2}$$

which satisfies $s'(T) = 0$ only if $0.06T = 1.67$, equivalently $T \approx 27.83°C$. Evaluating $s(T)$ at this critical point and at the end points we obtain

$$s(20) \approx 0.129 \qquad s(27.83) \approx 0.104 \qquad s(30) \approx 0.106.$$

Hence, on the interval $20 \leq T \leq 30$, $s(T)$ has a minimum at $T \approx 27.83$ and a maximum at $T = 20$.

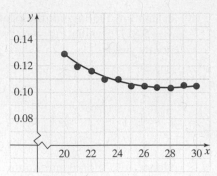

Figure 4.7 Codling moth search period.

Data Source: P. L. Shaffer, and H. J. Gold, "A Simulation Model of Population Dynamics of the Codling Moth, *Cydia pomonella*," *Ecological Modeling* 30 (1985): 247–274.

In many problems we need to find the global extrema on open intervals, half-closed intervals, or intervals involving infinity. For each of these cases, we deal with the limits as we approach the end points of the intervals, as illustrated with the open interval method. (The other cases appear as exercises in Problem Set 4.2.)

Open Interval Method

Let f be a continuous function defined on the open interval (a, b). Assume the limits $L = \lim\limits_{x \to a^+} f(x)$ and $M = \lim\limits_{x \to b^-} f(x)$ are well defined. Here we allow L and M to be $\pm\infty$, a to be $-\infty$, and b to be $+\infty$. To find the global extrema of f on (a, b), do the following:

Find critical points Find all the critical points on the interval (a, b).

Evaluate at critical points Evaluate f at all critical points.

Identify the extrema If L or M is greater than f evaluated at any critical point, then f has no global maximum on (a, b). Alternatively, if f evaluated at a critical point $x = c$ is greater than or equal to L, M, and f evaluated at any other critical point, then $f(c)$ is the global maximum. If L or M is less than f evaluated at any critical point, then f has no global minimum on (a, b). Alternatively, if f evaluated at a critical point $x = c$ is less than or equal to L, M, and f evaluated at any other critical point, then $f(c)$ is the global minimum.

Example 9   Using the open interval method

Use the open interval method to find the global extrema of the following functions on the indicated intervals.

a. $f(x) = \dfrac{1}{3x - x^2 - 2}$ on $(1, 2)$ **b.** $f(x) = \dfrac{x}{1 + x^2}$ on $(-\infty, \infty)$

Solution

a. We have $f(x) = \dfrac{1}{3x - x^2 - 2} = \dfrac{1}{(2 - x)(x - 1)}$ is continuous on $(1, 2)$. Note $f'(x)$ exists for all x on $(1, 2)$. Moreover,

$$\lim_{x \to 2^-} \frac{1}{(2 - x)(x - 1)} = +\infty$$

$$\lim_{x \to 1^+} \frac{1}{(2 - x)(x - 1)} = +\infty$$

Hence, $f(x)$ has no global maximum on $(1, 2)$. Solving for the critical points on $(1, 2)$, we get

$$f'(x) = 0$$
$$-\frac{3 - 2x}{(3x - x^2 - 2)^2} = 0$$
$$2x = 3$$
$$x = 1.5$$

Since f has only one critical point and $f(1.5) = 4$ is less than $\lim_{x \to 1^+} f(x)$ and $\lim_{x \to 2^-} f(x)$, the global minimum is 4 and occurs at $x = 1.5$.

b. Since $1 + x^2$ is positive for all x, $f(x) = \dfrac{x}{1 + x^2}$ is continuous on $(-\infty, \infty)$. Taking limits at infinity, we get

$$\lim_{x \to \infty} \frac{x}{1 + x^2} \frac{1/x}{1/x} = \lim_{x \to \infty} \frac{1}{1/x + x} = 0$$
$$\lim_{x \to -\infty} \frac{x}{1 + x^2} \frac{1/x}{1/x} = \lim_{x \to -\infty} \frac{1}{1/x + x} = 0$$

Solving for the critical points, we get

$$f'(x) = 0$$
$$\frac{1(1 + x^2) - x(2x)}{(1 + x^2)^2} = 0$$
$$\frac{1 - x^2}{(1 + x^2)^2} = 0$$
$$x = \pm 1$$

Since $f(1) = \dfrac{1}{2}$ and $f(1) = -\dfrac{1}{2}$ are greater than 0 and less than 0, respectively, these correspond to the global minimum and global maximum. ∎

PROBLEM SET 4.2

Level 1 DRILL PROBLEMS

In Problems 1 to 4, identify the local and global extrema.

1.

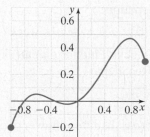

2.

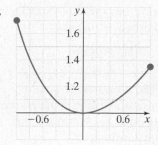

3.

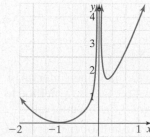

4.

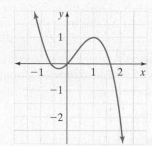

In Problems 5 to 12, find the critical points and use the first derivative test to classify them.

5. $y = 1 + 3x + 4x^2$ **6.** $f(x) = 10 + 6x - x^2$

7. $f(t) = t^2 e^{-t}$ **8.** $y = x^3 - \dfrac{3x^4}{4} + 5$

9. $f(x) = \dfrac{x}{1+x}$ **10.** $y = -3x - x^2 + \dfrac{x^3}{3}$

11. $y = -x + \dfrac{3x^2}{4} + \dfrac{x^3}{3}$ **12.** $y = e^{t^2 - 2t + 1}$

In Problems 13 to 16, find the critical points and use the second derivative test to classify them.

13. $y = -12x - \dfrac{9x^2}{2} + x^3$ **14.** $y = 1 - \exp(-x^2)$

15. $y = x + \dfrac{1}{2+x}$ **16.** $y = \dfrac{2x^2 - x^4}{4}$

In Problems 17 to 20, use the closed interval method to find the global extrema on the indicated intervals.

17. $f(x) = x^2 - 4x + 2$ on $[0, 3]$

18. $f(x) = x^3 - 12x + 2$ on $[-3, 3]$

19. $f(x) = x + \dfrac{1}{x}$ on $[0.1, 10]$

20. $f(x) = xe^{-x}$ on $[0, 100]$

In Problems 21 to 24, use the open interval method to find the global extrema on the indicated intervals.

21. $f(x) = x^2 - 4x + 2$ on $(-\infty, \infty)$

22. $f(x) = x^3 - 12x + 2$ on $(0, \infty)$

23. $f(x) = x + \dfrac{1}{x}$ on $(0, \infty)$

24. $f(x) = xe^{-x}$ on $(-\infty, \infty)$

25. Let f be continuous on the half-open interval $[a, b)$ with b possibly equal to $+\infty$. Devise a method to find the global extrema of f on this interval.

26. Let f be continuous on the half-open interval $(a, b]$ with a possibly equal to $-\infty$. Devise a method to find the global extrema of f on this interval.

In Problems 27 to 30, use the half-open interval methods you developed in Problems 25 and 26 to find the global extrema on the indicated intervals.

27. $f(x) = x^2 - 4x + 2$ on $[0, \infty)$

28. $f(x) = x^3 - 12x + 2$ on $[1, 10)$

29. $f(x) = x + \dfrac{1}{x+2}$ on $[-1, \infty)$

30. $f(x) = xe^{-x}$ on $(-\infty, -1]$

Level 2 APPLIED AND THEORY PROBLEMS

31. Let f be defined on (a, b) and $c \in (a, b)$. Prove that if $x = c$ is a local maximum and f is differentiable at $x = c$, then $f'(c) = 0$.

32. In Example 4 of Section 1.2, we examined how CO_2 concentrations (in ppm) have varied from 1974 to 1985. Using linear and periodic functions, we found that the following function gives an excellent fit to the data:

$$f(x) = 0.122463x + 329.253 + 3\cos\left(\frac{\pi x}{6}\right) \text{ ppm}$$

where x is months after April 1974. Using the closed interval method, find the global maximum and minimum CO_2 levels on the interval $[12, 24]$.

33. In the previous problem, use the closed interval method and find the global maximum and minimum CO_2 levels between April 2000 and April 2001.

34. A close relative of the codling moth is the pea moth, *Cydia nigricana*, which is a pest of cultivated and garden peas in several European countries. If its search period in one of the regions where it is a pest is given by the function

$$s(T) = \frac{1}{-0.04T^2 + 2T - 15} \quad \text{for} \quad 20 \le T \le 30,$$

then graph $s(T)$ using information about the first derivative over the domain $20 \le T \le 30$. Be sure that your graph indicates the largest and smallest value of s over this interval.

35. In Problem 30 of Problem Set 4.1, we saw that the weekly mortality rate during the outbreak of the plague in Bombay (1905–1906) can be reasonably well described by the function

$$f(t) = 890\,\mathrm{sech}^2(0.2t - 3.4) \text{ deaths/week}$$

where t is measured in weeks. Find the global maximum of this function. Recall that

$$\mathrm{sech}\, x = \frac{2}{e^x + e^{-x}}$$

36. Some species of plants (e.g., bamboo) flower once and then die. A well-known formula for the average growth rate r of a *semelparous species* (a species that breeds only once) that breeds at age x is

$$r(x) = \frac{\ln[s(x)n(x)p]}{x}$$

where $s(x)$ represents the proportion of plants that survive from germination to age x, $n(x)$ is the number of seeds produced at age x, and p is the proportion of seeds that germinate.

a. Find the age of reproduction that maximizes r in terms of the parameters a, b, c, and p where
$$s(x) = e^{-ax} \qquad a > 0$$
and
$$n(x) = bx^c \qquad b > 0$$
$0 < c < 1$.

b. Sketch the graph of $y = r(x)$ for the case where $a = 0.2$, $b = 3$, $c = 0.8$, and $p = 0.5$.

37. The production of blood cells plays an important role in medical research involving leukemia and other so-called *dynamical diseases*. In 1977, a mathematical model was developed by A. Lasota. (See W. B. Gearhart and M. Martelli, "A Blood Cell Population Model, Dynamical Diseases, and Chaos," in UMAP Modules 1990: *Tools for Teaching* [Arlington, MA: Consortium for Mathematics and Its Applications, 1991].) The model involved the cell production function
$$P(x) = Ax^s e^{-sx/r}$$
where A, s, and r are positive constants and x is the number of granulocytes (a type of white blood cell) present.

a. Find the granulocyte level x that maximizes the production function P. How do you know it is a maximum?

b. Graph this function.

38. When you cough, the radius of your trachea (windpipe) decreases, thereby affecting the speed of the air in the trachea. If r is the normal radius of the trachea, the relationship between the speed S of the air and the radius r of the trachea during a cough is given by a function of the form
$$S(r) = ar^2(r_0 - r)$$
where a is a positive constant. (Philip M. Tuchinsky, "The Human Cough," UMAP Modules 1976: *Tools for Teaching* [Lexington, MA: Consortium for Mathematics and Its Applications, 1977].) Find the radius r for which the speed of the air is the greatest.

39. Research indicates that the power P required by a bird to maintain flight is given by the formula
$$P = \frac{w^2}{2\rho Sv} + \frac{1}{2}\rho Av^3$$
where v is the relative speed of the bird, w is its weight, ρ is the density of air, and S and A are constants associated with the bird's size and shape. (See C. J. Pennycuick, "The Mechanics of Bird Migration," *IBIS III* (1969): 525–556.) What speed will minimize the power? You may assume that w, ρ, S, and A are all positive and constant.

40. An epidemic spreads through a community in such a way that t weeks after its outbreak, the number of residents who have been infected is given by a function of the form
$$f(t) = \frac{A}{1 + Ce^{kt}}$$
where A is the total number of susceptible residents. Show that the epidemic is spreading most rapidly when half the susceptible residents have been infected.

4.3   Optimization in Biology

One of the most important applications of calculus to biology in particular, and science and technology in general, involves finding the extrema of functions. Consider, for example, the problem of determining the most effective treatment regimen for a malignant tumor using chemo- or radiation therapy. Such treatments are toxic to the body, so physicians want to prescribe the minimum dosage that will do the job. Typically, a single treatment will not destroy the tumor. Instead, the tumor will initially shrink in size after treatment and sometime later will begin to regrow. Ideally, therapy should be readministered immediately—when the tumor is at its smallest—before this regrowth phase. Using calculus in conjunction with tumor growth data, we can estimate the time when therapy should be reapplied. As another example, think of a farmer planting a corn crop. The farmer might be interested in knowing the planting density of seeds that would maximize his or her profit. To find out, the farmer could formulate a function that describes how profits depend on planting density, and maximizing this function. In this section, we consider these problems as well as the behavior of dogs fetching balls, sustainable harvesting of arctic fin whales, and vascular branching. More examples appear in the problem set. We end this section with applications to finding best-fitting functions to empirically collected data.

Steps to solve an optimization problem

Optimization problems typically require that we develop an appropriate model for the problem being considered, then analyze this model to find the optimal solution for the problem. To tackle these problems successfully, use the following steps:

1. **Read, understand, and visualize.** Take the time to carefully read the problem so that you fully understand what is being asked. In particular, ask yourself: What am I trying to maximize or minimize? What information am I given? Is it sufficient to solve the problem at hand? When appropriate, draw a picture or figure that summarizes the problem.

2. **Identify key variables and quantities.** Ask yourself: What are the important quantities in the problem? What quantity is being optimized? This is the dependent variable. Which of the variables is the one whose value I can control to obtain my optimal solution? This is the independent variable. What additional quantities presented in the problem do I need to obtain the sought after relationship between the dependent and independent variable? Associate units with each of these variables.

3. **Write the function.** In this step, you need to determine how the dependent variable is determined by the quantities that you identified in the previous step. Think carefully about this crucial step. Make sure that units on both sides of your equation agree.

4. **Optimize.** Determine whether you need to minimize or maximize the function and over what interval you need to perform the optimization. To find the optimal value, it suffices to find the critical points and evaluate the function at these critical points and at the end points of the interval. Whichever of these values is the largest (smallest) yields the maximum (minimum).

5. **Interpret your answer.** Interpret the results of your optimization. Ask yourself whether your answer makes sense. If not, check your work.

In this next example, we demonstrate how these steps are applied to the problem of determining the density of seedlings that a farmer needs to plant to optimize profits from a particular crop. In the remaining examples in this section, we do not stress the steps involved, but these steps are implicit in our approach. Remember to you these steps whenever you get stuck in solving an optimization problem. The examples relate to behavior, physiology, resource management, and fitting functions to data. You won't necessarily have time to study them all, so you can pick and choose those of greatest interest to you.

Example 1   Maximum economic yield

In an article titled "*A 'Cookbook' Approach for Determining the 'Point of Maximum Economic Return,'*" Gaspar and colleagues lay out procedures for farmers to conduct field trials to determine the impact of different planting densities on crop yields. In an experiment to assess the impact of planting densities on yield Y of a corn hybrid, they fitted the following function to the data depicted in Figure 4.8.

$$Yield: \quad Y(x) = -0.1181x^2 + 8.525x + 12.95 \text{ bushels per acre}$$

where x is thousands of seeds planted per acre.

In a particular year, suppose that the price a farmer can obtain for his corn is \$1.50 per bushel and that the cost of seed is \$3 per thousand seeds. Use this information to determine the density of seeds (i.e., seeds per acre) that maximizes the farmer's net profit per acre.

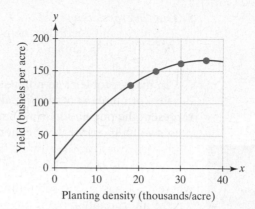

Figure 4.8 A yield curve (red) fitted to data (blue points) obtained from a corn hybrid experiment (since the curve does not pass through the point $(0, 0)$, it should not be used to estimate yield at densities close to zero).

Data Source: Gaspar, P.E., S. Paszkiewicz, P. Carter, M. McLeod, T. Doerge, and S. Butzen. 1999. Corn hybrid response to plant populations. Pioneer Hi-Bred International Inc. Northern Agronomic Research Summary. pp. 29–40.

Solution   We follow the five steps for solving an optimization problem.

1. From the statement of the problem, we realize that the farmer wants to maximize his net profit, measured in dollars per acre, rather than his yield, measured in bushels per acre. The planting density of seeds is the quantity that can be varied.

2. The independent variable is x, with units 1,000 seeds per acre; the dependent variable is net profit $P(x)$, with units dollars per acre.

3. Since $P(x)$ is determined by the total revenue $R(x)$ generated from the crops each season minus the cost $C(x)$ of the seeds each season, it follows that

$$\textit{Net profit:} \qquad P(x) = R(x) - C(x) \text{ dollars per acre}$$

where $R(x)$ is the yield $Y(x)$ in bushels per acre multiplied by the price $p = 1.5$ in dollars per bushel of corn; that is,

$$\textit{Total revenue:} \qquad R(x) = pY(x) \text{ dollars per bushel} \times \text{ bushels per acre}$$
$$= 1.5(-0.1181x^2 + 8.525x + 12.95) \text{ dollars per acre}$$

On the other hand, the cost per acre is \$3 for each thousand seeds, so that for x thousand seeds per acre the cost is

$$\textit{Cost:} \qquad C(x) = 3x \text{ dollars per acre}$$

Hence, we can *write the function*

$$P(x) = \overbrace{-0.1772\,x^2 + 12.7875\,x + 19.425}^{R(x)} - \overbrace{3x}^{C(x)}$$
$$= -0.1772\,x^2 + 9.7875\,x + 19.425$$

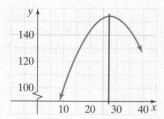

that we want to *maximize with respect to x on the interval* $[0, \infty)$. If we plot $P(x)$ verus x, then the red line in the figure on the left indicates optimum planting density.

4. The point at which $P(x)$ is maximized can be found by solving for the critical points of $P(x)$. Since

$$P'(x) = -0.3544\,x + 9.7875$$

it follows that $x = \dfrac{9.7875}{0.3544} = 27.617$ is the unique critical point. Since the graph of $P(x)$ is a downward facing parabola, the global maximum occurs at $x = 27.617$.

5. Our *interpretation* of 27.617 is that the farmer should plant approximately 27.6 thousand seeds per acre and, in doing so, should obtain a net profit of $P(27.6) \approx \$155$ per acre. ∎

In many problems in population biology, a variable x is used to represent the density (or number) of individuals in a population and the function $G(x)$ is used to represent the population growth rate. For instance, for the discrete logistic equation presented in Example 7 of Section 2.5, we modeled the growth using the function.

$$G(x) = rx\left(1 - \frac{x}{K}\right)$$

where $r > 0$ has the interpretation of the maximum per capita growth rate and $K > 0$ has the interpretation of the environmental carrying capacity. Here, we use the function $G(x)$ to determine the optimal rate at which a population of whales should be harvested, if not illegal to do so.

The Arctic fin whale *Balaenoptera physalus*, which at fifty to seventy tons for adults of both sexes is second in size to the blue whale, was a highly desirable catch during the whaling heydays of the nineteenth and twentieth centuries. As many as 30,000 individuals were slaughtered each year from 1935 to 1965. This level of exploitation could not be sustained for long, so today population levels are estimated to be an order of magnitude (i.e., a factor of 10) below historical highs of around a half million individuals. Some individuals are still taken each year for purposes of subsistence by aboriginal people in Greenland. A moratorium on whale hunting is needed, however, to allow this species to recover to levels where the populations can be safely exploited on a sustainable basis.

Mark Carwardine / Biosphoto

Figure 4.9 Arctic fin whale.

Example 2 Sustainable exploitation of the Arctic fin whale

Assume the Arctic fin whale growth rate is modeled by the logistic function $G(x) = rx(1 - x/K)$ with $r = 0.08$ (i.e., an 8% annual growth rate when the whale densities are low) and $K = 500,000$ (i.e., prior to exploitation, the Arctic fin whale population was estimated to be around half a million individuals). If the population is harvested at a constant rate of H individuals per year for an extended period of time, then this harvesting rate is *sustainable* if there exists a positive number x of whales at which the growth rate $G(x)$ equals the harvesting rate H. That is, it is possible for the growth to keep pace with the loss from harvesting. Determine the maximal sustainable harvesting rate.

Solution According to the statement of the problem, a harvesting rate H is sustainable if there is a positive x such that

$$H = G(x) = 0.08x\left(1 - \frac{x}{500,000}\right)$$

Hence, maximizing a sustainable harvesting rate is equivalent to finding $x > 0$ which maximizes $G(x)$. Since the graph of G is a downward facing parabola, we can find its maximum by taking the derivative of G, setting it equal to zero, and solving for x:

$$0 = G'(x) = 0.08 - 0.16\frac{x}{500,000}$$
$$0.16\frac{x}{500,000} = 0.08$$
$$x = 250,000$$

Hence, the maximum sustainable yield occurs at a harvesting rate of $H = G(250,000) = 10,000$ whales per year at which the whale population consists of 250,000 individuals. This maximum sustainable harvesting rate of 10,000 whales per year is *three times smaller* than the harvesting rate in the early twentieth century. Hence, the model reaffirms the statement that harvesting at 30,000 whales per year in the early twentieth century was not sustainable and may explain why the current population sizes are an order of magnitude lower than half a million. ◼

Sometimes when solving a problem it is useful to sketch a figure, as illustrated in the next example.

Example 3   Do dogs know calculus?

Professor Tim Pennings from Hope College wanted to determine whether his dog, Elvis, fetched balls thrown into Lake Michigan in an optimal way. (See T. Pennings, "Do Dogs Know Calculus?" *College Mathematics Journal* 34(2003):178–182.) Standing along the shoreline with Elvis at his side, Tim would throw the ball into the water. Elvis could choose to swim out directly from where Tim was standing to get the ball, hence taking a minimal-distance trajectory. Alternatively, he could run along the shore before he jumped into the water and swam to the ball. Because Elvis can only swim at an average speed of 0.91 meters per second whereas he can run at an average speed of 6.4 meters per second, it is likely that he ran for some distance along the shore. But how far along the shore should Elvis run?

Professor Pennings performed an experiment to assess what strategy Elvis was playing by throwing the ball repeatedly into the water and keeping track of where Elvis entered the water. For one throw, the ball landed 6 meters from the shore, as illustrated in Figure 4.10**a**.

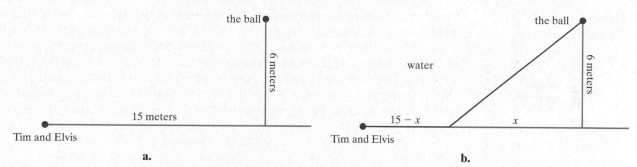

Figure 4.10 How will Elvis fetch a ball that landed 6 meters from shore?

What path would Elvis take if he were to minimize that amount of time it took him to retrieve the ball?

Solution   Let us begin by sketching a figure that indicates a hypothetical path Elvis could take (see Figure 4.10**b**). In this drawing, $15 - x$ is the distance Elvis runs along the shore. Assuming that Elvis wants to minimize the time to getting to the ball, we need to write a function that describes how the amount of time to get to the ball depends on x. Since he is running at a speed of 6.4 meters per second and runs a distance of $15 - x$ meters along the shore, we get that the time he spends running on the shore is

$$\frac{(15 - x) \text{ meters}}{6.4 \text{ meters per second}} = \frac{15 - x}{6.4} \text{ seconds}$$

By the Pythagorean theorem, the distance Elvis swims to the ball is $\sqrt{36 + x^2}$. Hence, the time he spends swimming is

$$\frac{\sqrt{36 + x^2} \text{ meters}}{0.91 \text{ meters per second}} = \frac{\sqrt{36 + x^2}}{0.91} \text{ seconds}$$

Hence, the total time T it takes him to get to the ball as a function of x is given by

$$T(x) = \frac{15 - x}{6.4} + \frac{\sqrt{36 + x^2}}{0.91} \text{ seconds}$$

We want to understand the graph of this function for $0 \leq x \leq 15$, so let us take the derivative of T.

$$T'(x) = -\frac{1}{6.4} + \frac{x}{0.91\sqrt{36 + x^2}}$$

To find the critical points, we need to solve $T'(x) = 0$. Doing so yields

$$\frac{x}{0.91\sqrt{36 + x^2}} = \frac{1}{6.4}$$

$$\frac{x^2}{0.91^2(36 + x^2)} = \frac{1}{6.4^2} \quad \textit{square both sides}$$

$$x^2 = 0.02(36 + x^2) \quad \textit{multiply both sides by common denominator}$$

$$0.98\,x^2 = 0.72 \quad \textit{and isolate } x^2$$

$$x^2 = 0.735$$

$$x \approx \pm 0.86$$

Hence, on the interval $[0, 15]$, T' vanishes only at $x = 0.86$. Since $T'(2) \approx 0.19 > 0$, we see that T is increasing on the interval $(0.86, 15]$. On the other hand, since $T'(0) \approx -0.16$, T is decreasing on $[0, 0.86]$. Thus, the minimum time is achieved at $x = 0.86$. Therefore, Elvis should run 14.1 meters along the shore before jumping into the water. ∎

So what was the outcome of Professor Pennings' experiment? When Professor Pennings measured the point at which Elvis entered the water, he found that Elvis ran $15 - x = 14.1$ meters along the shore (i.e., $x = 0.9$). Does this dog know calculus? Well, he could have been lucky on this one throw. So Professor Pennings performed thirty-five throws, with the ball landing different distances d from the shoreline. Professor Pennings measured the point x where Elvis entered the water on each throw. In the problems at the end of this section, you will be asked to show that the optimal place to enter the water as a function of the distance d the ball lands from the shore is

$$x = 0.144\,d \text{ meters}$$

A scatter plot of the data and the line is shown in Figure 4.11. This figure illustrates that Elvis is on average acting pretty optimal, which is quite remarkable.

The next example concerns treating tumors with chemotherapy (Figure 4.12), which was mentioned in the introduction to this section. Refer to Problems 44 and 45 in Problem Set 1.4 and Problem Set 44 in Problem Set 1.5 for further insights into modeling tumor growth.

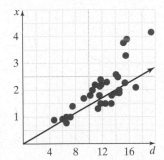

Figure 4.11 Scatter plot of distance of ball from shore (in the horizontal direction) and Elvis's point of entry $15 - d$ in the water (in the vertical direction). The best-fitting line $x = 0.144\,d$ passes through the center of the scatter plot.

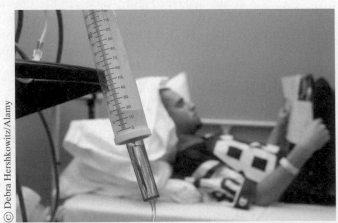

© Debra Hershkowitz/Alamy

Figure 4.12 Chemotherapy is a treatment for cancer patients in which powerful drugs are used to kill cancer cells. Often it is used in conjunction with other treatments, including radiation therapy and surgery.

Example 4   Tumor regrowth

In an experimental study performed at Dartmouth College (E. Demidenko. Mixed Models: Theory and Applications. Wiley 2004), two groups of mice with tumors were treated with the chemotherapeutic drug cisplatin. Before the therapy, the tumor consisted of proliferating cells (also known as *clonogenic cells*) that grew exponentially with a doubling time of approximately 2.9 days. Each of the mice was given a dose of 10 mg/kg of cisplatin. At the time of the therapy, the average tumor size was approximately 0.5 cm^3. After treatment, 99% of the proliferating cells became quiescent cells (also known as *nonproliferating* or *resting cells*). These quiescent cells do not divide and decay with a half-life of approximately 5.7 days.

a. Write a function $V(t)$ that represents the volume of the tumor t days after therapy. The tumor volume includes the volume of the proliferating cells and the quiescent cells.

b. Determine at what point in time the tumor starts to regrow and therapy should be readministered.

Solution

a. If $P(t)$ and $Q(t)$ represent the respective volumes of proliferating and quiescent cells in a tumor, then the total volume $V(t)$ of the tumor is given by

$$V(t) = P(t) + Q(t)$$

Assume the proliferating cells are increasing at an exponential rate and have an initial volume of $P(0) = 0.01 \times 0.5 = 0.005$ cm^3 (i.e., 1% of the previous untreated average size). Hence, $P(t) = 0.005e^{at}$ where we need to solve for a. Since the doubling time is 2.9 days, we can solve for a as follows:

$$P(2.9) = 2(0.005) \qquad \textit{tumor has doubled in size by } t = 2.9$$

$$0.005e^{2.9a} = 0.01$$

$$e^{2.9a} = 2 \qquad \textit{dividing by } 0.005$$

$$a = \ln 2/2.9 \approx 0.24$$

Hence, $P(t) = 0.005e^{0.24t}$.

Similarly, we have $Q(0) = 0.99(0.5) = 0.495$ and $Q(t) = 0.495e^{bt}$ where we have to solve for b. Since the half-life of quiescent cells is 5.7 days, we can solve for b as follows:

$$Q(5.7) = 0.5(0.495)$$

$$0.495e^{5.7b} = 0.5(0.495)$$

$$e^{5.7b} = 0.5$$

$$b = \ln 0.5/5.7 \approx -0.12$$

Hence, $Q(t) = 0.495e^{-0.12t}$.

Thus, it follows from the first equation that

$$V(t) = 0.005e^{0.24t} + 0.495e^{-0.12t}$$

b. To determine when $V(t)$ is increasing or decreasing, we need to compute its derivative:

$$V'(t) = 0.0012e^{0.24t} - 0.0594e^{-0.12t}$$

Since $V'(0) \approx -0.0582$, the volume of the tumor is initially decreasing after therapy. To see when $V'(t)$ changes sign, we solve

$$0.0012e^{0.24t} - 0.0594e^{-0.12t} = 0$$
$$0.0012e^{0.24t} = 0.0594e^{-0.12t}$$
$$0.0012e^{0.36t} = 0.0594$$
$$e^{0.36t} = 49.5$$
$$t = \frac{\ln 49.5}{0.36} \approx 10.84 \text{ days}$$

Hence, after 10.84 days, the tumor begins to regrow and therapy should be readministered. Indeed, this prediction is supported by the data shown in the left panel of Figure 4.13. (The data in the right panel are examined in Problem Set 4.3.)

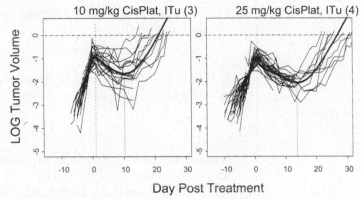

Figure 4.13 Plots of tumor size (log scale; each line comes from a different mouse) before and after a treatment of 10 mg/kg (left panel) and 25 mg/kg (right panel) of the drug cisplatin.

From E. Demidenko, Mixed Models: Theory and Applications, John Wiley & Sons, 2004. p. 544. Used with permission.

In the next example, we consider the vascular system, which consists of arteries and veins that branch in different directions to pump blood through all parts of the body. Ideally, the body is designed to minimize the amount of energy it expends in pumping blood. According to one of Poiseuille's laws, the resistance blood experiences by traveling down the center of a blood vessel with radius r and length L is proportional to

$$\frac{L}{r^4}$$

Without loss of generality, we assume that this proportionality constant equals one, and use this law to determine optimal branching angles in the vascular system of animals.

Example 5   Vascular branching

Consider a blood vessel that branches as illustrated below:

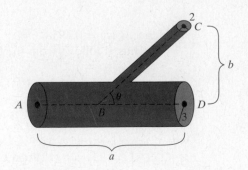

where a and b are positive constants. Given a and b, determine the angle θ which minimizes the total resistance in the blood flow from the point A to the point C.

Solution   We want to minimize the total resistance along the blood vessel from A to C. Let B be the point where the vessel branches. We need to determine the resistance from A to B and the resistance from B to C. To determine the resistance along the blood vessel from A to B, we need to determine how the distance from A to B depends on θ. Using the right triangle as shown below

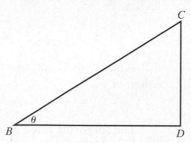

we get that the length from B to D is given by $b\cot\theta$. Hence, the distance from A to B is

$$a - b\cot\theta$$

and, as the radius of the vessel from A to B is 3, the resistance from A to B is

$$\frac{a - b\cot\theta}{3^4}$$

Using the same right triangle, we get that the distance from B to C is $b\csc\theta$. Hence, the resistance from B to C is

$$\frac{b\csc\theta}{2^4}$$

as the radius of the vessel from B to C is 2. Adding the resistance from A to B to the resistance from B to C, we get that the resistance from A to C is given by

$$R(\theta) = \frac{a - b\cot\theta}{81} + \frac{b}{16}\csc\theta$$

To minimize the resistance on the half-open interval $\left(0, \frac{\pi}{2}\right]$, we need to determine the critical points along this interval as follows:

$$R'(\theta) = 0$$

$$\frac{b}{81}\csc^2\theta - \frac{b}{16}\csc^2\theta\cos\theta = 0 \quad \textit{taking the derivative and using the fact that}$$
$$\textit{cot}\,\theta = \csc\theta\cos\theta$$

$$b\csc^2\theta\left(\frac{1}{16}\cos\theta - \frac{1}{81}\right) = 0 \quad \textit{common factor}$$

$$\textit{This implies} \quad \cos\theta = \frac{16}{81}$$

$$\theta \approx 1.37 \quad \textit{solving for } \theta$$

Hence, the optimal angle is given by ≈ 1.37 radians. Equivalently, since one radian is approximately 57.3 degrees, $\theta \approx 78.5$ degrees.

Regression

Regression, loosely speaking, is the process of drawing a graph through a set of data, where the graph represents the information in the data after process and measurement errors have been "averaged out." The term *regression* dates back to Sir Francis

Galton's work on measuring heritable traits in humans, such as height, and his publication of this work in 1886 under the title "Regression Towards Mediocrity in Hereditary Status" (*Journal of the Anthropological Institute of Great Britain and Ireland*, 15 (1886): 246–263). To simplify our notation in developing some of the ideas associated with regression theory, we introduce the summation notation, a notation that we revisit in Chapter 5 when developing integral calculus.

Summation Notation

The sum of a sequence of n real numbers $a_1, a_2, a_3, \cdots, a_n$ is

$$\sum_{i=1}^{n} a_i = a_1 + a_2 + a_3 + \cdots + a_n$$

The index i does not have to start at $i = 1$, but could start at any integer value less than or equal to n.

The following three properties of this summation notation are easily verified.

Properties of Summations

Let $a_1, a_2, a_3, \ldots, a_n, b_1, b_2, b_3, \ldots, b_n$, and c be real numbers. Then

$$\sum_{i=1}^{n}(a_i + b_i) = \sum_{i=1}^{n} a_i + \sum_{i=1}^{n} b_i$$

$$\sum_{i=1}^{n}(a_i - b_i) = \sum_{i=1}^{n} a_i - \sum_{i=1}^{n} b_i$$

$$\sum_{i=1}^{n} ca_i = c \sum_{i=1}^{n} a_i$$

In our development of regression, we use the following definition that we informally introduced in Section 1.2 and depicted in Figure 1.14 for the case of linear regression. Here we provide a definition for any function $f(x)$, not just the linear function $f(x) = mx + c$.

Residual Sum-of-Squares

The residual sum-of-squares S of the function $y = f(x)$ from a data set

$$D = \{(x_1, y_1), (x_2, y_2), \cdots, (x_n, y_n)\}$$

is the residuals

$$e_i = y_i - f(x_i)$$

squared and then summed to obtain

$$S = \sum_{i=1}^{n} e_i^2$$

If we interpret the residuals e_i as "errors" from the true value, then S is the **sum of squared errors**. To find a best-fitting function to a data set, regression aims to minimize the sum of these squared errors.

Example 6   Best line through the origin

In Example 7 of Section 2.2, we determined the slope a of how the rate (cells/hour) at which copepods feed on diatoms increases as a function of diatom density (cells/milliliter). In the first two rows of data in Table 4.1 we have extracted eleven representative points from these data, as diatom density increases from 0 to 200 cells/milliliter.

Table 4.1 Diatom densities x (cells/milliliter) and copepod feeding rates y (cells/hour)

i	1	2	3	4	5	6	7	8	9	10	11	$\sum_{i=1}^{11}$
x_i	19	20	39	68	73	100	106	149	155	174	193	
y_i	102	156	297	313	523	484	797	680	938	1328	914	
x_i^2	373	412	1494	4574	5392	10097	11295	22139	23898	30246	37340	147260
$x_i y_i$	1963	3170	11473	21135	38436	48617	84692	101132	144928	230978	176630	863210

a. Find the slope a of the line $f(x) = ax$ through the origin that minimizes the residual sum-of-squares through the data. This is the so-called *best-fitting* or *regression* line.

b. Plot these data and the regression line on the same graph.

Solution

a. From the definition of the residual sum-of-squares with $f(x) = ax$ and properties of summations, we have

$$S(a) = \sum_{i=1}^{11} (y_i - ax_i)^2$$

$$= \sum_{i=1}^{11} (y_i^2 - 2ax_i y_i + a^2 x_i^2) \quad \text{expanding the quadratic term}$$

$$= \sum_{i=1}^{11} y_i^2 - 2a \sum_{i=1}^{11} x_i y_i + a^2 \sum_{i=1}^{11} x_i^2 \quad \text{from summation properties}$$

S is a quadratic function of a. Since the coefficient of a^2 is positive, the graph of $S(a)$ is an upward facing parabola and, hence, has a unique minimum. To minimize $S(a)$ with respect to a simply requires finding where its derivative $S'(a)$ equals 0. Differentiating S yields

$$S'(a) = -2 \sum_{i=1}^{11} x_i y_i + 2a \sum_{i=1}^{11} x_i^2$$

Setting $S'(a) = 0$ and solving yields

$$a^* = \frac{\sum_{i=1}^{11} x_i y_i}{\sum_{i=1}^{11} x_i^2}$$

Using the values of the sums, as given in Table 4.1, we obtain that

$$a^* = \frac{863210}{147260} = 5.86$$

is the value of a that minimizes $S(a)$.

b. The line $f(x) = 5.86x$ and the data are plotted below

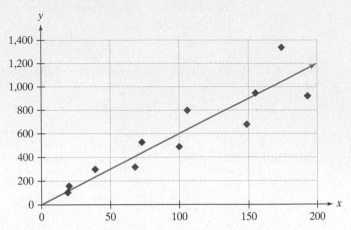

Recall Example 5 of Section 1.7. We plotted the data in Figure 1.46 of that example on the decline of the Glued gene in a species of fruit fly. The data, which exhibit an exponential decline from 0.5 to 0.0036 over six generations, are reported in Table 4.2. To model how rapidly the Glued gene is lost from the experimental fruit fly population in each generation, we used the function $0.5e^{-rt}$ for some appropriately chosen constant $r > 0$. Taking the logarithm of this equation yields the linear function $\ln 0.5 - rt$ of t. In the next example, we use this observation to find the best-fitting line to the logarithmically transformed data.

Table 4.2 Changes in the frequency x of the Glued gene in a species of fruit fly over seven (generation 0 is also experimentally generated) experimental generations t (values to two significant digits)

t (generation)	0	1	2	3	4	5	6
x_t	0.50	0.45	0.32	0.23	0.23	0.067	0.036

Example 7   Regression and lethal genes

a. Find the value of the decay rate parameter $r > 0$ of the function $\ln 0.5 - rt$ that minimizes the residual sum-of-squares of this function with respect to the data $(1, \ln x_1), (2, \ln x_2), \ldots, (6, \ln x_6)$ from Table 4.2.

b. Provide a semi-log plot of these data and the best-fitting, log-transformed, exponential decay function on the same graph.

Solution

a. Following the approach of Example 6, we want to minimize

$$S(r) = \sum_{t=1}^{6} (\ln x_t - (\ln 0.5 - rt))^2$$

Differentiating S with respect to r yields

$$S'(r) = \sum_{t=1}^{6} 2t \, (\ln x_t - \ln 0.5 + rt) = 2\sum_{t=1}^{6} t \ln x_t - 2\ln 0.5 \sum_{t=1}^{6} t + 2r \sum_{t=1}^{6} t^2$$

Setting $S'(r) = 0$ and solving for r yields

$$r = \frac{\ln 0.5 \sum_{t=1}^{6} t - \sum_{t=1}^{6} t \ln x_t}{\sum_{t=1}^{6} t^2}$$

Calculating relevant sums from Table 4.2, we obtain the following table.

t	0	1	2	3	4	5	6	$\sum_{t=1}^{6} t = 21$
x_t	0.50	0.45	0.32	0.23	0.23	0.067	0.036	
t^2	0	1	4	9	16	25	36	$\sum_{t=1}^{6} t^2 = 91$
$\ln x_t$	−0.69	−0.80	−1.14	−1.47	−1.47	−2.70	−3.32	
$t \ln x_t$	0	−0.80	−2.28	−4.41	−5.88	−13.5	−19.92	$\sum_{t=1}^{6} t \ln x_t = -46.8$

Thus, from the above equation for r we have

$$r = \frac{64.5}{182} = 0.35$$

We conclude that the best estimate of the rate at which the Glued gene is purged from the fruit fly population during the experiment is very close to 33% per generation.

b. The log-transformed data $\ln x_t$ and the best-fitting function $\ln 0.5 - rt$ are shown below.

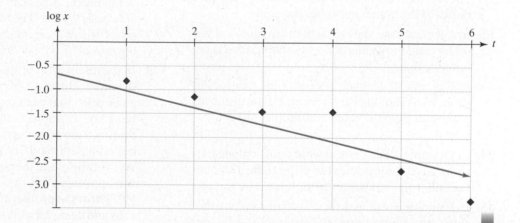

PROBLEM SET 4.3

Level 1 DRILL PROBLEMS

1. In Example 1, the selling price of corn was $1.50 per bushel and the seeds cost $3 per thousand seeds. Determine the density of seeds that maximize profit if the selling price and cost are both doubled.

2. In Example 1, determine the density of seeds that maximize profit if the selling price is $5 per bushel and the seeds cost $2 per thousand seeds.

3. In Example 1, determine the density of seeds that maximize profit if the selling price is $2.20 per bushel and the seeds cost $2.50 per thousand seeds.

4. In Example 2, we estimated that the maximum per capita growth rate of the Arctic fin whale is $r = 0.08$.

Suppose a better estimate is $r = 0.1$. Determine the maximum sustainable harvesting rate for this value of r.

5. In Example 2, we estimated that the long-term abundance of the Arctic fin whales in the absence of harvesting the Arctic fin whale is $K = 500{,}000$. Suppose a better estimate is $K = 400{,}000$. Determine the maximum sustainable harvesting rate for this value of K.

6. In a species of fish the growth rate function is given by $G(x) = 1.4x(1 - x/K)$, where $K = 5$ million metric tons (i.e., the population of fish is measured in metric tons rather than number of individuals). If

the harvest rate is a function of the harvesting effort h and the total amount of fish x, that is $H = hx$, find the harvesting effort value h that corresponds to the maximum sustainable yield.

7. In a species of fish, the growth rate function is given by $G(x) = 2.1x(1 - x/K)$, where $K = 8$ million metric tons. If the harvest rate is $H = hx$, find the harvesting effort value h that corresponds to the maximum sustainable yield.

In Problems 8 to 12, find the optimal angle for the following vascular branching problems, as considered in Example 5.

8. A larger artery has a radius of 0.05 mm, and a smaller artery of radius 0.025 mm branches from the larger artery with branching angle θ.

9. A larger artery has a radius of 0.06 mm, and a smaller artery of radius 0.04 mm branches from the larger artery with branching angle θ.

10. The radius of the main blood vessel is $r_1 = 2$ and the radius of the branching vessel is $r_2 = 1$.

11. The radius of the main blood vessel is $r_1 = 4$ and the radius of the branching vessel is $r_2 = 3$.

12. In a general case, the radius of the main blood vessel is r_1 and the radius of the branching vessel is r_2. Assume that $r_1 > r_2$.

13. In Example 3, calculate at what point x along the shore Elvis should enter the water if the distance of the ball from the shore is 20 meters rather than 6.

14. In Example 3, calculate at what point x along the shore Elvis should enter the water if the distance of the ball from the shore is 10 meters rather than 6.

15. In Example 3, calculate at what point x along the shore Elvis should enter the water if the distance of the ball from the shore is d meters.

In Problems 16 to 22, calculate the residual sum-of-squares of the listed function through this data set:

x_i	1	2	3	4	5
y_i	1	3	6	12	16

Draw a graph of the function and the data on the same plot. In Problems 21 to 22, plot the residual sum-of-squares graph as a function of the free parameter and selected range for this parameter. Note that you are not being asked to fit the exponential functions in the relevant examples below on a semi-log plot, but to use the function itself directly in the calculations of the residuals with respect to the data.

16. $f(x) = 2x$

17. $f(x) = 2x + 1$

18. $f(x) = 3x - 1$

19. $f(x) = 0.4e^{0.8x}$

20. $f(x) = 0.2e^{0.9x}$

21. $f(x) = mx, 0 \le m \le 4$

22. $f(x) = 0.2e^{rx}, 0 \le r \le 1$

Level 2 APPLIED AND THEORY PROBLEMS

23. Find a general formula for which Example 3 is a specific case that describes how to calculate at what point x along the shore Elvis should enter the water if the distance of the ball from the shore is d meters (rather than 6) and the point on the shore to which this distance d holds is k meters (rather than 15) from where Tim is standing.

24. In a species of fish, the growth rate function is given by $G(x) = 1.5x(1 - x/K)$, where $K = 6{,}000$ metric tons (i.e., the population of fish, x, is measured in metric tons rather than number of individuals). The price a fisher can get is $p = \$600$ per metric ton. If the amount the fisher can harvest is determined by the function $H = hx$, where each unit of h costs the fisher $c = \$100$, what is the maximum amount of money the fisher can expect to make on a sustainable basis? (Hint: The fisher's sustainable income is given by $pH - ch$, where H is a sustainable harvesting rate.)

25. In the tumor growth study described in Example 4, the tumor consisted of proliferating cells (clonogenic cells) that grew exponentially with a doubling time of approximately 2.9 days. Suppose that each mouse was given a dose of 25 mg/kg of cisplatin per treatment with the following results: At the time of the therapy, the average tumor size was approximately 0.44 cm^3. After treatment, 99.73% of the proliferating cells became quiescent cells and decayed with a half-life of approximately 6.24 days.

   a. Write a function $V(t)$ that represents the size of the tumor (proliferating plus quiescent cells) t days after therapy.

   b. Determine at what point in time the tumor starts to regrow and therapy should be readministered.

   c. Compare your answer to the data figure in Example 4.

26. In a follow-up study to the tumor growth study described in Example 4, mice were infected with a relatively aggressive line of proliferating clonogenic cells that grew exponentially with a doubling time of approximately 1.8 days. Each mouse was given a dose of 20 mg/kg of cisplatin per treatment with the following results: At the time of the therapy, the average tumor size was approximately 0.6 cm^3. After treatment, 99.10% of the proliferating cells became

quiescent cells and decayed with a half-life of approximately 4.4 days.

  a. Write a function $V(t)$ that represents the size of the tumor (proliferating plus quiescent cells) t days after therapy.

  b. Determine at what point in time the tumor starts to regrow and therapy should be readministered.

27. In certain tissues, cells exist in the shape of circular cylinders. Suppose such a cylinder has radius r and height h. If the volume is fixed (say, at v), find the value of r that minimizes the total surface area ($S = 2\pi r^2 + 2\pi rh$) of the cell.

28. Farmers regularly use fertilizers to enhance the productivity of their crops. Determining the appropriate amount of fertilizer to use requires balancing the costs of fertilization with the increases in yield. In a 2004 study, Baker and colleagues studied the relationship between nitrogen fertilization and yield of hard red spring wheat. (See Dustin A. Baker, Douglas L. Young, David R. Huggins, and William L. Pan, "Economically Optimal Nitrogen Fertilization for Yield and Protein in Hard Red Spring Wheat," *Agronomy Journal* 96 (2004): 116–123.) For conventional tillage practices in eastern Washington in the late 1980s, they found that the grain yield (in kilograms per hectare) as a function of nitrogen (in kilograms per hectare) is well approximated by

$$Y(N) = 1.86 + 0.02741N - 0.00009N^2$$

These researchers suggested that a high selling price for wheat would be \$191.1/kg and low cost for nitrogen would be \$0.49/kg. Determine the amount of nitrogen that maximizes profits per hectare.

29. Baker and colleagues suggested that a low selling price for wheat would be \$139.65/kg and a high cost for nitrogen would be \$0.71/kg. Using the same yield function as in the previous problem, determine the amount of nitrogen that maximizes profits per hectare.

30. If the effects of density dependence in a whale population set in less rapidly closer to the final carrying capacity, K, than the logistic equation used in Example 2, then the equation should be replaced by an asymmetric growth model

$$G(x) = 0.08x\left[1 - \left(\frac{x}{500,000}\right)^\alpha\right] \text{ whales per year}$$

for some $\alpha \in (0, 1)$. For the case $\alpha = 0.5$, calculate the stock level x that provides the maximum sustainable.

31. If the effects of density dependence in a whale population set in less rapidly or more rapidly closer to the final carrying capacity, K, than the logistic equation used in Example 2, then the equation should be replaced by an asymmetric growth model

$$G(x) = rx\left[1 - \left(\frac{x}{K}\right)^\alpha\right] \text{ whales per year}$$

For $\alpha > 0$, $r > 0$, and $K > 0$, calculate the stock level x that provides the maximum sustainable yield. Discuss whether rapid onset of density dependence (i.e., large α) or gradual onset of density dependence (i.e., small α) leads to larger sustainable yields.

32. During the winter, a species of bird migrates from the coast of a mainland to an island 500 miles southeast. If the energy the bird requires to fly one mile over the water is twice more than the amount of energy it requires to fly over the land, determine what path the species should fly to minimize the amount of energy used.

33. The Statue of Liberty is 92 meters high, including the 46 meter pedestal upon which it stands. How far from the base should an individual stand to ensure that the view angle, θ, is maximized?

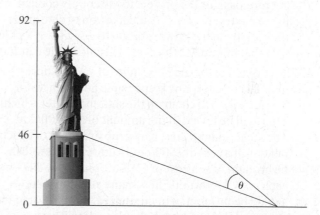

34. In the northeastern part of Sweet Water County, a large dam is being constructed on the Shuga River to produce *hydroelectricity* (i.e., the generation of electricity through water pressure). An important part of this project is running power lines from the power stations at the downstream side of the dam to various parts of the county, including Pickle City, the largest city in the county. On the recommendation of a number of other counties, Sweet Water County officials have hired you as a consultant to resolve cost issues for running these power lines.

    County officials have informed you that the Shuga River runs due east, and on its southern side lies an expanse of federally protected wetlands. Pickle City lies several miles to the south of these wetlands, as shown in the map below.

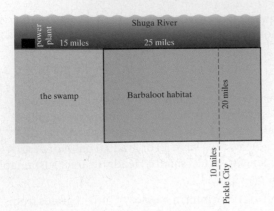

The federally protected wetlands are divided into two regions. In the western region, county officials expect that due to federal regulations it will cost 40% more to run conduit here than it does through non-wetland ground. The eastern region of the wetlands is a habitat for the endangered Brown Barbaloots. Consequently, federal law prevents the county from running conduits through this region.

As the county officials intend to submit a budget proposal for the project to the county council in the next week, they would like you to determine the path from the power station to downtown Pickle City that minimizes the cost of installing the conduit.

35. An oil spill has fouled 200 miles of Pacific shoreline. The oil company responsible has been given fourteen days to clean up the shoreline, after which a fine will be levied in the amount of $10,000/day. The local cleanup crew can scrub five miles of beach per day at a cost of $500/day. Additional crews can be brought in at a cost of $18,000, plus $800/day for each crew. Determine how many additional crews should be brought in to minimize the total cost to the company and how much the cleanup will cost.

36. Consider a spherical cell with radius r. Assume that the cell gains energy at a rate proportional to its surface area (i.e., nutrients diffusing in from outside of the cell) and that the cell loses energy at a rate proportional to its volume (i.e., all parts of the cell are using energy). If the cell is trying to maximize its net gain of energy, determine the optimal radius of the cell. Note: Your final expression will depend on your proportionality constants.

37. Consider a cylindrical cell with radius r and height $r/2$. Assume that the cell gains energy at a rate proportional to its surface area (i.e., nutrients diffusing in from outside of the cell) and that the cell loses energy at a rate proportional to its volume (i.e., all parts of the cell are using energy). If the cell

is trying to maximize its net gain of energy, determine the optimal value of r. Note: Your final expression will depend on your proportionality constants.

38. A dune buggy is in the desert at a point A located 40 km from a point B, which lies on a long, straight road, as shown in Figure 4.14. The driver can travel at 45 km/hour on the desert and 75 km/hour on the road. The driver will win a prize if she arrives at the finish line at point D, 50 km from B, in 85 minutes or less. Set up and analyze a model to help her decide on a route to minimize the time of travel. Does she win the prize?

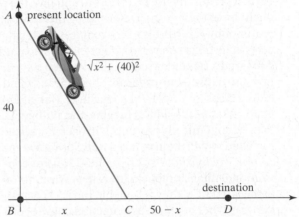

Figure 4.14 Path traveled by a dune buggy.

39. The question of whether an optimal body size exists for different kinds of animals is one that is of great interest to biologists. The reproductive power P of an individual can be modeled, following the ideas of ecologist James H. Brown (see his book *Macroecology*, University of Chicago Press, 1995), as the harmonic mean of two limiting rates. The harmonic mean of two numbers a and b is the reciprocals of the average of the inverses of the two numbers: $1/(1/a + 1/b) = ab/(a + b)$. The two rates are a per-unit-mass rate R_1 at which individuals acquire resources, and a per-unit-mass rate R_2 at which individuals convert those resources into new individuals; that is,

$$P = \frac{R_1 R_2}{R_1 + R_2}$$

Assuming both R_1 and R_2 are the following allometric functions of body mass measure in kilograms

$$R_1 = 2M^{3/4} \qquad \text{and} \qquad R_2 = 3M^{-1/4},$$

find the body mass M that maximizes the reproductive power P, and show that this extremum is a maximum for the case $b_1 = 0.75$ and $b_2 = -0.25$.

40. Suppose that we express the two rate functions in Problem 39 using the general form $R_1 = c_1 M^{b_1}$ and $R_2 = c_2 M^{-b_2}$. Show in this case that the maximum body size is given by the expression

$$M^* = \left(\frac{-c_2 b_1}{c_1 b_2}\right)^{1/(b_1 - b_2)}$$

41. In Example 6, we determined the slope a of the linear component of a type I functional response for the rate (cells/hour) at which copepods feed on diatoms as a function of diatom density (cells/milliliter). Suppose now that only the following partial set of data (Table 4.3) is available for estimating the parameter a.

Table 4.3 Diatom densities x (cells/milliliter) and copepod feeding rates y (cells/hour)

i	1	2	3	4	5	6
x_i	19	39	68	106	149	174
y_i	102	297	313	797	680	1328

a. Find the slope a of the line $f(x) = ax$ through the origin that minimizes the residual sum-of-squares through the data.

b. Plot these data and the best-fit line on the same graph.

42. In Table 1.4, the CO_2 concentrations at the Mauna Loa Observatory in Hawaii are listed by month, starting at month 1 (May 1974) until month 140 (December 1985). The data for the month of May, for the twelve periods spanning May 1974 to May 1985, are shown in Table 4.4.

a. Find the slope m of the regression line $f(x) = mx + 333.2$ through the point $(t, x_t) = (0, 333.2)$ that minimizes the residual sum-of-squares through the remaining eleven data points.

b. Plot these data and the regression line on the same graph.

43. In Example 7, we found the value of the decay rate parameter $r > 0$ using linear regression to fit a semi-log plot of the frequency of Glued genes in an experimental fruit fly population. This experiment was repeated by the same group of researchers and the following data were obtained:

t	0	1	2	3	4	5
x_t	0.59	0.37	0.21	0.16	0.047	0.015

Use these data to find the value r of the best-fitting exponential decay function $x_t = 0.59e^{-rt}$ through the starting point $(t, \ln x_t) = (0, -0.53)$. Present the solution on a semi-log plot together with a plot of the data.

44. In Section 1.4, we presented the following data on the growth of the United States from 1815 until 1895.

Year	Population (in millions)
1815	8.3
1825	11.0
1835	14.7
1845	19.7
1855	26.7
1865	35.2
1875	44.4
1885	55.9
1895	68.9

Use linear regression on a semi-log transformation of the above data to find the best estimate of the annual growth rate $r > 0$ in the population model $x(t) = 8.3e^{rt}$ (million individuals) for $t \in [0, 80]$ (years), with $t = 0$ corresponding to the year 1815.

Table 4.4 CO_2 x (cell/milliliter) in year t

t	0	1	2	3	4	5	6	7	8	9	10	11
x_t	333.2	333.9	334.8	336.8	338.0	339.0	341.5	343.0	344.3	345.8	347.5	348.7

4.4 Decisions and Optimization

Optimal decisions

The behavior of animals has been honed by natural selection to maximize the reproductive potential of individuals. Thus, from an individual point of view, individuals should act in ways that maximize the number of offspring they rear to sexual maturity. This number is referred to as an **individual's fitness**. From a genetic point of view, a gene encoding for a behavior that maximizes an individual's fitness will have a greater representation in the gene pool of future generations than a gene that encodes for a behavior that is detrimental to an individual's fitness (e.g., a gene that causes an individual to be excessively reckless, making it likely that the individual will die before reaching sexual maturity). Theories of optimal behavior are based on the premise that organisms maximize their fitness by behaving in a certain way. Using models, researchers can develop hypotheses about these optimal behaviors. These hypotheses can be tested experimentally or through comparative studies.

In our first example in this section, we obtain insights into the reason why Northwestern crows consistently drop whelks from a specific height to break them open. If the crows fly too low, the shells require too many drops to get them to break open. If they fly too high, the crows waste energy. Assuming that crows have evolved to minimize energy expenditures, scientists might be interested in testing this hypothesis by formulating a suitable function to minimize. This function would characterize the number of of drops, and hence work, required to break open a whelk as a function of the height from which the shell is dropped. In addition to modeling the dropping behavior of Northwestern crows, this section investigates optimal foraging in a patchy environment, optimal timing of seed production, and optimal time to harvest crops. In addition to these examples, we present a key theorem called the *marginal value theorem*, which has applications to problems maximizing or minimizing average rates of change.

The Northwestern crow, illustrated in Figure 4.15, feeds on whelks, a type of mollusk. To get the meat from inside the whelk's shell, individual crows lift whelks into the air and drop them onto a rock to break open the shell. The biologist Reto Zach (see Example 6 of Section 3.2) observed that individual crows typically drop the shells from a height of five meters. This led him to ask this question: Does the height from which Northwestern crows drop whelk shells minimize the amount energy required to open a shell?

Figure 4.15 A Northwestern crow.

Anthony Mercieca / Photo Researchers Inc

Example 1   Northwestern crows and whelks

After collecting data by dropping whelks from different heights, Zach found that, on average, the number of drops required to break a whelk dropped from h meters is modeled by the function

$$D(h) = 1 + \frac{20.4}{h - 0.84} \text{ drops}, \quad h > 0.84$$

This relationship implies $\lim\limits_{h \to 0.84^+} D(h) = \infty$, which in turn implies that if $h \leq 0.84$, the shell will never open. Use this equation

$$work = force \times distance$$

to find the optimal height from which a whelk should be dropped to minimize the amount of work required to break a whelk shell.

Solution   Since work is force times distance, the amount of work required to drop a shell of fixed weight is in proportion to the total height the crow flies when breaking a whelk. The total height is given by the number of drops times the height of the drop. In other words, up to a proportionality constant, the average amount of work that it will take a crow to break open a whelk shell is

$$W(h) = hD(h) = h + \frac{20.4h}{h - 0.84}$$

To determine where this function takes on its smallest value, we need to understand the graph of the function. It has a vertical asymptote at $h = 0.84$. Taking the derivative yields

$$W'(h) = 1 - \frac{0.84 \cdot 20.4}{(h - 0.84)^2}$$

$$= \frac{-16.4304 - 1.68h + h^2}{(h - 0.84)^2}$$

Since the denominator is positive wherever $h \neq 0.84$, we only need to understand when the numerator is positive or negative. Solving $-16.4304 - 1.68h + h^2 = 0$ for h yields $h \approx -3.3$ meters and $h \approx 4.98$ meters. Since this quadratic corresponds to an upward facing parabola, we get that the numerator of $W'(h)$ is positive when $h > 4.98$ and negative on the interval $(0.84, 4.98)$. Hence, $W(h)$ decreases on the interval $(0.84, 4.98)$ and increases on the interval $(4.98, \infty)$. The height $h \approx 4.98$ is a global minimum for $h > 0.84$.

Hence, the height that minimizes the amount of work is approximately five meters, the height observed by Reto Zach! ∎

The next example explores the optimal time for a plant to produce seeds. This is just one in a class of optimal time-to-reproduction problems. Others include the optimal time for a honey bee colony to swarm and the optimal time for salmon to return from the ocean to lay eggs upriver.

Example 2   Optimal time for producing seeds

A particular plant is known to have the following growth and seed production characteristics. At time of planting ($t = 0$), the seedling has a mass of 5 grams. At time ($t > 0$) days after planting, the seedling has grown into a plant that weighs $w(t) = 5 + 400t - t^2$ grams. The plant has a gene that can be manipulated to control the age t at which the plant matures. The number of seeds $S(t)$ produced by a plant maturing at age t is

$$S(t) = 0.1w(t) = 0.5 + 40t - 0.1t^2$$

A farmer asks the geneticists to genetically engineer a plant line that accounts for the fact that on his farm, because of losses from pests, drought, and disease, only a proportion

$$P(t) = \frac{100}{100 + t}$$

of germinating seeds develop and survive to age t. What age of maturity should the geneticist select for the plants to maximize the seed production of the mature crop for the farmer?

Solution   For every N seeds that the farmer plants on his land at time $t = 0$, $N \times P(t)$ will mature at time $t > 0$. The total yield from these plants is then

$$Y(t) = N \times P(t)S(t) = \frac{100N(0.5 + 40t - 0.1t^2)}{100 + t}$$

Since N is just a scaling factor that depends on the number of acres that farmer plants, we can set it to any convenient value such as $N = 1$. To find the germination time that maximizes this yield, we need to understand the first derivative:

$$Y'(t) = \frac{d}{dt}\left[\frac{50 + 4000t - 10t^2}{100 + t} \right]$$

$$= \frac{(100 + t)(4000 - 20t) - (50 + 4000t - 10t^2)}{(100 + t)^2}$$

$$= \frac{-10t^2 - 2000t + 399950}{(100 + t)^2}$$

Thus, $Y'(t)$ exists for $t > 0$ and the derivative vanishes at solutions to the equation

$$10t^2 + 2{,}000t - 399{,}950 = 0$$

We can use technology or the quadratic formula to obtain the roots

$$t^* = -323.6 \text{ and } 123.6$$

Since $Y'(0) > 0$ and $Y'(200) < 0$, we get that Y increases on the interval $(0, 123.6)$ and decreases on the interval $(123.6, \infty)$. Hence, $Y(t)$ is maximized at $t \approx 123.6$. We verify this directly by plotting Y as a function of time, as illustrated in Figure 4.16. The vertical line indicates the optimal maturation time $t^* = 123$ days.

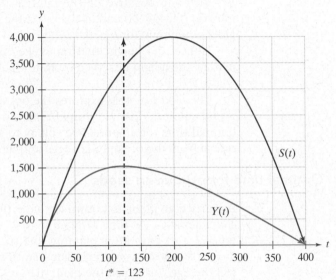

Figure 4.16 The red curve is the number of seeds $S(t)$ produced by a plant that survives to age t. The blue curve, which has its maximum value at t^*, is the expected number of seeds that plant will produce after taking into account the probability it may die before starting to seed.

Optimal foraging and marginal value

Very often, food is not distributed homogeneously over the environment; rather, it occurs in discrete patches in the environment. For fruit bats, a patch may correspond to a fruit tree or a stand of fruit trees. For a hummingbird, which feeds on the nectar of flowers, a patch may correspond to a single flower or a field of flowers. In *optimal foraging theory*, we want to know how long an animal should continue to collect resources in a patch when it has the choice of traveling to another resource-rich patch. The question of when to leave a patch as resources in the patch are being depleted is known as the *optimal residence time* problem.

Example 3   Optimal foraging in a multi-patch environment

Figure 4.17 House martin parent feeding its young

House martins make sorties from their nests to collect food to bring back to their young. In an experiment carried out in the early 1980s, two British scientists, D. M. Bryant and A. K. Turner, found that the travel time of martins from a particular nest to nearby foraging areas ranged from half a minute to several minutes, and the weight of the load of insects the martins collected and brought back to their nest to feed their chicks (see Figure 4.17) varied between 20 and 100 mg. (See Central place foraging by swallows (Hirundinidae): The question of load size. *Animal Behavior* 30 (1982): 845–856.) On an average foraging bout, Bryant and Turner observed that these martins collected insects at the rate of (roughly) 10 mg/minute from time of departure from the nest. Assume one of these martins encounters a patch three minutes after leaving its nest, and its cumulative load of insects after foraging for t minutes is given by the function

$$B(t) = \frac{200t}{6+t} \text{ mg}$$

If the martin is trying to maximize the average load accumulated per minute since leaving its nest, then what is the optimal time for the martin to quit foraging in this patch?

Solution   Since it takes three minutes for the martin to reach the patch, the average load accumulated per minute after t minutes in the patch is $R(t) = B(t)/(t+3)$. To determine the best time to leave the patch, we need to understand the graph of $R(t)$ for $t \geq 0$. Taking the first derivative of R we get

$$R'(t) = \frac{d}{dt}\left[\frac{200t}{(6+t)(t+3)}\right]$$

$$= \frac{d}{dt}\left[\frac{200t}{t^2 + 9t + 18}\right]$$

$$= \frac{200(t^2 + 9t + 18) - (2t + 9)200t}{(t^2 + 9t + 18)^2}$$

$$= \frac{200(18 - t^2)}{(t^2 + 9t + 18)^2}$$

We have $\frac{dR}{dt} = 0$ when $t^2 = 18$. Equivalently, $t = \pm\sqrt{18} \approx \pm 4.24$ minutes. Only the positive solution is relevant. Since $R'(0) > 0$ and $R'(\sqrt{18}) < 0$, R is increasing on the interval $(0, \sqrt{18})$ and decreasing on the interval $(\sqrt{18}, \infty)$. Hence, the maximum is achieved at $t = \sqrt{18}$ at which

$$R(\sqrt{18}) \approx 11.44 \text{ mg/minute}$$

which exceeds the background average rate of 10 mg/minute. This conclusion is reaffirmed by graphing $R(t)$ as follows:

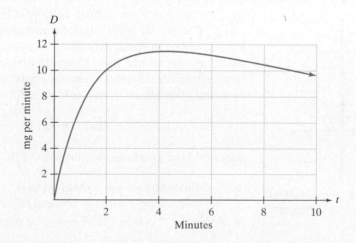

In Example 3, the average rate of change was being maximized over a time interval. A fundamental result for problems of this type is the *marginal value theorem*.

Theorem 4.2   Marginal value theorem

Let $V(t)$ be a function defined on an interval $[a, \infty)$. If $V(t)$ represents the accumulated value of the resource by time $t \geq a$, then the average rate of resource accumulation over

time $t - a$ is given by

$$A(t) = \frac{V(t) - V(a)}{t - a}$$

If a maximum or minimum of $A(t)$ occurs at $t = b > a$ and V is differentiable at $t = b$, then b satisfies the equation

$$V'(b) = \frac{V(b) - V(a)}{b - a}$$

In other words, this maximum or minimum occurs at a time where the average rate of change equals the instantaneous rate of change.

Proof. Since V achieves a maximum or minimum at $t = b > a$, we have $A'(b) = 0$. Computing the derivative yields

$$A'(b) = \frac{d}{dt}\Big|_{t=b} \left(\frac{V(t) - V(a)}{t - a} \right)$$

$$= \frac{V'(b)(b - a) - (V(b) - V(a))}{(b - a)^2} \qquad \textit{by the quotient rule}$$

Setting $A'(b) = 0$ and multiplying both sides of the equation by $(b - a)^2$, we get

$$V'(b)(b - a) - (V(b) - V(a)) = 0$$
$$V'(b)(b - a) = V(b) - V(a)$$

Equivalently, $V'(b) = \dfrac{V(b) - V(a)}{b - a}$. ∎

Example 4 Optimal foraging of great tits

Figure 4.18 Experimental tree in Cowie's experiments.

In a classic paper on animal behavior, biologist Richard Cowie studied the foraging behavior of great tits by constructing experimental trees in an aviary (see Figure 4.18). On these experimental trees, food was placed in plastic containers in a manner that would allow Cowie to manipulate the average travel time T between food containers. Through these experiments, Cowie estimated that the energy gained by a bird after eating from a container for $t \geq 0$ seconds is

$$E(t) = 6.3587(1 - e^{-0.0081\,t}) \text{ calories}$$

Assuming the great tits are maximizing their average energy gain, do the following:

a. Use the marginal value theorem to determine the relationship between the average travel time T and the optimal residence time t in a patch.

b. Solve for T in terms of the optimal residence time and plot it.

c. Discuss your findings with respect to data collected by Cowie, as depicted in Figure 4.19.

Solution

a. Assume that at $t = 0$ the bird arrives at a food container. Since it takes T seconds to get to a container, we are interested in the time interval $[-T, \infty)$ where $t = -T$ corresponds to the moment that the bird begins traveling to the container. Since we assume there is no energy gain during the flight, we define $E(t) = 0$ for $t \leq 0$.

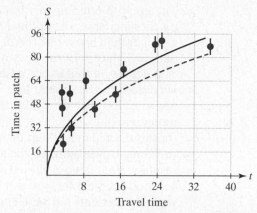

Figure 4.19 Amount of time spent by great tits in a foraging patch is plotted as the solid line through the data (solid points with measurement error bars) as a function of the amount of time taken by individuals to reach the patch. The dotted line is discussed in the solution to part **c** of Example 4.

Data Source: R. Cowie, "Optimal Foraging in Great Tits (*Parus major*)," *Nature* 268 (1977): 137–139.

Clearly, the maximum cannot occur during the interval $[-T, 0]$. By the marginal value theorem with $a = -T$, the time t at which the maximum occurs must satisfy

$$E'(t) = \frac{E(t) - E(-T)}{t + T}$$

$$0.0515e^{-0.0081t} = \frac{6.3587(1 - e^{-0.0081t})}{t + T}$$

b. Solving for the average travel time T in terms of the optimal residence time t yields

$$0.0515e^{-0.0081t} = \frac{6.3587(1 - e^{-0.0081t})}{t + T}$$

$$t + T = \frac{6.3587(1 - e^{-0.0081t})}{0.0515e^{-0.0081t}}$$

$$t + T = 123.5(e^{0.0081t} - 1)$$

$$T = 123.5(e^{0.0081t} - 1) - t$$

Plotting T as a function of t yields this graph:

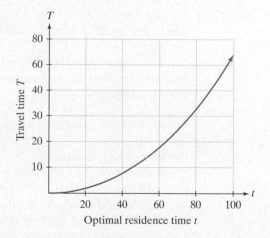

c. The graph plotted in part **b** implies that as the average travel time T between patches increases, the optimal residence time t within a patch should increase at a greater than linear rate (i.e., it is concave up). Notice that in the graph of Cowie's data in Figure 4.19 the axes are switched, so that the inverse of our function $T = f(t)$ is plotted as the dashed line. While five of the twelve data points are very close to the dashed line, the remaining seven data points lie significantly above it. In other words, for these seven experiments, the birds were spending more time in the patches than predicted by the model. One possible explanation for this discrepancy is that the model does not account for the energetic costs of traveling. Cowie adjusted the model to account for these energetic costs and the resulting prediction is plotted as a solid curve in Figure 4.19. In Problem 4 of this section's problem set, you will be asked to redo the analysis in this example in a way that accounts for these energetic costs. ▪

The marginal value theorem has a simple graphical interpretation, which is explored in the next example.

Example 5   Optimal time to harvest

Over a six-decade period, a forestry company has collected data on the profit $P(t)$ of stands of trees harvested at various ages t (units are decades). Initially, $P(t)$ is zero because the costs required to bring in the heavy equipment needed to harvest the trees exceeds the value of the harvest itself, and the company will not harvest before making a profit. Once the trees reach a certain size, a profit is possible and it steadily increases as the stand of trees ages. The company found that the function that best fit its data has the following graph:

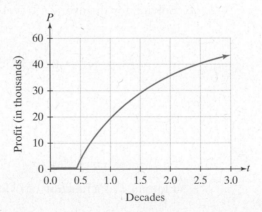

where the profit P is measured in thousands of U.S. dollars and t is measured in decades.

The company wants to maximize the profits it makes per unit time. Write the function $A(t)$ that the company wants to maximize and illustrate where the maximum occurs graphically.

Solution   The company wants to maximize the average rate of accumulation of profit $P(t)$, which is $A(t) = P(t)/t$. Since $P(0) = 0$, we can apply the marginal value theorem, which asserts that the average rate of change $P(a)/a$ equal the instantaneous rate of change $P'(a)$, where $t = a$ is the optimal harvesting time. Graphically, this corresponds to the line from $(0, 0)$ to $(a, P(a))$ being tangent to the graph of $P(t)$ at $t = a$. To find the optimal value a, we take a ruler and place one end at $(0, 0)$ and rotate the ruler until the line determined by its edge is tangent to the graph of $P(t)$. Doing so yields the following plot:

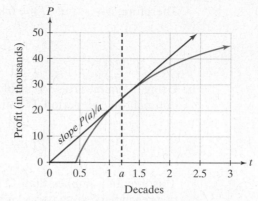

where the optimal time to harvest is approximately $a \approx 1.2$ decades, about 12 years.

Our final example in this section introduces the concept of discounting when optimizing a *sustainable stream of revenue* calculated for all time in the future. Discounting arises if someone promises to pay you D dollars next year, and the current interest rate compounded continuously is $r\%$; then this person should be willing to receive $De^{-r/100}$ dollars now. Namely, if this person took the $De^{-r/100}$ dollars now and invested it, then a year later the investment would yield $e^{r/100}$ dollars for each dollar invested. Hence, a year later the person would have $e^{r/100} De^{-r/100} = D$ dollars. As a result of this reasoning, economists use the *discount factor* $e^{-\delta t}$, where $\delta = r/100$ to reduce D dollars needed at time t in the future to their current value $De^{-\delta t}$ dollars now.

Example 6   Optimal rotation period for a plantation

In the mid-nineteenth century, a German forester by the name of Faustmann developed a theory for the optimal rotation period of a plantation. He calculated that if one planted a stand and harvested it every T years, and received the same value $V(T)$ each time, then the sum of all the discounted amounts (i.e., the sum of $V(T)e^{-\delta T}$ obtained after T years, $V(T)e^{-\delta 2T}$ obtained after $2T$ years, $V(T)e^{-\delta 3T}$ obtained after $3T$ years, and so on for all time into the future) constitutes the so-called *present value* $P(T)$ of the stand given by the formula

$$P(T) = \frac{V(T)}{e^{\delta T} - 1}$$

Now continue his analysis by doing the following:

a. Using his formula for $P(T)$, find a general expression for the optimal stand rotation period T^* that is defined to be the value of $T > 0$ that maximizes the present value $P(T)$ of the stand.

b. What does the expression in part **a** imply as $\delta \to 0$?

c. Use your technology to find the optimal rotation period when

$$V(T) = \left(\frac{2T^{5/2}}{1 + T^2} - 1 \right)$$

and the discount rate is $\delta = 0.1$.

Solution

a. The optimal rotation period T^* is an extremum of $P(T)$. If $T^* > 0$, then, T^* will satisfy the equation $P'(T) = 0$ where

$$P'(T) = \frac{d}{dT}\left[\frac{V(T)}{e^{\delta T} - 1} \right]$$

$$= \frac{V'(T)(e^{\delta t} - 1) - V(T)\delta e^{\delta T}}{(e^{\delta T} - 1)^2} \qquad \text{\textit{quotient rule}}$$

Therefore, if $\delta > 0$, $P'(T) = 0$ implies that

$$V'(T) = \frac{\delta e^{\delta T}}{e^{\delta T} - 1} V(T)$$

b. Using l'Hôpital's rule to calculate the limit as δ approaches 0, we obtain that the optimal rotation period T^* satisfies the equation

$$V'(T) = V(T) \lim_{\delta \to 0} \frac{\delta e^{\delta T}}{e^{\delta T} - 1} = \frac{V(T)}{T}$$

This equation implies that T^* maximizes the average profit accumulation rate over each harvesting period in the limit $\delta = 0$.

c. From part **a** and the specific form for $V(T)$, the optimal rotation period when $\delta = 0.1$ is the solution to

$$\left(\frac{2T^{3/2}}{2} \right) \frac{T^2 + 5}{(1 + T^2)^2} = \left(\frac{2T^{5/2}}{1 + T^2} - 1 \right) \left(\frac{0.1e^{0.1T}}{e^{0.1T} - 1} \right)$$

$$T^{3/2}(T^2 + 5)(e^{0.1T} - 1) = 0.1(1 + T^2)(2T^{5/2} - T^2 - 1)e^{0.1T}$$

$$T^* = 2.68361 \quad \textit{using technology}$$

PROBLEM SET 4.4

Level 1 DRILL PROBLEMS

In Problems 1 to 6, the amount of energy a hummingbird gains after remaining in a patch for t seconds is given. For each problem, find how long a hummingbird should stay in a patch if it wants to maximize its average energy intake rate.

1. The travel time between patches is 15 seconds and

$$f(t) = \frac{180t}{1 + 0.15t} \quad \text{calories}$$

2. The travel time between patches is 5 seconds and

$$f(t) = \frac{180t}{1 + 0.15t} \quad \text{calories}$$

3. The travel time between patches is 10 seconds and

$$f(t) = \frac{360t}{1 + 0.5t} \quad \text{calories}$$

4. The travel time between patches is 5 seconds and

$$f(t) = \frac{360t}{1 + 0.5t} \quad \text{calories}$$

5. The travel time between patches is 5 seconds and

$$f(t) = \frac{360t}{1 + 0.3t} \quad \text{calories}$$

6. The travel time between patches is 10 seconds and

$$f(t) = \frac{360t}{1 + 0.3t} \quad \text{calories}$$

In Problems 7 to 10, rework Example 5 with the given graphs.

7.

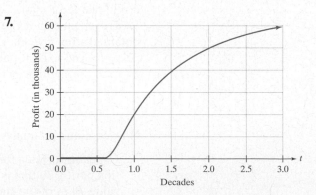

8.

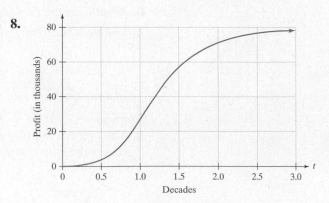

9.

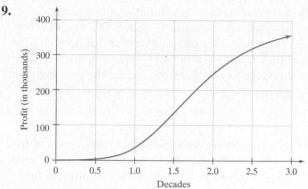

10.

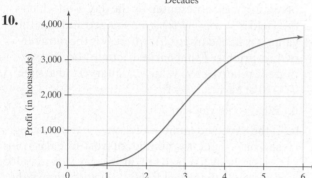

Assume the house martins in Example 3 can choose between two patches. In Problems 11 to 16, the time to fly to a patch and the energy yield as a function of patch residence time (t minutes) are given for two patches. If an individual can visit only one patch and wants to maximize the average amount of calories it receives, then which patch of each pair should it choose?

11. $B(t) = \dfrac{150t}{3+t}$ calories with a travel time of 2 minutes or $B(t) = \dfrac{250t}{5+t}$ calories with a travel time of 3 minutes.

12. $B(t) = \dfrac{150t}{3+t}$ calories with a travel time of 1 minute or $B(t) = \dfrac{250t}{5+t}$ calories with a travel time of 2 minutes.

13. $B(t) = \dfrac{150t}{3+t}$ calories with a travel time of 3 minutes or $B(t) = \dfrac{150t}{4+t}$ calories with a travel time of 2 minutes.

14. $B(t) = \dfrac{150t}{3+t}$ calories with a travel time of 2 minutes or $B(t) = \dfrac{150t}{4+t}$ calories with a travel time of 3 minutes.

15. $B(t) = \dfrac{250t}{5+t}$ calories with a travel time of 2 minutes or $B(t) = \dfrac{150t}{4+t}$ calories with a travel time of 3 minutes.

16. $B(t) = \dfrac{250t}{4+t}$ calories with a travel time of 2 minutes or $B(t) = \dfrac{150t}{4+t}$ calories with a travel time of 15 seconds.

In Example 5 (optimal time to harvest), assume that the profit function P(t) has the form specified in Problems 17 to 22. For these profit functions, find the optimal age at which to harvest the stands of trees to maximize profit where t is measured in decades.

17. $P(t) = \dfrac{2t^{5/2}}{1+t^2} - 1$ whenever $\dfrac{2t^{5/2}}{1+t^2} - 1$ is positive and 0 otherwise.

18. $P(t) = \dfrac{3t^{5/2}}{1+t^2} - 1$ whenever $\dfrac{3t^{5/2}}{1+t^2} - 1$ is positive and 0 otherwise.

19. $P(t) = \dfrac{2t^{5/2}}{1+2t^2} - 1$ whenever $\dfrac{2t^{5/2}}{1+2t^2} - 1$ is positive and 0 otherwise.

20. $P(t) = \dfrac{3t^{5/2}}{1+2t^2} - 1$ whenever $\dfrac{3t^{5/2}}{1+2t^2} - 1$ is positive and 0 otherwise.

21. $P(t) = \dfrac{5t^{5/2}}{1+2t^2} - 2$ whenever $\dfrac{5t^{5/2}}{1+2t^2} - 2$ is positive and 0 otherwise.

22. $P(t) = \dfrac{4t^{5/2}}{1+2t^2} - 3$ whenever $\dfrac{4t^{5/2}}{1+2t^2} - 3$ is positive and 0 otherwise.

23. Find the optimal rotation period for a forest stand that has a value $V(T) = \left(\dfrac{2T^{5/2}}{1+T^2} - 1 \right)$ when $\delta = 0.2$.

24. Find the optimal rotation period for a forest stand that has a value $V(T) = \left(\dfrac{2T^{5/2}}{1+T^2} - 3/2 \right)$ when $\delta = 0.1$.

25. Find the optimal rotation period for a forest stand that has a value $V(T) = \left(\dfrac{(7/3)T^{5/2}}{1+T^2} - 1 \right)$ when $\delta = 0.15$.

26. Find the optimal rotation period for a forest stand that has a value $V(T) = \left(\dfrac{(5/3)T^{5/2}}{2/3+T^2} - 1 \right)$ when $\delta = 0.1$.

Level 2 APPLIED AND THEORY PROBLEMS

27. At the National Council of Teachers of Mathematics (NCTM) illuminations website, students are encouraged to collect data on how many drops are required to break a blanched peanut in two pieces. The sample data provided at this website are shown in the following graph.

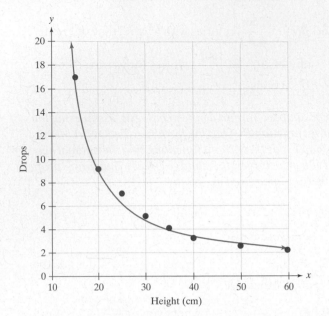

The data can be modeled by the function

$$f(h) = 0.8 + \frac{80}{h - 10} \text{ drops}$$

where h is the height in centimeters. Suppose that the "peanut hummingbird" collects peanuts and wants to minimize the amount of work required to break a peanut into two halves. Determine the height which minimizes the amount of work to break open the peanuts.

28. In Example 4, we found how the optimal residence time in a patch for a great tit depended on the travel time between patches. Although our prediction described the data reasonably well, more than half of the data points lay above the optimal curve. Cowie proposed that part of the reason for this result was that the birds expend energy traveling between patches and searching for food within a patch. In this problem, determine how these expenditures of energy influence the optimal residence time. Let

$$E(t) = 6.3587(1 - e^{-0.0081\,t}) \text{ calories}$$

denote the amount of energy gained by a bird after residing in a patch for t seconds. Assume that the bird requires T seconds to travel the patch. Cowie found that great tits expend approximately 0.697 calories per second while traveling between patches and expend approximately 0.155 calories per second while searching for food in a patch.

a. Write a function $V(t)$ that represents the average gain in energy in a patch after residing there for $t \geq 0$ seconds.

b. Use the marginal value theorem to find an expression relating the optimal residence time t to the travel time T.

c. Compare your solution to the solution found in Example 4.

29. Suppose the crop developed by plant geneticists, as discussed in Example 2—that is, the weight of the crop t days after planting satisfies the growth equation $w(t) = 5 + 400t - t^2$—is grown in a location that is relatively pest free, so that the proportion of germinating seeds is

$$P(t) = \frac{900}{900 + t}$$

The crop, however, must be harvested on or before the first frosty day of fall. Suppose the crop has relative value 1 when harvested at its optimum time of maturity, as represented by the day T on which the yield $Y = aw(t)P(t)$ is maximized and that this value is reduced by $10t_e\%$, where t_e is the number of days prior to T that harvest actually occurs, so that the relative crop value is 0 if harvested at time $T - 10$ or earlier.

a. What is the value of T?

b. If the expected number of days t_s in the growing season—that is the number of frost-free days plus 1—is equally likely to fall on any day from day 165 to day 190, then what is the expected value of the harvest in any year?

30. When a codling moth larva hatches from its egg, it goes looking for an apple. The period between hatching and finding an apple is called the *search period*. The search period S seems to be a function of the temperature, as shown in Table 4.5.

Table 4.5 Search period for the codling moth

Temperature	S, in days
20° C	0.129
21° C	0.122
22° C	0.116
23° C	0.112
24° C	0.109
25° C	0.106
26° C	0.105
27° C	0.104
28° C	0.104
29° C	0.105
30° C	0.106

Source: P.L. Shaffer and H.J. Gold, 1985. "A simulation model of population dynamics of the codling moth, *Cydia pomonella*" *Ecological Modeling* 30:247–274.

Following the lead of Shaffer and Gold (see Section 4.2, Example 8), find $1/S$ for each data value and then use technology to fit a quadratic function to the data. Find the largest and smallest value of this fitted function S.

31. In a plantation of a particular species of trees, a forest economist estimated the number of board feet that can be harvested as a function of the age of the plantation. Data are given in Table 4.6. By using your technology to fit a quadratic function to the data, estimate at what age the plantation should be harvested to maximize the yield of board per acre.

Table 4.6 Harvest yield for a lumber crop

Age (years)	Yield (board feet per acre)
15	6013
20	7021
25	8793
30	9411
35	9786
40	9958
45	9921
50	9766

32. By using your technology to fit a cubic equation to the data in Problem 31, find the age in [15, 50] at which the plantation represented by the data should be harvested to maximize the yield.

33. By using your technology to fit a quartic equation to the data in Problem 31, find the age in [15, 50] at which the plantation represented by this data should be harvested to maximize the yield.

34. By using your technology to fit a quintic equation to the data in Problem 31, find the age in [15, 50] at which the plantation represented by the data should be harvested to maximize the yield.

4.5 Linearization and Difference Equations

As we have seen in earlier chapters, difference equations $x_{n+1} = f(x_n)$ are useful for describing biological dynamics. The simplest dynamics occur at an equilibrium, because by definition, these equilibria are the solutions of the difference equation that remain constant for all time. Specifically, if a given first value x_0 satisfies $x_0 = f(x_0)$, then our difference equation implies $x_1 = x_0$. Repeated application results in $x_{n+1} = x_n = \cdots = x_0$ for all $n > 0$.

While equilibria may be easily identified, as discussed in Section 1.7, by solving the equation $x = f(x)$, their biological relevance depends on their *stability*. Many biological systems, when perturbed, naturally return to their equilibrium state. The temperature of the human body is a case in point. If a person's temperature is perturbed because of an infection, it returns to its equilibrium value of 98.6 °F once he or she is well again. Not all equilibria, however, are stable. If we stand up a six-month-old child, the child may stay upright for a second or two, but until the child is around a year old, he or she will soon fall over. Standing vertically without feedback control from muscles constantly moving to correct the tendency to fall over is an unstable situation.

Thus, when a biological system is perturbed away from equilibrium, it may do one of two things. A system may return to the equilibrium state, in which case the equilibrium is considered *stable*. Alternatively, even if the perturbation is small, a system may continue to drift away from the equilibrium. In this case, the equilibrium is *unstable*. In this section, we make the notion of stability precise and provide a simple algebraic method for checking stability—a method that relies on linearizing the difference equation near the equilibrium. These ideas and methods are applied to models of population growth and population genetics.

We conclude the section by considering another application of linearization and difference equations, namely, numerically solving for the roots of a nonlinear equation. This numerical method is an important alternative to the bisection method presented in Example 10 of Section 2.3.

Equilibrium stability

We begin with the following example, which motivates the notion of a stable equilibrium.

Example 1   Logistic equation

In Example 7 of Section 2.5, we introduced the *discrete logistic equation*, which is a simple model of population growth. If x_n denotes the population density (e.g., average number of individuals per acre) in the nth generation, then the model is given by

$$x_{n+1} = x_n + r x_n \left(1 - \frac{x_n}{K}\right) \qquad x_0 \text{ specified}$$

where r is the per capita growth rate at low densities and K is the carrying capacity of the population. The only two equilibrium solutions are $x = 0$ and $x = K$. For $K = 100$ and $r = 0.5, 1.5, 3.0$, simulate the model for the initial condition $x_0 = 99$. Discuss what you find.

Solution   Simulating the model with $x_0 = 99$ for twenty-five iterations yields the following figures:

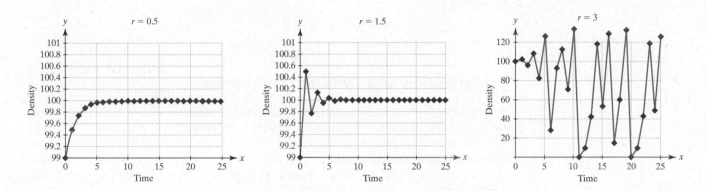

When $r = 0.5$, the population density gradually increases from the density 99 to the equilibrium density 100. When $r = 1.5$, the population density exhibits oscillations that dampen toward the equilibrium density 100. When $r = 3$, the population exhibits irregular oscillations that never approach the equilibrium density 100, despite having started near this equilibrium density.

Example 1 illustrates that some solutions starting near the equilibria approach the equilibrium, while other solutions starting near an equilibrium move away. These observations suggest the following definitions.

Equilibrium Stability

An equilibrium to $x_{n+1} = f(x_n)$, that is a solution satisfying $x^* = f(x^*)$, is:

stable provided that there exists an open interval (a, b) containing x^* such that $\lim_{n \to \infty} x_n = x^*$ and x_1 lies in (a, b); whenever x_0 lies in (a, b);

unstable provided that there is an interval (a, b) containing x^* such that all solutions x_n eventually leave (a, b) whenever x_0 lies in (a, b) but $x_0 \neq x^*$;

Note: In the definition of stability, the second condition that "x_1 lies in (a, b) whenever x_0 lies in (a, b)" is equivalent to the image of (a, b) under f lies in (a, b).

Stated more simply, stability of an equilibrium means that if the solution starts near the equilibrium, then it remains nears the equilibrium and asymptotically approaches the equilibrium. Alternatively, solutions starting near (but not at) an unstable equilibrium eventually move further away from the unstable equilibrium.

Example 2   Stability the hard way

Find the equilibria of the following difference equations and verify stability using the definitions of stable and unstable.

a. $x_{n+1} = \dfrac{x_n}{2}$

b. $x_{n+1} = x_n^2$

Solution

a. The equilibria are given by solutions of $x = x/2$. Hence, the only equilibrium is $x = 0$. Given any x_0, and using the methods developed in Section 1.7, it follows that $x_n = \dfrac{1}{2^n} x_0$. Therefore, given any $a > 0$, we get that $\lim\limits_{n \to \infty} x_n = 0$ for any x_0 in $(-a, a)$. In addition, the image of $(-a, a)$ under f is $\left(-\dfrac{a}{2}, \dfrac{a}{2}\right)$. Therefore, x^* is stable.

b. The equilibria of $x_{n+1} = x_n^2$ must sastisfy $x = x^2$. Hence, the equilibria are given by $x = 0$ and $x = 1$. For any x_0, we have that $x_1 = x_0^2, x_2 = x_1^2 = x_0^4, x_3 = x_2^2 = x_0^8$. Hence, $x_n = x_0^{2n}$. If x_0 lies in the interval $(-1, 1)$, then $\lim\limits_{n \to \infty} x_0^{2n} = 0$. Moreover, the image of $(-1, 1)$ under the function $f(x) = x^2$ is $[0, 1)$, which lies in $(-1, 1)$. Hence, 0 is a stable equilibrium for this difference equation. For any $x_0 > 1$ or $x_0 < -1$, $x_n = x_0^{2n}$ approaches $+\infty$ as n approaches ∞. Hence, for any initial condition near 1, the solution moves away from 1 so that the equilibrium 1 is unstable.

Example 3   Stability of linear difference equations

Consider the linear difference equation

$$x_{n+1} = r x_n$$

For this difference equation, the origin, $x = 0$, is always an equilibrium. Determine for which r values, the origin is stable or unstable.

Solution
The solution of this difference equation is given by $x_n = r^n x_0$. Suppose $x_0 \neq 0$. If $|r| < 1$, then $|x_n| = |r|^n |x_0|$ is decreasing to zero at a geometric rate. Therefore, if $|r| < 1$, then $x = 0$ is stable. Alternatively, if $|r| > 1$, then $|x_n| = |r|^n |x_0|$ is increasing without bound. Hence, $x = 0$ is unstable when $|r| > 1$. If $r = 1$, then $x_n = x_0$ for all n. Hence, x_n neither approaches or moves away from 0, so that 0 is neither stable or unstable when $r = 1$. Similarly, if $r = -1$, you can show that $x = 0$ is neither stable nor unstable.

Stability via linearization

Stability of an equilibrium can be verified directly using the definition, but this method can be challenging. To make things easier, we take advantage of linearization and our work in Example 3.

Suppose x^* is an equilibrium of $x_{n+1} = f(x_n)$ and f is differentiable at x^*. If we approximate f by its tangent line at $x = x^*$, we get

$$f(x) \approx f(x^*) + f'(x^*)(x - x^*)$$
$$= x^* + f'(x^*)(x - x^*) \quad \text{since } f(x^*) = x^*$$

Using this linear approximation and setting $r = f'(x^*)$, we get

$$x_{n+1} \approx x^* + r(x_n - x^*)$$

Equivalently,

$$(x_{n+1} - x^*) \approx r(x_n - x^*)$$

Using the change of variables $y_n = x_n - x^*$, we get

$$y_{n+1} \approx r y_n$$

Example 3 suggests that if $|r| < 1$ and x_0 is sufficiently close to x^*, then we expect that y_n approaches zero at a geometric rate. Equivalently, since we defined $y_n = x_n - x^*$, it follows that x_n approaches x^* at a geometric rate. Alternatively, if $|r| > 1$ and x_0 is sufficiently close to x^*, then we expect that y_n increases initially at a geometric rate. Equivalently, x_n initially moves away from x^* at a geometric rate. As it turns out, all of these statements hold provided that x_n is sufficiently close to x^*, as the following theorem states.

Theorem 4.3   Stability via linearization theorem

If $x_{n+1} = f(x_n)$ has an equilibrium at $x = x^*$ and $r = f'(x^*)$ exists, then x^* is stable if $|r| < 1$ and unstable if $|r| > 1$.

Theorem 4.3 is inconclusive about stability if $|f'(x^*)| = 1$.

Example 4   Logistic revisited

Consider the logistic difference equation

$$x_{n+1} = x_n + r x_n \left(1 - \frac{x_n}{100}\right)$$

with $r > 0$. Determine for which r values $x^* = 100$ is stable.

Solution   Let $f(x) = x + r x \left(1 - \frac{x_n}{100}\right)$. To determine whether an equilibrium is stable or not, we need to compute

$$f'(x) = 1 + r - \frac{rx}{50} = 1 + r\left(1 - \frac{x}{50}\right)$$

and evaluate at $x = 100$

$$f'(100) = 1 + r(1 - 2) = 1 - r$$

For stability, we need that $|1 - r| < 1$. Equivalently,

$$-1 < 1 - r < 1$$
$$-2 < -r < 0$$
$$2 > r > 0$$

Hence, the equilibrium $x^* = 100$ is stable provided that $0 < r < 2$ and unstable provided that $r > 2$. This conclusion is consistent with the simulations in Example 1. Indeed, for $r = 0.5$ and $r = 1.5$, the simulations approached the equilibrium $x^* = 100$. However, for $r = 3$, the simulation oscillated irregularly and never converged to any density. ∎

Example 5   Stability of insect population dynamics

Biology professor T. S. Bellows investigated the ability of several different difference equation models to describe the population dynamics of various insect species. (See T. S. Bellows, "The Descriptive Properties of Some Models for Density Dependence."

The Journal of Animal Ecology 50(1) (Feb. 1981), 139–156) He found that the so-called generalized Beverton-Holt model provided the best mathematical description for the insect species that he studied. If x_n denotes the population density in the nth generation, then the model is of the form

$$x_{n+1} = \frac{r x_n}{1 + x_n^b}$$

where r is the intrinsic fitness of the population and b measures the abruptness of density dependence. For three insect species, Bellows made the following parameter estimates:

- Budworm moth: $r = 3.5$ and $b = 2.7$
- Colorado potato beetle: $r = 75$ and $b = 4.8$
- Meadow plant bug: $r = 2.2$ and $b = 1.4$

These insects are shown in Figure 4.20.

Figure 4.20 Photos of the meadow plant bug (left), Colorado potato beetle (center), and the budworm moth (right).

a. Use these parameter estimates to determine which population, according to the model, supports a stable equilibrium.

b. For the species that do not support a stable equilibrium, simulate their dynamics.

Solution

a. To begin with, we need to find the equilibria of the model that must satisfy $x = r\dfrac{x}{1 + x^b}$. Equivalently, $x = 0$ or

$$1 = \frac{r}{1 + x^b}$$
$$1 + x^b = r$$
$$x^b = r - 1$$
$$x = (r - 1)^{1/b}$$

Hence, for the budworm moth, the equilibria are given by

$$x = 0 \text{ and } x = 2.5^{1/2.7} \approx 1.40$$

For the Colorado potato beetle, the equilibria are given by

$$x = 0 \text{ and } x = 74^{1/4.8} \approx 2.45$$

For the meadow plant bug, the equilibria are given by

$$x = 0 \text{ and } x = 1.2^{1/1.4} \approx 1.14$$

Let $f(x) = \dfrac{rx}{1 + x^b}$. To determine the stability of these equilibria, we need to evaluate the derivative

$$f'(x) = r\frac{1 + x^b - bx^{b-1}x}{(1 + x^b)^2}$$

$$= r\frac{1 + (1 - b)x^b}{(1 + x^b)^2}$$

at the equilibria. Since $f'(0) = r$ and $r > 1$ for all three species, 0 is an unstable equilibrium for all three species.

For the budworm moth, $f'(1.4) \approx -0.93$. Since $|-0.93| = 0.93 < 1$, the equilibrium $x \approx 1.4$ is stable for the budworm moth model. For the Colorado potato beetle, $f'(2.45) \approx -3.75$. Since $|-3.75| > 1$, the equilibrium $x \approx 2.45$ is unstable for the Colorado potato beetle model. Therefore, the Colorado potato beetle model has no stable equilibria. For the meadow plant bug, $f'(1.14) \approx 0.24$. Since $0.24 < 1$, the equilibrium $x \approx 1.14$ is stable for the meadow plant bug model.

b. Since all of the equilibria for the Colorado potato beetle model are unstable, we can ask the question: What is the long-term behavior of a nonequilibrium solution? Simulating the model with $x_0 = 2.4$ (a value "close to" the equilibrium 2.45) yields the following numerical solution.

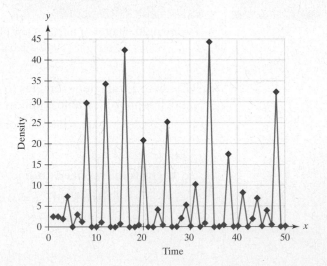

This figure suggests that the Colorado potato beetle is subject to episodic population outbreaks, which is a characteristic associated with agricultural insect pests. ■

Stability of monotone difference equations

A special, and important, class of difference equations $x_{n+1} = f(x_n)$ was introduced in Section 2.5. For these difference equations, f is a continuous and increasing function over some domain of interest. Within this domain, solutions to this difference equation are *monotone* (i.e., either increasing for all n or decreasing for all n). As a consequence of this monotonicity, it is possible to provide a simple graphical approach to stability for these difference equations.

Theorem 4.4   Stability of monotone difference equations theorem

Let f be a continuous, increasing function on an interval (a, b). Let x^* be an equilibrium for $x_{n+1} = f(x_n)$ that lies in (a, b). Then x^* is:

Stable *if $f(x) > x$ for x in (a, x^*) and $f(x) < x$ for x in (x^*, b). In particular,* $\lim_{n\to\infty} x_n = x^*$ *whenever x_0 lies in (a, b).*

Unstable *if $f(x) < x$ for x in (a, x^*) and $f(x) > x$ for x in (x^*, b). In particular, x_n leaves (a, b) for some n whenever x_0 lies in (a, x^*) or (x^*, b).*

Combined with the monotone convergence theorem in Section 2.5, this stability theorem allows us to determine the fate of solutions to difference equations where f is a continuous, increasing function.

Example 6   Graphical approach to stability

Consider the difference equation

$$x_{n+1} = f(x_n)$$

where the graph of f is given by

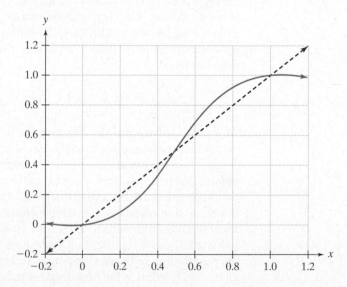

Assuming f is increasing on $[0, 1]$, identify the equilibria and determine their stability.

Solution   The equilibria correspond to points where the graph of $y = x$ intersects with the graph of $y = f(x)$. These intersections occur at $x = 0, \frac{1}{2}$, and 1. Inspection of the graph of f yields that $f(x) > x$ for x in $(-0.2, 0)$ and $(0.5, 1)$. Alternatively, $f(x) < x$ for x in $(0, 0.5)$ and $(1, 1.2)$. Applying the stability theorem for monotone difference equations implies that 0 and 1 are stable, while 0.5 is unstable. Moreover, x_n converges to 0 whenever x_0 lies in $(-0.2, 0.5)$ and x_n converges to 1 whenever x_0 lies in $(0.5, 1.2)$.

We can verify this stability with cobwebbing. Cobwebbing with an x_0 slightly above 0.5 and an x_0 slightly below 0.5 leads to the following figures:

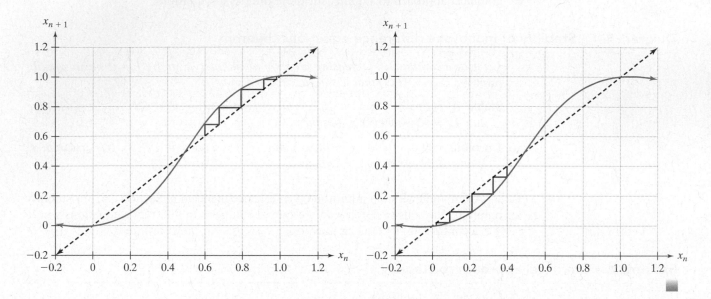

As we saw in Example 5 of Section 1.7, we can construct models that allow us to trace the fate of alleles that code for genes affecting the biological fitness (i.e., the ability to survive and reproduce) of individuals. Recalling our discussion in Section 1.7 regarding genetic models of diploid organisms, we consider an allele that codes for a particular trait. If the frequency of this allele in the population is $x \in [0, 1]$, then a well-known model describing how the frequency of this trait changes from generation n to generation $n + 1$ in a very large (essentially infinite) randomly mating population is

$$x_{n+1} = f(x_n) \quad \text{with} \quad f(x) = \frac{w_1 x^2 + x(1 - x)}{w_1 x^2 + 2x(1 - x) + w_2(1 - x)^2}$$

where w_1 and w_2 are the fitness of individuals who, respectively, have two or no copies, relative to individuals who have only one copy, of the allele in question. Referring back to the equation given in Example 5 of Section 1.7 regarding the spread of a deleterious mutant allele, we see that the equation is the same as the above equation for the special case $w_1 = 0$ and $w_2 = 1$. This case is equivalent to the statement that the allele a in question is recessive (heterozygous Aa individuals are not affected) and lethal (aa individuals die before reproducing). Despite the drastic effect of this lethal allele a, we found that it can take a very long time for it to be eliminated from the population. In the next example, we consider a variant of this model in which the allele that is lethal in the homozygous state actually confers a benefit on an individual when combined with the other allele (i.e., when in the heterozygous state).

Example 7   Fate of the sickle cell allele

In areas of the world where malaria occurs, it is known that individuals who have one sickle cell allele are more resistant to malaria than those who do not have the allele. On the other hand, individuals who have two sickle cell alleles suffer from sickle cell anemia, which can cause premature death. Let x denote the frequency of the allele that does not cause sickle cell anemia. Assume when malaria is prevalent that individuals not protected by the sickle cell allele will, on average, have 10% fewer progeny than individuals who have one sickle cell allele—that is, $w_1 = 0.9$. For the sake of simplicity, we assume that individuals with sickle anemia die before they reproduce (even though, in reality, this assumption is too extreme)—that is, $w_2 = 0$.

a. Write and simplify the difference equation, $x_{n+1} = f(x_n)$, under the assumption that $x \neq 0$.

b. Verify that f is increasing on the interval $(0, 1]$.

c. Find the equilibria on the interval $(0, 1]$ and determine their stability.

d. Interpret your results.

Solution

a. Under the assumption that $w_1 = 0.9$, $w_2 = 0$, and $x \neq 0$, we get

$$f(x) = \frac{0.9\,x^2 + x(1-x)}{0.9\,x^2 + 2x(1-x) + 0 \cdot (1-x)^2}$$

$$= \frac{-0.1\,x^2 + x}{-1.1\,x^2 + 2x}$$

$$= \frac{-0.1\,x + 1}{-1.1\,x + 2} \qquad \text{assuming } x \neq 0$$

b. To verify that $f(x)$ is increasing on the interval, we compute the derivative of $f(x)$:

$$f'(x) = \frac{d}{dx}\left[\frac{-0.1\,x + 1}{-1.1\,x + 2}\right]$$

$$= \frac{-0.1(-1.1\,x + 2) + 1.1(-0.1\,x + 1)}{(2 - 1.1\,x)^2}$$

$$= \frac{0.11\,x - 0.2 - 0.11\,x + 1.1}{(2 - 1.1\,x)^2}$$

$$= \frac{0.9}{(2 - 1.1\,x)^2}$$

Hence, $f'(x) > 0$ on $(0, 1]$ and f is increasing on this interval.

c. To find the equilibria in $(0, 1]$, we solve $x = f(x)$:

$$x = \frac{-0.1\,x + 1}{-1.1\,x + 2} \qquad \text{by definition of equilibrium}$$

$$-1.1\,x^2 + 2x = -0.1\,x + 1$$

$$0 = 1.1\,x^2 - 2.1\,x + 1 = (1.1\,x - 1)(x - 1)$$

Hence, the equilibria are given by $x = 1$ and $x = 1/1.1 \approx 0.91$.

To determine their stability, we can use the derivative calculated in part **b**. We have $f'(1) = \dfrac{0.9}{0.9^2} = \dfrac{1}{0.9} \approx 1.11$. Hence, $x = 1$ is unstable. Alternatively, $f'(1/1.1) = 0.9$ so that $x = 0.91$ is stable. In fact, these calculations imply that $f(x) < x$ on the interval $(0.91, 1)$ and $f(x) > x$ on the interval $(0, 0.91)$. Hence, the stability theorem for monotone difference equations implies that $\lim_{n \to \infty} x_n = 0.91$ whenever x_0 lies in $(0, 1)$.

d. The results imply that as long as both alleles are present in the population, they will persist, and the frequency of the sickle cell anemia allele will approach a value of $1 - 1/1.1 \approx 0.09$. Hence, under the assumptions made, we expect approximately 9% of this population to have the sickle cell allele. ∎

Newton's method

The final application of linearization to difference equations is to illustrate the inner workings of an algorithm called **Newton's method**. This method is used to find the roots of nonlinear algebraic equations of the form $f(x) = 0$ that are too difficult or impossible to solve algebraically. The algorithm is based on the *Newton-Raphson* difference equation, which we now describe. Suppose our initial guess for the solution of $f(x) = 0$ is $x = x_0$. Assuming that this guess is not the solution, we need to find an improved guess for the root. Since the nonlinear function in question is too hard to manipulate by hand, we consider the linear approximation of $y = f(x)$ at $x = x_0$:

$$y = f(x_0) + f'(x_0)(x - x_0)$$

To get our next guess, x_1, for the solution to $f(x) = 0$, we set $x = x_1$ and $y = 0$ in the linear approximation and solve for x_1:

$$0 = f(x_0) + f'(x_0)(x_1 - x_0)$$
$$-f(x_0)/f'(x_0) = x_1 - x_0 \qquad \text{assuming that } f'(x_0) \neq 0$$
$$x_1 = x_0 - f(x_0)/f'(x_0)$$

To get the next guess, x_2, we can proceed similarly to get the equation

$$x_2 = x_1 - \frac{f(x_1)}{f'(x_1)}.$$

Proceeding inductively, we get the following difference equation:

$$x_{n+1} = F(x_n) \quad \text{where} \quad F(x) = x - \frac{f(x)}{f'(x)} \quad \text{and} \quad f'(x) \neq 0.$$

This difference equation is illustrated in Figure 4.21. In this figure, r is a root of the equation $f(x) = 0$.

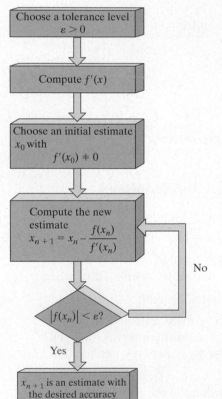

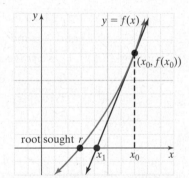
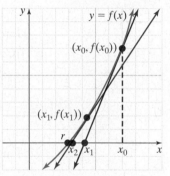

a. Estimating a root, r, of $y = f(x)$ **b.** First and second estimates

Figure 4.21 Graphical representation of Newton's method.

One of the key requirements of the method is to start with a reasonable guess x_0 for the root x^* because the closer x_0 is to x^* the more likely the solution will converge to x^*. The following theorem implies that if the sequence converges, then it converges to a root.

Theorem 4.5   Newton's method

Let $f(x)$ be a continuously differentiable function with $f'(x) \neq 0$. Any solution to

$$x_{n+1} = x_n - \frac{f(x_n)}{f'(x_n)} \qquad f'(x_n) \neq 0$$

will approach a limit that is a root of the equation or else will not have a finite limit.

When applying Newton's method, we choose a number $\epsilon > 0$ that determines the allowable tolerance for estimated solutions. Given an appropriate initial guess, x_0, we iteratively compute x_n until $\left| f(x_n) \right| < \epsilon$. This procedure is shown in the following flowchart.

Example 8   Time to tumor regrowth

In Example 4 of Section 4.3, we considered the growth of a tumor in a mouse after the mouse was given a drug treatment. To model the volume of the tumor, we used the function (renaming the variable x rather than t)

$$V(x) = 0.005e^{0.24x} + 0.495e^{-0.12x} \text{ cm}^3$$

where x is measured in days after the drug was administered. Using Newton's method, solve for the time $x > 0$ at which the tumor regrows to its starting volume of 0.5 cm^3. Find this time within a tolerance of 0.01.

Solution   We want to find a root of

$$f(x) = V(x) - 0.5 = 0.005e^{0.24x} + 0.495e^{-0.12x} - 0.5$$

To use Newton's method, we need to compute the first derivative:

$$f'(x) = 0.0012e^{0.24x} - 0.0594e^{-0.12x}$$

We will see what happens if we start with $x_0 = 20$, although other start values close to the anticipated solution can be chosen. To find x_1, we compute

$$x_1 = x_0 - \frac{f(x_0)}{f'(x_0)} = 20 - \frac{f(20)}{f'(20)} \approx 18.914$$

Since $|f(x_1)| \approx 0.02 > 0.01$, our stopping criterion has not been met and we compute

$$x_2 = x_1 - \frac{f(x_1)}{f'(x_1)} = 18.914 - \frac{f(18.914)}{f'(18.914)} \approx 18.732$$

Since $f(18.732) \approx 0.0004$, the stopping criterion has been met and our answer is $x = 18.732$.

Implementation of Newton's method for finding roots is widespread, as a quick search of the Web will reveal. Several websites will turn up that allow you to input a function, an initial condition, and the number of iterations, and you will obtain the corresponding sequence from Newton's method.

Newton's method may not converge to a solution, as shown by the following example.

Example 9   Nonconvergence of Newton's method

Consider the function $f(x) = e^x - 2x$. Use Newton's method with $x_0 = 1$ to find a solution to $f(x) = 0$. Discuss what you find.

Solution   Note that $f'(x) = e^x - 2$ so that

$$x_{n+1} = x_n - \frac{f(x_n)}{f'(x_n)}$$

$$= x_n - \frac{e^{x_n} - 2x_n}{e^{x_n} - 2}$$

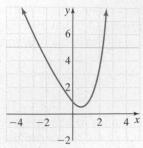

Figure 4.22 Graph of
$y = e^x - 2x$.

If we let $x_0 = 1$, then we find

$$x_1 = 1 - \frac{e^1 - 2(1)}{e^1 - 2} = 0$$

$$x_2 = 0 - \frac{e^0 - 2(0)}{e^0 - 2} = 1$$

$$x_3 = 1 - \frac{e^1 - 2(1)}{e^1 - 2} = 0$$

Note that the values simply alternate, and the method does not converge to a solution. The graph in Figure 4.22 shows why there can be no solution: the graph does not intersect with the x-axis and hence does not have any roots. ■

PROBLEM SET 4.5

Level 1 DRILL PROBLEMS

In Problems 1 to 4, find the equilibria of $x_{n+1} = f(x_n)$ and determine their stability using cobwebbing.

1.

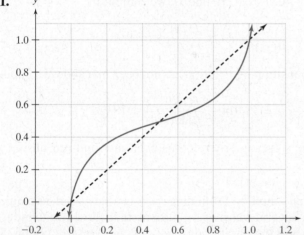

3.

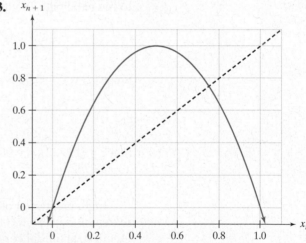

2.

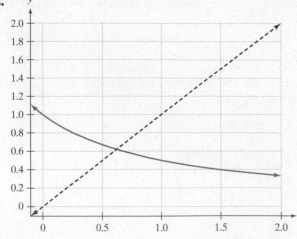

4.

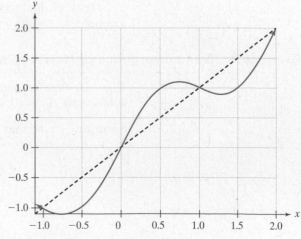

In Problems 5 to 10, find the equilibria of the difference equation. Moreover, use the definitions of an unstable equilibrium and a stable equilibrium to determine their stability.

5. $x_{n+1} = 2x_n$

6. $x_{n+1} = x_n^{1/3}$

7. $x_{n+1} = x_n^{1/2}$

8. $x_{n+1} = x_n^2$

9. $x_{n+1} = 4\sqrt{x_n}$

10. $x_{n+1} = \dfrac{x_n}{2 + 2x_n}$

In Problems 11 to 16, find the equilibria of the difference equation. Moreover, use linearization to determine their stability.

11. $x_{n+1} = x_n^2$

12. $x_{n+1} = \dfrac{x_n}{2 + 2x_n}$

13. $x_{n+1} = \dfrac{2x_n}{1 + 2x_n}$

14. $x_{n+1} = 2x_n(1 - x_n)$

15. $x_{n+1} = 4x_n(1 - x_n)$

16. $x_{n+1} = \dfrac{1}{1 + x_n}$

17. Consider the following alternative formulation of the logistic difference equation, which is different from the formulation presented in Example 1: $x_{n+1} = rx_n(1 - x_n/100)$ with $r > 0$.

   a. Find the equilibria.

   b. Determine under what conditions the origin is stable.

   c. Determine under what conditions the nonzero equilibrium is positive.

   d. Determine under what conditions the nonzero equilibrium is stable.

18. Consider the logistic difference equation $x_{n+1} = rx_n(1 - x_n/50)$ with $r > 0$.

   a. Find the equilibria.

   b. Determine under what conditions the origin is stable.

   c. Determine under what conditions the nonzero equilibrium is positive.

   d. Determine under what conditions the nonzero equilibrium is stable.

19. Consider the Beverton-Holt difference equation $x_{n+1} = \dfrac{rx_n}{1 + x_n}$ with $r > 0$.

   a. Find the equilibria.

   b. Determine under what conditions the origin is stable.

   c. Determine under what conditions the nonzero equilibrium is positive.

   d. Determine under what conditions the nonzero equilibrium is stable.

20. Consider the Beverton-Holt difference equation $x_{n+1} = \dfrac{rx_n}{1 + 2x_n}$ with $r > 0$.

   a. Find the equilibria.

   b. Determine under what conditions the origin is stable.

   c. Determine under what conditions the nonzero equilibrium is positive.

   d. Determine under what conditions the nonzero equilibrium is stable.

Following the approach laid out in Example 7 (i.e., graphing and using Theorem 4.4) investigate the fate of an allele in a large, randomly mating population when the fitnesses of individuals with two and zero copies of the allele relative to those that have one copy are given in Problems 21 to 24.

21. $w_1 = 1/2$ and $w_2 = 1/2$

22. $w_1 = 2$ and $w_2 = 1$

23. $w_1 = 1/2$ and $w_2 = 2$

24. $w_1 = 2$ and $w_2 = 2$

Use Newton's method to estimate a root of the equations in Problems 25 to 32. Use x_0 as a starting value and iterate twenty times.

25. $x^2 - 2 = 0$, $x_0 = 1$

26. $x^2 + 2 = 0$, $x_0 = 1$

27. $x^3 - x + 1 = 0$, $x_0 = -1$

28. $x^4 + 2x - 1 = 1$, $x_0 = 1$

29. $\cos x = x$, $x_0 = 1$

30. $\sin x + 0.1 = x^2$, $x_0 = 0$

31. $e^x - 5x = 0$, $x_0 = 0$

32. $e^x + x = 0$, $x_0 = -1$

33. Let $f(x) = -2x^4 + 3x^2 + \dfrac{11}{8}$

   a. Show that the equation $f(x) = 0$ has at least two solutions.

   b. Use $x_0 = 2$ and Newton's method to estimate a root of the equation $f(x) = 0$.

   c. Show that Newton's method fails if you choose $x_0 = \dfrac{1}{2}$ as the initial estimate.

34. Let $f(x) = x^6 - x^5 + x^3 - 3$

   a. Show that the equation $f(x) = 0$ has at least two solutions.

   b. Use $x_0 = 2$ and Newton's method to find a root of the equation $f(x) = 0$.

   c. Show that Newton's method fails if you choose $x_0 = 0$ as the initial estimate.

Level 2 APPLIED AND THEORY PROBLEMS

35. For the beetle species *Lasioderma serricorne*, Bellows found that the fraction $f(x)$ of eggs surviving as a function of their initial density x is well described by

$$f(x) = \frac{0.806x}{1 + (0.0114x)^{7.53}}$$

A graph of this function and the corresponding data are shown below:

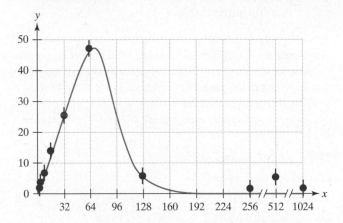

If we assume that each adult produces two eggs, then the dynamics of the population is given by

$$x_{n+1} = 2f(x_n)$$

a. Find the equilibria and determine their stability.

b. Simulate the model with $x_0 = 0.1$

36. For the flour beetle species *Tribolium castaneum*, Bellows found that the fraction $f(x)$ of eggs surviving as a function of their initial density x is well described by

$$f(x) = \frac{0.8x}{1 + (0.0149x)^{4.21}}$$

A graph of this function and the corresponding data are shown below:

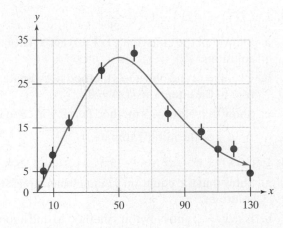

If we assume that each adult produces r eggs, then the dynamics of the population is given by

$$x_{n+1} = rf(x_n)$$

a. Find the equilibria and determine their stability for $r = 2, 4, 6$.

b. Simulate the model with $x_0 = 0.1$ for $r = 2, 4, 6$.

37. Consider the genetic model

$$p_{n+1} = \frac{w_1 p^2 + p(1 - p)}{w_1 p^2 + 2p(1 - p) + w_2(1 - p)^2}$$

Show that the following statements are true:

a. This model has three equilibrium solutions:
$$p = 0, \ p = 1, \text{ and } p^* = \frac{w_2 - 1}{w_1 + w_2 - 2}.$$

b. $p = 1$ is stable and $p = 0$ is unstable when $w_1 > 1 > w_2 > -1$.

c. (Harder problem) p^* as defined in part **a** is the only stable equilibrium when $w_1 < 1$ and $w_2 < 1$ (a condition known as *heterozygote superiority*).

d. (Harder problem) p^* as defined in part **a** is the only unstable equilibrium when $w_1 > 1$ and $w_2 > 1$ (a condition known as *inbreeding depression*).

38. It can be shown that the volume of a spherical cap is given by

$$V = \frac{\pi}{3} H^2 (3R - H)$$

where R is the radius of the sphere and H is the height of the cap, as shown in Figure 4.23.

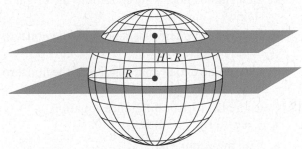

Figure 4.23 Spherical segment is the portion of a sphere between two parallel planes

If $V = 8$ and $R = 2$, use Newton's method to estimate the corresponding H.

39. ℌistorical Quest

Courtesy of Karl Smith

Archimedes
287–212 BC

The Greek geometer Archimedes is acknowledged to be one of the greatest mathematicians of all time. Ten treatises (as well as traces of some lost works) of Archimedes have survived the rigors of time and are masterpieces of mathematical exposition. In one of these works, *On the Sphere and Cylinder*, Archimedes asks where a sphere should be cut in order to divide it into two pieces whose volumes have a given ratio.

   Show that if a plane at distance d from the center of a sphere with $R = 1$ divides the sphere into two parts, one with volume five times that of the other, then

$$3H^3 - 9H^2 + 2 = 0$$

where $H = 1 - d$. Find d by using the Newton-Raphson method to estimate H. (*Hint:* see Problem 38.)

40. Suppose the plane in Problem 39 is located so that it divides the sphere in the ratio of 1:3. Find an equation for d, and estimate the value of d using Newton's method.

41. In Example 4 of Section 4.3, we considered the growth of a tumor in a mouse after the mouse was given a drug treatment. To model the volume of the tumor, we used the function

$$V(x) = 0.005e^{0.24x} + 0.495e^{-0.12x} \text{ cm}^3$$

where x is measured in days after the drug was administered. Using Newton's method, solve within a tolerance of 0.01 for the time x at which the tumor volume has doubled in volume. For an initial guess, use $x = 25$ days.

42. In Example 4 of Section 4.3, we considered the growth of a tumor in a mouse after the mouse was given a drug treatment. To model the volume of the tumor, we used the function

$$V(x) = 0.005e^{0.24x} + 0.495e^{-0.12x} \text{ cm}^3$$

where x is measured in days after the drug was administered. Using Newton's method, solve within a tolerance of 0.01 for the time x at which the tumor volume has quadrupled in volume. For an initial guess, use $x = 30$ days.

43. In Problem 25 in Problem Set 4.3, you found that the volume of a tumor for mice under a different drug regimen was

$$V(x) = 0.0044\,e^{0.239x} + 0.4356e^{-0.111x} \text{ cm}^3$$

where x is days after treatment. Using Newton's method, solve within a tolerance of 0.01 for the time x at which tumor volume has regrown to its original volume. For an initial guess, use $x = 20$ days.

44. In Problem 25 in Problem Set 4.3, you found that the volume of a tumor for mice under a different drug regimen was

$$V(x) = 0.0044\,e^{0.239x} + 0.4356e^{-0.111x} \text{ cm}^3$$

where x is days after treatment. Using Newton's method, solve within a tolerance of 0.01 for the time x at which tumor volume has doubled in volume. For an initial guess, use $x = 25$ days.

45. Show that for different initial values Newton's method converges to a unique solution for the function

$$y = x^3 - 3x^2 + 2x + 0.4$$

but yet converges to one of three solutions for the function

$$y = x^3 - 3x^2 + 2x + 0.3$$

Why is this the case?

46. According to an online article in the *New Scientist* (Catherine Brahic, "Carbon Emissions Rising Faster Than Ever," p. 9), recent research suggests that stabilizing carbon dioxide concentrations in the atmosphere at 450 parts per million (ppm) could limit global warming to 2°C. In Section 1.2, we modeled carbon dioxide concentrations in the atmosphere with the following function (which we now present to higher precision to make more transparent the numerical details of the convergence process):

$$f(x) = 0.122463x + 329.253 + 3\cos\frac{\pi x}{6} \text{ ppm}$$

where x is months after April 1974. In Example 11 of Section 2.3, we used the bisection method to estimate the first time that the model predicts carbon dioxide levels of 450 ppm. Use Newton's method to estimate this time with a stopping value of $\epsilon = 0.001$.

47. Repeat Problem 46 except estimate the first time until reaching 400 ppm.

CHAPTER 4 REVIEW QUESTIONS

1. Use the first derivative test and the second derivative test to find and classify all the extrema of $g(x) = x^3 - 3x - 4$.

2. Find the global maximum and global minimum of $f(x) = \sqrt{x}e^{-x}$ on $[0, 6]$.

3. Using asymptotes and first derivatives, graph $f(x) = \dfrac{x^3 + 3}{x(x + 1)(x + 2)}$ by hand and then check it using a calculator.

4. Consider the family of curves
$$y^2 = x^3 + x^2 + bx + 2b$$
Using calculus, graph the curves for the given values of b.

   a. $b = 0$

   b. $b = 0.05$

   c. $b = 0.01$

   d. $b = -0.05$

   e. $b = -0.1$

5. The canopy height (in meters) of a tropical grass may be modeled by (for $0 \le t \le 30$)
$$h(t) = 0.0000071t^3 - 0.0015852t^2 + 0.1419159t + 3.14$$
where t is the number of days after mowing.

   a. Sketch the graph of $h(t)$.

   b. When was the canopy height growing most rapidly? Least rapidly?

6. Find the value of r in the function $f(x) = e^{-rx}$ that provides the best fit through a semi-log plot of the points $\{(0, 1), (1, 0.6), (2, 0.4), (3, 0.3)\}$.

7. A travel company plans to sponsor a tour to Africa. There will be accommodations for no more than forty people, and the tour will be canceled if no more than ten people book reservations. Based on past experience, the manager determines that if n people book the tour, the profit (in dollars) may be modeled by the function
$$P(n) = -n^3 + 27.6n^2 + 970n - 4{,}235$$
What is the maximum profit?

8. Two towns, A and B, are 12.0 miles apart and are located 5.0 and 3.0 miles, respectively, from a long, straight highway, as shown in Figure 4.24.

Figure 4.24 Highway construction project.

A construction company has a contract to build a road from A to the highway and then to B. Analyze a model to determine the length (to the nearest tenth of a mile) of the shortest road that meets these requirements.

9. A particular plant is known to have the following growth and seed production characteristics: At time of planting ($t = 0$), the seedling has a mass of 3 grams. At time $t > 0$ days after planting, the seedling has grown into a plant that weighs
$$w(t) = 3 + 450t - t^2$$

grams. The plant has a gene that can be manipulated to control the age t at which the plant matures. At maturity the number of seeds $S(t)$ produced by the plant is given by
$$S(t) = \frac{w(t)}{3}$$
A farmer asks a geneticist to genetically engineer a plant line that accounts for the fact that on his farm, because of losses from pests, drought, and disease, a proportion
$$P(t) = \frac{100}{100 + t}$$
of germinating seeds can be expected to develop and survive to age t as plants. What age of maturity should the geneticist select for the plants to maximize the seed production of the mature crop for the farmer?

10. If the value of a forest stand (units are dollars per square meter) is given by the function
$$V(t) = \frac{100}{1 + e^{-0.1(t-40)}} - 1.8$$
where t represents the number of years after the stand has been clear-cut, and the discount is 0.02 per year, solve for the optimal rotation period, assuming that $V(t)$ applies every time the stand is clear-cut.

11. On a particular island in the tropics, scientists have determined that individuals who are not protected by the sickle cell allele will have, on average, 20% fewer progeny than individuals who have one sickle cell allele, while individuals who have two sickle cell alleles will not reproduce.

   a. Write a difference equation for the proportion x_n of non-sickle-cell alleles in the population in generation n, assuming that initially $x_0 > 0$, under assumptions of random mating and random segregation of alleles.

   b. Find the equilibria proportions for x on the interval $(0, 1]$ and determine if they are stable.

   c. Interpret your results.

12. Let $C(t)$ denote the concentration in the blood at time t of a drug injected into the body intramuscularly. In a now classic paper by E. Heinz ("Probleme bei der Diffusion kleiner Substanzmengen innerhalb des menschlichen Körpers," *Biochem. Z*, 319 (1949): 482–492), the concentration was modeled by the function
$$C(t) = \frac{k}{b - a}(e^{-at} - e^{-bt}) \qquad t \ge 0$$
where a, b (with $b > a$), and k are positive constants that depend on the drug. At what time does the largest concentration occur? What happens to the concentration as $t \to +\infty$?

13. Consider a bird that has arrived at a wooded patch with two trees. If the bird spends x minutes foraging for insects on the first tree, she gains $E_1(x) = 200(1 - e^{-x})$ Calories from insects. If the bird spends x minutes on the second tree, she gains $E_2(x) = 100(1 - e^{-x})$ Calories of insects. Assuming the bird has five minutes to spend in the patch, determine the time she should spend on each tree to optimize her energy intake.

14. For the flour beetle species *Tribolium confusum*, Bellows found that the fraction $f(x)$ of eggs surviving as a function of their initial density x is well described by

$$f(x) = \frac{0.61x}{1 + (0.0116x)^{3.12}}$$

A graph of this function and the corresponding data are shown below:

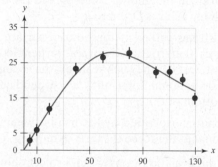

Data Source: T.S. Bellows, "Descriptive Properties of Some Models for Density Dependence," *Journal of Animal Ecology* 50(1)(1981): 139–156.

If we assume that each adult produces r eggs, then the dynamics of the population is given by

$$x_{n+1} = r f(x_n)$$

a. Find the equilibria and determine their stability for $r = 2, 4, 6$.

b. Simulate the model with $x_0 = 0.1$ for $r = 2, 4, 6$.

15. A model for the population growth rate of Eastern Pacific yellowfin tuna is given by

$$G(N) = 2.61N\left(1 - (N/148)^{\theta}\right)$$

where N is measured in thousands of tons, t is measured in years, and $\theta > 0$ is a parameter that determines the strength of density dependence. Find the population size at the maximum sustainable harvesting rate. Discuss what effect θ has on this population size.

16. The energy gain from nectar for a hummingbird flying at $t = 0$ to a flower is shown in the graph below:

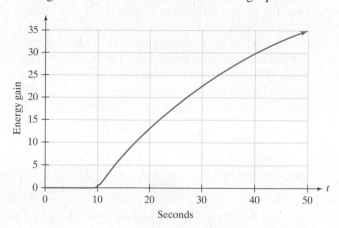

Ignoring the cost of flying constantly, find the time at which the hummingbird's energy gain per unit time is maximal and, consequently, the bird should fly to another flower. Discuss what effect including a fixed energetic cost per unit time would have on your answer.

17. A simple model of a biochemical switch is given by

$$x_{n+1} = \frac{3x_n^2}{1 + x_n^2}$$

where x_n is the concentration of a biochemical at time n.

a. Find the equilibria of this model.

b. Verify that $f(x) = \dfrac{3x^2}{1 + x^2}$ is an increasing function of x for $x > 0$.

c. Discuss what you can say about all nonnegative solutions to this difference equation.

18. Public awareness of a new drug is modeled by

$$P(t) = \frac{5.2t}{0.015t^2 + 0.342} + 0.18$$

where t is the number of months after FDA approval and $P(t)$ is the fraction of people who are aware of the drug and its possible uses.

a. Find the critical points for $P(t)$.

b. Sketch the graph of $P(t)$.

c. At what time, t, during the time interval $0 \le t \le 36$ is $P(t)$ the largest?

19. Suppose that systolic blood pressure of a patient t years old is modeled by

$$P(t) = 38.52 + 21.8\ln(0.98t + 1)$$

for $0 \le t \le 60$, where $P(t)$ is measured in millimeters of mercury.

a. Sketch the graph of $P(t)$.

b. At what rate is $P(t)$ increasing at age t?

20. During the time period 1905–1920, hunters virtually wiped out all large predators on the Kaibab Plateau near the Grand Canyon in northern Arizona. This, in turn, resulted in a rapid increase in the deer population $P(t)$, until food supplies were exhausted and famine led to a steep decline in $P(t)$. A study of this ecological disaster determined that during the time period 1905–1920, the rate of change of the population, $P'(t)$, could be modeled by the function

$$P'(t) = \frac{1}{8}(100 - 5t)t^3$$

$0 \le t \le 20$, where t is the number of years after the base year of 1905.

a. In what year during this period was the deer population the largest?

b. In what year did the rate of growth $P'(t)$ begin to decline?

GROUP PROJECTS

Seeing a project through on your own, or working in a small group to complete a project, teaches important skills. The following projects provide opportunities to develop such skills.

Project 4A   Optimal swimming patterns

In getting from one spot to another, fish have to contend with drag forces and gravity. Drag forces are much greater when a fish is swimming than when it is merely gliding. To reduce the amount of time spent swimming, fish that are heavier than water engage in burst swimming in which they alternate between gliding and swimming upward. This burst swimming leads to a vertical zigzag motion of the fish in the water as shown below:

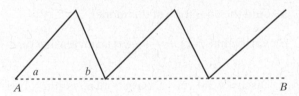

where a is the angle of the upward burst and b is the angle of the downward glide.

In this project, you will investigate the optimal swimming pattern under the following assumptions:

- Throughout its swim, the fish maintains a constant speed s to the right.
- The forces acting on the fish are its weight W relative to the water and drag forces.

- The drag on the gliding fish is D and the drag on the swimming fish is kD where $k \ge 1$.
- The fish has sufficient top/bottom surface area (e.g., a skate) that frictional forces perpendicular to the top/bottom of the fish cancel the component of the gravitational force that is perpendicular to the top/bottom of the fish.
- The energy expended by the fish in swimming is proportional to the force it exerts in moving.

Under these assumptions, your project should do the following:

- Find the ratio of energy in the burst mode to the energy for continuous horizontal swimming from A to B.
- It has been found empirically that $\tan a \approx 0.2$. Given this information, find the optimal value of b for the fish.
- Determine how much energy the fish saves by swimming with this b instead of swimming horizontally.
- Determine how sensitive the amount of energy used is to b, and how sensitive the optimal b is to the estimate of a.

Project 4B   Stability and bifurcation diagrams

Consider the normalized version of the logistic model introduced in the first example in Section 4.5; that is, we set $K = 1$ or, equivalently, interpret the units of x in terms of multiples of K to obtain the equation

$$x_{n+1} = f(x_n) \quad \text{where} \quad f(x) = rx(1 - x)$$

Now explore the behavior of this equation as follows:

1. Solve for the equilibrium solutions as a function of r and determine the stability properties of these equilibria for $r \in [0, 5]$. You will notice that as r increases, an equilibrium solution jumps at some point r_b from being stable on one side of r_b to unstable on the other side of r_b. The value r_b is called a *bifurcation point*.

2. Plot the equilibria in the rx plane (r is the horizontal axis spanning $[0, 5]$) using a solid line to denote where the nontrivial equilibrium solution $\hat{x}$ to $x = f(x)$ is stable and a dotted line where it is unstable.

3. Now consider the equilibria of the iterated logistic map $x_{n+2} = f(f(x_n))$ by constructing (see Section 1.6) the composite map $(f \circ f)(x)$. Use the terminology $f^2 \equiv (f \circ f)$. Find all the equilibria of $f^2(x)$ as a function of r and plot them on the same rx plane as above, but this time plot only the nontrivial stable solutions using a solid line (if you plot where they are unstable, your diagram will become too busy). Note that the equation $x = f^2(x)$ has many more solutions than the equation $x = f(x)$: It has both all the solutions to equation $x = f(x)$ (demonstrate this) and additional solutions that come in pairs, say x^* and x^{**}, such that the sequence $\{x^*, x^{**}, x^*, x^{**}, ...\}$ is a two-cycle solution of the original equation $x = f(x)$ (demonstrate this). Further, if for a particular value of r, x^* and x^{**} are stable equilibrium solutions of $x = f^2(x)$, then the two-cycle $\{x^*, x^{**}, x^*, x^{**}, ...\}$ is

a stable attractor of the equation $x_{n+1} = f(x_n)$. By this, we mean that for any initial condition x_0 starting close to x^* or x^{**}, the resulting sequence generated by our original equation will oscillate between two values that get closer and closer to x^* and x^{**} as time progresses.

4. You have now reached the limit of what you can probably do analytically. Research the literature (a good source is J. D. Murray's book *Mathematical Biology: I, An Introduction*, 3rd ed., New York: Springer-Verlag, 2001). Then discuss what happens as r increases on $[0, 5]$, focusing in terms of bifurcation values at which stable equilibrium solutions of the logistic equation are replaced by stable two-cycles, as well as stable n-cycles for $n > 2$.

5. If you have command of an appropriate technology, use it to summarize graphically your discussion in what is called a *bifurcation diagram*. (Instructions on how to do this are available in textbooks and on the Web, so locate a set of instructions and see if you can follow them.)

Project 4C   Economic production versus ecological welfare

Economic activities, such as extraction and processing of raw materials and the manufacture of finished goods, always result in some damage to the ecosystem. Because of pollution and the destruction of natural habitats, such activities may even severely degrade the ecosystem's delivery of clean water and clean air. Activities may also compromise the ecosystem's ability to produce food and provide a place for relaxation and recreation. In this project, you are asked to use optimization to explore the trade-off between economic production and ecological welfare. (This problem follows Problem II.5 in J. Harte, *Consider a Cylindrical Cow*. Sausalito, University Science Books, 2001.)

Figure 4.25 Industrial pollution.

Suppose the level of economic activity is measured by a variable X, the value of goods and services produced by this activity (also known as *economic output*) is measured by a variable Y, and the value of ecosystem services is measured by an environmental quality variable Z.

A very simple model of human welfare W is based on these assumptions:

- Welfare is proportional to both economic output Y and environmental quality Z.

- Economic output Y is itself proportional to economic activity X and and environmental quality Z. (The first assumption is self-evident and the second arises from the notion that it is much more difficult to produce the same unit of economic output in a poor environment where resources are depleted than in a pristine environment where resources are plentiful.)

- The environment declines from a pristine level linearly with activity X.

These assumptions are equivalent to the following mathematical statement: For positive constants a, b, c, and Z_0, our variables satisfy the equations

$$W = aYZ$$
$$Y = bXZ$$
$$Z = Z_0 - cX$$

1. Demonstrate that human welfare W is maximized at $X^* = Z_0/3c$ and has the maximum value

$$W^* = \frac{4abZ_0^3}{27c}$$

2. Show that the value of economic activity $\widehat{X}$ that maximizes production Y is 1.5 times larger than X^*; that is, $\rho = \widehat{X}/X^* = 3/2$. Further, show that if $\widehat{W}$ is the welfare obtained when production is maximized, then the "cost of greed" (defined to be the ratio $\gamma = \widehat{W}/W^*$) is $\gamma = 27/32$. Discuss the implications of the fact that $\rho > 1$ and $\gamma < 1$.

3. Assume that the economic production level Y has the more general Cobb-Douglas form $Y = bX^\alpha Z^\beta$ than assumed in Equation 4.1, where α and β are nonnegative, empirically determined constants with values that depend on the economic sector under consideration. If, in addition, welfare has the general form $W = aX^\mu Y^\nu$, then find the values of X that maximize both economic output and welfare. Calculate the ratios ρ and γ for this more general case. What do you conclude?

4. Show in the case of the given equations for W, Y, and Z that the level of economic output that maximizes welfare-per-unit-output—that is, the ratio W/Y—is $X = 0$. Does this hold true for the more general case when α, β, μ, and ν are not necessarily 1?

5. Look through the literature and see how many Cobb-Douglas functions you can find and what values of α and β are associated with various sectors of the world economy. Also, see if you can find a real problem where most of the parameters a, b, c, Z_0, α, β, μ, and ν are known. Describe the problem and the values of the parameters. (If one of or more of α, β, μ, and ν are not known, then set them equal to 1, and it is fine if relative rather than global values of the other constants are known or guessed at.) Now calculate the optimum production levels $\widehat{X}$ and X^* with respect to economic output and welfare, respectively, and elaborate in anyway you think appropriate.

CHAPTER 5

Integration

Figure 5.1 The peregrine falcon (*Falco peregrinus*) feeds primarily on pigeons, doves, mice, and shorebirds.

© Marcus Siebert/imagebroker/AGE Fotostock

Preview

"Nature laughs at the difficulties of integration."

Pierre-Simon de Laplace (1749–1827)

Calculus has two parts—*differential calculus*, the topic of the previous chapters, and *integral calculus*. At the core of differential calculus is the concept of the instantaneous rate of change of a function. We have seen how this concept can be used to locally approximate functions and to identify maxima and minima. Integral calculus, on the other hand, deals with accumulated change and, thereby, recovering a function from a mathematical description of its instantaneous rate of change. This recovery process, interestingly enough, is related to the concept of finding the area under a curve.

Using integral calculus, we can calculate the velocity of a stooping peregrine falcon (see Figure 5.1) as it dives toward Earth at great speed to catch prey; calculate the blossom date of a tree as a function of anticipated temperature patterns, so that an orchard can be stocked with bees to facilitate pollination; or estimate the amount of a drug in the bloodstream of a patient connected to an IV.

A systematic method for estimating the area under the curve was devised by Riemann, one of the great mathematicians of the nineteenth century. This method is commonly known as the Riemann sum and yields in the limit an object called the definite integral. The fathers of calculus, Newton and Leibniz, proved a connection between the problem of finding antiderivatives and finding areas under a curve. This connection, the *fundamental theorem of calculus*, which is presented in Section 5.4, helps make calculus one of the most powerful mathematical tools for understanding biological and physical processes.

In Sections 5.5 through 5.7, we provide a short apprenticeship in various techniques used to compute and approximate integrals. Armed with these techniques, the chapter concludes with applications to cardiac output, survival and renewal equations, and the scientific notion of work.

5.1

Antiderivatives

Many mathematical operations have an inverse. For example, to undo the addition of b to a we subtract b: $a + b - b = a$. To undo division of a by b we multiply by b: $\frac{a}{b} b = a$. To undo exponentiation, we take logarithms: $\ln e^a = a$. The process of differentiation can be undone by a process called *antidifferentiation*.

To motivate antidifferentiation, consider how long it takes an organism to develop when the rate of development depends on environmental factors such as heat, light, and humidity. For example, the developmental rate of plants and insects, which lack internal thermal regulation mechanisms, depend critically on ambient temperature. For ambient temperatures within a range defined by *developmental thresholds*, a plant's or insect's organismal developmental rate can often be approximated by an increasing linear function of temperature. For example, Eileen Cullen, a doctoral student at the University of California, collected data shown in Table 5.1 on the developmental rate of a particular species of stinkbug (Figure 5.2) reared in the laboratory.

Figure 5.2 A green stinkbug

Panoramic Images/Getty Images

Table 5.1 Developmental rates of stinkbugs

Temperature (°F)	Developmental rate (1/days)
64.4	1/89
69.8	1/58
80.6	1/37
89.6	1/25

We see from this table that a stinkbug kept at 64.4°F completes $\frac{1}{89}$ th of its development in one day and all of its development in eighty-nine days. Performing linear regression on these data (i.e., to find the "best-fitting" line as discussed in Section 4.3) yields

DEVELOPMENTAL RATE $= -0.06075 + 0.00112 \, T$

where T is temperature in degrees Fahrenheit. This relationship is illustrated in Figure 5.3. If $T(x)$ is the temperature at time x and $F(x)$ denotes the fraction of development completed by the stinkbug at time x, then the preceding equation yields the rate at which $F(x)$ changes with time; that is,

$$F'(x) = -0.06075 + 0.00112 \, T(x)$$

Figure 5.3 Graph of the developmental rate of stinkbugs. The red dots represent the actual data, and the line is the best-fitting line.

Thus, if we know $T(x)$ and want to know how long it takes the stinkbug to complete development, we need to "solve" for $F(x)$. More generally, if we are given that $f(x)$ is the developmental rate at time x, then $F(x)$ must satisfy

$$F'(x) = f(x)$$

Understanding solutions of this equation is the main goal of this section.

Antiderivative	Given a function f, an **antiderivative** F of f is a function F that satisfies $$F'(x) = f(x)$$

For example, x^3 is an antiderivative of $3x^2$ since $\frac{d}{dx} x^3 = 3x^2$. Is x^3 the only antiderivative of $3x^2$? The answer is no. For example x^3, $x^3 + 1$, and $x^3 + \pi$ all have the same derivative $3x^2$. Consequently, all are antiderivatives of $3x^2$. Luckily, all antiderivatives of a function are related. Suppose $F(x)$ and $G(x)$ are antiderivatives of $f(x)$ on some interval. Since $F'(x) = f(x) = G'(x)$, the function

$$H(x) = F(x) - G(x)$$

has derivative

$$H'(x) = f(x) - f(x) = 0$$

on this interval. What functions have a derivative equal to zero on an interval? The mean value theorem implies only the constant function. Hence, there must be a constant C such that $F(x) = G(x) + C$, and we have the following result.

General Form of an Antiderivative

If F is an antiderivative of f on an interval I, then every antiderivative of f on I has the form

$$F(x) + C$$

where C is a constant. For this reason, we call $F(x) + C$ the **general form of the antiderivative**.

Because of this general form, finding the general form of an antiderivative amounts to finding an antiderivative of f and adding an arbitrary constant C.

Example 1   Finding general forms of antiderivatives

Find the general forms of the antiderivatives of

a. e^x **b.** $\cos x$ **c.** x^5

Solution

a. Recall that $\dfrac{d}{dx}e^x = e^x$. Thus, the general form of the antiderivative is $e^x + C$.

b. Recall that $\dfrac{d}{dx}\sin x = \cos x$. Hence, the general form of the antiderivative is $\sin x + C$.

c. Recall that $\dfrac{d}{dx}x^6 = 6x^5$. This is not quite what we want, because we are off by a factor of 6. If we divide both sides of the equation by 6, then

$$\frac{d}{dx}\left(\frac{1}{6}x^6\right) = x^5$$

Thus, the general form of the antiderivative of x^5 is $\dfrac{x^6}{6} + C$. ∎

Warning! What we did in part **c**, namely, divide by 6 because we were off by a factor of 6, worked only because 6 is a constant. It doesn't work in general. For example, suppose we wanted to find an antiderivative of e^{x^2}. It would be incorrect to argue that since $\dfrac{d}{dx}e^{x^2} = 2xe^{x^2}$, we are off by a factor of $2x$ and the antiderivative is $\dfrac{1}{2x}e^{x^2}$. Indeed, $\dfrac{d}{dx}\dfrac{1}{2x}e^{x^2}$ does *not* equal e^{x^2} as you should verify for yourself.

Example 2　Antiderivative of cos(ax)

Find the general form of the antiderivative for $\cos(ax)$, where $a \neq 0$.

Solution　We know $\dfrac{d}{dx}\sin(ax) = a\cos(ax)$, but this is not quite what we want, since we are off by a factor of a. If we divide both sides by a, then

$$\frac{1}{a}\frac{d}{dx}\sin(ax) = \frac{1}{a}a\cos(ax)$$

$$\frac{d}{dx}\frac{\sin(ax)}{a} = \cos(ax)$$

Thus, the general form of the antiderivative of $\cos(ax)$ is $\dfrac{1}{a}\sin(ax) + C$.　■

Corresponding to the many rules of differentiation are rules of antidifferentiation. For instance, if $F(x)$ and $G(x)$ are antiderivatives of $f(x)$ and $g(x)$, respectively, then $H(x) = F(x) + G(x)$ is an antiderivative of $h(x) = f(x) + g(x)$. Indeed, since the derivative of a sum is the sum of the derivatives, we obtain

$$H'(x) = F'(x) + G'(x) = f(x) + g(x) = h(x)$$

In a similar manner, we can show that antiderivatives have the following properties.

Properties of Antiderivatives

Let $F(x)$ and $G(x)$ be antiderivatives of $f(x)$ and $g(x)$, respectively.

Addition $F(x) + G(x)$ is an antiderivative of $f(x) + g(x)$.

Subtraction $F(x) - G(x)$ is an antiderivative of $f(x) - g(x)$.

Scalar multiplication For any constant c, $cF(x)$ is an antiderivative of $cf(x)$.

Combining the antidifferentiation properties and formulas allows us to compute even more antiderivatives, as the following examples illustrate.

Example 3　Using antiderivative rules

Find the general antiderivative of $3x^2 + 3x + 7$.

Solution　Since an antiderivative of a sum is a sum of antiderivatives, an antiderivative of $3x^2 + 3x + 7$ is the sum of antiderivatives of $3x^2$, $3x$, and 7. Antiderivatives of $3x^2$, $3x$, and 7 are x^3, $\dfrac{3}{2}x^2$, and $7x$. Hence, an antiderivative of $3x^2 + 3x + 7$ is $x^3 + \dfrac{3}{2}x^2 + 7x$, and the general form of the antiderivative is $x^3 + \dfrac{3}{2}x^2 + 7x + C$ where C is an arbitrary constant.　■

To find a particular antiderivative $F(x)$ of $f(x)$ on an interval, we need to know a value of $F(x)$ at a particular value of x to determine the particular value of the arbitrary constant C. If we have this information, then finding the antiderivative is known as an *initial value problem*.

Example 4　An initial value problem

Find $F(x)$ such that $F(2) = 1$ and $F'(x) = 3x^2 + 3x + 7$.

Solution　From Example 3, the general form of the antiderivative is $F(x) = x^3 + \dfrac{3}{2}x^2 + 7x + C$. To solve for C, we solve the equation $F(2) = 1$

as follows:

$$F(2) = 2^3 + \frac{3}{2} \times 2^2 + 7 \times 2 + C = 1$$

This implies $28 + C = 1$ or $C = -27$. Thus,

$$F(x) = x^3 + \frac{3}{2}x^2 + 7x - 27$$

Example 5   Stinkbug development

Consider the development of the stinkbug from egg to adult, as summarized in Table 5.1. Suppose the stinkbugs are reared under a temperature $T(x)$ that sinusoidally oscillates between 60° and 80°F each day

$$T(x) = 70 + 10\cos(2\pi x)$$

where x is measured in days. Solve the following:

a. Assuming $F(0) = 0$, find the function $F(x)$ corresponding to the proportion of development completed after x days.

b. Use technology to find the number of days for development to be completed; that is, find x such that $F(x) = 1$.

Solution

a. Recall that we have $F'(x) = -0.06075 + 0.00112\,T(x)$. Substituting $T(x)$ into the expression for $F'(x)$ yields

$$F'(x) = -0.06075 + 0.00112\,(70 + 10\cos(2\pi x))$$
$$= 0.01765 + 0.0112\cos(2\pi x)$$

Since an antiderivative of 0.01765 is $0.01765\,x$ and an antiderivative of $0.0112\cos(2\pi x)$ is $\dfrac{0.0112}{2\pi}\sin(2\pi x)$, it follows that

$$F(x) = 0.01765\,x + \frac{0.0112}{2\pi}\sin(2\pi x) + C$$

To find C, we solve

$$F(0) = 0.01765(0) + \frac{0.0112}{2\pi}\sin(2\pi \times 0) + C = 0$$

which implies $C = 0$. Hence,

$$F(x) = 0.01765\,x + \frac{0.0112}{2\pi}\sin(2\pi x)$$

b. Plotting $F(x)$ as shown below suggests that development is completed in just under fifty-seven days.

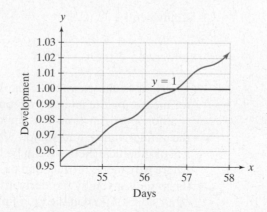

Differential equations and slope fields

An equation that involves derivatives is called a **differential equation**. Consider a function $y = F(x)$. Then any equation of the form $y' = f(x)$, or in Leibniz's notation

$$\frac{dy}{dx} = f(x)$$

is a differential equation, and solving for the antiderivative $y = F(x)$ of $f(x)$ (i.e., $F'(x) = f(x)$) corresponds to solving this differential equation.

In the next chapter, we discuss differential equations in greater detail. Here, we introduce the topic by considering a physiological phenomenon known as the *Weber-Fechner law*. This law describes the expected response of an animal or human subject to a stimulus, such as light or sound. More particularly, the Weber-Fechner law in physiological psychology asserts that when a subject is exposed to a stimulus, y, the rate of change of the response x with respect to y is inversely proportional to x. This statement can be written mathematically as

$$\frac{dy}{dx} = \frac{k}{x}$$

where k a positive constant to be determined through experiment. One can interpret this equation as saying if the stimulus x is small, then small changes in stimulus cause large changes in the response. Alternatively, if stimulus x is large, then small changes in the stimulus do not change the response much.

Example 6   Solving the Weber-Fechner differential equation

Find the solution to the Weber-Fechner equation

$$\frac{dy}{dx} = \frac{k}{x} \qquad k > 0$$

assuming that a threshold stimulus, $x_0 > 0$, is the lowest level for which a response can be detected. In other words, find $y(x)$ subject to the initial condition $y(x_0) = 0$.

Solution   This problem requires us to find a function $y(x)$ such that $y' = k/x$ and $y(x_0) = 0$. Taking the general antiderivative of k/x with respect to x yields

$$y(x) = k \ln x + C$$

where C is a constant. Since $y(x_0) = 0$, solving

$$y(x_0) = k \ln x_0 + C = 0$$

for C yields

$$C = -k \ln x_0$$

Hence, we obtain

$$y(x) = k \ln x - k \ln x_0$$

Equivalently,

$$y(x) = k \ln \left(\frac{x}{x_0} \right) \qquad \text{for } x_0 > 0$$

A particular example of the Weber-Fechner law is the logarithmic decibel scale for measuring the intensity of sound.

For some functions it is impossible to come up with an explicit expression for the antiderivative: for example, $f(x) = e^{-x^2}$ and $f(x) = \sin x^2$. In such cases, numerical or graphical methods can be used. One graphical approach involves slope fields, where we can use the fact that the slope of a function $y = F(x)$ at any point (x, y) on its graph is given by the derivative $F'(x)$. We exploit this fact to obtain a "picture" of all slopes $F'(x)$ on the (x, y)-plane, as illustrated in the next example.

Example 7   Antiderivatives with slope fields

Use technology to sketch the slope field for the equation

$$F'(x) = \sin(x^2)$$

for $0 \le x \le 3$ and $-2 \le y \le 2$. Sketch by hand antiderivatives $F(x)$ satisfying $F(0) = 0$ and $F(0) = -2$, respectively.

Solution   Using technology, we draw small line segments of slope $F'(x) = \sin(x^2)$ at regular intervals in the xy plane, for $0 \le x \le 3$ and $-2 \le y \le 2$, to obtain Figure 5.4a. The collection of these line segments is called a **slope field** or **direction field** of the function f.

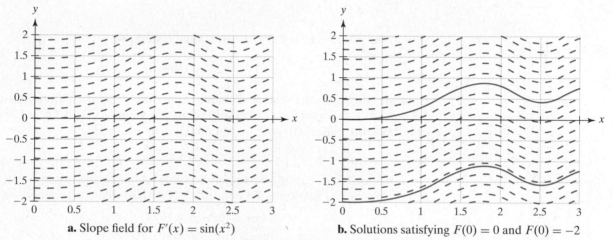

a. Slope field for $F'(x) = \sin(x^2)$ **b.** Solutions satisfying $F(0) = 0$ and $F(0) = -2$

Figure 5.4 Slope field with particular solutions

We see that all the line segments anchored at points $(0, y)$ are horizontal lines, because $F'(0) = \sin(0) = 0$. These horizontal line segments correspond to tangent lines of the antiderivatives at $x = 0$. Line segments anchored at other points (x, y) have slope $F'(x) = \sin x^2$, which varies between -1 and 1, and again are independent on the value of y. Sketching an antiderivative $F(x)$ satisfying $F(0) = 0$ corresponds to sketching a curve that passes through the point $(0, 0)$ and remains tangent to the line segments in Figure 5.4b. This sketch yields the upper curve in the right panel of the Figure 5.4b. The lower curve corresponds to the antiderivative $F(x)$ that satisfies $F(0) = -2$, since this curve passes through the point $(0, -2)$. Notice that the graphs of each of these antiderivatives are vertical translations of one another, because the slopes for each value of x do not depend on the location of y. ∎

Rectilinear motion

We can use antiderivatives to understand the motion of an object along a straight line. We have previously defined velocity $v(t)$ at time t to be the rate of change of position $s(t)$ of an object—that is, $s'(t) = v(t)$, and acceleration $a(t)$ to be the rate of change of velocity $v(t)$ of an object—that is, $v'(t) = a(t)$. Thus, it follows that position is the antiderivative of velocity which, in turn, is the antiderivative of acceleration.

These definitions allow us to study the stooping behavior of the peregrine falcon shown in Figure 5.1. The peregrine falcon is arguably the fastest animal in the world. This long-winged raptor favors direct pursuit of other birds, and its level speed exceeds the speed of most birds upon which it preys. Peregrines gain speed by launching attacks from high and then stooping (steep diving with feet back against the tail and wings close to the body) to attain speeds of well over 300 km/h.

Example 8   Stooping peregrines

Assume that the peregrine falcon's downward acceleration is due to gravity alone, which is 9.8 m/s², and that there is no air resistance. Determine how far a peregrine falcon would have to free-fall to achieve a speed of 300 km/h.

Solution   Let $v(t)$ denote the downward velocity at t seconds after a peregrine falcon has begun its stoop. Assuming acceleration is due purely to gravity, we have

$$\frac{dv}{dt} = 9.8 \text{ m/s}^2$$

To solve for v, we antidifferentiate to obtain $v(t) = 9.8\,t + C$ where C is a constant. Since the peregrine has no downward velocity at the beginning of its stoop, we have $v(0) = 0$. Hence, $0 = v(0) = C$ and

$$v(t) = 9.8\,t$$

To find the position $s(t)$ of the falcon at time t, we have

$$\frac{ds}{dt} = v(t) = 9.8\,t$$

Here, $s(t)$ describes the vertical distance (in meters) from the initial position of the falcon to its position at time t. Antidifferentiating yields $s(t) = 4.9\,t^2 + C$ where C is some constant. Since $s(0) = 0$, we obtain $0 = s(0) = C$ and

$$s(t) = 4.9t^2$$

Next, we need to determine how many seconds the peregrine falcon needs to fall to achieve a speed of 300 km/h. By converting 300 km/h to meters/second we obtain

$$\frac{300 \text{ km} \times 1000 \text{ m} \times 1 \text{ h}}{1 \text{ h} \times 1 \text{ km} \times 3600 \text{ s}} = 83\frac{1}{3} \text{ m/s}$$

Thus, to find the desired time, we solve

$$v = 83\frac{1}{3} = 9.8\,t$$

for t to obtain $t \approx 8.5$ seconds. Hence, the distance fallen to achieve a speed of 300 km/h is approximately

$$s(8.5) \approx 4.9(8.5)^2 \approx 354 \text{ m}$$

Thus, the peregrine falcon needs to free-fall about 350 meters to attain a speed of 300 km/h if it relies purely on the force of gravity.

$t=0$ s
$s=0$ m
$v=0$ m/s

stooping peregrine falcon falling under the influence of gravity

$t=8.5$ s
$s=354$ m
$v=83.33$ m/s

© Jim Zipp/ardea.com

Recently, scientists have accurately measured speeds achieved by peregrine falcons during stooping. One falcon was logged by radar at 183 km/h $\approx$ (114 mph) after a dive of 305 m $\approx$ (1,000 ft). This is considerably slower than the 300 km/h our current model would predict. One reason for this discrepancy is that we ignored air resistance in our calculations. This shortcoming can be addressed by formulating a differential equation model that includes the effects of air resistance.

PROBLEM SET 5.1

Level 1 DRILL PROBLEMS

Find the general antiderivative of the functions f shown in Problems 1 to 22.

1. 2

2. $f(x) = 4$

3. $f(x) = 2x + 3$

4. $f(x) = 4 - 5x$

5. $f(x) = 6x^4$

6. $f(x) = 2x^{-4}$ for $x > 0$

7. $f(x) = 2x^2 - 5$

8. $f(t) = 4t + 4t^2$

9. $f(t) = 8t^3 + 15t$

10. $f(x) = \dfrac{1}{2x}$ for $x > 0$

11. $f(x) = \dfrac{5}{x^2}$ for $x > 0$

12. $f(x) = \dfrac{2}{5x}$ for $x > 0$

13. $f(x) = \cos x$

14. $f(x) = 4\sin(5x)$

15. $f(x) = 3\sin(2\pi x)$

16. $f(x) = 14e^x$

17. $f(x) = 3e^x$

18. $f(\theta) = \sec^2 \theta$ for $-\pi/2 < x < \pi/2$

19. $f(x) = x^{3/2} + x^{1/2} + x^{-1}$ for $x > 0$

20. $f(u) = u^3 - 2u + \sqrt{u}$

21. $f(u) = 6u + 3\cos u$

22. $f(x) = 5x - 4\sin x$

Find the antiderivative $F(x)$ of the functions shown in Problems 23 to 28 satisfying the indicated initial condition.

23. $f(x) = 2$ with $F(0) = 1$

24. $f(x) = 4$ with $F(1) = -1$

25. $f(x) = 2x + 3$ with $F(-3) = 0$

26. $f(x) = 4 - 5x$ with $F(0) = 4$

27. $f(x) = 6x^4$ with $F(1) = -2$

28. $f(x) = 2x^{-4}$ for $x > 0$ with $F(2) = 0$

29. a. If $F'(x) = 1 - 4x$, find F so that $F(1) = 0$.

b. Sketch the graphs of $y = F(x)$, $y = F(x) - 2$, and $y = F(x) + 4$.

c. Find a constant C so that the largest value of $G(x) = F(x) + C$ is 0.

30. a. If $F'(x) = 2x - 1$, find F so that $F(2) = 0$.

b. Sketch the graphs of $y = F(x)$, $y = F(x) - 2$, and $y = F(x) + 4$.

c. Find a constant C so that the smallest value of $G(x) = F(x) + C$ is 0.

The slope $F'(x)$ at each point on a graph is given in Problems 31 to 34 along with one point (x_0, y_0) on the graph. Use this information to find F graphically.

31. $F'(x) = x^2 + 3x$ with point $(0, 0)$

32. $F'(x) = (2x - 1)^2$ with point $(1, 3)$

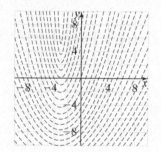

33. $F'(x) = x + e^x$ with point $(0, 2)$

34. $F'(x) = \dfrac{x^2 - 1}{x^2 + 1}$ with point $(0, 0)$

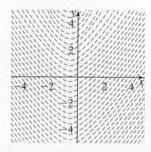

35. Sketch a slope field for

$$\frac{dy}{dx} = x$$

for $-2 \le x \le 2$ and $0 \le y \le 5$. Over this slope field, sketch the antiderivative of $F(x)$ of x which satisfies $F(0) = 1$.

36. Sketch a slope field for

$$\frac{dy}{dx} = 3x^2$$

for $-5 \leq x \leq 5$ and $-5 \leq y \leq 5$. Over this slope field, sketch the antiderivative $F(x)$ of x which satisfies $F(0) = 0$.

37. Sketch a slope field for

$$\frac{dy}{dx} = \cos x$$

for $-\pi \leq x \leq \pi$ and $-2 \leq y \leq 2$. Over this slope field, sketch the antiderivative $F(x)$ of x which satisfies $F(0) = 1$.

38. Sketch a slope field for

$$\frac{dy}{dx} = x \sin(\pi x)$$

for $-1 \leq x \leq 1$ and $-2 \leq y \leq 2$. Over this slope field, sketch the antiderivative $F(x)$ of x which satisfies $F(-1) = 0$.

39. Find the general antiderivative of $\sin(ax)$ where $a \neq 0$.

40. Find the general antiderivative of e^{kx} where $k \neq 0$.

Level 2 APPLIED AND THEORY PROBLEMS

41. As discussed in Example 5, the developmental rate of a stinkbug as a function of temperature T is $-0.06075 + 0.00112T$. Assume that the temperature of a typical spring in Davis, California, x days after the start of the stinkbug development period is adequately modeled by the function

$$T(x) = 80 + 10\cos(2\pi x)$$

a. Find the function $F(x)$ describing the amount of development completed by day x assuming that $F(0) = 0$.

b. Estimate at what time a stinkbug has completed development.

42. Recall that the developmental rate of a stinkbug as a function of temperature T is $-0.06075 + 0.00112T$. Assume that the temperature of an atypical day in Davis, California, x days after the start of the stinkbug development period is adequately modeled by the function

$$T(x) = 80 + x + 10\cos(2\pi x)$$

a. Find the function $F(x)$ describing the amount of development completed by day x assuming that $F(0) = 0$.

b. Estimate at what time a stinkbug has completed development.

43. Entomologists Godfrey and Anderson* studied the developmental rates of the hydrilla tuber weevil, which is a species that consumes a weed found in ponds and waterways. As illustrated in Figure 5.5, the developmental rate as a function of temperature (in degrees Celsius) is given by

$$F'(t) = -0.0582211 + 0.00417376\, T(t)$$

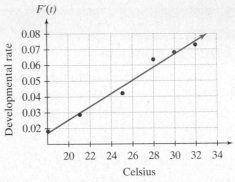

Figure 5.5 Developmental rate as a function of temperature

a. Suppose that the temperature in $C°$ is given by the function $T(t) = 30 + 10\sin(2\pi t)$ where t is measured in days. Find the fraction of development $F(t)$ that has been completed at time t for eggs laid at time 0.

b. Estimate how many days it takes the weevil to develop to adulthood.

44. Assume that the temperature in Problem 43 is given by the function $T(t) = 50 + \dfrac{1}{1+t}$. Using the developmental rate for the tuber weevil presented in that problem, estimate how many days it takes the weevil to complete half of its development.

45. A peregrine falcon stoops for 305 meters. Assuming a constant acceleration of 9.8 m/s^2, find its speed at the end of the stoop.

46. Suppose a food package is dropped out of a balloon that is 100 ft above the ground and ascending at a rate of 10 ft/s. Determine how long it takes the package to hit the ground, assuming a constant gravitational acceleration of 9.8 m/s^2.

47. Apollo 15 astronaut David Scott dropped a hammer and a feather on the moon to demonstrate that in a vacuum all objects fall at the same rate. He dropped both items from a height of approximately 4 ft. How long did it take each object to hit the ground? (Acceleration on the moon due to gravity is -5.2 ft/s^2.) How long would it take for a hammer to hit the ground on Earth if dropped from a height of 4 ft? (Gravitational acceleration on Earth is -32 ft/s^2.)

*K. E. Godfrey and L. W. J. Anderson, "Developmental Rates of *Bagous affinis* at Constant Temperatures," *Florida Entomologist* 77 (1994); 516–519.

48. Assume the brakes of a certain automobile produce a constant deceleration of 22 ft/s^2. If the car is traveling at 60 mi/h (88 ft/s) when the brakes are applied, how far will it travel before coming to a complete stop?

49. It is estimated that t months from now, the population of Ferndale, California, will be changing at the rate of $4 + 5t^{2/3}$ people per month. If the current population is 2,000, what will the population be eight months from now?

50. A hypothetical study of a community suggests that t years from now the level of carbon monoxide in the air will be changing at the rate of $0.1t + 0.1$ ppm/year. If the current level of carbon monoxide in the air is 3.4 ppm, what will be the level three years from now?

51. One of Poiseuille's laws for the flow of blood in an artery says that if $v(r)$ is the velocity of flow r centimeters from the central axis of the artery, then the velocity decreases at a rate proportional to r. That is, $v'(r) = ar$ where a is a negative constant. Find an expression for $v(r)$ assuming that $v(R) = 0$, where R is the radius of the artery.

52. Suppose that a silviculturist finds that a certain type of tree grows in such a way that its height $h(t)$ t years after planting is changing at the rate of

$$h'(t) = 0.2t^{2/3} + t \text{ ft/yr}$$

If the tree was two feet tall when it was planted, how tall will it be in twenty-seven years?

53. Suppose that a woman driving a sports car down a straight road at 60 mi/h (88 ft/s) sees a cow start to cross the road 200 feet ahead. She takes 0.7 seconds to react to the situation before hitting the brakes, which decelerates the car at the rate of 28 ft/s^2. Does she stop in time to avoid hitting the cow?

54. A hypothetical population, N, grows in such a way that at time t (years), the growth rate is given by

$$\frac{dN}{dt} = 0.15t + \cos t + 0.7 \sin t$$

where $N(t)$ is measured in thousands of individuals and $N(0) = 5$.

a. Find $N(t)$.

b. What is the minimum population size? When does it occur?

5.2

Accumulated Change and Area under a Curve

In this section, we deal with the problem of finding accumulated change, which can be interpreted as the area under a curve. The early Greeks, particularly Archimedes (287–212 BC), estimated the areas of geometrical objects using the "method of exhaustion," a precursor to integral calculus. They found increasingly better approximations by filling in areas with increasingly smaller elements of known area (much as we do later in this section in taking Riemann sums).

In elementary school you learned formulas to find areas of squares, triangles, and other polygons. You also are familiar with the formula for the area of a circle with radius r: $A = \pi r^2$. The Egyptians were the first to use this formula more than 5,000 years ago, but the Greeks derived the area of a circle by drawing inscribed polygons or circumscribed polygons, as shown in Figure 5.6, and then using triangles to find the area of those polygons as an approximation. This method, called the *method of exhaustion*, involves finding the area of a circle by inscribing polygons with increasing numbers of sides (Archimedes stopped at a ninety-six-sided polygon). The area of the circle is the limit of the areas of the inscribed polygons as the number of polygonal sides increases.

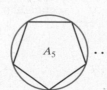

Figure 5.6 Using the limit of a sequence of inscribed polygons to find the area of a circle

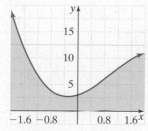

Figure 5.7 Area under a curve $y = f(x)$

In this section, we focus on estimating the area under a curve $y = f(x)$ over an interval a to b. As illustrated in Figure 5.7, this means estimating the area defined by the region bounded by the curves $y = f(x)$ (with $f(x) \geq 0$ on $[a, b]$), $x = a$, $x = b$, and $y = 0$. Similar to the method of exhaustion, we find these areas by approximating them with collections of finer and finer rectangles.

To motivate finding area under a curve, we show that area under a curve corresponds to accumulated change. We do this in the context of two biological processes: physiological time for insects and plants and disease incidence as studied by epidemiologists.

Physiological time: degree-days

Plants and insects often require a certain amount of heat to develop from one stage in their life cycle to another stage in their life cycle. This measure of accumulated heat is known as **physiological time**, and the units used are called **degree-days**—the accumulated product of time and temperature between the organism's lower and upper developmental thresholds. The **lower developmental threshold** is the temperature below which the insect or plant cannot develop, while the **upper development threshold** is the temperature above which it cannot develop.

To simplify the presentation, we initially assume that the temperature remains between the lower and upper developmental thresholds. The more general case is considered later. **One degree-day** is one day (24 hours) with the temperature one degree above the lower developmental threshold. For example, if the lower developmental threshold of the organism is $47°F$ and the temperature remains at $48°F$ for one day or $47.2°F$ for five days, then in each case, one degree-day is accumulated (i.e., $1 \times (48 - 47) = 5 \times (47.2 - 47) = 1$).

The concept of degree-days is used widely in agriculture and developmental biology. In agricultural publications we may come across the following types of statements: "Around 1539 degree-days are required for sweet corn to ripen, assuming a lower development threshold of $50°F$"; or, "corn earworms (pests of corn) have a lower developmental threshold of $54.7°F$ and require 760 degree-days to develop from egg to adult". These statements allow us to estimate the time it takes sweet corn to mature or corn earworms to develop in geographical regions with different temperature profiles, as well as to anticipate how these times may change in response to global warming.

Example 1   Degree-days under constant temperature

According the University of California at Davis Integrated Pest Management Program website, the lower developmental threshold of Thompson seedless grapevines is $50°F$, and this variety of grape requires approximately 3,000 degree-days for its fruit to mature. If the temperature were to remain constant at $70°F$, how long would it take for the fruit to mature?

Solution   The amount of degree-days accumulated in x days is

$$(70 - 50)x = 20x$$

Solving $20x = 3,000$ yields $x = 150$ days. Therefore, it would take 150 days for the grapes to mature. Notice that this answer can be interpreted as the following shaded rectangular area:

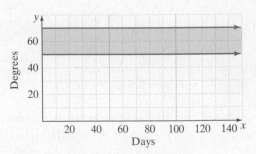

Unlike the preceding example, temperatures in fields vary over time. Consequently, computing the accumulation of degree-days, as we shall see, requires finding the area of an appropriate region defined by the temperature curve and the lower developmental threshold.

For example, the temperature in Lincoln, Nebraska, on June 23, 2006 is given by the function $f(t)°$F, illustrated in Figure 5.8, where t going from 0 to 1 represents one day from midnight to midnight. If an organism of interest (say, sweet corn) has a lower developmental threshold of 50°F, then, as we explore in the next example, the total accumulated degree-days is given by the area between the curves $y = f(t)$ and $y = 50$ from $t = 0$ to $t = 1$. This area corresponds to the shaded area in Figure 5.8.

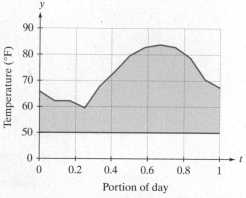

Figure 5.8 Temperature for Lincoln, Nebraska, on June 23, 2006. The shaded area corresponds to the accumulated degree-days for an organism with a lower developmental threshold of 50°F.

Example 2   Sweet corn in Nebraska

Estimate the accumulation of degree-days for sweet corn in Lincoln, Nebraska, on June 23, 2006 using the data in Table 5.2.

Table 5.2 Temperature in Lincoln, Nebraska, starting at midnight of June 23, 2006 and reported at two-hour intervals

Hour	Temperature (°F)	Excess above 50°F
0	65.8	15.8
2	62.2	12.2
4	62.2	12.2
6	59.5	9.5
8	67.8	17.8
10	73.4	23.4
12	79.7	29.7
14	82.8	32.8
16	83.8	33.8
18	82.8	32.8
20	78.8	28.8
22	70.5	20.5
24	67.3	17.3
Total		286.6

Solution   To approximate the number of degree-days that have accumulated, break up the day into two-hour intervals (i.e., intervals of width $\frac{1}{12}$ day). Within each interval, assume that the temperature is relatively constant. Then, accumulated degree-days within the first interval [0, 1/12] is given by

$$(65.8 - 50)\frac{1}{12} = \frac{15.8}{12}$$

This quantity simply corresponds to the area of a rectangle with height 15.8 and width $\frac{1}{12}$ days as illustrated in Figure 5.9. The total 286.6 in Table 5.2, when divided by 12 yields $\frac{286.6}{12} \approx 23.9$ degree-days, which corresponds to the sum of the areas of rectangles depicted in Figure 5.9. This sum of areas is an approximation for the shaded area in Figure 5.8.

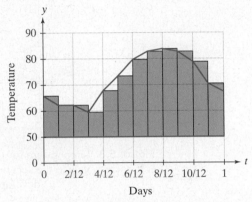

Figure 5.9 Accumulated degree-days approximated by the area of rectangles whose width is $\frac{1}{12}$ days

The Bombay plague epidemic

In epidemiology, scientists keep track of various rates associated with disease, including the incidence rate, which measures the number of new disease cases per unit of time (e.g., day or week), and the mortality rate, which reports the number of deaths due to the disease per unit of time. For instance, during the outbreak of the plague in Bombay in 1905–1906, the weekly mortality rates due to the plague were recorded, and the values obtained are plotted in Figure 5.10.

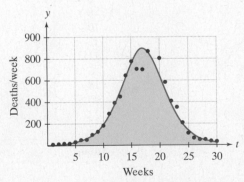

Figure 5.10 Mortality rate from plague in Bombay (now called Mumbai) from December 17, 1905 to July 21, 1906. Data shown in red and the fitted function in blue.

In a landmark paper,* two mathematicians, W. O. Kermack and A. G. McKendrick, showed that these data could be reasonably well fitted by the function

$$f(t) = 890 \, \text{sech}^2 (0.2t - 3.4) \text{ deaths/week}$$

where t is measured in weeks and the *hyperbolic secant* function $\text{sech} \, x$ equals $\dfrac{2}{e^x + e^{-x}}$. If we want to estimate the total number of deaths using this function, what do we need to compute? Consider a small interval of time, from t to $t + \Delta t$. Since the mortality rate over this interval is given approximately by $f(t)$, the number of deaths over this time interval is approximately $f(t)\Delta t$, that is, the area of a rectangle of width Δt and height $f(t)$. Notice how the units work out in this product: $f(t)$ has units of deaths/week and Δt has units of weeks. The product $f(t)\Delta t$ has units of deaths. These arguments suggest that the area under the curve $f(t)$ from $t = 0$ to $t = 30$ equals the total number of deaths, as discussed in the next example. Since we know the actual number of deaths—we simply add up all the weekly data—the next example examines how the area under $f(t)$ approximates the actual number of deaths.

Example 3   Mortality due to the plague

Use the function $f(t) = 890 \sec h^2(0.2t - 3.4)$ to approximate the total number of deaths in Bombay from $t = 0$ to $t = 30$ using intervals of five weeks.

Solution   Begin by breaking the interval from $t = 0$ to $t = 30$ into six subintervals of length 5, as shown in Figure 5.11. For the mortality rate in each interval, we can evaluate $f(t)$ at the right endpoint of each interval. This yields the following table (entries rounded to one decimal place):

Interval	Deaths/week (height of rectangle)	Deaths (area of rectangle)
[0, 5]	$f(5) \approx 28.8$	144
[5, 10]	$f(10) \approx 192.4$	962
[10, 15]	$f(15) \approx 761.5$	3,807.5
[15, 20]	$f(20) \approx 633.3$	3,166.5
[20, 25]	$f(25) \approx 134.0$	670
[25, 30]	$f(30) \approx 19.4$	97

Summing up the deaths yields 8,847 deaths, which is approximately 2% short of the actual 9,043 recorded number of deaths. This approximation is illustrated in Figure 5.11.

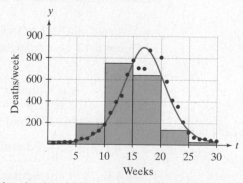

Figure 5.11 Using the data in Figure 5.10 to approximate the number of deaths

*W. O. Kermack and A. G. McKendrick, "A Contribution to the Mathematical Theory of Epidemics," *Proceedings of the Royal Statistical Society* 115 (1927): 700–721.

In Figure 5.11 notice that when the curve is on the rise, as in the first three rectangles, the area is overestimated (the green area above the curve), and when the curve is on the decline, the area is underestimated (the white area below the curve). This is a result of the height of the rectangles being defined by the value of the function on the right side of each interval. The reverse would be true if the height of the rectangles were defined by the value of the function on the left side of each interval.

The area problem

The previous examples illustrate the importance of finding areas under curves. These examples also show that we can approximate areas by approximating the region with rectangles, computing the area of each rectangle, and summing up the areas. This observation is the key that unlocks the area problem. We pursue this approach further in the following example.

Example 4　Estimating the area under a curve

Consider the function $f(x) = x^2$ over the interval $[0, 1]$. Use rectangles to find upper and lower bounds for the area under x^2, above the y axis, between $x = 0$ and $x = 1$.

Solution　Let A denote the area under $y = f(x)$, above $y = 0$, and between the lines $x = 0$ and $x = 1$, as shown in Figure 5.12.

Now, we find the area A by taking successive approximations. Notice that the largest value of x^2 on the interval $[0, 1]$ is 1 at $x = 1$. Hence, the region under x^2 is contained in a rectangle of height 1 and width 1. Thus, $A < 1 \times 1 = 1$. On the other hand, A is clearly greater than 0. To obtain a better estimate, subdivide the interval $[0, 1]$ into two subintervals $[0, 1/2]$ and $[1/2, 1]$, each with width $\Delta x = 1/2$, as shown in Figure 5.13.

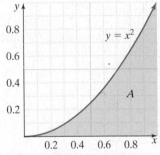

Figure 5.12 Area A under $y = x^2$ on $[0, 1]$

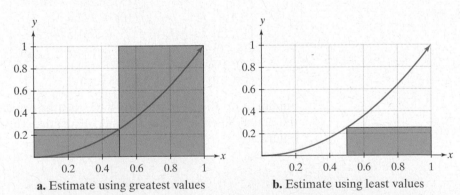

a. Estimate using greatest values　　**b.** Estimate using least values

Figure 5.13 Left and right sum approximations of the area under $y = x^2$

The greatest values that x^2 takes on these subintervals are $f(1/2) = 1/4$ and $f(1) = 1$. Hence, the two rectangles over the intervals $[0, 1/2]$ and $[1/2, 1]$ with heights, $1/4$ and 1, respectively, enclose our region. Therefore, $A < \dfrac{1}{4} \times \dfrac{1}{2} + 1 \times \dfrac{1}{2} = \dfrac{5}{8}$. Alternatively, the minimum values of x^2 on $[0, 1/2]$ and $[1/2, 1]$ are 0 and $1/4$, respectively. Therefore, $A > 0 \times \dfrac{1}{2} + \dfrac{1}{4} \times \dfrac{1}{2} = \dfrac{1}{8}$.

Subdividing the interval once improved our estimates, so more subdivisions should also improve our estimates. Suppose we divide the interval into n subintervals $\left[0, \dfrac{1}{n}\right]$, $\left[\dfrac{1}{n}, \dfrac{2}{n}\right]$, ..., $\left[\dfrac{n-1}{n}, 1\right]$ of width $\dfrac{1}{n}$. Since x^2 is an increasing function on the interval $[0, 1]$, the maximum values of $f(x) = x^2$ on these subintervals are

$\frac{1}{n^2}, \frac{2^2}{n^2}, \ldots, \frac{n^2}{n^2}$. The area of the n rectangles determined by these heights is given by

$$R_n = f\left(\frac{1}{n}\right)\frac{1}{n} + f\left(\frac{2}{n}\right)\frac{1}{n} + \cdots + f\left(\frac{n}{n}\right)\frac{1}{n}$$

$$= \frac{1^2}{n^2} \times \frac{1}{n} + \frac{2^2}{n^2} \times \frac{1}{n} + \cdots + \frac{n^2}{n^2} \times \frac{1}{n}$$

which is greater than A. Since the minimum values of x^2 on these subintervals of width $\frac{1}{n}$ are $0, \frac{1^2}{n^2}, \ldots, \frac{(n-1)^2}{n^2}$, A is greater than

$$L_n = f\left(\frac{0}{n}\right)\frac{1}{n} + f\left(\frac{1}{n}\right)\frac{1}{n} + \cdots + f\left(\frac{n-1}{n}\right)\frac{1}{n}$$

$$= 0 \times \frac{1}{n} + \frac{1^2}{n^2} \times \frac{1}{n} + \cdots \frac{(n-1)^2}{n^2} \times \frac{1}{n}$$

Thus,

$$L_n < A < R_n$$

This relationship is illustrated with $n = 4$ in Figure 5.14 where

$$L_4 < A < R_4$$

$$\frac{0^2}{4^2} \times \frac{1}{4} + \frac{1^2}{4^2} \times \frac{1}{4} + \frac{2^2}{4^2} \times \frac{1}{4} + \frac{3^2}{4^2} \times \frac{1}{4} < A < \frac{1^2}{4^2} \times \frac{1}{4} + \frac{2^2}{4^2} \times \frac{1}{4} + \frac{3^2}{4^2} \times \frac{1}{4} + \frac{4^2}{4^2} \times \frac{1}{4}$$

$$0.21875 < A < 0.46875$$

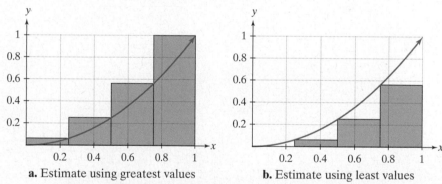

a. Estimate using greatest values   **b.** Estimate using least values

Figure 5.14 Estimating the area of A using four subintervals

Computing these estimates by hand for large n is tedious. Using technology, we calculated the entries in the following table to a precision of ten decimal places, but zeros are included only when necessary.

n	L_n	R_n
1	0.0	1.0
2	0.125	0.625
3	0.1851851852	0.5185185185
4	0.21875	0.46875
5	0.24	0.44
10	0.285	0.385
100	0.32835	0.33835
1,000	0.3328335	0.3338335
10,000	0.333283335	0.333383335
100,000	0.3333283334	0.3333383334

The sums suggest that as n becomes large, L_n and R_n both converge to $\frac{1}{3}$.

Example 4 suggests that area under x^2 over the interval $[0, 1]$ is $\frac{1}{3}$. But how can we *really* be sure that these numbers converge to $\frac{1}{3}$? We address this question in the next example.

Example 5   Finding the exact area under x^2 on the interval $[0, 1]$

Use the formula (which can be proved inductively)

$$1^2 + 2^2 + 3^2 + \cdots + n^2 = \frac{n(n+1)(2n+1)}{6}$$

to prove that

$$\lim_{n \to \infty} R_n = \frac{1}{3}$$

Solution   We have that

$$
\begin{aligned}
R_n &= \frac{1^2}{n^2} \times \frac{1}{n} + \frac{2^2}{n^2} \times \frac{1}{n} + \cdots + \frac{n^2}{n^2} \times \frac{1}{n} \qquad \textit{from Example 4} \\
&= \frac{1}{n}\left(1^2 \frac{1}{n^2} + 2^2 \frac{1}{n^2} + \cdots + n^2 \frac{1}{n^2}\right) \\
&= \frac{1}{n^3}(1^2 + 2^2 + \cdots + n^2) \\
&= \frac{1}{n^3}\frac{n(n+1)(2n+1)}{6} \qquad \textit{using the stated summation} \\
&\phantom{= \frac{1}{n^3}\frac{n(n+1)(2n+1)}{6} \qquad} \textit{formula} \\
&= \frac{(n+1)(2n+1)}{6n^2} \\
&= \frac{2n^2 + 3n + 1}{6n^2} \\
&= \frac{1}{3} + \frac{1}{2n} + \frac{1}{6n^2}
\end{aligned}
$$

Thus,

$$
\begin{aligned}
\lim_{n \to \infty} R_n &= \lim_{n \to \infty} \frac{1}{3} + \frac{1}{2n} + \frac{1}{6n^2} \\
&= \frac{1}{3}
\end{aligned}
$$

Similarly (see Problem 19 in Problem Set 5.2), it can be shown that $\lim_{n \to \infty} L_n = \frac{1}{3}$. Since $L_n \le A \le R_n$ for all $n \ge 1$, it follows that

$$A = \lim_{n \to \infty} R_n = \lim_{n \to \infty} L_n = \frac{1}{3}$$

Examples 4 and 5 provide the core idea of how to *define* area under a nonnegative function $y = f(x)$ from $x = a$ to $x = b$. First, we divide the interval $[a, b]$ into n equally spaced subintervals of width $\Delta x = \frac{b-a}{n}$. Let

$$a_0 = a, a_1 = a + \Delta x, a_2 = a + 2\Delta x, a_3 = a + 3\Delta x, \ldots, a_n = a + n\Delta x = b$$

To approximate the height of f over a subinterval $[a_i, a_{i+1}]$, choose a point x_i on the interval $[a_i, a_{i+1}]$. The points x_i are called *sample points*. In our examples, we chose left or right endpoints as our sample points, but we could have picked any point in each interval. The height of f over $[a_i, a_{i+1}]$ is approximately $f(x_i)$. The area of f over $[a_i, a_{i+1}]$ is approximately $f(x_i)\,\Delta x$. Adding all these rectangular areas up yields

$$\text{area} \approx f(x_1)\Delta x + f(x_2)\Delta x + \ldots f(x_n)\Delta x$$

This sum is known as a *Riemann sum* after the brilliant mathematician Georg Friedrich Bernhard Riemann (1826–1866; see the 𝔥𝔦𝔰𝔱𝔬𝔯𝔦𝔠𝔞𝔩 𝔔𝔲𝔢𝔰𝔱 in Problem Set 5.2). Now we write this sum succinctly, using the summation notation presented in the regression subsection of Section 4.3.

| Riemann Sum | Suppose a continuous function f is defined on the interval $[a, b]$. If the interval is divided into n subintervals so that $\Delta x = \dfrac{b - a}{n}$ and |

$$a = a_0, a_1 = a + \Delta x, a_2 = a + 2\Delta x, \ldots, a_n = a + n\Delta x = b$$

then a **Riemann sum** associated with f is the sum

$$\sum_{i=1}^{n} f(x_i)\Delta x = f(x_1)\Delta x + f(x_2)\Delta x + \ldots f(x_n)\Delta x$$

where x_i is any point we chose to select in the interval $[a_{i-1}, a_i]$.

We have seen how area can be approximated by a Riemann sum and how this approximation improves as n become large, approaching the true area as $n \to \infty$. Therefore, we write

$$\text{area} = \lim_{n \to \infty} \left[f(x_1)\Delta x + f(x_2)\Delta x + f(x_3)\Delta x + \cdots + f(x_n)\Delta x \right]$$

$$= \lim_{n \to \infty} \sum_{i=1}^{n} f(x_i)\Delta x$$

We cannot know that the method really works unless we have a theorem that tells us that a limit exists and that this limit is *independent* of the way we choose the sample points in the subintervals. For continuous functions such a theorem does exist, but its proof is a topic for a course in *real analysis*. (The "real" refers to real-valued functions in contrast to the "complex" of complex numbers and complex-valued functions.)

Theorem 5.1   Limit of a Riemann sum theorem

If $f(x)$ is continuous on $[a, b]$, then

$$\lim_{n \to \infty} \sum_{i=1}^{n} f(x_i)\Delta x$$

exists and is independent of the choice of sample points x_i.

In Problem Set 5.2, some of the problems require one or more of the following summation formulas. These formulas can be verified using mathematical induction.

| Summation Formulas | The following formulas can be verified using mathematical induction. You may use these formulas to find certain Riemann sums. |

$$\sum_{k=1}^{n} 1 = \overbrace{1 + 1 + \cdots + 1}^{n \text{ times}} = n$$

$$\sum_{k=1}^{n} k = 1 + 2 + 3 + \cdots + n = \frac{n(n + 1)}{2}$$

$$\sum_{k=1}^{n} k^2 = 1^2 + 2^2 + 3^2 + \cdots + n^2 = \frac{n(n + 1)(2n + 1)}{6}$$

$$\sum_{k=1}^{n} k^3 = 1^3 + 2^3 + 3^3 + \cdots + n^3 = \frac{n^2(n + 1)^2}{4}$$

PROBLEM SET 5.2

Level 1 DRILL PROBLEMS

First sketch the region under the graph of $y = f(x)$ on the interval $[a, b]$ in Problems 1 to 12. Then approximate the area of each region by using right endpoints and the formula

$$R_n = f(a + \Delta x)\Delta x + f(a + 2\Delta x)\Delta x + \cdots + f(a + n\Delta x)\Delta x$$

for $\Delta x = \dfrac{b - a}{n}$ and the indicated values of n.

1. $f(x) = 2x + 1$ on $[0, 1]$ for $n = 4$
2. $f(x) = 4x + 1$ on $[0, 1]$ for $n = 8$
3. $f(x) = x^2$ on $[0, 2]$ for $n = 4$
4. $f(x) = x^2$ on $[0, 2]$ for $n = 6$
5. $f(x) = x^3$ on $[1, 3]$ for $n = 4$
6. $f(x) = 4x^2 + 2$ on $[0, 1]$ for $n = 4$
7. $f(x) = x^2 + x^3$ on $[0, 1]$ for $n = 4$
8. $f(x) = e^x$ on $[0, 1]$ for $n = 4$
9. $f(x) = x^{-1}$ on $[1, 2]$ for $n = 4$
10. $f(x) = \sqrt{x}$ on $[1, 4]$ for $n = 4$
11. $f(x) = \cos x$ on $\left[-\dfrac{\pi}{2}, 0\right]$ for $n = 4$
12. $f(x) = x + \sin x$ on $\left[0, \dfrac{\pi}{4}\right]$ for $n = 3$

Use a calculator to estimate the area under the curve $y = f(x)$ on each interval given in Problems 13 to 18 as a sum of ten terms evaluated at right endpoints.

13. $f(x) = 4x$ on $[0, 1]$
14. $f(x) = x^2$ on $[0, 4]$
15. $f(x) = \cos x$ on $\left[-\dfrac{\pi}{2}, 0\right]$
16. $f(x) = x + \sin x$ on $\left[0, \dfrac{\pi}{4}\right]$
17. $f(x) = \ln(x^2 + 1)$ on $[0, 3]$
18. $f(x) = e^{-3x^2}$ on $[0, 1]$

Use a summation formula in Problems 19 to 24.

19. Prove that

$$\lim_{n \to \infty} L_n = \frac{1}{3}$$

for L_n as defined in Example 4.

20. Use Riemann sums and left endpoints to prove that the area under $y = x$ from $x = 0$ to $x = 2$ equals 2.

21. Use Riemann sums and right endpoints to prove that the area under $y = x$ from $x = 0$ to $x = 4$ equals 8.

22. Use Riemann sums and right endpoints to prove that the area under $y = x^3$ from $x = 0$ to $x = 4$ is 64.

23. Use Riemann sums and left endpoints to prove that the area under $y = x^3$ from $x = 0$ to $x = 2$ is 4.

24. Use Riemann sums and right endpoints to prove that the area under $y = x + 3x^2$ from $x = 0$ to $x = 2$ is 10.

Level 2 APPLIED AND THEORY PROBLEMS

25. The lower developmental threshold of sweet corn is $50°$F and requires 1,587 degree-days for maturing. If the temperature were to remain a constant $75°$F, how long would it take for the corn to mature?

26. The pistachio has a lower developmental threshold of $50°$F and requires 1,197 degree-days for shell hardening. If the temperature were to remain a constant $72°$F, how long would it take for the pistachio's shell to harden?

27. The black turtle bean has a lower developmental threshold of $41°$F and requires 1,365.5 degree-days for 50% anthesis (i.e., until 50% of all the flowers have blossomed). If the temperature were to remain a constant $68.5°$F, how long would it take to reach the required 50% anthesis?

28. Estimate mortality due to the plague in Bombay (Mumbai) by approximating the region under

$$f(t) = 890 \operatorname{sech}^2(0.2t - 3.4)$$

deaths per week from $t = 0$ to $t = 30$ with rectangles of width 15 weeks. Would you expect your answer to be more or less accurate than the result of Example 3?

29. Estimate mortality due to the plague by approximating the region under

$$f(t) = 890 \operatorname{sech}^2(0.2t - 3.4)$$

deaths per week from $t = 0$ to $t = 30$ with rectangles of width three weeks and height given by their right endpoints. Would you expect your answer to be more of less accurate than the result of Example 3?

30. The weekly rate of cases of influenza A (strain unknown) studied by WHO/NREVSS during the 2003–2004 season is plotted in Figure 5.15. Estimate the total number of cases (i.e., the area under the curve) over the interval $[40, 56]$ using the right endpoints of two-week intervals. Sketch the corresponding rectangles in the figure.

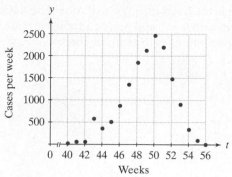

Figure 5.15 Weekly rate of cases of influenza A

31. Repeat Problem 30 using left endpoints.

32. The daily maximum (max) and minimum (min) temperatures at Westmoreland, California, for the period from April 1 to 9 are given in Table 5.3.

Table 5.3 Early morning minimum and mid-afternoon maximum temperatures at Westmoreland, California, for nine days (minimum on April 10 included for later use)

Data	Min temp	Max temp
April 1, 2012	53°F	76°F
April 2, 2012	44°F	79°F
April 3, 2012	40°F	88°F
April 4, 2012	46°F	82°F
April 5, 2012	49°F	85°F
April 6, 2012	44°F	77°F
April 7, 2012	41°F	85°F
April 8, 2012	39°F	89°F
April 9, 2012	48°F	91°F
April 10, 2012	48°F	NA

The lower developmental threshold for cotton is $k = 60°F$. If M_i and m_i are used to denote the max and min temperatures on day i, respectively, then estimate the degree-day accumulation over the nine-day period using this formula:

degree-days accumulated on day i

$$= \begin{cases} \dfrac{M_i + m_i}{2} - k & \text{if } k \le m_i \\[2mm] \dfrac{(M_i - k)^2}{2(M_i - m_i)} & \text{if } m_i < k < M_i \\[2mm] 0 & \text{if } k > M_i \end{cases}$$

Note that the logic behind this formula is that on days for which $k \le m_i$ we use the average temperature above the threshold, and on days for which $m_i < k < M_i$ we estimate that the temperature is above threshold for a proportion $\dfrac{M_i - k}{M_i - m_i}$ of the

day, and then use the average $\dfrac{M_i + k}{2}$ as applying over this period of time to obtain the middle expression in the equation.

33. The lower developmental threshold for the elm leaf beetle is 52°F. Use the formula in Problem 32 to estimate the degree-day accumulation for the elm leaf beetle in Stockton, California, over a two-week period of time using the data listed in Table 5.4.

Table 5.4 Early morning minimum and mid-afternoon maximum temperatures at Stockton, California (fire station # 4) for 14 days (minimum on March 15 included for later use)

Data	Min temp	Max temp
March 1, 2012	38°F	60°F
March 2, 2012	34°F	62°F
March 3, 2012	32°F	68°F
March 4, 2012	36°F	73°F
March 5, 2012	38°F	74°F
March 6, 2012	37°F	58°F
March 7, 2012	36°F	64°F
March 8, 2012	34°F	68°F
March 9, 2012	34°F	73°F
March 10, 2012	38°F	65°F
March 11, 2012	41°F	61°F
March 12, 2012	34°F	64°F
March 13, 2012	52°F	56°F
March 14, 2012	52°F	59°F
March 15, 2012	55°F	NA

34. In the figure that follows, m_i and m_{i+1} are the minimum daily temperatures on consecutive days i and $(i + 1)$, respectively, which are assumed to be exactly one day apart. M_i is the maximum daily temperature on day i and is assumed to occur exactly half way between the two minimum temperatures. The parameter k is the lower developmental threshold for a particular species.

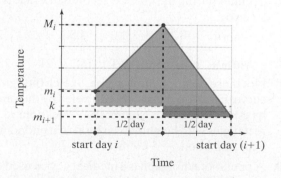

If these values satisfy $M_i > m_i > k$ and $k > m_{i+1}$, then show that the degree-day accumulation over the one-day period, as depicted by the green area in

the figure, is given by the expression

degree-days accumulated on day i

$$= \left(\frac{M_i - m_i}{4} + \frac{m_i - k}{2} \right) + \frac{(M_i - k)^2}{4(M_i - m_{i+1})}$$

Alternatively, if $M_i > m_i > k$ and $M_i > m_{i+1} > k$ then show that the degree-day accumulation over day i is given by

degree-days accumulated on day i

$$= \frac{2M_i - m_i - m_{i+1}}{4} + \frac{m_i + m_{i+1} - 2k}{2}.$$

35. The obliquebanded leafroller, is an agricultural pest with a lower developmental threshold of 43°F. Estimate the accumulated degree-days for this insect growing in Westmoreland, California, over the period from the morning of April 2nd to the morning of April 3rd using the data in Table 5.3 (see Problem 32), and the formula presented in Problem 34.

36. The codling moth, is an agricultural pest with a lower developmental threshold of 50°F. Estimate the accumulated degree-days for this insect growing in Stockton, California, over the period from the morning of April 1st to the morning of April 2nd using the data in Table 5.3 (see Problem 32), and the formula presented in Problem 34.

37. Assume the temperature in degrees Fahrenheit is given by

$$T(t) = 50 + 20 \cos(2\pi t/365) + 10 \sin(2\pi t)$$

where t is time in days. Assume the lower development threshold is 40°F and estimate the number of degree-days that accumulate from $t = 0$ to $t = 10$ days using time intervals of width two.

38. Use the temperature variation model in Problem 37 to estimate the number of degree-days accumulated from $t = 0$ to $t = 20$ for citrus flower which has a lower developmental threshold of 49°F. Use time intervals of width four days.

39. Suppose the velocity v (in meters per second) of a runner during the first few seconds of a race is given by

t in s	0	0.5	1.0	1.5	2.0	2.5
v in m/s	0	5	9.5	15.1	21	25

Plot these points in the tv plane. Sketch the velocity curve. Estimate the distance traveled by the runner by estimating the area under the velocity curve; use rectangles with heights given by a right endpoint approximation.

40. A pneumotachograph is a medical device used to measure the rate at which air is exhaled by a patient's lungs. Suppose Figure 5.16 shows the rate of exhalation for a particular patient. The area under the graph provides a measure of the total volume of air in the lungs during exhalation. Use a Riemann sum with $n = 8$ and right-endpoint subinterval representatives to estimate the volume.

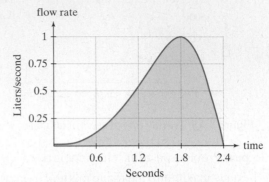

Figure 5.16 Rate of exhalation

41. An industrial plant spills pollutant into a lake. Suppose that the pollutant spread out to form the pattern shown in Figure 5.17. All distances are in feet.

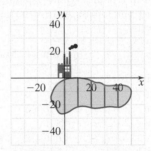

Figure 5.17 Pollutant spill

Use a Riemann sum with $n = 6$ and right-endpoint subinterval representatives to estimate the area of the spill.

42. Historical Quest

Georg Friedrich Bernhard Riemann (1826–1866)

In this section, we saw that history honored Georg Riemann by naming an important process after him. In his personal life Riemann was frail, bashful, and timid; but in his professional life, he was one of the all-time giants in mathematics. In his book, *Space Through the Ages*, Cornelius Lanczos wrote,

"Although Riemann's collected papers fill only one single volume of 538 pages, this volume weighs tons if measured intellectually. Every one of his many discoveries was destined to change the course of mathematical science." One of these discoveries is the Riemann zeta function

$$\zeta(s) = \sum \frac{1}{n^s} = \prod \frac{1}{1 - p^{-s}}$$

which had already been considered by Euler. In this function, the sum is over all natural numbers n, and the product is over all prime numbers. While Euler considered the zeta function in the context of a real variable z, Riemann considered the zeta function

in the context of a complex variable z. Except for a few trivial exceptions, the roots of $\zeta(s)$ all lie between 0 and 1. Riemann conjectured that the zeta function had infinitely many nontrivial roots, all with their real parts equal to 1/2. This is the famous Riemann hypothesis, which remains one of the most important unsolved problems in mathematics. The Clay Mathematics Institute, for example, has offered a million dollar prize for solving this conjecture. (For more information about this institute, visit http://www.claymath.org.) So you can become a millionaire doing mathematics!

Write a paper on Georg Riemann; in particular, discuss this million dollar prize.

5.3   The Definite Integral

Previously we defined area under a nonnegative function as the limit of a Riemann sum. In this section we define this limit for any continuous function (positive or negative) and develop its geometrical meaning as well as its properties.

For a nonnegative continuous function $f(x)$ from $x = a$ to $x = b$, we defined the area under the curve as

$$\text{area} = \lim_{n \to \infty} [f(x_1)\Delta x + f(x_2)\Delta x + \cdots + f(x_n)\Delta x] = \lim_{n \to \infty} \sum_{i=1}^{n} f(x_i)\Delta x$$

where $\Delta x = \dfrac{b - a}{n}$ and x_i is a point from the interval $[a + (i - 1)\Delta x, a + i\Delta x]$. Theorem 5.1 from the previous section implies that

$$\lim_{n \to \infty} \sum_{i=1}^{n} f(x_i)\Delta x$$

exists and is independent of the sample points x_i whenever f is continuous. When f takes on negative values, the integral no longer corresponds to the area under the curve, but the signed area as we soon shall see. The existence of the limit is so important that Leibniz (see 𝕳istorical 𝕼uest, page 538) developed the following special notation for it.

Definite Integral	Let f be continuous on $[a, b]$. Then the **definite integral** of f from a to b is defined to be

$$\int_a^b f(x)\,dx = \lim_{n \to \infty} \sum_{i=1}^{n} f(x_i)\Delta x$$

In the definition of the definite integral, the function f that is being integrated is called the **integrand**; the interval $[a, b]$ is the **interval of integration**; and the endpoints a and b are called, respectively, the **lower** and the **upper limits of integration**. The variable x is called the **variable of integration**. Notice that in taking the limit the Greek letters are supplanted by the Roman letters: Δ becomes a d and Σ becomes an elongated S.

Example 1 From sums to integrals

Write the sum

$$\lim_{n\to\infty} \sum_{i=1}^{n} \sin\left(\frac{2\pi i}{n}\right)\frac{2\pi}{n}$$

as a definite integral.

Solution   There are several ways we can answer this problem depending on how we view the Riemann sum. For instance, we can view this Riemann sum corresponding to an integrand $\sin x$ with sample points $x_i = \dfrac{2\pi i}{n}$ and $\Delta x = \dfrac{2\pi}{n}$. Since the first sample point $x_1 = \dfrac{2\pi}{n}$ approaches 0 as n increases, the lower limit of integration must be 0. Since the last sample point $x_n = \dfrac{2\pi n}{n} = 2\pi$ for all n, the upper limit of integration must be 2π. Hence, the definite integral is

$$\int_0^{2\pi} \sin x \, dx$$

Alternatively, we can always represent the limit of the Riemann sum as an integral from $x = 0$ to $x = 1$. (In fact, we can choose the limits of integration to be any $a < b$ and still get things to work out!) With this view, our sample points need to be $x_i = \dfrac{i}{n}$ and $\Delta x = \dfrac{1}{n}$. Hence, the argument of the sum is equal to $\sin(2\pi x_i)2\pi \Delta x$ and the Riemann sum converges to

$$\int_0^1 \sin(2\pi x)2\pi \, dx$$

Note that the two expressions obtained for the integrals in Example 1 must be the same for the theory of integration to be consistent. This will be demonstrated in Section 5.5, after we consider how to change the variable of integration.

Example 2 From integrals to sums

Write the integral

$$\int_1^4 \frac{dx}{x}$$

as a limit of a Riemann sum.

Solution   Our integral $\int_a^b f(x)\,dx$ has integrand $f(x) = \dfrac{1}{x}$ and the limits of integration $a = 1$ and $b = 4$. If we break up the interval $[1, 4]$ into n subintervals of equal width, then

$$\Delta x = \frac{4-1}{n} = \frac{3}{n}$$

Choosing the right endpoints of the intervals as sample points gives

$$x_1 = 1 + \frac{3}{n}, \quad x_2 = 1 + 2 \times \frac{3}{n}, \quad \cdots, \quad x_n = 1 + n \times \frac{3}{n}$$

Hence, the definite integral equals

$$\lim_{n\to\infty} \sum_{i=1}^{n} f(x_i)\Delta x = \lim_{n\to\infty} \sum_{i=1}^{n} \frac{1}{1+\frac{3i}{n}}\left(\frac{3}{n}\right)$$

Example 3   Approximating integrals with sums

Approximate the integral

$$\int_{-1}^{0.5} \tan x \, dx$$

by the sum

$$\sum_{i=1}^{6} \tan x_i \, \Delta x$$

where the x_i correspond to right endpoints.

Solution   Since the integrand, $\tan x$, is continuous on the interval $[-1, 0.5]$, the integral is well defined. The summation expression in the problem statement implies that $n = 6$. Thus, we choose $\Delta x = \dfrac{0.5 - (-1)}{6} = \dfrac{1.5}{6} = 0.25$, in which case $x_0 = -1$, $x_1 = -0.75$, $x_2 = -0.5$, $x_3 = -0.25$, $x_4 = 0$, $x_5 = 0.25$, and $x_6 = 0.5$. The Riemann sum with right end points is

$$\sum_{i=1}^{6} \tan x_i \, \Delta x = [\tan(-0.75) + \tan(-0.5) + \tan(-0.25) + \tan(0) + \tan(0.25) + \tan(0.5)]\, 0.25 \approx -.2329$$

A graphical representation of this sum is shown in green in Figure 5.18. Notice that we got a negative number, as the areas of the rectangles below the x axis were greater than the areas of the rectangles above the x axis.

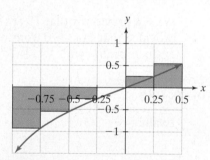

Figure 5.18 Graph of $y = \tan x$ with approximating rectangles

Example 4   Computing an integral using a summation formula

Use a summation formula to compute

$$\int_{0}^{2} (1 - x^2) \, dx$$

Solution   Break the interval $[0, 2]$ into n subintervals whose endpoints are $0, \dfrac{2}{n}, \dfrac{4}{n}, \ldots, \dfrac{2n}{n}$. Choose $x_i = \dfrac{2i}{n}$. The corresponding Riemann sum is

$$\sum_{i=1}^{n} \left[1 - \left(\frac{2i}{n} \right)^2 \right] \Delta x$$

with $\Delta x = \dfrac{2}{n}$. Expanding and rearranging terms yields

$$\sum_{i=1}^{n}\left[1-\left(\frac{2i}{n}\right)^2\right]\Delta x = \left[\left(1-\frac{4\times 1^2}{n^2}\right)+\left(1-\frac{4\times 2^2}{n^2}\right)+\cdots+\left(1-\frac{4\times n^2}{n^2}\right)\right]\frac{2}{n}$$

$$= \left[\underbrace{1+1+\cdots+1}_{n\text{ times}}\right]\frac{2}{n}-\left[\frac{4\times 1^2}{n^2}+\frac{4\times 2^2}{n^2}+\cdots+\frac{4\times n^2}{n^2}\right]\frac{2}{n}$$

$$= n\times\frac{2}{n}-\left[1^2+2^2+\cdots+n^2\right]\frac{4\times 2}{n^3}$$

$$= 2-\left[\frac{n(n+1)(2n+1)}{6}\right]\frac{8}{n^3} \qquad \textit{using a summation formula}$$

$$= 2-\frac{(n+1)(2n+1)4}{3n^2}$$

$$= 2-\frac{8n^2+12n+4}{3n^2}$$

$$= 2-\frac{8}{3}-\frac{4}{n}-\frac{4}{3n^2}$$

Taking the limit of this expression as $n\to\infty$ yields

$$\int_0^2 (1-x^2)\,dx = \lim_{n\to\infty}\left(2-\frac{8}{3}-\frac{4}{n}-\frac{4}{3n^2}\right) = 2-\frac{8}{3}-0-0 = -\frac{2}{3}$$

Again, we have an integral that is negative.

Geometrical meaning of the definite integral

We saw previously that $\int_a^b f(x)\,dx$ corresponds to the area under the curve $y = f(x)$ provided that $f(x) \geq 0$ from $x = a$ to $x = b$. The following example uses this fact to evaluate an integral.

Example 5   Integral of *dx* rule

Evaluate

$$\int_a^b 1\,dx$$

Solution   Let $f(x) = 1$ with limits of integration $x = a$ and $x = b$.

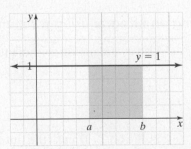

If we plot f over $[a, b]$, we can see this is the area of a rectangle of height 1 and width $(b-a)$. Thus,

$$\int_a^b 1\,dx = 1(b-a) = b-a$$

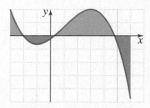

Figure 5.19 Geometry of the definite integral

What happens if $f(x)$ changes sign on the interval? In this case, $\int_a^b f(x)\,dx$ is the *signed area* of the region R determined by the curve $y = f(x)$ and the lines $y = 0$, $x = a$, and $x = b$. More specifically, if f changes sign on the interval $[a, b]$, then the region R breaks up into two pieces: one piece, call it R^-, that lies below the x axis as illustrated by the red region in Figure 5.19 and another piece, call it R^+, that lies above the x axis as illustrated by the green region in Figure 5.19.

If A^+ and A^- denote the areas of R^+ and R^-, respectively, then

$$\int_a^b f(x)\,dx = A^+ - A^-$$

Example 6   Evaluating integrals using signed areas

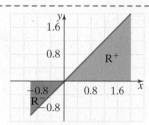

Figure 5.20 Graph of f

Use the signed area interpretation of integrals to find

a. $\displaystyle\int_{-1}^{2} x\,dx$ **b.** $\displaystyle\int_{-3}^{3} \sqrt{9 - x^2}\,dx$ **c.** $\displaystyle\int_{-3}^{3} x^5\,dx$

Solution

a. Let $f(x) = x$ on $[-1, 2]$, as shown Figure 5.20. The graph forms two triangles, R^+ and R^-, that lie above and below the x axis, respectively. The area of R^+ is 2 and the area of R^- is $\frac{1}{2}$. Hence,

$$\int_{-1}^{2} x\,dx = 2 - \frac{1}{2} = \frac{3}{2}$$

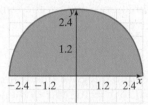

Figure 5.21 Graph of g

b. Let $g(x) = \sqrt{9 - x^2}$ on $[-3, 3]$, as shown in Figure 5.21. The graph forms a semicircle of radius 3. The graph is always above the axis; consequently, we need its area. Using the formula for the area of a circle,

$$\int_{-3}^{3} \sqrt{9 - x^2}\,dx = \frac{1}{2}(\pi 3^2) = \frac{9\pi}{2}$$

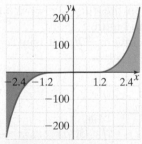

Figure 5.22 Graph of h

c. Let $h(x) = x^5$ on $[-3, 3]$, as shown in Figure 5.22. Notice that this graph is symmetric with respect to the origin; consequently, it has the same area above and below the x axis. Therefore,

$$\int_{-3}^{3} x^5\,dx = 0$$

Properties of definite integrals

Integrals satisfy several useful properties, some of which are summarized in the following box.

Properties of the Definite Integral: Part I

Let f and g be continuous functions on the interval $[a, b]$.

Sum rule $\displaystyle\int_a^b [f(x) + g(x)]\,dx = \int_a^b f(x)\,dx + \int_a^b g(x)\,dx$

Difference rule $\displaystyle\int_a^b [f(x) - g(x)]\,dx = \int_a^b f(x)\,dx - \int_a^b g(x)\,dx$

Scalar rule $\displaystyle\int_a^b c\,f(x)\,dx = c\int_a^b f(x)\,dx$

Opposite rule $\displaystyle\int_a^b f(x)\,dx = -\int_b^a f(x)\,dx$

These properties can be proved using Riemann sums and limit laws (see Problem Set 5.3).

Example 7   Using the properties of definite integrals

Evaluate $\int_{-3}^{3} [2\sqrt{9-x^2}-5]\,dx$.

Solution

$$\int_{-3}^{3} [2\sqrt{9-x^2}-5]\,dx = \int_{-3}^{3} 2\sqrt{9-x^2} - \int_{-3}^{3} 5\,dx \qquad \textit{difference rule}$$

$$= 2\int_{-3}^{3} \sqrt{9-x^2} - 5\int_{-3}^{3} 1\,dx \qquad \textit{scalar rule}$$

$$= 2\left(\frac{9\pi}{2}\right) - 5(3-(-3)) \qquad \textit{from Examples 5 and 6}$$

$$= 9\pi - 30$$

Combining the properties of integrals with the geometrical interpretation of the integral allows one to quickly compute certain integrals.

Example 8   Growing grapes

Thompson seedless grapes have a lower developmental threshold of $50°F$ and require approximately 3,000 degree-days to ripen after pollination. Suppose the temperature in the fields over a particular ten-day period is given by

$$T(x) = 70 + 10\sin(2\pi x)$$

where x is time in days. Write an expression involving definite integrals that represents the number of degree-days accumulated from $x = 0$ to $x = 10$, and evaluate this expression. Use the expression to determine the percent of development that takes place during this ten-day period.

Solution   We are interested in finding the area between the curves $y = 50$ and $y = 70 + 10\sin(2\pi x)$ from $x = 0$ to $x = 10$, as illustrated in Figure 5.23.

Since this area is computed by finding the area below the curve $y = 70 + 10\sin(2\pi x)$ and then subtracting the area below the curve $y = 50$, the result is

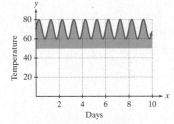

Figure 5.23 Degree-days accumulated for ten days

$$\text{ACCUMULATED DEGREE-DAYS} = \int_{0}^{10} [70 + 10\sin(2\pi x)]\,dx - \int_{0}^{10} 50\,dx$$

$$= \int_{0}^{10} 70\,dx + 10\int_{0}^{10} \sin(2\pi x)\,dx - \int_{0}^{10} 50\,dx \qquad \textit{sum rule}$$

$$= \int_{0}^{10} 20\,dx + 10\int_{0}^{10} \sin(2\pi x)\,dx \qquad \textit{difference rule}$$

$$= 20\int_{0}^{10} dx + 10\int_{0}^{10} \sin(2\pi x)\,dx \qquad \textit{scalar rule}$$

$$= 200 + 10\int_{0}^{10} \sin(2\pi x)\,dx \qquad \textit{integral of dx rule}$$

Since the integral of $\sin(2\pi x)$ has equal area above and below the x axis on the interval $[0, 10]$, its value is zero. Hence, the number of degree-days accumulated is 200. This area could be found by noticing that the "hills" of the temperature functions fit in the valleys, yielding a 20 by 10 rectangle.

Since 200 of 3000 degree-day units accumulated over the ten-day period, this period represents $p = 200/3000 \approx 0.0667$ of the total number of degree-days needed to ripen. Hence, 6.67% of needed ripening took place during this ten-day period.

We conclude this section with some additional properties of the definite integral.

Properties of the Definite Integral: Part II

Assuming all integrals exist, we have the following properties.

Nonnegativity   If $f(x) \geq 0$ from $x = a$ to $x = b$, then

$$\int_a^b f(x)\, dx \geq 0$$

Dominance   If $f(x) \geq g(x)$ from $x = a$ to $x = b$, then

$$\int_a^b f(x)\, dx \geq \int_a^b g(x)\, dx$$

Bounding   If $m \leq f(x) \leq M$ from $x = a$ to $x = b$, then

$$m(b - a) \leq \int_a^b f(x)\, dx \leq M(b - a)$$

Splitting

$$\int_a^b f(x)\, dx = \int_a^c f(x)\, dx + \int_c^b f(x)\, dx$$

Definite integral at a point

$$\int_a^a f(x)\, dx = 0$$

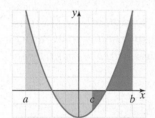

Figure 5.24 Geometrical depiction of the splitting property where the signed area $\int_a^b f(x)\, dx$ from $x = a$ to $x = b$ equals the signed area $\int_a^c f(x)\, dx$ from $x = a$ to $x = c$ plus the signed area $\int_c^b f(x)\, dx$ from $x = c$ to $x = b$

Nonnegativity can be proved using the definition of a definite integral, and nonnegativity, in turn, can be used to prove dominance and bounding. For example, to prove dominance, suppose that $f(x) \geq g(x)$ from $x = a$ to $x = b$. Then, $f(x) - g(x) \geq 0$ from $x = a$ to $x = b$. Applying the property of differences and nonnegativity yields

$$\int_a^b f(x)\, dx - \int_a^b g(x)\, dx = \int_a^b [f(x) - g(x)]\, dx \geq 0$$

Hence,

$$\int_a^b f(x)\, dx \geq \int_a^b g(x)\, dx$$

If we set M and m to be the maximum value and minimum value, respectively, of f on the interval $[a, b]$, then the bounding property provides crude estimates for the value of a definite integral. When working through detailed computations by hand, these crude estimates allow us to see whether our work has resulted in a reasonable answer. Finally, a proof of the splitting property is somewhat subtle, but geometrically intuitive, as illustrated in Figure 5.24.

Example 9   Using bounds

Show that

$$24 \leq \int_3^6 [10 + 2 \sin x^2]\, dx \leq 36$$

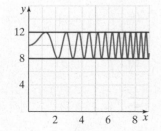

Solution   Since the sine function is bounded between -1 and 1, it follows that $8 \leq 10 + 2 \sin x^2 \leq 12$ for all x as illustrated in the left margin. The bounding property implies that

$$24 = 8(6 - 3) \leq \int_3^6 [10 + 2 \sin x^2]\, dx \leq 12(6 - 3) = 36$$

which yields the desired result.

Example 10 Using the splitting property

Suppose that $\int_4^9 f(x)\,dx = 100$ and $\int_{-3}^9 f(x)\,dx = 125$; find $\int_{-3}^4 f(x)\,dx$.

Solution By the splitting property,

$$
\begin{aligned}
125 &= \int_{-3}^9 f(x)\,dx \\
&= \int_{-3}^4 f(x)\,dx + \int_4^9 f(x)\,dx \\
&= \int_{-3}^4 f(x)\,dx + 100
\end{aligned}
$$

Thus,

$$
\int_{-3}^4 f(x)\,dx = 125 - 100 = 25
$$

PROBLEM SET 5.3

Level 1 DRILL PROBLEMS

Express the limits in Problems 1 to 6 as definite integrals of the form $\int_0^1 f(x)\,dx$.

1. $\displaystyle \lim_{n\to\infty} \sum_{i=1}^n \frac{i}{n^2}$

2. $\displaystyle \lim_{n\to\infty} \sum_{i=1}^n \frac{i^2}{n^3}$

3. $\displaystyle \lim_{n\to\infty} \sum_{i=1}^n \left(-2 + \frac{3i}{n}\right)\left(\frac{3}{n}\right)$

4. $\displaystyle \lim_{n\to\infty} \sum_{i=1}^n \left(1 - \frac{2i}{n}\right)\left(\frac{2}{n}\right)$

5. $\displaystyle \lim_{n\to\infty} \sum_{i=1}^n \left(1 - \frac{i^2}{n^2}\right)\frac{1}{n}$

6. $\displaystyle \lim_{n\to\infty} \sum_{i=1}^n \sin\left(\frac{\pi i}{n} - \pi\right)\frac{\pi}{n}$

Express the definite integrals in Problems 7 to 12 as limits of Riemann sums.

7. $\displaystyle \int_1^2 x^4\,dx$

8. $\displaystyle \int_{-1}^1 (x^2 - x)\,dx$

9. $\displaystyle \int_0^1 e^x\,dx$

10. $\displaystyle \int_{-1}^4 e^x\,dx$

11. $\displaystyle \int_{-1}^1 |x|\,dx$

12. $\displaystyle \int_{-1}^1 |\cos x|\,dx$

First sketch the region under the graph of $y = f(x)$ on the interval $[a, b]$. Then use the interpretation of the definite integral $\int_a^b f(x)\,dx$ as a signed area to evaluate the integrals in Problems 13 to 16.

13. $\displaystyle \int_{-4}^3 (1 - 2x)\,dx$

14. $\displaystyle \int_0^{2\pi} \cos x\,dx$

15. $\displaystyle \int_0^4 \sqrt{16 - x^2}\,dx$

16. $\displaystyle \int_{-1}^3 |x|\,dx$

Evaluate each of the integrals in Problems 17 to 22 by using the following information together with the sum rule and the splitting property:

$$
\int_{-1}^2 f(x)\,dx = 3; \quad \int_{-1}^0 f(x)\,dx = \frac{1}{3};
$$

$$
\int_{-1}^2 g(x)\,dx = \frac{3}{2}; \quad \int_0^2 g(x)\,dx = 2
$$

17. $\displaystyle \int_0^{-1} f(x)\,dx$

18. $\displaystyle \int_{-1}^2 [f(x) + g(x)]\,dx$

19. $\displaystyle \int_{-1}^2 [2f(x) - 3g(x)]\,dx$

20. $\displaystyle \int_0^2 f(x)\,dx$

21. $\displaystyle \int_{-1}^0 g(x)\,dx$

22. $\displaystyle \int_{-1}^0 [3f(x) - 5]\,dx$

Use integral properties to establish the statements in Problems 23 to 26.

23. $\displaystyle\int_0^\pi \sin x \, dx \le \pi$

24. $\displaystyle\frac{9}{10} \le \int_1^{10} \frac{dx}{x} \le 9$

25. $\displaystyle 2 \le \int_{-1}^1 \sqrt{1 + x^2} \, dx \le 2\sqrt{2}$

26. $\displaystyle\int_0^1 x^3 \, dx \le \frac{1}{2}$ Hint: $x^3 \le x$ on $[0, 1]$

27. Use the fact that $\int_0^1 x^2 \, dx = \dfrac{1}{3}$ and the geometrical interpretation of the integral to find

$$\int_{-1}^1 x^2 \, dx$$

28. Use the graph of $y = \cos x$ to evaluate

$$\int_a^b \cos x \, dx$$

on the indicated interval.
   a. $[0, 2\pi]$
   b. $\left[\dfrac{\pi}{2}, \dfrac{5\pi}{2}\right]$
   c. If $a = 0$, for what values of $b > 0$ does the integral take on its largest value?

29. Given $\int_{-2}^4 [5 f(x) + 2g(x)] \, dx = 7$ and $\int_{-2}^4 [3 f(x) + g(x)] \, dx = 10$, find

   a. $\displaystyle\int_{-2}^4 f(x) \, dx$ **b.** $\displaystyle\int_{-2}^4 g(x) \, dx$

30. Suppose $\int_0^2 f(x) \, dx = 3$, $\int_0^2 g(x) \, dx = -1$, and $\int_0^2 h(x) \, dx = 3$.
   a. Evaluate $\int_0^2 [2 f(x) + 5g(x) - 7h(x)] \, dx$
   b. Find the value of s so that
   $\int_0^2 [5 f(x) + sg(x) - 6h(x)] \, dx = 0$

31. Evaluate $\int_{-1}^2 f(x) \, dx$ given that $\int_{-1}^1 f(x) \, dx = 3$, $\int_2^3 f(x) \, dx = -2$, and $\int_1^3 f(x) \, dx = 5$.

32. If $\int_0^1 f(x) \, dx = 1$, $\int_0^2 f(x) \, dx = 3$, and $\int_1^2 g(x) \, dx = 4$, then find $\int_1^2 [f(x) - g(x)] \, dx$.

Using right endpoints with $n = 5$, approximate the definite integrals in Problems 33 to 36. Indicate whether each approximation is greater than or less than the actual definite integral.

33. $\displaystyle\int_{-2}^0 x^2 \, dx$ **34.** $\displaystyle\int_1^2 x^3 \, dx$ **35.** $\displaystyle\int_1^4 \frac{dx}{x}$

36. $\int_{-1}^0 \sqrt{1 + x^2} \, dx$ Hint: Is $\sqrt{1 + x^2}$ increasing or decreasing on the interval $[-1, 0]$?

Use Riemann sums with right endpoints, along with a summation formula (see Section 5.2) to evaluate the integrals in Problems 37 and 38.

37. $\displaystyle\int_0^3 (x^3 - 3) \, dx$ **38.** $\displaystyle\int_0^1 (2x^2 - 4) \, dx$

Show that each statement about area in Problems 39 to 42 is true in general, or if not, provide a counterexample. It will probably help to sketch the indicated region for each problem.

39. If $C > 0$ is a constant, the region under the line $y = C$ on the interval $[a, b]$ has area $A = C(b - a)$.

40. If $C > 0$ is a constant and $b > a \ge 0$, the region under the line $y = Cx$ on the interval $[a, b]$ has area $A = \dfrac{1}{2} C(b - a)$.

41. Let f be a function that satisfies $f(x) \ge 0$ for x in the interval $[a, b]$. Then the area under the curve $y = [f(x)]^2$ on the interval $[a, b]$ must always be greater than the area under $y = f(x)$ on the same interval.

42. A function f is said to be *even* if $f(-x) = f(x)$. If f is even and $f(x) \ge 0$ throughout the interval $[-a, a]$, then the area under the curve $y = f(x)$ on this interval is *twice* the area under $y = f(x)$ on $[0, a]$.

Level 2 APPLIED AND THEORY PROBLEMS

43. We saw in Example 8 that Thompson seedless grapes have a lower developmental threshold of $50°$F and require approximately 3,000 degree-days to ripen. Suppose the temperature in the field is given by

$$T(x) = 60 + 10 \sin(2\pi x)$$

over a 15 day period where x is time in days. Write an expression involving definite integrals that represents the number of degree-days accumulated from $x = 0$ to $x = 15$, and evaluate this expression. What percentage of ripening took place during this 15 day period?

44. Assume for a particular variety of grape that the lower developmental threshold is $30°$F and that it takes 2,500 degree-days to ripen. If the temperature in degrees Fahrenheit over a particular 15 day period is given by

$$T(x) = 50 + 20 \cos(2\pi x)$$

where x is the time in days, then write an expression involving definite integrals that represents the number of degree-days accumulated from $x = 0$ to $x = 15$. Evaluate this expression and calculate the percentage ripening the occurred over this 15 day period.

45. Assume that over a 50 day period the rate at which individuals in a population die from disease is given by the function $f(x)$, where x is days and the units of $f(x)$ are individuals per day. If $f(x)$ has the form

$$f(x) = \begin{cases} 4x & 0 \le x \le 25 \\ 200 - 4x & 25 < x \le 50 \end{cases}$$

then write an expression involving definite integrals that represents the number of deaths over this 50 day period and evaluate this expressing using your knowledge of how to calculate areas of geometrical objects.

46. Assume that over a 60 day period the rate at which individuals in a population die from disease is given by the function $f(x)$, where x is days and the units of $f(x)$ are individuals per day. If $f(x)$ has the form

$$f(x) = \begin{cases} 3x & 0 \le x \le 20 \\ 60 & 20 < x \le 40 \\ 180 - 3x & 40 < x \le 60 \end{cases}$$

then write an expression involving definite integrals that represents the number of deaths over this 60 day period and evaluate this expressing using your knowledge of how to calculate areas.

47. A function f is said to be *odd* if $f(-x) = -f(x)$. Show that if f is odd on the interval $[-a, a]$, then $\int_{-a}^{a} f(x)\, dx = 0$.

48. Prove the sum rule for integrals using the definition of a definite integral and properties of limits and summations.

49. Generalize the splitting property by showing that for $a \le c \le d \le b$

$$\int_{a}^{b} f(x)\, dx = \int_{a}^{c} f(x)\, dx + \int_{c}^{d} f(x)\, dx + \int_{d}^{b} f(x)\, dx$$

whenever all these integrals exist.

50. Prove the *bounding rule* for definite integrals: If f is continuous on the closed interval $[a, b]$ and $m \le f(x) \le M$ for constants, m, M, and all x in the closed interval, then

$$m(b - a) \le \int_{a}^{b} f(x)\, dx \le M(b - a)$$

51. 𝕳istorical 𝕼uest Gilles de Roberval (1602–1675) started his study of mathematics at the age of 14 years. He had a distinguished career and was a founding member of the *Académie Royale des Sciences*. In 1669 he invented the Roberval balance (Figure 5.25), which is still widely used today.

Figure 5.25 Roberval's balance variations

Roberval had a chair position as professor of mathematics at the *Collége Royale*. Every three years there was a contest by competitive examination (written by the incumbent!) to determine who would occupy this position. It is said for this reason, Roberval kept many of his techniques of integration secret until his death. We do know, however, that he developed powerful methods in the early study of integration. These methods are described in his treatise *Traité des indivisibles*. For instance, in this treatise, he computed the definite integral of $\sin(x)$ using obscure trigonometric identities. It is these identities that the computer algebra system uses to simplify the sum and take the limit. For this quest, you may stand on the shoulders of Roberval and use technology to write a Riemann sum for

$$\int_{0}^{\pi} \sin x\, dx$$

using right endpoints. Then use technology to simplify this sum to evaluate this definite integral.

5.4 The Fundamental Theorem of Calculus

In this section, we discuss the evaluation theorem and the fundamental theorem of calculus. These theorems link antiderivatives, which we can compute relatively easily, with definite integrals and Riemann sums. We show that antiderivatives and Riemann sums, when they both exist, are the same thing.

Evaluation theorem and net change

Theorem 5.2   Evaluation theorem

Let f be a continuous function on $[a, b]$ and F be any antiderivative of f. Then,

$$\int_a^b f(x)\, dx = F(b) - F(a)$$

The proof of this theorem is a corollary of the fundamental theorem of calculus (FTC), which is discussed later in this section. Why is the evaluation theorem useful? Well, as we saw in the beginning of this chapter, finding antiderivatives is much easier than taking limits of Riemann sums. This theorem allows us to evaluate definite integrals by finding and evaluating an antiderivative. This fact is so important that *this* theorem is sometimes itself called the fundamental theorem of calculus, even though it is only a corollary to the more powerful theorem of the same name.

Example 1   Using the evaluation theorem

Evaluate the following definite integrals.

a. $\int_0^1 x^7\, dx$ **b.** $\int_1^2 x^{-1}\, dx$ **c.** $\int_{-2}^{-1} x^{-1}\, dx$ **d.** $\int_0^\pi \sin x\, dx$

Solution

a. Since an antiderivative of $f(x) = x^7$ is $F(x) = \frac{1}{8}x^8$, the evaluation theorem tells us

$$\int_0^1 x^7\, dx = F(1) - F(0) = \frac{1}{8} - 0 = \frac{1}{8}$$

Notice that if we took another antiderivative, say $F(x) = \frac{1}{8}x^8 + 14$, we still get

$$F(1) - F(0) = \frac{1}{8} + 14 - (0 + 14) = \frac{1}{8}$$

as the constant term 14 cancels out.

b. Since an antiderivative of $f(x) = x^{-1}$ is $F(x) = \ln|x|$, the evaluation theorem tells us

$$\int_1^2 x^{-1}\, dx = F(2) - F(1) = \ln 2 - 0 = \ln 2$$

Notice that we used the fact that $\frac{1}{x}$ is continuous on the interval $[1, 2]$. The fundamental theorem would not apply to $\int_{-1}^1 \frac{dx}{x}$, as $\frac{1}{x}$ is not continuous on the interval $[-1, 1]$.

c. Since an antiderivative of $f(x) = x^{-1}$ is $F(x) = \ln|x|$, the evaluation theorem tells us

$$\int_{-2}^{-1} x^{-1}\, dx = F(-1) - F(-2) = \ln|-1| - \ln|-2| = \ln 1 - \ln 2 = -\ln 2$$

Compare this with the previous answer, which is $+\ln 2$.

d. Since the antiderivative of $f(x) = \sin x$ is $F(x) = -\cos x$, the evaluation theorem tells us

$$\int_0^\pi \sin x \, dx = F(\pi) - F(0) = -\cos \pi - (-\cos 0) = 2$$

To appreciate the power of the evaluation theorem, compare the work of Example 1 with work required to find the limit of the Riemann sum for each of these functions.

To simplify our work, we introduce the following notation:

$$F(b) - F(a) = F(x)\Big|_{x=a}^{x=b}$$

Or, when there is no ambiguity,

$$F(b) - F(a) = F(x)\Big|_a^b$$

Example 2   Using evaluation notation

Evaluate $\int_\pi^{5\pi/4} \sec^2 x \, dx$.

Solution   Since the antiderivative of $\sec^2 x$ is $\tan x$, the evaluation theorem tells us

$$\int_\pi^{5\pi/4} \sec^2 x \, dx = \tan x \Big|_\pi^{5\pi/4} = 1 - 0 = 1$$

Notice that we used the fact that $\sec^2 x$ is continuous on the interval from $x = \pi$ to $5\pi/4$. The fundamental theorem does not apply on the interval from $x = \pi$ to 2π, as $\sec^2 x$ is not defined at $x = 3\pi/2$.

As illustrated in Example 3 of Section 5.2, an important interpretation of the evaluation theorem is that it relates the accumulated change of a function over an interval to the area under its derivative. Thus, we have the following procedure for calculating accumulated change.

Accumulated Change

If F is an antiderivative of f, then by the evaluation theorem we have

$$\text{ACCUMULATED CHANGE OF } F \text{ OVER } [a, b] = \int_a^b f(x) \, dx = F(b) - F(a)$$

For instance, if $F(x)$ represents the total number of births in the world by year x, then $F(b) - F(a)$ is the number of births that occurred between years a and b. Now suppose that $f(x) = F'(x)$, which by definition implies that $F'(x)$ is the instantaneous birthrate x. The evaluation theorem asserts that the area under the instantaneous birthrate equals the accumulated change in births. Does this make sense? At the level of units it certainly does. The instantaneous birthrate has units births per year. Hence, the area over an interval of time has units births per year multiplied by years. This equals births, which has the same units as the accumulated change. Moreover, if we broke up the interval $[a, b]$ into small subintervals, then the number of births in a given subinterval would be approximately the instantaneous birthrate at some point in the subinterval in question multiplied by the length of the subinterval, that is, the area of a rectangle lying above this subinterval. Adding up all these little rectangular areas would give us simultaneously an approximation for the number of births over $[a, b]$ and the area under F'. More generally, if one integrates the rate of change over an interval $[a, b]$, then one gets the accumulated change over $[a, b]$.

Example 3 Horn increase for the bighorn ram

CR Courson/Flickr/Getty Images

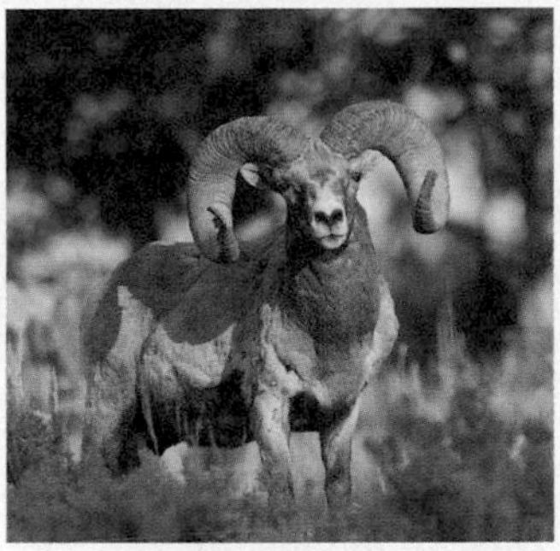

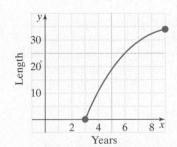

Figure 5.26 Bighorn ram

Bighorn sheep (*Ovis canadensis*) (see Figure 5.26) inhabit remote mountain and desert regions. They are restricted to semi-open, precipitous terrain with rocky slopes, ridges, and cliffs or rugged canyons. Forage, water, and escape terrain are the most important components of their habitat. Jon Jorgenson and colleagues* found that the rate of increase of a bighorn ram's horn is approximated by the function

$$0.1762\,x^2 - 3.986\,x + 22.68 \text{ cm per year}$$

for x between three and nine years.

Find the accumulated change in the length of a bighorn ram's horn from age $x = 3$ to $x = 9$.

Solution Let $F(x)$ denote the length of a ram's horn at age x years. Then,

$$\text{ACCUMULATED CHANGE OVER } [3, 9] = F(9) - F(3) = \int_3^9 (0.1762\,x^2 - 3.986\,x + 22.68)\,dx$$

The evaluation theorem implies

$$\begin{aligned}\int_3^9 (0.1762\,x^2 - 3.986\,x + 22.68)\,dx &= \left[\frac{0.1762}{3}x^3 - \frac{3.986}{2}x^2 + 22.68\,x\right]\Bigg|_3^9 \\ &= \left[\frac{0.1762}{3}(9)^3 - \frac{3.986}{2}(9)^2 + 22.68(9)\right] \\ &\quad - \left[\frac{0.1762}{3}(3)^3 - \frac{3.986}{2}(3)^2 + 22.68(3)\right] \\ &= 85.5036 - 51.6888 \\ &= 33.8148 \text{ cm}\end{aligned}$$

Fundamental theorem of integral calculus

The answer to Example 3 gives us the *net* increase in the length of the ram's horn over the whole six-year period. Suppose, though, that we want to find the net increase at any age x during the six-year period from age 3 to age 9. The net increase is given by the function

$$F(x) = \int_3^x (0.1762\,u^2 - 3.986\,u + 22.68)\,du$$

for $3 \leq x \leq 9$. For example, we found $F(9) \approx 33.8148$ cm. In writing the integral defining $F(x)$, we took advantage of the fact that the variable of integration is a dummy variable (it disappears once the integration has been performed). Consequently, to avoid confusion, we choose the variable u of integration to be different from our time variable x. By the evaluation theorem,

$$\begin{aligned}F(x) &= \left[\frac{0.1762}{3}u^3 - \frac{3.986}{2}u^2 + 22.68\,u\right]\Bigg|_3^x \\ &= 0.0587\,x^3 - 1.993\,x^2 + 22.68\,x - 51.6888\end{aligned}$$

Figure 5.27 Estimated growth of a ram's horn in centimeters

Plotting $F(x)$ from $x = 3$ to $x = 9$ as shown in Figure 5.27 illustrates how the net increase of the length of the horn changes over this time interval. Notice, as we might expect, the length of the horn is increasing at a decreasing rate.

*Jon Jorgenson et al., "Effects of Population Density on Horn Development in Bighorn Rams," *Journal of Wildlife Management* 62 (1998): 1011–1020.

We can now generalize this idea for any continuous function f defined on the interval $[a, b]$ by considering the function

$$F(x) = \int_a^x f(u)\, du \quad \text{for} \quad a \le x \le b$$

If we interpret $f(x)$ as a rate, then $F(x)$ describes how the accumulated change varies as a function of x. Alternatively, $F(x)$ describes how the signed area under f confined to the interval $[a, x]$ varies as a function of x.

Theorem 5.3   Fundamental theorem of calculus (FTC)

Consider a continuous function f on the interval $[a, b]$. Then, F defined by

$$F(x) = \int_a^x f(u)\, du$$

is an antiderivative of $f(x)$ on (a, b). In other words,

$$\frac{d}{dx} \int_a^x f(u)\, du = f(x) \qquad \text{on } (a, b)$$

Why should this be true? The idea of the proof is as follows. The splitting property of integrals implies that

$$F(x + \Delta x) - F(x) = \int_a^{x+\Delta x} f(u)\, du - \int_a^x f(u)\, du$$

$$= \int_a^x f(u)\, du + \int_x^{x+\Delta x} f(u)\, du - \int_a^x f(u)\, du$$

$$= \int_x^{x+\Delta x} f(u)\, du$$

On the other hand, continuity of f implies that $f(u) \approx f(x)$ for u between x and $x + \Delta x$. Furthermore, this approximation gets better and better as Δx approaches zero. Thus,

$$F(x + \Delta x) - F(x) \approx \int_x^{x+\Delta x} f(x)\, du = f(x) \int_x^{x+\Delta x} du = f(x)[x + \Delta x - x] = f(x)\Delta x$$

Dividing both sides by Δx and letting Δx go to zero suggests that

$$F'(x) = f(x)$$

To really show that this final statement is true requires a bit more care using ϵ-δ arguments of the type introduced in Section 2.2.

An important consequence of the fundamental theorem of calculus is that it proves that every continuous function f has an antiderivative F, even though F cannot always be expressed using a combination of elementary functions (e.g., polynomial, exponential, and trigonometric functions). A corollary of the fundamental theorem of calculus is the evaluation theorem, as discussed in Problem 33 in Problem Set 5.4.

Example 4   Derivatives via the fundamental theorem

Compute the following derivatives.

a. $\dfrac{d}{dx} \displaystyle\int_1^3 \sqrt{u + u^3}\, du$

b. $\dfrac{d}{dx} \displaystyle\int_0^x \sqrt{u - u^3}\, du$ on the interval $[0, 1]$

Solution

a. Since $\int_1^3 \sqrt{u + u^3}\, du$ is a number, we are taking the derivative of a constant and

$$\frac{d}{dx} \int_1^3 \sqrt{u + u^3}\, du = 0.$$

b. By the fundamental theorem of calculus, $\dfrac{d}{dx} \int_0^x \sqrt{u - u^3}\, du = \sqrt{x - x^3}$.

Example 5   From integrals to integrands

Suppose

$$\int_3^x f(u)\, du = \sqrt{x} + a$$

Find f and a.

Solution   Let $F(x) = \int_3^x f(u)\, du$. By the fundamental theorem of calculus,

$$f(x) = F'(x) = \frac{d}{dx}[\sqrt{x} + a] = \frac{1}{2\sqrt{x}}$$

To find a, notice that

$$\sqrt{3} + a = F(3)$$
$$= \int_3^3 f(u)\, du = 0$$

Therefore, $a = -\sqrt{3}$ and $f(x) = \dfrac{1}{2\sqrt{x}}$.

Using the fundamental theorem, we can easily compute the accumulation of degree-days.

Example 6   Seedless grapes

Ed Young/© AgStock Images, Inc./ Alamy

Figure 5.28 Seedless grapes

Thompson seedless grapes (see Figure 5.28) have a lower developmental threshold of 50°F and require approximately 3,000 degree-days to ripen.

Suppose the temperature (degrees Fahrenheit) in the fields is given by

$$T(t) = 70 + 10 \sin(2\pi t)$$

where t is time in days. Write an expression involving definite integrals that represents the number of degree-days accumulated from day 0 to day x, evaluate this expression, and find the time x at which 3,000 degree-days have accumulated.

Solution   Since $T(t) \geq 50$ for all t (i.e., the lower developmental threshold does not require consideration), the number of degree-days accumulated by day x is given by $F(x) = \int_0^x (T(t) - 50)\, dt$. Integrating yields

$$F(x) = \int_0^x (T(t) - 50)\, dt$$
$$= \int_0^x (20 + 10 \sin(2\pi t))\, dt$$
$$= \left[20t - \frac{10}{2\pi} \cos(2\pi t) \right] \Bigg|_0^x$$
$$= 20x - \frac{5}{\pi} \cos(2\pi x) + \frac{5}{\pi}$$

Since $F'(x) = 20 + 10\sin(2\pi x) > 0$, F is an increasing function. Therefore, if we find a positive solution, then it is the only solution. Notice that if x is an integer, then

$$F(x) = 20x - \frac{5}{\pi} + \frac{5}{\pi} = 20x$$

Solving

$$20x = 3{,}000$$
$$x = 150$$

The grapes will be ready for picking in 150 days. Although the first part of this solution required the fundamental theorem of calculus, the latter part of the problem could be answered using the geometrical interpretation of the definite integral. ■

Indefinite integrals

Since the fundamental theorem of calculus ensures the existence of antiderivatives via integrals, it is appropriate to introduce a notation for the general antiderivative.

Indefinite Integral

If $f(x)$ is a continuous function, then

$$\int f(x)\,dx$$

is called the **indefinite integral** of f and is equal to the general antiderivative of f.

The fact that the indefinite integral has no upper limit of integration or lower limit of integration distinguishes it from a definite integral. It is important to remember that the indefinite integral represents a family of functions (a particular function plus an arbitrary constant), whereas the definite integral of a specified function represents a value.

Example 7   Finding indefinite integrals

Compute the following indefinite integrals.

a. $\displaystyle\int e^x\,dx$ **b.** $\displaystyle\int x^3 + 2\,dx$ **c.** $\displaystyle\int \sec^2 x\,dx$

Solution
a. $\int e^x\,dx = e^x + C$ where C is an arbitrary constant.

b. $\int (x^3 + 2)\,dx = \dfrac{x^4}{4} + 2x + C$ where C is an arbitrary constant.

c. $\int \sec^2 x\,dx = \tan(x) + C$ where C is an arbitrary constant. ■

PROBLEM SET 5.4

Level 1 DRILL PROBLEMS

In Problems 1 to 8, evaluate the definite integral.

1. a. $\displaystyle\int_{-10}^{10} 6\,dx$

b. $\displaystyle\int_{-3}^{5} (2x + a)\,dx$

2. a. $\displaystyle\int_{-5}^{7} (-3)\,dx$

b. $\displaystyle\int_{-2}^{2} (b - x)\,dx$

3. a. $\displaystyle\int_{0}^{4} (x^2 - 1)\,dx$

b. $\displaystyle\int_{0}^{\pi} (\sin x + x)\,dx$

4. a. $\int_{-1}^{1} (x^3 + bx^2)\, dx$ **b.** $\int_{-2}^{-1} \frac{b}{x^2}\, dx$

5. a. $\int_{0}^{9} \sqrt{x}\, dx$ **b.** $\int_{0}^{1} (5u^7 + \pi^2)\, du$

6. a. $\int_{0}^{27} \sqrt[3]{x}\, dx$ **b.** $\int_{0}^{1} (7u^8 + \sqrt{\pi})\, du$

7. a. $\int_{1}^{2} (2x)^\pi\, dx$ **b.** $\int_{-1}^{1} e^{x+1}\, dx$

8. a. $\int_{1}^{2} x^{2a}\, dx, a \neq -\frac{1}{2}$ **b.** $\int_{1}^{2} x^{2a}\, dx, a = -\frac{1}{2}$

Find the indefinite integrals in Problems 9 to 16.

9. a. $\int (4t^3 + 3t^2)\, dt$ **b.** $\int (-8t^3 + 15t^5)\, dt$

10. a. $\int \frac{dx}{2x}$ **b.** $\int 14e^x\, dx$

11. a. $\int (-3\cos u)\, du$ **b.** $\int (5t^3 - \sqrt{t})\, dt$

12. a. $\int 2\sin\theta\, d\theta$ **b.** $\int \frac{\cos\theta}{3}\, d\theta$

13. a. $\int \sqrt{x}(x+1)\, dx$ **b.** $\int \sqrt{t}(t - \sqrt{t})\, dt$

14. a. $\int \frac{x^2 + 1}{x^2}\, dx$ **b.** $\int \frac{x^2 + x - 1}{\sqrt{x}}\, dx$

15. a. $\int \frac{x^2 - 4}{x - 2}\, dx$ **b.** $\int \frac{x^2 - 1}{x + 1}\, dx$

16. a. $\int (\sin^2 x + \cos^2 x)\, dx$ **b.** $\int (\sec^2 t - \tan^2 t)\, dt$

Compute $F'(x)$ for the functions given in Problems 17 to 22.

17. $F(x) = \int_{-1}^{x} \sqrt{1 + u^2}\, du$

18. $F(x) = \int_{\pi/3}^{x} (\sec^2 t)\tan t\, dt$

19. $F(x) = \int_{4}^{x} \frac{dt}{2 + \sin(t^2)}$

20. $F(x) = \int_{1}^{x} \frac{\sin u}{u}\, du$

21. $F(x) = \int_{x}^{2} \frac{e^u}{u}\, du$

22. $F(x) = \int_{x}^{3} e^{t^2}\, dt$

23. Find a function f and a number a such that
$$\int_{0}^{x} f(t)\, dt = \cos(2x) + a$$

24. Find a function f and a number a such that
$$\int_{a}^{x} f(t)\, dt = \ln(x) + 4$$

25. a. If $F'(x) = \frac{1}{\sqrt{x}} - 4$ and $F(1) = 0$, find F.

   b. Sketch the graphs of $y = F(x)$, $y = F(x) + 3$, and $y = F(x) - 1$.

26. Let $F(x) = \int_{-2}^{x} f(u)\, du$ where the graph of f is shown in Figure 5.29.

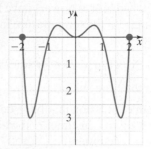

Figure 5.29 Graph of f

a. For what values of x does $F(x)$ have a local maximum or minimum?

b. For what values of x is F concave up? concave down?

c. At what values of x does $F(x)$ achieve a global maximum? global minimum?

d. Sketch the graph of $F(x)$.

27. Let $G(x) = \int_{0}^{x} g(u)\, du$ where the graph of g is shown in Figure 5.30.

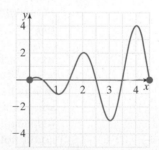

Figure 5.30 Graph of g

a. For what values of x does $G(x)$ have a local maximum or minimum?

b. For what values of x is G concave up? concave down?

c. At what values of x does $G(x)$ achieve a global maximum? global minimum?

d. Sketch the graph of $G(x)$.

Level 2 APPLIED AND THEORY PROBLEMS

28. Use the model for bighorn rams formulated by Jorgenson and colleagues (see Example 3) to find the net increase in length of a ram's horn from $x = 3$ to $x = 7$.

29. Use the model for bighorn rams formulated by Jorgenson and colleagues (see Example 3) to find the net increase in length of a ram's horn from $x = 5$ to $x = 9$.

30. Citrus flowers in Tulare County, California, have a lower developmental threshold of 49°F and require approximately 767 degree-days to reach petal-fall (i.e., 50% of citrus flowers have lost their petals). Suppose the temperature in the fields is given by

$$T(x) = 74 + 14\sin(2\pi x)$$

where x is the time in days.

   a. Write an expression involving definite integrals that represents the time, x, at which 767 degree-days have accumulated.

   b. Use technology to estimate the time x.

31. Sweet corn in western Oregon has a lower developmental threshold of 50°F and requires

approximately 1,597 degree-days to reach maturity. Suppose the temperature in the fields is given by

$$T(x) = 68 + 17\sin(2\pi x)$$

where x is the time in days.

   a. Write an expression involving definite integrals that represents the time, x, at which 1,597 degree-days have accumulated.

   b. Use technology to estimate the time x.

32. The rate of change of ant diversity along an elevational gradient is given by

$$F'(x) = 24.9 - 15.4\,x \text{ species on average per km}$$

where x is elevation above sea level measured in kilometers. If $F(1) = 6.9$, find an expression for $F(x)$, the number of ant species at an elevation of x km. Compare your answer to Example 4 of Section 2.7.

33. Prove the evaluation theorem (Theorem 5.2) using the fundamental theorem of calculus (Theorem 5.3).

5.5   Substitution

In the next two sections, we discuss three techniques of integration: substitution, integration by parts, and partial fractions. The first two of these techniques are counterparts to rules of differentiation. Unlike differentiation, techniques of integration are incomplete; not every function has an elementary representation of its indefinite integral. Consequently, part of the skill you need to acquire is knowing which integration techniques apply to which functions. Since it is possible to compute integrals using technology, you may be asking: Why bother with these techniques? There are several reasons. First, by learning integration techniques and determining when and in which order to use them, you gain insight into how these techniques work. Second, sometimes technology needs a helping hand; implementing an integration technique (especially substitution) by hand may allow technology to complete a calculation it could not otherwise do. Third, computing integrals builds important mathematical skills: it's like lifting weights to improve your strength for sports.

Substitution for indefinite integrals

Our integration efforts begin with an antidifferentiation form of the chain rule. Consider the integral

$$\int \frac{2x}{x^2 + 5}\, dx$$

Basic rules of antidifferentiation provide no direct way of computing this integral. However, look carefully at this integral, and note that the derivative of the denominator equals the numerator. This observation suggests *introducing a new variable*, $u = x^2 + 5$, that, on differentiation, yields $\dfrac{du}{dx} = 2x$. In this next step we replace $2x\,dx = \dfrac{du}{dx}dx$ in the integrand with du. A mathematical justification for this step is dealt with in more advanced texts. Finally, integrating with respect to the variable u, we obtain

$$\int \frac{2x}{x^2 + 5}\, dx = \int \frac{2x\, dx}{x^2 + 5}$$

$$= \int \frac{du}{u}$$

$$= \ln |u| + C$$

$$= \ln(x^2 + 5) + C$$

Although we have not justified our steps rigorously, we can verify our answer by differentiating. Since

$$\frac{d}{dx} \ln(x^2 + 5) = \frac{2x}{x^2 + 5}$$

it follows that $\ln(x^2 + 5) + C$ is the general antiderivative of $\dfrac{2x}{x^2 + 5}$.

To see why this approach worked, consider the integral

$$\int f[g(x)]g'(x)\, dx$$

for functions f and g. In our example $\int \dfrac{2x}{x^2 + 5}\, dx$, we have $f(x) = 1/x$ and $g(x) = x^2 + 5$. If F is an antiderivative of f, then

$$\int f[g(x)]g'(x)\, dx = \int F'[g(x)]g'(x)\, dx \qquad \textit{definition of } F$$

$$= \int \frac{d}{dx} F[g(x)]\, dx \qquad \textit{chain rule}$$

$$= F[g(x)] + C \qquad \textit{fundamental theorem of calculus}$$

Equivalently, if we make the change of variables $u = g(x)$, then

$$\int f[g(x)]g'(x)\, dx = F[g(x)] + C$$

$$= F(u) + C \qquad \textit{substitution}$$

$$= \int F'(u)\, du \qquad \textit{fundamental theorem of calculus}$$

$$= \int f(u)\, du \qquad \textit{definition of } F$$

We summarize this procedure in the following box.

Integration by Substitution

Over any interval of x for which $u = g(x)$ is continuously differentiable and f is continuous on the range of this function, the relationship

$$\int f(\underbrace{g(x)}_{u})\, \underbrace{g'(x)\, dx}_{du} = \int f(u)\, du$$

holds.

Example 1   Integration by substitution

Find

$$\int 9(x^2 + 3x + 5)(2x + 3)\, dx$$

Solution   For the procedure of substitution, we need to identify the appropriate change of variables.

$$\int 9(x^2 + 3x + 5)(2x + 3)\, dx$$

Let $u = x^2 + 3x + 5$. Then, $du = (2x + 3)\, dx$ and

$$\int 9(x^2 + 3x + 5)(2x + 3)\, dx = \int 9u\, du$$

$$= \frac{9}{2}u^2 + C \qquad\qquad power\ rule$$

$$= \frac{9}{2}(x^2 + 3x + 5)^2 + C \qquad return\ to\ original\ variable$$

This procedure may seem difficult now, because you may not be sure of what to let the variable u represent. Just remember, at least initially, you are looking for one part of the integrand that is the derivative of another part of the integrand. If you practice enough, things will get easier!

Example 2   Substitution with a radical function

Find $\int \sqrt{3x + 7}\, dx$.

Solution   Let $u = 3x + 7$, so $du = 3dx$ or $dx = \dfrac{du}{3}$. Substituting and integrating yields

$$\int \sqrt{3x + 7}\, dx = \int \sqrt{u}\, \frac{du}{3} \qquad substitute$$

$$= \frac{1}{3}\int u^{1/2}\, du \qquad simplify$$

$$= \frac{1}{3} \times \frac{2}{3}u^{3/2} + C \qquad power\ rule$$

$$= \frac{2}{9}(3x + 7)^{3/2} + C \qquad return\ to\ original\ variable$$

As the previous two examples illustrate, after making a substitution and simplifying, there should be no x values in the integrand. Sometimes eliminating all the x's requires additional work.

Example 3   Substitution with leftover x values

Find $\int x\sqrt{4x+5}\,dx$.

Solution   Let $u = 4x+5$. Then, $du = 4dx$ and $dx = \dfrac{du}{4}$.

$$\int x\sqrt{4x+5}\,dx = \int x\sqrt{u}\,\frac{du}{4} \qquad \textit{substitute}$$

$$= \frac{1}{4}\int x\sqrt{u}\,du \qquad \begin{array}{l}\textit{leftover x value: since } u = 4x+5,\\ \textit{it follows that } x = \dfrac{u-5}{4}\end{array}$$

$$= \frac{1}{4}\int \frac{u-5}{4}\sqrt{u}\,du$$

$$= \frac{1}{16}\int (u\sqrt{u} - 5\sqrt{u})\,du \qquad \textit{simplify}$$

$$= \frac{1}{16}\int u^{3/2}\,du - \frac{5}{16}\int u^{1/2}\,du \qquad \textit{difference rule}$$

$$= \frac{1}{16}\frac{u^{5/2}}{\frac{5}{2}} - \frac{5}{16}\frac{u^{3/2}}{\frac{3}{2}} + C \qquad \textit{power rule}$$

$$= \frac{1}{40}(4x+5)^{5/2} - \frac{5}{24}(4x+5)^{3/2} + C \qquad \begin{array}{l}\textit{simplify and return to}\\ \textit{the original variable}\end{array}$$

Example 4   Substitution with a trigonometric function

Find
$$\int \tan x\,dx$$

Solution   Recall that $\tan x = \dfrac{\sin x}{\cos x}$ and $\dfrac{d}{dx}\cos x = -\sin x$. Hence, using $u = \cos x$ gives

$$\int \tan x\,dx = \int \frac{\sin x}{\cos x}\,dx$$

$$= \int \frac{1}{\cos x}(\sin x\,dx) \qquad \textit{let } u = \cos x, \textit{ so } du = -\sin x\,dx$$

$$= \int \frac{1}{u}(-1)\,du \qquad \textit{substitution}$$

$$= -\ln|u| + C \qquad \textit{antiderivative of } 1/u$$

$$= -\ln|\cos x| + C \qquad \textit{return to the original variable}$$

Some of you will, no doubt, have access to a calculator or a computer program that can assist in the process of integration. Although technology is very useful, there are times when evaluating an integral by hand will result in a simpler form of the answer. In other cases, technology will give an incomplete form that needs to be adjusted.

Example 5   Using technology to integrate

Use technology to find

$$\int \tan x \, dx$$

and compare the result with the results shown in Example 4.

Solution   Some calculators or software programs will return the answer $-\ln(|\cos(x)|)$ and others may return $-\ln\cos x$. In either case, compared with the previous example, the constant, C, is missing.   ∎

Lest you think that if you purchase a calculator you will not need to study and master techniques of integration, consider the following example.

Example 6   Use substitution with the help of technology

Find a closed-form algebraic expression for

$$\int (1 + \ln x)\sqrt{1 + (x \ln x)^2} \, dx$$

Solution   If we apply technology to solve this problem using the form of the integrand as it stands, we will likely not obtain a satisfactory solution. However, if we first substitute $u = x \ln x$ and $du = (1 + \ln x) \, dx$ then we have

$$\int (1 + \ln x)\sqrt{1 + (x \ln x)^2} \, dx = \int \sqrt{1 + u^2} \, du$$

Using technology on this transformed integrand will yield one of the following two particular antiderivatives:

Expression 1

$$\frac{1}{2}\left(u\sqrt{1 + u^2} + \sinh^{-1} u\right)$$

where $\sinh\theta = (e^\theta - e^{-\theta})/2$ is called the hyperbolic sine function and $\sinh^{-1}$ is its inverse.

Expression 2

$$\frac{\ln(|\sqrt{u^2 + 1} + u|)}{2} + \frac{u\sqrt{u^2 + 1}}{2}$$

Despite looking quite different, these two expressions must be algebraically equivalent up to a constant. If you cannot show this algebraically, you can graph the expressions to demonstrate their equivalence.   ∎

Substitution for definite integrals

We have two methods for dealing with definite integrals. The first is to return to the original variable (as we did with indefinite integrals). The second is to keep track of the change of variables in the limits of integration. We illustrate both of these methods by revisiting Example 4.

Method I: Return to the original variable.

$$\int_0^{\pi/4} \tan x \, dx = \int_{x=0}^{x=\pi/4} -\frac{du}{u} \qquad \text{where } u = \cos x$$

$$= -\ln|u| \Big|_{x=0}^{x=\pi/4}$$

$$= -\ln|\cos x| \Big|_0^{\pi/4}$$

$$= -\ln|\cos(\pi/4)| - (-\ln|\cos(0)|)$$

$$= -\ln\left(\frac{1}{\sqrt{2}}\right) + \ln 1$$

$$= \ln\sqrt{2} + 0 = \ln\sqrt{2}$$

Method II: Keep track of the change of variables in the limits of integration.

$$\int_0^{\pi/4} \tan x \, dx = \int_1^{1/\sqrt{2}} \frac{-du}{u} \qquad \begin{array}{l} \text{If } x = 0 \text{ (lower limit), then } u = \cos 0 = 1, \text{ and} \\ \text{if } x = \pi/4 \text{ (upper limit), then } u = \cos(\pi/4) = 1/\sqrt{2}. \end{array}$$

$$= -\ln|u| \Big|_1^{1/\sqrt{2}} \qquad \begin{array}{l} \textit{Since limits of integration were changed, it is not} \\ \textit{necessary to return to the original variable.} \end{array}$$

$$= -\ln\left(\frac{1}{\sqrt{2}}\right) + \ln 1 \qquad \textit{evaluate}$$

$$= \ln\sqrt{2}$$

Consider the general case of this second method. Let $u = g(x)$ and $F(x)$ be an antiderivative of $f(x)$. Then,

$$\int_a^b f(g(x))g'(x)\,dx = \int_a^b F'(g(x))g'(x)\,dx \qquad \textit{definition of } F(x)$$

$$= \int_a^b \frac{d}{dx} F(g(x))\,dx \qquad \textit{chain rule for differentiation}$$

$$= F[g(b)] - F[g(a)] \qquad \textit{fundamental theorem of calculus}$$

$$= \int_{g(a)}^{g(b)} f(u)\,du \qquad \textit{fundamental theorem of calculus}$$

We summarize this observation in the following box.

Substitution with Definite Integrals

If $g'(x)$ is a continuous function on $[a, b]$ and f is continuous on the range of $u = g(x)$, then

$$\int_a^b f(g(x))g'(x)\,dx = \int_{g(a)}^{g(b)} f(u)\,du$$

Example 7 United States Population Growth

The logistic formula

$$P(t) = \frac{389.2e^{0.23t}}{e^{0.23t} + e^4}$$

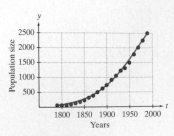

Figure 5.31 U.S. population growth (in millions)

provides a reasonably good fit to the population of the United States (in millions) during the period 1790–1990, as illustrated in Figure 5.31. The variable t is the time (in decades) after 1790. Thus, $t = 0$ for 1790, $t = 20$ for 1990. Suppose that each person eats food at a rate of one annual-person-ration (APR) per year. Find the total number of APRs of food eaten in the United States between 1790 and 1990.

Solution   Since each person consumes one APR per year, the rate at which food is being eaten in decade t is $10 P(t)$ APRs per decade. To find the number of APRs consumed over 20 decades, we integrate $10 P(t)$ from $t = 0$ to $t = 20$.

$$10 \int_0^{20} P(t)\, dt = 3892 \int_0^{20} \frac{e^{0.23t}}{e^{0.23t} + e^4}\, dt \qquad \begin{array}{l} \text{Let } u = e^{0.23t} + e^4, \text{ then } du = 0.23e^{0.23t}\, dt \\ \text{If } t = 0, \text{ then } u = 1 + e^4 \\ \text{If } t = 20, \text{ then } u = e^{4.6} + e^4 \end{array}$$

$$= 3892 \int_{1+e^4}^{e^{4.6}+e^4} \frac{\frac{du}{0.23}}{u}$$

$$\approx 16{,}922 \int_{1+e^4}^{e^{4.6}+e^4} \frac{du}{u}$$

$$= 16{,}922 \ln |u| \Big|_{1+e^4}^{e^{4.6}+e^4}$$

$$\approx 17{,}249 \text{ million}$$

Thus, between the years 1790 and 1990, 17,249 million APRs were consumed! ■

Example 8   Breathing

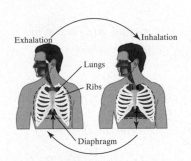

Figure 5.32 Breathing cycle

Breathing is a cyclic process, as illustrated in Figure 5.32. One cycle of breathing from the beginning of inhalation to the end of exhalation takes about five seconds. Since the maximum rate of airflow into the lungs during a typical breath is approximately $\frac{1}{2}$ liter each second, we can model the rate of airflow into the lungs by the function

$$f(t) = \frac{1}{2} \sin\left(\frac{2\pi}{5}t\right) \text{ liter/second}$$

where t is the time in seconds. Find the total amount of air inhaled in one cycle. This volume of is air is known as the *tidal volume*.

Solution   The time for inhalation is $\frac{5}{2}$ seconds, so total amount of air inhaled in one cycle is $\int_0^{5/2} f(t)\, dt$. Let $u = \frac{2\pi}{5}t$, in which case $dt = \frac{5\, du}{2\pi}$. When $t = 0$, $u = 0$, and when $t = 5/2$, $u = \pi$. Hence,

$$\int_0^{5/2} \frac{1}{2} \sin\left(\frac{2\pi}{5}t\right)\, dt = \frac{1}{2} \int_0^{\pi} (\sin u)\frac{5}{2\pi}\, du$$

$$= \frac{5}{4\pi} \int_0^{\pi} \sin u\, du$$

$$= \frac{5}{4\pi}(-\cos u)\Big|_0^{\pi}$$

$$= \frac{5}{4\pi}[-(-1) - (-1)]$$

$$= \frac{5}{2\pi}$$

$$\approx 0.80 \text{ liters}$$

PROBLEM SET 5.5

Level 1 DRILL PROBLEMS

Problems 1 to 8 present pairs of integration problems; one of the pair will use substitution and one will not. As you are working these problems, think about when substitution may be appropriate.

1. a. $\displaystyle\int_0^4 (2t + 4)\, dt$ **b.** $\displaystyle\int_0^4 (2t + 4)^{-1/2}\, dt$

2. a. $\displaystyle\int_0^{\pi/2} \sin\theta\, d\theta$ **b.** $\displaystyle\int_0^1 e^\theta \sin(e^\theta)\, d\theta$

3. a. $\displaystyle\int_0^{\pi/2} \cos t\, dt$ **b.** $\displaystyle\int_0^\pi t\cos(t^2)\, dt$

4. a. $\displaystyle\int_0^4 \sqrt{x}\,dx$ **b.** $\displaystyle\int_{-4}^0 \sqrt{-x}\, dx$

5. a. $\displaystyle\int_0^{16} \sqrt[4]{x}\, dx$ **b.** $\displaystyle\int_0^1 \sqrt[4]{x+2}\, dx$

6. a. $\displaystyle\int x(3x^2 - 5)\, dx$ **b.** $\displaystyle\int x(3x^2 - 5)^5\, dx$

7. a. $\displaystyle\int x^2\sqrt{2x^3}\, dx$ **b.** $\displaystyle\int 6x^2\sqrt{2x^3 - 5}\, dx$

8. a. $\displaystyle\int (2x + 1)\, dx$ **b.** $\displaystyle\int (2x + 1)^{1,000}\, dx$

Use substitution to find the indefinite integrals in Problems 9 to 16.

9. $\displaystyle\int (2x + 3)^4\, dx$ **10.** $\displaystyle\int (5x - 2)^{20}\, dx$

11. $\displaystyle\int x\sqrt{x^2 + 4}\,dx$ **12.** $\displaystyle\int \frac{x\, dx}{\sqrt{x^2 + 1}}$

13. $\displaystyle\int \cot x\, dx$ **14.** $\displaystyle\int \sin^3 t \cos t\, dt$

15. $\displaystyle\int \frac{\ln x}{x}\, dx$ **16.** $\displaystyle\int \frac{z^3\, dz}{\sqrt{z^4 + 12}}$

Use substitution to evaluate the definite integrals in Problems 17 to 24.

17. $\displaystyle\int_{-1}^2 (5x^2 - x)^2 (10x - 1)\, dx$

18. $\displaystyle\int_0^1 \frac{5x^2\, dx}{2x^3 + 1}$ **19.** $\displaystyle\int_1^2 \frac{e^{1/x}}{x^2}\, dx$

20. $\displaystyle\int_0^1 \frac{\ln(x + 1)}{x + 1}\, dx$ **21.** $\displaystyle\int_0^1 \frac{0.58e^{0.2x}}{1 + e^{0.2x}}\, dx$

22. $\displaystyle\int_0^{12} \frac{5{,}000e^{0.2t}\, dt}{e^{0.2t} + 10}$ **23.** $\displaystyle\int_1^2 x\sqrt{x - 1}\, dx$

24. $\displaystyle\int_0^2 (e^x - e^{-x})^2\, dx$

25. Find

$$\int e^{\sin x} \cos x\, dx$$

26. Assume that f is continuous and $\int_1^8 f(x)\, dx = 12$. Find

$$\int_1^2 f(x^3)x^2\, dx$$

27. Assume that f is continuous and $\int_1^6 f(2x)\, dx = -3$. Find

$$\int_2^{12} f(x)\, dx$$

Level 2 APPLIED AND THEORY PROBLEMS

28. In Example 7, U.S. population growth was modeled by

$$P(t) = \frac{389.2e^{0.23t}}{e^{0.23t} + e^4} \text{ millions of individuals}$$

where t is decades after 1790. If each person eats food at a rate of one ration per year, find the total number of rations of food eaten in the United States between 1800 and 1900.

29. Assume that a dust mite population starts with 10 dust mites and grows at a rate of $10e^{0.3t}$ dust mites per hour. How many dust mites will there be one day from now?

30. Suppose an environmental study indicates that the ozone level, L, in the air above a major metropolitan center is changing at a rate modeled by the function

$$L'(t) = \frac{0.24 - 0.03t}{\sqrt{36 + 16t - t^2}}$$

parts per million per hour (ppm/h) t hours after 7:00 A.M.

a. Express the ozone level $L(t)$ as a function of t if L is 4 ppm at 7:00 A.M.

b. Use the graphing utility of your calculator to find the time between 7:00 A.M. and 7:00 P.M. when the highest level of ozone occurs. What is the highest level?

31. *The Gompertz law* of tumor growth is given by the equation

$$\int \frac{dN}{N \ln(N/b)} = -\int at \, dt$$

where N is the size of the tumor, t is time (measured in days), b is the asymptotic size of the tumor, and a is a measurement of the tumor growth rate. Assume $a = 1$ and $b = 10$. Integrate both sides of the Gompertz equation and solve for N in terms of t. To get rid of the integration constant, assume that N equals 5 at time $t = 0$.

32. In Example 6 of Section 1.5, we modeled the uptake of glucose by bacterial populations off of the coast of Peru by the function

$$f(x) = \frac{1.2078x}{1 + 0.0506x} \text{ micrograms per hour}$$

where x is micrograms of glucose per liter. Suppose the concentration of glucose x is decaying exponentially in time: $x(t) = 100 \, e^{-0.01t}$ micrograms per liter where t is measured in hours.

a. Write a function $U(t)$ that describes how the uptake rate is changing in time.

b. Determine the net uptake of a cell from $t = 0$ to $t = 6$ hours.

33. In Example 5 of Section 2.4, we found that the rate at which wolves kill moose can be modeled by

$$f(x) = \frac{3.36x}{0.42 + x} \quad \begin{array}{l}\text{moose killed per wolf} \\ \text{per hundred days}\end{array}$$

where x is measured in number of moose per km^2. Suppose that the density of moose is increasing exponentially according to the function $x(t) = 0.1e^{0.2t}$ moose per km^2 where t is measured in hundreds of days. Determine the number of moose killed by a wolf from $t = 0$ to $t = 3$.

34. In Problem 42 in Problem Set 2.4, we examined how wolf densities in North America depend on moose densities. We found that the following function provides a good fit to the data:

$$f(x) = \frac{58.7(x - 0.03)}{0.76 + x} \text{ wolves per } 1000 \, km^2$$

where x is number of moose per km^2. Assume the moose density is increasing exponentially according the function $x(t) = 0.1e^{0.2t}$ moose per km^2 where t is measured in hundreds of days. Determine the change in the wolf density from $t = 0$ to $t = 3$.

5.6   Integration by Parts and Partial Fractions

In this section, we consider two important techniques of integration that will be useful in later chapters.

Integration by parts

Integration by parts is a procedure based on inverting the product rule for differentiation. To derive a formula for this procedure, we begin with the product rule for differentiating functions $f(x)$ and $g(x)$, assuming these derivatives exist.

$$\frac{d}{dx} f(x)g(x) = f'(x)g(x) + f(x)g'(x) \qquad \textit{product rule}$$

$$\int \frac{d}{dx} f(x)g(x) \, dx = \int [f'(x)g(x) + f(x)g'(x)] \, dx \qquad \textit{antidifferentiate both sides}$$

$$\int \frac{d}{dx} f(x)g(x) \, dx = \int f'(x)g(x) \, dx + \int f(x)g'(x) \, dx \qquad \textit{properties of integrals}$$

$$f(x)g(x) = \int f'(x)g(x) \, dx + \int f(x)g'(x) \, dx \qquad \textit{FTC}$$

$$f(x)g(x) - \int f'(x)g(x) \, dx = \int f(x)g'(x) \, dx \qquad \textit{subtract } \int f'(x)g(x)dx \textit{ from both sides}$$

If we let $u = f(x)$ and $v = g(x)$, then $du = f'(x) \, dx$, $dv = g'(x) \, dx$, and we obtain the following simplified formula.

Integration by Parts

$$\int u\, dv = uv - \int v\, du$$

To evaluate integrals using integration by parts, we want to choose u and dv so that the new integral is easier to integrate than the original integral.

Example 1 Integration by parts

Find

$$\int xe^x\, dx$$

Solution For this example, there are two ways we can choose u and dv. Suppose we choose $u = x$ and $dv = e^x\, dx$. We differentiate u and integrate dv. Thus, $du = dx$, and $v = e^x$. Now, substitute these values into the integration by parts formula.

$$\int u\, dv = uv - \int v\, du$$

$$\int xe^x\, dx = xe^x - \int e^x\, dx$$

$$= xe^x - e^x + C$$

We noted that there were two possible choices for u and dv in Example 1. The other choice is to let $u = e^x$ and $dv = x\, dx$. If we make this choice, and substitute into the formula for integration by parts, then we obtain

$$\int xe^x\, dx = \frac{1}{2}x^2 e^x - \int \frac{1}{2}x^2 e^x\, dx$$

(Try this yourself to practice the technique.) In this case, instead of simplifying the problem and solving it, we just made it more complicated! Thus, it is important to try different substitutions to see which works best.

Example 2 When the differentiable part is the entire integrand

Find

$$\int \ln x\, dx$$

assuming $x > 0$.

Solution Let $u = \ln x$ and $dv = dx$. Then, $du = \dfrac{dx}{x}$, $v = x$, and

$$\int u\, dv = uv - \int v\, du$$

$$\int \ln x\, dx = (\ln x)x - \int x\frac{dx}{x}$$

$$= x\ln x - \int dx$$

$$= x\ln x - x + C = x(\ln x - 1) + C$$

Example 3 Repeated use of integration by parts

Find

$$\int x^2 e^{2x}\, dx$$

Solution   Let $u = x^2$ and $dv = e^{2x}\,dx$. Then, $du = 2x\,dx$, $v = \frac{1}{2}e^{2x}$, and

$$\int x^2 e^{2x}\,dx = x^2\left(\frac{1}{2}e^{2x}\right) - \int \frac{1}{2}e^{2x}(2x\,dx)$$

To compute the rightmost integral, we need another application of integration by parts. Let $u = x$ and $dv = e^{2x}\,dx$. Then, $du = dx$, $v = \frac{1}{2}e^{2x}$, and

$$\int x^2 e^{2x}\,dx = x^2\left(\frac{1}{2}e^{2x}\right) - \int x e^{2x}\,dx$$

$$= \frac{1}{2}x^2 e^{2x} - \left[x\left(\frac{1}{2}e^{2x}\right) - \int \frac{1}{2}e^{2x}\,dx\right]$$

$$= \frac{1}{2}x^2 e^{2x} - \frac{1}{2}x e^{2x} + \frac{1}{4}e^{2x} + C$$

$$= \frac{1}{4}e^{2x}(2x^2 - 2x + 1) + C$$

In the next example, it is necessary to apply integration by parts more than once. As you will see, though, when we do so a second time, we return to the original integral.

Example 4   There and back again

Find

$$\int e^x \cos x\,dx$$

Solution   For this problem it will be useful to call the initial antiderivative I and ignore the constant of integration until the end of the calculation. That is, let

$$I = \int e^x \cos x\,dx$$

$$= e^x \sin x - \int \sin x\,(e^x\,dx) \qquad \begin{array}{l}\text{let } u = e^x \text{ and } dv = \cos x\,dx \\ \text{so that } du = e^x\,dx, \text{ and } v = \sin x; \\ \text{use integration by parts}\end{array}$$

$$= e^x \sin x - \left[-e^x \cos x - \int(-\cos x)e^x\,dx\right] \qquad \begin{array}{l}\text{let } u = e^x \text{ and } dv = \sin x\,dx \\ \text{so that } du = e^x\,dx, \text{ and } v = -\cos x; \\ \text{use integration by parts again}\end{array}$$

$$= e^x \sin x + e^x \cos x - \int e^x \cos x\,dx \qquad \text{simplify}$$

$$= e^x \sin x + e^x \cos x - I \qquad \text{notice the integral is } I$$

$$2I = e^x \sin x + e^x \cos x \qquad \text{add } I \text{ to both sides}$$

$$I = \frac{e^x}{2}(\cos x + \sin x) \qquad \begin{array}{l}\text{divide both sides by 2 and factor} \\ \text{the common factor on the right}\end{array}$$

Hence, the general form of the antiderivative is

$$\int e^x \cos x\,dx = \frac{e^x}{2}(\cos x + \sin x) + C$$

Sometimes you need to combine techniques to conquer an integral.

Example 5   Combining substitution and integration by parts

Find

$$\int x^3 e^{-x^2} \, dx$$

The function $x^3 e^{-x^2}$ arises in the context of considering the properties of the *Gaussian or normal distribution*, which is the most important distribution in statistics.

Solution

$$\int x^3 e^{-x^2} \, dx = \int t e^{-t} \frac{dt}{2} \qquad \qquad \textit{substitution: let } t = x^2, \textit{ so } dt = 2x \, dx$$

$$= \frac{1}{2} \int t e^{-t} \, dt \qquad \qquad \begin{array}{l} \textit{integration by parts:} \\ u = t, dv = e^{-t} \, dt; \\ \textit{so } du = dt, v = -e^{-t} \end{array}$$

$$= \frac{1}{2} \left[t(-e^{-t}) - \int (-e^{-t} \, dt) \right]$$

$$= \frac{1}{2} \left[-t e^{-t} + \int e^{-t} \, dt \right]$$

$$= \frac{1}{2} [-t e^{-t} - e^{-t} + C]$$

$$= -\frac{1}{2} e^{-t}(t + 1) + C \qquad \qquad \textit{rename C/2 just C (an arbitrary constant)}$$

$$= -\frac{1}{2} e^{-x^2}(x^2 + 1) + C \qquad \qquad \textit{return to the original variable x}$$

Integration by parts extends to definite integrals in a natural way.

Integration by Parts with Definite Integrals	If $f(x)$ and $g(x)$ are differentiable functions of x on the interval $[a, b]$, then $$\int_a^b f(x)g'(x) \, dx = f(x)g(x)\Big|_a^b - \int_a^b f'(x)g(x) \, dx$$

Example 6   Integration by parts with a definite integral

Evaluate

$$\frac{1}{4} \int_0^t s e^{-s/2} \, ds$$

Solution   Let $u = s$ and $dv = e^{-s/2} \, ds$, so that $du = ds$ and $v = -2e^{-s/2}$.

$$\frac{1}{4} \int_0^t s e^{-s/2} \, ds = \frac{1}{4} \left[s(-2e^{-s/2})\Big|_0^t - \int_0^t (-2e^{-s/2}) \, ds \right]$$

$$= \frac{1}{4} \left[-2t e^{-t/2} - \left(4e^{-s/2}\Big|_0^t \right) \right]$$

$$= \frac{1}{4} [-2t e^{-t/2} - (4e^{-t/2} - 4)]$$

$$= -\frac{1}{2} e^{-t/2}(t + 2 - 2e^{t/2})$$

Example 7   Survival to age t

Suppose a biologist found that for a particular population of monkeys, the proportion of individuals born each year who die before they are t years old is

$$p(t) = \frac{1}{4} \int_0^t s e^{-s/2} ds$$

a. What proportion of individuals dies before the age of 3?

b. What proportion of individuals dies between ages 3 and 4?

c. What proportion of individuals lives to be at least age 6?

d. At what rate is the proportion changing at age 1? age 4?

Solution   From Example 6, we found that $p(t) = -\frac{1}{2} e^{-t/2}(t + 2 - 2e^{t/2})$. In particular, the proportion of individuals dying at age 0 equals $p(0) = 0$.

a. $p(3) = -\frac{1}{2} e^{-3/2}(3 + 2 - 2e^{3/2}) \approx 0.442$; so about 44% of the population will die before the age of 3.

b. The proportion that will die before the age of 4 is

$$p(4) = -\frac{1}{2} e^{-4/2}(4 + 2 - 2e^{4/2}) \approx 0.594$$

Thus, the proportion dying between ages 3 and 4 is

$$p(4) - p(3) \approx 0.152$$

That is, about 15% of the population will die between the ages of 3 and 4.

c. The proportion living to be at least age 6 is one minus the number that die before the age of 6. We find

$$p(6) = -\frac{1}{2} e^{-6/2}(6 + 2 - 2e^{6/2}) \approx 0.800$$

Thus, the desired number is

$$1 - 0.80 = 0.20$$

Therefore, we would expect 20% of the individuals to live to at least the age of 6.

d. Using properties of integrals and the fundamental theorem of calculus, we have that

$$\frac{d}{dt} \frac{1}{4} \int_0^t s e^{-s/2} ds = \frac{1}{4} \frac{d}{dt} \int_0^t s e^{-s/2} ds$$

$$= \frac{1}{4} t e^{-t/2}$$

Hence, the proportion is changing at an instantaneous rate of $\frac{1}{4} e^{-1/2} \approx 0.1516$ at age 1 and $\frac{1}{4} 4 e^{-2} \approx 0.1353$ at age 4. ∎

Partial fractions

Partial fractions is an integration technique enable us to integrate any *rational function;* that is, functions of the form

$$f(x) = \frac{P(x)}{Q(x)} = \frac{a_0 + a_1 x + a_2 x^2 + \cdots + a_m x^m}{b_0 + b_1 x + b_2 x^2 + \cdots + b_n x^n}$$

Integration problems involving rational functions arise commonly in enzyme kinetics, evolutionary games, and population dynamics. For example, in describing the growth of a population of size $N(t)$ with a growth that is positively impacted by its own size, we may encounter an integral of the form

$$\int \frac{dN}{N(N-1)}$$

The appropriate integration procedure is to write (expand) the rational function $\frac{1}{N(N-1)}$ into a sum of two simpler functions that we can directly integrate. More specifically, we try to find constants A and B such that

$$\frac{1}{N(N-1)} = \frac{A}{N} + \frac{B}{N-1}$$

Equivalently, multiplying both sides by $N(N-1)$ gives

$$1 = A(N-1) + BN = (A+B)N - A$$

This equation holds for all N only if $A = -1$ and $A + B = 0$; in other words, $B = -A = -(-1) = 1$. Thus, we can write

$$\frac{1}{N(N-1)} = -\frac{1}{N} + \frac{1}{N-1}$$

and

$$
\begin{aligned}
\int \frac{1}{N(N-1)} \, dN &= \int \left[\frac{-1}{N} + \frac{1}{N-1} \right] dN \\
&= -\int \frac{dN}{N} + \int \frac{dN}{N-1} \\
&= -\ln|N| + \ln|N-1| + C \\
&= \ln \left| \frac{N-1}{N} \right| + C
\end{aligned}
$$

Although it is possible to deal with all rational functions, we confine our discussion to rational functions $f(x) = P(x)/Q(x)$, such that $Q(x)$ can be expressed as a product of n distinct linear factors:

$$Q(x) = (a_1 + b_1 x)(a_2 + b_2 x) \ldots (a_n + b_n x)$$

If the degree of $P(x)$ is less than the degree of $Q(x)$ (i.e., $n > m$), then we can always find constants $A_1, A_2, \ldots, A_n$ such that

$$\frac{P(x)}{Q(x)} = \frac{A_1}{a_1 + b_1 x} + \frac{A_2}{a_2 + b_2 x} + \cdots + \frac{A_n}{a_n + b_n x}$$

For integration methods for more general rational functions, go online or read another calculus text.

Alternatively, if the degree of $P(x)$ is greater than or equal to the degree of $Q(x)$, then we can perform long division and factor the remainder term. When $Q(x)$ can be decomposed into linear factors, there is a simple method to determine the coefficients. This method is the Heaviside "cover-up" method—named after Oliver Heaviside (see Problem 35, 𝔥istorical 𝔔uest, in Problem Set 5.6).

Example 8   Integrating a rational function

Find

$$\int \frac{x+2}{x^3 - x} \, dx$$

Solution   Since we can express $x^3 - x = x(x^2 - 1) = x(x - 1)(x + 1)$ as a product of distinct linear factors, there exist constants A_1, A_2, and A_3 such that

$$\frac{x + 2}{x(x - 1)(x + 1)} = \frac{A_1}{x} + \frac{A_2}{x - 1} + \frac{A_3}{x + 1}$$

Multiplying both sides of the equation by $x(x - 1)(x + 1)$ yields

$$x + 2 = A_1(x - 1)(x + 1) + A_2 x(x + 1) + A_3 x(x - 1)$$

Since this equation has to hold for all x, we solve for A_1 by simply setting $x = 0$ to get $2 = -A_1$. Hence, $A_1 = -2$. To solve for A_2, we set $x = 1$ to get $3 = 2A_2$. Hence, $A_2 = 3/2$. Finally, to solve for A_3 we set $x = -1$, to get $1 = 2A_3$. Hence, $A_3 = 1/2$. Notice that our choices of x ensured that all but one of the terms on the right-hand side of the equation vanished.

Thus, we can write

$$\int \frac{x + 2}{x^3 - x} = -2 \int \frac{1}{x} \, dx + \frac{3}{2} \int \frac{1}{x - 1} \, dx + \frac{1}{2} \int \frac{1}{x + 1} \, dx$$

$$= -2 \ln |x| + \frac{3}{2} \ln |x - 1| + \frac{1}{2} \ln |x + 1| + C$$

Example 9   Second-order chemical kinetics

Consider two compounds A and B that bind to form a third compound C. Assume a and b are the initial concentrations of A and B. If the rate at which C is produced is proportional to the product of concentrations of A and B, then it has been shown that the following integral equation holds when y is the concentration of C, k is a constant of proportionality, and t is time:

$$\int \frac{dy}{(a - y)(b - y)} = \int k \, dt$$

Integrate both sides of this equation and solve for y as a function of t, assuming that $a = 2$, $b = 1$, $k = 3$, and that $y = 0$ when $t = 0$. Sketch this function.

Solution   Solve

$$\int 3 \, dt = \int \frac{dy}{(2 - y)(1 - y)} \qquad \textit{given equation}$$

$$= \int \frac{dy}{(y - 2)(y - 1)} \qquad \textit{more familiar form}$$

To deal with the right-hand side, we need to solve for A_1 and A_2 such that

$$\frac{1}{(y - 2)(y - 1)} = \frac{A_1}{y - 2} + \frac{A_2}{y - 1}$$

for all y. Multiplying both sides by $(y - 2)(y - 1)$ yields

$$1 = A_1(y - 1) + A_2(y - 2)$$

Setting $y = 2$ gives $A_1 = 1$. Setting $y = 1$ gives $A_2 = -1$. Therefore, we can rewrite the equation as

$$\int 3 \, dt = \int \left[\frac{1}{y - 2} + \frac{(-1)}{y - 1} \right] dy$$

$$3t + C = \ln |y - 2| - \ln |y - 1| \qquad \textit{integrating}$$

$$3t + C = \ln \left| \frac{y - 2}{y - 1} \right|$$

$$\left| \frac{y - 2}{y - 1} \right| = e^{3t + C} \qquad \textit{definition of logarithm}$$

$$\frac{y - 2}{y - 1} = \pm e^{3t + C} \qquad \textit{definition of absolute value}$$

If $t = 0$, then $y = 0$; so we have

$$\frac{0-2}{0-1} = \pm e^{3\times 0 + C}$$

$$2 = \pm e^{C}$$

Hence, we need the positive solution $+e^{C}$ to equal 2. Thus,

$$\frac{y-2}{y-1} = (\pm e^{C})e^{3t}$$

$$\frac{y-2}{y-1} = 2e^{3t}$$

$$y - 2 = 2ye^{3t} - 2e^{3t}$$

$$y(1 - 2e^{3t}) = 2 - 2e^{3t}$$

$$y = \frac{2 - 2e^{3t}}{1 - 2e^{3t}}$$

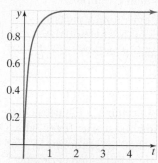

Figure 5.33 Graph of a second-order chemical process

The graph is shown in Figure 5.33. To obtain this graph by hand, it is not hard to check that $y'(t) > 0$ and $\lim\limits_{t\to\infty} y(t) = 1$.

PROBLEM SET 5.6

Level 1 DRILL PROBLEMS

Find each integral in Problems 1 to 10.

1. $\displaystyle\int xe^{-x}\,dx$ **2.** $\displaystyle\int e^{t}\sin t\,dt$

3. $\displaystyle\int x\ln x\,dx$ **4.** $\displaystyle\int x\sin(2x)\,dx$

5. $\displaystyle\int \frac{\ln\sqrt{x}}{\sqrt{x}}\,dx$ **6.** $\displaystyle\int x^{2}\ln x\,dx$

7. $\displaystyle\int e^{2x}\sin(3x)\,dx$ **8.** $\displaystyle\int x^{2}\sin x\,dx$

9. $\displaystyle\int x\sin x\cos x\,dx$

10. $\displaystyle\int \sin^{-1}x\,dx$. Hint: $\dfrac{d}{dx}\sin^{-1}x = \dfrac{1}{\sqrt{1-x^{2}}}$.

Find the exact value of the definite integrals in Problems 11 to 16 using integration by parts. Check your answer by using a calculator to find an approximate answer correct to four decimal places.

11. $\displaystyle\int_{0}^{4} xe^{-x}\,dx$ **12.** $\displaystyle\int_{1}^{e} (\ln x)^{2}\,dx$

13. $\displaystyle\int_{1/3}^{e} 3[\ln(3x)]^{2}\,dx$ **14.** $\displaystyle\int_{0}^{\pi} x\sin x\,dx$

15. $\displaystyle\int_{0}^{\pi} x(\sin x + \cos x)\,dx$ **16.** $\displaystyle\int_{1}^{e} x^{3}\ln x\,dx$

Find the indicated integrals in Problems 17 to 22.

17. $\displaystyle\int \frac{dN}{N(1{,}000 - N)}$ **18.** $\displaystyle\int \frac{x+1}{x(1-x)}\,dx$

19. $\displaystyle\int \frac{x}{x(x-1{,}000)}\,dx$ **20.** $\displaystyle\int \frac{(x+1)\,dx}{(x+2)(x+3)}$

21. $\displaystyle\int \frac{dx}{x(x+1)(x-2)}$

22. $\displaystyle\int \frac{4}{(x+1)(x+2)(x+3)}\,dx$

In Problems 23 to 28, first use an appropriate substitution and then use integration by parts or partial fractions to evaluate the integral. Remember to give your answers in terms of x.

23. $\displaystyle\int \cos(\ln x)\,dx$ **24.** $\displaystyle\int x^{3}\,e^{x^{2}}\,dx$

25. $\displaystyle\int \frac{\ln x\,\sin(\ln x)}{x}\,dx$ **26.** $\displaystyle\int [\sin x\ln(2+\cos x)]\,dx$

27. $\displaystyle\int \frac{e^{2x}\,dx}{e^{2x}+3e^{x}+2}$ **28.** $\displaystyle\int \frac{\cos x\,dx}{(1-\sin x)(2-\sin x)}$

29. a. Evaluate $\displaystyle\int \frac{x^{3}}{x^{2}-1}\,dx$ using integration by parts.

    b. Evaluate the integral using partial fractions.

30. a. Evaluate $\int \cos^2 x\, dx$. Hint: Use the trigonometric identity: $\cos^2 x = \frac{1}{2}(1 + \cos(2x))$.

b. Use part **a** to evaluate $\int x \cos^2 x\, dx$ using integration by parts.

Level 2 APPLIED AND THEORY PROBLEMS

Warner Bros./Photofest

31. The 1988 film *Stand and Deliver*, starring Edward James Olmos pictured above, provides an alternative perspective — tabular integration — on integration by parts. This technique involves writing a table with two columns, one labeled *D* for differentiation and another labeled *I* for integration. The first row of the *D* column contains *u*, the part to be differentiated in the original integral. The second row in the *D* column contains $\frac{du}{dx}$. The third row in the *D* column contains $\frac{d^2u}{dx^2}$. Proceed in this manner until the product of the functions in the last row either equals 0 or is a constant multiple of what you started with. The first row of the *I* column contains *dv*, the part to be integrated. For the second, third, and so on rows in the *I* column, place the successive integrals. For example, rework Example 1, namely $\int xe^x\, dx$, with $u = x$ and $dv = e^x$:

D	*I*
x	e^x
1	e^x
0	e^x

Now, draw diagonal lines from the first element of the *D* column to the second element of the *I* column, from the second element of *D* to the third element of *I* and so forth. Multiply the elements at the ends of each of the diagonal lines, take an alternating sum of these products, and add the integral of the product of terms in the last row. For Example 1, and the table shown here we have

$$xe^x - 1 \times e^x + \int 0\, dx = e^x(x - 1) + \int 0\, dx$$

$$= e^x(x - 1) + C$$

which is the same result shown in Example 1. Use this method to find the integral in Example 2.

32. Use the tabular method from *Stand and Deliver* (Problem 31) on the integral in Example 3.

33. Use the tabular method from *Stand and Deliver* (Problem 31) on the integral in Example 4.

34. Contrast the method of integration by parts as illustrated by the examples in the text and the tabular method from *Stand and Deliver* (Problem 31).

35. 𝕳𝕚𝕤𝕥𝕠𝕣𝕚𝕔𝕒𝕝 𝕼𝕦𝕖𝕤𝕥* Oliver Heaviside (1850–1925) was born in the same London slums as Charles Dickens. When young, Heaviside had scarlet fever, which left him partly deaf. He finished his schooling in 1865 at age 16. He was a top student but failed geometry. Heaviside went to work as a telegrapher, which drew him into the study of electricity. He read Maxwell's *Treatise on Electricity and Magnetism* and managed to reduce Maxwell's whole field theory into two equations.

Next Heaviside picked up the new ideas of vector analysis, which inspired him to develop a theory he called operational calculus, a powerful tool. However, its foundation was not rigorous, so his methods were not accepted by many mathematicians of his day. Only people like Kelvin, Rayleigh, and Hertz saw the brilliance that was driving Heaviside faster than method could follow. Heaviside knew what he was doing. He growled at his detractors, "Shall I refuse my dinner because I do not fully understand . . . digestion?"

Like vector analysis, Heaviside's calculus stood the test of time. So did the rest of his work. For example, he gave us the theory for long distance telephones. Ultimately, Heaviside grew sick of fighting his detractors and retired to Torquay in southwest England. A sketch of Oliver Heaviside is shown in Figure 5.34.

Figure 5.34 Oliver Heaviside (1850–1925)

* Data from *The Engines of Our Ingenuity* public radio program produced by KUHF-FM Houston, Episode No. 426.

Now let us consider Heaviside's so-called cover-up method for determining coefficients with partial fractions. Consider the antiderivative from Example 8. Find A_1, A_2, and A_3 such that

$$\frac{x+2}{x(x+1)(x-1)} = \frac{A_1}{x} + \frac{A_2}{x+1} + \frac{A_3}{x-1}$$

The coefficients are found, one at a time, by "covering" that factor and evaluating the remaining expression by the value that causes the "covered" factor to be zero. That is, first cover x:

$$\frac{x+2}{\boxed{x}(x+1)(x-1)}$$

The covered factor is 0 when $x = 0$, so evaluate the noncovered portion at $x = 0$:

$$\frac{0+2}{(0+1)(0-1)} = -2$$

Thus, $A_1 = -2$. Next, cover the factor under the A_2 term:

$$\frac{x+2}{x\boxed{x+1}(x-1)}$$

Evaluate for $x = -1$:

$$\frac{-1+2}{-1(-1-1)} = \frac{1}{2}$$

Finally, cover the factor under the A_3 term:

$$\frac{x+2}{x(x+1)\boxed{x-1}}$$

Evaluate for $x = 1$:

$$\frac{1+2}{1(1+1)}$$

Thus,

$$\frac{x+2}{x(x+1)(x-1)} = \frac{-2}{x} + \frac{1/2}{x+1} + \frac{3/2}{x-1}$$

Explain why this cover-up method of Heaviside works.

36. Assume that after t hours on the job, a factory worker can produce $100te^{-0.5t}$ units per hour. How many units does the worker produce during the first three hours?

37. After t weeks, suppose that contributions in response to a fund-raising campaign were coming in

at the rate of $2{,}000te^{-0.2t}$ dollars per week. How much money was raised during the first five weeks?

38. An actuary measures the probability that a person in a certain population will die at age x by the formula

$$P(x) = \lambda^2 xe^{-\lambda x}$$

where λ is a parameter such that $0 < \lambda < e$.

a. For a given λ, find the maximum value of $P(x)$.

b. Sketch the graph of $P(x)$.

c. Find the area under the probability curve $y = P(x)$ for $0 \le x \le 10$, and interpret your result.

39. A population P, grows at the rate

$$P'(t) = 5(t+1)\ln\sqrt{t+1}$$

thousand individuals per year at time t (in years). By how much does the population change during the eighth year?

40. Suppose that a drug is assimilated into a patient's bloodstream at a rate modeled by

$$A(t) = 2te^{-0.31t}$$

where t is the number of minutes since the drug was taken. Find the total amount of drug assimilated into the patient's bloodstream during the second minute.

41. Recovering from an environmental perturbation, a (hypothetical) population exhibits dampened oscillations of the form

$$N(t) = 100 + 50\sin(2\pi t)\,e^{-0.01t} \text{ individuals per acre}$$

where t is measured in days. As a part of a sampling effort, a scientist captures and releases individuals from this population at a rate of $0.1N(t)$ individuals per acre per day. If the scientist is sampling one acre, determine the number of individuals she captures and releases in seven days.

42. Recovering from an environmental perturbation, a (hypothetical) population exhibits dampened oscillations of the form

$$N(t) = 50 + 50\cos(\pi t)\,e^{-0.2t} \text{ individuals per acre}$$

where t is measured in days. As a part of a sampling effort, a scientist captures and releases individuals from this population at a rate of $0.01N(t)$ individuals per acre per day. If the scientist is sampling one acre, determine the number of individuals she captures and releases in ten days.

5.7   Numerical Integration

We have seen that integration is, in general, a more difficult task than differentiation. In differentiation, knowing the derivatives of several elementary functions (e.g., $\sin x$, e^x, and x^n) and a set of basic rules (e.g., product rule, chain rule, and quotient rule) allows us to differentiate rather complex-looking functions. In contrast, integration is more complicated. The number of rules and special cases, and uncertainty about which rule to apply, make integration more of an art than a science. An optimist would argue that armed with enough rules, and a great deal of practice, we could express the integral of any reasonable continuous function in terms of familiar functions. Unfortunately, this is not true.

You may have encountered some of these functions. For instance, some calculators return the same integral when asked to compute $\int e^{-x^2}\,dx$, while others return the answer $\frac{1}{2}\sqrt{\pi}\,\mathrm{erf}(x)$. This answer becomes circular when you find that the definition of the erf function, which is short for **error function**, is

$$\mathrm{erf}(x) = \frac{2}{\sqrt{\pi}} \int_0^x e^{-t^2}\,dt \quad \text{for all } x \geq 0$$

Why is technology of no help when it comes to integrating this function analytically?

To answer this question, recall that an "elementary function" is a function that can be expressed in terms of power functions, exponentials, sines, cosines, and logarithms via the usual algebraic processes, including the solving (with or without radicals) of polynomials. Thus, *elementary functions* are all the "precalculus functions," including polynomials and trigonometric and logarithmic functions. A theorem in mathematics asserts that there are an infinite number of elementary functions without an elementary antiderivative. Here are some examples of these functions:

$$\int e^{x^2}\,dx, \quad \int \frac{\sin x}{x}\,dx, \quad \int \sin x^2\,dx, \quad \int \sqrt{1 + x^3}\,dx, \quad \int x^x\,dx, \quad \int \frac{dx}{\ln x}$$

The more common ones get their own names. For instance, we saw that up to some scaling factors, "erf" is the antiderivative of e^{-x^2}. We can also find out using technology that "Si" is the antiderivative of $\sin x / x$. Unfortunately, these functions are not exceptional.

What can we do when we need to integrate functions that are integrable but do not have elementary derivatives? If they are definite integrals, we could approximate them using Riemann sums, or we could use one of several other approximation schemes. In this section, we discuss five numerical schemes for approximating definite integrals. Three of these schemes, *left endpoint rule*, *right endpoint rule*, and *midpoint rule*, differ only in the manner that the sample points x_i are chosen. A fourth scheme, *the trapezoidal rule*, and a fifth scheme, *Simpson's rule*, involve approximating the functions with piecewise linear and piecewise quadratic segments, respectively. These five schemes, three of which are depicted in Figure 5.35, differ in how rapidly they converge (as $n \to \infty$) to the true value of the definite integral. These rates of convergence are described via error estimates.

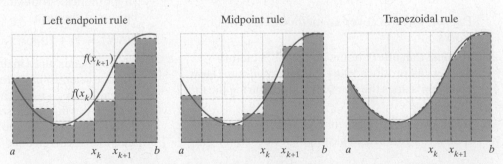

Figure 5.35 Numerical estimates of the area under a curve on the interval $[a, b]$ using the left endpoint, midpoint, and trapezoidal rules, as labeled.

Left and right endpoint approximations

We begin with the simplest approximation schemes, **left endpoint approximation** and **right endpoint approximation**. For presentation purposes, our discussion focuses on left endpoint approximation (see first panel in Figure 5.35). By analogy, similar statements apply to the right endpoint approximation.

Suppose f is a continuous function from $x = a$ to $x = b$ and we want to estimate $\int_a^b f(x)\,dx$. By definition, $\int_a^b f(x)\,dx$, is a limit of Riemann sums. Consequently, given n, we partition the interval $[a, b]$ into n equal subintervals with endpoints:

$$a = a_0 < a_1 < a_2 < \cdots < a_n = b$$

where $a_1 = a + \Delta x$, $a_2 = a + 2\Delta x, \ldots, a_n = a + n\Delta x$, and $\Delta x = \dfrac{b - a}{n}$. Taking the left endpoints, $x_1 = a_0, x_2 = a_1, \ldots, x_n = a_{n-1}$ as our sample points, we have

$$\int_a^b f(x)\,dx \approx \sum_{k=1}^{n} f(x_k)\Delta x = L_n$$

As a first example, we begin with a simple function, so that we can examine the error generated by taking an approximation.

Example 1 Using technology for a left endpoint approximation

We know

$$\int_1^4 \frac{3\sqrt{x}}{2}\,dx = 7$$

Use the left endpoint rule with $n = 5, 10, 25, 50,$ and 100 to approximate this integral.

Solution We have $f(x) = 3\sqrt{x}/2$, $a = 1$, and $b = 4$. We will show the detail for $n = 5$ and then use technology to generate other values. For $n = 5$, we have $\Delta x = \dfrac{4 - 1}{5} = 0.6$ and

n	x_n	$f(x_n)$	$f(x_n)\,\Delta x$
1	1	1.5	0.9
2	1.6	1.897	1.1382
3	2.2	2.225	1.3350
4	2.8	2.510	1.5060
5	3.4	2.766	1.6596
			Sum = 6.5388

Thus,

$$\int_1^5 \frac{3\sqrt{x}}{2}\,dx \approx \sum_{k=1}^{n} f(x_k)\Delta x \approx 6.539$$

When we use technology to generate the approximation L_n, the other values for n are

n	L_n
5	6.539
10	6.772
25	6.910
50	6.955
100	6.977

Hence, the approximation given by technology appears to converging to the known value of 7.

We might have added another column to our answer for Example 1. What is the error of the approximation? That is,

$$\text{ERROR} = \left| \int_a^b f(x)\,dx - \text{NUMERICAL APPROXIMATION} \right|$$

We calculate the error for each of the n values in the 2nd table of Example 1 and find that the approximations are underestimates by 0.461, 0.228, 0.090, 0.045, and 0.023 for the respective entries in the table. As we would expect, the errors tend to decrease as n increases. This leads to two questions: First, how quickly do the errors decrease with n? Second, if we don't know the true value of the definite integral, how can we estimate the error?

To answer both questions requires introducing error bounds: upper bounds for the magnitude of the error. These upper bounds often involve understanding the derivatives of the integrand. For example, suppose we know for some constant $K_1 > 0$ that $|f'(x)| \le K_1$ for all x between a and b. The evaluation theorem implies that

$$f(x) - f(a) = \int_a^x f'(u)\, du$$

for any point x in $[a, a + \Delta x]$. Since $f'(u) \le K_1$, the dominance property of integrals implies that

$$f(x) - f(a) \le K_1(x - a)$$

for x between a and $a + \Delta x$. Equivalently,

Inequality I: $f(x) \le f(a) + K_1(x - a)$ for x in $[a, a + \Delta x]$

Similarly, since $f'(u) \ge -K_1$, the dominance property of integrals implies

Inequality II: $f(x) \ge f(a) - K_1(x - a)$ for x in $[a, a + \Delta x]$

A graphical interpretation of these two inequalities is shown in Figure 5.36. The graph of $f(x)$ above the interval $[a, a + \Delta x]$ lies in a triangular wedge with area $K_1(\Delta x)^2$. To estimate the error, we can consider three cases: the graph lies entirely in the upper half of triangular wedge (as illustrated in Figure 5.36), lies entirely in the lower half of the triangular wedge, or pass through both halves of the triangular wedge. When the graph lies entirely within one half of the wedge, it lies in a region with area $K(\Delta x)^2/2$. Hence, the error in approximating $\int_a^{a+\Delta x} f(x)dx$ with $f(x_1)\Delta x$ is less than or equal to $K_1(\Delta x)^2/2$. On the other hand, if the graph of $f(x)$ passes through both the upper and lower half of the triangular wedge, then over- and under-estimates partially cancel one another, and the error is less than if the graph lied entirely in either the upper or lower half of the triangular wedge.

Similarly, over any of the subintervals, the error in approximating the actual value with $f(x_k)\Delta x$ is at most $K_1(\Delta x)^2/2$. Let E_L be the error by using a left endpoint approximation. Then, summing the error estimates over the n subintervals yields

$$E_L \le nK_1(\Delta x)^2 = \frac{K_1(b-a)^2}{2n}$$

where

$$E_L = \left| \int_a^b f(x)\, dx - L_n \right|$$

is the error of the left endpoint approximation.

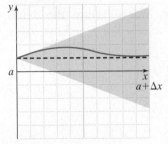

Figure 5.36 Estimating errors for the left endpoint rule

Example 2   Using the left endpoint rule

Consider

$$\int_0^\pi \sin(x^2)\, dx$$

a. Use technology to evaluate this integral.

b. Use the left endpoint rule with $n = 10$ to approximate this integral.

c. Give an error estimate for the approximation found for $n = 10$.

d. Find the smallest value of n that ensures the left endpoint rule approximation has an error no larger than 0.001.

Solution

a. Using technology, we get an estimate of 0.77265.

b. Setting up a table of values for $f(x) = \sin(x^2), a = 0, b = \pi, n = 10$, and $\Delta x = \dfrac{\pi - 0}{10} = \dfrac{\pi}{10}$ (we leave the details for you), we obtain an estimate of 0.78997.

c. To find an upper bound to the error, we need an upper bound to the derivative of $f(x) = \sin(x^2)$ on the interval $[0, \pi]$. Since

$$f'(x) = 2x \cos(x^2)$$

and $|\cos(x^2)| \leq 1$,

$$|f'(x)| \leq 2x \leq 2\pi$$

for x on $[0, \pi]$. Setting $K_1 = 2\pi, n = 10, a = 0$, and $b = \pi$ into the error bound yields

$$E_L \leq \frac{2\pi(\pi - 0)^2}{2 \times 10} \approx 3.101$$

Hence, our estimate of 0.78997 is not guaranteed high-assured accuracy, despite it being reasonably close to the answer we found in part **a**.

d. We want n such that $E_L \leq 0.001$:

$$E_L \leq \frac{K_1(b - a)^2}{2n} \qquad \textit{left endpoint error formula}$$

Thus, we find n such that

$$\frac{K_1(b - a)^2}{2n} \leq 0.001$$

$$K_1(b - a)^2 \leq 0.002n$$

$$2\pi(\pi - 0)^2 \leq 0.002n \qquad \textit{substituting known values}$$

$$\frac{2\pi^3}{0.002} \leq n$$

$$31{,}006.3 \leq n$$

Since n is an integer, we need to choose $n = 31{,}007$. Implementing $n = 31{,}007$ with technology, we obtain

$$I = \int_0^\pi \sin x^2 \, dx \approx 0.77267$$

Thus,

$$0.77267 - 0.001 \leq I \leq 0.77267 + 0.001$$

$$0.77167 \leq I \leq 0.77367$$

The original calculator answer is within this range of accuracy.

The preceding example illustrates several important points. First, even though we initially chose a reasonably large n (say, 10), the error bound was so large that we could not be certain about any of the digits. Second, an extremely large n is needed to ensure an estimate with accuracy to 0.001. Third, the approximation for the large n value was not very different from the estimate for the smaller n value of 10.

Midpoint rule

An alternative numerical approximation scheme is the **midpoint rule** in which the sample points chosen are the midpoints of each of the subintervals (see center panel in Figure 5.35). We begin, as we did with the left endpoints, by partitioning the interval $[a, b]$ into n subintervals with endpoints:

$$a = a_0 < a_1 < a_2 < \cdots < a_n = b$$

where $a_1 = a + \Delta x$, $a_2 = a + 2\Delta x$, ..., $a_n = a + n\Delta x$, and $\Delta x = \dfrac{b-a}{n}$. This time, we take the midpoints,

$$x_1 = \frac{a_0 + a_1}{2}, x_2 = \frac{a_1 + a_2}{2}, \cdots, x_n = \frac{a_{n-1} + a_n}{2}$$

as our sample points to get

$$\int_a^b f(x)\,dx \approx \sum_{k=1}^{n} f(x_k)\Delta x = M_n$$

Associated with the midpoint rule is an error estimate. If $|f''(x)| \le K_2$ for x on $[a, b]$, then the *midpoint error bound* is

$$E_M \le \frac{K_2(b-a)^3}{24n^2}$$

where

$$E_M = \left| \int_a^b f(x)\,dx - M_n \right|$$

Example 3   Using the midpoint rule

Consider

$$\int_0^\pi \sin(x^2)\,dx$$

a. Use the midpoint rule with $n = 10$ to approximate this integral and compare it to the technology solution in the previous example.

b. Provide an error estimate for the approximation found for $n = 10$.

c. Find the smallest value of n that ensures the midpoint rule approximation will have an error no larger than 0.001.

Solution

a. Setting up a table of values for the case $f(x) = \sin(x^2)$, $a = 0$, $b = \pi$, $n = 10$, and $\Delta x = \dfrac{\pi - 0}{10} = \dfrac{\pi}{10}$ (we leave the details to you), we obtain the estimate 0.79918141, which is approximately 0.027 larger than the technology solution 0.77265 in the previous example.

b. To find an upper bound to the error, we need an upper bound to the second derivative of $f(x) = \sin x^2$ on the interval $[0, \pi]$. Since

$$f'(x) = 2x \cos x^2, \quad f''(x) = 2\cos x^2 - 4x^2 \sin x^2$$

and $|\sin x^2| \le 1$ as well as $|\cos x^2| \le 1$,

$$|f''(x)| \le 2 + 4|x|^2 \le 2 + 4\pi^2$$

for x on $[0, \pi]$. Setting $K_2 = 2 + 4\pi^2$, $n = 10$, $a = 0$, and $b = \pi$ into the error bound yields

$$E_M \le \frac{(2 + 4\pi^2)(\pi - 0)^3}{24 \times 10^2} \approx 0.53587$$

Hence, our estimate is accurate to within 0.54, which is substantially less than our error estimate of 6.2 with the left endpoint rule.

c. We want n such that $E_M \le 0.001$.

$$E_M \le \frac{K_2(b-a)^3}{24n^2} \qquad \textit{midpoint error formula}$$

Thus, we find n such that

$$\frac{K_2(b-a)^3}{24n^2} \leq 0.001$$

$$K_2(b-a)^3 \leq 0.001(24n^2)$$

$$(2+4\pi^2)(\pi-0)^3 \leq 0.001(24n^2) \qquad \textit{substituting known values}$$

$$\frac{2\pi^3 + 4\pi^5}{(0.001)(24)} \leq n^2$$

$$231.5 \leq n \qquad\qquad \textit{since } n > 0$$

Since n is an integer, we need to choose $n = 232$. Compare this to the $n = 31,007$ that we needed for the left endpoint rule! Using midpoint rule with $n = 232$, we obtain

$$I = \int_0^\pi \sin(x^2)\, dx \approx 0.7727$$

which we know is within 0.001 of the true answer.   ∎

If the midpoint rule differs from the left and right endpoint rules only by shifting the sample points by $\Delta x/2$, why does it do so much better? First, observe that both the midpoint rule and left endpoint rule integrate constant functions perfectly. Indeed, L_n and M_n for $\int_a^b c\, dx$ equal $c(b-a)$. Second, for the case of linear functions $f(x) = cx + d$ on an interval $[a, b]$, it is simple to show (verify this for yourselves!) that $E_M = 0$ while $E_L = |c|(b-a)/2$. Hence, the midpoint rule integrates linear functions perfectly; consequently, it introduces error only for functions with nonzero, second-order derivatives. This explains why the error bound for the midpoint rule involves second-order terms (i.e., $|f''(x)|$ and n^2). In contrast, since the left endpoint rule introduces errors for functions whose derivatives are nonzero, the error bound for the left endpoint rule involves first-order terms (i.e., $|f'(x)|$ and n).

Another rule that perfectly estimates linear functions is the trapezoidal rule. This rule also has error bounds in terms of second-order rather than first-order derivatives. This rule is presented in the definition box at the end of this section and can be studied in more detail in Problem 38 of Problem Set 5.7.

Simpson's rule

As Example 3 illustrated, the midpoint rule was a significant improvement over the left endpoint rule because the error bound decreased like $1/n^2$ instead of like $1/n$. The small price we had to pay for this improvement was that bounds are calculated in terms of the second rather than first derivative. With this sweet taste of success, can we do better? As it turns out, yes! There is another rule, Simpson's rule, for approximating integrals using parabolas to approximate the curve. This method requires breaking the interval into *an even number n of subintervals*. Let $a = x_0 < x_1 = a + \Delta x < \cdots < x_n = b$ be the endpoints of these subintervals of width $\Delta x = (b-a)/n$. The approximation is given by

$$\int_a^b f(x)\, dx \approx [f(x_0) + 4f(x_1) + 2f(x_2) + 4f(x_3) + \cdots + 2f(x_{n-2}) + 4f(x_{n-1}) + f(x_n)]\frac{\Delta x}{3} = S_n$$

Given $|f^{(4)}(x)| \leq K_4$ for x on $[a, b]$, the *Simpson error bound* is

$$E_S \leq \frac{K_4(b-a)^5}{180n^4}$$

where

$$E_S = \left| S_n - \int_a^b f(x)\, dx \right|$$

Note that throughout this section we used the notation K_n to represent the bound of the nth derivative of $f(x)$ for the cases $n = 1, 2,$ and 4; that is,

$$\left| f^{(n)}(x) \right| \le K_n$$

This convention stresses the fact that each derivative has its own bound.

Example 4   How good is Simpson's rule?

Compare Simpson's rule with the left endpoint and midpoint rules in calculating the value of

$$I = \int_0^\pi \sin x^2 \, dx$$

Consider:

a. The case where the interval is partitioned into $n = 10$ subintervals

b. The smallest value of n that ensures the approximation is no larger than 0.001

From the results, what do you conclude?

Solution

a. The approximation given by Simpson's rule for the case $n = 10$ can be calculated as outlined in the following table:

i	x_i	Simpson's weighting $\times \sin x_i^2$
0	0	1×0
1	$\pi/10$	4×0.09854
2	$\pi/5$	2×0.38461
3	$3\pi/10$	4×0.77598
4	$2\pi/5$	2×0.99997
5	$\pi/2$	4×0.62427
6	$3\pi/5$	2×-0.39995
7	$7\pi/10$	4×-0.99236
8	$4\pi/5$	2×0.03336
9	$9\pi/10$	4×0.99016
10	π	1×-0.43030

Weighted total $\times \dfrac{\pi/10}{3} = 0.79503$

From technology, we know that the correct answer (to five decimal places) is $I = 0.77265$, and for $n = 10$:

- The left endpoint rule approximation is $I = 0.78997$, yielding an error of 0.01732.
- The midpoint rule approximation is $I = 0.79918$, yielding an error of 0.02653.
- The Simpson's rule approximation is $I = 0.79503$, yielding an error of 0.02238.

Thus, Simpson's rule, as expected, yields a smaller error than the midpoint rule. Rather surprisingly, however, both these rules are outperformed by the left endpoint rule, which in general is the least accurate of the three (see part **b** below)! This outcome is purely fortuitous in that positive and negative errors in the estimates obtained on each of the subintervals can sometimes cancel out to give a very good result for a generally inaccurate method.

b. To find n such that $E_S \le 0.001$, we need to solve for n in the inequality

$$\frac{K_4(b-a)^5}{180n^4} \le 0.001$$

where K_4 is a bound on the fourth derivative of $\sin x^2$ on the interval $[0, \pi]$. Repeated differentiation yields

$$\frac{d^4}{dx^4} \sin x^2 = (16x^4 - 12)\sin x^2 - 48x^2 \cos x^2$$

from which we conclude that $K_4 = (16\pi^4 - 12) + 48\pi^2$ for x ranging over the interval $[0, \pi]$. Since in this case $(b - a) = \pi$, it follows from the above inequality that the lower bound for n to ensure an error of less than 0.001 is

$$n^4 \geq \frac{((16\pi^4 - 12) + 48\pi^2)\pi^5}{180 \times 0.001}$$

Solving this, we obtain $n \geq 43.05$. Thus, selecting a value $n \geq 44$ (remember for Simpson's rule n must be even) ensures that the accuracy of the estimate of I is better than 0.001.

Again, from the two previous examples and the calculations here, we can see the requirements to ensure an accuracy of at least 0.001:

- $n \geq 31,007$ for the left endpoint rule
- $n \geq 232$ for the midpoint rule
- $n \geq 44$ for Simpson's rule

As the error estimate for Simpson's rule suggests from the foregoing example, Simpson's rule has much better convergence properties (i.e., the rate at which the error decreases with increasing n) than the midpoint rule. Why? Well, the Simpson's rule integrates a cubic function (i.e., third-order polynomials) perfectly. Hence, only nonzero fourth-order derivatives result in errors, and the error bound involves fourth-order terms (i.e., $|f^{(4)}(x)|$ and n^4).

Example 5   Estimating crab harvests

The Dungeness crab (*Cancer magister*) is an important element of commercial fishing along the northern Pacific coast (California to Alaska). The data depicted in Figure 5.37 show the commercial harvest level of Dungeness crabs (in millions of pounds and excluding sport fishery and nontreaty landings) off the coast of Washington State for the period 1950 to 1999.

© George Ostertag/Pur/ AGE Fotostock

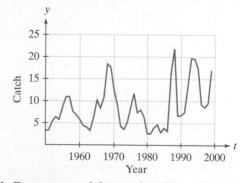

a. Pacific Coast Dungeness crab   **b.** Dungeness crab harvest in millions of pounds

Figure 5.37 Crab harvests along the Pacific coast

A subset of these catch data are reported in the table below (unit of catch is in millions of pounds).

Year	Catch	Year	Catch
1950	3.3	1975	8.5
1955	8.5	1980	2.7
1960	5.9	1985	3.9
1965	10.2	1990	6.8
1970	12.6		

Data source: The Quantitative Environmental Learning Project (http://www.seattlecentral.edu/qelp/ sets/062/062.html)

Use Simpson's rule to estimate the total amount of Dungeness crabs caught between 1950 and 1990.

Solution   Applying Simpson's rule with $n = 8$ yields

$$(3.3 + 4 \times 8.5 + 2 \times 5.9 + 4 \times 10.2 + 2 \times 12.6 + 4 \times 8.5 + 2 \times 2.7 + 4 \times 3.9 + 6.8)\frac{5}{3} \approx 294.8$$

For convenience, we next summarize the four methods of numerical integration described in this section. We include the trapezoidal rule (see the third panel in Figure 5.35 and Problem 38 in Problem Set 5.7).

Numerical Integration

Let f be continuous on $[a, b]$. Divide this interval into n equal parts:

$$a = a_0 < a_1 < a_2 < \cdots < a_n = b$$

Define $\Delta x = \dfrac{b - a}{n}$. The following rules provide estimates of $\int_a^b f(x)\,dx$ with error bounds given in terms of the bounding parameters $|f^{(i)}(x)| \le K_i$ for all $a \le x \le b$.

LEFT ENDPOINT RULE

$$L_n = [f(a_0) + f(a_1) + \cdots + f(a_{n-1})]\Delta x$$

and

$$E_L \le \frac{K_1(b - a)^2}{2n}$$

RIGHT ENDPOINT RULE

$$R_n = [f(a_1) + f(a_2) + \cdots + f(a_n)]\Delta x$$

and

$$E_R \le \frac{K_1(b - a)^2}{2n}$$

MIDPOINT RULE

$$M_n = \left[f\left(\frac{a_0 + a_1}{2}\right) + f\left(\frac{a_1 + a_2}{2}\right) + \cdots + f\left(\frac{a_{n-1} + a_n}{2}\right)\right]\Delta x$$

and

$$E_M \le \frac{K_2(b - a)^3}{24n^2}$$

TRAPEZOIDAL RULE

$$T_n = \left[\frac{f(a_0)}{2} + f(a_1) + \cdots + f(a_{n-1}) + \frac{f(a_n)}{2}\right]\Delta x$$

and

$$E_T \le \frac{K_2(b - a)^3}{12n^2}$$

SIMPSON'S RULE (n is even)

$$S_n = \frac{\Delta x}{3}[f(a_0) + 4f(a_1) + 2f(a_2) + 4f(a_3) + \cdots$$
$$+ 2f(a_{n-2}) + 4f(a_{n-1}) + f(a_n)]$$

and

$$E_S \le \frac{K_4(b - a)^5}{180n^4}$$

A simple example reinforces how efficient Simpson's rule is compared with the left endpoint and midpoint rules in converging to a solution.

Example 6   Comparing the efficiency of the different methods

Consider

$$\int_1^3 \frac{dx}{x}$$

What is the smallest value of n needed to ensure that

a. $E_L \leq 0.0001$?   **b.** $E_M \leq 0.0001$?   **c.** $E_S \leq 0.0001$?

Solution   We have $f(x) = x^{-1}$, $f'(x) = -x^{-2}$, $f''(x) = 2x^{-3}$, $f'''(x) = -6x^{-4}$, and $f^{(4)} = 24x^{-5}$.

a. The maximum value of $|f'(x)|$ on $[1, 3]$ is $K_1 = 1$. Since $|E_L| \leq \dfrac{K_1(b-a)^2}{2n}$, we need

$$\frac{1 \times (3-1)^2}{2n} \leq 0.0001$$

$$\frac{2}{0.0001} \leq n$$

$$20,000 \leq n$$

Hence, $n = 20,000$ will suffice.

b. The maximum value of $|f''(x)|$ on $[1, 3]$ is $K_2 = 2$. Since $|E_M| \leq \dfrac{K_2(b-a)^3}{24n^2}$, we need

$$\frac{2 \times (3-1)^3}{24n^2} \leq 0.0001$$

$$\frac{16}{24(0.0001)} \leq n^2$$

$$6,666\frac{2}{3} \leq n^2$$

$$81.6 \leq n$$

Hence, $n = 82$ will suffice.

c. The maximum value of $|f^{(4)}(x)|$ on $[1, 3]$ is $K_4 = 24$. Since $|E_S| \leq \dfrac{K_4(b-a)^5}{180n^4}$, we need

$$\frac{24(3-1)^5}{180n^4} \leq 0.0001$$

$$\frac{768}{180(0.0001)} \leq n^4$$

$$14.4 \leq n$$

Since Simpson's rule requires an even number of intervals, $n = 16$ will suffice.

PROBLEM SET 5.7

Level 1 DRILL PROBLEMS

Approximate the integrals in Problems 1 to 12 using these four rules:

a. *left endpoint rule*   **b.** *right endpoint rule*

c. *midpoint rule*   **d.** *Simpson's rule*

1. $\int_1^2 x^2 \, dx$ with $n = 4$

2. $\int_0^4 \sqrt{x} \, dx$ with $n = 6$

3. $\int_0^1 \cos 2x \, dx$ with $n = 4$

4. $\int_1^2 x^{-1} \, dx$ with $n = 6$

5. $\int_0^1 \frac{1}{1 + x^2} \, dx$ with $n = 4$

6. $\int_{-1}^0 \sqrt{1 + x^2} \, dx$ with $n = 4$

7. $\int_0^2 x \cos x \, dx$ with $n = 6$

8. $\int_0^2 xe^{-x} \, dx$ with $n = 6$

9. $\int_0^1 \frac{1}{1 + x^3} \, dx$ with $n = 4$

10. $\int_0^\pi \sin x \, dx$ with $n = 4$

11. $\int_{-2}^2 \cos x^2 \, dx$ with $n = 6$

12. $\int_0^2 e^{-x} \, dx$ with $n = 6$

Estimate the value of the integrals in Problems 13 to 22 to within the prescribed accuracy; use the rule indicated by the subscript on the error bound.

13. $\int_0^2 x\sqrt{4 - x} \, dx, |E_L| < 0.01$

14. $\int_1^4 \sqrt{x} \, dx, |E_M| < 0.01$

15. $\int_{-2}^0 e^x \, dx, |E_S| < 0.01$

16. $\int_0^{\pi/2} \cos^2 \theta \, d\theta, |E_R| < 0.1$

17. $\int_0^{\pi/2} \cos^2 \theta \, d\theta, |E_M| < 0.01$

18. $\int_0^\pi \sin(2\theta) \, d\theta, |E_S| < 0.01$

19. $\int_0^1 e^{\sin x} \, dx, |E_R| < 0.001$

20. $\int_0^1 e^{\sin x} \, dx, |E_S| < 0.001$

21. $\int_0^1 \sin(x^3) \, dx, |E_S| < 0.02$

22. $\int_1^5 \frac{1}{x} \, dx, |E_S| < 0.001$

In Problems 23 to 28, determine how many subintervals are required to guarantee accuracy to within 0.00005 using these two rules:

a. *midpoint rule*   **b.** *Simpson's rule*

23. $\int_1^4 x^{-1} \, dx$

24. $\int_1^4 \frac{dx}{\sqrt{x}}$

25. $\int_0^2 \cos x \, dx$

26. $\int_0^1 e^{-2x} \, dx$

27. $\int_{-1}^4 (x^3 + 2x^2 + 1), \, dx$ **28.** $\int_1^3 \ln \sqrt{x} \, dx$

29. Estimate the area in the graph in Figure 5.38 using the left endpoint rule, right endpoint rule, and Simpson's rule with $n = 6$.

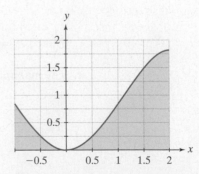

Figure 5.38 Estimate shaded area

30. Estimate the area in the graph in Figure 5.39 using the left endpoint rule, right endpoint rule, and Simpson's rule with $n = 4$.

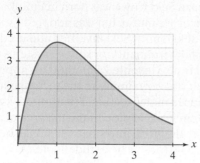

Figure 5.39 Estimate shaded area

Level 2 APPLIED AND THEORY PROBLEMS

31. Governmental health agencies throughout the world monitor cases of human diseases. For example, the Office of Population Censuses and Surveys in the United Kingdom published weekly case reports about measles in England and Whales. Theoretical ecologist Ben Bolker published the data from 1948 to 1966 on the Web at http://ms.mcmaster.ca/bolker/measdata/ewmeas.dat The first 89 weeks at four week intervals is shown in the table below.

week	cases	week	cases	week	cases
1	3762	33	4606	65	15875
5	5604	37	2610	69	11498
9	8828	41	4529	73	9644
13	10901	45	6986	77	6899
17	10210	49	10133	81	5044
21	13511	53	13279	85	2151
25	10670	57	13872	89	668
29	8521	61	18694		

Use Simpson's rule and right endpoint rule to estimate the total number of cases over these 89 weeks.

32. Using inverse trigonometric functions, it can be shown that

$$\int_0^1 \frac{dx}{1 + x^2} = \frac{\pi}{4}$$

Use this result to estimate π correct to four decimal places by applying Simpson's rule to this integral and using the appropriate error estimate.

33. **Black Plague revisited** Recall that for the outbreak of the plague in Bombay in 1905–1906, the mortality rate due to the plague was approximated by Kermack and McKendrick with the function

$$f(t) = 890 \operatorname{sech}^2(0.2t - 3.4)$$

deaths per week.

a. Write a definite integral that represents the number of deaths that accumulated from $t = 0$ to $t = 30$.

b. Estimate the definite integral using Simpson's rule with $n = 10$.

34. The data set for 80 hours of the discharge (in m^3/s) for the Raging River is shown here:

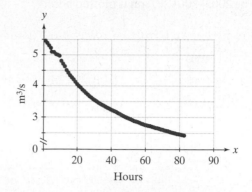

The data set for the first 24 hours is summarized in the following table:

h	m^3/s	h	m^3/s	h	m^3/s
0	5.45	10	4.81	20	4.06
2	5.41	12	4.67	22	3.97
4	5.25	14	4.49	24	3.83
6	5.10	16	4.29		
8	5.00	18	4.19		

Use Simpson's rule to estimate the total amount of discharge in the first 24 hours.

35. Sweet corn has a lower development threshold of $50°F$ and requires 1587 degree-days to complete development. On July 3, 2006, the temperatures in northern Illinois were as follows (measurements performed by the Northern Illinois Agronomy Research Center).

Hour	Temperature	Hour	Temperature
0	66.7	13	72.1
1	66.7	14	75.4
2	65.3	15	77.9
3	66.0	16	79.7
4	69.6	17	80.8
5	70.2	18	81.0
6	69.1	19	80.6
7	68.7	20	79.2
8	68.7	21	76.5
9	68.5	22	75.6
10	68.7	23	73.9
11	69.3	24	73.0
12	70.2		

a. Using the right endpoint rule, estimate the number of degree-days that elapsed on this summer day.

b. If the temperatures on July 3rd typified the temperatures throughout the summer, estimate how many days it would take sweet corn to mature.

36. Repeat Problem 35 using Simpson's rule. Do you expect your answer to be more or less accurate than the answer to Problem 35?

37. The weekly rate of cases of influenza A (strain unknown) studied by WHO/NREVSS during the 2003–2004 season is plotted here:

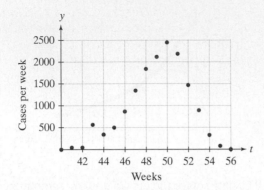

Estimate the total number of cases (i.e., the area under the curve) from week 40 to week 56 using Simpson's rule on two-week intervals.

38. *Trapezoidal rule for numerical integration:* Assume that the area under a curve $f(x)$ on an interval $[a, b]$ is approximated by n trapezoids made up from rectangles topped by right-angle triangles so that the left- and right-hand sides of the $(k + 1)^{th}$ trapezoid over the interval $[x_k, x_{k+1}]$, are of heights $f(x_k)$ and $f(x_{k+1})$, respectively, as illustrated in Figure 5.35. Let $a = x_0 < x_1 = a + \Delta x < \cdots < x_n = b$ and $\Delta x = (b - a)/n$. Show that the approximate area is given by

$$T_n = [f(x_0) + 2f(x_1) + 2f(x_2) + 2f(x_3) + \ldots$$

$$+ 2f(x_{n-2}) + 2f(x_{n-1}) + f(x_n)]\frac{\Delta x}{2} \approx \int_a^b f(x)\, dx$$

For completeness, we note that the error

$$E_T = \left| T_n - \int_a^b f(x)\, dx \right|$$

satisfies

$$E_T \leq \frac{K_2(b - a)^3}{12n^2}$$

where K_2 satisfies $\left| f''(x) \right| \leq K_2$ on the interval $[a, b]$.

39. 𝕳istorical 𝕼uest

Takakazu Seki Kōwa (1642–1708)

Takakazu Seki Kōwa was born in Fujioka, Japan, the son of a samurai, but was adopted by a patriarch of the Seki family. Seki invented and used an early form of determinants for solving systems of equations. He also invented a method for approximating areas that is very similar to the rectangular method introduced in this section. His method, known as the *yenri* (circle principle), found the area of a circle by dividing the circle into small rectangles, as shown in Figure 5.40 (drawn by a student of Seki).

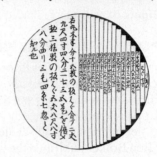

Figure 5.40 Early Asian calculus

Data Source: From Sawguchi Kazuyuki kokon Sampoki 1670. A History of Japanese Mathematics David Eugene Smith and Yoshio Mikami

Draw a circle with radius of 10 cm. Draw vertical chords through each centimeter on a diameter (you should have 18 rectangles). Measure the heights of the rectangles and approximate the area of the circle by adding the areas of the rectangles. Compare this method with the formula for the area of this circle.

40. 𝕳istorical 𝕼uest

Roger Cotes (1682–1716)

Isaac Newton invented a preliminary version of Simpson's rule. In 1779, Newton wrote an article to an addendum to *Methodus Differentials* (1711) in which he gave the following example: If there are four ordinates at equal intervals, let A be the sum of the first and fourth, B the sum of the second and third, and R the interval between the first and fourth; then the area between the first and fourth ordinates is approximated by $\frac{1}{8}(A + 3B)R$. This is known today as the Newton-Cotes three-eighths rule, which can be expressed in the form

$$\int_{x_0}^{x_3} f(x)\, dx \approx \frac{3}{8}(y_0 + 3y_1 + 3y_2 + y_3)\Delta x$$

Roger Cotes and James Stirling (1692–1770) both knew this formula, as well as what we call in this section Simpson's rule. In 1743, this rule was rediscovered by Thomas Simpson (1710–1761).

Estimate the integral

$$\int_0^3 \tan^{-1} x\, dx$$

using the Newton-Cotes three-eights rule, then compare with an approximation made using left endpoints (rectangles) with $n = 8$. Which of the rules gives the most accurate estimate?

5.8   Applications of Integration

In the preceding sections of this chapter, we motivated definite integration with area under a curve and accumulated change. In this section, we consider some additional applications that use Riemann sums to formulate integrals for survival and renewal processes, estimate cardiac output, and compute the amount of work required to perform a task.

Survival and renewal

Survival and renewal is the study of a population, or group of individuals, with the goal of predicting the size of the group at some future time. In the following example, a **survival function** gives the fraction of individuals in a group, or population, that can be expected to remain in the group for any specified period of time. In addition, a **renewal function** gives the rate at which new members arrive. Survival and renewal problems arise in many areas of study, including sociology, ecology, demography, and even finance—where the "population" is the number of dollars in an investment account, and "survival and renewal" refer to results of an investment strategy.

Example 1   Survival and renewal in a clinic

A new county mental health clinic has just opened. Statistics from similar facilities suggest that the fraction of patients who will still be receiving treatment at the clinic t months after their initial visit is given by the *survival function* $s(t) = e^{-t/20}$. The clinic initially accepts 300 people for treatment and plans to accept new patients at the rate of 10 per month. Approximately how many people will be receiving treatment at the clinic 15 months from now?

Solution   Since $e^{-15/20}$ is the fraction of patients whose treatment we expect to continue for at least 15 months, it follows that of the current 300 patients, only $300e^{-15/20} \approx 141.7$ will still be receiving treatment 15 months from now.

Each month, however, 10 new patients enter, and some of these will also still be around at month $t = 15$. To account for this, we divide the 15-month time interval $[0, 15]$ into n equal subintervals, each of length $\Delta t = \dfrac{15}{n}$ months. Let $t_k = (k-1)\Delta t$ denote the beginning of the kth subinterval for $k = 1, \ldots, n$. Since new patients are accepted at the rate of 10 per month, the number of new patients accepted during the kth subinterval is $10\,\Delta t$. When Δt is small, we can estimate $15 - t_k$ to be the time that elapses for all of these patients by the 15th month. Consequently, approximately

$$e^{-(15-t_k)/20}10\,\Delta t$$

of these patients will still be receiving treatment 15 months from now. Thus, the total number of patients arriving at times $t_1, t_2, \ldots, t_n$ who are still receiving treatment at time $t = 15$ is approximated by the sum

$$\sum_{k=1}^{n} e^{-(15-t_k)/20}10\Delta t$$

As $n \to \infty$, we obtain the integral

$$\lim_{n \to \infty} \sum_{k=1}^{n} e^{-(15-t_k)/20} 10\Delta t = \int_0^{15} 10e^{(t-15)/20}\, dt$$

which is also referred to as a *renewal function*.

Adding this integral to the 141.7 original patients still receiving treatment after 15 months, the total number of patients receiving treatment at time $t = 15$ is

$$141.7 + \int_0^{15} 10e^{(t-15)/20}\, dt = 141.7 + 200e^{(t-15)/20}\Big|_0^{15}$$
$$\approx 141.7 + 105.5$$
$$= 247.2$$

That is, 15 months from now, the clinic will be treating approximately 247 patients.

This example provides a guide to developing a more general formulation for survival and renewal processes. Suppose a population initially has N_0 individuals and receives new individuals (renews) at a rate $r(t)$, and suppose the fraction of individuals remaining (surviving) in the population after t units of time after entering the population is $s(t)$. If we want to determine the number of individuals in the population at time T, we can divide the interval $[0, T]$ into subintervals of width $\Delta t = T/n$. The number of individuals arriving into the population during the kth time interval is approximately $r(t_k)\Delta t$. The fraction of these $r(t_k)\Delta t$ individuals surviving to time T is approximately $s(T - t_k)$. Hence, the number of individuals entering during the kth time interval and surviving to time T is approximately $s(T - t_k)r(t_k)\Delta t$. Summing up over all these time intervals yields

$$\text{NEW INDIVIDUALS SURVIVING TO TIME T} \approx \sum_{k=1}^{n} s(T - t_k)r(t_k)\Delta t$$

Taking the limit as $n \to \infty$ yields

$$\text{NEW INDIVIDUALS SURVIVING TO TIME T} = \int_0^T s(T - t)r(t)dt$$

Of the N_0 individuals who were initially present, $s(T)N_0$ of them survive to time T. Hence, the number of individuals in the population at time T is given by the following *survival and renewal function*.

Survival and Renewal Equation

Suppose there are N_0 individuals initially present, a fraction of $s(t)$ individuals survive a period of length t, and individuals arrive at a rate of $r(t)$ individuals per unit time at time t. Then the total number of individuals present at time T is given by the **survival renewal equation**

$$s(T)N_0 + \int_0^T s(T - t)r(t)\, dt$$

Example 2   Fire ants

Figure 5.41 Fire ants were imported from South America

The imported fire ant (*Solenopsis invecta*) (Figure 5.41) is a pest in both urban and rural areas. Damage estimates for the United States range in the millions of dollars.

The fire ant has colonies in which workers live approximately 10 to 70 weeks and queens survive for about seven years. A single colony can have from 10 to 100 or more queens, each producing 1,000 to 1,500 eggs per year for 7 years. Suppose a colony is formed with 100 queens, in which each queen produces workers at a rate of $1{,}250 + 250 \sin(2\pi t)$ per year and in which the fraction of workers living t years after their birth is given by $s(t) = e^{-1.25t}$. Find the number of workers in the colony seven years from now, assuming all 100 queens survive the seven years under consideration.

Solution Initially there are no workers and $N_0 = 0$. The rate at which workers are renewed is

$$r(t) = 100(1{,}250 + 250 \sin(2\pi t))$$
$$= 125{,}000 + 25{,}000 \sin(2\pi t) \text{ workers/year}$$

The survival function is given by $s(t) = e^{-1.25t}$. Setting $T = 7$ into the renewal equation

$$\int_0^T s(T - t) r(t)\, dt = \int_0^7 [125{,}000 + 25{,}000 \sin(2\pi t)] e^{-1.25(7-t)}\, dt$$
$$\approx 96{,}157 \qquad \textit{using technology}$$

The renewal equation predicts that we should expect the colony to have around 100,000 workers seven years from now. It is possible to do the prior integral without technology: separate the integral into two integrals, use substitution with $u = -1.25(7 - t)$ in the first integral, and use integration by parts (twice) for the second integral. ∎

Another application of the renewal equation is in the area of finances, where we are concerned with the growth of economies or our personal fortunes. The key difference in this application is that instead of calculating how capital (population of dollars) decays, we are interested in how capital grows.

Although it is not common practice in banking systems to compound continuously, it is a reasonable approximation for daily compounding. Recall from Example 6 of Section 1.4 that compounding continuously at a rate of $c\%$ per year implies that if you put N dollars into an account, then t years later there will be $e^{ct/100} N$ dollars in the account.

Example 3 Saving for retirement

Starting at age 20, Peggy Sue puts money into a retirement account at a rate of $2,000 per year. The money in this account is compounded continuously at a rate of 10% per year. How much money will be in her account when she turns 60? How much would she have if she starts at the age of 30?

Solution To determine the total amount in Peggy Sue's account, let us break up the time interval $[0, 40]$ into n subintervals of width $\Delta t = 40/n$. The amount of money she puts into the account during the kth time interval $[(k - 1)\Delta t, k\Delta t]$ is approximately $2{,}000\Delta t$. Over the $40 - k\Delta t$ year period, this money grows to approximately $e^{0.1(40 - t_k)} 2{,}000\Delta t$ where $t_k = k\Delta t$. Hence, the total amount of money she has at age 60 is approximately

$$\sum_{k=1}^{n} e^{0.1(40 - t_k)} 2{,}000\Delta t$$

Taking the limit as $n \to \infty$ yields

$$\int_0^{40} e^{0.1(40 - t)} 2{,}000\, dt$$

Integrating yields

$$\int_0^{40} e^{0.1(40 - t)} 2{,}000\, dt \approx \$1{,}071{,}960$$

Peggy Sue will be millionaire! Alternatively, if she started saving at age 30, then at age 60 she would have

$$\int_0^{30} e^{0.1(30 - t)} 2{,}000\, dt \approx \$381{,}711$$

Not even close to being a millionaire! ∎

The solution to Example 3 shows that if money is being added to an account at a rate of $r(t)$, the account is continually compounded at an interest rate of $c\%$, and there is initially N_0 dollars in the account, then the total amount in the account T years from now is

$$N_0 e^{cT/100} + \int_0^T e^{c(T-t)/100} r(t)\, dt$$

This is just another survival-renewal equation with $s(t) = e^{cT/100}$.

Cardiac output

Cardiac output is the volume of blood pumped by the heart in a specified interval of time. Estimating cardiac output is important, because it is an indicator of certain heart diseases.

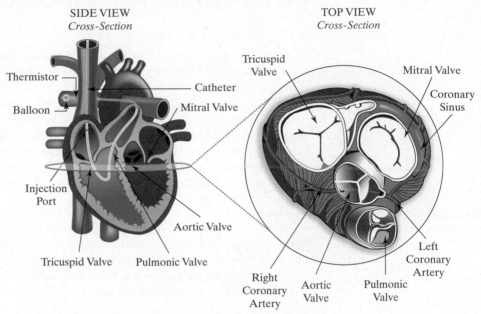

Figure 5.42 Schematic of the heart

Cardiac output can be measured using *dye dilution*. A known quantity of dye, say D milligrams (mg), is injected into a main vein near the heart. This dye circulates with the blood through the body (from the right ventricle of the heart to the lungs to the left ventricle and into the arteries) and returns to the left ventricle (see schematic in Figure 5.42). The concentration of the dye, $c(t)$ milligrams per liter (mg/l), passing through an artery is monitored. To compute cardiac output from these recorded concentrations, assume that the cardiac output (i.e., blood flow) remains at a constant rate, F liters per second (l/s), during the experiment. The rate at which dye is passing through the artery at time t seconds is given by $Fc(t)$ milligrams per second (mg/s). Notice how the units work out here: $c(t)$ has units mg/L and F has units L/s. Hence $c(t)F$ has units

$$\frac{mg}{L} \times \frac{L}{s} = \frac{mg}{s}$$

Assume that the entire amount of dye passes through the artery between time $t = 0$ and $t = T$. The net amount of dye passing through the artery over the time interval from 0 to T is

$$\int_0^T Fc(t)\, dt = F \int_0^T c(t)\, dt$$

By conservation of mass, the net amount of dye observed must equal the initial amount of dye, D; that is,

$$F \int_0^T c(t)\,dt = D$$

Solving for the cardiac output, F, yields

$$F = \frac{D}{\int_0^T c(t)\,dt}$$

Example 4   Dye dilution

A (hypothetical) patient is given an injection of 5 mg of dye. The measured concentrations of dye are recorded in the following table.

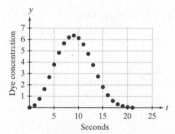

Figure 5.43 Dye concentrations in the heart after an injection

t	c(t)	t	c(t)	t	c(t)	t	c(t)
0	0.00	6	4.84	12	4.74	18	0.30
1	0.20	7	5.67	13	3.76	19	0.15
2	0.77	8	6.19	14	2.75	20	0.05
3	1.63	9	6.35	15	1.82	21	0.00
4	2.69	10	6.13	16	1.10		
5	3.81	11	5.57	17	0.60		

A plot of these data is given in Figure 5.43. Use Simpson's rule to estimate the cardiac output of the patient.

Solution   We want to use Simpson's rule with $\Delta t = 1$. However, we have 22 data points and consequently 21 (an odd number) of intervals. Since Simpson's rule requires an even number of intervals and $c(21) = 0$, we omit the last data point and make the following approximation:

$$\int_0^{20} c(t)\,dt \approx \frac{1}{3}(0 + 4 \times 0.2 + 2 \times 0.77 + \cdots + 4 \times 0.15 + 0.05)$$

$$\approx 59.1$$

Since the initial amount of dye is 5 mg, we get the following result for cardiac output:

$$F = \frac{D}{\int_0^T c(t)\,dt}$$

$$\approx \frac{5}{59.1}$$

$$\approx 0.085 \text{ L/s}$$

$$\approx 5.1 \text{ L/min}$$

Example 5   Thermodilution

Another approach to measuring cardiac output involves use of a pulmonary artery catheter that allows rapid, easy measurements of cardiac output using *thermodilution*. The principle of thermodilution is the same as that of dye dilution. Instead of injecting dye, doctors inject 10 ml of a cold dextrose solution. As the cold solution mixes with the blood in the heart, the temperature variations in the blood leaving the

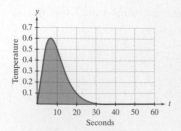

Figure 5.44 Temperature variation in the heart resulting from an injection of cold dextrose (The interpretation here is that the peak represents 0.6°C *below* normal body temperature.)

heart are measured. A hypothetical temperature variation curve may be described by the function

$$f(t) = 0.1t^2 e^{-0.3t} \text{ degrees Celsius}$$

This curve is plotted in Figure 5.44, with this interpretation: for a cold solution, $f(t)$ is the amount the temperature drops below normal body temperature.

Assuming that the temperature of a body is 37°C and the temperature of the dextrose solution is 0°C, estimate the cardiac output of a patient over a one-minute time interval.

Solution   This example is just like the previous example but replacing dye concentration with temperature variation. The initial "amount of cold" (the equivalent of the initial amount of dye) is given by

$$10 \text{ mL} \times (37 - 0)°C = 370 \text{ mL-}°C$$

If the cardiac output is F, then the rate of "cold" passing by at time t is

$$F \times f(t) \text{ mL-}°C/s$$

The accumulated change in cold is $F \times \int_0^{60} f(t)\, dt$ mL-°C which must equal 370 mL-°C. Hence,

$$F = \frac{370}{\int_0^{60} 0.1t^2 e^{-0.3t}\, dt}$$

Applying integration by parts (twice) and evaluating yields

$$F = \frac{370}{\int_0^{60} 0.1t^2 e^{-0.3t}\, dt} \approx 49.95$$

We can convert units to get 49.95 ml/s $\approx$ 3.00 L/min.

Work

How much pasta should a person eat to dig a posthole? How much energy should a grizzly bear expend to dig out a pocket gopher? To answer these questions, we need to understand the relationship between work and energy. To complete any work, we need energy. A standard unit of energy is a *calorie*. A **calorie** is the amount of energy required to heat one gram of water one degree Celsius. What does this mean?

In the article "What is Energy?" in Exploring Food Magazine (vol. 14 #4 1990), Paul Doherty writes about his undergraduate days as a biophysics student at MIT and describes an experiment where the professor set a peanut on fire. The energy from this peanut was sufficient to heat a test tube filled with 10 grams of water from room temperature to 100° Celsius, and then boil away 2 grams of water.

Doherty calculated that amount of energy required to heat and then boil the water. To heat 10 grams of water from 20° C to boiling (100° C) requires $(100 - 20) \times 10 = 800$ calories. Further, since 540 calories are needed to boil away a gram of water (at sea level), another 1080 calories were needed to vaporize 2 grams of water. Thus, in the demonstration, the burning peanut delivered 1880 calories of heat flow to the test tube of water. This might seem like a lot of calories for one peanut. The reason for this surprising number is that a single dietary calorie (i.e., the type of calorie that is reported with food), which is abbreviated Cal or simply C, is actually one kilocalorie, abbreviated kcal. Hence, a peanut typically represents 1.5 to 2 Cal of energy.

What can we do with all this energy once ingested? We can turn it into work.

| Work Done by a Constant Force | If a body moves a distance d in the direction of an applied constant force F, the **work**, W, done is $$W = Fd$$ |

From Newtonian physics we know that

$$\text{force} = \text{mass} \times \text{acceleration}$$

which implies that units of force are mass × distance × time^{-2}. Hence, the units of work are mass × distance^2 × time^{-2}. A *standard work unit*, called a *Joule*(J), is defined as follows:

$$1 \text{ joule (J)} = 1 \text{ kilogram-meter}^2/\text{second}^2 \text{ (kg-m}^2/\text{s}^2)$$

The conversion between joules and dietary calories is given by the formula

$$1 \text{ Cal} = 4{,}184 \text{ J}$$

Example 6   Calories used while working out

How much work is done lifting 30 kilograms 20 meters? (This is equivalent to 40 arm curls lifting about 67 pounds). Give your answer in joules and in calories.

Solution   To solve this, we make use of the fact that on the surface of Earth, the force of gravity on 1 kilogram of mass is approximately 9.81 kg-m/s^2. Thus,

$$
\begin{aligned}
W &= Fd \\
&= (30 \times 9.81) \times 20 \quad \textit{given } 30 \textit{ kg lifted } 20 \textit{ m} \\
&= 5{,}886 \text{ J} \\
&= \frac{5{,}886}{4{,}184} \text{ Cal} \\
&\approx 1.41 \text{ Cal}
\end{aligned}
$$

All that work, and so little to show for it? Well, actually, we are not 100% efficient at translating calories from food to work. Roughly, humans have 10% efficiency (all that overhead from maintaining body temperature, etc.). Thus, we might estimate the number of calories being burnt off as 14 Cal. That is, we need about 7 to 10 peanuts to carry out this work. ■

Example 7   Climbing mountains with a candy bar

The article "What is Energy?" in Exploring Food Magazine (vol. 14 #4 1990) reveals that a Milky Way bar contains more energy than a stick of dynamite. The candy bar contains 270 Cal. If the energy from the Milky Way bar is used with 100% efficiency, determine how high (in meters) a 70 kg human could be lifted with the energy from the Milky Way bar.

Solution   First find the number of joules:

$$270 \text{ Cal} \times 4{,}184 = 1{,}129{,}680 \text{ J}$$

Figure 5.45 El Capitan, Yosemite

The amount of work required to lift a 70 kg person x meters is

$$70 \times 9.81 \times x = 686.7x$$

Thus,

$$686.7x = 1{,}129{,}680$$
$$x \approx 1{,}645$$

This is almost twice the height of the cliff face of Yosemite's El Capitan (see Figure 5.45). No stick of dynamite can do that! In fact, an ounce of dynamite produces only one quarter as many calories when it explodes as an ounce of sugar does when it is burnt!

Often people try to achieve great things by moving huge loads of rocks and dirt, such as digging canals, cutting through mountains to build roads and railways, or hollowing out deep cellars to build skyscrapers. Calculus can be used to compute the amount of work required to accomplish such feats, as evidenced in the following example.

Example 8   Digging a cellar

Consider the problem of digging a cellar that it is 7 m deep, 100 m long, and 50 m wide. How much work, in theory, does it take just to remove all the dirt out of the hole being dug?

Solution   To answer this question, we use the fact that the density of soil is approximately the same density as water. In this case, one cubic centimeter (i.e., milliliter) has a mass of one gram. Since our problem is phrased in meters and kilograms, we need to translate this statement into these units. Since one cubic meter equals 100^3 cm^3, one cubic meter of soil has a mass of approximately $1{,}000{,}000$ g $= 1{,}000$ kg (one metric ton).

The amount of work required to lift one scoop of dirt to ground level depends on the depth of that scoop of dirt. Dirt at the bottom of the cellar has to be lifted higher than dirt at the top of the cellar. To find the amount of work, envision cutting the cellar up into n thin horizontal slices of thickness $\Delta x = 7/n$ (see Figure 5.46). Let x denote the depth of a slice in meters.

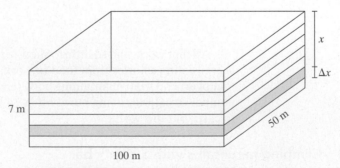

Figure 5.46 Slicing a cellar into horizontal slices

The volume of a slice with thickness Δx is

$$(100 \, \text{m}) \times (50 \, \text{m}) \times (\Delta x \, \text{m}) = 5{,}000 \times \Delta x \, \text{m}^3$$

The mass of this slice is given by

$$(1{,}000 \, \text{kg/m}^3)(5{,}000 \Delta x \, \text{m}^3) = 5{,}000{,}000 \Delta x \, \text{kg}$$

The weight of this slice is given by

$$(9.81 \, \text{m/s}^2)(5{,}000{,}000 \Delta x \, \text{kg}) = 49{,}050{,}000 \Delta x \, \text{kg-m/s}^2$$

If this slice is at depth x meters, then the work required to lift the slice is

$$\approx 49{,}050{,}000\, x\, \Delta x \ \text{J}$$

If the depths of the slices are $x_1, x_2, \ldots, x_n$, then the work to dig the cellar is the sum of the work to lift all of the slices, approximately given by

$$\sum_{k=1}^{n} 49{,}050{,}000\, x_k \Delta x \ \text{J}$$

Letting Δx get smaller and smaller should yield better and better approximation. Consequently, taking the limit as $n \to \infty$ yields

$$\text{AMOUNT OF WORK} \ = \int_0^7 49{,}050{,}000 x \, dx = 1{,}201{,}730{,}000 \ \text{J}$$

This is equivalent 287,219 Cal—at last a way to burn off a considerable number of calories!

The next example computes the amount of work required to remove soil from a geometrically more complicated region.

Example 9   A hungry grizzly

Figure 5.47 A gopher treat for Spirit? Spirit is the first Montana grizzly to reside at the Grizzly & Wolf Discovery Center.

Since the early 1980s, Steve and Marilynn French, founders of the Yellowstone Grizzly Foundation, have been watching grizzlies in Yellowstone National Park (see Figure 5.47). They report watching a bear digging a trench twenty feet long to get a little gopher as a tasty treat. Is the effort worth all the energy expended?

Assuming that the trench has a semicircular cross-section with radius 1 m and that the density of soil is 1,000 kg/m^3, find the amount of work performed by the grizzly bear.

Solution   We convert 20 feet to meters to obtain ≈ 6.1 m. To determine the approximate amount of work, we slice the trench into n slices of thickness Δx meters (see Figure 5.48).

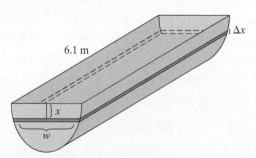

Figure 5.48 Grizzly's trench

To determine the width w of a slice at depth x meters, we use the fact that the cross-sectional profile of the trench is a semicircle of radius 1. Thus, $(w/2)^2 + x^2 = 1$ so that

$$w = 2\sqrt{1 - x^2}$$

The volume of a slice at depth x meters is approximately

$$\underbrace{2\sqrt{1 - x^2}}_{width} \ \underbrace{6.1}_{length} \ \underbrace{\Delta x}_{height} = 12.2\sqrt{1 - x^2}\, \Delta x \ \text{m}^3$$

The weight of the slice is approximately

$$(9.81\text{m/s}^2)(1{,}000\,\text{kg/m}^3)(12.2\sqrt{1 - x^2}\,\Delta x\,\text{m}^3) = 119{,}682\sqrt{1 - x^2}\,\Delta x\text{kg-m/s}^2$$

The amount of work to lift a slice at depth x is

$$119,682\, x\sqrt{1-x^2}\,\Delta x \ \text{J}$$

If $x_1, x_2, \ldots, x_n$ are depths of the n slices, then the total work is approximately

$$\sum_{k=1}^{n} 119,682 x_k \sqrt{1-x_k^2}\,\Delta x \ \text{J}$$

Taking the limit as $n \to \infty$ yields

$$\text{AMOUNT OF WORK} = 119,682 \int_0^1 x\sqrt{1-x^2}\,dx$$

$$\approx 39,894 \ \text{J} \qquad \text{by calculator or substitution} \\ \text{where } u = 1 - x^2$$

$$\approx 39,894/4,184$$

$$\approx 9.5 \ \text{Cal}$$

The answer of nine to ten calories seems surprisingly few calories! Of course all this presumes that the grizzly was able to perform the work 100% efficiently, which is certainly not the case. For instance, if the grizzly worked with 5% efficiency, then the bear used about 200 Calories—the amount of Calories in one Twinkie.

PROBLEM SET 5.8

Level 1 DRILL PROBLEMS

After reviewing the mental health clinic in Example 1, for Problems 1 to 6 calculate the number of patients in the clinic after 15 months if the patient survival rate $s(t)$ and the renewal rate $r(t)$ are as given.

1. $s(t) = e^{-t/20}$ and $r(t) = 20$ per month. How did doubling the renewal rate change the answer from what we found in Example 1?

2. $s(t) = e^{-t/40}$ and $r(t) = 10$ per month. How did halving the survival rate change the answer from what we found in Example 1?

3. $s(t) = e^{-t/10}$ and $r(t) = 20$ per month

4. $s(t) = e^{-t/40}$ and $r(t) = 20$ per month

5. $s(t) = e^{-t/20}$ and $r(t) = 10 + t$ per month

6. $s(t) = \dfrac{1}{1+t}$ and $r(t) = 10$ per month

After reviewing the fire ants in Example 2, for Problems 7 to 12 calculate the number of workers after five years if the worker survival rate $s(t)$ is as given. Use technology to numerically evaluate the integrals.

7. $s(t) = e^{-0.625t}$, that is, survival rate is doubled

8. $s(t) = e^{-2.5t}$, that is, survival rate is halved

9. $s(t) = \dfrac{1}{0.25 + t^2}$

10. $s(t) = e^{-1.25t}$ and, in addition, the proportion of queens alive at time t is $q(t) = e^{-0.1t}$

11. $s(t) = e^{-1.25t}$ and, in addition, the proportion of queens alive at time t is $q(t) = e^{-0.2t}$

12. $s(t) = e^{-1.25t}$ and, in addition, the proportion of queens alive at time t is $q(t) = e^{-0.05t}$

For Problems 13 to 16, reconsider Example 3. Calculate the amount of money Peggy Sue will have in her account by age 60 if she adds A dollars per year to her account, she opens her account at age B years, and the annual interest rate is 10%.

13. $A = 4,000$ and $B = 20$

14. $A = 4,000$ and $B = 30$

15. $A = 1,000$ and $B = 10$ (she starts really young!)

16. $A = 10,000$ and $B = 40$ (she starts very late!)

Answer Problems 17 to 19 by finding the work done; express your answer using the unit of foot-pounds (ft-lb).

17. Lifting a 90 lb bag of concrete 3 ft

18. Lifting a 50 lb bag of salt 5 ft

19. Lifting a 850 lb billiard table 15 ft

Level 2 APPLIED AND THEORY PROBLEMS

20. Analysts speculate that patients will enter a new clinic at a rate of $300 + 100 \sin \dfrac{\pi t}{6}$ individuals per month. Moreover, the likelihood an individual is in the clinic t months later is e^{-t}. Find the number of patients in the clinic one year from now.

21. A patient receives a continuous drug infusion at a rate of 10 mg/h. Studies have shown that t hours

after injection, the fraction of drug remaining in a patient's body is e^{-2t}. If the patient initially has 5 mg of drug in her bloodstream, then what is the amount of drug in the patient's bloodstream 24 hours later?

22. Consider a mental health clinic that initially has 300 patients, and accepts 100 new patients per month; the fraction of patients receiving treatment for t or more months is given by $f(t)$.

t (in months)	$f(t)$
0	1
3	0.5
6	0.3
9	0.2
12	0.1

Using Riemann sums with left endpoints, estimate the number of patients in the clinic after 12 months.

23. Consider the following two scenarios involving an IRA account that yields 9% continuous interest.

   a. You graduate from college at age 22, get a job, and open an IRA account. You deposit $1,000 per year until age 65. How much money is the account at age 65? How much money did you pay into this account?

   b. You graduate from college at age 22 and do not bother to start an IRA account until you reach 32. Then you deposit $2,000 per year into the IRA account until you reach age 65. How much money is in your IRA account at age 65? How much money did you pay into this account?

24. The administrators of a town estimate that the fraction of people who will still be residing in the town t years from now is given by the function $S(t) = e^{-0.04t}$. The current population is 20,000 people and new people are arriving at a rate of 500 per year.

   a. What will be the population size 10 years from now?

   b. What will be the population size 100 years from now?

25. After 5 mg of dye is injected into a vein, we obtain the concentration levels shown in the following table. The variable t is in seconds and $c(t)$ is in mg/liter. Using Simpson's rule, compute the cardiac output.

t	$c(t)$	t	$c(t)$	t	$c(t)$	t	$c(t)$
0	0.00	6	4.8	12	4.5	18	0.50
1	0.20	7	5.5	13	3.5	19	0.2
2	0.7	8	6.0	14	2.5	20	0.1
3	1.6	9	6.3	15	1.8	21	0.00
4	2.5	10	6.3	16	1.10	22	0.00
5	3.5	11	5.5	17	0.60		

26. *Sediment flow.** Methods used in the measuring of cardiac output can be applied to other situations. For example, ecologists and scientists are interested in how much sediment is moved by a river. Data on the water flow and suspended sediment in the Des Moines River near Saylorville Lake, Iowa, is given in the following table. Using Simpson's rule, compute the total amount (kilograms) of suspended sediment that passed the measurement point for the period ending December 15, 1993.

Des Moines River Basin Water Discharge Records (December 1993)

Day	Discharge (ft³/s)	Suspended Sediment (mg/l)
1	1300	8
2	1590	35
3	2000	58
4	2200	64
5	2350	66

Note: One cubic foot equals 28.3 liters. One kilogram equals 1,000,000 milligrams.

27. *Kety-Schmidt technique.** Seymour Kety and Carl Schmidt described a widely acknowledged and accurate method for the determination of cerebral blood flow and cerebral physiological activity (e.g., metabolic rate of oxygen). For example, a patient breathes 15% nitrous oxide (N_2O). After the start of administration, the arterial concentration, A, is measured in the radial artery. This is the concentration before the blood enters the brain. The venous concentration, V, is measured at the base of the skull in the superior bulb of the internal jugular vein (at the point of exit of the jugular vein from the brain). This process for measuring the blood flow of cerebral physiological activity is commonly referred to as the *Kety-Schmidt technique*. A sample table is shown below.

Time (minute)	A (cc N_2O per cc blood)	V (cc N_2O per cc blood)
0.0	0.000	0.000
2.5	0.031	0.012
5.0	0.039	0.027
7.5	0.041	0.034
10.0	0.044	0.042

   a. Initially, the rate of N_2O flow into the brain is greater than the flow out of the brain. Moreover, after approximately ten minutes, the concentrations flowing into the brain and from the brain are approximately equal. The brain has become saturated with N_2O. Assuming a constant cerebral blood flow rate, F, use Simpson's rule to estimate the total amount of N_2O accumulated in the brain during ten minutes. Your answer will depend on F.

*Problem 26 and 27 were found at http://illuminations.nctm.org/LessonDetail.aspx?ID=L461 on February 25, 2013.

b. Through other means, the maximum amount of N_2O in the brain can be measured. Suppose that the maximum amount is determined to be 58.8 cc. Determine F.

28. A bucket weighing 75 lb when filled and 10 lb when empty is pulled up the side of a 100 ft building. How much more work in foot-pounds is done in pulling up the full bucket than the empty bucket?

29. A 20 ft rope weighing 0.4 lb/ft hangs over the edge of a building 100 ft high. How much work in foot-pounds is done in pulling the rope to the top of the building? Assume that the top of the rope is flush with the top of the building and that the lower end of the rope is swinging freely.

30. How much ice water do you need to ingest to burn off 300 calories? Assume your body temperature is 37°C and the energy required to digest ice water is the energy needed to raise the ice to body temperature.

31. A children's book, a steam shovel claims that she can do as much work in one day as 100 men can do in seven days. In a modern society we assume that the work needed to dig the cellar considered in Example 8 is done by a machine. But how many calories would 100 men produce in seven days, if

we assume that each ate two pounds of pasta a day and worked with 10% efficiency? How does this compare with the work done (i.e., calories needed) to dig the cellar considered in Example 8? Assume a serving of pasta is two ounces and contains 200 Cal.

32. Determine the length of a rectangular trench you can dig with the energy gained from eating one Milky Way bar (270 Cal). Assume that you convert the energy gained from the food with 10% efficiency and that the trench is 1 meter wide and 1 meter deep. Assume the density of soil is $1,000 \text{ kg/m}^3$.

In the next two problems, use the fact that a serving of pasta contains 200 Cal and the density of soil is 1,000 kg/m³.

33. How much work does it take to dig up a conical hole of depth 5 meters and diameter 6 meters? How many servings of pasta are required to complete this work, assuming the energy from the pasta is converted with 5% efficiency to work?

34. How much work does it take to dig a hemispherical pit with radius 10 meters? How many servings of pasta are required to complete this work, assuming the energy from the pasta is converted with 5% efficiency to work?

CHAPTER 5 REVIEW QUESTIONS

1. Find the general antiderivative of $f(x) = \dfrac{1}{\sqrt{x}}$.

Evaluate the definite integrals in Problems 2 to 5.

2. $\displaystyle\int_0^4 (x^2 - 1)\, dx$

3. $\displaystyle\int_0^{\pi/2} t^2 \sin(2t)\, dt$

4. $\displaystyle\int_{-2}^2 xe^{-x^2}\, dx$

5. $\displaystyle\int_{-1}^1 \dfrac{x+1}{(x+3)(x+2)}\, dx$

6. Find the area under the curve $y = \dfrac{x+1}{x}$ over $[1, 2]$.

7. The slope $F'(x) = \dfrac{x+1}{x^2}$ at each point is shown in Figure 5.49.

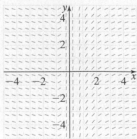

Figure 5.49 Slope field

Find F passing through $(1, -2)$ both graphically and analytically.

8. The Royalty rose has a lower developmental threshold of 41.4°F and requires 473 degree-days for harvesting time. If the temperature were to remain a constant 72°F, how long would it take for this rose to mature?

9. Find $\dfrac{dy}{dx}$ where
$$y = \int_1^{2x} \sin(t^2)\, dt$$

10. Evaluate the following integral
$$\int \frac{dN}{N(100 - N)}$$

11. Consider a mental health clinic that initially has 450 patients and accepts 150 new patients per month; the fraction of patients receiving treatment for t or more months is given by $f(t)$:

t (in months)	$f(t)$
0	1
3	0.5
6	0.3
9	0.2
12	0.1

Using Riemann sums with left endpoints, estimate the number of patients in the clinic after 12 months.

12. Find an upper bound for

$$\int_{-2}^{2} 5\sin(x^3)\,dx$$

13. The rate of infection of a disease in a population of 10,000 is given by the function

$$R(t) = 10,000\,te^{-t}\ \text{people per month}$$

where t is the time in months since the disease broke out.

   a. Use technology to plot $R(t)$. Why is this a reasonable description of a disease spreading in a population?

   b. Compute the number of people infected by the disease by time T.

   c. Use technology to approximate the time when 50% of the population have the disease.

14. In a wild week of temperature fluctuations, the temperature in Corvallis, Oregon, is given by

$$T(t) = 75 + t\cos(2\pi t)^{\circ}F$$

where t is measured in days. Find the number of degree-days that have elapsed for a beet armyworm over the first week. Note: The lower developmental threshold of a beet armyworm is 54°F.

15. Express $\int_{3}^{7} \tan x\,dx$ as the limit of a Riemann sum using right endpoints.

16. Express $\displaystyle\lim_{n\to\infty}\sum_{i=1}^{n}\left(\frac{1}{5+2i/n}\right)\frac{2}{n}$ as a definite integral.

17. Find $\int_{1}^{2}\dfrac{x}{\sqrt{x+1}}\,dx$.

18. Suppose $\int_{1}^{3} f(x)\,dx = 4$, $\int_{2}^{3} f(x)\,dx = 5$, and $\int_{1}^{2} g(x)\,dx = 6$. Find $\int_{1}^{2} f(x) - 2g(x)\,dx$.

19. Use the geometrical interpretation of the definite integral to find $\int_{-1}^{2}(1 - |x|)\,dx$. Be sure to provide a sketch.

20. A stone was dropped off a tower and hit the ground at a speed of 200 ft/s. What was the height of the tower?

GROUP PROJECTS

Seeing a project through on your own, or working in a small group to complete a project, teaches important skills. The following projects provide opportunities to develop such skills.

Project 5A Physiological Time

Section 5.1 introduced the concept of *developmental thresholds* that defined the range of temperatures over which plants and poikilothermic animals (those without an internal mechanism for maintaining their body temperature within a narrow range of values, as in homeothermic birds and mammals) grow and develop. In Section 5.2 this idea was articulated further through the concept of *physiological time*, as measured through the accumulation of heat units called *degree-days*. The number of degree-days that accumulate over time for a given temperature profile is the area under the curve of this profile between the lower and upper thresholds, as illustrated in Figure 5.50. Note that a lower threshold is always needed, to bound the area from below, but the calculated area is either bounded above by the temperature curve itself or by an upper threshold, depending on which is the minimum for the time in question. Thus, a lower threshold is always needed, but an upper threshold is only included as a refinement of physiological time as a model for estimating the growth and developmental rates of plants and poikilotherms.

What if we could continuously measure the temperature in an orchard, for example, from the time of *bud break* (the first buds appear on otherwise bare trees), and we knew how many degree-days between minimum and maximum thresholds were needed until

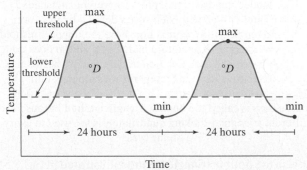

Figure 5.50 Solid line represents the continuous temperature that a plant or poikilotherm experiences, and shaded area represents the accumulated degree-days that the organism in question will experience subject to development being arrested above and below the upper and lower thresholds, respectively.

the trees come into blossom? If so, we could predict— using anticipated weather patterns—the expected date for the occurrence of blossoms and make sure we have honeybee hives in the orchard in sufficient time to anticipate this event. Thus, the calculation of degree-days can help growers optimize their use of honeybee pollinators, schedule harvest activities, and so on.

It is generally not possible, or even desirable, to continuously monitor the temperature of an orchard. Most growers have temperature gauges that record only the maximum (max) and minimum (min) temperatures each day. These data can be used to generate a degree-day calculation based on the assumption that the maximum and minimum temperatures occur 12 hours apart (as idealized in Figure 5.50) using an appropriate function (i.e., model) for interpolating the temperature between each consecutive pair of max-min and min-max temperatures. If a linear function is used, and only a lower threshold is assumed, the method is equivalent to constructing a sequence of right-angle triangles with either a rectangular piece added below when the minimum temperature is above the threshold; or, the base of the triangle is raised for the case when the minimum temperature is below a lower threshold (as illustrated in Figure 5.51).

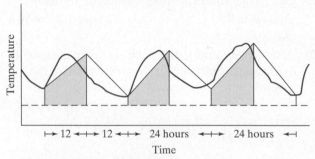

Figure 5.51 The thick, irregular red line represents a hypothetical temperature profile that oscillates like a distorted sine wave, so that in every 24-hour period it has a maximum and a minimum value. The thin line is a linear interpolation between these maximum and minimum values. The shaded quadrilaterals, plus intervening nonshaded quadrilaterals—all with their bases defined by the lower threshold temperature (dotted line: note the upper threshold is above the max temperature in all cases and so does not apply)—are the accumulation of degree-days between consecutive min-max temperatures and max-min temperatures, respectively. This method of accumulating degree-days is called the **double triangle method** because two different triangular-looking quadrilaterals are used in every 24-hour cycle.

1. Use the double triangle method illustrated in Fig. 5.51 to calculate the number of degree-days accumulated over a three-day period in which the minimum and maximum temperatures in degrees Celsius are $T = \{(5, 23), (7, 22), (4, 26), (5, \text{not measured})\}$ and the lower threshold is $0°C$ with no upper threshold assumed to exist.

2. Using the double triangle method, recalculate the number of degree-days accumulated when the lower threshold is $5°C$, first for the case when there is no upper threshold, then when the upper threshold is $30°C$, and finally when the upper threshold is $25°C$.

3. Instead of using a line to interpolate between min and max temperatures, use the rising first quarter phase of a sine function to interpolate between the given min and max and the falling second quarter of a sine function to interpolate between the max and min temperatures. This method is referred to as the **double sine method** (Figure 5.52).

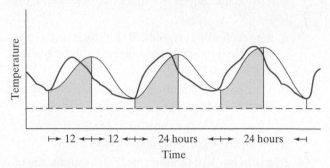

Figure 5.52 The thick irregular red line represents a hypothetical temperature profile. The thin line, instead of being a linear interpolation as depicted in the double triangle method, is a quarter sine wave interpolation (of 12 hour duration) between each min-max and max-min pair of temperatures. This method of accumulating degree-days, is called the **double sine method**.

4. Repeat step 2 using the double sine method and compare your results with the double triangle method.

5. Use your precalculus knowledge of algebra and geometry to write a general expression for the number of accumulated degree-days under the double triangle method when the temperature profile is $T = \{(m_1, M_1), (m_2, M_2), \cdots, (m_n, M_n)\}$ and the minimum and maximum developmental thresholds are k and K, respectively.

6. Use your knowledge of integral calculus to repeat the exercises and write a general expression for the double sine method.

7. Find a real data set on the Web of daily max and min temperatures that spans a period of several months (if you find a longer data set, select a subset); use double triangle and double sine formulas, implemented in your favorite technology (e.g., a spreadsheet application, Mathematica, Maple, or some other programming language) to calculate the number of degree-days progressively accumulating each day from the start to end date of your data, if the lower and upper thresholds, respectively, are equal to the average min and average max over the data. Plot these results on a graph of accumulated degree-days to date to provide a visual sense of how much the two methods differ over time.

Project 5B   Life Histories and Population Growth (challenging!)

Every biological species has a life history characterized by two functions: the mortality function $\ell(x)$ and natality function $b(x)$. The interpretation of the first function is that $\ell(x)$ represents the proportion of individuals in a large population that survive to age x or, in a small population the probability that an individual survives to age x. As we will see in Chapter 7, when we look at the relationship between integration and probability theory, $\ell(x)$ also represents the probability that any given individual will survive until age x (which can be a fractional number). Thus, $\ell(30) = 0.2$ implies that only 20% of individuals in a population will survive until age 30 or that any particular individual has a 20% chance of surviving until age 30. Note that we don't have to use years as our unit of time. In the case of fruit flies, for example, a more appropriate measure of age is weeks or days.

The function $b(x)$ represents a *force of natality* which only has a clear meaning in terms of being integrated over some nonzero age interval x. For example, $\int_2^3 b(x)dx = 3.5$ implies that each individual in its third period of life (i.e., from age 2 to age 3) is expected (on average) to produce 3.5 offspring. If these are sexually reproducing organisms, then this implies that any male-female pair in its third year of life is expected to produce 7 offspring.

The theory we are about to explore assumes that either the species is clonal, or males and females have the same life histories, or males and females have different gender-specific life histories but only females are considered. In the latter case, $b(x)$ is interpreted as the force of natality of female progeny per reproducing female of age x—that is, the statement $\int_2^3 b(x)dx = 3.5$ implies that each female in her third period of life is expected to have 3.5 daughters. Of course some might have 0 and others might have 10, but the average for the age range in question is 3.5.

Demographers have shown, under assumptions of *stationarity* (a technical term that requires more advanced concepts than we have to define it, but can be loosely thought of as a population that has an unchanging age-structure over time), that the quantity

$$R_0 = \int_0^{x_{max}} \ell(x)b(x)dx$$

represents the number of individuals being born for every individual that dies, given that no individual lives beyond age x_{max}. This implies that the population is growing if $R_0 > 1$ and declining if $R_0 < 1$. Further, demographers have shown that this rate of growth or decline is equivalent to the mathematical statement that $N_{t+G} = R_0 N_t$, where G is the length of a generation which is given by the integral

$$G = \frac{\int_0^{x_{max}} x\ell(x)b(x)dx}{R_0}$$

Thus, if we rescale time so $G = 1$, then this model implies that the population will have grown from an initial size N_0 to a size $N_m = R_0^m N_0$ after m generations.

1. If the proportion of individuals that die each time period in an age-specific cohort (i.e., a group of individuals of the same age) is independent of their age, then the mortality schedule (curve, function) for the species in question is said to be Type II. Demonstrate that the form

$$\ell(x) = \begin{cases} e^{-rx} & 0 \leq x \leq x_{max} \\ 0 & x \geq x_{max} \end{cases}$$

is a Type II mortality curve on $[0, x_{max}]$ for some constant $r > 0$.

2. Species are said to have Type I mortality schedules if mortality rates are much higher in immature than mature individuals (except, of course, for the very old). Type III is the reverse case. By searching the Internet or other reference sources, identify three to five species conforming to each of the three mortality schedule types.

3. Over long periods of time, ecological processes ensure that most populations either stay the same size or go extinct, since the finiteness of our world does not permit them to grow without bound. In the former case, we expect in the long run (i.e., on average over time) that $R_0 = 1$, which implies:

$$\int_0^{x_{max}} \ell(x)b(x)dx = 1$$

If a species has the mortality schedule given in part 1, and a natality schedule $b(x)$ of the form

$$b(x) = \begin{cases} 0 & 0 \leq x < m \\ b & x \geq m \end{cases},$$

where $m < x_{max}$ (i.e., individuals start reproducing at age m beyond which reproduction is the age-independent rate of b progeny per time period), then for $x_{max} = 100$ explore the trade-offs in the values of r, b, and m that correspond to a stable population (i.e., $R_0 = 1$) and provide an expression for the corresponding generation time. (Hint: Use the condition $R_0 = 1$ to get a relationship among r, b, and m; then express one of the parameters in terms of the other two. For selected values of one of the parameters, you can then graph relationships between the other two. What general statements can you make about these relationships?)

4. The mortality schedule

$$\ell(x) = \begin{cases} \dfrac{1}{1 + (x/d)^2} & 0 \leq x \leq x_{max} \\ 0 & x \geq x_{max} \end{cases}$$

is of Type III on $[0, x_{max}]$, provided $d > x_{max}$, because mortality rates are relatively low until individuals approach age d, around which mortality rates increase strongly. Repeat the previous exercise with this mortality schedule instead of the Type II schedule; look at trade-offs in the values of d, b, and m.

CHAPTER 6

Differential Equations

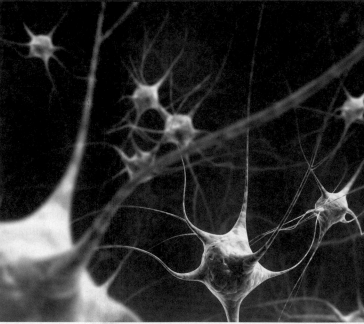

Figure 6.1 We are able to think, move, and eat because of collections of neurons in our bodies. This photograph shows the neurons of a ground squirrel.

Photo Researchers, Inc.

Preview

"Among all the mathematical disciplines the theory of differential equations is the most important . . . It furnishes the explanation of all those elementary manifestations of nature which involve time."

Sophus Lie, *Leipziger Berichte*, 47 (1805)

Equations containing one independent and one dependent variable, as well as the derivative of the dependent variable with respect to the independent variable, are known as **first-order ordinary differential equations** (or **first-order ODEs** for short). An example of a first-order ODE where the independent variable is x and the dependent variable y is:

$$\frac{dy}{dt} = 3 + \sin(t)y(1 - y)$$

The right hand side of this differential equation depends on t and y and, consequently, is known as a **non-autonomous** differential equation. When the right hand side only depends on the dependent variable, the equation is **autonomous**. In this Chapter, we develop quantitative and qualitative methods to understand solutions of these equations. A **solution** is a function $y(t)$ that satisfies the equation in question over a specified interval of time.

Despite their moniker, ODEs are anything but "ordinary." They have been used to describe extraordinary things such as world population growth, the dynamics of nerve impulses, and the spread of diseases. For instance, in Example 5 of Section 6.6, we use differential equations to explore how populations of neurons can store memories. In this chapter, we study these extraordinary equations three ways. First, after introducing some basic terminology and models, we will derive analytical solutions for special types of ODEs using integration techniques. Second, as ODEs often cannot be solved explicitly, we will introduce techniques that shed light on the qualitative behavior of ODEs. Third, we will use technology to generate and visualize numerical solutions to these ODEs. To this end, we will discuss a numerical method, namely *Euler's method*, to generate specific solutions to ODEs. By no means will this discussion be exhaustive. This chapter and Chapter 8 provide only a taste of this powerful mathematical formulation, which has been used extensively to understand the dynamics of biological systems.

6.1 A Modeling Introduction to Differential Equations

Differential equations describe how quantities change continuously over time. One of the first applications of differential equations to biology was to population growth: understanding how the sizes of populations—whether they be viruses, plants, or animals—vary in time. We begin our study of differential equations by examining some of these applications.

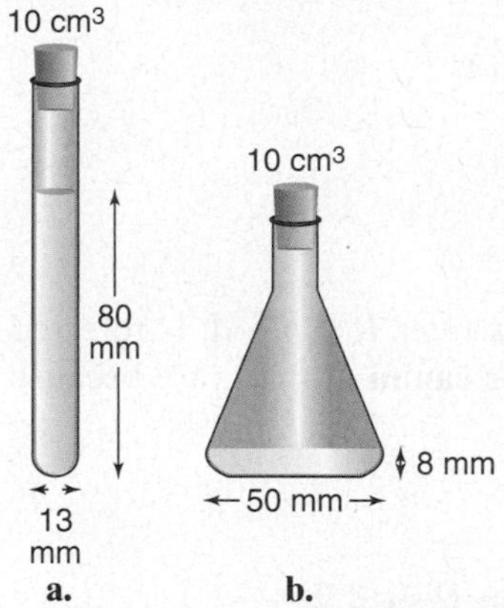

Exponential population growth and decay

At the beginning of the twentieth century, several notable biologists, including Georgyi F. Gause and T. Carlson, studied population growth of yeast. Both biologists grew yeast under constant environmental conditions in test tubes and flasks as illustrated in Figure 6.2, and they regularly monitored the densities. The resulting data from Carlson's experiment are shown in Figure 6.3.

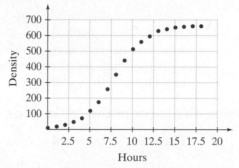

Figure 6.3 Carlson yeast data plotted hourly over an eighteen-hour period.

Data Source: "Über Geschwindigkeit und Grösse der Hefevermehrung in Würze,"
Biochem. Z.57 (1913): 313–334.

One of the goals of this section is to come up with a model that describes Carlson's yeast data. In developing the model, we adhere to the *principle of parsimony*, an operational principle used in science that requires beginning with the simplest model and only add more complexity as needed. It is also known as *Occam's razor* or more simply as the KISS principle—Keep It Simple, Stupid! Thus, we begin by formulating and analyzing the simplest possible model and introduce elaborations only as necessary. We start by modeling the initial growth phase of a relatively simple population: the number of cells in a yeast culture.

Figure 6.2 sidebar:

10 cm³

10 cm³

80 mm

13 mm

50 mm **8 mm**

a.   **b.**

Figure 6.2 Density of yeast growing in a test tube **a** and a flask **b** configured so that the same amount of medium can have two different levels of exposure to the air.

Data Source: G. F. Gause, "The Struggle for Existence." This is a book published in 1934 by Williams and Wilkins, Baltimore.

The principle *entia non sunt multiplicanda praeter necessitatem* (i.e., entities must not be multiplied beyond necessity) is attributed to Franciscan friar William of Ockham (ca. 1287–1347).

Example 1   Constant per capita growth rate model

The growth of a population of size N at time t, denoted by the function $N(t)$, is determined by four processes: birth, death, immigration, and emigration. The simplest model follows from these assumptions:

- **The system is closed:** There is no immigration or emigration.
- **Birth rates are proportional to the population density:** The more individuals there are, the more births there are. If b is the proportionality constant, then the birth rate is $bN(t)$ and b is called the **per capita birth rate**.
- **Death rates are proportional to the population density:** The more individuals there are, the more deaths there are. If d is the proportionality constant, then the total death rate is $dN(t)$ and d is called **the per capita death rate**.

Write a differential equation model that embodies these assumptions.

Solution   Let N denote the population density and t time. Under the stated assumptions, the model is as follows:

$$\frac{dN}{dt} = \text{birth rates} - \text{death rates}$$
$$= bN - dN$$
$$= (b-d)N$$
$$= RN \qquad \text{setting } R = b - d$$

Here, the per capita birth rate minus per capita death rate, $R = b - d$, is referred to as the **intrinsic growth rate** or the **instantaneous per capita growth rate** because $R = \frac{1}{N}\frac{dN}{dt}$ for $N > 0$.

A solution of the equation, $\frac{dN}{dt} = RN$, is a function N such that

$$N'(t) = RN(t)$$

We will analyze the solution of this differential equation qualitatively and analytically. A **qualitative analysis** involves discovering the qualitative behavior of solutions. In other words, it involves determining whether the solutions are increasing, decreasing, remaining constant, or even oscillating, without worrying about the exact form of the solution.

Example 2   Qualitative behavior of the constant per capita growth rate model

Consider the growth of a population modeled by

$$\frac{dN}{dt} = RN$$

where the *initial value* of the population $N(0)$ at time $t = 0$ is positive. Discuss how the behavior of the population depends on the sign of R.

Solution
Case $R = 0$. Since $N'(t) = RN = 0$ for all t, the rate of change in N is identically zero. Hence, the population density $N(t)$ stays equal to its initial value $N(t) = N(0)$ for all $t > 0$.

Case $R > 0$. In this case $N'(t) = RN > 0$ for all t. Therefore, the population density increases indefinitely.

Case $R < 0$. In this case $N'(t) = RN < 0$ for all t. Therefore, the population density decreases indefinitely.

The three qualitative cases in this example correspond to three regimes of population behavior:

Constancy: $R = 0$ and the per capita birth rate b equals the per capita death rate d. The population neither grows nor declines.

Growth: $R > 0$ and the per capita birth rate b exceeds the per capita death rate d. The population increases over time.

Decay: $R < 0$ and the per capita death rate d exceeds the per capita birth rate b. The population decreases over time.

We made all of these qualitative predictions by looking at the sign of the right-hand side of $N' = RN$. General qualitative methods for making these predictions are discussed further in Sections 6.4 and 6.5.

In contrast to a qualitative analysis, an **analytical approach** involves finding explicit solutions to differential equation models. For this constant per capita growth rate model, finding an analytical solution means finding a function $N(t)$ such that its derivative is R times itself; that is, $N'(t) = RN(t)$.

Example 3   Exponential growth model

Verify that the function

$$N(t) = Ce^{Rt}$$

is a solution to the differential equation

$$\frac{dN}{dt} = RN$$

for any value C. Provide an interpretation for C and for the behavior of $N(t)$ when $R > 0$ and $R < 0$.

Solution   To verify that $N(t) = Ce^{Rt}$ is a solution, it suffices to plug $N(t)$ into both sides of the equation $\dfrac{dN}{dt} = RN$:

$$\frac{dN}{dt} = CRe^{Rt} = R(Ce^{Rt}) = RN(t)$$

Thus, $N(t) = Ce^{Rt}$ is a solution for any constant C. Evaluating $N(t)$ at $t = 0$ gives $N(0) = Ce^{R \times 0} = C$. Hence, C represents the initial population density and, population models must satisfy $C \geq 0$. Furthermore, because $N(t)$ is an exponential function, we have:

Exponential growth: For $R > 0$, the population density $N(0)e^{Rt}$ increases exponentially to infinity over time.

Exponential decay: For $R < 0$, the population density $N(0)e^{Rt}$ declines at an exponential rate, asymptotically approaching zero.

Using the solution $N(t) = Ce^{Rt}$, we can determine how well our simple model $N'(t) = RN$ fits the initial phase of Carlson's yeast data. We can either fit the model directly or use a semi-log plot to fit a linear model through the data since, by taking logarithms, our exponential growth model $N(t) = N(0)e^{rt}$ can be written as

$$\ln N(t) = Rt + k$$

where $k = \ln C = \ln N(0)$.

Example 4   Carlson's data: exponential growth

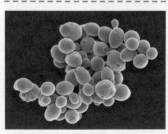

Figure 6.4 Yeast cells, as seen under a microscope.

Yeast provide an ideal organism for population growth studies because they are easy to culture in liquid media, and the number of individuals per unit volume can be counted under a microscope, as depicted in Figure 6.4 where some cells can be seen reproducing by budding. The data that Carlson gathered in his yeast culturing study are show in Table 6.1.

Table 6.1 Population densities (number/unit volume) for a growing yeast culture at one-hour intervals

Time	Population	Time	Population	Time	Population
0	9.6	6	174.6	12	594.8
1	18.3	7	257.3	13	629.4
2	29.0	8	350.7	14	640.8
3	47.2	9	441.0	15	651.1
4	71.1	10	513.3	16	655.9
5	119.1	11	559.7	17	659.6

Data source: See Figure 6.3.

As illustrated in Figure 6.3, the initial phase of population growth appears to be exponential.

a. Use the data to estimate the parameters C and R in the growth model $N(t) = Ce^{Rt}$, where t is measured in hours, over the first three hours; then plot the model together with the data over the first eight hours to see how well the model continues to fit.

b. Use the logarithms of the data to estimate the value of R that provides the best-fitting line $\ln N(t) = Rt + \ln N(0)$ to a semi-log plot of the first eight hours of data.

Solution

a. Recall, C represents the initial population density, so $C = N(0)$. Hence, $C = 9.6$. To estimate R over the first three hours, we choose the data point $N(3) = 47.2$ (bearing in mind that a different data point would yield a similar, but different graph) and solve

$$N(t) = Ce^{Rt}$$
$$N(3) = 9.6e^{3R} \qquad \textit{substitute known value for } C$$
$$47.2 = 9.6e^{3R} \qquad \textit{substitute known value for } N(3)$$
$$3R = \ln(47.2/9.6) \qquad \textit{definition of logarithm}$$
$$R \approx 0.53$$

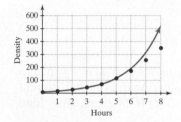

Figure 6.5 Exponential model (solid line) fitted to the first three hours of the Carlson yeast data (dots).

Since the time is in hours, $R \approx 0.53$ per hour over the first three hours.

A plot of $N(t) = 9.6\, e^{0.53\,t}$ against the data is shown in Figure 6.5. The equation we derived fits the data well until $t = 6$. After $t = 6$, the equation overestimates the population size.

b. Using the approach laid out in Example 7 of Section 4.3, we obtain the following table:

t (hours)	0	1	2	3	4	5	6	7	8	$\sum_{t=1}^{8} t = 36$
$N(t)$	9.6	18.3	29.0	47.2	71.1	119.1	174.6	257.3	350.7	
t^2	0	1	4	9	16	25	36	49	64	$\sum_{t=1}^{8} t^2 = 204$
$\ln N(t)$	2.26	2.91	3.37	3.85	4.26	4.78	5.16	5.55	5.86	
$t \ln N(t)$	0	2.91	6.73	11.56	17.06	23.90	30.97	38.85	46.88	$\sum_{t=1}^{8} t \ln N(t) = 178.86$

Now we seek to minimize the function

$$S(R) = \sum_{t=1}^{8} \left(\ln N(t) - (\ln 9.6 + Rt) \right)^2$$

Differentiating S with respect to R yields

$$S'(R) = 2\sum_{t=1}^{8} t \ln N(t) - 2\ln 9.6 \sum_{t=1}^{8} t - 2R \sum_{t=1}^{8} t^2$$

Setting $S'(R) = 0$ and solving for R yields

$$R = \frac{\displaystyle\sum_{t=1}^{8} t \ln N(t) - \ln 9.6 \sum_{t=1}^{8} t}{\displaystyle\sum_{t=1}^{8} t^2} \approx 0.48$$

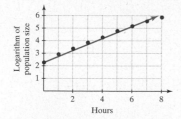

Figure 6.6 Semi-log regression plot (solid line) of exponential growth model through the first eight hours of the Carlson yeast data (dots).

which is somewhat lower than the $R = 0.53$ that we estimated directly from the data over the first three hours. The semi-log regression line is

$$\ln N(t) = 2.26 + 0.48t$$

A plot of this line is provided in Figure 6.6.

It is not surprising that we obtained a lower value for R using a semi-log regression for the first eight hours of data, as our exponential model in part **a** overestimates the growth rate from hours 6 to 8. ∎

Using the exponential model of growth, we can estimate the doubling time for a yeast population during the early stages of colony growth.

Example 5   Doubling time for yeast

For a population satisfying the equation

$$\frac{dN}{dt} = 0.53\,N$$

find the time in hours for the population to double.

Solution   The solution of this differential equation is of the form

$$N(t) = Ce^{0.53t}$$

where C corresponds to the initial population density. To find the doubling time in hours, we need to find t such that $N(t) = 2N(0) = 2C$. Hence, we need to solve

$$
\begin{aligned}
2C &= Ce^{0.53t} &&\textit{solve equation for } t \\
2 &= e^{0.53t} &&\textit{divide both sides by } C \\
0.53t &= \ln 2 &&\textit{definition of logarithm} \\
t &= \frac{\ln 2}{0.53} \approx 1.31 &&\textit{divide both sides by } 0.53
\end{aligned}
$$

The doubling time is about 1 hour and 18 minutes. This conclusion is consistent with the Carlson data during the first few hours of growth. For instance, we can see in Figure 6.5 that after $3 \times 1.31 \approx 4$ hours, the yeast density has increased approximately by a factor of $2^3 = 8$. ∎

As you will see in Problem Set 6.1 and in Section 6.3, the exponential growth model $dN/dt = RN$ is used to model radioactive decay, decay of drugs in the bloodstream, and decay of the number of viral particles in the blood of an individual treated with antiviral drugs.

Logistic growth

Although the exponential model provides a reasonable fit for the initial growth of the yeast population considered in the previous example, it substantially overestimates the population density during the seventh and eighth hours. Moreover, beyond eight hours, the data plotted in Figure 6.7 indicate that the yeast culture asymptotically approaches a density of around 660, while the exponential growth model exhibits unbounded growth. This phenomenon of decreasing per capita growth rate with increasing population density was first elaborated by Thomas Malthus (1766–1834) in his 1798 treatise "An Essay on the Principle of Population Growth." Malthus recognized that as populations get larger, their per capita growth rate declines due to limited resources and interference among individuals. To deal with these limitations, we modify our model using the principle of parsimony.

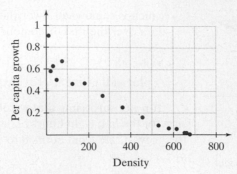

Figure 6.7 Per capita growth rate as a function of density for the Carlson yeast data.

Using the Carlson data in Table 6.1, we estimated the per capita growth rate, R, of yeast as a function of population density, N. These estimates are plotted in Figure 6.7 and show that the per capita growth rate is a decreasing function. The exact form of R as a function of N is not uniquely determined. Following the parsimony principle, we begin with the simplest decreasing function of N with positive intercept on the R axis, which is the linear function. Let K denote the horizontal intercept and r the vertical intercept of this linear function. In other words, we choose the per capita growth rate $R(N)$ in the model

$$\frac{dN}{dt} = R(N)N$$

to be the linear function

$$R(N) = r(1 - N/K)$$

This equation is the *logistic equation*, which is arguably the single most important equation in population ecology. From the data plotted in Figure 6.7, we might guess that the vertical intercept r lies somewhere between 0.5 and 0.6 and that the horizontal intercept K lies somewhere between 600 and 700. For reasons that become clearer in the next example, the value $N = K$ is called the *carrying capacity* for the population. The parameter r is called the *intrinsic growth rate*.

Logistic Equation

If in the equation

$$\frac{dN}{dt} = R(N)N$$

the instantaneous per capita growth rate function $R(N)$ has the form

$$R(N) = r(1 - N/K)$$

then we obtain the **logistic equation**

$$\frac{dN}{dt} = r\left(1 - \frac{N}{K}\right)N$$

The parameter $r > 0$ is the **intrinsic growth rate** and the parameter $K > 0$ is the **carrying capacity**. The parameters r and K are, respectively, the vertical and horizontal intercepts of the graph of the function $R(N)$.

What can we say about the behavior of the solutions to the logistic equation? We partially answer this question with a qualitative analysis. Finding explicit solutions have to wait until the next section.

Example 6   Qualitative analysis of the logistic equation

Assuming that $r > 0$ and $K > 0$, describe qualitatively how solutions to the logistic equation depend on the initial value of N.

Solution   Qualitatively there are three types of solutions when initially $N \geq 0$.

Equilibrium solutions: If $N = 0$ initially, then the population growth rate $\frac{dN}{dt} = r \times 0(1 - 0/K) = 0$ initially and the population density continues to remain at zero. We verify in the next section that $N = 0$ for all t is a solution to this differential equation. Similarly, if $N = K$ initially, then $\frac{dN}{dt} = rK(1 - K/K) = 0$ initially and the population density continues to remain at K. These unchanging (i.e., constant) solutions are called **equilibrium solutions** and are illustrated in Figure 6.8a.

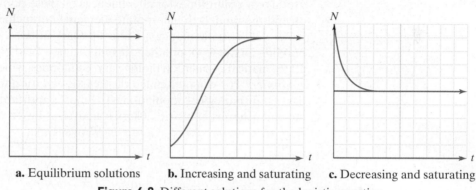

a. Equilibrium solutions   **b.** Increasing and saturating   **c.** Decreasing and saturating

Figure 6.8 Different solutions for the logistic equation.

Increasing and saturating: If $0 < N < K$, then $rN(1 - N/K) > 0$ and the population growth rate $\frac{dN}{dt}$ is positive. For a population starting between 0 and K, we expect the population density to increase. However, since $\frac{dN}{dt}$ gets close to zero as N gets close to K, we expect the population to increase less rapidly as it approaches K and to asymptotically saturate at K, as illustrated in Figure 6.8b. We will show this formally in the next section where we find explicit solutions to the logistic equation.

Decreasing and saturating: If $N > K$, then $rN(1 - N/K) < 0$ and the population growth rate $\frac{dN}{dt}$ is negative. In this case the population density declines over time. As $\frac{dN}{dt}$ becomes less negative as N approaches K from above, we expect population density to decline less rapidly as it approaches K and the population density to asymptotically level off at K. This is illustrated in Figure 6.8c.

Hence, as long as $N > 0$ initially, we expect the population density to approach the carrying capacity K of the environment, as will be seen to be true once we have solved the logistic equation in the next section. ∎

Example 7   Logistic model for the yeast data

Parameterize the logistic model for the Carlson yeast data given in Example 4.

Solution   The Carlson data in Table 6.1 suggest that the population density is approaching an asymptotic value of 660. Hence, we choose $K = 660$. To estimate r, notice that when N is small

$$\frac{dN}{dt} = rN\left(1 - \frac{N}{660}\right) \approx rN$$

In other words, at low densities we expect to see approximately exponential growth. Using our work from Example 4, we set $r = 0.53$. Thus, the specific logistic equation in this case is

$$\frac{dN}{dt} = 0.53N\left(1 - \frac{N}{660}\right)$$

In the next section we derive the solution to the logistic equation that we can compare directly to the data. Problem Set 6.1 has examples in which the logistic model describes the spread of AIDS and the use of the iPhone in the United States.

External influences on populations

In addition to understanding the feedbacks of populations on their own growth, it is useful to account for external influences on these populations. To do this, we need to extend the models to incorporate elements not included in the initial model. In the 1970s the biomathematician Colin W. Clark at the University of British Columbia, building on the work of others, invented a new field of research at the intersection of mathematical population biology and economics, which he called *mathematical bioeconomics*.* His work extended the logistic equation to account for the economics of harvesting biological populations. The most important applications were in the whaling and fisheries industries. Clark's analysis is based on logistically growing populations that are harvested at a rate $h(N)$ over time:

$$\frac{dN}{dt} = rN\left(1 - \frac{N}{K}\right) - h(N)$$

Two cases are of particular interest are:

- **Constant harvesting:** In this case, the harvest rate is $h(N) = h$ for all $N > 0$. If harvesting drives N to 0, as it has in some real populations, then $h(N) = 0$ for $N = 0$.

- **Proportional harvesting:** In this case, the harvest rate is $h(N) = vN$, where the constant of proportionality $v > 0$ is called the *harvesting effort variable*.

Example 8   Harvesting queen conchs

- -

Figure 6.9 Conchs are harvested for home reef aquariums, as well as for their beautiful shells.

© blickwinkel/Alamy

Consider a population of queen conchs in the Bahamas that, in the absence of harvesting, exhibit logistic growth. Let N represent the number of conchs in a well-defined area and t be measured in years. For ease of computation, assume that the intrinsic growth rate of this population is $r = 10$ and that the carrying capacity of the area in which the conchs are located is $K = 10,000$ individuals.

a. Write a logistic harvesting model for the case where 21,000 individuals are removed from the population every year.

b. Determine qualitatively the fate of the population and how it depends on the initial number of conchs in the population.

c. Discuss what happens if the harvesting rate is $h = 30,000$ conchs per year.

*Colin W. Clark, *Mathematical Bioeconomics: The Optimal Management of Renewable Resources,* (New York: John Wiley & Sons, 1976).

Solution

a. Since we are harvesting at a constant rate of 21,000 individuals per year, the model is

$$\frac{dN}{dt} = 10N\left(1 - \frac{N}{10,000}\right) - 21,000$$

b. The qualitative analysis reduces to understanding for what values of N is

$$\frac{dN}{dt} = 0, \quad \frac{dN}{dt} < 0, \quad \text{or} \quad \frac{dN}{dt} > 0$$

Case I: $\dfrac{dN}{dt} = 0$.

$$\frac{dN}{dt} = 0$$

$$10N\left(1 - \frac{N}{10,000}\right) - 21,000 = 0$$

$$N^2 - 10,000N + 21,000,000 = 0 \qquad \textit{expanding and multiplying by } 1,000$$

$$(N - 3,000)(N - 7,000) = 0$$

$$N = 3,000 \text{ or } 7,000$$

These are the equilibrium values, that is, values where $\dfrac{dN}{dt} = 0$.

Case II: $\dfrac{dN}{dt} < 0$. From our work in case I, we see that $\dfrac{dN}{dt} < 0$ if $N < 3,000$ or $N > 7,000$. Hence, if $N > 7,000$, then the population would decrease, but decrease more slowly as N approaches 7,000 (i.e., $\dfrac{dN}{dt}$ is close to zero for N near 7,000). Consequently, if $N > 7,000$, we would expect the population to decrease and to saturate at 7,000 (we say *expect*, because the notions expressed here can only be made more precise once we have considered additional theory). Alternatively, if $N < 3,000$, the population would continually decrease to 0 (i.e., extinction) as $\dfrac{dN}{dt}$ becomes more and more negative as N continues to decrease.

Case III: $\dfrac{dN}{dt} > 0$. From our work in case I, we see that $\dfrac{dN}{dt} > 0$ if $3,000 < N < 7,000$. We would expect the population to increase, increase more slowly as it approaches 7,000, and to saturate at 7,000.

c. First we note that the conch growth rate

$$F(N) = 10N\left(1 - \frac{N}{10,000}\right)$$

has a maximum at $N = 5,000$ with

$$F(5,000) = 50,000\left(1 - \frac{1}{2}\right) = 25,000$$

Thus, the right-hand side of the conch growth equation subject to a harvesting rate of 30,000 conchs per year satisfies

$$\frac{dN}{dt} = 10N\left(1 - \frac{N}{10,000}\right) - 30,000 \leq -5,000 \text{ for all } N$$

Hence, harvesting the population at this rate will drive it rapidly to extinction.

PROBLEM SET 6.1

Level 1 DRILL PROBLEMS

Write a differential equation to model the situation in Problems 1 to 8. Do not solve.

1. The number of bacteria in a culture grows at a rate that is proportional to the number of bacteria present.

2. A sample of radium decays at a rate that is proportional to the amount of radium present in the sample.

3. In Section 6.3, we will introduce Newton's law of cooling. Newton's law states that the rate at which the temperature of a body changes is proportional to the difference between the body's temperature T and the ambient temperature A.

4. In Section 6.3, we will study the von Bertalanffy growth equation. As part of that study, we will formulate a differential equation which states that the rate at which the mass, M, of a healthy critter grows through absorption of food is directly proportional to its surface area L^2 and declines through respiration at a rate proportional to its mass L^3. If M is proportional to L^3 — that is, $M = kL^3$ for some positive constant k — then write the model in terms of the unknown function M.

5. According to Benjamin Gompertz (1779–1865), the growth rate of a population is proportional to the number of individuals present, where the factor of proportionality is an exponentially decreasing function of time.

6. When a person is asked to recall a set of N facts, the rate at which the facts are recalled is proportional to the number of relevant facts in the person's memory that have not yet been recalled.

7. The rate at which an epidemic spreads through a community of P susceptible people is proportional to the product of the number of people y who have caught the disease and the number $P - y$ who have not.

8. The rate at which people are implicated in a government scandal is proportional to the product of the number N of people already implicated and the number of people involved who have not yet been implicated.

For Problems 9 to 14 find the time for the population to double or halve its initial level, as appropriate.

9. $\dfrac{dN}{dt} = 0.64N$ Units of t are years.

10. $\dfrac{dx}{dt} = 1.2x$ Units of t are weeks.

11. $\dfrac{dN}{dt} = -0.41N$ Units of t are months.

12. $\dfrac{dx}{dt} = -0.8x$ Units of t are decades.

13. $\dfrac{dy}{dt} = -0.03y$ Units of t are centuries.
    Give answer in years.

14. $\dfrac{dy}{dx} = 3.5y$ Units of x are millennia.
    Give answer in years.

15. How long does it take for a population to quadruple if its growth over years is modeled by
$$\frac{dN}{dt} = 0.2N$$

16. How long does it take for a population to quintuple if its growth over years is modeled by
$$\frac{dN}{dt} = 0.04N$$

17. How long does it take for the level of radioactivity to decay to 10% of its current level, if the level of radioactivity over years is modeled by
$$\frac{dx}{dt} = -0.0025x$$

18. How long does it take for the level of radioactivity to decay to 1% of its current level, if the level of radioactivity over years is modeled by
$$\frac{dx}{dt} = -0.03x$$

A population model for Problems 19 to 22 is given by
$$\frac{dP}{dt} = P(100 - P)$$
where $P(t)$ denotes population density at time t.

19. For what values is the population at equilibrium?

20. For what values is $\dfrac{dP}{dt} > 0$?

21. For what values is $\dfrac{dP}{dt} < 0$?

22. Describe how the fate of the population depends on the initial density.

A population model for Problems 23 to 26 is given by
$$\frac{dP}{dt} = P(P - 1)(100 - P)$$
where $P(t)$ denotes population density at time t.

23. For what values is the population at equilibrium?

24. For what values is $\dfrac{dP}{dt} > 0$?

25. For what values is $\dfrac{dP}{dt} < 0$?

26. Describe how the fate of the population depends on the initial density.

27. Find the equilibria to the model
$$\frac{dP}{dt} = -p^3 + 3p^2 - 2p$$

Level 2 APPLIED AND THEORY PROBLEMS

Radioactive decay: *Certain types of atoms (e.g., carbon-14, xenon-133, lead-210) are inherently unstable. They exhibit random transitions to a different atom while emitting radiation in the process. Based on experimental evidence, the number, N, of atoms in a radioactive substance can be described by the equation*

$$\frac{dN}{dt} = -\lambda N$$

where t is measured in years and $\lambda > 0$ is known as the **decay constant.** *The decay constant is found experimentally by measuring the half-life, τ, of the radioactive substance (i.e., the time it takes for half of the substance to decay). Use this information in Problems 28 to 32.*

28. Find a solution to the decay equation assuming that $N(0) = N_0$.

29. For xenon-133, the half-life is five days. Find λ. Assume t is measured in days.

30. For carbon-14, the half-life is 5,568 years. Find the decay constant λ. Assume t is measured in years.

31. How old is a piece of human bone which contains just 60% of the amount of carbon-14 expected in a sample of bone from a living person? Assume the half-life of carbon-14 is 5,568 years.

32. The Dead Sea Scrolls were written on parchment at about 100 BC. Given that the half-life of carbon-14 is 5,568 years, what percentage of carbon-14 originally contained in the parchment remained when the scrolls were discovered in 1947?

33. King Arthur's Round Table. In Winchester castle there hangs a wooden round table, 18 feet in diameter and divided into twenty-five sections, one for the king and twenty-four for the knights. Some speculate that the Winchester round table is King Arthur's round table from the fifth century.* We know that the round table has been at Winchester since the fifteenth century. John Harding says in his chronical (1484) that the round table "ended at Winchester, and there it hangs still." To put an end to the speculation regarding the Winchester round table, in 1976 it was taken down from the wall and tests were employed to determine the date of origin. The rate of decay of carbon-14

in the table (i.e., in dead wood) was found to be 6.08 atoms per minute per gram of sample. Estimate the age of the table to determine whether the Winchester table was King Arthur's round table. Hint: Use the facts that the half-life of carbon-14 in dead wood is 5,568 years and that in living wood the rate of decay of carbon-14 is 6.68 atoms per minute per gram of wood.

34. Historical Quest

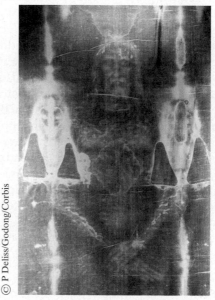

© P Deliss/Godong/Corbis

Shroud of Turin

The Shroud of Turin is a rectangular linen cloth kept in the Chapel of the Holy Shroud in the cathedral of St. John the Baptist in Turin, Italy. It shows the image of a man whose wounds correspond with the biblical accounts of the crucifixion of Jesus Christ. In 1389, Pierre d'Arcis, the Bishop of Troyes, wrote a memo to the pope accusing a colleague of passing off a certain cloth, cunningly painted, as the burial shroud of Jesus Christ. Despite this early testimony of forgery, the Shroud of Turin has survived as a famous relic. In 1988, a small sample of the Shroud of Turin was taken and scientists from Oxford University, the University of Arizona, and the Swiss Federal Institute of Technology were permitted to test it. Suppose the cloth contained 92.3% of the original amount of carbon. Use this information to determine the age of the shroud.

35. Consider the queen conch logistic growth model presented in Example 8 with a general harvesting function $h(N)$:

$$\frac{dN}{dt} = 10N\left(1 - \frac{N}{10,000}\right) - h(N)$$

---

*From D. N. Burghes, I. Huntley, and J. Mc-Donald, *Applying Mathematics: A Course in Mathematical Modeling* (Halsted Press, 1982).

a. Describe the qualitative behavior of solutions to this equation (i.e., the long-term abundance of the population) when $h(N) = 25,000$ for all N.

b. Describe the qualitative behavior of solutions to this equation (i.e., the long-term abundance of the population) when $h(N) = 5N$ for all N.

c. Describe the qualitative behavior of solutions to this equation (i.e., the long-term abundance of the population) when $h(N) = 12N$ for all N.

36. The cane toad (*Bufo marinus*) was introduced to Australia by the sugar cane industry to control two pests of sugar cane: the grey-backed cane beetle and the frenchie beetle.* Just over a hundred toads were brought to Gordonvale (near Edmonton, North Queensland) in 1935 and released into the cane fields. Unfortunately, due to an asynchrony between the life cycles of the cane toad and the sugar cane pests, the cane toad did not help suppress the cane beetle and the frenchie beetle. However, the cane toads ate almost everything else and grew at a tremendous pace. Now the cane toad is a major pest in Australia. The data in Table 6.2 describe the extent of the area occupied by the cane toads as a function of time.

A simple model to describe these data is given by

$$\frac{dA}{dt} = RA$$

where $A(t)$ is the area occupied at time t (years).

a. Find a solution to this model such that $A(0) = 32,800$ and $A(10) = 73,600$. Estimate the parameter R.

b. Estimate the area occupied by cane toads in 2004.

c. Modify this model to account for removing cane toads at a rate H km²/yr beginning in the year 2004. Determine how large H needs to be to ensure that A starts decreasing.

Table 6.2 Area occupied (in square kilometers) by the cane toad in Australia

Year	Area
1939	32,800
1944	55,800
1949	73,600
1954	138,000
1964	257,000
1969	301,000
1974	584,000

37. Consider the following problem of historical curiosity. The percentage of U.S. households that own a videocassette recorder (VCR) rose steadily from the time of their introduction in the late 1970s to the point at which other technologies displaced them. Let $y(t)$ denote the percentage of U.S. households with a VCR where t is measured in years from 1980 to 1991.

Year	1978	1979	1980	1981	1982	1983	1984
%	0.3	0.5	1.1	1.8	3.1	5.5	10.6

Year	1985	1986	1987	1988	1989	1990	1991
%	20.8	36	48.7	58	64.6	71.9	71.9

Data source: The Veronis, Suhler and Associates Communications Industry Forecast.

Assume that

$$\frac{dy}{dt} = ry\left(1 - \frac{y}{K}\right)$$

can be used to describe the data.

a. Use the first and third data points and the approximation $\frac{dy}{dt} \approx ry$ when y is small compared with K to estimate r.

b. Use the fact that the data are saturating to estimate the value of K.

38. In the previous example, compare an estimate obtained for r using growth from 1981 to 1982 versus 1981 to 1984.

39. The Ohio Department of Health released the following data tallying the number of newly diagnosed cases of AIDS in the state from the initial stages of the epidemic to the early 1990s when the first antiretroviral drugs began to become widely available: (*Cincinnati Enquirer*, December 11, 1994).

Year	1981	1982	1983	1984	1985	1986
Cases	2	8	27	58	121	209

Year	1987	1988	1989	1990	1991	1992
Cases	394	533	628	674	746	725

Let $y(t)$ denote the number of AIDS cases in Ohio in year t. Assume that

$$\frac{dy}{dt} = ry\left(1 - \frac{y}{K}\right)$$

can be used to describe the data.

a. Using the first few data points and the fact that $\frac{dy}{dt} \approx ry$ when y is small, estimate r.

b. Estimate the value of K.

Data source: M. D. Sabbath, W. C. Boughton, and S. Easteal, "Expansion of the Range of the Introduced Toad *Bufo marinus* in Australia from 1935 to 1974, *Copeia* 3 (1981): 676–680.

40. Hyperthyroidism is caused by growth of tumor-like cells that secrete thyroid hormones in excess of normal amounts. If untreated, an individual with hyperthyroidism may experience extreme weight loss, anorexia, muscle weakness, heart disease, and intolerance to stress and eventually may die. The most successful and least invasive treatment option is generally considered to be radioactive iodine-131 therapy. This involves the injection of a small amount of radioactive iodine into the body. For the type of hyperthyroidism called Graves' disease, it is usual for about 40% to 80% of the administered radioactive iodine to concentrate in the thyroid gland. For functioning adenomas ("hot nodules"), the uptake is closer to 20% to 30%. Excess iodine-131 is excreted rapidly by the kidneys. The quantity of radioiodine used to treat hyperthyroidism is not enough to injure any tissue except the thyroid, which slowly shrinks over a matter of weeks to months. Radioactive iodine is either swallowed in a capsule or sipped in solution through a straw. A typical dose is 5 to 15 millicuries. The half-life of iodine-131 is eight days.

a. Suppose that it takes forty-eight hours for a shipment of iodine-131 to reach a hospital. How much of the initial amount shipped is left once it arrives at the hospital?

b. Suppose a patient is given a dose of 10 millicuries, of which 30% concentrates in the thyroid gland. How much is left one week later?

c. Suppose a patient is given a dose of 10 millicuries, of which 30% concentrates in the thyroid gland. How much is left thirty days later?

6.2   Solutions and Separable Equations

In this chapter, we consider differential equations in which the right-hand side may explicitly depend on both the dependent and independent variables. If t is the independent variable and y is the dependent variable, then the equations we are considering have the general form

$$\frac{dy}{dt} = h(t, y)$$

where h is an expression involving t and y. Although this moves us into the realm of functions of two variables, we postpone a more formal development of the theory for such functions to Chapter 8. Even though $h(t, y)$ is a function of two variables, the solution $y(t)$ to the differential equation is a function of the single variable t.

In this section, we discuss what it means to be a solution to these differential equations and develop an important method, *separation of variables*, for solving differential equations of the form $\dfrac{dy}{dt} = \dfrac{f(t)}{g(y)}$. Using this method, we explore super-exponential growth and logistic growth of populations.

Solutions to differential equations

A function $y(t)$ is a **solution** to a differential equation if, when the function $y(t)$ is substituted into both sides of the differential equation, the differential equation is satisfied.

Example 1   Verifying a function is a solution to a differential equation

Consider the differential equation

$$\frac{dy}{dt} = \frac{y - 6}{t + 1}$$

which is defined for all $t \neq -1$. Which of the following functions are solutions for all $t \neq -1$?

a. $y(t) = t + 7$ **b.** $y(t) = 3t + 21$ **c.** $y(t) = 3t + 9$

Solution

a. To verify whether $y(t) = t + 7$ is a solution, we substitute this expression for $y(t)$ into the differential equation and simplify both sides.

$$y'(t) = \frac{y(t) - 6}{t + 1}$$

$$\text{left-hand side (LHS)} = \frac{d}{dt}(t + 7)$$

$$= 1$$

$$\text{right-hand side (RHS)} = \frac{t + 7 - 6}{t + 1}$$

$$= \frac{t + 7 - 6}{t + 1}$$

$$= \frac{t + 1}{t + 1}$$

$$= 1$$

Therefore: LHS = RHS for all $t \neq -1$

Since the equation is satisfied for all $t \neq -1$, we deduce that $y(t) = t + 7$ is a solution.

b. To verify whether $y(t) = 3t + 21$ is a solution, we substitute this expression for $y(t)$ into the differential equation and simplify both sides.

$$y'(t) = \frac{y(t) - 6}{t + 1}$$

$$\text{LHS} = \frac{d}{dt}(3t + 21)$$

$$= 3$$

$$\text{RHS} = \frac{3t + 21 - 6}{t + 1}$$

$$= \frac{3t + 15}{t + 1}$$

$$\neq 3 \text{ for all } t \neq -1$$

Therefore: LHS $\neq$ RHS

Since this equation is not satisfied for all $t \neq -1$, we deduce that $y(t) = 3t + 21$ is not a solution.

c. To verify whether $y(t) = 3t + 9$ is a solution, we substitute this expression for $y(t)$ into the differential equation and simplify both sides.

$$y'(t) = \frac{y(t) - 6}{t + 1}$$

$$\text{LHS} = \frac{d}{dt}(3t + 9)$$

$$= 3$$

$$\text{RHS} = \frac{3t + 9 - 6}{t + 1}$$

$$= \frac{3t + 3}{t + 1}$$

$$= 3\frac{t + 1}{t + 1}$$

$$= 3 \quad \textit{provided } t \neq -1$$

Therefore: LHS = RHS provided $t \neq -1$

Since this equation is satisfied for all $t \neq -1$, we deduce that $y(t) = 3t + 9$ is a solution.

Example 2   Verifying an implicit solution to a differential equation

Verify that if y satisfies the relationship

$$x^2 + y^2 = 4$$

then it is a solution to the differential equation

$$\frac{dy}{dx} = -\frac{x}{y} \quad provided\ y \neq 0$$

Solution   From the given equation, we find $\frac{dy}{dx}$:

$$x^2 + y^2 = 4$$

$$2x + 2y\frac{dy}{dx} = 0 \quad using\ implicit\ differentiation$$

$$\frac{dy}{dx} = -\frac{x}{y} \quad provided\ y \neq 0$$

Example 3   From solutions to differential equations

Find a function $g(t)$ such that $y(t) = \cos t$ is a solution to

$$\frac{dy}{dt} = \frac{y}{g(t)}$$

on some interval of time.

Solution   For $y(t) = \cos t$ to be a solution to the given differential equation, we need

$$\frac{dy}{dt} = \frac{y}{g(t)} \quad provided\ g(t) \neq 0$$

$$\frac{d}{dt}(\cos t) = \frac{\cos t}{g(t)}$$

$$-\sin t = \frac{\cos t}{g(t)} \quad differentiating\ left\text{-}hand\ side$$

$$g(t) = -\frac{\cos t}{\sin t} \quad solving\ for\ g(t)\ assuming\ \sin t \neq 0$$

Therefore, if we chose $g(t) = -\cot t$, then $y(t) = \cos t$ is a solution to the differential equation $\frac{dy}{dt} = \frac{y}{g(t)}$ on a time interval for which both $\sin t \neq 0$ and $\cos t \neq 0$.

Separation of variables

As with the case of integration, solving differential equations requires specialized techniques, and there is no guarantee that in general we can find an elementary solution. A special class of differential equations for which we often can find solutions are *separable equations*—differential equations that can be written in this form:

$$\frac{dy}{dt} = \frac{f(t)}{g(y)}$$

To solve such an equation on an interval of time for which $g(y) \neq 0$, we separate the variables to obtain

$$g(y)\frac{dy}{dt} = f(t)$$

and then integrate both sides separately to obtain

$$\int g(y)\frac{dy}{dt}dt = \int f(t)dt$$

The expression we derived for integration by substitution in Section 5.5 implies that the left-hand side can be expressed purely in terms of y (i.e., without reference to t) to obtain the equation

$$\int g(y)dy = \int f(t)dt$$

which we can then integrate to solve for y in terms of t, as illustrated in the next example.

Example 4   Solving a separable differential equation

Solve

$$\frac{dy}{dt} = -\frac{t}{y}$$

Solution   In this case $g(y) = y$ and $f(t) = -t$. Hence, separating variables and integrating yields

$$\int y\,dy = -\int t\,dt$$

$$\frac{1}{2}y^2 + C_1 = -\frac{1}{2}t^2 + C_2$$

$$y^2 = -t^2 + C \qquad \textit{where } C = 2(C_2 - C_1) \textit{ is an arbitrary constant}$$

Solving for y yields $y = \pm\sqrt{C - t^2}$ for any nonnegative C and for $-\sqrt{C} < t < \sqrt{C}$.

Notice the treatment of constants in the solution to Example 4. Because all constants can be combined into a single constant, it is customary not to write $C = 2(C_1 - C_2)$, but rather to simply replace all the arbitrary constants in the problem by a single arbitrary constant after the last integral is found.

Example 5   Finding and plotting solutions

a. Solve the differential equation

$$\frac{dy}{dt} = ty^2$$

b. Find and plot a solution of this equation that satisfies $y(0) = 1$.

Solution
a. First observe that $y = 0$ is a solution, but it is not the solution that passes through the point $y(0) = 1$. To find this latter solution, we define $g(y) = y^{-2}$ and $f(t) = t$ and use the separation of variables technique to obtain the integrals.

$$\int y^{-2}dy = \int t\,dt$$

$$-y^{-1} = \frac{t^2}{2} + C \qquad \textit{integrating}$$

$$y = \frac{-1}{t^2/2 + C} \qquad \textit{solving for y}$$

To check our work, we can substitute this solution into the differential equation $y'(t) = t[y(t)]^2$ to ensure that both sides are the same.

$$\text{LHS:} \quad \frac{d}{dt}\left(\frac{-1}{t^2/2 + C}\right) = \frac{t}{(t^2/2 + C)^2}$$

$$\text{RHS:} \quad t\frac{(-1)^2}{(t^2/2 + C)^2} = \frac{t}{(t^2/2 + C)^2} = \text{LHS}$$

b. To satisfy $1 = y(0)$, we need

$$1 = \frac{-1}{0^2/2 + C}$$

$$1 = \frac{-1}{C}$$

$$C = -1 \quad \textit{solving for C}$$

Thus,

$$y(t) = \frac{-1}{t^2/2 - 1} = \frac{-2}{t^2 - 2} \quad \textit{for all } t \neq \pm\sqrt{2}$$

The solution is plotted in Figure 6.10 on the interval $(-\sqrt{2}, \sqrt{2})$. ◼

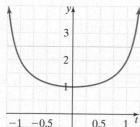

Figure 6.10 Solution to $\dfrac{dy}{dt} = ty^2$ on $(-\sqrt{2}, \sqrt{2})$ with $y(0) = 1$

Sometimes separation of variables leads to integrals we cannot compute or leads to expressions for which y is only implicitly defined.

Example 6   Implicitly defined solutions

Consider

$$\frac{dy}{dt} = \frac{2t}{y + \sin y}$$

a. Use separation of variables to solve for y implicitly in terms of t. Use technology to graph this solution.

b. Find an implicit solution of this equation that satisfies $y(-1) = \pi$. Use technology to graph this particular solution.

Solution

a. In this case, $g(y) = y + \sin y$ and $f(t) = 2t$. Hence, separating variables and integrating yields

$$\int (y + \sin y)\, dy = \int 2t\, dt$$

$$\frac{y^2}{2} - \cos y = t^2 + C$$

We use technology to plot the solutions as implicitly defined functions yielding the family of solutions shown in Figure 6.11.

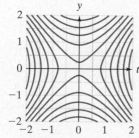

Figure 6.11 Plots of solutions $y(t)$ to the equation $\dfrac{dy}{dt} = \dfrac{2t}{y + \sin y}$ (the horizontal axis $y = 0$ must be excluded)

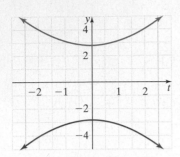

Figure 6.12 Plots of the solutions $y(t)$ to $\dfrac{dy}{dt} = \dfrac{2t}{y + \sin y}$ with $y(-1) = \pi$ (in blue) and $y(-1) = -\pi$ (in red).

b. To find the constant C from our answer in part **a**, we substitute $y = \pi$ and $t = -1$ into our implicit solution.

$$\frac{y^2}{2} - \cos y = t^2 + C$$

$$\pi^2/2 - (-1) = (-1)^2 + C \quad \text{substituting } y = \pi \text{ and } t = -1$$

$$\pi^2/2 = C \quad \text{solving for } C$$

Thus, the particular solution is

$$\frac{y^2}{2} - \cos y = t^2 + \frac{\pi^2}{2}$$

and this solution is shown graphically (using technology) in Figure 6.12. Notice that we get two curves in our plot. Only one of them (the blue curve) corresponds to the desired solution. The other curve (in red) corresponds to another solution satisfying the initial condition $y(-1) = -\pi$.

Super-exponential and logistic population growth

In Section 1.4 we modeled human population growth in the United States during the nineteenth century with an exponential growth model. Although this model worked reasonably well, the human population on the entire globe appears to be growing faster than exponentially. In the next example, we develop a model of this super-exponential growth.

Example 7   Super-exponential growth and a doomsday prediction

Using data on human abundance across the globe from 1000 AD to 2010 AD, we can model human population growth by a differential equation of the form

$$\frac{dN}{dt} = aN^{1+b}$$

where N is population size in billions, t is years after 0 AD, b is estimated to be 1.144, and a is estimated to be 0.0035. Assuming that there were 159 million people on Earth in 0 AD, solve for $N(t)$ and estimate in what year the human population will "blow up."

Solution   We want to solve

$$\frac{dN}{dt} = 0.0035 N^{2.144}$$

where $N(0) = 0.159$. Separating the variables and integrating gives

$$\int N^{-2.144} dN = \int 0.0035 dt \quad \text{separating variables}$$

$$N^{-1.144}/(-1.144) = 0.0035t + C \quad \text{integrating}$$

Before solving for N, we can solve for C by setting $N = 0.159$ and $t = 0$, which yields

$$C = 0.159^{-1.144}/(-1.144) = -7.164$$

Solving for N we get

$$N^{-1.144}/(-1.144) = 0.0035t - 7.164 \quad \text{equation with } C = -7.164$$

$$N^{-1.144} = -0.004t + 8.2$$

$$N = (8.2 - 0.004t)^{-1/1.144}$$

Plotting the solution $N(t)$ of our differential equation gives a remarkably good fit to the data, as shown in Figure 6.13.

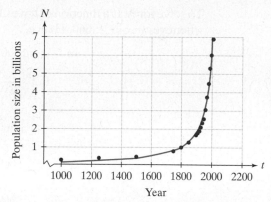

Figure 6.13 Human population growth data from 1000 AD until 2010 AD (as red points) plotted against the solution of the differential equation $\dfrac{dN}{dt} = 0.0035 N^{2.144}$ with $N(0) = 0.159$.

There is a discontinuity of $N = (8.2 - 0.004t)^{-0.874}$ when $8.2 - 0.004t = 0$, equivalently $t = 2050$. Hence, our model predicts that the population will become infinitely large in 2050 AD. Since no planet can sustain an infinite number of individuals, one can view this solution as a prediction that at the current rate of population growth a *doomsday* will occur sometime prior to the year 2050.

We know that populations cannot increase to arbitrarily large sizes. The logistic equation, which we introduced in Section 6.1, accounts for limits on population growth. Recall, the logistic equation is

$$\frac{dN}{dt} = rN\left(1 - \frac{N}{K}\right)$$

where N is the population abundance, r is the intrinsic rate of growth, and K is the carrying capacity. For many data sets, we were able to estimate the parameters r and K. In Example 4 of Section 6.1, we estimated $r \approx 0.53$ and $K \approx 660$ for the yeast data set of Carlson. Using the method of separation of variables, we can find an analytic solution to this equation and see how well it describes the yeast data of Carlson.

Example 8   Logistic growth of Carlson's yeast data
- -

Find a solution to

$$\frac{dN}{dt} = 0.53\, N(1 - N/660) \qquad N(0) = 9.6$$

and compare the solution to the Carlson yeast data set from Example 4 in Section 1.

Solution   In this case $g(N) = \dfrac{1}{N(1 - N/660)}$ and $f(t) = 0.53$. Hence, separating variables and integrating yields

$$\int \frac{1}{N(1 - N/660)}\, dN = \int 0.53\, dt \qquad \textit{integrate both sides}$$

$$\int \left[\frac{1}{N} + \frac{1}{660 - N}\right] dN = 0.53 \int dt \qquad \textit{method of partial fractions}$$

$$\ln|N| - \ln|660 - N| = 0.53t + C \qquad \textit{integrate}$$

To find C corresponding to the solution that passes through the point $t = 0$ and $N = 9.6$, we solve

$$\ln 9.6 - \ln(660 - 9.6) = 0 + C$$

$$C \approx -4.2158$$

To solve for N as a function of t, we use the fact that $|N| = N$ and $|660 - N| = 660 - N$ whenever $0 < N < 660$. Hence,

$$\ln N - \ln(660 - N) = 0.53t - 4.2158$$

$$\ln\left(\frac{N}{660 - N}\right) = 0.53t - 4.2158$$

$$\frac{N}{660 - N} = e^{(0.53t - 4.2158)}$$

$$\frac{N}{660 - N} = e^{-4.2158}e^{0.53t}$$

$$\frac{N}{660 - N} = 0.01476e^{0.53t}$$

$$N = \frac{9.7416e^{0.53t}}{1 + 0.01476e^{0.53t}}$$

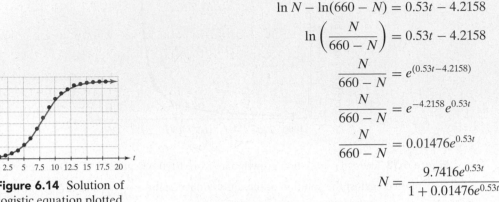

Figure 6.14 Solution of logistic equation plotted against the Carlson yeast data

A plot of the solution against the Carlson yeast data is shown in Figure 6.14 and illustrates a very good fit.

Note that if $N(0)$ had exceeded 660 in Example 8, the equation to solve for $N(t)$ would be $\ln N - \ln(N - 660) = 0.53t + C$, rather than $\ln N - \ln(660 - N) = 0.53t + C$ as was done above.

PROBLEM SET 6.2

Level 1 DRILL PROBLEMS

Verify in Problems 1 to 8 that if y satisfies the prescribed relationship with t

1. If $t^2 + y^2 = 7$, then $\dfrac{dy}{dt} = -\dfrac{t}{y}$.

2. If $5t^2 - 2y^2 = 3$, then $\dfrac{dy}{dt} = \dfrac{5t}{2y}$.

3. If $y = C/t$ for $t \neq 0$, then $\dfrac{dy}{dt} = \dfrac{-y}{t}$.

4. If $t^2 - 3ty + y^2 = 5$, then $\dfrac{dy}{dt} = \dfrac{2t - 3y}{3t - 2y}$.

5. If $y = e^{\sin t}$, then $\dfrac{dy}{dt} = y\cos t$.

6. If $y = \dfrac{1}{1 + t}$, then $\dfrac{dy}{dt} = -y^2$.

7. If $y = 100 - 2e^{-t}$, then $\dfrac{dy}{dt} = 100 - y$.

8. If $y = 100 - 2e^{-3t}$ then $\dfrac{dy}{dt} = 300 - 3y$.

Determine whether the functions given in Problems 9 to 12 are solutions of

$$\frac{dy}{dt} = \sin t - y$$

9. $y(t) = \dfrac{1}{2}(\sin t - \cos t)$

10. $y(t) = \dfrac{1}{2}(10 + \sin t - \cos t)$

11. $y(t) = \sin t - \cos t$

12. $y(t) = e^{-t} + \dfrac{1}{2}(\sin t - \cos t)$

Determine whether the functions given in Problems 13 to 16 are solutions of

$$\frac{dy}{dt} = \frac{1}{2}(y^2 - 1)$$

13. $y(t) = \dfrac{1 + e^t}{1 - e^t}$

14. $y(t) = \dfrac{1 - e^t}{1 + e^t}$

15. $y(t) = 2 - e^t$

16. $y(t) = \dfrac{2 + e^t}{2 - e^t}$

Solve the differential equations in Problems 17 to 28.

17. $\dfrac{dy}{dt} = y^3$

18. $\dfrac{dy}{dt} = y\sin t$

19. $\dfrac{dy}{dt} = \cos t$

20. $\dfrac{dy}{dt} = \dfrac{t}{y}$

21. $\dfrac{dy}{dt} = e^{-y}$

22. $\dfrac{dy}{dt} = y - 1$

23. $\dfrac{dy}{dx} = 3xy$

24. $\dfrac{dy}{dx} = \dfrac{x}{y}\sqrt{1 + x^2}$

25. $\dfrac{dy}{dx} = \dfrac{2xy}{\sqrt{1 + x^2}}$

26. $\dfrac{dy}{dx} = \dfrac{\sin x}{\cos y}$

27. $\dfrac{dy}{dx} = \sqrt{xy}$

28. $\dfrac{dy}{dx} = -\sec y/x^2$

Find the solutions to Problems 29 to 36.

29. $\dfrac{dy}{dt} = (1 + y)^2$ with $y(0) = 2$

30. $\dfrac{dy}{dt} = yt$ with $y(1) = -1$

31. $\dfrac{dy}{dt} = te^{-t}/y$ with $y(0) = 3$

32. $\dfrac{dy}{dt} = e^{-y}t$ with $y(-2) = 0$

33. $\dfrac{dy}{dt} = \dfrac{t+1}{y + e^y}$ with $y(3) = 4$

34. $\dfrac{dy}{dt} = ty^2 + 3t^2y^2$ with $y(-1) = 2$

35. $\dfrac{dy}{dt} = y(y - 1)$ with $y(0) = 1/2$

36. $\dfrac{dy}{dt} = y(y - 1)$ with $y(0) = 2$

37. Create a differential equation of the form

$$\frac{dy}{dt} = 5 - t + g(y)$$

such that $y(t) = e^t$ is a solution.

38. Create a differential equation of the form

$$\frac{dy}{dt} = yh(t)$$

such that $y(t) = \sin t$ is a solution.

Level 2 APPLIED AND THEORY PROBLEMS

39. Doomsday prediction. In 1960, three electrical engineers at the University of Illinois published a paper in *Science* titled "Doomsday." Based on world population growth data from 1000 AD to 1960 AD, the engineers found that population growth was faster than proportional to the population size. Using the data, they modeled the growth of the population as

$$\frac{dP}{dt} = 0.4873\,P^2 \qquad P(0) = 0.2$$

where P is the population size in billions and t is centuries after 1000 AD. Solve this differential equation and sketch the solution. What year is doomsday according to their model? Compare to the prediction found in Example 7.*

40. The logistic equation did a remarkable job in describing the number of new cases of AIDS in the United States from 1980 until the early 1990s, as seen in Figure 6.15. (see the website http://www.nlreg.com/aids.htm). Let $y(t)$ denote the number of new cases t years after 1980.

Nonlinear regression techniques used to fit the data resulted in the equation

$$\frac{dy}{dt} = 0.8y(1 - y/50{,}000) \quad y(0) = 334$$

a. Find the solution to this differential equation.

b. Plot this solution. What happens as $t \to \infty$?

c. Check the Web to see how this compares with the incidence of AIDS in the United States today. What do you conclude and how do you explain the discrepancy? (There is no right or wrong answer to this part.)

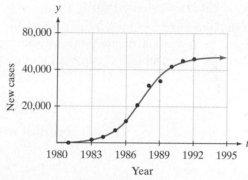

Figure 6.15 New cases of AIDS in the United States
Permission of Phill Sherrod, http://www.nlreg.com and http://www.dtreg.com

41. A model for tumor growth is the Gompertz function that is a solution to the differential equation

$$\frac{dy}{dt} = ay \ln \frac{K}{y}$$

where y is the weight of tumor in milligrams, t is measured in days, a is a constant, and K is the limiting size of the tumor. Assume that $a = 0.5$ and $K = 100$.

a. Find a solution to this differential equation that satisfies $y(0) = 1$ mg.

b. Plot this solution.

42. The 1984 U.S. census recorded a population of 15,757,000 Hispanics; in 1990, the size was 16,098,000. Assuming that the rate of population growth is proportional to the population, predict the population of Hispanics in the United States in the year 2000. Use the Web to find the actual Hispanic population in 2000. How does your prediction compare with the actual number? What do you think can account for any differences?

43. Consider a chemical reaction involving two reactants, A and B, that form a product C. Let $[A]$, $[B]$, and $[C]$ denote the concentrations of A, B, and C. If a molecule of A encounters molecules of B at a rate

*H. von Foerster, P. M. Mora, L. W. Amiot, 1960. "Doomsday: Friday, 13 November, A.D. 2026," *Science*, Vol. 132 no. 3436 pp. 1291–1295 DOI:10.1126/science.132.3436.1291.

proportional to their concentration, then the *law of mass action* states that

$$\frac{d[C]}{dt} = k[A] \cdot [B]$$

where k is a positive constant. If the initial concentration of A is a, the initial concentration of B is b, and we set $y = [C]$, then $[A] = a - [C]$, $[B] = b - [C]$, and

$$\frac{dy}{dt} = k(a - y)(b - y)$$

Assume that $a = b$. Find and plot the solution to this differential equation satisfying $y(0) = 0$.

44. Populations may exhibit seasonal growth in response to seasonal fluctuations in resource availability. A simple model accounting for seasonal

fluctuations in the abundance N of a population is

$$\frac{dN}{dt} = (R + \cos t)N$$

where R is the average per capita growth rate and t is measured in years.

a. Assume $R = 0$ and find a solution to this differential equation that satisfies $N(0) = N_0$. What can you say about $N(t)$ as $t \to \infty$?

b. Assume $R = 1$ (more generally $R > 0$) and find a solution to this differential equation that satisfies $N(0) = N_0$. What can you say about $N(t)$ as $t \to \infty$?

c. Assume $R = -1$ (more generally $R < 0$) and find a solution to this differential equation that satisfies $N(0) = N_0$. What can you say about $N(t)$ as $t \to \infty$?

6.3   Linear Models in Biology

An important class of models is described by the **linear differential equation**

$$\frac{dy}{dt} = a + cy$$

where the constants a and c are model parameters that have specific physical or biological interpretations. For example, in Section 6.1 we saw how models with $a = 0$ and $c = r$ were used to describe exponential population growth ($c = R$) and radioactive decay ($c = -\lambda$). In this section, we discuss more applications where the constant coefficient a is nonzero.

Mixing models

Mixing models formulated on the premise that the density of individuals or concentration of molecules, which are generically characterized in terms of a number of *objects* per unit area or volume, form a homogeneous *pool*. The flow of objects into the pool is controlled by an external constant rate, while the flow of objects out of the pool is in proportion to the density of objects in the pool. This latter assumption implies that the greater the density of objects in the pool, the faster the total flow of objects out of the pool.

Mixing Model

Let $y(t)$ represent the density of objects in a pool at time t. If objects flow into this pool at a constant *total rate* $a > 0$ and out of this pool at a rate $by(t) > 0$, that is, at a constant *per capita clearance rate* $b > 0$, then the density of objects in the pool over time is governed by the equation

$$\frac{dy}{dt} = \text{RATE IN} - \text{RATE OUT} = a - by$$

Our first application of mixing models is to obtain a better understanding of within-host pathogen dynamics of HIV. Human immunodeficiency virus-type 1 (HIV-1) has many puzzling quantitative features. For instance, most persons with HIV undergo a ten-year period during which concentration of the virus in plasma is very low. It is only after this quiescent period that a person experiences the onset of AIDS. The reason for this quiescent period is unknown, and it had been presumed that during this period the virus was relatively inactive. Using models, as described in the next example, Perelson and colleagues quantified viral levels in the blood of infected individuals during this quiescent period.*

Example 1   Modeling HIV levels

In building a model of HIV levels in infected hosts, Perelson and colleagues used the variable $V(t)$ to represent the concentration of viral particles in a host's blood plasma (Figure 6.16). They assumed that HIV viral particles infused into the blood, from production sites in lymphatic tissue, at a constant rate $a > 0$ and were cleared from the blood at a rate bV. From these assumptions, they obtained the mixing model

$$\frac{dV}{dt} = a - bV$$

where t is measured in days. Both a and b are unknown, positive constants.

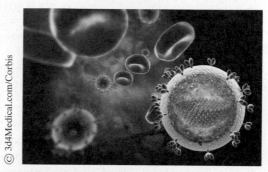

Figure 6.16 An Illustration of HIV in a host's bloodstream

a. Data showed that after a host received a potent antiviral drug, viral concentration in the blood fell exponentially. Assuming that the drug rapidly fell to zero, Perelson and colleagues estimated the half-life of the viral particles to be 0.2 days. Use this information to estimate the clearance rate constant, b.

b. Perelson and colleagues estimated that prior to administration of the drug, the mean plasma viral level was $2.16 \cdot 10^5$ viral particles per milliliter (ppmL). Assume that before the administration of the drug the system was at equilibrium (i.e., V is such that $dV/dt = 0$). Using the estimate of b from part **a**, estimate the rate a of production of viral particles from the lymphatic tissue prior to administration of the drug.

Solution

a. To estimate the clearance rate of all the viral particles currently in the blood of an individual, we set the production parameter a to zero, which yields

$$\frac{dV}{dt} = -bV$$

We solved this equation previously and found the general solution $V(t) = V(0)e^{-bt}$ where $V(0)$ is the initial viral load. Since the half-life is 0.2 days, we know that $\frac{V(0.2)}{V(0)} = \frac{1}{2}$ when $t = 0.2$. Thus,

$$\frac{1}{2} = \frac{V(0)e^{-b(0.2)}}{V(0)} = e^{-b(0.2)}$$

which gives $b = 5\ln 2 \approx 3.47$.

*A. S. Perelson, A. U. Neumann, M. Markowitz, J. M. Leonard, and D. D. Ho, "HIV-1 Dynamics In Vivo: Virion Clearance Rate, Infected Cell Lifespan, and Viral Generation Time," *Science* 271 (1996): 1582–1586, and A. S. Perelson and P. W. Nelson, "Mathematical Analysis of HIV-1 Dynamics In Vivo," *SIAM Review* 41 (1999): 3–44.

b. If the viral density in a person's blood is at equilibrium prior to administration of the drug, then $\dfrac{dV}{dt} = 0$ and we can solve for a.

$$\dfrac{dV}{dt} = a - bV \qquad \text{given equation}$$

$$0 = a - 3.47\,V \qquad \text{substitute known values}$$

$$0 = a - 3.47\,(2.16 \cdot 10^5) \qquad \text{substitute given plasma level}$$

$$a = 749{,}520 \qquad \text{solve for } a$$

Hence, we estimate about 75×10^4 viral particles per milliliter per day. The typical individual has approximately 5.6 liters of blood, which means during the quiescent phase

$$75 \times 10^4 \times 5.6 \times 10^3 \approx 4.2 \times 10^9$$

viral particles are being created per day. Thus, this dormant phase still exhibits "the raging fire of active HIV replication." Consequently, Perelson and colleagues suggested that "early and aggressive therapeutic intervention is necessary if a marked clinical impact is to be achieved." ∎

Hospital patients often receive drugs by intravenous infusion. For drugs to be administered effectively and safely, correct infusion rates must be determined. Differential equation models are a basic tool used by doctors to determine these infusion rates. These models are known as *pharmacokinetics* or *biopharmaceutics* models. (Check out **http://www.boomer.org/c/p1/index.html** for a whole course on this topic.)

Example 2   Determining an infusion rate

A patient who has asthma is given a continuous infusion of theophylline to relax and open the air passages in his lungs. The desired steady-state level of theophylline in the patient's bloodstream is 15 mg/L. The average half-life of theophylline is about four hours, and the patient has 5.6 L of blood in his body.

a. Find the necessary infusion rate.

b. Determine how long it takes for the concentration of theophylline to be 10 mg/L.

Solution

a. First, we write a differential equation model. Let $y(t)$ be the amount (milligrams) of theophylline in the blood plasma at time t, in hours. Let a denote the rate (milligrams per hour) at which theophylline enters the bloodstream via infusion. Let b denote the clearance rate constant of the theophylline. Then,

$$\dfrac{dy}{dt} = a - by$$

To determine b, we use the fact that the half-life of theophylline is about four hours. What this means is that in the absence of the infusion (i.e., when $a = 0$), half of the theophylline leaves the blood plasma in four hours. Solving $y'(t) = -by(t)$ yields $y(t) = y(0)e^{-bt}$. Since $\dfrac{y(4)}{y(0)} = \dfrac{1}{2}$, we can solve for b as follows:

$$\dfrac{1}{2} = e^{-b4} \qquad \text{half-life is four hours}$$

$$b = \dfrac{1}{4}\ln 2 \approx 0.17 \qquad \text{solve for the clearance rate } b$$

To find a, we want the equilibrium (i.e., the y value for which $\frac{dy}{dt} = 0$) to hold at $y = 15$ mg/L $\times 5.6$ L $= 84$ mg:

$$\frac{dy}{dt} = a - by$$
$$0 = a - 0.17 \cdot 84$$
$$a \approx 14.28$$

The desired infusion rate is approximately 14.3 mg/h.

b. To determine how long it takes to reach a concentration of 10 mg/L, we need to solve the differential equation subject to the initial condition $y(0) = 0$ (i.e., initially, there is no drug in the patient). First, using the separation of variables method, the solution for any value of a and b is

$$\frac{dy}{dt} = a - by$$

$$\int \frac{dy}{a - by} = \int dt \quad \textit{provided } y \neq a/b, \textit{ which we note is also a solution}$$

$$\frac{\ln|a - by|}{-b} = t + C_1$$

$$\ln|a - by| = -bt - bC_1 \quad \textit{we use} \pm \textit{ in the next step to account for the two cases}$$
$$\textit{a > by and a < by}$$

$$a - by = C_2 e^{-bt} \quad \textit{where } C_2 = \pm e^{-bC_1} \textit{ is still an arbitrary constant}$$

$$y = \frac{a}{b} - C_3 e^{-bt} \quad \textit{where } C_3 = \frac{C_2}{b} \textit{ is still an arbitrary constant}$$

Now we solve for C_3 corresponding to the solution that passes through $y = 0$ at $t = 0$. This implies $C_3 e^0 = a/b$ or simply $C_3 = a/b$. Since $a = 14.28$ and $b = 0.17$, $\frac{a}{b} = \frac{14.28}{0.17} = 84$, and the particular solution is

$$y = 84(1 - e^{-0.17t})$$

Finally, since a concentration of 10 mg/L corresponds to having 10×5.6 mg $= 56$ mg in the blood, we need to solve

$$84(1 - e^{-0.17t}) = 56$$
$$e^{-0.17t} = \frac{1}{3}$$
$$-0.17t = \ln\frac{1}{3}$$
$$t = 6.46$$

Thus, it will take about 6.5 hours for the concentration of theophylline to reach 10 mg/L.

In the solution to Example 2**b**, we found the general solution to the mixing model, which we restate for future reference.

General Solution of the Mixing Model

The general solution of the differential equation

$$\frac{dy}{dt} = a - by$$

is given by

$$y(t) = \frac{a}{b} - Ce^{-bt}$$

where C is an arbitrary constant.

In Example 1, we determined the clearance rate constant when the half-life of the viral particles in the patient's blood was known. In Example 2, we determined the infusion rate necessary to maintain a desired concentration of drug in the patient's blood. Next, in Example 3, we determine the clearance rate constant when the half-life of the viral particles in the patient's blood is not known.

Example 3   Determining a clearance rate constant

Consider a patient who is receiving a drug intravenously at a rate of 10 mg/h. An hour later, the concentration of drug in the patient's body is 1 mg/L. Assume that the patient has 5 L of blood and that the drug is lost at a rate proportional to the amount of drug in the body; find the clearance rate constant of the drug. Finally, determine the limiting concentration of drug in the patient's body.

Solution   Let y denote the amount of drug in the patient's body. Then, because the patient has 5 L of blood, $y/5$ is the concentration of drug per liter in the patient's body. The rate of change of y is given by

$$\frac{dy}{dt} = 10 - by$$

where 10 is the infusion rate and $b > 0$ is the clearance rate constant. Either redoing our separation of variables work or directly using the general solution of the mixing model, we get

$$y(t) = \frac{10}{b} - Ce^{-bt}$$

To solve for C, we use the initial condition $y(0) = 0$, which gives $0 = y(0) = \frac{10}{b} - C$. Therefore, $C = \frac{10}{b}$ and we have

$$y(t) = \frac{10}{b}(1 - e^{-bt})$$

To find the clearance rate constant b, we can use the fact that

$$y(1) = 1\frac{mg}{L} \cdot 5\,L = 5\,mg$$

Now, we need to solve

$$5 = \frac{10}{b}(1 - e^{-b})$$

We cannot solve for b analytically in this equation. Hence, we use technology to find $b \approx 1.6$ is a solution to this equation.

To find the limiting concentration of the drug in the patient's body, we must find the limit of $y(t)$ as t approaches ∞.

$$\lim_{t \to \infty} \frac{10}{1.6}(1 - e^{-1.6t}) = \frac{10}{1.6} = 6.25$$

The limiting quantity is 6.25 mg in 5 L of blood, or 1.25 mg/L.   ∎

Another application of the mixing model is the following analysis of the concentration of pollution in a lake.

Example 4   Lake pollution

A well-mixed lake with a constant volume of 100 km^3 is fed by rivers and tributaries at a rate of 48 km^3/year. Factories are dumping polluted water, as depicted in Figure 6.17, into the lake at a rate of 2 km^3/year. Environmental studies have shown that after mixing, if the percentage of water in the lake from polluted sources exceeds 2%, then the water becomes a hazardous environment for fish.

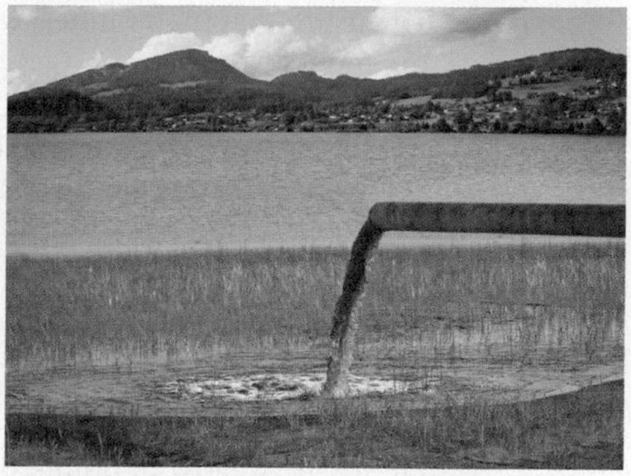

Figure 6.17 A pipe dumping polluted water into a lake

a. Does this lake ever reach the 2% level of polluted water mixed in with fresh water? If so, when?

b. How much would the polluted water input into the lake have to be decreased to reduce the long-term mixture of water from polluted sources to 2%?

Solution

a. Let y (units in cubic kilometers) denote the total amount of polluted water mixed into the lake. Assume that $y(0) = 0$. The rate polluted water enters the lake is $2\,\text{km}^3/\text{year}$. Finding the rate at which pollutants are leaving is more difficult. Since the lake volume is assumed to be constant, the rate at which the well-mixed water leaves the lake is equal to the rate at which the fresh and polluted water enter the lake; that is, $48 + 2 = 50\,\text{km}^3/\text{year}$. The proportion of polluted water in the lake is $y/100$. Hence, the rate at which polluted waters leave the lake is

$$\frac{y}{100} \times 50 = \frac{y}{2}\frac{\text{km}^3}{\text{year}}$$

Thus, our initial value problem is

$$\frac{dy}{dt} = \text{ENTERING RATE} - \text{LEAVING RATE}$$

$$= 2 - \frac{y}{2}$$

with initial value $y(0) = 0$. Using the general solution for the mixing model with $a = 2$ and $b = 1/2$, we get

$$y(t) = 4 - Ce^{-t/2}$$

The initial condition $y(0) = 0$ implies that $C = 4$, and

$$y(t) = 4(1 - e^{-t/2})$$

Since $\lim_{t\to\infty} y(t) = 4$, the eventual amount of polluted water in the well-mixed lake is $4\,\text{km}^3$, which is 4% of the lake's $100\,\text{km}^3$ volume. Thus, the lake will reach the hazardous level. To find the time at which it reaches the 2% hazardous level, we solve

$$2 = 4 - 4e^{-t/2}$$

$$e^{-t/2} = 0.5$$

$$t = 2\ln 2 \approx 1.39$$

This is about 1 year, 5 months.

b. To ensure that the polluted water never exceeds 2% of the total, we reformulate the model as

$$\frac{dy}{dt} = p - \frac{y}{2}$$

where p is the rate polluted water is dumped into the lake, and the initial value is $y(0) = 0$. Polluted water levels in the lake will approach the equilibrium level given by

$$0 = p - \frac{y}{2}$$
$$y = 2p \text{ km}^3$$

Hence, if the flow of polluted water is reduced to 1 km^3/year, then the long-term (i.e., equilibrium) proportion of polluted water in the lake is 2%.

Newton's cooling law and forensic medicine

In forensic medicine, determining the time of death of a victim can be critical in convicting the murderer. Mathematics can aid this process by using **Newton's law of cooling**. This law states that the rate at which the temperature of a body changes is proportional to the difference between the body's temperature T and the ambient temperature A. Mathematically, this statement corresponds to the following differential equation:

$$\frac{dT}{dt} = k(A - T)$$

where k a positive constant proportionate to the thermal conductivity of the body. A large k means the body readily conducts heat and quickly adjusts to the ambient temperature. A small k means the body is well insulated and slowly adjusts to the ambient temperature. Notice that this differential equation is also a mixing model where $y = T$, $a = kA$, and $b = k$. Hence, the general solution of this cooling model is

$$T(t) = A - Ce^{-kt}$$

where C is an arbitrary constant.

Example 5   Put on your detective caps

On a dark and stormy night, Sherlock Holmes and Dr. Watson were called to investigate the shocking murder of Jacob Marley. The main suspects of this crime were three people who would benefit from his death. First, there was Marley's business partner, Ebenezer Scrooge, who was having strong disagreements with Marley about how to run the business. Scrooge spent the evening alone working late at his office. His household staff confirmed that he arrived home at 9 P.M. and remained home for the rest of the night. Second, there was Marley's wife, Claudia, who was having an affair with another man. Claudia stood to inherit Marley's fortune. Claudia was at the theater from 8 P.M. to 9:30 P.M., as verified by several people at the theater. Finally, there was Marley's client, Sam Wise Gange, whom Marley swindled out of a large sum of money. Sam was at a local pub until 11:00 P.M., which was verified by several people. Marley's body was found in an alley at 1:30 A.M. The alley temperature was a nippy 55 degrees and the body temperature was 87°F. One hour later, the body temperature had cooled to 85°F. Given this information, determine who has a good alibi.

Solution   Let T denote the temperature of the body. By Newton's law of cooling

$$\frac{dT}{dt} = k(55 - T)$$

Using the general solution of this differential equation, we get

$$T(t) = 55 - Ce^{-kt}$$

To determine C and k, let us identify $t = 0$ with 1:30 A.M. Since at this time the body temperature was 87°F, it follows that

$$87 = T(0) = 55 - C$$
$$C = -32$$

Thus,

$$T(t) = 55 + 32e^{-kt}$$

To find k, we use the information that the body temperature was 85°F an hour after 1:30 A.M., which implies

$$85 = T(1) = 55 + 32e^{-k}$$
$$30 = 32e^{-k}$$
$$-k = \ln\frac{15}{16}$$
$$k = \ln\frac{16}{15} \approx 0.065$$

Finally, to determine the time of death, we need to solve backwards in time to the point where the body temperature was a normal 98.6°F, that is,

$$98.6 = T(t) = 55 + 32e^{-0.065t}$$
$$43.6 = 32e^{-0.065t}$$
$$-0.065t = \ln\frac{43.6}{32}$$
$$t \approx -4.76$$

Thus, Marley died approximately four hours and forty-five minutes before the body was found, that is, at approximately 8:45 P.M. Thus, Claudia and Sam have alibis for the murder, while Scrooge does not.

Organismal growth

Figure 6.18 Cubical critter

The study of growth involves determining body size as a function of age. Various measurements of body size exist, including weight, length, and girth. A famous equation, the *von Bertalanffy growth equation*, which describes growth of an organism, can be derived from first principles using scaling laws. To derive this equation, consider a cubical critter with length L as illustrated in Figure 6.18.

The surface area of this critter is $6L^2$ and the volume is L^3. If we assume that length is measured in centimeters and that the critter is mostly made of water, which has a density of 1 g/cm^3, then the critter's mass M is L^3 grams. If a critter ingests food at a rate proportional to its surface area and respires at a rate proportional to its mass, then

$$\frac{dM}{dt} = aL^2 - bL^3$$

where a and b are positive proportionality constants. Since $M = kL^3$, where the value of k depends on units of measurement and so can be set to 1 without affecting the form of the solution (the values of a and b would be adjusted according to the selected units), we obtain

$$\frac{dM}{dt} = 3L^2\frac{dL}{dt}$$

Combining the previous two equations yields

$$\frac{dL}{dt} = \frac{1}{3}L^{-2}\frac{dM}{dt}$$

$$= \frac{1}{3}L^{-2}(aL^2 - bL^3)$$

$$= \frac{a}{3} - \frac{b}{3}L$$

Defining $k = b/3$ and $L_\infty = a/b$ yields the two-parameter von Bertalanffy growth equation

$$\frac{dL}{dt} = k(L_\infty - L)$$

The next example clarifies why biologists use the notation L_∞ for one of the parameters.

Example 6   von Bertalanffy growth equation

Find a general solution to the von Bertalanffy growth equation

$$\frac{dL}{dt} = k(L_\infty - L)$$

with initial value $L(t_0) = 0$; that is, the organism was "born" at time t_0.

Solution   This differential equation is just another mixing model with $y = L$, $a = kL_\infty$, and $b = k$. Hence, the general solution is

$$L(t) = L_\infty - Ce^{-kt}$$

Next, we use the initial condition $L(t_0) = 0$ to find C

$$0 = L_\infty - Ce^{-kt_0}$$

$$Ce^{-kt_0} = L_\infty$$

$$C = e^{kt_0}L_\infty$$

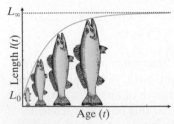

Thus,

$$L(t) = L_\infty - e^{kt_0}L_\infty e^{-kt}$$

$$= L_\infty(1 - e^{kt_0}e^{-kt})$$

$$= L_\infty(1 - e^{-k(t-t_0)})   \quad laws\ of\ exponents$$

Figure 6.19 von Bertalanffy growth equation.

As shown in Figure 6.19. This time t_0 is sometimes thought of as the *theoretical time of conception*, but it is only a meaningful concept if the same growth equation applies at all stages of development (which is not generally a reasonable assumption).

PROBLEM SET 6.3

Level 1 DRILL PROBLEMS

In Example 1 we modeled HIV levels using the data of Perelson and colleagues. In that example we assumed that the half-life of the viral particles was 0.2 days (or 4.8 hours) and that the mean plasma viral level was $2.16 \cdot 10^5$ viral particles per milliliter (ppmL). Estimate the clearance rate constant b for the half-life given

in Problems 1 to 6 and then estimate the daily rate of production of HIV particles for the specified mean plasma viral level.

1. 2.4 h; $1.89 \cdot 10^5$ viral ppmL
2. 3 h; $2.15 \cdot 10^5$ viral ppmL
3. 4 h; $2.25 \cdot 10^5$ viral ppmL

4. 5 h; $2.35 \cdot 10^5$ viral ppmL

5. 6 h; $3.15 \cdot 10^5$ viral ppmL

6. 7.2 h; $2.75 \cdot 10^5$ viral ppmL

Using the information from Example 2, determine the length of time it takes for the concentration of theo-phylline to be the quantity given in Problems 7 to 12.

7. 5 mg/L   **8.** 7 mg/L   **9.** 12 mg/L

10. 14 mg/L   **11.** 14.5 mg/L   **12.** 14.99 mg/L

Find the amount of drug in a patient's body as a function of time t, given the infusion rate and the concentration of the drug one hour later, as given in Problems 13 to 18. Assume the patient has 5 L of blood.

13. 10 mg/h; 1.6 mg/L   **14.** 12 mg/h; 1 mg/L

15. 12 mg/h; 2 mg/L   **16.** 20 mg/h; 1 mg/h

17. 20 mg/h; 2 mg/h   **18.** 20 mg/h; 3 mg/h

Rework Example 4 using all of the given information and changing only the lake size and outflow as shown in Problems 19 to 24.

19. Lake size of 50 km^3 and outflow of 23 km^3/year

20. Lake size of 100 km^3 and outflow of 62 km^3/year

21. Lake size of 100 km^3 and outflow of 23 km^3/year

22. Lake size of 120 km^3 and outflow of 10 km^3/year

23. Lake size of 50 km^3 and outflow of 18 km^3/year

24. Lake size of 80 km^3 and outflow of 10 km^3/year

Level 2 APPLIED AND THEORY PROBLEMS

In Problems 25 to 28, set up an appropriate model to an-swer the given question. These problems use a special case of the linear model in which the relative rate of change remains constant.

25. In 1990 the gross domestic product (GDP) of the United States was $5,464 billion. Suppose the growth rate from 1989 to 1990 was 5.08%. Predict the GDP in 2003. Check your answer by finding the actual 2003 GDP.

26. In 1980 the gross domestic product (GDP) of the United States in constant 1972 dollars was $1,481 billion. Suppose the growth rate from 1980 to 1984 was 2.5% per year. Predict the GDP in 2003. Check your answer by finding the actual 2003 GDP.

27. According to the Department of Health and Human Services, the annual growth rate in the number of divorces per year in 1990 in the United States was 4.7% and there were 1,175,000 divorces that year. How many divorces will there be in 2004 if the annual growth rate in the number of divorces per year remains constant?

28. According to the Department of Health and Human Services, the annual growth rate in the number of marriages per year in 1990 in the United States was 9.8% and there were 2,448,000 marriages that year. How many marriages will there be in 2004 if the annual growth rate in the number of marriages per year is constant?

29. The rate at which a drug is absorbed into the blood system is given by

$$\frac{dx}{dt} = a - bx$$

where $x(t)$ is the concentration of the drug in the bloodstream at time t. What does $x(t)$ approach in the long run (that is, as $t \to \infty$)? At what time is $x(t)$ equal to half this limiting value? Assume $x(0) = 0$.

30. Calculate the infusion rate in milligrams per hour required to maintain a long-term drug concentra-tion of 50 mg/L (i.e., the rate of change of drug in the body equals zero when the concentration is 50 mg/L). Assume that the half-life of the drug is 3.2 hours and that the patient has 5 L of blood.

31. Calculate the infusion rate in milligrams per hour required to maintain a desired drug concentration of 2 mg/L. Assume the patient has 5.6 L of blood and the half-life of the drug is 2.7 h.

32. Calculate the infusion rate required to achieve a desired drug concentration of 2 mg/L in 1 h. Assume the elimination rate constant of the drug is 5 mg/L per hour and the patient has 6 L of blood.

33. Calculate the infusion rate required to achieve a desired drug concentration of 12 mg/L in 20 min. Assume the clearance rate constant is 2 mg/L per hour and the patient has 5 L of blood.

34. A drug is given at an infusion rate of 50 mg/h. The drug concentration value determined at 3 h after the start of the infusion is 8 mg/L. Assuming the patient has 5 L of blood, estimate the half-life of this drug.

35. A drug is given at an infusion rate of 250 mg/h. The drug concentration determined at 4 h after the start of the infusion is 50 mg/L. Assuming the patient has 5.5 L of blood, estimate the elimination rate con-stant of this drug.

36. A lake with a constant volume of 10,000 m^3 is ini-tially clean and pristine. Water flows into the lake from two streams, Babbling Brook and Raging Rapids, at rates of 250 m^3 per day and 750 m^3 per day, respectively. At time $t = 0$, road salt from a nearby road contaminates Babbling Brook with concentration of 2 kg/m^3. Find an equation that describes the amount of salt in the lake for all $t \geq 0$ and find the limiting amount of salt in the lake.

37. After one hydrodynamic experiment, a tank contains 300 L of a dye solution with a dye concentration of 2 g/L. To prepare for the next experiment, the tank is to be rinsed with water flowing in at a rate of 2 L/min, with the well-stirred solution flowing out at the same rate. Write an equation that describes the amount of dye in the container. Be sure to identify variables and their units.

38. At midnight the coroner was called to the scene of the brutal murder of Casper Cooly. The coroner arrived and noted that the air temperature was 70°F and Cooly's body temperature was 85°F. At 2 A.M., she noted that the body had cooled to 76°F. The police arrested Cooly's business partner Tatum Twit and charged her with the murder. She has an eyewitness who said she left the theater at 11:00 P.M. Does her alibi help?

39. A cup of coffee at a cafe is served at 95°C and left on the counter. The cafe is air-conditioned with an ambient temperature of 20°C. After five minutes, the coffee's temperature is 45°C. Determine how long before the coffee loses its taste quality; that is, it cools down to the temperature of 22°C.

40. In 1986 the Chernobyl nuclear disaster in the Soviet Union contaminated the atmosphere. The buildup of radioactive material in the atmosphere satisfies the differential equation

$$\frac{dM}{dt} = r\left(\frac{k}{r} - M\right)$$

where M = mass of radioactive material in the atmosphere after time (in years); k is the rate at which the radioactive material is introduced into the atmosphere; r is the annual decay rate of the radioactive material. Find the solution, $M(t)$, of this differential equation in terms of k and r.

41. The von Bertalanffy curve was used to examine growth patterns in both body length and mass of female and male polar bears (*Ursus maritimus*) captured live near Svalbard, Norway (see Figure 6.20).

A longer growth period in males resulted in pronounced sexual dimorphism in both body length and mass. Males were 1.16 times longer and 2.10 times heavier than females. For females, $L_\infty = 194$ cm, $k = 0.75$/year, and $t_0 = -0.27$ is the theoretical age at which the polar bear would have no length ($L_0 = 0$). For males, $L_\infty = 225$ cm, $k = 0.537$/year, and $t_0 = -0.395$ is the theoretical age at which the polar bear would have no length. Use the von Bertalanffy curve to determine at what age males and females achieve half of their limiting size.

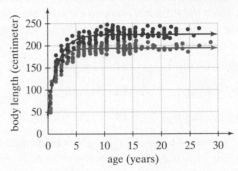

Figure 6.20 The von Bertalanffy curve fitted to age and body length: data for female (blue dots and curve) and male (red dots and curve) polar bears.

Data Source: A. E. Derocher and Ø Wiig, "Postnatal Growth in Body Length and Mass of Polar Bears (*Ursus maritimus*) at Svalbard," *J. Zool., Lond.* 256 (2002): 343–349.

42. In Chapter 5 Section 8, we introduced survival-renewal equations to describe how populations (e.g. money, organisms) change when individuals arrive at rate $r(t)$ and survive with probability $s(t)$ for t units of time. If initially the population size is $y(0)$, then the survival-renewal equation is

$$y(t) = y(0)s(t) + \int_0^t s(t - x)r(x)dx$$

If $s(t) = e^{at}$, then verify that $y(t)$ solves the linear differential equation:

$$\frac{dy}{dt} = r(t) - ay$$

6.4   Slope Fields and Euler's Method

Not all equations are separable, and many separable equations do not lead to explicit solutions. Furthermore, even when you find a solution, it may be so complex that it is nearly impossible to interpret. To address these issues, we discuss a qualitative method, *slope fields*, and a numerical method, *Euler's method*, for studying solutions of differential equations.

Slope fields

Consider a differential equation of the general form

$$\frac{dy}{dt} = f(t, y)$$

where $f(t, y)$ denotes an expression involving t and y. Since a solution $y(t)$ to this differential equation satisfies $y'(t) = f(t, y(t))$, it follows that the slope of all solutions at time t are given by the right-hand side $f(t, y(t))$ of the differential equation. Equivalently, $f(t, y(t))$ is the slope of the tangent line of $y(t)$ at time t. A qualitative way to investigate the behavior of solutions to $y'(t) = f(t, y)$ is to sketch its **slope field**, a figure in the ty plane with infinitesimal line segments of slope $f(t, y)$ at (t, y).

Consider, for example, sketching the slope field for $\dfrac{dy}{dt} = \dfrac{1}{t}$. Since for $t = 1$ the slope is $\dfrac{1}{1} = 1$, we draw short line segments at $t = 1$, each with slope 1, for different y values, as shown in Figure 6.21a. For $t = -3$, the slope is $-1/3$ and we draw short line segments at $t = -3$, each with slope $-1/3$, also shown in Figure 6.21a. If we continue to plot these slope points for different values of t, we obtain many little slope lines. The resulting graph, shown in Figure 6.21b, is the slope field for the equation $y'(t) = 1/t$.

We can also examine the relationship between the slope field for $y'(t) = 1/t$ and its solutions $y = \ln|t| + C$. If we choose particular values for C, say $C = 0$, $C = -\ln 2$, or $C = 2$, and draw these particular solutions as shown in Figure 6.21c, we see that these solutions are tangent to the slope field.

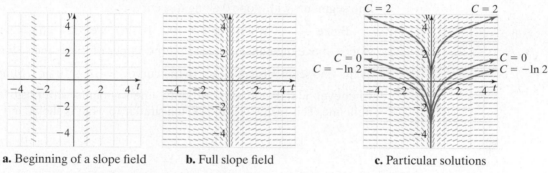

a. Beginning of a slope field   **b.** Full slope field   **c.** Particular solutions

Figure 6.21 Solution of the differential equation $y' = \dfrac{1}{y}$ using a slope field

The slope field for $y'(t) = 1/t$ was relatively straightforward to sketch, but such sketches can be more challenging for differential equations with a more complicated right-hand side. In the next example, we illustrate how to obtain a qualitative understanding using nullclines.

Nullclines

The set of points (t, y) for which the equation

$$\frac{dy}{dt} = f(t, y) = 0$$

holds are called the **nullcline(s)**. For most models, these nullclines correspond to a finite number of curves. In general, however, they are a closed subset of the ty-plane.

Example 1   Sketching solutions using a slope field

Consider a drug that continuously infuses into a patient at a periodic rate. One possible differential equation modeling such a scenario is

$$\frac{dy}{dt} = 10 + 10\sin t - y$$

where y is the amount of drug (in milligrams) and t is the time in units of hours. Notice that this is just a mixing problem (see Example 2 of Section 3.4) where the input is now the time-dependent infusion rate $10 + 10 \sin t$ mg/h and the elimination rate constant is 1 mg/h. Sketch the slope field for this differential equation, and sketch the particular solution that satisfies $y(0) = 0$.

Solution   To sketch the slope field by hand, it often suffices to determine

$$\text{the nullclines:} \quad \frac{dy}{dt} = 0$$

and

$$\text{the regions where} \quad \frac{dy}{dt} < 0 \quad \text{versus} \quad \frac{dy}{dt} > 0$$

In this example, the nullclines satisfies

$$\frac{dy}{dt} = 10 + 10 \sin t - y = 0$$

which implies

$$y = 10 + 10 \sin t$$

Since everywhere along the curve $y = 10 + 10 \sin t$ the slope $\frac{dy}{dt}$ is 0, we draw small line segments with slope 0 along this curve.

Since $\frac{dy}{dt} < 0$ whenever $y > 10 + 10 \sin t$, we draw line segments with negative slopes above the nullcline such that the slopes of these line segments get closer to 0 as the line segments get closer to the curve $y = 10 + 10 \sin t$. Similarly, since $\frac{dy}{dt} > 0$ whenever $y < 10 + 10 \sin t$, we draw line segments with positive slopes below the nullcline. This recipe yields a sketch similar to what is shown in Figure 6.22a.

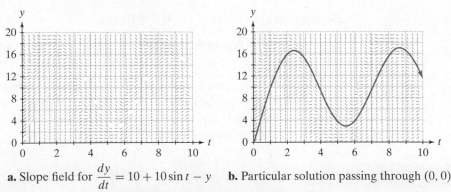

a. Slope field for $\dfrac{dy}{dt} = 10 + 10 \sin t - y$ **b.** Particular solution passing through $(0, 0)$

Figure 6.22 Slope field and a particular solution

To sketch the solution satisfying $y(0) = 0$, we sketch a curve starting at $t = 0$, $y = 0$ that remains tangent to the slope field, which leads to a sketch similar to Figure 6.22b. This qualitative analysis correctly suggests that this solution eventually exhibits well-defined oscillations. In fact, the conclusion can be verified by solving this differential equation using the solution to Problem 42 in Problem Set 6.3. ∎

Slope fields can provide qualitative insights when solutions are only implicitly defined or are very complicated. In the next example, the equations are separable and an explicit solution can be found. The solution, however, requires solving for the roots of a cubic equation, yielding complicated expressions that are hard to interpret. The behavior of the solution, however, is quite easy to infer using slope fields. The example is inspired by the work of Warder Clyde Allee, an American ecologist who argued that the per capita growth rate of populations may be negative when population densities are low, despite being positive at higher densities. Reasons for this so-called **Allee effect** can be due to cooperative hunting or the difficulty of finding mates at low population densities.

Example 2   The Allee effect

Consider the population model

$$\frac{dN}{dt} = rN\left(\frac{N}{A} - 1\right)\left(1 - \frac{N}{K}\right)$$

where N is the population density, t is time, $r > 0$ and $0 < A < K$. Use the parameters $r = 1$, $K = 200$, and $A = 50$.

a. Sketch the per capita growth rate function.

b. Sketch the slope field.

c. Sketch and compare solutions satisfying $N(0) = 49$ and $N(0) = 55$.

Solution

a. The per capita growth rate function for the specified parameters is

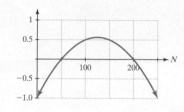

$$f(N) = \left(\frac{N}{50} - 1\right)\left(1 - \frac{N}{200}\right)$$

and is illustrated on the left.

Consistent with the Allee effect, the per capita growth rate is negative at low densities yet positive at intermediate densities. At high densities it is negative again due to competitive effects.

b. To sketch the slope field, we first solve for the nullclines. This corresponds to the set of points in the tN-plane for which

$$\frac{dN}{dt} = N\left(\frac{N}{50} - 1\right)\left(1 - \frac{N}{200}\right) = 0$$

Hence, the nullclines are given by the lines $N = 0$, $N = 50$, and $N = 200$ in the tN-plane. Along the lines, we sketch horizontal line segments.

For $0 < N < 50$ and $N > 200$, we have $\dfrac{dN}{dt} < 0$ (shaded red). Hence, between the lines $N = 0$ and $N = 50$ and above the line $N = 200$, we sketch line segments with negative slope. Moreover, the slope of these line segments gets closer to zero as the line segments get closer to $N = 0$, $N = 50$, or $N = 200$.

For $50 < N < 200$, we have $\dfrac{dN}{dt} > 0$ (shaded blue). Hence, between the lines $N = 50$ and $N = 200$, we draw line segments with positive slope. Moreover, the slope of these line segments gets closer to zero as the line segments get closer to $N = 50$ or $N = 200$. This work yields a sketch similar to Figure 6.23a.

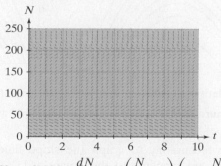

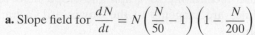

a. Slope field for $\dfrac{dN}{dt} = N\left(\dfrac{N}{50} - 1\right)\left(1 - \dfrac{N}{200}\right)$

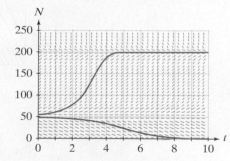

b. Solutions with initial values (0, 49) and (0, 55)

Figure 6.23 Slope field with two particular solutions

c. To sketch a solution satisfying $N(0) = 49$, we sketch a curve passing through the point $t = 0$, $N = 49$ that remains tangent to the slope field (i.e., "go with the flow"). This curve is shown in Figure 6.23b; it starts at $(0, 49)$ and becomes asymptotic to the t axis.

Similarly, we sketch a solution satisfying $N(0) = 55$, which is shown in Figure 6.23b starting at $(0, 55)$ and becoming asymptotic to the line $N = 200$.

These solutions suggest that whenever $0 < N(0) < A$, the population declines to extinction. Whenever $N(0) > A$, the population converges to K. ∎

Any model of the form $y' = f(y)$ is called **autonomous** because the associated slope field is independent of time (i.e., the function $f(y)$ does not explicitly depend on time). We discuss these equations in much greater detail in the next two sections. The counterparts to these autonomous models are purely time dependent equations $\dfrac{dy}{dt} = f(t)$—that is, the slope field does not depend at all on the variable y. In this case, a solution $y(t)$ is an antiderivative of $f(t)$.

Example 3   Purely time-dependent slope field

Consider

$$\frac{dy}{dt} = -t$$

Find all solutions to this differential equation; then sketch several members of the family of curves representing the solutions. Finally, use technology to compare the slope field with this family of solutions.

Solution   We begin by separating the variables and integrating.

$$\frac{dy}{dt} = -t$$

$$\int dy = -\int dt$$

$$y(t) = -\frac{t^2}{2} + C$$

If we sketch the solutions for a variety of C values, we obtain a family of downward facing parabolas as illustrated in Figure 6.24a.

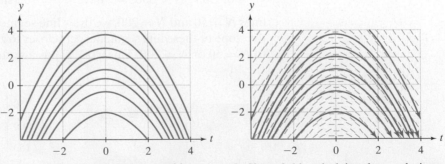

a. Plot of solutions to equation $y'(t) = -t$ **b.** Slope field underlying these solutions

Figure 6.24 Family of solutions compared with slope-field solution

Now, use technology to graph the slope field for $y'(t) = -t$. As expected, the slope lines are tangents to the downward facing parabolas, as illustrated in Figure 6.24b. ∎

Using slope fields, we sometimes can quickly answer questions about the long-term behavior of solutions to a differential equation.

Example 4   Lake pollution revisited

In Example 4 of Section 6.3 we showed that the amount of polluted water $y(t)$ (in cubic kilometers) at time t (years) from factories dumping effluent at a rate of 2 km^3/year into a well-mixed lake of constant volume 100 km^3, fed by rivers and tributaries at a rate of 50 km^3/year, is given by the equation

$$\frac{dy}{dt} = 2 - \frac{y}{2}$$

Sketch a slope field for this equation and use it to find the limiting values as $t \to \infty$.

Solution   We used technology to draw the slope field in Figure 6.25. Note that the nullcline is

$$2 - \frac{y}{2} = 0 \qquad \text{which implies} \qquad y = 4$$

and that the slopes above and below this nullcline are negative and positive, respectively.

Sketches of several solutions on this slope field are shown in Figure 6.25. Hence, as we had previously shown analytically, the volume of polluted water in the well-mixed lake approaches a limiting value of $y = 4$ km^3 (or equivalently, an asymptotic concentration of 4% polluted water).

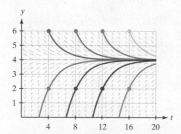

Figure 6.25 Slope field and solutions to the equation $\dfrac{dy}{dt} = 2 - \dfrac{y}{2}$

Euler's method

Sometimes it is not possible to solve for the solution of a differential equation analytically; nevertheless, we want more than a qualitative sense of the solution. Such situations require numerical methods. The simplest numerical method is **Euler's method**, which roughly corresponds to sliding short linear segments along the slope field. More precisely, Euler's method for the differential equation

$$\frac{dy}{dt} = f(t, y)$$

involves choosing a step size $h > 0$ and making the approximation

$$\frac{y(t + h) - y(t)}{h} \approx y'(t) = f(t, y(t))$$

Rearranging this expression gives

$$y(t + h) \approx y(t) + f(t, y(t))h$$

Therefore, if we are given y at time t, we can approximate y at time $t + h$. Applying this approximation iteratively yields an approximation to the solution of the differential equation.

Euler's Method

Consider the equation

$$\frac{dy}{dt} = f(t, y) \qquad y(t_0) = y_0$$

To estimate the solution of this equation, follow these steps:

Step 1: Choose a step size $h > 0$ and define $t_1 = t_0 + h$, $t_2 = t_0 + 2h, \ldots$

Step 2: Estimate $y(t_1)$ by y_1 where

$$y_1 = y_0 + h f(t_0, y_0)$$

In other words, approximate $y(t_1)$ using the tangent line to $y(t)$ at t_0, as illustrated in Figure 6.26a.

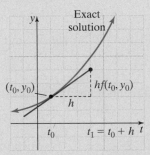 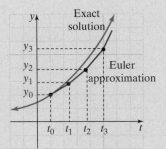

a. First Euler approximation   **b.** Graphical representation of Euler's method

Figure 6.26

Step 3: Estimate $y(t_2)$ by y_2 where

$$y_2 = y_1 + h f(t_1, y_1)$$

In other words, approximate $y(t_2)$ following the slope field at (t_1, y_1).

Step 4: Repeat this process and at the i-th step estimate $y(t_i)$ by y_i where

$$y_i = y_{i-1} + h f(t_{i-1}, y_{i-1})$$

as illustrated in Figure 6.26b.

Euler's method is illustrated in the following example.

Example 5   Euler's method

Use Euler's method with $h = 0.1$ to estimate the solution of the initial value problem

$$\frac{dy}{dt} = t + y^2 \qquad y(0) = 1$$

over the interval $[0, 0.5]$.

Solution   To use Euler's method for this example, note:

$$f(t, y) = t + y^2 \qquad t_0 = 0 \qquad y_0 = 1$$

For $h = 0.1$, the Euler approximation (correct to four decimal places) is:

$$t_0 = 0.0: \quad y_0 = y(0) = 1$$
$$t_1 = 0.1: \quad y_1 = y_0 + h f(t_0, y_0) = 1 + 0.1(0 + 1^2) = 1.1$$
$$t_2 = 0.2: \quad y_2 = y_1 + h f(t_1, y_1) = 1.1 + 0.1(0.1 + 1.1^2) = 1.2310$$
$$t_3 = 0.3: \quad y_3 = y_2 + h f(t_2, y_2) = 1.2310 + 0.1(0.2 + 1.2310^2) \approx 1.4025$$
$$t_4 = 0.4: \quad y_4 = y_3 + h f(t_3, y_3) = 1.4025 + 0.1(0.3 + 1.4025^2) \approx 1.6292$$
$$t_5 = 0.5: \quad y_5 = y_4 + h f(t_4, y_4) = 1.6292 + 0.1(0.4 + 1.6292^2) \approx 1.9346$$

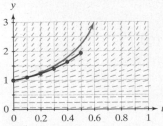

Figure 6.27 Comparison of an analytically derived solution (blue) and the approximate solution due to Euler's method (red).

These points can be plotted to approximate the solution, as shown by the red line in Figure 6.27—a reasonable approximation of the analytically derived solution shown in blue.

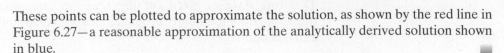

Example 6 Comparing Euler's method for two time steps h

Use Euler's method to approximate the solution to

$$\frac{dy}{dt} = \sin \pi t - y \qquad y(0) = 0$$

on the interval $[0, 2]$ with $y(0) = 0$.

a. Solve by hand, with $h = 0.5$.   **b.** Solve using technology, with $h = 0.1$.

Solution

a. We have $f(t, y) = \sin \pi t - y$ with $h = 0.5$, $t_0 = 0$, and $y_0 = 0$. Thus, by Euler's method we obtain

$$t_1 = 0.5: \quad y_1 = y_0 + hf(t_0, y_0) = 0 + 0.5[\sin(\pi \times 0) - 0.0] = 0$$

$$t_2 = 1.0: \quad y_2 = y_1 + hf(t_1, y_1) = 0 + 0.5[\sin(\pi \times 0.5) - 0.0] = 0.5$$

$$t_3 = 1.5: \quad y_3 = y_2 + hf(t_2, y_2) = 0.5 + 0.5[\sin(\pi \times 1.0) - 0.5] = 0.25$$

$$t_4 = 2.0: \quad y_4 = y_3 + hf(t_3, y_3) = 0.25 + [\sin(\pi \times 1.5) - 0.25] = -0.375$$

To visualize the solution, we plot these points in the ty-plane, and connect them with line segments, as shown in the left panel of Figure 6.28.

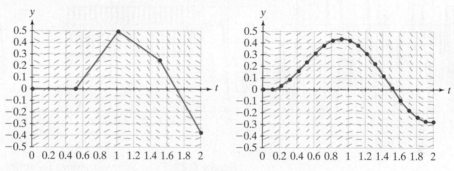

Figure 6.28 Euler approximation of $\dfrac{dy}{dt} = \sin \pi t - y$ for $h = 0.5$ (left panel)
and $h = 0.01$ (right panel)

b. We use technology to graph a slope field, along with the solution using Euler's method for $h = 0.1$, as shown in the right panel of Figure 6.28.

The one thing that is obvious in comparing the solutions graphed in the left and right panels of Figure 6.28 is that approximate solutions with larger h values can be crude. They may even go horribly wrong, as in the next example. However, as the step size decreases, the time it takes to complete the calculation increases. Thus it is important, as with any numerical scheme, to have error bounds to determine how small h needs to be to get an answer to within a desired level of accuracy. The ultimate trade-off involves accuracy versus time to complete the calculation. Although we do not discuss error bounds for differential equations in this book, you can learn about them in any introductory numerical analysis course.

Example 7 Effect of the choice of h

Consider the logistic equation

$$\frac{dy}{dt} = 30y\left(1 - \frac{4y}{3}\right) \qquad y(0) = 0.1$$

which has this solution (using separation of variables):

$$y(t) = \frac{3e^{30t}}{26 + 4e^{30t}}$$

Compare the plots on $[0, 5]$ of the numerical solution and actual solution for the given values of h.

a. $h = 0.1$ b. $h = 0.08$ c. $h = 0.05$

Solution

a. Using Euler's method for $h = 0.1$ on $[0, 5]$ yields 51 values for t and y. Plotting these values in the ty-plane yields the red curve shown in Figure 6.29a. The actual solution is shown in blue. As you can see, the numerical solution acts quite wildly.

b. Using Euler's method for $h = 0.08$ on $[0, 5]$ yields 63 values for t and y. Comparing these values (red) with the actual solution (blue). As we can see, Figure 6.29b reveals that the numerical solution is not a good approximation, even though it is not quite as wild as that shown in part a.

c. Using Euler's method for $h = 0.05$ on $[0, 5]$ yields a 101 values for t and y. Comparing these values (red) with the actual solution (blue). Note that Figure 6.29c reveals that the numerical and actual solutions are almost identical.

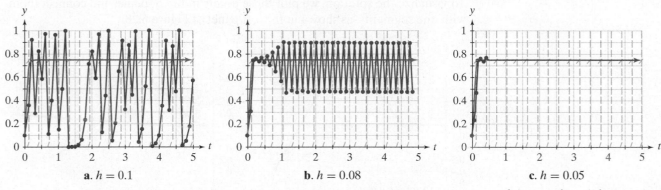

a. $h = 0.1$ b. $h = 0.08$ c. $h = 0.05$

Figure 6.29 Euler approximations for solution of $\dfrac{dy}{dt} = 30y\left(1 - \dfrac{4y}{3}\right)$

The moral of the previous example is that with the Euler method, as with other numerical methods for solving differential equations, one must be careful in choosing the appropriate step size h.

PROBLEM SET 6.4

Level 1 DRILL PROBLEMS

Sketch at least three particular solutions for each of the slope fields shown in Problems 1 to 6.

1.   2.   3.   4.

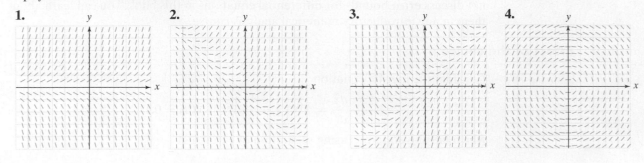

5.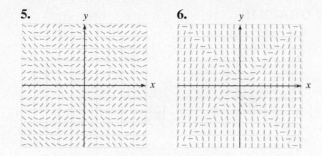

6.

Sketch a solution satisfying the specified initial conditions shown over the slope field in Problems 7 to 10.

7. $y(0) = 0.3$

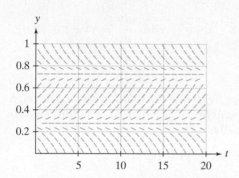

8. $y(0) = 2$

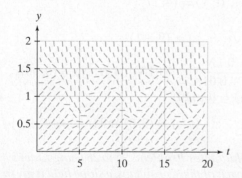

9. $y(6) = 0$

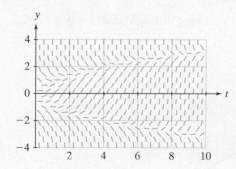

10. $y(8) = 2$

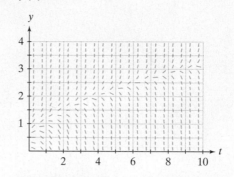

11. Match the following four equations with the four slope fields.

a. $\dfrac{dy}{dt} = \sin t$ **b.** $\dfrac{dy}{dt} = t \sin y$

c. $\dfrac{dy}{dt} = \sin y$ **d.** $\dfrac{dy}{dt} = y \sin t$

GRAPH A

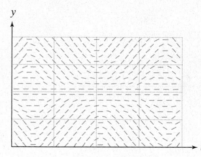

GRAPH B

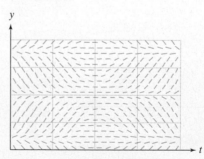

GRAPH C

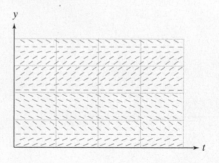

GRAPH D

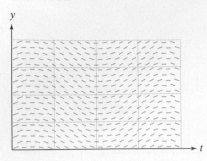

12. Match the following four equations with the four slope fields.

a. $\dfrac{dy}{dt} = y(1-y)(1+y)$ **b.** $\dfrac{dy}{dt} = \sin t$

c. $\dfrac{dy}{dt} = \sin(t+y)$ **d.** $\dfrac{dy}{dt} = t/10 + y$

GRAPH E

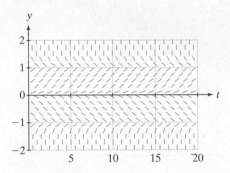

GRAPH F

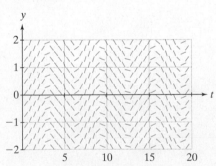

GRAPH G

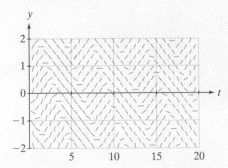

GRAPH H

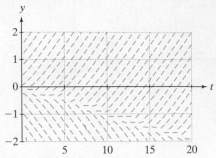

Sketch the slope fields and sketch a few solutions for the differential equations given in Problems 13 to 18.

13. $\dfrac{dy}{dt} = y(4-y)(y-2)$ **14.** $\dfrac{dy}{dt} = t^2 - y$

15. $\dfrac{dy}{dt} = \sin t$ **16.** $\dfrac{dy}{dt} = y^2 + t^2 - 1$

17. $\dfrac{dy}{dx} = -\dfrac{y}{x}$ **18.** $\dfrac{dy}{dx} = e^{x+y}$

Sketch the slope fields and the solution passing through the specified point for the differential equations given in Problems 19 to 24.

19. $\dfrac{dy}{dt} = t^2 - y^2, (t, y) = (0, 0)$

20. $\dfrac{dy}{dt} = 1.5y(1-y), (t, y) = (0, 0.1)$

21. $\dfrac{dy}{dt} = \sqrt{\dfrac{t}{y}}, \ (t, y) = (4, 1)$

22. $\dfrac{dy}{dt} = y^2\sqrt{t}, \ (t, y) = (9, -1)$

23. $\dfrac{dN}{dt} = \dfrac{0.1N}{1 + 0.01N} - 0.01N - 4, (t, N) = (0, 90)$
and $(t, N) = (0, 110)$

24. $\dfrac{dz}{dt} = 4(z - z^3), (t, z) = (0, 0)$ and $(t, z) = (0, 0.1)$

Estimate a solution for Problems 25 to 28 using Euler's method. For each of these problems, a slope field is given with an actual solution. Superimpose the segments from Euler's method on the given slope field and assess how well your solution approximates the actual solution as drawn.

25. $\dfrac{dy}{dt} = \dfrac{t}{y} - t$ passing through $(0, 4)$ for $0 \le t \le 7$,
$h = 1$

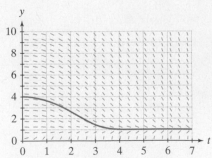

26. $\dfrac{dy}{dt} = \dfrac{t}{y} + \dfrac{t}{4} - 2$ passing through $(0, 5)$ for

$0 \le t \le 4,\ h = 1$

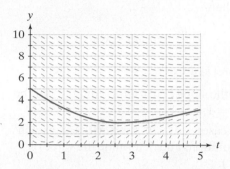

27. $\dfrac{dy}{dt} = 2t(y - t^2)$ passing through $(0, 1)$ for

$0 \le t \le 3, h = 0.5$

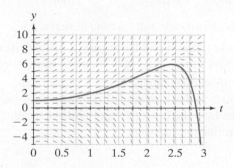

28. $\dfrac{dy}{dt} = \dfrac{4t - 2ty}{1 + t^2}$ passing through $(0, 1)$ for

$0 \le t \le 5, h = 0.5$

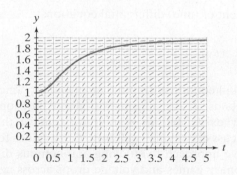

Use Euler's method to approximate the solution to

$\dfrac{dy}{dt} = f(t, y)$ *and sketch the approximate solution in*

Problems 29 to 32 over the specified interval.

29. Over the interval $0 \le t \le 2$ with
$f(t, y) = (4 - y)(y + 2),\ y(0) = 0.1, h = 0.5$

30. Over the interval $0 \le t \le 1$ with
$f(t, y) = y - t,\ y(0) = 2, h = 0.2$

31. Over the interval $0 \le t \le 3.5$ with
$f(t, y) = \sin \pi t - 2y,\ y(0) = 0, h = 0.5$

32. Over the interval $-1 \le t \le 0$ with
$f(t, y) = (4 - y)(y + 2),\ y(-1) = 0, h = 0.2$

33. Consider the differential equation

$$\frac{dy}{dt} = \frac{1}{t}$$

a. Verify that $y(t) = \ln t$ is a solution to this differential equation satisfying $y(1) = 0$.

b. Use Euler's method to approximate $y(2) = \ln 2$ with $h = 0.5$.

34. Consider the differential equation

$$\frac{dy}{dt} = y$$

a. Verify that $y(t) = e^t$ is a solution to this differential equation satisfying $y(0) = 1$.

b. Use Euler's method to approximate e with $h = 0.2$.

Level 2 APPLIED AND THEORY PROBLEMS

35. A patient receives a continuous drug infusion of 100 mg/h. The half-life of the drug is two hours.

a. Write a differential equation for the amount of drug in the body. (Hint: Review Example 2 in Section 6.3.)

b. Sketch the slope field for this differential equation.

c. Determine the limiting amount of the drug in the patient's body.

36. A patient receives a continuous drug infusion of 50 mg/h. The half-life of the drug is one hour.

a. Write a differential equation for the amount of drug in the body. (Hint: Review Example 2 in Section 6.3.)

b. Sketch the slope field for this differential equation.

c. Determine the limiting amount of the drug in the patient's body.

37. A population subject to seasonal fluctuations can be described by the logistic equation with an oscillating carrying capacity. Consider, for example,

$$\frac{dP}{dt} = P\left(1 - \frac{P}{100 + 50 \sin 2\pi t}\right)$$

Although it is difficult to solve this differential equation, it is easy to obtain a qualitative understanding.

a. Sketch a slope field over the region $0 \leq t \leq 5$ and $0 \leq P \leq 200$.

b. Sketch solutions that satisfy $P(0) = 0$, $P(0) = 10$, and $P(0) = 200$.

c. Use technology to obtain a better rendition of the slope field and solutions.

d. Comment on your solutions and compare to your work using different methods.

38. The velocity $v(t)$ of a skydiver is governed by the equation

$$m\frac{dv}{dt} = mg - kv^2$$

where m is the mass of the skydiver, g is gravitational acceleration, and k is a dampening constant (i.e., accounts for air friction).

a. Sketch the slope field for this equation assuming that $m = 70\,\text{kg}$, $g = 9.8\,\text{m/s}^2$, and $k = 110\,\text{kg/s}$.

b. Using the slope field, determine the value of $\lim_{t \to \infty} v(t)$ for the solution $v(t)$ satisfying $v(0) = 0$. Note that this limiting value is known as the *terminal velocity*.

39. An autocatalytic chemical reaction involves two molecules, A and B. Let a denote the concentration of A and assume that the concentration b of B remains constant throughout the experiment

(e.g., B is added to the mixture in such a way to keep b constant). If A combines with a molecule of B to form two molecules of A and in a backward reaction, two molecules of A form a molecule of A and B, then

$$\frac{da}{dt} = k_1 a b - k_2 a^2$$

where k_1 and k_2 are positive rate constants.

a. Sketch the slope field for this equation for the case $k_1 = 1$, $b = 1$, $k_2 = 0.5$.

b. For the cases $a(0) = 0.2$ and $a(0) = 3$, sketch in the solutions and determine the value of $\lim_{t \to \infty} a(t)$.

40. A population, in the absence of harvesting, exhibits the following growth

$$\frac{dN}{dt} = N\left(\frac{N}{100} - 1\right)\left(1 - \frac{N}{1{,}000}\right)$$

where N is abundance and t is time in years.

a. Write an equation that corresponds to harvesting the population at a rate of 0.5% per day.

b. Sketch the slope field for the differential equation you found in part **a**; by sketching solutions, describe how the fate of the population depends on its initial abundance.

6.5   Phase Lines and Classifying Equilibria

Now we focus on autonomous (independent of time) differential equations:

$$\frac{dy}{dt} = f(y)$$

The slope fields for autonomous differential equations are time independent. Since the slope field at any point in time contains all the information about the slope field, an infinite amount of redundancy exists. In this section we remove this redundancy using *phase lines*, and we discuss classifying **equilibria**, the y values for which $f(y) = 0$. With this new qualitative approach to studying autonomous differential equations, we examine evolutionary games and voltage drops across cell membranes.

Phase lines

In the previous section, we sketched slope fields by determining where the slope is zero (nullcline), where it is positive, and where it is negative. In this section, we consider a *phase-line* diagram that collapses the two-dimensional slope field to the y axis without losing any information about the qualitative behavior of solutions to $\frac{dy}{dt} = f(y)$ as illustrated Figure 6.30.

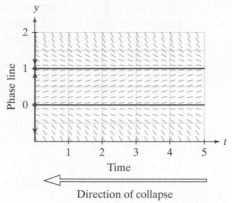

Figure 6.30 Illustration of how the slope field for the logistic equation $\dfrac{dy}{dt} = y(1 - y)$, can be collapsed on the y axis by removing (or projecting down) the time axis t.

Phase Lines

To draw a phase line for $\dfrac{dy}{dt} = f(y)$, follow these steps:

Step 1. Draw a vertical line corresponding to the y axis.

Step 2. Draw solid circles on this line corresponding to the equilibria of $\dfrac{dy}{dt} = f(y)$, that is, y values where $f(y) = 0$.

Step 3. Draw an upward arrow on intervals where $f(y) > 0$. On these intervals, solutions of the differential equation are increasing.

Step 4. Draw a downward arrow on intervals where $f(y) < 0$. On these intervals, the solutions of the differential equation are decreasing.

In the next example we consider competition between two clonal populations of the same species. A clonal population consists of genetically identical individuals who only have mothers (reproducing asexually) and are all descendants of the same female. Clonal populations include many kinds of plants, fungi, and bacteria, and some insects, fish, amphibians, and reptiles such as the whip-tail lizard shown in Figure 6.31.

Figure 6.31 Certain species of whiptail lizards are clonally reproducing; that is, all individuals are female and all progeny are identical genetic copies of their mothers, except for changes due to mutations.

Example 1   Competition of clonal genotypes

Consider two clonally reproducing lines of the same species that exhibit two genotypes denoted by a and A. Suppose the growth rate of each line satisfies an exponential growth model, with the per capita growth rates of lines a and A being r_a and r_A, respectively. Further, suppose these two clonal lines are growing together in the same geographic area, and let y denote the proportion of genotype a in this population. The solution to Problem 39 in Problem Set 6.5 shows that the variable y satisfies the equation

$$\frac{dy}{dt} = (r_a - r_A)y(1 - y)$$

a. Draw the phase line for this equation when $r_a > r_A$.

b. Draw the phase line for this equation when $r_a < r_A$.

c. Discuss why your answers to parts **a** and **b** make sense.

Solution

a. We begin by drawing the y axis. The equilibria are determined by the solutions of

$$0 = (r_a - r_A)y(1 - y)$$

Since the equilibria are $y = 0$ and $y = 1$, we draw solid circles on the y axis at these y values. Since $r_a > r_A$, we have $\frac{dy}{dt} > 0$ for $0 < y < 1$ and we draw an upward arrow on this interval. Since $\frac{dy}{dt} < 0$ for $y > 1$ and $y < 0$, we draw downward arrows on these intervals. This results in the phase line illustrated in Figure 6.32a.

b. Again begin by drawing the y axis. The equilibria are determined, as before, by the solutions of

$$0 = (r_a - r_A)y(1 - y)$$

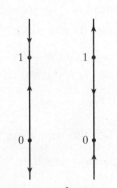

a. $r_a > r_A$ **b.** $r_a < r_A$

Figure 6.32 Phase lines for $\frac{dy}{dt} = (r_a - r_A)y(1 - y)$

Since the equilibria are $y = 0$ and $y = 1$, we draw solid circles on the y axis at these y values. Since $r_a < r_A$, we have $\frac{dy}{dt} < 0$ for $0 < y < 1$ and we draw a downward arrow on this interval. Since $\frac{dy}{dt} > 0$ for $y > 1$ and $y < 0$, we draw upward arrows on these intervals. This results in the phase line illustrated in Figure 6.32b.

c. If the per capita growth rate of genotype a is greater than the per capita growth rate of genotype A, then we would expect genotype a to become more and more prevalent in the population. Hence, provided that $y > 0$ initially, y approaches 1 as seen in the phase line for part **a**. Conversely, if the per capita growth rate of genotype a is less than the per capita growth rate of genotype A, then we would expect a to become less and less prevalent in population. Hence, y should approach 0, as seen in the phase line for part **b**.

In the previous example, we found phase lines from an equation, but sometimes we have a graph of $f(y)$ and not an equation. The next example shows how to find phase lines in such a case.

Example 2   From graphs to phase lines to solutions

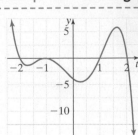

Figure 6.33 Graph of $\frac{dy}{dt} = f(y)$

Let the graph of $f(y)$ defined on the domain $(-\infty, \infty)$ be as shown in Figure 6.33.

a. Draw a phase line for $\frac{dy}{dt} = f(y)$.

b. Sketch solutions for this differential equation that satisfy $y(0) = -1.1$, $y(0) = 1.1$, and $y(0) = 0.9$.

Solution

a. Since the graph of $f(y)$ intersects the y axis at the points $-2, -1, 1,$ and 2, these y values are the equilibria of $y' = f(y)$. We draw solid circles at these points of the phase line. Since $f(y) > 0$ on the intervals $(-\infty, -2)$ and $(1, 2)$, we draw upward arrows on these intervals, as shown in Figure 6.34. For all the other intervals, $(-2, -1), (-1, 1),$ and $(2, \infty)$, we draw downward arrows.

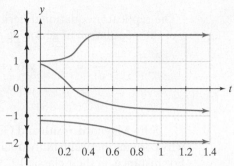

Figure 6.34 Phase line (left) and solutions (right) to the differential equation depicted in Figure 6.33

b. According to the phase line, a solution initiated at $y = -1.1$ initially decreases slowly (as it is near the equilibrium $y = -1$), decreases more rapidly, and asymptotes at the equilibrium $y = -2$. A solution initiated at $y = 1.1$ initially increases slowly, increases more rapidly, and asymptotes at the equilibrium $y = 2$. A solution initiated at $y = 0.9$, initially decreases slowly, decreases more rapidly, and asymptotes at the equilibrium $y = -1$. These solutions are shown in Figure 6.34. ∎

The equation $y = (r_a - r_A)y(1 - y)$ in Example 1 has a special name in the context of *evolutionary game theory*. It is called the **replicator equation**. Evolutionary games were formalized into a coherent mathematical theory in the late 1970s by the theoretical evolutionary biologist John Maynard Smith (1920–2004). (For more information about one of the world's greatest evolutionary biologists, see the 𝔥𝔦𝔰𝔱𝔬𝔯𝔦𝔠𝔞𝔩 𝔔𝔲𝔢𝔰𝔱 in Problem Set 6.5.) Perhaps the best known of his games is the Hawk–Dove game, which describes under what conditions aggressive versus nonaggressive behaviors persist in populations.

In general, for any two inherited contrasting strategies, the growth rates r_a and r_A for genotype a (e.g., doves) and genotype A (e.g., hawks), respectively, in the replicator equation are constructed from a two-by-two table. This table is known as the **payoff matrix**; it tells how much payoff (benefit if positive, cost if negative) an individual gets after a pairwise interaction with another individual. The payoffs when a meets a are denoted by P_{aa}. Similarly, P_{aA}, P_{Aa}, and P_{AA} denote the payoffs when a meets A, A meets a, and A meets A, respectively. This information is summarized in Table 6.3.

Table 6.3 Payoff Matrix that specifies the payoff to individuals in the rows interacting with individuals in the columns.

	Type a	Type A
Type a (proportion y)	P_{aa}	P_{aA}
Type A (proportion $(1 - y)$)	P_{Aa}	P_{AA}

If the payoff affects an individual's reproductive rate, we determine the per capita (i.e., proportional) growth rate of a genotype. We find the **expected payoff** by calculating the product of the chance of meeting an individual playing a particular strategy and the corresponding payoff.

For genotype a, this expected payoff is

$$r_a = yP_{aa} + (1 - y)P_{aA}$$

and for genotype A the expected payoff is

$$r_A = yP_{Aa} + (1 - y)P_{AA}$$

Substituting these expressions for r_a and r_A into $y' = y(1 - y)(r_a - r_A)$, we get the two-strategy replicator equations.

The replicator equation describing the dynamics of the proportions y and $(1 - y)$ of the population of genotype a and A, with payoff matrix P_{ij}, $(i, j = a$ and $A)$ is

$$\frac{dy}{dt} = y(1 - y)(r_a - r_A)$$
$$= y(1 - y)\left[P_{aA} - P_{AA} + y(P_{aa} + P_{AA} - P_{aA} - P_{Aa})\right]$$

John Maynard Smith and George Price used replicator equations to understand why conflicts between individuals within species rarely escalate to fights to the death. For instance, in their classic 1973 paper "The Logic of Animal Conflict," *Nature* 246: 15–18, they wrote with regard to mule deer (Figure 6.35):

"In mule deer (Odocoileus hemionus) *the bucks fight furiously but harmlessly by crashing or pushing antlers against antlers, while they refrain from attacking when an opponent turns away, exposing the unprotected side of its body."*

Figure 6.35 Two male mule deer engaging in a contest for a mate. Such furious conflicts rarely lead to either individual being seriously hurt.

To model animal conflicts, they considered a population of individuals competing for a limiting resource such as mates, food, or shelter. To win this resource, individuals engage in pairwise contests and play one of two strategies, hawk or dove. Individuals playing the **hawk strategy** constantly escalate the intensity of the contest until either they get the resource or they get injured. Individuals playing the **dove strategy** leave the contest whenever their opponent escalates the conflict. This game is illustrated in the next two examples.

Example 3   The Hawk-Dove replicator equation

Suppose a hawk gets a payoff of $V > 0$ every time it meets a dove and the dove gets 0. Every time two doves meet they share the payoff V, while if two hawks meet they escalate the contest until one gets the net payoff V and the other pays a cost $C > 0$. What are the payoff matrix entries for this contest and the replicator equation that describes the frequency of doves in the population?

Solution   Let a represent doves and A represent hawks. In this game, the payoffs are as follows:

$$P_{aa} = \frac{V}{2} \qquad \text{two doves sharing the payoff}$$
$$P_{aA} = 0 \qquad \text{dove gets nothing}$$
$$P_{Aa} = V \qquad \text{hawk gets } V$$
$$P_{AA} = \frac{V}{2} - \frac{C}{2} \qquad \begin{array}{l}\text{half of the time the hawk gets } V,\\ \text{half of the time is gets } -C\end{array}$$

If we substitute these values in the two-strategy replicator equation we obtain

$$\frac{dy}{dt} = y(1-y)(r_a - r_A)$$

$$= y(1-y)\left[0 - \frac{V-C}{2} + y\left(\frac{V}{2} + \frac{V-C}{2} - 0 - V\right)\right]$$

$$= y(1-y)\left[\frac{C-V}{2} - \frac{Cy}{2}\right]$$

Example 4   Dynamics of a Hawk-Dove game

Consider a Hawk-Dove game with the payoff $V = 2$ and the cost $C = 3$. Sketch the phase line and discuss the evolutionary implications.

Solution   When $V = 2$ and $C = 3$, we obtain from the previous example the specific replicator equation

$$\frac{dy}{dt} = y(1-y)\left(\frac{1}{2} - \frac{3y}{2}\right)$$

The equilibria solutions are values for which $dy/dt = 0$:

$$y = 0, \ y = 1, \ y = 1/3$$

For $0 < y < 1/3$, $\dfrac{dy}{dt} > 0$. To see this, choose a value in the interval, say $y = \dfrac{1}{6}$, and calculate

$$\frac{dy}{dt} = \frac{1}{6}\left(1 - \frac{1}{6}\right)\left(\frac{1}{2} - \frac{1/2}{2}\right) > 0$$

For $1/3 < y < 1$, $\dfrac{dy}{dt} < 0$. To see this, choose a representative value, say $y = \dfrac{1}{2}$, and calculate

$$\frac{dy}{dt} = \frac{1}{2}\left(1 - \frac{1}{2}\right)\left(\frac{1}{2} - \frac{3/2}{2}\right) < 0$$

Finally, for $y > 1$, $\dfrac{dy}{dt} > 0$. To see this, choose some representative value, say $y = 2$, and calculate

$$\frac{dy}{dt} = 2(1-2)\left(\frac{1}{2} - \frac{6}{2}\right) > 0$$

The phase line is shown in Figure 6.36. The phase line implies that if initially hawks and doves are present, then the population approaches an equilibrium consisting of one-third doves and two-thirds hawks. This approach to equilibrium is illustrated in Figure 6.36.

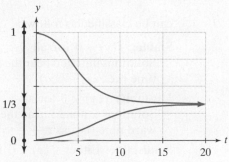

Figure 6.36 Phase line (left) and solutions (right) to the equation $\dfrac{dy}{dt} = y(1-y)(1-3y)/2$

The equilibrium value $y = 1/3$ in Example 4 supports both the hawk and dove strategies in the proportion of one individual playing the dove strategy for every two individuals playing the hawk strategy. Such an equilibrium is called a **polymorphic** strategy, as opposed to a **monomorphic** strategy where all individuals play either hawk or dove. In Example 4, we can understand the growth rates of hawks and doves at low frequencies as follows. Imagine the population consists mainly of doves and only a few hawks. In this case, individuals are most likely to have a contest with a dove. For a dove this means getting a payoff of $V/2 = 2 \times (1/2) = 1$. For a hawk this means getting a payoff of $V = 2$. Therefore, hawk numbers will grow at twice the rate of doves. Alternatively, consider a population consisting mostly of hawks and only a few doves. In this case, individuals are most likely to encounter a hawk. For a hawk this means an average payoff of $(V - C)/2 = (2 - 3)/2 = -1/2$. For a dove this means a payoff of 0. Therefore, hawk numbers will decline until a balance is reached at the polymorphic equilibrium.

Classifying equilibria

When a system starts at an equilibrium, it remains there for all time. In the real world, however, biological systems are constantly subject to environmental perturbations. Thus, if a system starting at equilibrium is slightly perturbed from equilibrium, we need to ask if it tends to return to the equilibrium or not. When the system tends to return to the equilibrium, we call the equilibrium *stable*. Otherwise, we call it *unstable*. Precise definitions follow; see Figure 6.37 for a graphical representation.

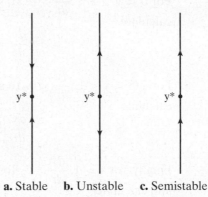

a. Stable   **b.** Unstable   **c.** Semistable

Figure 6.37 Graphical representation of the classification of equilibria

Classification of Equilibria

An equilibrium y^* of the equation

$$\frac{dy}{dt} = f(y)$$

that is a solution satisfying

$$f(y*) = 0$$

can be classified as follows:

Stable: $f(y) > 0$ for all $y < y^*$ near y^* and $f(y) < 0$ for all $y > y^*$ near y^*. Solutions initiated near the equilibrium tend toward the equilibrium in forward time (i.e., as $t \to \infty$).

Unstable: $f(y) < 0$ for all $y < y^*$ near y^* and $f(y) > 0$ for all $y > y^*$ near y^*. Solutions initiated near the equilibrium tend toward the equilibrium in backward time (i.e., as $t \to -\infty$).

Semistable: Either $f(y) < 0$ for all $y \neq y^*$ near y^* or $f(y) > 0$ for all $y \neq y^*$ near y^*. Solutions initiated near one side (respectively, other side) of the equilibrium tend toward the equilibrium in backward (respectively, forward) time.

Example 5   Classifying equilibria

Classify the equilibria for $\dfrac{dy}{dt} = f(y)$ where the graph of $f(y)$ is the graph given in Figure 6.33 in Example 2, which we repeat here for convenience.

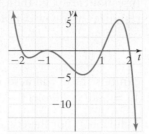

Solution   Previously, we sketched the phase line for $\dfrac{dy}{dt} = f(y)$ and found four equilibria: $y = -2$, $y = +2$, $y = -1$, $y = +1$. From the phase line sketch in Example 2, Figure 6.34, we classify the equilibria as follows: $y = -2$ and $y = 2$ are stable, $y = 1$ is unstable, and $y = -1$ is semistable.

The voltage V across the membrane of a typical cell is maintained by voltage-gated (i.e., controlled) protein channels embedded in the cell membrane. These channels (see Figure 6.38) regulate the flow of positively charged potassium ions and negatively charged organic molecules out of the cell, and negatively charged chlorine ions and positively charged sodium ions into the cell.

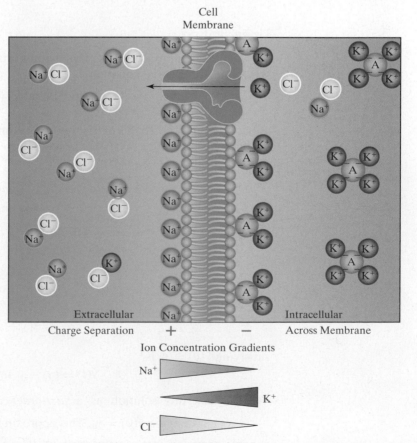

Figure 6.38 An illustration of a voltage-gated channel depicting the electrical field gradients and ions that flow across cell membranes

Example 6   Membrane potential

If a cell membrane is perturbed from its resting potential V_0 by a small input current (e.g., coming from another neuron), it will return to its resting potential. However, if this perturbing current is sufficiently large to cause $V(t)$ to drop below a critical threshold level V_c, then the sodium ions flow across the membrane until the voltage stabilizes at a new *depolarized* equilibrium level V_d. Show that model

$$\frac{dV}{dt} = -k(V - V_0)(V - V_c)(V - V_d)$$

exhibits these characteristics by finding and classifying its equilibria for the values $V_0 = -70$ millivolts (mV), $V_c = -30$ mV, $V_d = 55$ mV, and $k = 1$.

Solution   For the constants in question, the right-hand side of the equation is

$$f(V) = -(V + 70)(V + 30)(V - 55)$$

This function is cubic in the variable V with roots at $V = -70, -30$, and 55. The graph of this cubic is given by the graph shown in Figure 6.39.

Since $f(V) > 0$ for V less than but close to -70 and $f(V) < 0$ for V less than but close to -70, $V_0 = -70$ mV is stable. Similarly, $V_d = 55$ mV is stable. In contrast, since $f(V) < 0$ for V less than but close to -30 and $f(V) > 0$ for V more than but close to -30, $V_c = -30$ mV is unstable. Hence, the membrane returns to its resting potential, -70 mV, whenever perturbations are small. However, if there is a sufficiently large perturbation to cause the membrane potential to be above -30 mV, then the membrane approaches the depolarized state of 55 mV. Moreover, small perturbations from this depolarized state will not return the membrane to its resting potential. Only a sufficiently large perturbation will.

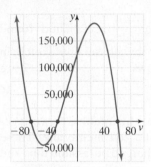

Figure 6.39 Graph of $F(V) = -(V + 70) \times (V + 30)(V - 55)$

Linearization

An analytical approach to classifying equilibria involves linearizing the right-hand side of the differential equation

$$\frac{dy}{dt} = f(y)$$

about its equilibria. To this end, suppose y^* is an equilibrium of this equation and we evaluate the derivative of $f(y)$ at y^* to obtain $a = f'(y^*)$. A linear approximation to $f(y)$ for y near y^* is given by

$$f(y) \approx f(y^*) + f'(y^*)(y - y^*) \qquad \text{\textit{linear approximation}}$$
$$= a(y - y^*) \qquad \text{\textit{since } } f(y^*) = 0 \text{ \textit{and} } f'(y^*) = a$$

Hence,

$$\frac{dy}{dt} \approx a(y - y^*)$$

for values of y near y^*. As you are asked to show in Problem 37 of Problem Set 6.5, the solution to

$$\frac{dy}{dt} = a(y - y^*)$$

satisfying $y(0) = y_0$ is

$$y(t) = (y_0 - y^*)e^{at} + y^*$$

We can use this solution as a *first-order approximation* for the solution to $\frac{dy}{dt} = f(y)$ satisfying $y(0) = y_0$. This approximation $y(t) = (y_0 - y^*)e^{at} + y^*$ is reasonable provided $y(t)$ remains near y^*. If $a < 0$, then this approximation suggests that $\lim_{t \to \infty} y(t) = y^*$ whenever y_0 is sufficiently close to y^*. Alternatively, if $a > 0$, then this approximation suggests that $\lim\limits_{t \to -\infty} y(t) = y^*$ whenever y_0 is sufficiently close to y^*. These observations can be made mathematically precise and can be used to prove the following theorem.

Theorem 6.1 Stability of a first-order autonomous differential equation

Suppose $f(y)$ is differentiable at $y = y^*$ and $y = y^*$ is an equilibrium of the equation

$$\frac{dy}{dt} = f(y)$$

Then y^* is

Stable *if $f'(y^*) < 0$,*

Unstable *if $f'(y^*) > 0$, and*

Indeterminate *if $f'(y^*) = 0$ (no conclusion is possible without looking at higher-order derivatives)*

Beyond the stated result, we can use the approximation

$$y(t) \approx (y_0 - y^*)e^{at} + y^* \quad \text{where } a = f'(y^*)$$

to gauge at what "speed" solutions move toward or away from y^*. The more positive a is, the more rapidly solutions move away from y^*. The more negative a is, the more rapidly solutions approach y^*.

Example 7 Population resilience

Consider two populations whose dynamics are described by

$$\frac{dN}{dt} = N\left(1 - \frac{N}{10,000}\right) \quad \text{and} \quad \frac{dP}{dt} = 0.5\, P\left(1 - \frac{P}{10,000}\right)$$

a. Find the equilibria and use linearization to classify their stability properties.

b. Describe in what ways the populations are similar and dissimilar.

Solution

a. For both populations, the equilibria are given by 0 and 10,000. For the first model, we have $f(N) = N(1 - N/10,000)$ and

$$f'(N) = \left(1 - \frac{N}{10,000}\right) - \frac{N}{10,000}$$

Evaluating this derivative at the equilibria, we find:

Equilibria	Evaluate	Classification
$N = 0$	$f'(0) = 1$	unstable
$N = 10,000$	$f'(10,000) = -1$	stable

For the second model, we have $g(P) = 0.5\, P\left(1 - \dfrac{P}{10,000}\right)$ and can find:

Equilibria	Evaluate	Classification
$P = 0$	$g'(0) = \dfrac{1}{2}$	unstable
$P = 10,000$	$g'(10,000) = -\dfrac{1}{2}$	stable

b. The populations are similar in that both populations have equilibria at 0 and 10,000, which are unstable and stable, respectively. Hence, both populations tend to approach the equilibrium value of 10,000.

The populations differ in that population P (the second model) tends to grow less rapidly at low densities; that is, $g'(0) = \dfrac{1}{2} < f'(0) = 1$. Moreover, if the populations are at the equilibrium of 10,000, the P population recovers less rapidly from a perturbation; that is, $g'(10,000) = -\dfrac{1}{2} > f'(10,000) = -1$.

When one population recovers more rapidly from environmental perturbations than another population (as with P versus N in Example 7), it is said to be more *resilient*.

Example 8   Hawk-Dove game revisited

Consider the Hawk-Dove game

$$\frac{dy}{dt} = y(1-y)\left(\frac{1}{2} - \frac{3y}{2}\right)$$

where y is the frequency of doves in the population.

a. Use linearization to classify each of the equilibria.

b. Use your work from part **a** to determine whether the hawks increase more rapidly at low frequencies or the doves increase more rapidly at low frequencies.

Solution
a. Let

$$f(y) = y(1-y)\left(\frac{1}{2} - \frac{3y}{2}\right)$$

As we have seen, the equilibria are $y = 0$, $y = 1$, and $y = 1/3$. To linearize, we need this derivative:

$$f'(y) = (1-y)\left(\frac{1}{2} - \frac{3y}{2}\right) - y\left(\frac{1}{2} - \frac{3y}{2}\right) - y(1-y)\frac{3}{2} \qquad \textit{by product rule}$$

Evaluated at $y = 0$, we obtain

$$f'(0) = \frac{1}{2} > 0$$

Hence, the equilibrium $y = 0$ is unstable.
Since

$$f'(1) = 1 > 0$$

the equilibrium $y = 1$ is unstable.
Since

$$f'\left(\frac{1}{3}\right) = -\frac{1}{3} < 0$$

the equilibrium $y = \dfrac{1}{3}$ is stable.

b. Since $f'(1) = 1 > f'(0) = \dfrac{1}{2}$, we see that hawks increase more rapidly at low frequency than doves do.

Example 9   Linearization of membrane voltage model

Consider the membrane potential model from Example 6:

$$\frac{dV}{dt} = -k(V - V_0)(V - V_c)(V - V_d)$$

Assume that $V_0 < V_c < V_d$ and $k > 0$.

a. Use linearization to classify the equilibria.

b. Discuss the resilience of the two stable equilibria.

Solution

a. Define

$$f(V) = -k(V - V_0)(V - V_c)(V - V_d)$$

The roots of the function $f(V)$ are the equilibria $V^* = V_0$, V_c, and V_d.

Two applications of the product rule imply

$$f'(V) = -k[(V - V_c)(V - V_d) + (V - V_0)(V - V_d) + (V - V_0)(V - V_c)]$$

Thus, at $V = V_0$, we have

$$f'(V_0) = -k[(V_0 - V_c)(V_0 - V_d) + 0 + 0] = -k(V_0 - V_c)(V_0 - V_d)$$

Since $V_0 < V_c < V_d$, $f'(V_0) < 0$ which implies V_0 is stable.

At $V = V_c$, we have

$$f'(V_c) = -k(V_c - V_0)(V_c - V_d)$$

Since $V_0 < V_c < V_d$, $f'(V_c) > 0$ which implies V_c is unstable.

At $V = V_d$, we have

$$f'(V_d) = -k(V_d - V_0)(V_d - V_c)$$

Since $V_0 < V_c < V_d$, $f'(V_d) < 0$ which implies $V = V_d$ is stable.

b. The resting state V_0 is more resilient than the depolarized state V_d whenever

$$f'(V_0) < f'(V_d)$$

$-k(V_0 - V_c)(V_0 - V_d) < -k(V_d - V_0)(V_d - V_c)$ *substituting work from part **a***

$-k(V_c - V_0)(V_d - V_0) < -k(V_d - V_0)(V_d - V_c)$ *rearranging left-hand side*

$V_c - V_0 > V_d - V_c$ *dividing both sides by a negative term*

Hence, provided the difference between the resting potential and the critical potential is greater than the difference between the depolarized potential and the critical potential, the resting state is more resilient. ∎

PROBLEM SET 6.5

Level 1 DRILL PROBLEMS

Draw phase lines, classify the equilibria, and sketch a solution satisfying the specified initial value for the equations in Problems 1 to 10.

1. $\frac{dy}{dt} = 1 - y^2$, $y(0) = 0$

2. $\frac{dy}{dt} = 2 - 3y$, $y(0) = 2$

3. $\frac{dy}{dt} = -7$, $y(0) = -2$

4. $\frac{dy}{dt} = 10$, $y(0) = 5$

5. $\frac{dy}{dt} = y(y - 10)(20 - y)$, $y(0) = 9$

6. $\frac{dy}{dt} = y(y - 5)(25 - y)$, $y(0) = 7$

7. $\frac{dy}{dt} = \sin y$, $y(0) = 0.1$

8. $\frac{dy}{dt} = 1 - \sin y$, $y(0) = -0.6$

9. $\dfrac{dy}{dt} = y^2 - 2y + 1,\ y(0) = 0$

10. $\dfrac{dy}{dt} = y^3 - 4y,\ y(0) = 0.1$

Draw a phase line for $\dfrac{dy}{dt} = f(y)$ for the graphs shown in Problems 11 to 14. Sketch the requested solutions.

11. $y(0) = -1.1,\ y(0) = 1.1,\ y(0) = 0.9$

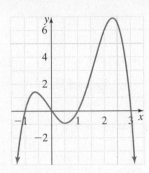

12. $y(0) = -0.1,\ y(0) = 0.9,\ y(0) = 1.1$

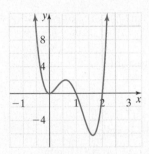

13. $y(0) = -2,\ y(0) = 1,\ y(0) = 1.5$

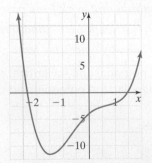

14. $y(0) = -0.1,\ y(0) = 1.9,\ y(0) = 3.9$

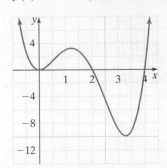

Linearize about the equilibrium in Problems 15 to 20 and classify stability properties.

15. $\dfrac{dy}{dt} = 4 - y^2,\ y^* = 2$

16. $\dfrac{dy}{dt} = \cos y,\ y^* = \dfrac{\pi}{2}$

17. $\dfrac{dy}{dt} = \dfrac{1}{\sqrt{2}} - \cos y,\ y^* = \dfrac{\pi}{4}$

18. $\dfrac{dy}{dt} = 2y - y^2 - y^{10},\ y^* = 0$

19. $\dfrac{dy}{dt} = 3 - y,\ y^* = 3$

20. $\dfrac{dy}{dt} = y(10 - y)(100 - y),\ y^* = 100$

Sketch the phase line and classify the equilibria for the Hawk-Dove game with the values V and C given in Problems 21 to 24.

21. $V = 2, C = 2$ **22.** $V = 4, C = 2$

23. $V = 3, C = 2$ **24.** $V = 2, C = 4$

Sketch the phase line and classify the equilibria for the replication equations with the indicated payoffs in Problems 25 to 28.

25. $P_{aa} = 2, P_{aA} = 1, P_{Aa} = 1,$ and $P_{AA} = 2$

26. $P_{aa} = 1, P_{aA} = 2, P_{Aa} = 3,$ and $P_{AA} = 4$

27. $P_{aa} = -1, P_{aA} = 2, P_{Aa} = 1,$ and $P_{AA} = -1$

28. $P_{aa} = 2, P_{aA} = -1, P_{Aa} = -1,$ and $P_{AA} = 3$

Level 2 APPLIED AND THEORY PROBLEMS

29. Consider a pack of wolves that can either jointly hunt a stag or individually on their own hunt hares. Suppose during the course of a stag hunt, a hare comes along and an individual wolf considers either remaining with the pack to hunt the stag or going off on its own to hunt the hare. Suppose the payoffs for these two strategies (remain with the pack or go off and hunt the hare) are given by the following payoff matrix in the context of an evolutionary game:

	Hunt stag	Hunt hare
Hunt stag	7.5	4
Hunt hare	7	5

    a. Find the replicator equation, assuming that in any one hunt only one wolf in the pack can make the decision to go or not go after the hare.

    b. Sketch the phase line and classify the equilibria.

    c. Discuss how the outcome of the evolutionary game depends on the initial strategy composition of the population.

30. Consider two scenarios based on Problem 29:

- In a population of stag hunters, a few individuals decide to hunt hares.
- In a population of hare hunters, a few individuals decide to hunt stags.

Use linearization to determine in which of these scenarios the "defecting" individuals are more rapidly excluded.

31. Evolution of cooperation, part 1. Consider a population with two strategies, cooperate and defect. Individuals that cooperate provide a benefit B to their opponents and pay a cost C for providing this benefit. Defectors provide no benefits to their opponents and pay no cost. Under these assumptions, we get the following payoff matrix:

	Cooperate	Defect
Cooperate	$B - C$	$-C$
Defect	B	0

a. Write a replicator equation for this payoff matrix.

b. Assuming $B > 0$ and $C > 0$, sketch the phase line for the replicator equation.

c. Discuss the implications of your phase line.

32. Evolution of cooperation, part 2. In Problem 31, cooperation could not evolve. However, cooperation occurs in natural populations. In this problem, we investigate how individuals that interact frequently and respond to the strategy of their opponents can promote the evolution of cooperation. Imagine that each time two opponents meet they interact an average n times. Individuals can play one of two strategies: defect always or tit-for-tat, in which case an individual initially cooperates but switches to defecting if its opponent defected.

a. If each time individuals interact the payoffs are as in Problem 31, then discuss why the payoff matrix should be this:

	Tit-for-tat	Defect
Tit-for-tat	$n(B - C)$	$-C$
Defect	B	0

b. Write a replicator equation for this game.

c. Assume $B = 3$ and $C = 2$. Sketch phase lines for $n = 2, 3, 4$.

d. Discuss the implications for the evolution of cooperation.

33. To account for the effect of a generalist predator (with a type II functional response) on a population,

ecologists often write differential equations of the form

$$\frac{dN}{dt} = 0.1N\left(1 - \frac{N}{1{,}000}\right) - \frac{10N}{1 + N}$$

where N is the population abundance and t is time (in years). The first term of the equation corresponds to logistic growth and the second term corresponds to saturating predation.

a. Sketch the phase line for this system.

b. Discuss how the fate of the population depends on its initial abundance.

34. Construct the phase line for the model

$$\frac{dV}{dt} = -2V^3 - 20V^2 + 3000V$$

and demonstrate that this equation belongs to the class of membrane voltage models presented in Example 6.

35. Use a phase line diagram to discuss the behavior of the membrane voltage models presented in Example 6 with constants $k = 3$, $V_0 = -65$ mV, $V_c = 40$ mV, and $V_d = 40$ mV. Does this membrane have the property that it is able to switch between two states when perturbed by a current?

36. 𝔥𝔦𝔰𝔱𝔬𝔯𝔦𝔠𝔞𝔩 𝔔𝔲𝔢𝔰𝔱

John Maynard Smith (1920–2004)

John Maynard Smith, or JMS as he was almost always known, was a professor at the University of Sussex (United Kingdom), and one of the world's great evolutionary biologists. JMS introduced mathematical modeling from game theory into the study of mathematical biology and completely revolutionized the way that biologists think about behavioral evolution. Jonathan Weiner in his book titled *A Conversation with John Maynard Smith* describes JMS as a classical geneticist and leading theorist in evolutionary biology. Weiner also relates how JMS applied game theory to explain jousting matches among many species, including sticklebacks, sea lions, stag beetles, and stags.

JMS received many prestigious awards. For this 𝔥𝔦𝔰𝔱𝔬𝔯𝔦𝔠𝔞𝔩 𝔔𝔲𝔢𝔰𝔱, research and write a few words about each of the following awards received by JMS, or, in the case of the last one, established in his honor.

a. Balzan Prize

b. Crafoord Prize

c. Kyoto Prize

d. John Maynard Smith Prize

37. Verify that the solution to

$$\frac{dy}{dt} = a(y - y^*)$$

satisfying $y(0) = y_0$ is given by

$$y(t) = (y_0 - y^*)e^{at} + y^*$$

38. Show that the linearization theorem is inconclusive when the derivative equals zero at the equilibrium by considering the stability of the equilibria to these equations:

$$\frac{dy}{dt} = y^3, \qquad \frac{dy}{dt} = -y^3, \qquad \frac{dy}{dt} = y^2$$

39. Consider a population of clonally reproducing individuals consisting of two genotypes, a and A, with per capita growth rates, r_a and r_A, respectively. If N_a and N_A denote the densities of genotypes a and A, then

$$\frac{dN_a}{dt} = r_a N_a \qquad \frac{dN_A}{dt} = r_A N_A$$

Also, let $y = \dfrac{N_a}{N_a + N_A}$ be the fraction of individuals in the population that are genotype a. Show that y satisfies

$$\frac{dy}{dt} = (r_a - r_A)y(1 - y)$$

40. In the Hawk-Dove replicator equation

$$\frac{dy}{dt} = \frac{y}{2}(1 - y)(C(1 - y) - V)$$

if the value $V > 0$ is specified, then find the range of values of C (in terms of V) that will ensure a polymorphism exists (i.e., find conditions that ensure the existence of an equilibrium $0 < y^* < 1$ that is stable).

41. Production of pigments or other protein products of a cell may depend on the activation of a gene. Suppose a gene is autocatalytic and produces a protein whose presence activates greater production of that protein. Let y denote the amount of the protein (say, micrograms) in the cell. A basic model for the rate of this self-activation as a function of y is

$$A(y) = \frac{ay^b}{k^b + y^b} \text{ micrograms/minute}$$

where a represents the maximal rate of protein production, $k > 0$ is a "half saturation" constant, and $b \geq 1$ corresponds to the number of protein molecules required to active the gene. On the other hand, proteins in the cell are likely to degrade at a rate proportional to y, say cy. Putting these two components together, we get the following differential equation model of the protein concentration dynamics:

$$\frac{dy}{dt} = \frac{ay^b}{k^b + y^b} - cy$$

a. Verify that $\lim\limits_{y \to \infty} A(y) = a$ and $A(k) = a/2$.

b. Verify that $y = 0$ is an equilibrium for this model and determine under what conditions it is stable.

42. Consider the model of an autocatalytic gene in Problem 41 with $b = 1, k > 0, a > 0,$ and $c > 0$.

a. Sketch the phase line for this model when $ck > a$.

b. Sketch the phase line for this model when $ck < a$.

43. Sketch the phase line for the autocatalytic gene model in Problem 41 with $b = 2, k = 1,$ and $c = 1$.

a. Sketch the phase for this model when $a = 1$.

b. Sketch the phase for this model when $a = 2$.

c. Sketch the phase for this model when $a = 2.5$.

d. Discuss what you found.

6.6   Bifurcations

Biological systems can exhibit a range of dynamic behaviors that change abruptly or gradually in response to external perturbations. The term **bifurcation** is used in the context of difference or differential equation models to denote a change in the qualitative behavior of the solutions that occurs as a parameter in the model is varied— that is, as one of the constants in the model changes its value. In this section, we introduce *bifurcation theory*. We will use the notation

$$\frac{dy}{dt} = f(y, a)$$

to represent an expression in y and a where y is a model variable and a is a parameter. The goal is to understand how the behavior of the solutions of this differential equation depends on a. More precisely, we will study how the phase line varies with the parameter a.

We will examine bifurcation theory in populations subjected to harvesting, protein dynamics associated with an autocatalytic gene, and the firing rates of neural populations.

Sudden population disappearances

Example 1   **Harvesting queen conchs**

- -

Consider a population of queen conchs in the Bahamas whose dynamics are given by

$$\frac{dy}{dt} = 10\, y \left(1 - \frac{y}{10{,}000}\right) - a$$

where t is time in years, y is number of conchs, and a is the constant annual harvesting rate.

a. Draw phase lines for $a = 0$, $a = 21{,}000$, and $a = 30{,}000$.

b. Discuss the biological implications of these phase lines.

c. Discuss how the number of equilibria depends on the harvesting rate a.

Solution   The growth rate function $f(y) = 10\, y\,(1 - y/10{,}000)$ is plotted in Figure 6.40 (blue curve) along with the values of a (broken horizontal lines) and associated equilibria (vertical dotted lines).

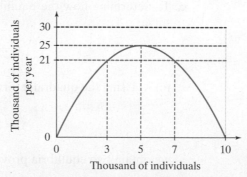

Figure 6.40 Annual growth rate of conchs is plotted as a function of population density (blue curve) subject to harvesting at three levels: 0, $a = 21{,}000$ (broken green line), $a = 25{,}000$ (broken brown lines) and $a = 30{,}000$ (broken red line). The vertical dotted lines indicate the associated equilibria, as discussed in the text.

a. Consider $a = 0$. The equilibria are given by the solutions of

$$0 = 10y\left(1 - \frac{y}{10{,}000}\right) - 0$$

Solving this equation yields the equilibria $y = 0$ and $y = 10{,}000$. Since $\dfrac{dy}{dt} > 0$ for $0 < y < 10{,}000$ and $\dfrac{dy}{dt} < 0$ for the other intervals, we obtain the phase line on the left in Figure 6.41. Now consider $a = 21{,}000$. The equilibria are given by solutions to

$$0 = 10\, y \left(1 - \frac{y}{10{,}000}\right) - 21{,}000$$

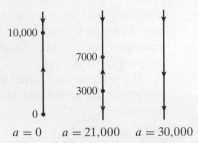

Figure 6.41 Phase lines for the density (y) of conchs for the three harvesting levels a, as labeled, inserted into the conch harvesting equation. The direction arrows indicate which points are stable (i.e., approached) and which are unstable.

which yields $y = 3,000$ and $y = 7,000$ (where the green dotted line intersects the blue curve in Figure 6.40). Since $\dfrac{dy}{dt} > 0$ for $3,000 < y < 7,000$ and $\dfrac{dy}{dt} < 0$ elsewhere, we get a phase line as shown in the center of Figure 6.41. Finally, consider $a = 30,000$ (the red dotted line in Figure 6.40). In this case there are no equilibria because

$$10y\left(1 - \frac{y}{10,000}\right) - 30,000$$

is negative for all $y \geq 0$ and we get a phase line as shown on the right-hand side of Figure 6.41.

b. The phase lines in Figure 6.41 show that as a increases, the number of equilibria goes from two to zero. In particular, at sufficiently high harvesting rates, the population is unable to persist at an equilibrium.

c. To determine how the equilibria depend on the harvesting rate a, we need to solve

$$0 = 10\,y\left(1 - \frac{y}{10,000}\right) - a$$

for y. Using the quadratic formula

$$y = 5,000 \pm 100\sqrt{2,500 - \frac{a}{10}}$$

we obtain two equilibria provided that

$$2,500 - a/10 > 0$$

which occurs if and only if $a < 25,000$. If $a = 25,000$ (brown dotted line in Figure 6.40), then we get only one equilibrium given by $y = 5,000$. Finally, if $a > 25,000$, then $2,500 - \dfrac{a}{10}$ is negative and there are no equilibria. Therefore, a change in the number of equilibria occurs at $a = 25,000$. ∎

Example 1 illustrates that the phase line of $\dfrac{dy}{dt} = f(y, a)$ can vary substantially as we vary the parameter a. Moreover, it shows that at certain parameter values (i.e., $a = 25,000$ in Example 1) there is a qualitative change in the phase line. These values are important enough to have their own name: **bifurcation values**. A bifurcation value is defined as the value of a parameter in an equation where either the number of equilibrium solutions changes or the stability of these equilibria undergo a transition from stable to unstable. A simple way to graphically summarize how the behavior of the system depends on a is to create a **bifurcation diagram**.

Bifurcation Diagram

A **bifurcation diagram** summarizes the behavior of the differential equation $\frac{dy}{dt} = f(y, a)$ in the ay plane and is created as follows:

Step 1. Draw the a axis (horizontal) and the y axis (vertical).

Step 2. Sketch the set of equilibria in the ay plane, that is, the set of points (a, y) that satisfy $0 = \frac{dy}{dt} = f(y, a)$.

Step 3. Determine in which regions of the ay plane, $\frac{dy}{dt}$ is positive or negative.

Step 4. For a collection of a values, draw a phase line. In particular, draw phase lines at bifurcation values of a and at values of a that lie between bifurcation values.

Example 2   Sudden disappearances of queen conchs

Sketch a bifurcation diagram for Example 1:

$$\frac{dy}{dt} = 10\, y \left(1 - \frac{y}{10{,}000} \right) - a$$

with $a \geq 0$ and $y \geq 0$. Discuss the implications for harvesting queen conchs.

Solution   We begin by solving

$$0 = 10\, y \left(1 - \frac{y}{10{,}000} \right) - a$$

for a and graphing $a = 10y(1 - y/10{,}000)$ in the ay plane. The graph is a parabola as shown in Figure 6.42a.

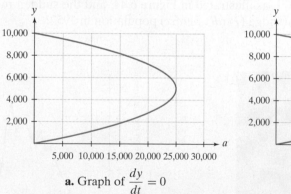
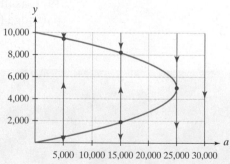

a. Graph of $\frac{dy}{dt} = 0$

b. Bifurcation diagram

Figure 6.42 The curve of equilibria and bifurcation diagram for the differential equation

$$\frac{dy}{dt} = 10y \left(1 - \frac{y}{10{,}000} \right) - a$$

Choosing a point inside the parabola, say $(0, 5{,}000)$, we obtain $\frac{dy}{dt} = 10 \times 5{,}000(1 - 1/2) = 25{,}000 > 0$. Hence, $\frac{dy}{dt} > 0$ inside of the parabola. Choosing a point outside of the parabola, say $(10, 0)$, we obtain $\frac{dy}{dt} = -10 < 0$. Hence $\frac{dy}{dt} < 0$ outside of the parabola.

Next, we sketch phase lines for several a values, say $a = 5{,}000$, $a = 15{,}000$, $a = 25{,}000$, and $a = 30{,}000$. For each of these values of a, we draw a vertical line. Where the line intersects the parabola, we draw a solid circle (in red in Fig. 6.42b) as this corresponds to points where $\frac{dy}{dt} = 0$. Where the line lies inside the parabola, we draw an upward arrow. Where the line lies outside the parabola, we draw a downward arrow. The resulting bifurcation diagram is illustrated in Figure 6.42b. Notice that for $a = 0$, $a = 21,000$, and $a = 30,000$, we get the same phase lines as in Example 1.

This bifurcation diagram indicates that for $0 < a < 25,000$ there are two equilibria. The lower equilibrium is unstable and the upper equilibrium is stable. When the two equilibria coalesce at $a = 25,000$, the resulting equilibrium is semistable—that is, solutions starting above the density $y = 5,000$ decrease to asymptotically approach the equilibrium $y = 5,000$, while solutions that start below 5,000 also decrease to 0.

It follows that for harvesting rates over the range $0 < a < 25,000$, the population can persist provided that its initial population abundance is sufficiently large (e.g., above 5,000 for the case $a = 25,000$). Moreover, the stable population equilibrium is always greater than 5,000. On the other hand, if the population is harvested at a rate $a > 25,000$, it will eventually be driven to 0 regardless of the initial abundance $y(0)$. Note that from a modeling point of view, harvesting must necessarily be set to 0 at $y(t) = 0$, since a population that has 0 individuals can no longer be harvested. From a purely mathematical point of view, though, the equation remains valid for $y(t) < 0$.

An important implication of the bifurcation diagram in Example 2 is that gradual changes in harvesting can bring about discontinuous changes in population abundance. More specifically, when the harvesting rate is ever so slightly increased beyond the bifurcation value ($a = 25,000$ in Example 1), the population begins slowly at first, but then more rapidly, to decline from high abundance to extinction. Such population disappearances have been observed in natural populations. Dramatic examples include the precipitous drop of blue pike (*Stizostedion vitreum glaucum*) from annual catches of 10 million pounds to less than a thousand pounds in the mid-1950s, or the unexpected collapse of the Peruvian anchovy (see Figure 6.43) population in 1973, as illustrated in Figure 6.44, and the sudden reduction of Great Britain's grey partridge (*Perdix perdix*) population in 1952.

Figure 6.43 The Peruvian anchovy (*Engraulis ringens*) once supported the biggest fishery of all time with a catch of 13.1 million metric tons in 1970.

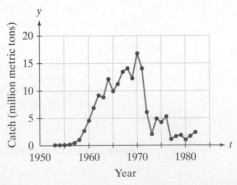

Figure 6.44 Catch data for Peruvian anchovies in the twentieth century
Data source: Table 3 in: Castillo, S., & Mendo, J. (1987). Estimation of unregistered Peruvian anchoveta (*Engraulis ringens*) in official catch statistics, 1951–1982. In D. Pauly, & I. Tsukayama (Eds.), The Peruvian anchoveta and its upwelling ecosystem: Three decades of change. ICLARM studies and reviews (Vol. 15, pp. 109–116)

The bifurcation occurring at $a = 25,000$ in the queen conch example is a *saddle node* bifurcation, because the transition from two equilibria to zero equilibria is preceded by the appearance of a semistable (or saddle) equilibrium. A more colorful name for this bifurcation is *blue sky catastrophe*, as two equilibria vanish into the blue sky as a increases past the value 25,000. Another important type of bifurcation is the *pitchfork bifurcation* illustrated by the next example. A look ahead at Figure 6.45 indicates the source of this name: one equilibrium bifurcates into three to create a pitchfork.

Example 3   Pitchfork bifurcation

Sketch a bifurcation diagram for

$$\frac{dy}{dt} = ay - y^3$$

Solution   The equilibria are given by

$$0 = y(a - y^2)$$

Hence, at an equilibrium, either $y = 0$ or $y^2 = a$. The sketches of the curves $y = 0$ for all a and $a = y^2$ for all y in the ay plane yields the blue lines in Figure 6.45. These curves determine four regions in the ay plane: the regions above and below the pitchfork and the upper and lower parabolic wedges of the pitchfork. Using the point $(a, y) = (0, 1)$, we obtain $\frac{dy}{dt} = -1 < 0$. Hence, $\frac{dy}{dt} < 0$ in the region above the pitchfork. Using the point $(a, y) = (0, -1)$, we obtain $\frac{dy}{dt} = 1 > 0$. Hence, $\frac{dy}{dt} > 0$ in the region below the pitchfork. Using the point $(a, y) = (2, 1)$, we obtain $\frac{dy}{dt} = 1 > 0$. Hence, $\frac{dy}{dt} > 0$ in the upper parabolic wedge of the pitchfork. Using the point $(a, y) = (2, -1)$, we obtain $\frac{dy}{dt} = -1 < 0$. Hence, $\frac{dy}{dt} < 0$ in the lower parabolic wedge of the pitchfork.

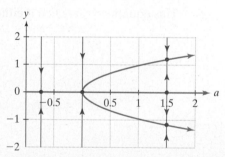

Figure 6.45 Diagram of the pitchfork bifurcation associated with the differential equation $\frac{dy}{dt} = y - y^3$ obtained by plotting the equilibria $\frac{dy}{dt} = 0$ (blue lines) and using vertical phase lines to (red lines) illustrate how the dynamics vary with a.

To complete the bifurcation diagram, it suffices to sketch phase lines for a negative a value (i.e., only one equilibrium), a positive a value (i.e., three equilibria), and the bifurcation value $a = 0$. Drawing vertical lines at these a values, solid circles at the equilibria, upward arrows where $\frac{dy}{dt} > 0$, and downward arrows where $\frac{dy}{dt} < 0$, results in the completed bifurcation diagram illustrated in Figure 6.45.

Biological switches

In order to respond appropriately to changes in their environment, living cells receive environmental signals and respond to these signals in variety of ways, including gene expression and cell activity. Similar to your computer, information processing is carried out by a complex network of switches. Unlike your computer, these switches are not made from electronic parts; rather, they consist of interacting genes and the proteins activated by these genes. Here, we examine two forms of biological switches, an autocatalytic gene introduced in Problem 41 of Section 6.5 and a neural switch for memory.

Example 4   An autocatalytic gene

As discussed in Problem 41 of Section 6.5, production of pigments or other protein products of a cell may depend on the activation of a gene. The gene can be *autocatalytic*; that is, it produces a protein whose presence activates the production of more protein. If the gene requires two proteins to bind to its receptor region to activate production of the protein, and if y denotes the concentration of the protein, then a simple model of the dynamics of y is given by

$$\frac{dy}{dt} = \frac{ay^2}{k^2 + y^2} - cy$$

where a represents the maximal rate of protein production, $k > 0$ is a "half saturation" constant, and c is the per capita degradation rate of the protein. For $k = 1$ and $c = 1$, sketch a bifurcation diagram of the dynamics for $a > 0$ and $y \geq 0$.

Solution   Setting the parameters $k = 1$ and $c = 1$ leaves us with the following model:

$$\frac{dy}{dt} = \frac{ay^2}{1 + y^2} - y$$

To solve for the equilibria as a function of a, we set this differential equation to zero and factor out a y term

$$0 = \frac{dy}{dt} = y\left(\frac{ay}{1 + y^2} - 1\right)$$

This equation is satisfied if either $y = 0$ or

$$\frac{ay}{1 + y^2} - 1 = 0$$

$$\frac{ay}{1 + y^2} = 1$$

$$ay = 1 + y^2$$

$$a = \frac{1}{y} + y$$

Sketching the equilibrium curves $y = 0$ and $a = \dfrac{1}{y} + y$ in the ay plane yields the blue curves in Figure 6.46a. Note that there are two regions in the nonnegative

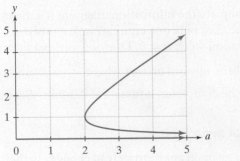

a. Graph of the solutions to $0 = \dfrac{ay^2}{1 + y^2} - y$ **b.** Bifurcation diagram for $\dfrac{dy}{dt} = \dfrac{ay^2}{1 + y^2} - y$

Figure 6.46 Constructing a bifurcation diagram for a model of an autocatalytic gene

quadrant: the region where $a < \dfrac{1}{y} + y$ and the region where $a > \dfrac{1}{y} + y$. At the point $(a, y) = (1, 2)$ in the region $a < \dfrac{1}{y} + y$, we have $\dfrac{dy}{dt} = 2(1/5 - 1) < 0$. Hence, in this region, $\dfrac{dy}{dt} < 0$. At the point $(a, y) = (4, 1)$ in the region $a > \dfrac{1}{y} + y$, we have $\dfrac{dy}{dt} = 1(4/2 - 1) > 0$. Hence, in this region, $\dfrac{dy}{dt} > 0$.

Since there is only one equilibrium for $a < 2$, two equilibria at $a = 2$, and three equilibria for $a > 2$, we sketch a phase line for $a = 1, 2,$ and 3. At $a = 1$, there is only equilibrium at $y = 0$ and $\dfrac{dy}{dt} < 0$ for all $y > 0$. We get the left phase line in Figure 6.46b. At $a = 2$, there are equilibria at $y = 0$ and near $y = 1$. Since $\dfrac{dy}{dt} < 0$ for all other y, we get the center phase line in Figure 6.46b. At $a = 3$, there are equilibria at $y = 0$, near $y = \dfrac{1}{2}$, and near $y = 2.5$. We have $\dfrac{dy}{dt} < 0$ for y between the first two equilibria, $\dfrac{dy}{dt} > 0$ for y between the latter two equilibria, and $\dfrac{dy}{dt} < 0$ for larger y values. Hence, we get the right phase line in Figure 6.46b. ■

Behind the motions, sensations, and thoughts of every animal lies a vast network of cells, the nervous system. The network comprises billions of cells called *neurons*. A typical neuron is illustrated in Figure 6.47, although neurons with various shapes and sizes make up the total neural system of any animal.

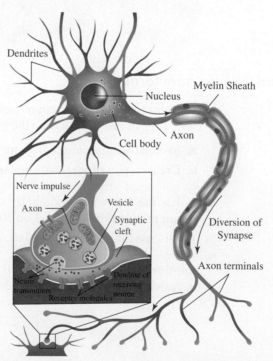

Figure 6.47 Sketch of a neuron with inset showing details of a synapse

Neurons specialize in carrying "messages" from one part of the body to another through an electrochemical process that typically causes a voltage spike to travel along the membrane of the neural cell. Messages are received by *dendrites*, which look like tentacles attached to the cell body. The chemical messages pass through these tentacles into the cell body and then out through one main, long *axon*. The end of this axon then communicates with dendrites of neurons further down the neural chain, thereby passing messages along from one neuron to the next. Messages between two neurons are usually passed in the form of a chemical flux of so-called *neurotransmitters*. Excitatory neurotransmitters trigger "go" signals that allow messages to be passed to the next neuron in the communication line. *Inhibitory neurotransmitters* produce "stop" signals that prevent messages from being forwarded. A single neuron "integrates" incoming signals to determine whether or not to pass the information along to other cells. The activity within a single neuron is typically measured by the rate at which it "fires" voltage spikes.

The simplest model of a population of neurons is the **Wilson-Cowan model**. It assumes that the entire population of neurons fire at the same rate y (units are number of spikes per millisecond) and are of the same type (i.e., release the same type of neurotransmitters).

Wilson-Cowan Neural Population Model

Let a be the rate at which an external source produces neurotransmitters that stimulate the dendrites of a population of neurons. If a is positive or negative, then the external neurotransmitters are, respectively, excitatory or inhibitory.

Let b be the rate at which each individual neuron releases neurotransmitters when it fires. If b is positive or negative, then the internal neurotransmitters are, respectively, excitatory or inhibitory.

Let c be the rate at which the firing of an active neuron decays exponentially in the absence of external stimulation.

Then the **Wilson-Cowan model** models the firing rate y (measured in spikes per unit time) of each neuron in the network by the equation

$$\frac{dy}{dt} = -cy + \frac{1}{1 + e^{-a-by}}$$

Example 5   Modeling memory formation

Consider the Wilson-Cowan model with $b = 6$ (i.e., the neurons are excitatory) and $c = 1$ (i.e., in one unit of time the firing rate has dropped by a factor $1/e$).

a. Sketch the bifurcation diagram with respect to parameter a.

b. Create a plot of $y(t)$ that corresponds to a population of neurons that are initially quiescent—that is, $y(0) = 0$—and are then subject to an external stimulus that has the following "switching" characteristics: $a = -3$ for $0 \le t < 20$ (units of t are milliseconds), $a = -1$ on $20 \le t \le 40$, and $a = -3$ on $40 < t \le 100$.

c. Discuss the implications of the results.

Solution

a. Solving for a in terms of y under equilibrium conditions yields:

$$\frac{1}{1 + e^{-a-6y}} = y$$

$$1 + e^{-a}e^{-6y} = \frac{1}{y}$$

$$e^{-a} = e^{6y}\left(\frac{1}{y} - 1\right)$$

$$e^{a} = e^{-6y}\left(\frac{y}{1 - y}\right)$$

$$a = -6y - \ln\left(\frac{1 - y}{y}\right)$$

Using technology, we obtain the blue curve plotted in Figure 6.48. At the point $(a, y) = (-5, 1)$, we obtain $\dfrac{dy}{dt} < 0$. Hence, $\dfrac{dy}{dt} < 0$ in the left region. At the point $(a, y) = (0, 0)$, we obtain $\dfrac{dy}{dt} > 0$. Hence, $\dfrac{dy}{dt} > 0$ in the right region. To complete the bifurcation diagram, we draw five phase lines (in red): one at each bifurcation value (i.e., $a \approx -2.5$ and $a \approx -3.5$) and one to either side of the bifurcation values. Doing so, we obtain the bifurcation diagram illustrated in Figure 6.48.

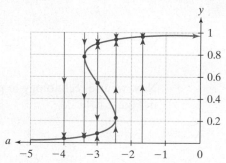

Figure 6.48 A bifurcation diagram for the Wilson-Cowan model, where the blue S-shaped curve is the equilibrium solution $\dfrac{dy}{dt} = 0$ as a function of the parameter a. The red vertical lines are selected phase lines that illustrate how the dynamics depend on a.

b. We use technology to solve the differential equation ($a = -3$):

$$\frac{dy}{dt} = -y + \frac{1}{1 + e^{3-6y}}$$

for $0 \le t < 20$ with $y(0) = 0$. This solution is shown below:

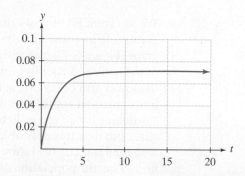

Then, as the domain shifts to $20 \le t \le 40$, we are given that $a = -1$, so we again use technology to graph a solution of

$$\frac{dy}{dt} = -y + \frac{1}{1 + e^{1-6y}} \qquad y(20) \approx 0.07$$

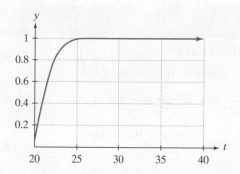

Finally, a returns to -3 for the domain $40 < t \leq 100$, and we use from the previous graph an initial value of $y(40) = 1$, to give the following graph:

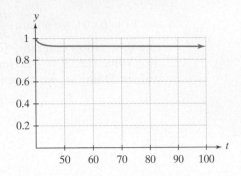

Now we use technology to put these parts together into a single graph, as shown in Figure 6.49.

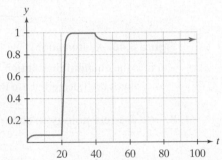

Figure 6.49 Graph of how a population of neurons records that it has been subject to a change in background firing rate a on the interval $t \in [0, 100]$ (milliseconds)

c. We see from Figure 6.49 that the population of neurons initially rises from 0 to asymptote at a low firing rate around $y = 0.07$. In terms of the bifurcation diagram (Figure 6.48), the activity rises from 0 to the equilibrium on the lower arm of the S-shaped equilibrium curve in ya space. When a increases from -3 to -1, the lower arm of equilibria in the bifurcation diagram ceases to exist and the activity rises and approaches an equilibrium close to 1, as seen in Figure 6.49. After "switching back" at $t = 40$ to the value $a = -1$, the neural firing rate starts to decline, but now it is only able to drop to the equilibrium on the upper arm of the S-shaped curve in the bifurcation diagram (Figure 6.48). By remaining at the high firing rate, the population of neurons is effectively "remembering" that the background stimulus was in a higher activity state ($a = -1$) for some period of time before switching back to the lower activity state ($a = -3$). In this way, the neuron remembers that it was once "switched on." To clear the memory, the background stimulus a would need to drop below approximately -3.5 (see Figure 6.48). ∎

PROBLEM SET 6.6

Level 1 DRILL PROBLEMS

Draw the phase lines requested in Problems 1 to 6.

1. $\dfrac{dy}{dt} = y\left(1 - \dfrac{y}{100}\right) - a; a = 0, a = 9, a = 25$

2. $\dfrac{dy}{dt} = 2y\left(1 - \dfrac{y}{1{,}000}\right) - a; a = 0, a = 180,$
$a = 600$

3. $\dfrac{dy}{dt} = 450 - ay; a = -10, a = 0, a = 10$

4. $\dfrac{dy}{dt} = (100 - y)(y - 250) - a; a = 0, a = 2{,}000,$
$a = 5{,}000$

5. $\dfrac{dy}{dt} = y^2 - ay + 1; a = 0, a = 2, a = 4$

6. $\dfrac{dy}{dt} = 2y^2 - ay + 240; \; a = 0, a = 50, a = 200$

Sketch bifurcation diagrams for the equations in Problems 7 to 12.

7. $\dfrac{dy}{dt} = ay - y^2$ 8. $\dfrac{dy}{dt} = y^2 - a$

9. $\dfrac{dy}{dt} = 1 + ay$ 10. $\dfrac{dy}{dt} = 1 - ay^2$

11. $\dfrac{dy}{dt} = \sin y + a$ 12. $\dfrac{dy}{dt} = y^2 - ay + 2$, for $a > 0$

Consider the model

$$\frac{dy}{dt} = \frac{ay^2}{k^2 + y^2} - cy$$

of an autocatalytic gene from Example 4. In Problems 13 to 16, two of the parameters are specified. Sketch a bifurcation diagram with respect to the third parameter.

13. $k = 1, c = 2$ with a as the bifurcation parameter

14. $k = 2, c = 1$ with a as the bifurcation parameter

15. $a = 10, k = 1$ with c as the bifurcation parameter

16. $a = 10, c = 1$ with k^2 as the bifurcation parameter

Consider the Wilson-Cowan model for the values of b and c specified in Problems 17 to 22. Sketch the bifurcation diagram with respect to parameter a.

17. $b = 5, c = 1$ 18. $b = 4, c = 1$

19. $b = 8, c = 1$ 20. $b = 4, c = 2$

21. $b = 8, c = 2$ 22. $b = 12, c = 2$

Level 2 APPLIED AND THEORY PROBLEMS

SIS model in Epidemiology

Mathematical epidemiologists often use these symbols: S to denote the number of individuals in a population who are susceptible to a disease, I to denote the number of people infected with the disease, and R to denote the number of individuals who have recovered and are now immune. If no individuals die then the total number of individuals in the population is $N = S + I + R$. This kind of model is called a SIR model. In the case that all individuals who recover are immediately susceptible, then $R = 0$ for all time and the model is called an SIS model. Many sexually transmitted infections (STIs)—for example, gonorrhea—do not confer immunity and are best described by SIS models. This is one reason why a single round of antibiotics, even if administered widely on a population basis, will not have a long-term effect in lowering the incidence of STIs.

Let us assume that a susceptible individual encounters and gets infected by infected individuals at a rate proportional to the densities of the product of susceptible and infected individuals in the population. Call this proportionality constant $b \geq 0$. The constant b is known as the *transmission rate* in the epidemiological literature. Also assume that individuals infected with the disease recover from the disease at a constant rate $r \geq 0$. Under these assumptions, we obtain this SIS model:

$$\frac{dI}{dt} = bIS - rI$$

Since $I + S = N$, we know $S = N - I$, so

$$\frac{dI}{dt} = bI(N - I) - rI$$

Rearranging terms yields

$$\frac{dI}{dt} = I(bN - r - bI)$$

In Problems 23 to 26 sketch a bifurcation diagram with respect to b for r = 1 and with respect to r for b = 1. Discuss under what conditions the disease persists in a population confined to living in a group of indicated size.

23. $N = 1,000$ (boarding school)

24. $N = 10,000$ (army camp)

25. $N = 100,000$ (isolated town)

26. $N = 1,000,000$ (isolated city)

Habitat destruction

Consider a population living in a patchy environment. Let y be the fraction of patches occupied by the species of interest. Let $c \geq 0$ denote the colonization rate (i.e., the rate at which individuals from one patch colonize an empty patch), $d \geq 0$ the rate at which individuals clear out of a patch, and $0 \leq D \leq 1$ the fraction of patches destroyed by humans. Then we get the following model (developed by Harvard biologists, Richard Levins and David Culver*):

$$\frac{dy}{dt} = cy(1 - D - y) - dy$$

Sketch the bifurcation diagram for this differential equation for the information given in Problems 27 to 32.

27. Assume that $D = 0$. Sketch bifurcation diagrams for d when $c = 1$ and for c when $d = 0$. Under what conditions does the population persist?

28. Assume that $D = 0$. Sketch bifurcation diagrams for d when $c = 2$ and for c when $d = 1$. Under what conditions does the population persist?

29. Assume that $D = 0.5$. Sketch bifurcation diagrams for d when $c = 1$ and for c when $d = 0.5$. Under what conditions does the population persist?

30. Assume that $D = 0.5$. Sketch bifurcation diagrams for d when $c = 2$ and for c when $d = 2$. Under what conditions does the population persist?

---

*R. Levins and D. Culver, Regional coexistence of species and competition between rare species *Proceedings of the National Academy of Sciences* 68 (1971), pp. 1246–1248.

31. Assume that $c = 3/2$. Sketch bifurcation diagrams for D when $d = 1/2$ and for d when $D = 0$. Under what conditions does the population persist?

32. Assume that $d = 2$. Sketch bifurcation diagrams for D when $c = 4$ and for c when $D = 1/2$. Under what conditions does the population persist?

Lotka-Volterra predation

In the 1920s two mathematicians, Vito Volterra (1860–1940) and Alfred Lotka (1880–1949) considered models of the density of a prey species, denoted here by the variable x, predated by a species at a density denoted here by the variable y. They wrote two differential equations, one for the prey and one for the predator, in which the prey equation included the predator density and the predator equation included the prey density. In Chapter 8 we will explore how to analyze a system of two interdependent differential equations. Here, however, with the methods covered in this chapter, we can analyze the behavior of the prey equation *or* the predator equation where the density of the other species appears as a parameter. The general form of the **prey equation** is

$$\frac{dx}{dt} = xg(x) - yh(x)$$

where $g(y)$ is a per capita growth rate of the prey species and $h(x)$ is the rate at which each unit of predator is able to extract prey. Note it is assumed that both $g(0) = 0$ and $h(0) = 0$. The general form of the **predator equation** is

$$\frac{dy}{dt} = ybh(x) - yf(y)$$

where $h(x)$ is the prey extraction rate per predator appearing in the prey equation, $0 < b < 1$ is the efficiency with which predators can convert a unit of consumed prey into their own biomass (ingestion, digestion, metabolism, etc.), and $f(y)$ is the rate at which predators die when they have no prey species to feed upon.

In Problems 33 to 35 sketch the bifurcation diagram for the specified growth and extraction functions in the prey equation in which the density y of the predators is regarded as a parameter in the prey equation, and in Problems 36 to 37 sketch the bifurcation diagram for the specified extraction and mortality functions in the predator equation in which the density x of the prey species is regarded as a parameter.

33. In one form of the Lotka-Volterra model, $g(x)$ is a decreasing linear function and $h(x)$ is proportional to x. Consistent with these assumptions, let $g(x) = 0.5\left(1 - \frac{x}{3}\right)$ and $h(x) = x$. Determine under what conditions the population persists.

34. In another form of the Lotka-Volterra model, $g(x)$ is constant and $h(x)$ is an increasing and saturating function of prey density x. Consistent with these

assumptions, let $g(x) = 0.5$ and $h(x) = \frac{x}{x + 2}$. Under what conditions is the prey population extinction bound?

35. Assume $g(x) = 0.5\left(1 - \frac{x}{3}\right)$ and $h(x) = \frac{x}{x + 2}$. Under what conditions is the prey population driven to extinction by predation?

36. Assume $b = 0.2$ and $f(y) = 1$. Under what conditions does the predator population persist when $h(x) = x$?

37. Assume $b = 0.2$ and $f(y) = y$. Under what conditions does the predator population persist when $h(x) = x$?

38. **A self-regulatory genetic network.** Smolen and colleagues[*] investigated a model of a single transcription factor, TF-A, that activates its own transcription TF-A by first forming a homodimer, which then activates transcription by binding to enhancers (TF-REs). A rapid equilibrium is assumed between monomeric and dimeric TF-A. The transcription rate saturates with TF-A dimer concentration to a maximal rate a, which is proportional to TF-A phosphorylation. Responses to stimuli are modeled by varying the degree of TF-A phosphorylation. A basal synthesis rate d is present, as well as a first-order process for degradation, $-cy$. If y denotes the concentration of TF-A then the model is given by

$$\frac{dy}{dt} = \frac{ay^2}{b + y^2} - cy + d$$

Assume that $b = 1, c = 1,$ and $d = 0.1$. Sketch a bifurcation diagram over the region $1 \le a \le 3$ and $0 \le y \le 3$. Discuss when you expect to see two stable equilibria.

39. Evolution of cooperation, part 3. Problem 31 in Section 6.5 investigated how individuals who interact frequently and respond to the strategy of their opponents can promote the evolution of cooperation. If opponents interact on average n times and cooperation gives a benefit B to the opponent and a cost C to the cooperator, then the payoff matrix for the tit-for-tat and defect strategies are as shown here:

	Tit-for-tat	Defect
Tit-for-tat	$n(B - C)$	$-C$
Defect	B	0

a. Write a replicator equation for this game.

b. Assume $B = 4$ and $C = 3$ and sketch a bifurcation diagram with respect to the parameter n.

c. Discuss the implications for the evolution of cooperation.

[*]P. Smolen, D. A. Baxter, and J. H. Byrne. Frequency, selectivity, multistability, and oscillations emerge from models of gene networks, *Am. J. Physiol.*, 274 (1998), pp. C531–C542.

40. Suppose the growth rate of a whale population at density N (individuals per million square kilometers of ocean), harvested at a rate h, is given by

$$\frac{dN}{dt} = 0.07N \left(\frac{N}{10} - 1\right)\left(1 - \frac{N}{80}\right) - h$$

where the units of t are years.

a. Sketch a bifurcation diagram with respect to the parameter h as it varies over the interval $[0, 8]$.

b. If $h = vN$, then sketch a bifurcation diagram with respect to the parameter v as it varies over the interval $[0, 0.12]$.

CHAPTER 6 REVIEW QUESTIONS

1. Determine for what values of a, b, and c,
$y(t) = e^t + at^2 + bt + c$ is a solution to $t^2 + \dfrac{dy}{dt} = y$.

2. Find the general solution to the differential equation
$\dfrac{dy}{dt} = y \tan t$.

3. In Ludwig von Bertalanffy's classic paper on modeling the growth of individuals, he modeled the growth of a species of guppy (*Lebistes reticulatus*) using the differential equation

$$\frac{dL}{dt} = k(L_\infty - L)$$

where $L(t)$ is the length of the guppy in millimeters in week t. The data and the best-fitting version of the model are shown in Figure 6.50. Given that the length at one week is 12.5 mm and at two weeks is 15.6 mm, and that the limiting length is 26.1, find $L(t)$.

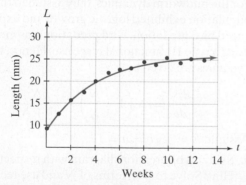

Figure 6.50 Growth of the guppy species *Lebistes reticulatus* with the best-fitting von Bertalanffy growth curve.

4. Let $\dfrac{dy}{dt} = f(y)$ where the graph of $f(y)$ is as shown here:

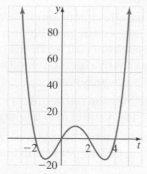

Sketch the phase line for this differential equation and sketch solutions of this differential equation with initial conditions $y(0) = -0.1$ and $y = 0.1$.

5. Find the implicit solution to the initial value problem $y\dfrac{dy}{dt} = e^{t+2y} \sin t$ where $y(0) = 0$.

6. Cooperation can be observed in natural populations. To better understand under what conditions it may evolve, imagine that each time two opponents meet they interact two times. During these interactions, individuals play one of two strategies: defect always or tit-for-tat, in which case an individual initially cooperates but switches to defecting if the opponent defected. If the benefit of interacting with a cooperating individual is B and the cost of cooperating is seven, then the payoff matrix is given by

	Tit-for-tat	Defect
Tit-for-tat	$2(B-7)$	-7
Defect	B	0

a. Write a replicator equation for this game where y is the frequency of individuals playing the tit-for-tat strategy.

b. Linearize the equation at the $y = 1$ equilibrium and determine for what values of B this equilibrium is stable.

c. Discuss the implications for the evolution of cooperation.

7. Sketch the bifurcation diagram for the differential equation $\dfrac{dy}{dt} = ry + y^3$ with respect to the parameter r.

8. Estimate a solution to the differential equation $\dfrac{dy}{dt} = 2t(t^2 - y)$ with $y(0) = 3$ over the interval $0 \le t \le 2$ using Euler's method with step size $h = 0.4$.

9. The radioactive substance gallium-67 (symbol ^{67}Ga) used in the diagnosis of malignant tumors has a half-life of 46.6 hours. If we start with 100 mg of ^{67}Ga, what percentage is lost between the 30th and 35th hours? Is this the same as the percentage lost over any other five-hour period?

10. Sketch the solution to the initial value problem $\dfrac{dy}{dt} = f(t, y)$ with $y(0) = 0.25$ for the slope field shown here:

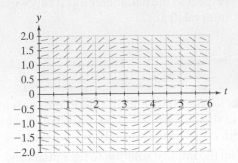

11. A population of animals on Catalina Island is limited by the amount of food available. Suppose it is shown that there were 1,800 individuals present in 1980 and 2,000 in 1986, and studies also suggest that 5,000 individuals can be supported by the conditions present on the island. Use the logistic model $\dfrac{dN}{dt} = rN(1 - N/K)$ to predict the size of the population in the year 2000.

12. In the 1960s, research scientist Anna Laird for the first time successfully used the Gompertz model to fit data of growth of tumors. Recall the Gompertz model is given by

$$\frac{dy}{dt} = -\alpha y \log(y/K)$$

where y is the size of the tumor, t is measured in hours, K is the theoretical limiting size of the tumor (never achieved), and α is a positive constant. For one of the rat data sets used by Laird, the growth of the rat's tumor was modeled using the parameter estimates $\alpha = 0.02$, $y(0) = 0.4$ grams, and $K = 9,600$ grams. Find the size of the tumor after t hours.

13. A lake has a volume of 6 billion ft^3, and its initial pollutant content is 0.22%. A river whose waters contain only 0.06% pollutants flows into the lake at the rate of 350 million ft^3/day, and another river flows out of the lake also carrying 350 million ft^3/day. Assume that the water in the two rivers and the lake is always well mixed. How long does it take for the pollutant content to be reduced to 0.15%?

14. The velocity v of a skydiver falling to the ground is governed by the equation $m\dfrac{dv}{dt} = mg - kv^2$, where m is the mass of the skydiver, g is acceleration due to gravity, and $k > 0$ is a constant proportional to air resistance. Plot the phase line for this equation and find the terminal velocity $\lim\limits_{t \to \infty} v(t)$.

15. Consider the differential equation $\dfrac{dy}{dt} = ay - y^2$. Verify that $y = 0$ is an equilibrium and use linearization to determine for what a values 0 is stable or unstable equilibrium.

16. A refined model of harvesting a logistically growing population is given by

$$\frac{dN}{dt} = rN\left(1 - \frac{N}{K}\right) - H\frac{N}{a + N}$$

where N is the population abundance in thousands of individuals, r is the population's intrinsic rate of growth in the absence of harvesting, K is the carrying capacity of the population in the absence of harvesting, H is the harvesting effort, and $a > 0$ is the half saturation constant for harvesting. Assume $r = 0.05$, $K = 100$, and $a = 1$.

   a. Sketch a bifurcation diagram for $N \geq 0$ and $H \geq 0$.

   b. What is the maximal harvesting effort H that is sustainable?

17. Spruce budworm is a serious insect pest in eastern Canada. In years with large outbreaks, these insects can defoliate and kill a major portion of balsam firs in a forest within four years. Mathematical biologist Donald Ludwig and colleagues developed an elegant model of spruce budworm and forest dynamics. For the budworm dynamics, they assumed that the population exhibited logistic growth and experienced bird predation. Bird predation was modeled using a type III functional response. Hence, the dynamics are given by

$$\frac{dN}{dt} = rN\left(1 - \frac{N}{K}\right) - \frac{bN^2}{a^2 + N^2}$$

Assume $K = 15, b = a = 1$.

   a. Sketch a bifurcation diagram with respect to r. Hint: Solve for r in terms of N and use technology to graph the resulting curve.

   b. Experimental observation suggests that for a young forest $r < 0.2$. As the forest matures, r slowly increases to 1. Discuss what increasing r slowly implies about the abundance of spruce budworms.

18. A patient requires a concentration of 2 mg per liter of a drug in his bloodstream. The clearance rate of the drug is 0.2 per hour (i.e., a half-life of ≈ 3.5 hours). The patient has five liters of blood and initially has no drug in his bloodstream.

   a. Find the infusion rate required to get a concentration of 2 mg per liter in two hours.

   b. Find the infusion rate required to maintain a concentration of 2 mg per liter in the long term.

19. Consider

$$\frac{dy}{dt} = (y - 1)(y - 4)$$

a. Sketch the slope field for this differential equation.

b. Use Euler's method with $\Delta t = 0.5$ to approximate the solution satisfying $y(0) = 0$ over the interval $0 \leq t \leq 1.5$.

c. Sketch the numerical solution over your slope field and *briefly* discuss whether the approximation is reasonable.

20. Consider the initial value problem

$$\frac{dy}{dt} = (y - 1)(y - 4) \qquad y(0) = 0$$

a. Find the solution to the initial value problem.

b. Find $\lim\limits_{t \to \infty} y(t)$ for the answer you found in part **b.**

GROUP PROJECTS

Seeing a project through on your own, or working in a small group to complete a project, teaches important skills. The following projects provide opportunities to develop such skills.

Project 6A   Modeling Diseases

In Problem Set 6.6 we explored the behavior of epidemics in a population of size N individuals. We separated these individuals, into susceptible persons, with numbers denoted by S, and infected persons, with numbers denoted by I. In our differential equation model for epidemics, infected individuals once recovered were immediately susceptible, and the relationship $N = S + I$ held throughout; that is, no new individuals entered the population (births or immigration) and no individuals left (deaths or emigration). We also raised but did not explore the possibility that individuals can be removed from the epidemic process either by recovering and becoming immune or by dying. Let the number of individuals in this removed class be represented by R.

The goal of this project is to address epidemiological questions such as these: If an infected individual is introduced into a population of susceptible individuals will an epidemic occur? If it does occur, how many people will ultimately catch the disease?

Begin with the assumptions that the disease under consideration confers permanent immunity on any individual who has completely recovered from it, and that the disease has a negligibly short incubation period. This latter assumption implies that an individual who contracts the disease becomes infected immediately afterward.

To complete this project, perform the following tasks:

• Write a system of three first-order differential equations based on the following additional assumptions:

Assumption 1: The total population remains fixed at a level N in the time interval of consideration.

Assumption 2: The rate of change of the susceptible population is proportional to the product of the number of susceptible and the number of infected.

Assumption 3: Individuals from the infected class are removed and enter the removed class at a rate proportional to the number of infected.

• Assume that $R(0) = 0$. Use the fact that $S(t) + I(t) + R(t) = N$, and special features of the $\dfrac{dS}{dt}$ and $\dfrac{dR}{dt}$ equations, to show that there exists a function $F(R)$ such that $\dfrac{dR}{dt} = F(R)$—that is, the system reduces to **one** first-order equation. Hint: Use chain rule to write dS/dR and solve the resulting differential equation to express S as a function of R.

• Show that $\dfrac{dR}{dt} = F(R)$ can be rewritten as

$$\frac{dx}{d\tau} = a - bx - e^{-x} \qquad (6.1)$$

with $a \geq 1$ and $b > 0$ by appropriately rescaling R and t to x and τ.

• Determine the number of fixed points of equation 6.1 and classify their stability.

• Show that if $b < 1$ and $x(0) = 0$, then $x'(\tau)$ is increasing from $\tau = 0$ until it reaches a maximum at some time $\tau_{\max} > 0$. Show that if $b > 1$ and $x(0) = 0$, then $x'(\tau)$ is decreasing. What do these facts imply about $I(t)$? Discuss the biological implications.

Project 6B   Save the Perch Project*

Happy Valley Pond is currently populated by yellow perch. A map of the pond is shown in Figure 6.51. This map shows at each grid point the depth of the pond in feet when the dam is at spillover level. Assume that each grid cell is 5×5 ft^2. Use this map to estimate the number of gallons of water in the pond when the water level is exactly even with the top of the spillover dam. This information will be needed to construct your model to account for the following additional facts.

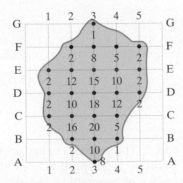

Figure 6.51 Happy Valley Pond is fed by two springs

Water flows into the pond from two springs and evaporates from the pond at rates given in the following table.

Spring	Dry season (6 months)	Rainy season (6 months)
A	50 gal/h	60 gal/h
B	60 gal/h	75 gal/h
Evaporation	110 gal/h	75 gal/h

At all times the pond is well mixed by the inflows, outflows and wind. Recently, spring B became contaminated by an underground salt deposit so that its water is a 10% salt solution, which means that 10% of a gallon of water from spring B is salt. Assume that the salt does not evaporate but is instead well mixed with the water in the pond so that the rate of salt lost is determined by the outflow rate of water and the well-mixed concentration of salt in the pond at the time of outflow.

The yellow perch in the pond are salt intolerant and start to die when the concentration of salt exceeds 1%. There was no salt in the pond before the contamination

of spring B. You and the members of your group (if you have one) have been called upon by the Happy Valley Bureau of Fisheries to try to save the perch. Unfortunately, spring B is underground and cannot be capped off, but fresh water can be piped from other sources to help dilute the salt concentration in the pond.

Assume the salt contamination in spring B started at the beginning of the dry season in 2004 ($t = 0$), when the pond was exactly even with the top of the spillover dam. Selecting the units of t to be hours, formulate a differential equation for the amount of salt in the pond at any time t after the start of the dry season in 2004. Remembering to take into account the seasonal nature of the flows. Using differential equation solving technology, draw a graph of the amount of salt in the pond over the dry season in 2004 and over the following wet season.

Now use the model to address the following questions (assume no management interventions unless specifically asked to consider them):

1. What is the equilibrium solution under persistent dry season conditions?

2. What is the equilibrium solution under persistent wet season conditions?

3. What will the salt profile look like in the long run over a combined dry and wet season? (Use your model to produce the profile for the first dry-wet season, the second dry-wet season, and so on until all consecutive profiles are identical to the desired number of decimal places.)

4. From this long-term profile, identify the periods when the perch are threatened (i.e., salt concentrations exceed 1%) and calculate how much water will need to be piped in to ensure that the perch remain safe. Whenever fresh water is piped in, it is always at the rate of 100 gallons of pure water per hour (all the Happy Valley Bureau of Fisheries Management Council can decide is when to switch the spigot on and off).

In your report include the design and analysis of a plan that can be used to ensure that the water in the pond never gets too salty for the perch. Can you come up with any interesting innovations that might help manage the salinity of the pond?

---

*Reprinted with permission from Stephen Hilbert, Diane D. Schwartz, Stan Seltzer, John Maceli and Eric Robinson, *Calculus: An Active Approach with Projects* (MAA 2010). Copyright the Mathematical Association of America 2013. All rights reserved.

CHAPTER 7

Probabilistic Applications of Integration

7.1 Histograms, PDFs, and CDFs

7.2 Improper Integrals

7.3 Mean and Variance

7.4 Bell-Shaped Distributions

7.5 Life Tables

Figure 7.1 Albert Einstein (1879–1955), best known for his formula $E = MC^2$ and for his statement "God does not play dice with the world," was awarded the Nobel Prize in Physics in 1921.

Bettmann/CORBIS

Preview

"Lest men suspect your tale untrue, Keep probability in view."*

John Gay, English poet and dramatist (1685–1732)

At the beginning of the nineteenth century, the prevailing scientific view was a clockwork universe in which everything was determinable, if not actually determined. An asteroid hitting Earth was not a chance event; rather, it could have been anticipated if the position and velocity of all asteroids in the solar system were known. This was the view of Pierre-Simon Laplace (1749–1827) whose methods on how to compute future positions of planets and comets from observations of their past positions were published in a five-volume treatise titled *Méchanique céleste*. A century later, the clockwork view of classical mechanics was shattered by twentieth century quantum mechanics, built on Werner Heisenberg's (1901–1976) uncertainty principle. This principle implies that the precision with which both the position and velocity of any object can be known at a given point in time is limited. Einstein's difficulty in accepting this principle is encapsulated in one of his most quoted statements (translated from the German): "God does not play dice with the world." Whether we believe that the time of death of an individual is preordained by God or is subject to the throw of a cosmic die is not relevant. In either case, we know that many kinds of computations in biology are intrinsically predictions of the frequencies of the occurrence of particular events. Think of flipping coins: heads can be predicted to occur only 50% of the time.

The mathematical theory of chance started when a French writer, Chevalier de Méré, with a penchant for gambling and an interest in mathematics, challenged the French mathematician Blaise Pascal (1623–1662) to solve a betting problem. Pascal teamed up with another French mathematician, Pierre de Fermat (1601–1665), to solve the problem; in the process, they laid the foundations of *probability theory*. It is no exaggeration to say that without probability theory, the biological sciences would not exist as we know them today. All concepts, ideas, and calculations relating to the theory and practice of mathematical statistics and its indispensable application to experimental biology would not exist. Thus, students in the biological sciences need to become immersed in ideas that relate to chance and probabilities as early as possible and have these ideas

*This version of the statement can be found in William Hermann's published conversations with Einstein (W. Hermann, 1983. *Einstein and the Poet: In Search of Cosmic Man*. Branden Press, Inc., Brookline Village, MA 02147, p. 58.).

reinforced as often as possible. Certain foundational aspects of probability theory rely on integration; this chapter provides an elementary introduction to these aspects of probability.

We begin with organizing data into histograms, a topic reminiscent of how we used discrete rectangles to approximate the area under a continuous curve. Integration allows us to make the transition from histograms to continuous probability density functions and cumulative distribution functions. These functions often have infinite domains (e.g., heights theoretically can take on any positive value) and, consequently, require improper integrals. Next we consider means and variances of these functions, fundamental concepts for all sciences. We then focus on applications: the probability of getting infected as a disease sweeps through a population, identification of underweight children, time to extinction of marine species, life expectancy of humans and dinosaurs, and probabilities of tumor regrowth after medical treatment.

7.1   Histograms, PDFs, and CDFs

Histograms and probabilities

In previous chapters we saw how gathering data can lead to uncovering and ultimately understanding many interesting phenomena. In the simplest case, data represent single values, each associated with an object or an individual. Some examples of sets of data we considered (or, in the case of a coin or die, familiar to us all) are listed in Table 7.1.

Table 7.1 Examples of random variables with expected ranges

Object	Variable	Type*	Range	Reference
Coin	side after flip	discrete	$H (= 1)$ or $T (= 0)$	
Die	side after throw	discrete	$\{1, 2, 3, 4, 5, 6\}$	
Power plant	CO_2 output	continuous	$[0, 3.5]$ (million tons)	Example 3, Section 1.2
Individual animal	daily calories required	continuous	$[0, 10]$ (thousand kCal/day)	Example 12, Section 1.6
Individual trout	mass (g)	continuous	$[0, 1000]$ (g)	Problem 45, Section 1.6
Recruits (salmon)	numbers	discrete	$\{0, 1, \ldots, 499, 500\}$ (individuals)	Figure 2.43
Species	biodiversity index	discrete	$[0, 14]$ (number of species)	Figure 2.58, Section 2.7
Fruit fly population	proportion Glued genes	continuous	$[0, 1]$	Example 7, Section 4.3
Temperature gauge	max temperature[†]	continuous	$[32, 74]$ (°F)	Table 5.4
Crab fishery	annual catch	continuous	$[0, 25]$ (million pounds)	Figure 5.37

* Continuous is essentially also discrete when measures are rounded to a fixed number of decimal points.

[†] Any day in March in Stockton, California.

One way of visualizing large data sets is to organize values by the proportion that fall into subintervals, known as *bins*. For example, consider the fourth data set listed in Table 7.1: If we have data on 100 animals no larger than a cow, we can ask what proportion of these 100 individuals require, respectively, $(0, 1]$, $(1, 2]$, ..., and $(9, 10]$ thousand kilocalories per day. We can then plot the proportions as a bar graph, referred to as a *histogram*.

> **Histograms**
>
> A **histogram** is a graphical representation of a data set shown as adjacent rectangles, erected over nonoverlapping intervals (bins), with the area of each rectangle equal to the fraction of data points in the interval.

Given a data set, various types of histograms can be created, depending on how we divide up the interval over which the data are defined. The most common type is obtained by splitting the range of the data into equal-sized bins, as illustrated in the following example.

Example 1   Bird diversity in oak woodlands

In spring 1994, the number of bird species in forty different oak woodland sites in California were collected. Each site was around 5 hectares in size—the equivalent of about 12 acres, or 0.019 square mile—and the sites were situated in a relatively homogeneous habitat. The numbers of bird species found in these sites are listed here:

$$
\begin{array}{cccccccccc}
37, & 21, & 26, & 27, & 21, & 21, & 28, & 22, & 22, & 26 \\
47, & 26, & 29, & 34, & 28, & 25, & 19, & 32, & 32, & 29 \\
29, & 16, & 21, & 24, & 37, & 38, & 30, & 20, & 23, & 30 \\
27, & 32, & 17, & 24, & 32, & 29, & 40, & 31, & 38, & 35
\end{array}
$$

a. Construct a histogram with two intervals corresponding to 0 to 25 species and 25 to 50 species.

b. Construct a histogram with intervals of width 10 over the domain [0, 50).

c. Use technology to construct a histogram with intervals of width 5 over the domain [0, 50).

d. Determine the units on the vertical axis of the histograms.

For this problem, assume the intervals include the left endpoint but not the right endpoint.

Solution

a. Since 13 of the 40 data points are between 0 and 24, the fraction of data points in the first interval is $13/40 = 0.325$. Since the remaining 27 data points are in the second interval, the fraction of data points in the second interval is $27/40 = 0.675$. To draw the histogram, we sketch a rectangle over the right interval that is approximately twice as high as the rectangle over the left interval. More precisely, we want the area of the left rectangle to equal 0.325. Since the base of the rectangle is of length 25, its height must be $0.325/25 = 0.013$. We want the area of the right rectangle to equal 0.675. Therefore, the height of this rectangle is $0.675/25 = 0.027$. The resulting histogram is illustrated in Figure 7.2a. Note that, by construction, the total shaded area is 1.

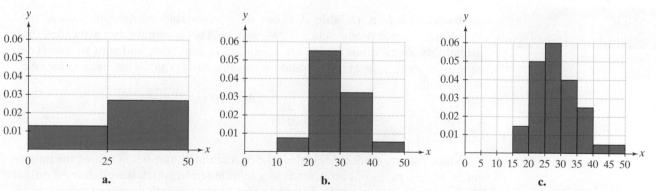

Figure 7.2 Frequencies y of species richness x in oaklands binned into histograms with **a.** two bins, **b.** five bins (first is empty), and **c.** 10 bins (first three are empty).

b. Intervals of width 10 correspond to [0, 10), [10, 20), [20, 30), [30, 40), and [40, 50). The number of data points in [0, 10) interval is 0. Hence, we draw no rectangle over this interval. The number of data points in the [10, 20) interval is 3. Hence, the fraction of data in this interval is $3/40 = 0.075$. Since the width of the interval is 10, the height of the rectangle over the [10, 20) interval should be $0.075/10 = 0.0075$. Similarly, the number of data points in the [20, 30) interval is 22, implying that the fraction of data in this interval is $22/40 = 0.55$ and that the height of the rectangle over this interval should be $0.55/10 = 0.055$. For the interval $[30, 40)$, the number of data points is 13; the fraction of data is thus $13/40 = 0.325$, so the height of the rectangle over this interval should be $0.325/10 = 0.0325$. Finally, for the interval $[40, 50)$ the number of data points is 2; the fraction of data is thus $2/40 = 0.05$, so the height of the rectangle over this interval should be $0.05/10 = 0.005$. The resulting histogram is illustrated in Figure 7.2b. Again, by construction, the total shaded area is 1.

c. Many programs exist (e.g., most spreadsheet and statistical software) that create histograms for which one can specify the size of the intervals to be plotted. Specifying intervals of width 5 in one of these programs yields Figure 7.2c.

d. Since the areas of the rectangles are unitless, the units on the vertical axes of the histogram have to be the reciprocal of the units on the horizontal axes. For the histograms in Figure 7.2, the units on the horizontal axis are number of species. Hence, the units on the vertical axes are $\dfrac{1}{\text{number of species}}$.

As we have seen in Example 1, histograms provide a sense of whether there is a center to a data set (i.e., where most of the data points lie), how much spread there is to the data set (i.e., is it even over the whole range, peaked in more than one region, or concentrated around a center), and how skewed the data set is (i.e., whether there are more data points to the left or right of the center). For example, in Figure 7.2c, the data are centered around the bin of 25–30 species and the histogram is slightly skewed to the right.

One interpretation of the area of a rectangle in a histogram is the proportion of the data in the interval. An alternative interpretation is in terms of random variables. A **random variable**, usually written X, is a variable whose possible values are outcomes of a random phenomenon, such as the rolling of a die, this evening's winning lottery numbers, or the height of a randomly chosen individual. Two important types of random variables are *discrete random variables*, which we discuss next, and *continuous random variables*, which we examine later in this section. In the discussion of random variables, we provide only simple, intuitive definitions to facilitate our presentation of probabilistic applications of integration. Introductory statistics and probability classes provide more in-depth studies of these concepts.

Discrete Random Variables

A **discrete random variable** X takes on a countable number of values, say $x_1, x_2, x_3, \ldots$, with probabilities $p_1, p_2, p_3, \ldots$. These probabilities are called the **probability distribution** for X, are nonnegative numbers, and sum to one (i.e., $p_1 + p_2 + p_3 + \cdots = 1$) as the random variable always takes on some value. We write

$$P(X = x_i) = p_i$$

to denote "$X = x_i$ with probability p_i."

When discrete random variables take on an infinite number of values, the infinite sum $p_1 + p_2 + p_3 + \cdots$ corresponds to a *infinite series*, which is well-defined only if $\lim_{n \to \infty} p_1 + p_2 + \cdots + p_n$ exists. As with the *improper integrals* discussed in Section 7.2, understanding when these limits are well-defined is a delicate issue. However, all of our examples only deal with finite number of values.

Example 2   Playing roulette

Roulette is a casino game named after the French diminutive for "little wheel." In American casinos, the roulette wheel has 38 colored and numbered pockets of which 18 are black, 18 are red, and 2 are green; a small ball moves as the wheel rotates. A player betting one dollar on red wins one dollar if the ball lands in a red pocket and loses otherwise. Let X be the earning of a player betting one dollar on red. Find the probability distribution for X.

Solution   Assuming that the ball is equally likely to land in each pocket (i.e., the wheel is fair), the probability of the ball landing in a red pocket is $18/38 = 9/19$ and the probability of landing in a black or green pocket is $20/38 = 10/19$. Hence, X equals 1 with probability $9/19$ and -1 with probability $10/19$. As with all casino games, betting on red slightly favors the house.

To see how random variables relate to data sets and their histograms, imagine writing down each data value on a slip of paper, putting all the slips in a hat, drawing one slip at random from the hat, and calling this value X. For an interval of data values $[a, b)$, we define

$$P(a \le X < b) = \text{ the probability that } X \text{ lies in the interval } [a, b)$$

For a histogram of this data set, the area of the rectangle above the interval $[a, b)$ equals $P(a \le X < b)$.

Example 3   Species diversity in a randomly chosen site

An oak woodland site is randomly selected from the forty sites presented in Example 1. Let X denote the number of bird species in that site.

a. Find and interpret the probability $P(0 \le X < 25)$.

b. Find $P(20 \le X < 30)$.

c. Find $P(20 \le X < 40)$.

Solution
a. Since the proportion of sites with less than 25 bird species is $13/40 = 0.325$, $P(0 \le X < 25) = 0.325$. In other words, there is a 32.5% chance that a randomly chosen site has fewer than 25 bird species. This value corresponds to the area over the interval $[0, 25)$ in Figure 7.2a.

b. Since the proportion of sites with at least 20 species and fewer than 30 species is $22/40 = 0.55$, $P(20 \le X < 30) = 0.55$. In other words, there is a 55% chance that a randomly chosen site has between 20 and 30 bird species. This corresponds to the area over the interval $[20, 30)$ in Figure 7.2b.

c. Since the proportion of sites with at least 20 species and fewer than 40 species is $(22 + 13)/40 = 0.875$, $P(20 \le X < 40) = 0.875$. In other words, there is a 87.5% chance that a randomly chosen site has between 20 and 40 bird species. This corresponds to the total area over the intervals $[20, 30)$ and $[30, 40)$ in Figure 7.2b.

Example 4   From histograms to probabilities

In a study involving 252 men, Garath Fisher estimated the percentage of body fat by weighing individuals underwater and taking various body circumference measurements. A histogram for the data is shown in Figure 7.3. Assume a man is randomly

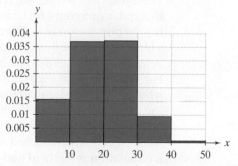

Figure 7.3 Frequencies y of percentage body fat x in a study of 225 men

selected from this study. Let X denote the percentage of body fat of this randomly selected man.

    a. Estimate $P(X < 10)$ **b.** Estimate $P(10 \leq X < 30)$ **c.** Estimate $P(X \geq 30)$

Solution

a. The area of the rectangle above the interval $[0, 10]$ of the histogram depicted in Figure 7.3 is approximately $0.015 \times 10 = 0.15$. Hence, we approximate $P(X < 10) = 0.15$. Equivalently, we estimate that 15% of the men have less than 10% body fat.

b. The area of the rectangle above the interval $[10, 20]$ of the histogram depicted in Figure 7.3 is approximately $0.037 \times 10 = 0.37$. The area of the rectangle above the interval $[20, 30]$ is approximately $0.038 \times 10 = 0.38$. Hence, we approximate $P(10 \leq X < 30) = 0.37 + 0.38 = 0.75$. Equivalently, we estimate that 75% of the men have between 10% and 30% body fat.

c. Since the sum of the areas of the rectangles must be one (i.e., the total fraction of data is one), the area of the rectangles over the intervals $[30, 40]$ and $[40, 50]$ must equal $1 - 0.75 - 0.15 = 0.10$. Hence, we approximate $P(X \geq 30) = 0.10$ or, in words, we estimate that 10% of the men in the study have greater than 30% body fat. ∎

Probability density functions

Some data sets are naturally discrete, such as the distribution of litter size among female cats of a particular age. Other data sets involving physical measurements—such as height, weight, or time—can take on a continuum of values. In the latter case, when sufficiently many measurements are taken, the histogram may be well approximated by a continuous function, as seen in the next example.

Example 5   Like father, like son

    a. A demographer decides he wants to understand the heights of fathers and their sons. Discuss how he might conduct this study.

    b. Karl Pearson (1857–1936), one of the founders of mathematical statistics, collected data on the heights of 1078 father-son pairs. Use his data (readily available on the Internet) and compare the histograms of heights for fathers and sons using 5, 10, and 20 bins. Discuss what you find.

Solution

a. The demographer would first identify the population for which he wants to understand the heights of fathers and sons. It is well known that individuals from different nationalities and groups differ on average with respect to height. For example, Tutsi men of Burundi and Rwanda are regarded as the tallest humans, averaging over 6 feet, whereas Pygmy men and women of central Africa are the shortest, averaging 4 feet 5 inches and 4 feet 6 inches, respectively. After identifying the population (e.g., Pygmies in central Africa), the demographer would take a large, randomized sample of fathers and their sons within the target population. A large sample ensures that he is unlikely to get a misleading result due to chance. For example, if he only chooses two fathers, he might get by chance two fathers who are much taller than their sons, even though this does not accurately reflect the population. Choosing individuals randomly (e.g., via a telephone directory or census data) prevents biases in data collecting. For instance, a demographer who selects only the tallest fathers (e.g., only professional basketball players) in a population is likely to find their sons are typically shorter.

b. In 1903, Pearson measured the heights of 1,078 fathers and their sons in England. Histograms of these data with 5, 10, and 20 bins are shown in Figure 7.4. Despite sons having a slightly wider range of heights, these histograms suggest that fathers and sons have quite similar distributions of heights. Later we will see that the sons are about an inch taller on average than their fathers. However, this slight difference is not readily apparent in the histograms. Figure 7.4 illustrates, quite unexpectedly, that as the number of bins is increased, the histogram is well approximated by a continuous curve.

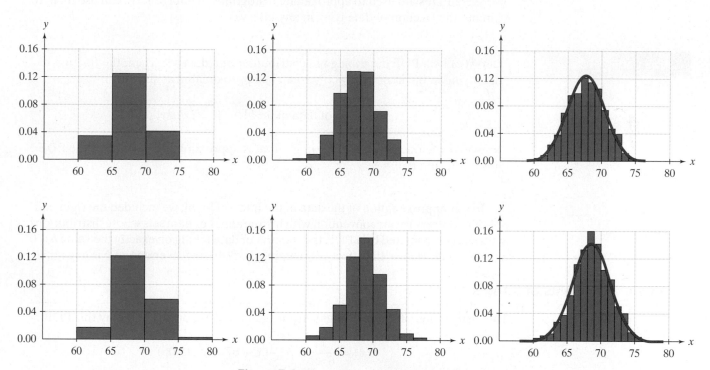

Figure 7.4 Histograms of heights x of fathers (top row) and sons (bottom row) with 5, 10, and 20 equally sized bins. The red curve is a probability density function that approximates the histogram.

The continuous function approximating the histogram in Figure 7.4 is an example of a probability density function. The graphs of these functions are the continuous analogues of histograms. Since probability density functions are used to describe distributions from data sets, they need to satisfy certain natural properties.

Probability Density Function (PDF)	A **probability density function (PDF)** is a piecewise continuous function $f(x)$ such that **1.** $f(x) \geq 0$ for all x; that is, probabilities are nonnegative **2.** total area under $f(x)$ equals one; that is, the area under the histogram equals one

Example 6   Constructing a PDF from a nonnegative function

Let a be a constant. Consider the function defined by $f(x) = ax$ for $0 \leq x \leq 5$ and $f(x) = 0$ otherwise. Determine for what value of a the function f is a PDF.

Solution   In order for f to be a PDF, f needs to be nonnegative. Hence, a must be nonnegative. The area under f must equal one. Since $f(x) = 0$ outside the interval $[0, 5]$, the area under f is given by

$$\int_0^5 ax\,dx = \frac{ax^2}{2}\bigg|_0^5 = \frac{25a}{2} = 1$$

Solving for a yields $a = 2/25 = 0.08$.

Since PDFs are used to approximate histograms of data sets, we can use them to estimate the fraction of data lying in any interval.

> ### Area Under a PDF
>
> Let $f(x)$ be a PDF describing the distribution of a data set. Then, the fraction of data lying in the interval $a \le x \le b$ can be approximated as shown:
>
> $$\text{fraction of data in } [a, b] = \int_a^b f(x)\,dx$$
>
> In words, the fraction of data in an interval is approximately the area of the PDF above this interval.

In our approximation of the data in the interval $[a, b]$, we included the right endpoint b, contrary to our convention with histograms. For data sets whose distributions are well approximated by a PDF, the fraction of data taking on exactly the value b is 0 or very small. Hence, the effect of including or excluding an endpoint b is negligible.

Example 7   Heights of fathers

The distribution of fathers' heights from Example 5 is approximated by this PDF:

$$f(x) = \frac{1}{6.88} \exp\left(-(x - 67.69)^2/15.06\right)$$

Use numerical integration to approximate the fraction of fathers whose heights are between 6 and 7 feet.

Solution   Since 6 and 7 feet correspond to 72 and 84 inches, we can estimate the fraction of fathers with heights between 6 and 7 feet by the integral

$$\int_{72}^{84} \frac{1}{6.88} \exp\left(-(x - 67.69)^2/15.06\right) dx$$

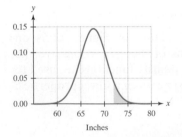

which is given approximately by 0.058. Hence, approximately 6% of the fathers in Pearson's study were between 6 and 7 feet. A graphical representation of this fraction is shown in the blue shaded region in the graph at the left.

An alternative interpretation of the area under a PDF is in terms of continuous random variables. Imagine we have a hat that contains an infinite number of (infinitely thin) slips of paper, each with different numbers such that the proportion of slips with numbers in the interval $[a, b]$ is given by

$$\int_a^b f(x)\,dx$$

where f is a PDF. Now shake this hat and, with your eyes closed, grab a slip of paper. Let X denote the value on this slip. Since X can assume any real value, it is called a *continuous random variable with PDF* $f(x)$, where the probability that X takes on a value in the interval $[a, b]$ equals $\int_a^b f(x)\,dx$.

Continuous Random Variables

A **continuous random variable** X with PDF $f(x)$ satisfies

$$P(a \leq X \leq b) = \int_a^b f(x)\,dx$$

In other words, the probability of lying in an interval equals the area of the PDF above this interval.

The simplest continuous random variable is a uniform random variable that places equal weight on all values in a given interval. For example, one might expect the birth time of babies to be approximately uniformly distributed over the year, as illustrated in Figure 7.5. However, some interesting patterns emerge, including high peaks in September and late December (could the latter be a possible synergism of tax considerations and induced/Cesarean births?) and a low trough in late April/early May. The peaks and troughs are surely climate and culture dependent.

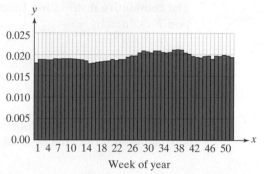

Figure 7.5 Fraction of births each week in 1978 in the United States

Example 8   Birth times

In this example, ignore the subtleties of birth dates indicated in Figure 7.5. Let's assume that the time of birth X in days after January 1st for a randomly chosen individual is equally likely to be any time of the year. Ignoring leap years, our birth-time distribution has the following PDF:

$$f(x) = \begin{cases} \dfrac{1}{365} & 0 \leq x \leq 365 \\ 0 & otherwise \end{cases}$$

a. Show that f is a PDF.

b. Compute the probability of a randomly chosen individual having a birth date in January.

Solution

a. Since $f(x) = 0$ outside of the interval $[0, 365]$, the area under $f(x)$ is

$$\int_0^{365} \frac{dx}{365} = \frac{365}{365} = 1$$

Since $f(x) \geq 0$ for all x, f is a PDF.

b. Since January comprises the first thirty-one days of the year, we obtain

$$P(0 \leq X \leq 31) = \int_0^{31} \frac{dx}{365} = \frac{31}{365} \approx 0.0849315$$

In other words, there is approximately a 8.5% chance that a randomly chosen student from your calculus class was born in January.

The PDF in Example 8 is an example of the following general class of density functions:

Uniform PDF

The **uniform PDF** on the interval $[a, b]$ is given by

$$f(x) = \begin{cases} \dfrac{1}{b-a} & \text{if } a \leq x \leq b \\ 0 & \text{elsewhere} \end{cases}$$

Cumulative distribution functions

Another way of describing a discrete or continuous random variable is with a cumulative distribution function.

Cumulative Distribution Function (CDF)

The **cumulative distribution function (CDF)** of a random variable X is the function F defined by

$$F(x) = P(X \leq x)$$

If X describes a data set, then $F(x)$ equals the fraction of data in the interval $(-\infty, x]$.

If X is a continuous random variable with PDF $f(x)$, then $F(x)$ corresponds to the area under f over the interval $(-\infty, x]$. Formally, we write this as follows:

$$F(x) = \int_{-\infty}^{x} f(t)\, dt$$

If there exists an a such that $f(t) = 0$ for $t \leq a$, then this expression simplifies to

$$F(x) = \int_{a}^{x} f(t)\, dt$$

The case for which there is no such a results in an improper integral (i.e., an integral over an infinite range) which is discussed in Section 7.2.

There are several nice properties of CDFs (opposed to PDFs). For example, if X is a random variable with a CDF F, then

$$P(a < X \leq b) = F(b) - F(a)$$

Thus, when you are given a CDF, computing probabilities is much easier as *no integration* needs to be performed. If you are only given the PDF, you are stuck with doing the integration one way or the other.

Example 9   From PDF to CDF

Consider the random variable X corresponding to the birth time of a randomly chosen individual with the PDF

$$f(x) = \begin{cases} \dfrac{1}{365} & 0 \leq x \leq 365 \\ 0 & \text{otherwise} \end{cases}$$

a. Find and plot the CDF for X.

b. Use the CDF to find the probability of X lying in January. Compare your answer to what was found in Example 8.

Solution

a. Since $f(x) = 0$ for $x \leq 0$, we obtain $F(x) = P(X \leq x) = 0$ whenever $x \leq 0$. Alternatively, for $0 < x < 365$, we obtain

$$F(x) = P(X \leq x) = P(0 \leq X \leq x) = \int_0^x \frac{dx}{365} = \frac{x}{365}$$

Finally, we have

$$F(x) = P(X \leq x) = 1$$

for $x \geq 365$. Thus,

$$F(x) = \begin{cases} 0 & \text{if } x \leq 0 \\ \dfrac{x}{365} & \text{if } 0 < x < 365 \\ 1 & \text{if } x \geq 365 \end{cases}$$

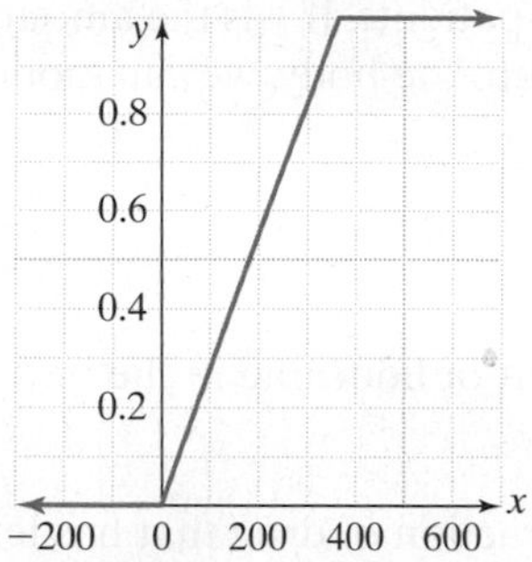

Figure 7.6 CDF for the birth time distribution

Plotting the CDF yields Figure 7.6.

b. January corresponds to the interval $[0, 31]$. The fraction of data in this interval is given by $F(31) - F(0) = \dfrac{31}{365} - 0 \approx 0.0849315$. This answer agrees with what was found in Example 8.

A CDF $F(x)$ for a random variable X is characterized by the following properties.

CDF Properties

A nonnegative function $F(x)$ is a CDF if and only if it has these properties:

1. $F(y) \geq F(x)$ whenever $y \geq x$; that is, F is a nondecreasing function
2. $\lim\limits_{x \to \infty} F(x) = 1$ and $\lim\limits_{x \to -\infty} F(x) = 0$
3. $F(x)$ is right continuous for all x; that is, $\lim\limits_{y \to x^+} F(y) = F(x)$ for all x

The first property requires that the fraction data with values less than x are nondecreasing with x. Intuitively, as x increases, one is only including more data in the interval $(-\infty, x]$; hence, the fraction cannot get smaller. The second property ensures that all the data lie somewhere on the real line. The third property ensures that there can be no jump discontinuities in $F(x)$ from the right. For discrete random variables, there always are jump discontinuities from the left. For example, if $X = 0$ with probability $\dfrac{1}{2}$ and $X = 1$ with probability $\dfrac{1}{2}$, then the CDF for X is

$$F(x) = \begin{cases} 0 & \text{if } x < 0 \\ \dfrac{1}{2} & \text{if } 0 \leq x < 1 \\ 1 & \text{if } x \geq 1 \end{cases}$$

Clearly, $F(x)$ has jump discontinuities at $x = 0, 1$. However, as expected, $\lim\limits_{x \to 0^+} F(x) = \dfrac{1}{2}$ and $\lim\limits_{x \to 1^+} = 1$. It is worth noting, as we saw in Example 9, that the CDFs for continuous random variables are continuous and, therefore, easily satisfy the third property.

CDFs can arise quite naturally from differential equation models, as the following example illustrates.

Example 10   Drug decay and the exponential CDF

Lidocaine is a common local anesthetic and antiarrhythmic drug. The elimination rate constant for lidocaine is $c = 0.43$ (per hour) for most patients. If y is the amount of drug in the body and there is no further input of drug into the body, we can model the drug dynamics by

$$\frac{dy}{dt} = -0.43\, y \qquad y(0) = y_0$$

where t denotes times in hours and y_0 is the initial amount of lidocaine in the body.

a. Solve for $y(t)$.

b. Write an expression, call it $F(t)$, that represents the fraction of drug that has left the body by time $t \geq 0$.

c. If we define $F(t) = 0$ for $t \leq 0$, verify that $F(t)$ satisfies the three properties of a CDF.

d. What is the probability that a randomly chosen molecule of drug leaves the body in the first two hours? What is the probability that a randomly chosen molecule of drug leaves the body between the second hour and fourth hour?

Solution

a. Separating and integrating yields

$$\int \frac{dy}{y} = -\int 0.43 \, dt$$
$$\ln|y| = -0.43t + C_1$$
$$y = C_2 e^{-0.43t}$$

Since $y_0 = y(0) = C_2$, we obtain $y(t) = y_0 e^{-0.43t}$.

b. The fraction of drug in the body at time t is $\dfrac{y(t)}{y_0} = e^{-0.43t}$. Hence, the fraction that has left by time t is $F(t) = 1 - e^{-0.43t}$ for $t \geq 0$.

c. Let $F(t) = 0$ for $t \leq 0$. Since $F'(t) = 0.43e^{-0.43t} > 0$ for $t \geq 0$, F is nondecreasing for all t. Since $F(t) = 0$ for $t \leq 0$, $\lim\limits_{t \to -\infty} F(t) = 0$. Since $\lim\limits_{t \to \infty} e^{-0.43t} = 0$, $\lim\limits_{t \to \infty} F(t) = 1$. Finally, since

$$\lim_{t \to 0^+} F(t) = \lim_{t \to 0^+} 1 - e^{-0.43t} = 0$$

and $F(0) = 0$ for all $t \leq 0$, $F(t)$ is continuous at $t = 0$ and, therefore, continuous for all t. Hence, F is a CDF.

d. The likelihood that a particular molecule of drug is eliminated in the first two hours is given by $F(2) \approx 0.58$. The likelihood that a particular molecule of drug is eliminated between the second and fourth hour is $F(4) - F(2) \approx 0.24$. Hence, a randomly chosen molecule of drug is much more likely to be eliminated in the first two hours than in the second two hours.   ■

Example 10 is a particular instance of the *exponential distribution* that arises in many applications. The general exponential distribution and additional applications are discussed in this chapter's problem sets, but to emphasize the importance of exponential distribution, we define it here.

Exponential PDF and CDF

The **exponential PDF** on the interval $[0, \infty)$, with rate parameter $c > 0$, is given by

$$f(x) = \begin{cases} ce^{-cx} & x \geq 0 \\ 0 & x < 0 \end{cases}$$

The corresponding **exponential CDF** is

$$f(x) = \begin{cases} 1 - e^{-cx} & x \geq 0 \\ 0 & x < 0 \end{cases}$$

Since CDFs are nondecreasing continuous functions, it is often easier to fit functions to empirically derived CDFs than to empirically derived PDFs, which might be less smooth. When fitting CDFs, however, we need to be careful—as the following example illustrates.

Example 11   Survivorship histograms and CDFs for the Mediterranean fruit fly

The Mediterranean fruit fly is one of the world's most destructive pests of deciduous fruits—such as apples, pears, and peaches—and of citrus fruits as well. Adults of both sexes may live six months or more under favorable conditions. University of California scientist James Carey and his colleagues reared Mediterranean fruit flies under laboratory conditions and daily recorded the number of adults surviving a given number of days after emerging from the pupal stage. This resulted in the following data:

Interval number	1	2	3	4	5	6	7	8	9
Interval in days	0–10	10–20	20–30	30–40	40–50	50–60	60–90	70–80	>80
Proportion that die	0.03	0.19	0.08	0.11	0.08	0.10	0.11	0.25	0.05
Cumulative proportion of dead	0.03	0.22	0.30	0.41	0.49	0.59	0.70	0.95	1.00

Data Source: James R. Carey: *Applied Demography for Biologists, with Special Emphasis on Insects.* Oxford University Press, New York and Oxford, 1993.

The histogram associated with the data is illustrated in Figure 7.7a. The cumulative proportion of dead individuals at times $0, 10, 20, \ldots 80$ is shown in Figure 7.7b. The experiment was stopped after 85 days when 3% of individuals were still alive, so we don't know at what time all the individuals died. Using technology to fit a quartic equation $F(x)$ representing the cumulative data yields (each coefficient rounded to five significant figures)

$$F(x) = 0.00059818\, x + 0.00069088\, x^2 - 0.000015411\, x^3 + 1.0672 \cdot 10^{-7}\, x^4$$

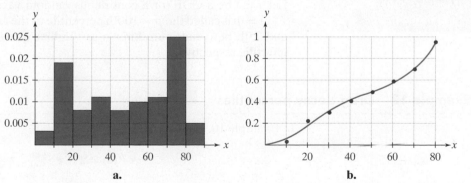

Figure 7.7 Mortality histogram and cumulative mortality distribution for the Mediterranean fruit fly

a. Use the fitted CDF $F(x)$ to estimate the probability that an individual dies before age 18 days.

b. What is the probability that an individual survives at least until age 46 days?

c. What is the probability that an individual lives beyond 100 days?

Solution

a. The probability that an individual dies before reaching 18 days is, by definition of the CDF,

$$F(18) - F(0) = 0.156$$

Hence, there is a 15.6% chance that a randomly chosen fruit fly dies before its 18th day of life.

b. The probability that an individual survives at least until age 46 days is equal to 1 minus the probability that the individual dies before reaching age 46 days:

$$1 - F(46) = 1 - 0.467 = 0.533$$

In other words, 53.3% of the fruit flies survived at least 46 days.

c. The probability that an individual lives beyond 100 days is $1 - F(101)$ (i.e., 1 minus the probability of dying by age 101). However, $F(101) = 2.335$, which clearly violates the requirement that $F(x) \leq 1$. The reason is that we only fitted the data up to day 80. We do not have sufficient data to know how to construct $F(x)$ beyond 80 days because, in the original data set, not all individuals had died at the termination of the experiment.

Example 11 suggests that for some populations, we could have a problem constructing $F(x)$ if we do not have an estimate of the maximum life span of individuals. At the beginning of this millennium, for example, the Guinness *Book of Records* reported that the oldest fully authenticated age to which any human has ever lived is attributed to a French woman, Jeanne-Louise Calment, who was born on February 21, 1875 and died at age 122 years and 164 days. Individuals who appear to be older than this are alive today, but authentication of their birth date is required for them to be listed in the Guinness book of records.

Percentiles

Using the CDF, we can define quantities called *percentiles*, which play a special role in statistics and probability.

Medians, Quartiles, and Percentiles	Let $F(x)$ be a CDF for a continuous random variable. The value of x such that $F(x) = p$ is called the $p \times 100$th percentile of the random variable. The 25th, 50th, and 75th percentiles are known as the **first quartile**, the **median**, and the **third quartile**, respectively.

Example 12   Drug decay percentiles

In Example 10, we found the CDF

$$F(t) = 1 - e^{-0.43t} \quad t \geq 0$$

which describes the fraction of lidocaine that has left the body after t hours. Find the median and 90th percentile for this CDF. Discuss what these numbers mean.

Solution To find the median, we need to solve $F(t) = 0.50$ as follows:

$$0.5 = 1 - e^{-0.43t}$$
$$e^{-0.43t} = 0.5$$
$$-0.43t = \ln 0.5$$
$$t = \ln 2/0.43 \approx 1.61$$

The median of 1.61 hours corresponds to the time when 50% of the drug has left the body.

To find the 90th percentile, we need to solve $F(t) = 0.9$ as follows:

$$0.9 = 1 - e^{-0.43t}$$
$$e^{-0.43t} = 0.1$$
$$-0.43t = \ln 0.1$$
$$t = \ln 10/0.43 \approx 5.35$$

The 90th percentile of 5.35 hours corresponds to the time when 90% of the drug has left the body. ◼

Example 13   Birth time quartiles

In Example 9, we presented

$$f(x) = \begin{cases} \dfrac{1}{365} & 0 \le x \le 365 \\[2mm] 0 & \textit{otherwise} \end{cases}$$

as the PDF of the birth time X (in days) of a randomly chosen individual in a population where births are equally likely at any time of the year. In such a population, compute the birthdays of individuals falling on the median and first and third quartiles of f.

Solution The median and first and third quartiles of f are solutions to, respectively,

$$P(0 \le X \le c) = \int_0^c \frac{dx}{365} = \frac{c}{365} = 0.5, 0.25, \text{ and } 0.75$$

which are $c = 182.5, 91.25$, and 273.75. For a non-leap year, these correspond to the first of July, the second of April, and the first of October. ◼

Example 14   An overweight baby

A medical practitioner examines a young boy of 24 months and finds that the child is 83 cm tall and weighs 14.1 kg. Use the percentile charts of the Centers for Disease Control and Preventions (CDC) to decide if the boy is heavier than normal for his height and how his height and weight relate to boys of other ages.

Solution Reading off the CDC percentile charts presented in Figure 7.8 for length and weight of boys aged 0 to 24 months, we see that 83 cm corresponds to the 10th percentile for height of a 24-month-old boy, while 14.1 kg corresponds to the 90th percentile for weight. Thus, the boy is well above the median weight for his height. His height equals the median for boys aged 19 months; the median age for his weight is not on the 0 to 24 month chart but it is about the 98th percentile for a 19th-month-old-boy.

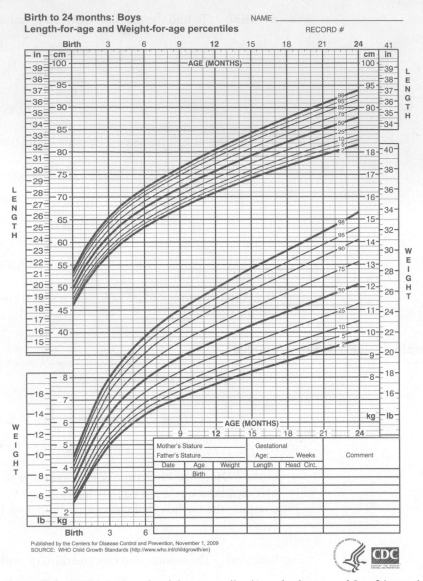

Birth to 24 months: Boys
Length-for-age and Weight-for-age percentiles

NAME _____

RECORD # _____

Published by the Centers for Disease Control and Prevention, November 1, 2009
SOURCE: WHO Child Growth Standards (http://www.who.int/childgrowth/en)

Figure 7.8 CDC length and weight percentile charts for boys aged 0 to 24 months

PROBLEM SET 7.1

Level 1 DRILL PROBLEMS

In Problems 1 to 4 construct a histogram for the given data sets.

1.

Score	Frequency
50–59	3
60–69	0
70–79	8
80–89	4
90–99	1

2.

Score	Frequency
1–10	5
11–20	8
21–30	6
31–40	10
41–50	17
51–60	15

3.

Score	Frequency
1–35	10
36–70	20
71–105	35
106–140	20
141–175	10

4.

Score	Frequency
0–99	50
100–199	45
200–299	65
300–399	75
400–499	60
500–599	50
600–699	80
700–799	75
800–899	30

5. If X denotes a score in Problem 1, find

    a. $P(50 \leq X \leq 59)$ **b.** $P(50 \leq X < 69)$

    c. $P(70 \leq X \leq 89)$ **d.** $P(90 \leq X < 100)$

6. If X denotes a score in Problem 2, find

    a. $P(1 \leq X \leq 10)$ **b.** $P(1 \leq X < 21)$

    c. $P(31 \leq X < 41)$ **d.** $P(51 \leq X \leq 60)$

7. If X denotes a score in Problem 3, find

    a. $P(X < 71)$ **b.** $P(1 \leq X < 141)$

8. If X denotes a score in Problem 4, find

    a. $P(X < 500)$ **b.** $P(X \geq 500)$

In Problems 9 to 12, find a constant a so that the given function is a PDF and find the values of x that correspond to the median, the first quartile, and the third quartile.

9. $f(x) = 2ax, 0 \leq x \leq 2$

10. $f(x) = 5ax, 1 \leq x \leq 5$

11. $f(x) = ax^2, 0 \leq x \leq 1$

12. $f(x) = 3ax^2, 1 \leq x \leq 4$

In Problems 13 to 16, use the CDC charts in Figure 7.8 to estimate the length for age and weight for age percentiles for the following boys of age a months, w kg, and l cm.

13. $a = 21$, $w = 12.5$, $l = 85$ cm

14. $a = 20$, $w = 12.2$, $l = 87$ cm

15. $a = 15$, $w = 13.4$, $l = 87$ cm

16. $a = 18$, $w = 10.9$, $l = 83$ cm

17. If $f(x) = \begin{cases} \dfrac{1}{20} & \text{if } 0 \leq x \leq 20 \\ 0 & otherwise \end{cases}$

Find and plot the CDF for this PDF.

18. If $f(x) = \begin{cases} \dfrac{1}{100} & \text{if } 0 \leq x \leq 100 \\ 0 & otherwise \end{cases}$

Find and plot the CDF for this PDF.

19. If $\dfrac{dy}{dt} = -0.25y$, $y(0) = y_0$

    a. Find $y(t)$.

    b. If we define $F(t) = 0$ for $t \leq 0$ and $F(t) = 1 - y(t)/y_0$, verify that $F(t)$ is a CDF.

    c. If X is a random variable whose CDF is given by $F(t)$, find $P(0 < X \leq 1)$.

20. If $\dfrac{dy}{dt} = -0.15y$, $y(0) = y_0$

    a. Find $y(t)$.

    b. If we define $F(t) = 0$ for $t \leq 0$ and $F(t) = 1 - y(t)/y_0$, verify that $F(t)$ is a CDF.

    c. If X is a random variable whose CDF is given by $F(t)$, find $P(0 < X \leq 1)$.

21. Consider the function $g(x)$ whose graph is shown below:

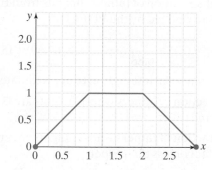

    a. For what value of c is $f(x) = cg(x)$ a PDF?

    b. For a continuous random variable with PDF $f(x)$, find $P(2 \leq X \leq 3)$.

22. Consider the function $g(x)$ whose graph is shown below:

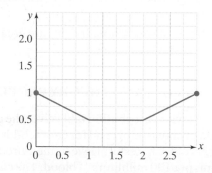

    a. For what value of c is $f(x) = cg(x)$ a PDF?

    b. For a continuous random variable with PDF $f(x)$, find $P(3 \leq X \leq 12)$.

23. For Problem 21, find an expression for the CDF and plot it.

24. For Problem 22, find an expression for the CDF and plot it.

25. Consider

$$F(x) = \begin{cases} \dfrac{x}{1+x} & \text{if } x \geq 0 \\ 0 & elsewhere \end{cases}$$

    a. Verify that $F(x)$ is a CDF.

    b. Assume X is a continuous random variable with CDF $F(x)$. Find $P(0 \leq X \leq 1)$, $P(2 \leq X \leq 10)$.

26. Consider

$$F(x) = \begin{cases} \dfrac{x^2}{1+x^2} & \text{if } x \geq 0 \\ 0 & \text{elsewhere} \end{cases}$$

a. Verify that $F(x)$ is a CDF.

b. Assume X is a continuous random variable with CDF $F(x)$. Find $P(0 \leq X \leq 1)$ and $P(2 \leq X \leq 10)$.

27. The distribution of fathers' heights from Example 5 is approximated by the PDF

$$f(x) = \frac{1}{6.88} \exp(-(x - 67.69)^2/15.06)$$

Use numerical integration to approximate:

a. Fraction of fathers whose heights are between 5 and 6 feet

b. Fraction of fathers whose heights are greater than 7 feet or less than 5 feet

28. The distribution of sons' heights from Example 5 is approximated by the PDF

$$f(x) = \frac{1}{7.06} \exp(-(x - 68.68)^2/15.84)$$

Use numerical integration to approximate:

a. Fraction of sons whose heights are between 5 and 6 feet

b. Fraction of sons whose heights are greater than 7 feet or less than 5 feet

Level 2 APPLIED AND THEORY PROBLEMS

29. The following distribution table gives the distribution of cholesterol level for 6,000 children, 4 to 19 years old. Cholesterol level is measured in milligrams per 100 milliliters of blood. The class intervals include the left endpoint but exclude the right endpoint.

Cholesterol (in mg)	Percentage
100–139	18
140–179	52
180–219	20
220–260	10

a. Sketch the histogram for the given intervals.

b. Find the probability that a randomly selected child in this group has a cholesterol level of ≥ 140.

c. Find the probability that a randomly selected child in this group has a cholesterol level between 100 and 220.

30. A study of grand juries compared the demographic characteristics of jurors with those of the general population to see if the jury panels were representative. Here are the results for age. Only persons 21 and older are considered; countywide age distribution is known from public health department data.

Age	Countywide percentage	Number of jurors
20 to 39	42	5
40 to 49	23	9
50 to 59	16	19
60 and older	19	33
Total	100	66

Sketch the histogram for countywide percentage and the number of jurors. What do you notice? For simplicity, assume that the last bin is $[60, 70)$.

31. According to the US Census Bureau's International Data Base, these were the life expectancies in 2000 for the following countries:

Country	Life expectancy	Country	Life expectancy
Argentina	75.1	Brazil	62.9
Canada	79.4	China	71.4
Colombia	70.3	Egypt	63.3
Ethiopia	45.2	France	78.8
Germany	77.4	India	62.5
Indonesia	68.0	Iran	69.7
Italy	79.0	Japan	80.7
Kenya	48.0	Korea, South	74.4
Mexico	71.5	Morocco	69.1
Pakistan	61.1	Peru	70.0
Philippines	67.5	Poland	73.2
Romania	69.9	Russia	67.2
South Africa	51.1	Spain	78.8
Turkey	71.0	Ukraine	66.0
United Kingdom	77.7	United States	77.1
Venezuela	73.1	Vietnam	69.3
Zambia	37.2		

a. Sketch a histogram with the following bins: less than 50 (plot as if on the interval $[45, 50)$), $[50, 55)$, $[55, 60)$, $[60, 65)$, $[65, 70)$, $[70, 75)$, $[75, 80)$, and greater than 80 (plot as if on the interval $[80, 85)$).

b. Say you are selecting one of the countries at random (i.e., each country is equally likely to be selected). Give the probability of getting a country with a life expectancy of

i. <60 **ii.** ≥ 70

32. Let $f(x)$ represent the PDF for the weight of a field mouse in Williamsburg, Virginia, where x is measured in grams. Express the following probabilities as integrals:

a. A randomly chosen field mouse weighs between 20 and 30 grams.

b. A randomly chosen field mouse weighs less than 40 grams.

33. Let $f(x)$ represent the PDF for the weight of a pigeon in New York City where x is measured in ounces. Express the following probabilities as integrals:

a. A randomly chosen pigeon does *not* weigh between 13 and 14 ounces.

b. A randomly chosen pigeon is in the weight class 12–15 ounces, but does not weigh between 13 and 14 ounces.

34. If you are really bad at darts, then the PDF for the distance x (in inches) that your dart is from the center of a 12-inch dartboard may be given by

$$f(x) = \begin{cases} \dfrac{x}{72} & \text{if } 0 \le x \le 12 \\ 0 & elsewhere \end{cases}$$

a. Verify that $f(x)$ is a PDF.

b. Compute the probability that your dart is more than 9 inches from the center.

c. Compute the probability that you dart is less than 3 inches from the center.

Note: This PDF assumes that you are equally likely to hit any point on the dartboard (a fact that you are asked to verify in Problem 25 of Problem Set 7.3).

35. Suppose you are a champion dart player with a PDF for the distance x (in inches) that your dart is from the center of a 12-inch dartboard given by

$$f(x) = \begin{cases} 1 - \dfrac{x}{2} & \text{if } 0 \le x \le 2 \\ 0 & elsewhere \end{cases}$$

a. Verify that $f(x)$ is a PDF.

b. Compute the probability that your dart is more than 1 inch from the center.

c. Compute the probability that your dart is between $1/4$ and $1/2$ inch from the center.

36. According to Thomson and colleagues,* the elimination constant for lidocaine for patients with hepatic impairment is 0.12 per hour. Hence, for a patient who has received an initial dose of y_0 mg, the lidocaine

level $y(t)$ in the body can be modeled by the differential equation

$$\frac{dy}{dt} = -0.12y \qquad y(0) = y_0$$

a. Solve for $y(t)$.

b. Write an expression, call it $F(t)$, that represents the fraction of drug that has left the body by time $t \ge 0$.

c. If $F(t) = 0$ for $t \le 0$, verify that $F(t)$ is a CDF.

d. What is the probability that a randomly chosen molecule of drug leaves the body in the first two hours?

e. What is the probability that a randomly chosen molecule of drug leaves the body between the second hour and fourth hour?

37. Consider a drug that has an elimination rate constant of c. If y is the amount of drug in the body and there is no further input of drug into the body, we can model the drug dynamics by

$$\frac{dy}{dt} = -cy \qquad y(0) = y_0$$

where t denotes times in hours and y_0 is the initial amount of drug in the body.

a. Solve for $y(t)$.

b. Write an expression, call it $F(t)$, that represents the fraction of drug that has left the body by time $t \ge 0$.

c. If $F(t) = 0$ for $t \le 0$, verify that $F(t)$ is a CDF.

d. Find an expression that allows one to calculate for any value $c > 0$ and times $0 < r < s$ what proportion of the drug is removed on the interval $[r, s]$.

38. In the early 1960s, Robert MacArthur of Princeton University and Edward O. Wilson of Harvard University developed a theory to explain why big islands generally have more species than smaller islands, and why the numbers of species on islands of similar sizes are inversely related to island distance from continental landmasses. They argued that the number of species on an island represents a dynamic balance between the rate at which new species arrive at that island and the rate at which species on the island go extinct. The simplest model of island biodiversity assumes that the rate of change of the number N of species is given by a constant rate I of immigration of new species from the mainland and that species on the island go extinct at a rate proportional to N.

*P. D. Thomson, K. L. Melmon, J. A. Richardson, et al., "Lidocaine Pharmacokinetics in Advanced Heart Failure, Liver Disease, and Renal Failure in Humans," *Ann Intern Med* 78 (1973): 499–508.

If the proportionality constant is c, then we obtain

$$\frac{dN}{dt} = I - cN$$

where t denotes time in years. To know what the species immigration rate I might be for a particular island, we need to know the number of species on the mainland that serve as a source for the colonization process. On the other hand, the extinction rate c on each island is a characteristic of the island alone, rather than of the surrounding mainlands and the distance of the island to these mainlands. To understand the likelihood a species already on the island has gone extinct by time t, we can ignore the immigration process (i.e., keep track only of the species currently on the island) and consider the model

$$\frac{dN}{dt} = -cN$$

a. Solve for $N(t)$.

b. Write an expression, $F(t)$, for the fraction of species that has gone extinct by year t.

c. Donald Levin, a botany professor at the University of Texas, Austin, was quoted in *Science Daily*: "Roughly 20 of the 297 known mussel and clam species and 40 of about 950 fishes have perished in North America in the last century."[*] Use these data to approximate the extinction constants c for mussel and clam species and for fish species.

d. Using your estimates from part **b**, estimate the probability that a specific clam or mussel species goes extinct in the next decade.

e. Using your estimates from part **b**, estimate the probability that a specific fish species goes extinct in the next decade.

[*]Posted January 10, 2002 at http://www.sciencedaily.com/releases/2002/01/020109074801.htm

7.2   Improper Integrals

As we saw in the previous section in defining cumulative distribution functions, we often encounter integrals in which the limits of integration are not finite. These *improper integrals* come in three varieties:

$$\int_a^\infty f(x)\,dx \qquad \int_{-\infty}^a f(x)\,dx \qquad \int_{-\infty}^\infty f(x)\,dx$$

where a is a real number. In this section, we discuss when these integrals are well defined.

One-sided improper integrals

What is the area under the graph of $y = e^{-x}$ for $x \geq 0$? At first, one might reason this way: Since the region under the curve goes on forever, the area is infinite. To evaluate this statement, define $A(t)$ to be the area under e^{-x} from $x = 0$ to $x = t$, as illustrated in Figure 7.9. In other words,

$$A(t) = \int_0^t e^{-x}\,dx$$

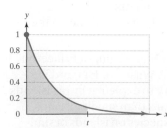

Figure 7.9 Area under e^{-x} from $x = 0$ to $x = t$

Computing $A(t)$ yields

$$A(t) = \int_0^t e^{-x}\,dx = -e^{-x}\Big|_0^t = 1 - e^{-t}$$

$A(t)$ is always less than 1 for any $t > 0$. Therefore, the area under e^{-x} for $x \geq 0$ cannot be infinite. In fact, it is natural to define the area under e^{-x} for $x \geq 0$ to be

$$\lim_{t\to\infty} A(t) = \lim_{t\to\infty} 1 - e^{-t} = 1$$

Thus, even though the curve is of infinite length, the area under this curve is finite. Our first guess was wrong!

Inspired by this example, we propose the following definition.

Convergent and Divergent Improper Integrals

For any given real number a define

$$\int_a^\infty f(x)\,dx = \lim_{t\to\infty} \int_a^t f(x)\,dx$$

When the limit exists and is finite, $\int_a^\infty f(x)\,dx$ is **convergent**, otherwise it is **divergent**. Similarly, define

$$\int_{-\infty}^a f(x)\,dx = \lim_{t\to-\infty} \int_t^a f(x)\,dx$$

When the limit exists and is finite, $\int_{-\infty}^a f(x)\,dx$ is **convergent**, otherwise it is **divergent**.

Example 1   Convergent versus divergent

Determine whether the following integrals are convergent or divergent. If convergent, determine their value.

a. $\displaystyle\int_2^\infty \frac{dx}{x^2}$ 　　　　 **b.** $\displaystyle\int_2^\infty \frac{dx}{x}$ 　　　　 **c.** $\displaystyle\int_0^\infty \sin x\,dx$

Solution

a. For any $t \geq 2$,

$$\int_2^t \frac{dx}{x^2} = -\frac{1}{x}\Big|_2^t = \frac{1}{2} - \frac{1}{t}$$

Taking the limit yields

$$\lim_{t\to\infty} \int_2^t \frac{dx}{x^2} = \lim_{t\to\infty} \frac{1}{2} - \frac{1}{t} = \frac{1}{2}$$

Hence, $\displaystyle\int_2^\infty \frac{dx}{x^2}$ is convergent and equals $\dfrac{1}{2}$.

b. For any $t \geq 2$,

$$\int_2^t \frac{dx}{x} = \ln x\Big|_2^t = \ln t - \ln 2$$

Since

$$\lim_{t\to\infty} \int_2^t \frac{dx}{x} = \lim_{t\to\infty} [\ln t - \ln 2] = \infty$$

$\displaystyle\int_2^\infty \frac{dx}{x}$ is divergent.

c. For any $t \geq 0$,

$$\int_0^t \sin x\,dx = -\cos x\Big|_0^t = 1 - \cos t$$

Since

$$\lim_{t\to\infty} \int_0^t \sin x\,dx = \lim_{t\to\infty} (1 - \cos t)$$

does not exist (i.e., the values oscillate between 0 and 2), $\int_0^\infty \sin x\,dx$ is divergent.

Example 1 shows that while the curves $\dfrac{1}{x}$ and $\dfrac{1}{x^2}$ are similar (i.e., both decreasing to zero as x goes to ∞), the areas under these curves are infinitely different: $\dfrac{1}{x}$ encloses an infinite area for $x \geq 2$, while $\dfrac{1}{x^2}$ encloses a finite area for $x \geq 2$. Figure 7.10 shows that $\dfrac{1}{x}$ decreases to zero much slower than $\dfrac{1}{x^2}$. This observation suggests the following question: How fast does the function have to approach zero to ensure convergence? The following example formulates a precise answer to this question for *p-integrals*.

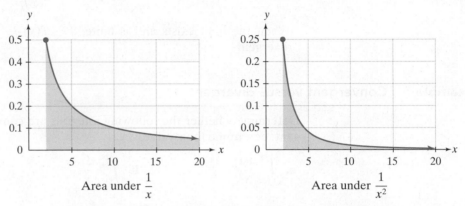

Figure 7.10 Area under curves that go to zero at different rates

Example 2   p-integrals

Determine for which $p > 0$, the integral

$$\int_1^\infty \frac{dx}{x^p}$$

is convergent and divergent.

Solution   Example 1 dealt with the case of $p = 1$ and found $\displaystyle\int_1^\infty \frac{dx}{x}$ to be divergent. Assume that $p \neq 1$. In which case

$$\int_1^t \frac{dx}{x^p} = \frac{1}{1-p} x^{1-p} \Big|_1^t = \frac{1}{1-p}(t^{1-p} - 1)$$

When $p > 1$, it follows that t^{1-p} has a negative exponent and

$$\lim_{t \to \infty} \frac{1}{1-p}(t^{1-p} - 1) = \frac{1}{p-1}$$

Hence, $\displaystyle\int_1^\infty \frac{dx}{x^p}$ is convergent if $p > 1$. When $p < 1$, it follows that t^{1-p} has a positive exponent and

$$\lim_{t \to \infty} \frac{1}{1-p}(t^{1-p} - 1) = \infty$$

Hence, $\displaystyle\int_1^\infty \frac{dx}{x^p}$ is divergent if $p < 1$.   ∎

Example 2 illustrates that convergence depends subtly on the speed at which $f(x)$ approaches zero as x approaches ∞. For instance, while $\dfrac{1}{x^{1.0001}}$ seems to go to zero only slightly faster than $\frac{1}{x}$, the integral $\int_1^\infty \frac{dx}{x^{1.0001}}$ is convergent (i.e.,

$p = 1.0001 > 1$) while the integral $\int_1^\infty \frac{dx}{x}$ is divergent. Although this might appear shocking at first, notice that the former integral converges to a very large value: $\frac{1}{1.0001 - 1} = 10{,}000$. More generally, as $p > 1$ approaches 1 from above, the area under $\frac{1}{x^p}$ approaches ∞ because $\lim_{p \to 1^+} \frac{1}{p - 1} = +\infty$.

The p-integrals are related to the *Pareto distribution*, named after the Italian economist Vilfredo Pareto (1848–1923). Pareto originally used this power distribution to describe the allocation of wealth among individuals; it also has been used to describe social, scientific, geophysical, and many other types of observable phenomena. In the next example, we examine the Pareto distribution and its use to describe the frequency of individuals visiting websites.

Example 3   The Pareto Distribution

The PDF for the Pareto distribution is of the form

$$f(x) = \begin{cases} 0 & \text{if } x < 1 \\ Cx^{-p} & \text{if } x \geq 1 \end{cases}$$

where $p > 1$ and C is a constant that you will determine.

a. Determine for what value of C, $f(x)$ is a PDF. Your answer will depend on $p > 1$.

b. Find the CDF for the Pareto distribution.

c. A scientist at Hewlett Packard's Information Dynamics Lab used the Pareto distribution to describe how many AOL users visited various websites on one day in 1997 (which is "ancient time" for the Internet). The data are shown in Figure 7.11 (from L. A. Acamic and B. A. Huberman (2002) Zipf's law and the internet. Glottometrics 3: 143–150) and conform to a Pareto distribution with $p = 2.07$. Estimate the fraction of websites that received visits from ten or fewer AOL users.

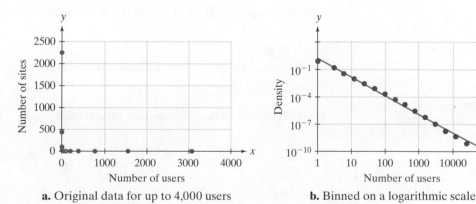

a. Original data for up to 4,000 users   **b.** Binned on a logarithmic scale

Figure 7.11 Number of websites visited by different numbers of AOL users.

Solution

a. We need to compute the area under $f(x)$. Since $f(x) = 0$ for $x \geq 1$, the area under f is given by

$$\int_1^\infty \frac{C}{x^p}\,dx$$

Since $k > 1$, Example 2 implies that

$$C \int_1^\infty \frac{1}{x^p} \, dx = \frac{C}{p-1}$$

Hence, in order for f to be a PDF, we need that $C = p - 1$ and we get

$$f(x) = \frac{p-1}{x^p}$$

for $x \geq 1$.

b. The CDF is given by

$$F(x) = \int_{-\infty}^x f(t) \, dt$$

Since $f(x) = 0$ for $x < 1$, we get $F(x) = 0$ for $x < 1$ and for $x > 1$,

$$F(x) = \int_1^x \frac{p-1}{t^p} \, dt$$

$$= -t^{1-p} \Big|_1^x$$

$$= 1 - x^{1-p}$$

Thus, the CDF is given by

$$F(x) = \begin{cases} 1 - x^{1-p} & \text{for } x \geq 1 \\ 0 & \text{otherwise} \end{cases}$$

c. If $p = 2.07$, then the fraction of websites visited by ≤ 10 AOL users can be approximated by

$$F(10) = 1 - 10^{-1.07}$$

$$\approx 0.9149$$

Hence, approximately 91.5% of the websites were visited by ten or fewer users.

Example 3 leads us to the following definition.

Pareto PDF and CDF

The PDF of the **Pareto distribution** for parameter $p > 1$ is defined to be

$$f(x) = \begin{cases} 0 & \text{if } x < 1 \\ \dfrac{p-1}{x^p} & \text{if } x \geq 1 \end{cases}$$

Its CDF is

$$F(x) = \begin{cases} 0 & \text{if } x < 1 \\ 1 - \dfrac{1}{x^{p-1}} & \text{if } x \geq 1 \end{cases}$$

Example 3 also illustrates how we go from PDFs to CDFs by integrating over the interval $(-\infty, x)$. Conversely, suppose that you are given a CDF for a continuous random variable. How do you find the associated PDF? The following theorem says all you have to do is differentiate. Hence, *integrate to go from PDF to CDF and dif-ferentiate to go from CDF to PDF.*

Theorem 7.1   Fundamental theorem of PDFs

Suppose that f is a probability density function. Then the CDF

$$F(x) = \int_{-\infty}^{x} f(s)\,ds$$

satisfies

$$F'(x) = f(x)$$

Proof.   The proof of this theorem follows from the fundamental theorem of calculus presented in Section 5.4 by letting $a \to -\infty$ in the statement of that theorem.   ∎

Example 4   Exponential distribution revisited

Recall that in Example 10 of Section 7.1 we considered a model of the decay of lido-caine in the human body. We found that the fraction of molecules of this drug that have been eliminated by $t \geq 0$ hours is given by

$$F(t) = 1 - e^{-0.43t} \quad \text{for } t \geq 0$$

and $F(t) = 0$ for $t \leq 0$.

a. If $F(t)$ represents the CDF of the random variable of how long a randomly chosen lidocaine molecule spends in the body, then find the corresponding PDF of this random variable.

b. Use the PDF in part **a** to find the probability that a randomly chosen molecule of this drug is eliminated in the first two hours. Compare your answer to what was found in Example 10 of Section 7.1.

Solution

a. The derivative of $F(t)$ for $t > 0$ is $F'(t) = 0.43e^{-0.43t}$. The derivative of $F(t)$ for $t < 0$ is $F'(t) = 0$. Hence, the PDF is given by

$$f(t) = \begin{cases} 0.43e^{-0.43t} & \text{if } t \geq 0 \\ 0 & \text{elsewhere} \end{cases}$$

b. Let X be the random variable whose CDF is given by $F(t)$. X corresponds to the time a randomly chosen molecule of drug gets eliminated. Using the PDF, we obtain

$$P(0 \leq X < 2) = \int_{0}^{2} f(t)\,dt$$

$$= \int_{0}^{2} 0.43e^{-0.43t}\,dt$$

$$= -e^{-0.43t}\Big|_{0}^{2}$$

$$= 1 - e^{-0.86}$$

$$\approx 0.58$$

Hence, there is a 58% chance that a randomly chosen molecule of drug gets elim-inated in the first two hours. This is the same answer we found in Example 10 of Section 7.1.   ∎

Convergence tests

As we have seen, the integral of a function cannot be always expressed in terms of elementary functions (e.g., $f(x) = e^{-x^2}$). One way to get around this issue is to numerically estimate $\int_a^b f(x)\, dx$. However, if the upper limit is $+\infty$, then numerical estimates only make sense if the integral converges because, informally,

$$\int_a^\infty f(x)\, dx \approx \int_a^b f(x)\, dx$$

if the left-hand side is convergent and b is sufficiently large. Consequently, it is important to have methods that determine whether an improper integral is convergent or not. A powerful yet simple test for convergence is the comparison test. The basic idea of this test is to compare the integral in question (the one for which convergence is not understood) to an integral for which convergence is understood.

Theorem 7.2   Comparison test

Suppose that $f(x) \geq g(x) \geq 0$ for $x \geq a$. Then:

Convergence If $\int_a^\infty f(x)\, dx$ is convergent, then $\int_a^\infty g(x)\, dx$ is convergent.

Divergence If $\int_a^\infty g(x)\, dx$ is divergent, then $\int_a^\infty f(x)\, dx$ is divergent.

The idea behind this theorem, as illustrated in Figure 7.12, is intuitive. If the area under f is finite and $f \geq g \geq 0$, then the area under g is finite. Conversely, if the area under g is infinite, then the area under f is infinite.

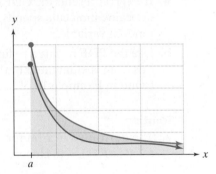

Figure 7.12 Comparing areas under $f(x)$ (in blue) and $g(x)$ (in red) when $f(x) \geq g(x) \geq 0$

Example 5   Using comparison test

Use the comparison test to determine whether the following integrals are convergent or divergent.

a. $\displaystyle\int_1^\infty \frac{2 + \sin x}{x^2}\, dx$ **b.** $\displaystyle\int_2^\infty \frac{dx}{x + \sqrt{x}}$ **c.** $\displaystyle\int_0^\infty e^{-x^2}\, dx$

Solution
a. Since $1 \leq 2 + \sin x \leq 3$ for all x,

$$0 \leq \frac{2 + \sin x}{x^2} \leq \frac{3}{x^2}$$

for all $x > 0$. Moreover, since $\displaystyle\int_1^\infty \frac{3}{x^2}\, dx = 3\int_1^\infty \frac{dx}{x^2}$ is convergent (i.e., a p-integral with $p > 1$), the comparison test implies that $\displaystyle\int_1^\infty \frac{2 + \sin x}{x^2}\, dx$ is convergent.

b. Since $x \geq \sqrt{x}$ for all $x \geq 1$, we have $x + \sqrt{x} \leq 2x$ for $x \geq 1$. Hence,

$$\frac{1}{x + \sqrt{x}} \geq \frac{1}{2x}$$

for $x \geq 1$. Since $\dfrac{1}{2} \displaystyle\int_2^\infty \dfrac{dx}{x}$ is divergent (i.e., a p-integral with $p = 1$), the comparison test implies that $\displaystyle\int_2^\infty \dfrac{dx}{x + \sqrt{x}}$ is divergent.

c. Since $x^2 \geq x$ for all $x \geq 1$, $e^{-x^2} \leq e^{-x}$ for all $x \geq 1$. Since $\displaystyle\int_1^\infty e^{-x}\,dx = \dfrac{1}{e}$ is convergent, the comparison test implies that $\displaystyle\int_1^\infty e^{-x^2}\,dx$ is convergent. Moreover, as $e^{-x^2} \leq 1$, we have that $\displaystyle\int_0^1 e^{-x^2}\,dx$ is finite. Hence, $\displaystyle\int_0^\infty e^{-x^2}\,dx = \int_0^1 e^{-x^2}\,dx + \int_1^\infty e^{-x^2}\,dx$ is convergent.

Improper integrals can lead to maddening paradoxes, as the following example illustrates.

Example 6   Torricelli's trumpet (or Gabriel's horn)

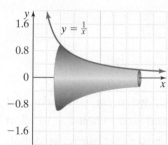

Figure 7.13 $y = \dfrac{1}{x}$ for $x \geq 1$ rotated about the x axis

Consider the surface created by rotating the curve $y = \dfrac{1}{x}$ about the x axis as illustrated in Figure 7.13.

This surface is sometimes called Torricelli's trumpet, named after the Italian mathematician Evangelista Torricelli (1608–1647). It can be shown that the volume of this infinite trumpet is given by the expression

$$\int_1^\infty \frac{\pi}{x^2}\,dx$$

and the surface area is given by the expression

$$\int_1^\infty \frac{2\pi}{x}\sqrt{1 + \frac{1}{x^2}}\,dx$$

a. Determine whether the surface area and volume are convergent or divergent.

b. Discuss how much paint it would take to paint the surface versus how much paint it would take to fill the trumpet.

Solution

a. Since the volume is determined by a p-integral with $p = 2 > 1$, we can say the volume is finite. In fact,

$$\int_1^\infty \frac{\pi}{x^2}\,dx = \lim_{t \to \infty} \int_1^t \frac{\pi}{x^2}\,dx = \lim_{t \to \infty}\left(\pi - \frac{\pi}{t}\right) = \pi$$

For the surface area, we have $\sqrt{1 + \dfrac{1}{x^2}} \geq 1$ for $x > 0$ and $\dfrac{2\pi}{x}\sqrt{1 + \dfrac{1}{x^2}} \geq \dfrac{2\pi}{x}$ for $x > 0$. Since $\displaystyle\int_1^\infty \dfrac{2\pi}{x}\,dx$ is a p-integral with $p = 1$, the comparison test implies that $\displaystyle\int_1^\infty \dfrac{2\pi}{x}\sqrt{1 + \dfrac{1}{x^2}}\,dx$ is divergent. Therefore, the surface area is infinite!

b. As the surface area is infinite, it would take an infinite amount of paint to paint the surface. On the other hand, if we plugged up the hole at the end of trumpet, then we could fill the trumpet with a finite amount of paint. After being poured out the same paint would cover the interior surface of the trumpet. How can this be? Paraphrasing the words of Thomas Hobbes: To make sense of this conundrum, one needs to be mad rather than a geometrician or a logician. ■

Two-sided improper integrals

We conclude this section by defining

$$\int_{-\infty}^{\infty} f(x)\, dx$$

A first attempt at this definition might be

$$\lim_{t \to \infty} \int_{-t}^{t} f(x)\, dx$$

Unfortunately this definition is flawed, as the following example illustrates.

Example 7   When definitions go wrong

Compute the integral $\int_{-\infty}^{\infty} 2x\, dx$ using the definition

$$\int_{-\infty}^{\infty} 2x\, dx = \lim_{t \to \infty} \int_{k-t}^{k+t} 2x\, dx$$

for any value of k and discuss any anomalies that arise.

Solution

$$\int_{k-t}^{k+t} 2x\, dx = x^2 \Big|_{k-t}^{k+t} = (k+t)^2 - (k-t)^2 = 4kt$$

Hence, if we believe that $\int_{-\infty}^{\infty} 2x\, dx$ is well defined, we must conclude that $\lim_{t \to \infty} \int_{-t}^{t} 2x\, dx$ equals 0 (when $k = 0$), ∞ (when $k > 0$), and $-\infty$ (when $k < 0$) all at the same time! Since this is clearly impossible, we have shown that our naïve definition of a doubly infinite integral is flawed. ■

To avoid the faulty path taken in Example 7, we need to avoid simultaneously confounding two computations involving ∞ (one in the positive direction and one in the negative direction). We avoid this by separating out the two computations, as presented in the following property box:

Doubly Infinite Integrals

$\int_{-\infty}^{\infty} f(x)\, dx$ is *convergent* if these limits exist:

$$\lim_{t \to -\infty} \int_{t}^{0} f(x)\, dx \quad \text{and} \quad \lim_{t \to \infty} \int_{0}^{t} f(x)\, dx$$

otherwise, $\int_{-\infty}^{\infty} f(x)\, dx$ is *divergent*. If convergent, we define

$$\int_{-\infty}^{\infty} f(x)\, dx = \lim_{t \to -\infty} \int_{t}^{0} f(x)\, dx + \lim_{t \to \infty} \int_{0}^{t} f(x)\, dx$$

In Problem Set 7.2, you will be asked to show that for convergent integrals

$$\int_{-\infty}^{\infty} f(x)\, dx = \lim_{t \to -\infty} \int_{t}^{a} f(x)\, dx + \lim_{t \to \infty} \int_{a}^{t} f(x)\, dx$$

for any a. Hence, for convergent integrals, we cannot make the infinite from nothing as in Example 7.

Example 8   Convergence of doubly improper integrals

Determine whether the following integrals are convergent or divergent.

a. $\displaystyle\int_{-\infty}^{\infty} 2x \, dx$ **b.** $\displaystyle\int_{-\infty}^{\infty} xe^{-x^2} \, dx$

The signed area for each of these curves is shown in Figure 7.14.

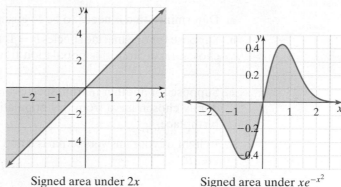

Signed area under $2x$ Signed area under xe^{-x^2}

Figure 7.14 Signed areas for Example 8

Solution

a. Since $\int_0^t 2x \, dx = x^2\big|_0^t = t^2$, $\int_0^{\infty} 2x \, dx = \lim_{t \to \infty} t^2 = \infty$ is divergent. Hence, $\int_{-\infty}^{\infty} 2x \, dx$ is divergent.

b. To compute $\int xe^{-x^2} \, dx$, we use the substitution $u = x^2$, $du = 2x \, dx$. Then,

$$\int xe^{-x^2} \, dx = \int e^{-u} \frac{du}{2} = -\frac{e^{-u}}{2} + C$$

Therefore,

$$\int_0^{\infty} xe^{-x^2} dx = \lim_{t \to \infty} \int_0^t xe^{-x^2} dx$$

$$= \lim_{t \to \infty} -\frac{e^{-x^2}}{2}\bigg|_0^t$$

$$= \lim_{t \to \infty} \left(\frac{1}{2} - \frac{e^{-t^2}}{2} \right) = \frac{1}{2}$$

Similarly,

$$\int_{-\infty}^0 xe^{-x^2} dx = \lim_{t \to -\infty} \left(\frac{e^{-t^2}}{2} - \frac{1}{2} \right) = -\frac{1}{2}$$

Hence,

$$\int_{-\infty}^{\infty} xe^{-x^2} dx = \int_{-\infty}^0 xe^{-x^2} dx + \int_0^{\infty} xe^{-x^2} dx = -\frac{1}{2} + \frac{1}{2} = 0$$

Many PDFs have bi-infinite tails (i.e., $f(x) > 0$ for all $x \in (-\infty, \infty)$). One such example is the Laplace distribution.

Example 9   The Laplace distribution

An important distribution discovered by the French mathematician and astronomer Pierre-Simon Laplace (1749–1827) is the double exponential or *Laplace distribution* whose probability density function is given by $f(x) = ae^{-b|x|}$ where $b > 0$ is a parameter and $a > 0$ is a constant that you will determine. As the Laplace distribution describes the random motion of a particle in a liquid with a constant settling rate, it has been used to describe dispersal of marine larvae along a coastline. Let X denote the distance (say, in kilometers) a larva has traveled northward from its birthplace. If X is negative, then the larva has traveled south.

a. Determine what a needs to be to ensure that f is a probability density function.

b. Suppose for one marine species $b = 1$. Determine the probability that a randomly chosen larva from this population travels more than 1 km north from its birthplace.

c. Suppose for another marine species $b = 2$. Determine the probability that a randomly chosen larva from this population travels more than 1 km north from its birthplace.

d. In light of your answers to parts **b** and **c**, provide an interpretation of the parameter b.

Solution

a. We need that $\int_{-\infty}^{\infty} ae^{-b|x|}\, dx = 1$. Computing the first half of this improper integral leads to

$$\int_{-\infty}^{0} ae^{bx}\, dx = \lim_{t \to -\infty} \int_{t}^{0} ae^{bx}\, dx \qquad \text{using the fact that } |x| = -x \text{ for } x \leq 0$$

$$= \lim_{t \to -\infty} \frac{a}{b} e^{bx}\Big|_{t}^{0} \qquad \text{using the substitution } u = bx$$

$$= \lim_{t \to -\infty} \frac{a}{b}(1 - e^{bt}) = \frac{a}{b}$$

Computing the second half of this improper integral leads to

$$\int_{0}^{\infty} ae^{-bx}\, dx = \lim_{t \to \infty} \int_{0}^{t} ae^{-bx}\, dx \qquad \text{using the fact that } |x| = x \text{ for } x \geq 0$$

$$= \lim_{t \to \infty} -\frac{a}{b} e^{-bx}\Big|_{0}^{t} \qquad \text{using the substitution } u = bx$$

$$= \lim_{t \to \infty} \frac{a}{b}(1 - e^{-bt}) = \frac{a}{b}$$

Hence, $\int_{-\infty}^{\infty} ae^{-b|x|}dx = 2\frac{a}{b}$. Since we need that $2\frac{a}{b} = 1$, we obtain $a = \frac{b}{2}$.

b. If $b = 1$ (the units of b are km^{-1}), then the fraction of larvae that travel at least 1 km north is given by

$$\int_{1}^{\infty} \frac{1}{2} e^{-x}\, dx = \lim_{t \to \infty} \int_{1}^{t} \frac{1}{2} e^{-x}\, dx$$

$$= \lim_{t \to \infty} -\frac{1}{2} e^{-x}\Big|_{1}^{t}$$

$$= \lim_{t \to \infty} -\frac{1}{2} e^{-t} + \frac{1}{2} e^{-1}$$

$$= \frac{1}{2} e^{-1} \approx 0.1839$$

Hence, there is approximately an 18% chance that a randomly chosen larvae travels at least 1 km north. The area corresponding to this integral is illustrated in the following figure:

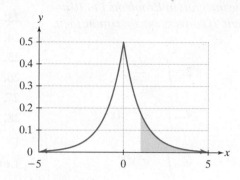

c. If $b = 2$, then the fraction of larva that travel at least 1 km north is given by

$$\int_1^\infty e^{-2x}\, dx = \lim_{t\to\infty} \int_1^t e^{-2x}\, dx$$

$$= \lim_{t\to\infty} -\frac{1}{2} e^{-2x}\Big|_1^t$$

$$= \lim_{t\to\infty} -\frac{1}{2} e^{-2t} + \frac{1}{2} e^{-2}$$

$$= \frac{1}{2} e^{-2} \approx 0.0677$$

Hence, there is approximately a 7% chance that a randomly chosen larva travels at least 1 km north.

d. The larger b is the more likely it is that a randomly chosen larva travels a shorter distance before settling. In fact, redoing parts **b** and **c** with an arbitrary b, we find that the chance of a randomly chosen larva moving at least 1 km north is $\frac{1}{2} e^{-b}$.

Given the importance of the Laplace distribution, we conclude by providing a general definition for this distribution.

Laplacian PDF and CDF

The **Laplacian PDF** on the interval $[0, \infty)$, with distance parameter $b > 0$, is given by

$$f(x) = \frac{be^{-|b|x}}{2} \qquad -\infty < x < \infty$$

The corresponding **Laplacian CDF** is

$$f(x) = \begin{cases} \dfrac{e^{bx}}{2} & x \le 0 \\[2mm] \dfrac{1}{2} + \dfrac{e^{-bx}}{2} & x \ge 0 \end{cases}$$

The derivation of the Laplace CDF is left to Problem 32 in Problem Set 7.2.

PROBLEM SET 7.2

Level 1 DRILL PROBLEMS

Determine whether the integrals in Problems 1 to 10 are convergent or divergent. If convergent, determine their value.

1. $\displaystyle\int_4^\infty \frac{dx}{x^2}$

2. $\displaystyle\int_{-\infty}^{-1} \frac{dx}{x^4}$

3. $\displaystyle\int_{-\infty}^0 \frac{dx}{1-x}$

4. $\displaystyle\int_0^\infty e^{-2x}\,dx$

5. $\displaystyle\int_0^\infty e^x\,dx$

6. $\displaystyle\int_{-\infty}^0 e^x\,dx$

7. $\displaystyle\int_0^\infty x^2 e^{-x}\,dx$

8. $\displaystyle\int_{-\infty}^0 x^2 e^{-x}\,dx$

9. $\displaystyle\int_{-\infty}^\infty x^2 e^{-x}\,dx$

10. $\displaystyle\int_{-\infty}^\infty \frac{e^x}{(1+e^x)^2}\,dx$

For Problems 11 to 14, use the comparison test to determine whether the integrals are convergent or divergent.

11. $\displaystyle\int_1^\infty \frac{dx}{1+e^x}$

12. $\displaystyle\int_2^\infty \frac{dx}{\sqrt{x^2-2}}$

13. $\displaystyle\int_1^\infty \frac{\cos^2 x}{1+x^2}\,dx$

14. $\displaystyle\int_1^\infty \frac{dx}{x^{1.01}+2}$

For Problems 15 to 18, find the CDF of the given PDF.

15. $f(x) = \dfrac{e^x}{(1+e^x)^2}$

16. $f(x) = \dfrac{1}{2}e^{-|x|}$

17. $f(x) = \dfrac{1}{x^2}$ for $x \geq 1$ and $f(x) = 0$ otherwise

18. $f(x) = \dfrac{1}{(1+x)^2}$ for $x \geq 0$ and $f(x) = 0$ otherwise

For Problems 19 to 22, find the PDF of the given CDF.

19. $F(x) = \dfrac{1}{1+e^{-x}}$

20. $F(x) = 0$ for $x \leq 1$ and $F(x) = 1 - \dfrac{1}{x}$ for $x \geq 1$

21. $F(x) = e^x$ for $x \leq 0$ and $F(x) = 1$ for $x \geq 0$

22. $F(x) = e^{-e^{-x}}$

In many species of birds, the age at which individuals die is assumed to follow an exponential distribution. In Problems 23 to 28, estimate the age of death corresponding to the first, second (median), and third quartiles of the distribution for the named species and given rate-of-death parameter b (units are per year rates: data from

D. B. Botkin and B. S. Miller, "Mortality Rates and Survival of Birds," American Naturalist, 108(1974): 181–192).

23. Blue tit, $b = 0.72$

24. Starling, $b = 0.52$

25. Grey heron, $b = 0.31$

26. Alpine swift, $b = 0.18$

27. Yellow-eyed penguin , $b = 0.10$

28. Royal albatross, $b = 0.06$ (an average of two different estimates)

Level 2 APPLIED AND THEORY PROBLEMS

29. Estimate the numerical value of $\int_0^\infty e^{-x^2}$ by writing it as the sum of $\int_0^4 e^{-x^2}\,dx$ and $\int_4^\infty e^{-x^2}\,dx$. Approximate the first integral using Simpson's rule with $n = 8$. Show that the second integral is smaller than 0.0000001. Hint: Compare to $\int_4^\infty e^{-4x}\,dx$.

30. Determine how large a needs to be to ensure that
$$\int_a^\infty \frac{dx}{1+x^3} < 0.01$$
Hint: Compare to $\displaystyle\int_a^\infty \frac{dx}{x^3}$.

31. If $\int_{-\infty}^\infty f(x)\,dx$ is convergent, show that
$$\int_{-\infty}^0 f(x)\,dx + \int_0^\infty f(x)\,dx = \int_{-\infty}^a f(x)\,dx + \int_a^\infty f(x)\,dx$$
for all a.

32. Show that
$$F(x) = \begin{cases} \dfrac{e^{bx}}{2} & x \leq 0 \\[2mm] 1 + \dfrac{e^{-bx}}{2} & x \geq 0 \end{cases}$$
is the CDF to PDF
$$f(x) = b\frac{e^{-|b|x}}{2} \qquad -\infty < x < \infty$$

33. Consider a marine species whose larvae disperse northward and southward according to the Laplace distribution $f(x) = e^{-2|x|}$. For all the individuals that travel in both directions:

  a. Determine the fraction of individuals that travel north at least 2 km.

  b. Determine the fraction of individuals that travel south at least 2 km.

  c. Determine the fraction of individuals that travel at most 2 km north.

34. Consider a large pine tree situated in a long, narrow valley. Suppose the seeds of this pine tree disperse up and down the valley according to the Laplace distribution $f(x) = \dfrac{1}{4}e^{-|x|/2}$.

a. What proportion of all seeds disperse up the valley at least 1 km.

b. What proportion of seeds disperse down valley at least 3 km.

c. Determine the fraction of seeds that disperse at most 2 km in either direction of the tree.

35. The following graph shows a plot on a log-log scale of the frequency of trips of different durations (in hours) made by albatrosses (data from A. M. Edwards et al., "Revisiting Levy Flight Search Patterns of Wandering Albatrosses, Bumblebees and Dear," *Nature* 449 (2007): 1044–1048).

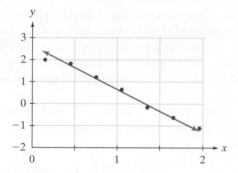

From this plot deduce the proportion of trips that lie between 5 and 20 hours.

36. Journal Problem *College Mathematics Journal* 24 (September 1993): 343. Peter Lindstrom reported that a student handled an ∞/∞ form as follows:

$$\int_1^{+\infty} (x-1)e^{-x}dx = \int_1^{+\infty} \frac{x-1}{e^x}dx$$

$$= \int_1^{+\infty} \frac{1}{e^x}dx \qquad l'Hopital's\ rule$$

$$= \frac{1}{e}$$

What is wrong, if anything, with this student's solution?

37. Historical Quest

Evangelista Torricelli (1608–1647)

Evangelista Torricelli studied in Galileo's home near Florence. When Galileo died, Torricelli succeeded his teacher as mathematician and philosopher for the Grand Duke of Tuscany, their friend and patron.

Torricelli described his amazement at discovering an infinitely long solid with a surface that calculates to have an infinite area, but a finite volume: "It may seem incredible that although this solid has an infinite length, nevertheless none of the cylindrical surfaces we considered has an infinite length but all of them are finite." (See http://curvebank.calstatela.edu/torricelli/torricelli.htm.)

In Example 6, we introduced Torricelli's trumpet and paraphrased Thomas Hobbes, that is, to solve Torricelli's conundrum one needs to be mad rather than a geometrician or a logician. Present an argument that resolves this paradox.

38. Historical Quest

Pierre-Simon Laplace (1749–1827)

Courtesy of Karl Smith

Newton and Leibniz have been credited with the discovery of calculus, but much of its development was due to the mathematicians Pierre-Simon Laplace, Joseph-Louis Lagrange, and Karl Gauss.

These three great mathematicians of calculus were contrasted by W. W. Rouse Ball*:

"The great masters of modern analysis are Lagrange, Laplace, and Gauss, who were contemporaries. It is interesting to note the marked contrast in their styles. Lagrange is perfect both in form and matter, he is careful to explain his procedure, and though his arguments are general they are easy to follow. Laplace, on the other hand, explains nothing, is indifferent to style, and, if satisfied that his results are correct, is content to leave them either with no proof or with a faulty one. Gauss is exact and as elegant as Lagrange, but even more difficult to follow than Laplace, for he removes every trace of the analysis by which he reached his results, and strives to give a

*W. W. Rouse Ball, *A Short Account of the History of Mathematics*, London: Macmillan, 1893.

proof which while rigorous will be as concise and synthetical as possible."

Pierre-Simon Laplace taught Napoleon Bonaparte, who appointed him for a short time as France's Minister of Interior. Today, Laplace is best known as a major contributor to probability, taking it from gambling to a true branch of mathematics. He was one of the earliest to evaluate the improper integral

$$I = \int_{-\infty}^{+\infty} e^{-x^2} dx$$

which plays an important role in the theory of probability. Use the web to find the value of this improper integral and its applications in mathematics, particularly probability theory.

7.3   Mean and Variance

As we have seen in Section 7.1, histograms provide visual summaries of large data sets. Sometimes these histograms are nicely approximated by the graph of a continuous function, the probability density function (PDF). When this occurs, a scientist can describe concisely his or her data set to another scientist by describing the PDF. Many important PDFs can be expressed in terms of families of functions whose parameters provide some basic information about the shape of the PDF. These parameters are often related to the *mean* and *variance* of the PDF. The mean is a measurement of the centrality of a data set. The variance, on the other hand, describes the spread of the data set around the mean; the greater the variance, the greater the spread in the data.

Means

There are numerous ways to characterize the central tendency in a set of data, including several ways of defining the concept of a mean—the most important being the **arithmetic mean**. For a data set $\{x_1, \ldots, x_n\}$ of real numbers, the **arithmetic mean**, or **average**, is given by the sum of the data values divided by the number of data values; namely, $\dfrac{x_1 + x_2 + \cdots + x_n}{n}$. It is useful to express this well-known expression in a slightly different manner, as illustrated in the following example.

Example 1   Computing the mean

In its May 1995 issue, the journal *Condor* published a study of competition for nest holes among collared flycatchers, a species of bird. Researchers collected data by periodically inspecting nest boxes located on the island of Gotland in Sweden. The accompanying data give the number of flycatchers breeding at fourteen distinct plots.

$$5 \quad 4 \quad 3 \quad 2 \quad 2 \quad 1 \quad 1 \quad 1 \quad 1 \quad 0 \quad 0 \quad 0 \quad 0 \quad 0$$

a. Find the arithmetic mean of this data set.

b. Consider the random variable X given by randomly selecting a data point. Find the probability distribution of X, that is, p_i such that $P(X_i = x_i) = p_i$ for $x_1 = 0$, $x_2 = 1$, $x_3 = 2$, $x_4 = 3$, $x_5 = 4$, and $x_6 = 5$.

c. Express the arithmetic mean in terms of the p_i and x_i in part **b**.

Solution

a. The arithmetic mean is given by

$$\frac{0+0+0+0+0+1+1+1+1+2+2+3+4+5}{14} = \frac{20}{14} = \frac{10}{7} \approx 1.42857$$

Hence, on average there are approximately 1.4 fly catchers breeding in a plot.

b. The data values are $x_1 = 0$, $x_2 = 1$, $x_3 = 2$, $x_4 = 3$, $x_5 = 4$, and $x_6 = 5$. The fraction of zeros is $\dfrac{5}{14}$. The fractions of ones and twos, respectively, are $\dfrac{4}{14}$ and $\dfrac{2}{14}$. The fractions of 3s, 4s, and 5s are all $\dfrac{1}{14}$. Hence, $p_1 = \dfrac{5}{14}$, $p_2 = \dfrac{4}{14}$, $p_3 = \dfrac{2}{14}$, and $p_4 = p_5 = p_6 = \dfrac{1}{14}$.

c. From part **a**, we rewrite the arithmetic mean as

$$\frac{(0+0+0+0+0)+(1+1+1+1)+(2+2)+3+4+5}{14}$$

$$=\frac{0\cdot 5+1\cdot 4+2\cdot 2+3\cdot 1+4\cdot 1+5\cdot 1}{14}$$

$$=0\cdot\frac{5}{14}+1\cdot\frac{4}{14}+2\cdot\frac{2}{14}+3\cdot\frac{1}{14}+4\cdot\frac{1}{14}+5\cdot\frac{1}{14}$$

From our work in part **b**, we get that the arithmetic mean equals

$$p_1 x_1 + p_2 x_2 + p_3 x_3 + p_4 x_4 + p_5 x_5 + p_6 x_6$$

Example 1 motivates the following more general definition.

Mean for a Discrete Random Variable	Consider a discrete random variable X that takes on the values $x_1, x_2, x_3, \ldots, x_k$ with probabilities $p_1, p_2, p_3, \ldots, p_k$. The **mean of X** equals $$\mu = p_1 x_1 + p_2 x_2 + \cdots + p_k x_k = \sum_{i=1}^{k} p_i x_i$$

One inspiration for this definition is Blaise Pascal's statement that the excitement felt by a gambler is equal to the amount he might win times the probability of winning it. From a gambling perspective, if $x_1, \ldots, x_k$ are the amounts you can win and $p_1, \ldots, p_k$ are the likelihoods of winning these amounts, then the mean is what you expect to win. Each term in the sum corresponds to the "amount you might win times the probability of winning it."

Example 2   How much do you expect to win playing roulette?

Recall from Example 2 of Section 7.1, the roulette wheel in American casinos has 38 colored and numbered pockets of which 18 are black, 18 are red, and 2 are green. A player betting one dollar on red wins one dollar if the ball lands in a red pocket and loses otherwise. Let X be the earning of a player betting one dollar on red. Find the mean of X and discuss its meaning.

Solution   We have X taking on the values $x_1 = 1$ and $x_2 = -1$ with probabilities $\dfrac{18}{38} = \dfrac{9}{19}$ and $\dfrac{20}{38} = \dfrac{10}{19}$, respectively. Hence, the mean of X equals

$$p_1 x_1 + p_2 x_2 = \frac{9}{19}\cdot 1 + \frac{10}{19}\cdot(-1) = -\frac{1}{19} \approx -0.052$$

Thus, an individual playing roulette on "average" loses about a nickel. One way to interpret this statement is to say that if 1,000 individuals bet 1 dollar on red, the casino will earn about $1{,}000 \cdot 0.052 = 52$ dollars.

Now suppose X is a continuous random variable with PDF $f(x)$. To find the mean of X, we may consider approximating X by a discrete random variable by dividing the real line into intervals of length Δx with endpoints:

$$\ldots,\ x_{-2} = -2\Delta x,\ x_{-1} = -\Delta x,\ x_0 = 0,\ x_1 = \Delta x,\ x_2 = 2\Delta x, \ldots$$

The probability of X taking on a value in the interval $[x, x + \Delta x)$ is approximately $f(x)\Delta x$. Using the definition of the mean for a discrete random variable as motivation, the mean of X should be approximately given by the sum of the values weighted by their probabilities, that is,

$$\cdots + x_{-2} f(x_{-2})\Delta x + x_{-1} f(x_{-1})\Delta x + x_0 f(x_0)\Delta x + x_2 f(x_2)\Delta x + \cdots = \sum_{k=-\infty}^{\infty} x_k f(x_k)\Delta x$$

Taking the limit as Δx goes to zero yields the integral $\int_{-\infty}^{\infty} x f(x)\, dx$, which suggests the following definition.

Mean for a Continuous Random Variable	For a continuous random variable X with PDF $f(x)$, the **mean** of X is given by $$\int_{-\infty}^{\infty} x\, f(x)\, dx$$ provided that the improper integral is convergent.

Example 3   Throwing darts

Sebastian is a terrible dart player. In his honor, the local pub has created a large dartboard with a radius of 2 feet. With this dartboard, Sebastian always hits the board, but his dart is equally likely to hit any point on the board. Let X be the distance from the center that the dart lands. In Problem 25 of Problem Set 7.3, you are asked to show that the PDF for X is given by

$$f(x) = \begin{cases} \dfrac{x}{2} & \text{if } 0 \leq x \leq 2 \\ 0 & elsewhere \end{cases}$$

a. Find the mean distance that Sebastian's darts land from the center.

b. Find the probability that a dart lands less than the mean distance from the center.

Solution

a. To find the mean, we compute

$$\int_{-\infty}^{\infty} x f(x)\, dx = \int_0^2 \frac{x^2}{2}\, dx$$
$$= \frac{x^3}{6}\Big|_0^2$$
$$= \frac{4}{3}$$

So on average, a dart thrown by Sebastian lands $\frac{4}{3}$ feet from the center.

b. To find $P\left(X \leq \dfrac{4}{3}\right)$, we compute

$$\int_{-\infty}^{4/3} f(x)\, dx = \int_0^{4/3} \frac{x}{2}\, dx$$
$$= \frac{x^2}{4}\Big|_0^{4/3}$$
$$= \frac{4}{9} \approx 0.444$$

Hence, there is less than a 50% chance that Sebastian's dart will land within $\frac{4}{3}$ feet of the center, even though $\frac{4}{3}$ is the mean distance of all shots from the center of the dartboard. ■

Example 4   Exponential means

Consider a drug with elimination constant c. The fraction of drug left after t hours has an exponential distribution with parameter c. As illustrated in Example 10 of Section 7.1, the PDF for this distribution is given by

$$f(t) = \begin{cases} 0 & \text{if } t < 0 \\ ce^{-ct} & \text{if } t \geq 0 \end{cases}$$

a. Find the mean of the exponential distribution. What is its interpretation in the context of drug decay?

b. For the typical patient, lidocaine has an elimination constant of 0.43 per hour. What is the mean time for a molecule to leave? What is the half-life in the body of a lidocaine molecule?

Solution

a. The mean of the exponential distribution is given by

$$\int_{-\infty}^{\infty} t f(t) \, dt = \int_{0}^{\infty} tce^{-ct} \, dt$$

$$= \lim_{s \to \infty} \int_{0}^{s} tce^{-ct} \, dt$$

$$= \lim_{s \to \infty} \left[-te^{-ct} \Big|_{0}^{s} + \int_{0}^{s} e^{-ct} \, dt \right] \qquad \textit{using integration by parts with } u = t \textit{ and } dv = ce^{-ct} dt$$

$$= \lim_{s \to \infty} \left[-te^{-ct} - \frac{1}{c}e^{-ct} \Big|_{0}^{s} \right]$$

$$= \lim_{s \to \infty} \left[-se^{-cs} - \frac{1}{c}e^{-cs} + \frac{1}{c} \right]$$

$$= \frac{1}{c} \text{ days}$$

Since c has units "per day", $\frac{1}{c}$ has units "days" and corresponds to the mean number of days it takes for a molecule of drug to be cleared from the body.

b. The mean elimination time for lidocaine is $\dfrac{1}{0.43} \approx 2.33$ days. On the other hand, the half-life is given by the solution to

$$\frac{1}{2} = e^{-0.43t}$$

$$\ln \frac{1}{2} = -0.43t$$

$$\frac{1}{0.43} \ln 2 = t$$

$$1.61 \approx t$$

Hence, half of the molecules are eliminated before the mean time to elimination. ■

As discussed in Problem Set 7.1, exponential distribution can be used to model extinction times for species.

Example 5   Extinction rates

In their article, "Extinction Rates of North American Freshwater Fauna" (*Conservation Biology* 13 (1999): 1220–1222), Anthony Ricciardi and Joseph B. Rasmussen showed that time to extinction of a species is exponentially distributed, with 0.1% of terrestrial and marine animals going extinct per decade.

a. What is the elimination constant c for this data set? What is the mean extinction time?

b. How long is it expected to take for half the species to go extinct?

c. Ricciardi and Rasmussen estimated future extinction rates by assuming all currently imperiled species (i.e., endangered or threatened) will not survive this century. Under this assumption, 0.8% of species would be going extinct per decade. Determine how this alters the answers to parts **a** and **b**.

Solution

a. If species extinctions are exponentially distributed and time x is measured in years, then the data of Ricciardi and Rasmussen tell us that

$$0.1\% = 0.001 = 1 - e^{-10c}$$

Solving for c yields

$$c = 0.00010005$$

From part **a** of Example 4, we get that the mean time to extinction for a species is $\frac{1}{c} = 9{,}995$ years.

b. To determine the half-life of the extinction process, we need to solve

$$0.5 = 1 - e^{-0.00010005\,t}$$

for t which yields $6{,}928.01$ years.

c. Solving

$$0.008 = 1 - e^{-10c}$$

for c yields $c = 0.000803217$. The mean to extinction shrinks by a factor of approximately 8 to $\frac{1}{c} \approx 1{,}245$ years. Solving

$$0.5 = 1 - e^{-0.000803217\,t}$$

for t yields a half-life of 863 years, which is the expected time it will take for half the currently extant species to go extinct. ■

Examples 3 and 4 illustrate that the fraction of data to the left of the mean can be significantly greater than 50%. This raises two questions. First, what is the geometrical interpretation of the mean? To answer this question, imagine that we take a (infinitely) long board and cut out the area lying under the PDF. If we placed this wooden PDF as shown in Figure 7.15 on a fulcrum at the mean, then the PDF would balance perfectly. Second, for what type of PDFs is 50% of the area to the left (and to the right) of the mean? A partial answer to this question is provided in the following example, using the concept of a symmetrical function and an odd function. Recall from Chapter 1 that an odd function $g(x)$ satisfies $g(x) = -g(-x)$ for all x.

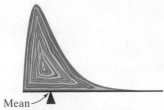

Mean

Figure 7.15 PDF with fulcrum at the mean

Example 6 Symmetrical PDFs

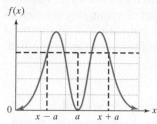

Figure 7.16 A symmetric PDF $f(x)$ around a satisfies $f(x + a) = f(x - a)$ for all x.

Let $f(x)$ be a PDF that is symmetric around $x = a$. In other words, $f(x)$ is a PDF and $f(a + x) = f(a - x)$ for all x as illustrated in Figure 7.16. If the mean associated with $f(x)$ is well defined, we expect it to equal a, as the PDF should balance at this point. To verify this assertion analytically, assume the mean is well defined (i.e., $\int_{-\infty}^{\infty} x f(x)\, dx$ is convergent) and do the following:

a. Show that the mean is zero if $a = 0$.

b. Show that $g(x) = f(x + a)$ is a PDF.

c. For $a \neq 0$, use parts **a** and **b** and the change of variables $u = a + x$ to find the mean.

Solution

a. Assume $a = 0$. Then, f is symmetric around 0; namely, $f(-x) = f(x)$ for all x. For any $b > 0$,

$$\int_{-b}^{b} x f(x)\, dx = \int_{-b}^{0} x f(x)\, dx + \int_{0}^{b} x f(x)\, dx$$

$$= \int_{b}^{0} u f(u)\, du + \int_{0}^{b} x f(x)\, dx \qquad \textit{change of variables } u = -x$$

$$= -\int_{0}^{b} u f(u)\, du + \int_{0}^{b} x f(x)\, dx \qquad \textit{integral properties}$$

$$= 0$$

Since we have assumed that $\int_{-\infty}^{\infty} x f(x)\, dx$ is convergent, the mean equals

$$\int_{-\infty}^{\infty} x f(x)\, dx = \lim_{b \to \infty} \int_{-b}^{b} x f(x)\, dx = 0$$

b. Let $g(x) = f(a + x)$. We have that $g(x)$ is nonnegative for all x as f is nonnegative for all x. Furthermore, using the change of variables $u = a + x$,

$$\int_{-\infty}^{\infty} f(a + x)\, dx = \int_{-\infty}^{\infty} f(u)\, du = 1$$

Therefore, $g(x)$ is a PDF. In Problem Set 7.3, you are asked to show that if X is a random variable with PDF $f(x)$, then the PDF of $X - a$ is $g(x)$.

c. Let $g(x) = f(a + x)$. By our assumption of symmetry around $x = a$, $g(-x) = f(a - x) = f(a + x) = g(x)$. In other words, g is symmetric around $x = 0$. Using the change of variables $u = x - a$, we get the mean of the PDF $f(x)$:

$$\int_{-\infty}^{\infty} x f(x)\, dx = \int_{-\infty}^{\infty} (u + a) f(u + a)\, du$$

$$= \int_{-\infty}^{\infty} (u + a) g(u)\, du$$

$$= \int_{-\infty}^{\infty} u g(u)\, du + a \int_{-\infty}^{\infty} g(u)\, du$$

$$= 0 + a \int_{-\infty}^{\infty} g(u)\, du \qquad \textit{part } \textbf{a} \textit{ applied to } g(x)$$

$$= a \qquad \textit{part } \textbf{b} \textit{ implies } \int_{-\infty}^{\infty} g(u)\, du = 1$$

In summary, the preceding example proves that for symmetric PDFs with a convergent mean, the mean corresponds to the point of symmetry of the PDF. Furthermore, by symmetry, the area under the PDF below the point of symmetry is equal to the area under the PDF above the point symmetry. Each contains half of the total area, so that, by definition, the point of symmetry is the 50th percentile, or median. Thus, we have the following result:

Means and Medians of Symmetrical Random Variables

A continuous random variable X with PDF $f(x)$ is **symmetric around** $x = a$ if $f(a - x) = f(a + x)$ for all x. A discrete random variable X is **symmetric around** $x = a$ if $P(X = a - x) = P(X = a + x)$ for all x. In either case, if the mean of X is well defined, then the mean of X and the median of X equal a.

Example 7 Means of symmetric PDFs

Assuming the means are well defined, find the means of the following probability densities.

a. Birthday PDF:

$$f(x) = \begin{cases} \dfrac{1}{365} & \text{if } 0 \leq x \leq 365 \\ 0 & elsewhere \end{cases}$$

b. Triangular PDF:

$$f(x) = \begin{cases} x & \text{if } 0 \leq x \leq 1 \\ 2 - x & \text{if } 1 \leq x \leq 2 \\ 0 & elsewhere \end{cases}$$

c. Laplacian PDF:

$$f(x) = \frac{b}{2} e^{-b|x|}$$

Solution

a. Since $f(x)$ is symmetric around $x = \dfrac{365}{2}$, the mean of the birthday distribution is $\dfrac{365}{2} = 182.5$.

b. Since the triangular distribution is symmetric around $x = 1$, $x = 1$ is the mean.

c. Since the Laplacian distribution is symmetric around $x = 0$, 0 is the mean. ∎

Sometimes the mean of a random variable can be infinite or not well defined, as the following example illustrates.

Example 8 Pareto distribution revisited

In Section 7.2, we introduced the Pareto PDF

$$f(x) = \begin{cases} 0 & \text{if } x < 1 \\ \dfrac{a - 1}{x^a} & \text{if } x \geq 1 \end{cases}$$

with parameter $a > 1$.

a. Find for what values of $a > 1$ the Pareto PDF has a finite mean; and when it is finite, find the mean.

b. In an article in *Contemporary Physics*, Michael Newman found that the distribution of population sizes (ten thousands) of cities in the United States in the year 2000 was well approximated by a Pareto PDF with $a = 2.18471$, as illustrated in Figure 7.17. Estimate the mean population size of a city.

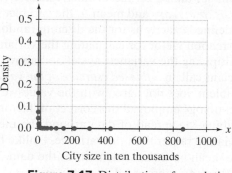

Figure 7.17 Distribution of population sizes for cities in the United States in the year 2000. In the right panel, the distribution is plotted on a log–log scale. The best-fitting Pareto PDF is shown as a solid line.

Data Source: M. W. Newman, "Power Laws, Pareto Distributions, and Zipf's Law," *Contemporary Physics* 46 (2005): 323–351.

Solution

a. Provided the integral is convergent, the mean of the Pareto PDF is given by

$$\int_1^\infty x \frac{a-1}{x^a}\,dx = (a-1)\int_1^\infty \frac{dx}{x^{a-1}}$$

Since $\int_1^\infty \frac{dx}{x^{a-1}}$ is a p-integral with $p = a - 1$ (see Example 2 of Section 7.2), it is convergent if $a > 2$ and divergent if $a \leq 2$. For $a > 2$, Example 2 of Section 7.2 with $p = a - 1$ implies that

$$(a-1)\int_1^\infty \frac{dx}{x^{a-1}} = \frac{a-1}{a-2}$$

is the mean of the Pareto distribution.

b. Since $a = 2.18471 > 2$, the mean is well defined and given by

$$\frac{a-1}{a-2} = \frac{1.18471}{0.18471} \approx 6.41389$$

Hence, according to the Pareto distribution approximation, the average population size in a U.S. city in the year 2000 was approximately sixty four thousand individuals.

Variance and standard deviation

The importance of going beyond the mean was captured by the English mathematician Sir Francis Galton (1822–1911). He famously queried why statisticians of his day typically limited their enquiries to computing averages; he commented (pp. 62–63 in Galton, F., 1889, *Natural Inheritance*. Macmillan & Co., London), that souls of these statisticians "seem as dull to the charm of variety as that of the native of one of our flat English counties, whose retrospect of Switzerland was that, if its mountains could be thrown into its lakes, two nuisances would be got rid of at once." One method to calculate the spread of data around the mean is to compute *variance*.

Variance for Discrete Random Variables

Let X be a random variable taking on the values $x_1, x_2, x_3, \ldots, x_k$ with probabilities $p_1, p_2, p_3, \ldots, p_k$. Let μ be the mean of X. The **variance**, which we denote σ^2, is defined by

$$\sigma^2 = p_1(x_1 - \mu)^2 + p_2(x_2 - \mu)^2 + \cdots + p_k(x_k - \mu)^2 = \sum_{i=1}^k p_i(x_i - \mu)^2$$

while its square root σ, which has the same units as the mean, is referred to as the **standard deviation**.

For a data set taking on the distinct values $x_1, x_2, x_3, \ldots, x_k$ with relative frequencies $p_1, p_2, p_3, \ldots, p_k$ and mean μ, the variance and standard deviation for the data set are defined exactly as for the discrete random variable. Some technologies employ a correction factor for calculating the variance associated with a sample of size n by multiplying the variance, as defined above, by the factor $n/(n-1)$ to get what statisticians call an *unbiased estimate of the variance*. If the value you obtain in solving a problem does not agree with the value we provide as solution, it may be that your technology employed this correction factor.

In describing the spread of data around the mean, scientists typically use the standard deviation rather than the variance. Unlike the variance, the standard deviation and the mean have the same units as the data. We will show, at the end of this section, that three-quarters of a data set always lies within two standard deviations of its mean. Equivalently, the probability that a random variable takes on a value within two standard deviations of its mean is at least three-quarters. In the next section, we will show that for certain bell-shaped distributions, approximately "two-thirds" of a data set lies within one standard deviation of its mean.

Example 9   Computing variances and standard deviations

Recall Example 1 which referred to a study of competition for nest holes among collared flycatchers. Researchers collected data by periodically inspecting nest boxes located on the island of Gotland in Sweden. The accompanying data give the number of flycatchers breeding at fourteen distinct plots:

$$5 \quad 4 \quad 3 \quad 2 \quad 2 \quad 1 \quad 1 \quad 1 \quad 1 \quad 0 \quad 0 \quad 0 \quad 0 \quad 0$$

Let X be the number of flycatchers in a randomly chosen plot. Find the variance and standard deviation of X.

Solution   In Example 1 we found that $\mu \approx 1.43$. Hence, the variance is given by

$$\sigma^2 = \frac{1}{14} \cdot (5 - 1.43)^2 + \frac{1}{14} \cdot (4 - 1.43)^2 + \frac{1}{14} \cdot (3 - 1.43)^2 + \frac{2}{14} \cdot (2 - 1.43)^2$$
$$+ \frac{4}{14} \cdot (1 - 1.43)^2 + \frac{5}{14} \cdot (0 - 1.43)^2$$
$$\approx 2.388$$

and the standard deviation is $\sigma \approx \sqrt{2.388} \approx 1.55$.   ∎

The following example illustrates that standard deviations measure the spread of the data around its mean.

Example 10   Seeing the spread

A person places multiple bets on three "fair" games. Her winnings for the games are as follows:

- **Game A** $-1, 0, 0, 0, 1$ (dollar values)
- **Game B** $-1, -1, 0, 1, 1$ (dollar values)
- **Game C** $-2, -1, 0, 1, 2$ (dollar values)

a. Plot the histograms of the frequencies of observed values for the each of these data sets.

b. Compute the variances for the each of these data sets.

c. Discuss what you find.

Solution
a. Plotting the histograms yields:

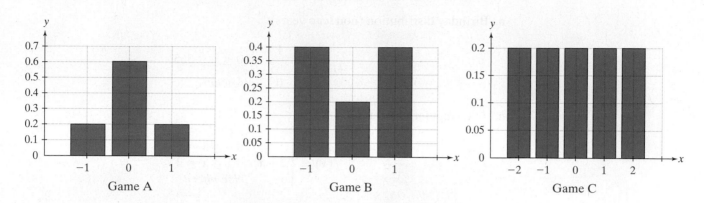

Game A   Game B   Game C

b. Since all the histograms balance at 0, the mean for all data sets is 0. The variances are as follows:

$$\text{Game A}: \quad \sigma^2 = (0+1)^2 \cdot 0.2 + 0^2 \cdot 0.6 + (0-1)^2 \cdot 0.2 = 0.4$$

$$\text{Game B}: \quad \sigma^2 = (0+1)^2 \cdot 0.4 + 0^2 \cdot 0.2 + (0-1)^2 \cdot 0.4 = 0.8$$

$$\text{Game C}: \quad \sigma^2 = (0+2)^2 \cdot 0.2 + (0+1)^2 \cdot 0.2 + 0^2 \cdot 0.2 + (0-1)^2 \cdot 0.2$$
$$+ (0-2)^2 \cdot 0.2 = 2.0$$

c. For game A, the variance is 0.4 as there is some variation about the mean 0. Since game B has more data points away from the mean than game A, the variance for this game is greater than that for game A. Since game C has the greatest spread in winnings, it has the largest variance.

Consider a continuous random variable X with PDF $f(x)$. Assume the mean $\mu = \int_{-\infty}^{\infty} x f(x)\, dx$ is well defined. To define the variance of X, we can approximate X by a discrete random variable by dividing the real line into intervals of length Δx with endpoints:

$$\ldots, \quad x_{-2} = -2\Delta x, \quad x_{-1} = -\Delta x, \quad x_0 = 0, \quad x_1 = \Delta x, \quad x_2 = 2\Delta x, \quad \ldots$$

Since the probability X takes on a value between x and $x + \Delta x$ is approximately $f(x)\Delta x$, the variance should approximately equal

$$\cdots + (x_{-2} - \mu)^2 f(x_{-2})\Delta x + (x_{-1} - \mu)^2 f(x_{-1})\Delta x$$
$$+ (x_0 - \mu)^2 f(x_0)\Delta x + (x_1 - \mu)^2 f(x_1)\Delta x + \cdots$$

Equivalently,

$$\sum_{k=-\infty}^{\infty} (x_k - \mu)^2 f(x_k)\Delta x$$

Taking the limit as Δx goes to zero yields $\int_{-\infty}^{\infty} (x - \mu)^2 f(x)\, dx$.

Variance and Standard Deviation for a Continuous Random Variable

For a continuous random variable X with PDF $f(x)$ and mean μ, the **variance of** X is given by

$$\sigma^2 = \int_{-\infty}^{\infty} (x - \mu)^2 f(x)\, dx$$

provided the improper integral converges. The **standard deviation of** X is given by σ, the square root of the variance.

Example 11   Integrating to get a variance

Find the variances of the following PDFs.

a. Birthday distribution (non leap years):

$$f(x) = \begin{cases} \dfrac{1}{365} & \text{if } 0 \le x \le 365 \\ 0 & elsewhere \end{cases}$$

b. A Triangular distribution:

$$f(x) = \begin{cases} x & \text{if } 0 \le x \le 1 \\ 2 - x & \text{if } 1 \le x \le 2 \\ 0 & elsewhere \end{cases}$$

Solution

a. Earlier, we found the mean of the non leap-year birthday PDF is $\mu = \dfrac{365}{2}$. Hence, the variance is given by

$$\int_0^{365} \left(x - \frac{365}{2} \right)^2 \frac{dx}{365} = \frac{1}{3} \left(x - \frac{365}{2} \right)^3 \frac{1}{365} \Bigg|_{x=0}^{365}$$

$$= \frac{365^2}{12} \approx 11{,}102$$

b. Earlier we found that the mean of the triangular distribution is $\mu = 1$. Hence, the variance is given by

$$\int_0^2 (x-1)^2 f(x)\, dx = \int_0^1 (x-1)^2 x\, dx + \int_1^2 (x-1)^2 (2-x)\, dx$$

$$= \frac{1}{6}$$

The following example studies the effect of the standard deviation on the shape of the distribution.

Example 12   Laplacian variance

Recall from Example 9 in Section 7.2, the Laplacian PDF $f(x) = \dfrac{b}{2} e^{-b|x|}$. Since this distribution is symmetric, its mean is 0.

a. Find the standard deviation of this PDF.

b. Use technology to plot the PDF for different b values and discuss how the standard deviation affects the shape of the PDF.

Solution

a. We need to compute

$$\frac{b}{2} \int_{-\infty}^{\infty} x^2 e^{-b|x|}\, dx = \frac{b}{2} \int_{-\infty}^{0} x^2 e^{bx}\, dx + \frac{b}{2} \int_0^{\infty} x^2 e^{-bx}\, dx$$

$$= b \int_0^{\infty} x^2 e^{-bx}\, dx \qquad by\ symmetry$$

Applying integration by parts twice yields

$$b \int x^2 e^{-bx}\, dx = -x^2 e^{-bx} + \int 2x e^{-bx}\, dx \qquad \text{using } u = x^2 \text{ and } dv = e^{-bx}\, dx$$

$$= -x^2 e^{-bx} - \frac{2x}{b} e^{-bx} + \frac{2}{b} \int e^{-bx}\, dx \qquad \text{using } u = 2x \text{ and } dv = e^{-bx}\, dx$$

$$= -x^2 e^{-bx} - \frac{2x}{b} e^{-bx} - \frac{2}{b^2} e^{-bx} + C$$

$$= -e^{-bx} \left(x^2 + \frac{2x}{b} + \frac{2}{b^2} \right) + C$$

Hence,

$$b \int_0^\infty x^2 e^{-bx}\, dx = \lim_{t \to \infty} b \int_0^t x^2 e^{-bx}\, dx$$

$$= \lim_{t \to \infty} -e^{-bx} \left(x^2 + \frac{2x}{b} + \frac{2}{b^2} \right) \Big|_0^t$$

$$= \lim_{t \to \infty} -e^{-bt} \left(t^2 + \frac{2t}{b} + \frac{2}{b^2} \right) + \frac{2}{b^2}$$

$$= \frac{2}{b^2}$$

Therefore, $\sigma = \dfrac{\sqrt{2}}{b}$.

b. Plotting the PDF for $b = 1, 5, 10$ yields this graph:

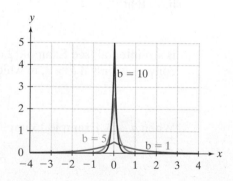

For larger b values, the standard deviation is smaller. The PDF tends to concentrate more around the mean of 0 when the standard deviation is smaller.

The following example provides an easier way of computing variances.

Example 13 Variance: mean-squared property

Let f be a PDF with mean μ. Assuming σ^2 is well defined, show that

$$\sigma^2 = \int_{-\infty}^\infty x^2 f(x)\, dx - \mu^2$$

Solution　The definition of variance and rules of integration imply

$$\sigma^2 = \int_{-\infty}^{\infty} (x - \mu)^2 f(x)\, dx$$

$$= \int_{-\infty}^{\infty} (x^2 - 2x\mu + \mu^2) f(x)\, dx$$

$$= \int_{-\infty}^{\infty} x^2 f(x)\, dx - 2\mu \int_{-\infty}^{\infty} x f(x)\, dx + \int_{-\infty}^{\infty} \mu^2 f(x)\, dx$$

$$= \int_{-\infty}^{\infty} x^2 f(x)\, dx - 2\mu^2 + \mu^2 \quad using \int_{-\infty}^{\infty} f(x)\, dx = 1 \text{ and definition of } \mu$$

$$= \int_{-\infty}^{\infty} x^2 f(x)\, dx - \mu^2$$

Example 14　Variance for the Pareto distribution

In Example 8, we found that the Pareto distribution with parameter $a > 0$ has a finite mean $\mu = \dfrac{a - 1}{a - 2}$ provided that $a > 2$. Determine when the variance for the Pareto distribution is finite and find the variance when it is finite.

Solution　Recall that the Pareto PDF is given by

$$f(x) = \begin{cases} 0 & \text{if } x < 1 \\ \dfrac{a - 1}{x^a} & \text{if } x \ge 1 \end{cases}$$

Assume that $a > 2$, in which case the mean equals $\mu = \dfrac{a - 1}{a - 2}$. Example 13 implies that the variance of this PDF is convergent only if $\int_1^{\infty} x^2 f(x)\, dx$ is convergent. We have that

$$\int_1^{\infty} x^2 f(x)\, dx = (a - 1) \int_1^{\infty} \frac{dx}{x^{a-2}}$$

is a p-integral (see Example 2 of Section 7.2) with $p = a - 2$. Therefore, this integral converges only if $a > 3$, in which case

$$(a - 1) \int_1^{\infty} \frac{dx}{x^{a-2}} = \frac{a - 1}{a - 3}$$

When $a > 3$, Example 13 implies that the variance of the Pareto distribution equals

$$\sigma^2 = \frac{a - 1}{a - 3} - \left(\frac{a - 1}{a - 2} \right)^2$$

Chebyshev's inequality

As we saw in the previous example, through its square root (i.e., standard deviation), the variance provides both a measurement and a sense of the spread around the mean. Larger standard deviations suggest greater spread around the mean. A basic inequality from probability theory provides a general method of estimating what fraction of the data is within a certain number of standard deviations of the mean. This inequality is *Chebyshev's inequality*, named after the mathematician Pafnuty Chebyshev (1821–1894), who first proved it.

Theorem 7.3　Chebyshev's inequality

Let X be a random variable with mean μ and standard deviation σ. Then,

$$P(\mu - k\sigma \le X \le \mu + k\sigma) \ge 1 - \frac{1}{k^2}$$

Proof. We provide a proof in the case of a continuous random variable with PDF $f(x)$. In which case, we obtain

$$\sigma^2 = \int_{-\infty}^{\infty} (x-\mu)^2 f(x)\,dx$$

$$= \int_{-\infty}^{\mu-k\sigma} (x-\mu)^2 f(x)\,dx + \int_{\mu-k\sigma}^{\mu+k\sigma} (x-\mu)^2 f(x)\,dx + \int_{\mu+k\sigma}^{\infty} (x-\mu)^2 f(x)\,dx$$

$$\geq \int_{-\infty}^{\mu-k\sigma} (x-\mu)^2 f(x)\,dx + \int_{\mu+k\sigma}^{\infty} (x-\mu)^2 f(x)\,dx$$

$$\geq (k\sigma)^2 \int_{-\infty}^{\mu-k\sigma} f(x)\,dx + (k\sigma)^2 \int_{\mu+k\sigma}^{\infty} f(x)\,dx$$

$$\text{since } (x-\mu)^2 > (k\sigma)^2 \text{ for all } x < \mu - k\sigma \text{ and } x > \mu + k\sigma$$

$$= (k\sigma)^2 P(X \leq \mu - k\sigma) + (k\sigma)^2 P(X \geq \mu + k\sigma)$$

Thus, we have shown that

$$\sigma^2 \geq (k\sigma)^2 P(X \leq \mu - k\sigma) + (k\sigma)^2 P(X \geq \mu + k\sigma)$$

$$\frac{1}{k^2} \geq P(X \leq \mu - k\sigma) + P(X \geq \mu + k\sigma)$$

$$\frac{1}{k^2} \geq 1 - P(\mu - k\sigma \leq X \leq \mu + k\sigma)$$

$$P(\mu - k\sigma \leq X \leq \mu + k\sigma) \geq 1 - \frac{1}{k^2}$$

Since this theorem holds for all PDFs, no matter how peaked or how flat, the bounds in some cases can be rather weak.

Example 15   Using Chebyshev's inequality

In 1998 in Hong Kong, the number of newborns was 52,955 whose mean birth weight was 3.2 kg and standard deviation was 0.5 kg.* Using only these data, estimate the following quantities.

a. Fraction of newborns weighing between 2.2 and 4.2 kg

b. Fraction of newborns weighing between 1.7 and 4.7 kg

Solution   To apply Chebyshev's inequality, let X be the weight of a randomly selected newborn.

a. Since $\mu = 3.2$ and $\sigma = 0.5$, we find that $2.2 = \mu - 2\sigma$ and $4.2 = \mu + 2\sigma$. By Chebyshev's inequality with $k = 2$, we find that at least $1 - \dfrac{1}{2^2} = \dfrac{3}{4}$ of the newborns weighed between 2.2 and 4.2 kg.

b. Since $\mu = 3.2$ and $\sigma = 0.5$, we find that $1.7 = \mu - 3\sigma$ and $4.7 = \mu + 3\sigma$. By Chebyshev's inequality with $k = 3$, we find that at least $1 - \dfrac{1}{3^2} = \dfrac{8}{9}$ of the newborns weighed between 1.7 and 4.7 kg.

Example 15 illustrates that Chebyshev's inequality states that at least $\dfrac{3}{4}$ of the data ($k = 2$) are within two standard deviations of the mean, at least $\dfrac{8}{9}$ of the data ($k = 3$) are with three standard deviations of the mean, and so on.

*See www.dh.gov.hk.

PROBLEM SET 7.3

Level 1 DRILL PROBLEMS

Compute the mean, variance, and standard deviation of the data set given in Problems 1 to 8.

1. 1, 1, 0, 1, 1

2. 2, 0, 2

3. 1, 1, 1, 1, 1

4. 1, 2, 3, 4, 5, 6 (a die)

5. 1, 5, 7

6. −1, −2, 1, 4

7. The set of numbers that contains 2 zeros, 6 ones, 17 twos, and 8 threes.

8. The set of numbers that contains 7 negative twos, 5 negative ones, 3 zeros, 8 ones, and 12 twos.

Compute the mean of the random variable with the indicated PDF in Problems 9 to 16.

9. $f(x) = \dfrac{1}{2}$ for $0 \le x \le 2$ and $f(x) = 0$ elsewhere

10. $f(x) = \dfrac{3}{x^4}$ for $x \ge 1$ and $f(x) = 0$ elsewhere

11. $f(x) = \dfrac{1.5}{x^4}$ for $|x| \ge 1$ and $f(x) = 0$ elsewhere

12. $f(x) = e^{-x}$ for $x \ge 0$ and $f(x) = 0$ elsewhere

13. $f(x) = \dfrac{1}{\sqrt{2\pi}} e^{-x^2/2}$

14. $f(x) = \dfrac{1}{1 + x^2} \dfrac{1}{\pi}$

15. $f(x) = xe^{-x}$ for $x \ge 0$ and $f(x) = 0$ elsewhere

16. $f(x) = \dfrac{4x^2 e^{-x^2}}{\sqrt{\pi}}$ for $x \ge 0$ and $f(x) = 0$ elsewhere

(Hint: See Example 5 of Section 5.6)

Compute the variance of the PDFs in Problems 17 to 20.

17. $f(x) = \dfrac{1}{2}$ for $0 \le x \le 2$ and $f(x) = 0$ elsewhere

18. $f(x) = \dfrac{3}{x^4}$ for $x \ge 1$ and $f(x) = 0$ elsewhere

19. $f(x) = \dfrac{1.5}{x^4}$ for $|x| \ge 1$ and $f(x) = 0$ elsewhere

20. $f(x) = xe^{-x}$ for $x \ge 0$ and $f(x) = 0$ elsewhere

21. Consider the following data set:

$$-1 \quad 0 \quad 0 \quad 0 \quad 0 \quad 0 \quad 0 \quad 0 \quad 1$$

　a. Find the mean and standard deviation.

　b. According to Chebyshev's inequality, what fraction of data (at the bare minimum) has to lie in the interval $[\mu - 2\sigma, \mu + 2\sigma]$? What fraction of the data does lie in this interval?

22. Consider the following data set:

$$-1 \quad -1 \quad 0 \quad 0 \quad 0 \quad 0 \quad 0 \quad 1 \quad 2$$

　a. Find the mean and standard deviation.

　b. According to Chebyshev's inequality, what fraction of data (at the bare minimum) has to lie in the interval $[\mu - 2\sigma, \mu + 2\sigma]$? What fraction of the data does lie in this interval?

23. Suppose that a random variable X has a PDF of

$$f(x) = 0.5 + x \quad \text{for} \quad 0 \le x \le 1$$

　a. Find $P(0.2 \le X \le 0.6)$. Sketch this probability on a graph of the PDF.

　b. Find and graph the CDF.

　c. Find the mean value.

24. Suppose that a random variable X has a PDF of

$$f(x) = \dfrac{1}{x} \quad \text{for} \quad 1 \le x \le e$$

　a. Find $P(1.2 \le X \le 2.4)$. Sketch this probability on a graph of the PDF.

　b. Find and graph the CDF.

　c. Find the mean value.

Level 2 APPLIED AND THEORY PROBLEMS

25. Let X be the distance from the center that a dart lands on a dartboard with a radius of 2 feet.

　a. Show that the PDF for x is given by

$$f(x) = \begin{cases} \dfrac{x}{2} & \text{if } 0 \le x \le 2 \\ 0 & elsewhere \end{cases}$$

　b. Find the mean and variance of this PDF.

26. In spring 1994, the number of bird species in forty different oak woodland sites in California were collected. Each site was around 5 hectares in size—the equivalent of about 12 acres, or 0.019 square miles—and the sites were situated in a relatively homogeneous habitat. The numbers of bird species found in these sites are listed below:

37, 21, 26, 27, 21, 21, 28, 22, 22, 26
47, 26, 29, 34, 28, 25, 19, 32, 32, 29
29, 16, 21, 24, 37, 38, 30, 20, 23, 30
27, 32, 17, 24, 32, 29, 40, 31, 38, 35

Let X be the number of species in a randomly chosen site. Using technology, compute the mean and standard deviation of X.

27. The following numbers are length in seconds of scenes showing tobacco use in six animation movies from a film studio:

0　223　0　176　0　548

Let X be the length of a scene from a randomly chosen movie.

a. Compute the mean for X.

b. Compute the standard deviation of X.

28. The following numbers are belly bristles per fruit fly in a sample size of six:

$$30 \quad 32 \quad 27 \quad 30 \quad 32 \quad 32$$

a. Find the relative frequency of 30.

b. Compute the mean number of belly bristles in this sample.

c. Compute the standard deviation of this data set.

d. According to Chebyshev's inequality, what is the minimum fraction of the data taking on values between 26.5 and 34.5? What is the actual fraction of data taking on values between 26.5 and 34.5?

29. Consider the exponential random variable X with PDF given by

$$f(t) = \begin{cases} ce^{-ct} & \text{if } t \geq 0 \\ 0 & elsewhere \end{cases}$$

Find the variance and standard deviation of X. Compare these numbers to the mean of the exponential distribution. What do you notice?

30. According to Thomson et al. (see Problem 36 in Problem Set 7.1), the elimination constant for lidocaine for patients with hepatic impairment is 0.12 per hour.

a. Determine the mean time μ for a lidocaine molecule to be eliminated.

b. Determine the fraction of lidocaine eliminated by time $t = \mu$.

31. Donald Levin was quoted in *Science Daily* (see Problem 38 in Problem Set 7.1) as stating: "Roughly 20 of the 297 known mussel and clam species and 40 of about 950 fishes have perished in North America in the last century."

a. Use these data to approximate the mean time to extinction constants for mussel and clam species and for fish species.

b. Determine the fraction of mussel and clam species and fish species that will be lost in the next century.

32. In Example 3 of Section 7.2, the following Pareto PDF was used to describe how many AOL users visited certain websites on one day in 1997:

$$f(x) = \begin{cases} 0 & \text{if } x < 1 \\ 1.07x^{-2.07} & \text{if } x \geq 1 \end{cases}$$

a. Find the mean of this PDF.

b. Compute the variance for this PDF. What does the variance suggest about the variability in the number of hits that a website can experience?

33. Let X denote the number of years a patient lives after receiving treatment for an acute disease such as cancer. Under appropriate conditions, X is exponentially distributed. Suppose that the probability that a patient will live at least five years after treatment is 0.85.

a. Find the mean value of X.

b. Find the probability a patient will live at least ten years.

34. Based on data from 1974 to 2000 in Humboldt and Del Norte counties in California, the mean time to the next earthquake of magnitude ≥ 4 is approximately 2.5 weeks. Assuming that the time to the next earthquake is exponentially distributed, find the probability there will be an earthquake of magnitude ≥ 4 in the next week.

35. According to a newspaper article titled "Babies by the Dozen for Christmas: 24-Hour Baby Boom," a record forty-four babies were born in one 24-hour period at the Mater Mothers' Hospital, Brisbane, Australia, on December 18, 1997. The article listed the times of birth for all of the babies.* The histogram of the times between births is as follows:

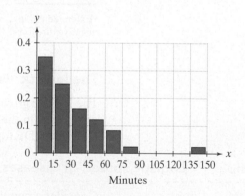

a. If this histogram has proportions {0.35, 0.25, 0.16, 0.12, 0.08, 0.02, 0.0, 0.0, 0.0, 0.02} centered on the values 7.5, 22.5, and so on every 15 minutes up to 142.5, then what is the mean time between births?

b. If this histogram is approximately exponentially distributed with a mean of 33.26 minutes between births, then what fraction of the times between births were less than 30 minutes? According to the histogram, what fraction of times between births were less than 30 minutes?

*See Peter K. Dunn, "A Simple Dataset for Demonstrating Common Distributions," *Journal of Statistics Education* 7 (no. 3, 1999).

c. According to the exponential distribution, what fraction of the times between births were more than 75 minutes? Compare this with the actual fraction of times between births that are more than 80 minutes, as depicted in the histogram.

36. Here is a fun but challenging exercise: Construct a data set so that only 75.1% of the data lie in the interval $[\mu - 2\sigma, \mu + 2\sigma]$.

37. Let X be a continuous random variable with PDF $f(x)$. Show that $g(x) = f(x + a)$ is a PDF for the continuous random variable $Y = X - a$.

7.4   Bell-Shaped Distributions

An important class of PDFs has graphs that are bell-shaped (see center panel in Figure 7.18). In this section, we investigate two PDFs with this property: the logistic PDF and the normal PDF. Both are symmetric about their means and nonzero on the interval $(-\infty, \infty)$. Although logistic and normal PDFs have similar shapes, each is used to represent biological data arising from different types of processes. In this section, we also study random variables that are normal on a log scale. This lognormal distribution is right skewed or right tailed (also referred to as positively skewed—see Figure 7.18) because it is zero on $(-\infty, 0]$.

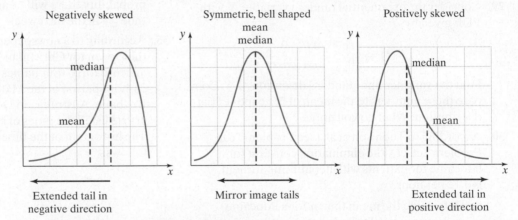

Figure 7.18 Symmetric and skewed PDFs

The logistic distribution

The logistic growth equation studied in Chapter 6 describes how populations change over time. In the next example, we show that solutions to the logistic equation can lead to a CDF for the *logistic distribution*.

Example 1   Logistic spread of diseases

Consider a population of individuals in which a disease is spreading, and individuals once infected remain infected (e.g., HIV/AIDS). Let y denote the fraction of infected individuals also known as the *prevalence of the disease*—and let t denote time in months. If the rate of increase of infected individuals is proportional to the product of the fraction of infected individuals and the fraction of uninfected individuals, then

$$\frac{dy}{dt} = ry(1 - y)$$

where r is a constant that describes how rapidly the disease spreads in the population.

a. Assuming that $r = 1$ and $y(0) = 0.5$, solve for $y(t)$.

b. Verify that $y(t)$ is a CDF on $-\infty < t < \infty$.

c. Determine the probability that a randomly chosen individual from this population will be infected within the next two months.

d. Find the PDF associated with the CDF $y(t)$ and show that it is symmetric.

Solution

a. Separating and integrating yields

$$\int \frac{dy}{y(1-y)} = \int dt$$

$$\int \left(\frac{1}{y} + \frac{1}{1-y} \right) dy = t + C$$

$$\ln |y| - \ln |1 - y| = t + C$$

$$\ln \left| \frac{y}{1-y} \right| = t + C$$

$$\frac{y}{1-y} = C_2 e^t \qquad \text{defining } C_2 = e^C > 0$$

$$y = C_2 e^t (1 - y)$$

$$y(1 + C_2 e^t) = C_2 e^t$$

$$y = \frac{C_2 e^t}{1 + C_2 e^t} \qquad \text{which is defined for } -\infty < t < \infty$$

Using the initial condition $y(0) = 0.5$, we solve for C_2:

$$0.5 = \frac{C_2}{1 + C_2}$$

$$0.5(1 + C_2) = C_2$$

$$0.5 = 0.5 C_2$$

$$C_2 = 1$$

Hence,

$$y(t) = \frac{e^t}{1 + e^t}$$

b. To verify that $y(t)$ is a CDF for $-\infty < t < \infty$, we need to check three things (see Section 7.1). First, to see that $y(t)$ is non-decreasing, we can take the derivative using the quotient rule

$$y'(t) = \frac{e^t(1 + e^t) - e^t e^t}{(1 + e^t)^2}$$

$$= \frac{e^t}{(1 + e^t)^2}$$

Since $y'(t) > 0$ for all t, $y(t)$ is increasing. Second, we need to verify that $\lim_{t \to \infty} y(t) = 1$ and $\lim_{t \to -\infty} y(t) = 0$. Indeed, dividing the numerator and denominator of $y(t)$ by e^t yields

$$\lim_{t \to \infty} y(t) = \lim_{t \to \infty} \frac{1}{e^{-t} + 1} = 1$$

Similarly, $\lim_{t \to -\infty} y(t) = 0$. Finally, since $y(t)$ is continuous for all t, it is right continuous for all t.

c. The probability that a randomly chosen individual will get infected before the second month is $P(X \leq 2) = y(2) = \frac{e^2}{1 + e^2}$. The probability that a randomly chosen individual will get infected before $t = 0$ is $P(X \leq 0) = y(0) = \frac{1}{2}$. Hence, the probability that a randomly chosen individual will get infected between $t = 0$ and $t = 2$ is $y(2) - y(0) = \frac{e^2}{1 + e^2} - \frac{1}{2} \approx 0.381$. Hence, there is an approximately 38% chance that a randomly chosen individual will get infected within two months.

d. To find the PDF, we can use the fundamental theorem of PDFs; namely, the PDF $f(t)$ is given by the derivative of the CDF:

$$f(t) = y'(t)$$
$$= \frac{e^t}{(1 + e^t)^2}$$

It follows that

$$f(-t) = \frac{e^{-t}}{(1 + e^{-t})^2}$$
$$= \frac{e^t}{(1 + e^t)^2} \qquad \textit{multiplying numerator and denominator by } e^{2t}$$
$$= f(t).$$

Hence, the PDF is symmetric around $t = 0$. ∎

In the previous example, we selected $y(0) = 0.5$, resulting in a symmetric PDF around zero. Thus, the mean is zero, provided that the associated improper integral is convergent. More generally, we can derive a logistic PDF for any initial condition $y(0)$, as well as for any arbitrary $r > 0$ in which case the PDF is symmetric around a value other than zero. In particular, in Problem Set 7.4 asks you to show the following:

Logistic PDF and CDF

A solution $y(t)$ to $\dfrac{dy}{dt} = ry(1 - y)$ with $y(0) \in (0, 1)$ and $r > 0$ gives a CDF of the following form:

$$y(t) = \frac{1}{1 + e^{a-rt}}$$

where $a = \ln(1/y(0) - 1)$. This CDF corresponds to the **logistic distribution**. The associated PDF is

$$f(t) = \frac{re^{a-rt}}{(1 + e^{a-rt})^2}$$

Note that $a > 0$ if $y(0) \in (0, 0.5)$ and $a < 0$ if $y(0) \in (0.5, 1)$.

Example 2   Playing with the logistic PDF
- -

Assume that the logistic PDF describes the distribution of infection times. Let r be the intrinsic rate of growth of the disease and $y(0)$ the fraction of individuals infected during week $t = 0$.

a. Consider a disease for which $r = 1$. Determine the fraction of people infected by the disease in the next two weeks if $y(0) = 0.25$ or $y(0) = 0.75$.

b. Use technology to plot the PDF for $r = 1$ and $y(0) = 0.25, 0.5$, and 0.75. Discuss what you find.

c. Consider a disease for which $y(0) = 0.1$. Determine the fraction of people infected by the disease in the next two weeks if the intrinsic rate of growth is $r = 0.5$ or $r = 5$.

d. Use technology to plot the PDF for $y(0) = 0.1$ and $r = 0.5, 1$, and 5. Discuss what you find.

Solution

a. If $r = 1$ and $y(0) = 0.25$, then $a = \ln(1/y(0) - 1) \approx 1.1$ and the CDF is given by

$$y(t) = \frac{1}{1 + e^{1.1-t}}.$$

The fraction of people infected in the next two weeks is given by

$$y(2) - y(0) \approx 0.71 - 0.25 = 0.46$$

Hence, 46% are infected in the next two weeks. If $r = 1$ and $y(0) = 0.75$, then $a = \ln(1/y(0) - 1) \approx -1.1$ and the CDF is given by

$$y(t) = \frac{1}{1 + e^{-1.1-t}}.$$

The fraction of people infected in the next two weeks is given by

$$y(2) - y(0) \approx 0.96 - 0.75 = 0.21$$

Hence, only 21% are infected in the next two weeks.

b. Using technology, we obtain the PDFs plotted in Figure 7.19. These plots illustrate that as we increase $y(0)$, the "center" of the PDF tends to move to the left. In other words, as the fraction of individuals infected at week 0 increases, the time-to-infection for all individuals decreases.

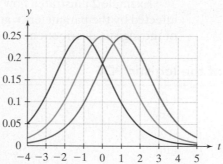

Figure 7.19 Plots of logistic PDFs for the case $r = 1$ and $a = \ln(1/y(0) - 1)$, where $y(0) = 0.75$ (red curve), $y(0) = 0.5$ (green curve), and $y(0) = 0.25$ (blue curve)

c. If $y(0) = 0.1$, then $a = \ln(1/y(0) - 1) \approx 2.2$. Hence, if $r = 0.5$, then the CDF is given by

$$y(t) = \frac{1}{1 + e^{2.2-0.5t}}.$$

The fraction of individuals infected in the next two weeks is given by

$$y(2) - y(0) \approx 0.23 - 0.1 = 0.13$$

Hence, 13% are infected in the next two weeks. Alternatively, if $r = 5$, then the CDF is given by

$$y(t) = \frac{1}{1 + e^{2.2-5t}}.$$

The fraction of individuals infected in the next two weeks is given by

$$y(2) - y(0) \approx 1.0 - 0.1 = 0.9$$

Hence, 90% are infected in the next two weeks.

d. Using technology, we obtain the PDFs plotted in Figure 7.20.

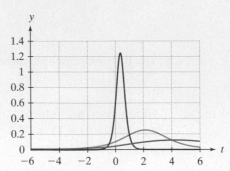

Figure 7.20 Plots of logistic PDFs for the case $a = 2.197$ (corresponding to $y(0) = 0.1$), and $r = 0.5$ (blue curve), $r = 1.0$ (green curve) and $r = 5.0$ (red curve)

These plots illustrate that as we increase r, the "center" of the PDF tends to move to the left and the spread around the center decreases. In other words, for diseases that spread quickly (i.e., r is larger), most people catch the disease quickly and around the same time. For diseases that spread slowly (i.e., r is smaller), there is greater variability in the time it takes for a person to get infected, although in the logistic model everyone gets infected in the end.   ■

Example 2 illustrated how the mean and variance of the logistic distribution are affected by the parameters r and $y(0)$. The following example determines the mean of the logistic distribution.

Example 3   Mean of the logistic PDF

Let

$$y(t) = \frac{1}{1 + e^{a-rt}}$$

and

$$f(t) = \frac{re^{a-rt}}{(1 + e^{a-rt})^2}$$

be the CDF and PDF of the logistic distribution. Do the following:

a. Find $t = T$ such that $y(T) = 0.5$. In other words, find T such that half of the data lie to the left of T and half of the data lie to the right of T.

b. Verify that $f(t)$ is symmetric around $t = T$.

c. Assuming the mean is well defined, find the mean.

In Problem Set 7.4, you will be asked to verify that the mean is well defined, that is, $\int_{-\infty}^{\infty} t f(t)\, dt$ is convergent.

Solution

a. Solving $y(T) = \dfrac{1}{2}$ yields

$$\frac{1}{1 + e^{a-rT}} = \frac{1}{2}$$

$$2 = 1 + e^{a-rT}$$

$$1 = e^{a-rT}$$

$$0 = a - rT$$

$$T = \frac{a}{r}$$

b. To check symmetry of $f(t)$ about $t = T$, we need to verify that $f(a/r + t) = f(a/r - t)$ for all t. Indeed, we have

$$f(a/r + t) = \frac{re^{a - r(a/r + t)}}{(1 + e^{a - r(a/r + t)})^2}$$

$$= \frac{re^{-rt}}{(1 + e^{-rt})^2}$$

$$= \frac{re^{-rt}}{(1 + e^{-rt})^2} \frac{e^{2rt}}{e^{2rt}}$$

$$= \frac{re^{rt}}{(e^{rt} + 1)^2}$$

$$= \frac{re^{a - r(a/r - t)}}{(1 + e^{a - r(a/r - t)})^2}$$

$$= f(a/r - t)$$

c. Since f is symmetric around a/r, the mean is given by $\mu = a/r$ provided that the integral $\int_{-\infty}^{\infty} t f(t)\, dt$ is convergent.

In addition to determining the mean, it is possible—but more challenging—to compute the variance of the logistic PDF.

Mean and Variance of the Logistic PDF

The mean of the logistic PDF $f(t) = \dfrac{re^{a - rt}}{(1 + e^{a - rt})^2}$ is

$$\mu = \frac{a}{r}$$

and the variance is

$$\sigma^2 = \frac{1}{3}\left(\frac{\pi}{r}\right)^2$$

The logistic distribution can also describe the spread of an organism across a landscape.

Example 4 Organismal spread

Pyura praeputialis is a large tunicate (a species of sea squirt reaching lengths of up to 35 cm) that, in Chile, is distributed exclusively along 60 to 70 km of coastline in and around the bay of Antofagasta. This tunicate is a sessile, dominant species, capable of forming extensive beds of barrel-like individuals tightly cemented together in rocky intertidal and shallow subtidal zones. Using experimental quadrats, biologist Jorge Alvarado and colleagues[*] investigated recolonization dynamics of *P. praeputialis* in Chile after removal of adult individuals. Alvardo and colleagues found that the fraction of occupied habitat is approximately

$$y(t) = \frac{1}{1 + e^{4 - 1.7t}}$$

where t is measured in hundreds of days. For a randomly chosen point in the habitat, what is the mean time for it to be occupied?

Solution Since $r = 1.7$ and $a = 4$, the mean time for a location being occupied is $\dfrac{a}{r} = \dfrac{4}{1.7} \approx 2.35$. Therefore, on average, it takes 235 days for a randomly chosen location to get occupied.

---

[*]J. L. Alvardo et al., 2001. "Patch Recolonization by the Tunicate *Pyura praeputialis* in the Rocky Intertidal of the Bay of Antofagasta, Chile: Evidence for Self-Facilitation Mechanisms", *Marine Ecology Progress Series* 224 (2001): 93–101.

An important type of regression analysis is associated with the logistic PDF. In Chapter 1, we demonstrated fitting linear models $y = ax + b$ to data and inferring values for y from values of x. Suppose we want to infer the probability p of a certain event occurring associated with some measurement t. For example, t could be the age of a healthy cow and p could be the probability that this cow will die during the next year. If we have data on the proportion $p(t)$ of cows that have died at or before t, then we may consider fitting the logistic CDF to data. A method for fitting the logistic CDF to data is illustrated in the following two examples.

Example 5   Transforming the logistic into a linear equation

Show the function $y = \ln \dfrac{p(t)}{1 - p(t)}$ is linear in t when $p(t)$ is a logistic CDF.

Solution   Since $p(t) = \dfrac{1}{1 + e^{a - rt}}$, we get the following:

$$
\begin{aligned}
y &= \ln p(t) - \ln(1 - p(t)) \\
&= \ln\left(\frac{1}{1 + e^{a-rt}}\right) - \ln\left(\frac{(1 + e^{a-rt}) - 1}{1 + e^{a-rt}}\right) \\
&= \ln 1 - \ln(1 + e^{a-rt}) - \ln e^{a-rt} + \ln(1 + e^{a-rt}) \\
&= rt - a
\end{aligned}
$$

Example 5 implies that if we have a set of n data points, $(t_1, p_1), \ldots, (t_n, p_n)$, whose distribution we want to describe with the logistic CDF, it suffices to find the best-fitting line through the transformed data, $\left(t_1, \ln \dfrac{p_1}{1 - p_1}\right), \ldots, \left(t_n, \ln \dfrac{p_n}{1 - p_n}\right)$. Finding the best-fitting parameters r and a in this manner is called *logistic regression*.

Example 6   Logistic regression

A medical researcher used chemicals to induce the growth of prostate tumors in several hundred male rats. He surgically removed the resulting tumors after 150 days. Then he measured the cumulative proportion of individuals that had the tumors return within 90 days as a function of the size of the original tumor that he removed. The results are given in the first two columns in Table 7.2. Find the best-fitting logistic equation to this data set.

Table 7.2 Cumulative proportion p of rates growing new prostate tumors as a function of weight t (grams) of original tumor removed

Weight t	Proportion p	Transformed variable $y = \ln\left(\dfrac{p}{1 - p}\right)$
0–1	0.01	−4.60
1–2	0.02	−3.89
2–3	0.05	−2.94
3–4	0.11	−2.09
4–5	0.18	−1.52
5–6	0.32	−0.75
6–7	0.56	0.24
7–8	0.76	1.15
8–9	0.83	1.59
9–10	0.95	2.94
10+	0.92	2.44

Solution We use the transformation $y = \ln\left(\dfrac{p}{1-p}\right)$ of the p values in the second column of Table 7.2 to obtain the third column of this table. For the t values, we select the midpoint values $t_1 = 0.5$, $t_2 = 1.5$, $\ldots t_{10} = 9.5$, and for the last bin we use $t_{11} = 10.5$, even though it represents all weights ≥ 10. Using technology to find the best-fitting line, we get $y = 0.76t - 4.9$—that is, $r = 0.76$ and $a = 4.9$. The transformed data and regression line are illustrated in Figure 7.21; we obtain a very good fit, except at the highest weight of 10 g.

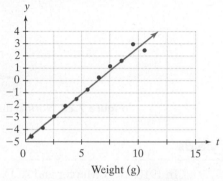

Weight (g)

Figure 7.21 Linear regression on transformed logistic data

Normal distribution

The most ubiquitous probability distribution in the natural and behavioral sciences is the *normal distribution*. For example, the normal distribution describes the distributions of heights (see Example 5 of Section 7.1) and weights of many organisms, crop yields (see Example 7), human IQs (see Example 9), and much more. Its ubiquity stems from this fact: if each data point is under the influence of many, independent, additive effects, then one can prove that the distribution of the data is well approximated by the normal distribution. The normal distribution is also known as the *Gaussian distribution*, after the German mathematician Karl Friedrich Gauss (1777–1855), who is shown in Figure 7.22 with the normal distribution in the background.

Figure 7.22 Deutsche mark showing Karl Gauss and the normal distribution in the background

PDF of the Normal Distribution

The PDF of the **normal distribution** is given by

$$f(x) = \frac{1}{\sqrt{2\pi}\,\sigma} e^{-\frac{(x-\mu)^2}{2\sigma^2}}$$

where μ is the mean of the distribution and σ is the standard deviation.

The effect of increasing μ on this distribution is to move the graph to the right. The standard deviation σ, on the other hand, controls the spread of the distribution about its center. For small σ, the distribution is more peaked or concentrated around the mean; for larger σ, the distribution is fatter with a lower, broader peak, as illustrated in Figure 7.23. As we discussed in Chapter 5, there is no elementary representation of the antiderivative of $f(x)$. Hence, we use numerical estimates.

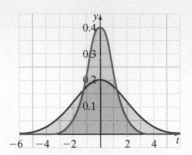

Figure 7.23 Normal distributions, with mean 0 and standard deviations of 1 (blue) and 2 (red)

Example 7   Wheat yields

In 1910, W. B. Mercer and A. D. Hall conducted a wheat yield experiment at Rothamsted Experimental Station in Great Britain. In 500 identical plots, wheat was grown and yield (in bushels) was recorded. The resulting histogram of the data is approximately normal, as illustrated in Figure 7.24.

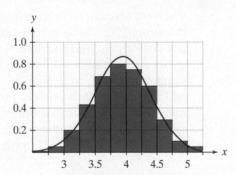

Figure 7.24 Histogram for the Rothamsted experiment

Data Source: Mercer, W. B. and Hall, A. D. (1911). The Experimental Error of Field Trials. *Journal of Agricultural Science* 4, 107–132.

The mean of the data is 3.95 bushels and the standard deviation is 0.45 bushels. Use numerical integration to approximate the following quantities:

a. The likelihood that the yield in a randomly chosen plot is between 3.5 and 4.5 bushels

b. The likelihood that the yield in a randomly chosen plot is at least 5 bushels

Solution   For this problem, we have

$$f(x) = \frac{1}{\sqrt{2\pi}0.45} e^{-\frac{(x-3.95)^2}{2(0.45)^2}}$$

a. Integrating $\int_{3.5}^{4.5} f(x)\,dx$ numerically using technology yields 0.731 (to three decimal places). Hence, there is approximately a 73% chance the yield will between 3.5 and 4.5 bushels.

b. Integrating $\int_{5}^{\infty} f(x)\,dx$ numerically using technology yields 0.00982 (to three significant digits). Hence, there is slightly less than 1% chance the yield will be at least 5 bushels.

Aside from using numerical integrators, we can use tables to estimate areas under normal densities. At first, you might think that we need an infinite number of tables to deal with all possible values of μ and σ. However, this is not the case. Using a simple substitution, we can reduce everything to a question about one normal distribution, the *standard normal distribution*.

Standard Normal Distribution

A random variable Z has a **standard normal distribution** if it is normally distributed with mean 0 and standard deviation 1; that is, it has the PDF

$$f(z) = \frac{1}{\sqrt{2\pi}} e^{-z^2/2} \quad \text{for} \quad z \in (-\infty, \infty)$$

The following example illustrates how all questions about normal distributions can be reformulated as a question about the standard normal distribution.

Example 8 From arbitrary normal to standard normal distributions

Let X be normally distributed with mean μ and standard deviation σ. Let Z be normally distributed with mean 0 and standard deviation 1. Show that for any a,

$$P(X \le a) = P(Z \le (a - \mu)/\sigma)$$

Solution   Since X has a normal distribution with mean μ and standard deviation σ, we have that

$$P(X \le a) = \int_{-\infty}^{a} \frac{1}{\sqrt{2\pi}\sigma} e^{-\frac{(x-\mu)^2}{2\sigma^2}} \, dx$$

Consider the change of variables, $z = (x - \mu)/\sigma$. Then $dz = \dfrac{dx}{\sigma}$, $z = (a - \mu)/\sigma$ when $x = a$, $\lim\limits_{x \to -\infty} z = -\infty$, and

$$P(X \le a) = \int_{-\infty}^{(a-\mu)/\sigma} \frac{1}{\sqrt{2\pi}\sigma} e^{-z^2/2} \sigma \, dz$$

$$= \int_{-\infty}^{(a-\mu)/\sigma} \frac{1}{\sqrt{2\pi}} e^{-z^2/2} \, dz$$

$$= P(Z \le (a - \mu)/\sigma)$$

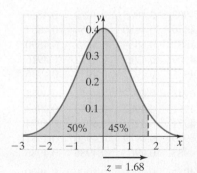

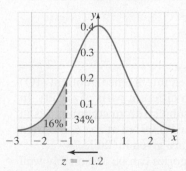

Figure 7.25 Using z scores to calculate area under a standard normal PDF

z Scores

Since questions about normally distributed random variables or data can be converted to questions about the standard normal distribution, we can use z **scores** to determine what fraction of normally distributed data lies between the data's mean and z standard deviations above the mean. Table 7.3 reports z scores where rows determine the first two digits of z and columns the third digit of z. For example, in the row labeled (at the left) 1.0 and in the column headed 0.00, the entry is 0.3413. Therefore, for data with a standard normal distribution, the fraction of data lying in the interval $[0, 1]$ is 34.13%. Alternatively, suppose the data have a standard normal distribution and we want to know what fraction of the data lies in the interval $(-\infty, 1.68]$. The z table tells us the fraction of data lying in the interval $[0, 1.68]$ is 0.4535. Since 50% of the data lies in $(-\infty, 0]$, the fraction in the interval $(-\infty, 1.68]$ is $0.4535 + 0.5 = 0.9535$, as shown in the top panel of Figure 7.25. By contrast for $z < 0$, the symmetry of the standard normal distribution around 0 implies that subtraction, rather than addition, is required. For example, if $z = -1.2$, as shown in the bottom panel of Figure 7.25, $50\% - 34.13\% = 15.87\%$ of the area under the standard normal PDF lies on the interval $(-\infty, -1.2]$.

Table 7.3 Standard normal distribution

z-scores

For a particular value, this table gives the percent of scores between the mean and the z-value of a normally distributed random variable.

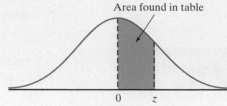

Area found in table

z	0.00	0.01	0.02	0.03	0.04	0.05	0.06	0.07	0.08	0.09
0.0	0.0000	0.0040	0.0080	0.0120	0.0160	0.0199	0.0239	0.0279	0.0319	0.0359
0.1	0.0398	0.0438	0.0478	0.0517	0.0557	0.0596	0.0636	0.0675	0.0714	0.0753
0.2	0.0793	0.0832	0.0871	0.0910	0.0948	0.0987	0.1026	0.1064	0.1103	0.1141
0.3	0.1179	0.1217	0.1255	0.1293	0.1331	0.1368	0.1406	0.1443	0.1480	0.1517
0.4	0.1554	0.1591	0.1628	0.1664	0.1700	0.1736	0.1772	0.1808	0.1844	0.1879
0.5	0.1915	0.1950	0.1985	0.2019	0.2054	0.2088	0.2123	0.2157	0.2190	0.2224
0.6	0.2257	0.2291	0.2324	0.2357	0.2389	0.2422	0.2454	0.2486	0.2517	0.2549
0.7	0.2580	0.2611	0.2642	0.2673	0.2704	0.2734	0.2764	0.2794	0.2823	0.2852
0.8	0.2881	0.2910	0.2939	0.2967	0.2995	0.3023	0.3051	0.3078	0.3106	0.3133
0.9	0.3159	0.3186	0.3212	0.3238	0.3264	0.3289	0.3315	0.3340	0.3365	0.3389
1.0	0.3413	0.3438	0.3461	0.3485	0.3508	0.3531	0.3554	0.3577	0.3599	0.3621
1.1	0.3643	0.3665	0.3686	0.3708	0.3729	0.3749	0.3770	0.3790	0.3810	0.3830
1.2	0.3849	0.3869	0.3888	0.3907	0.3925	0.3944	0.3962	0.3980	0.3997	0.4015
1.3	0.4032	0.4049	0.4066	0.4082	0.4099	0.4115	0.4131	0.4147	0.4162	0.4177
1.4	0.4192	0.4207	0.4222	0.4236	0.4251	0.4265	0.4279	0.4292	0.4306	0.4319
1.5	0.4332	0.4345	0.4357	0.4370	0.4382	0.4394	0.4406	0.4418	0.4429	0.4441
1.6	0.4452	0.4463	0.4474	0.4484	0.4495	0.4505	0.4515	0.4525	0.4535	0.4545
1.7	0.4554	0.4564	0.4573	0.4582	0.4591	0.4599	0.4608	0.4616	0.4625	0.4633
1.8	0.4641	0.4649	0.4656	0.4664	0.4671	0.4678	0.4686	0.4693	0.4699	0.4706
1.9	0.4713	0.4719	0.4726	0.4732	0.4738	0.4744	0.4750	0.4756	0.4761	0.4767
2.0	0.4772	0.4778	0.4783	0.4788	0.4793	0.4798	0.4803	0.4808	0.4812	0.4817
2.1	0.4821	0.4826	0.4830	0.4834	0.4838	0.4842	0.4846	0.4850	0.4854	0.4857
2.2	0.4861	0.4864	0.4868	0.4871	0.4875	0.4878	0.4881	0.4884	0.4887	0.4890
2.3	0.4893	0.4896	0.4898	0.4901	0.4904	0.4906	0.4909	0.4911	0.4913	0.4916
2.4	0.4918	0.4920	0.4922	0.4925	0.4927	0.4929	0.4931	0.4932	0.4934	0.4936
2.5	0.4938	0.4940	0.4941	0.4943	0.4945	0.4946	0.4948	0.4949	0.4951	0.4952
2.6	0.4953	0.4955	0.4956	0.4957	0.4959	0.4960	0.4961	0.4962	0.4963	0.4964
2.7	0.4965	0.4966	0.4967	0.4968	0.4969	0.4970	0.4971	0.4972	0.4973	0.4974
2.8	0.4974	0.4975	0.4976	0.4977	0.4977	0.4978	0.4979	0.4979	0.4980	0.4981
2.9	0.4981	0.4982	0.4982	0.4983	0.4984	0.4984	0.4985	0.4985	0.4986	0.4986
3.0	0.4987	0.4987	0.4987	0.4988	0.4988	0.4989	0.4989	0.4989	0.4990	0.4990

Note: For values of z above 3.09, use 0.4999.

Example 9 IQ score in socially disparate communities

Psychologists and sociologists use scores on standardized intelligence quotient (IQ) tests to predict performance outcomes of individuals in different parts of society. In a study conducted by Naomi Breslau and colleagues, subjects from communities in southeastern Michigan and the city of Detroit had their IQ tested at age 6 and then again five years later at age 11. The summary statistics are given in Table 7.4.

Assume the distribution of IQs in each of the categories can be reasonably well approximated by a normal distribution. Determine the proportion of six-year-olds who have an IQ less than or equal to 110 in each of the two normal birth-weight groups.

Table 7.4 Mean score with standard deviation in parentheses of IQ measurements at age 6 and 5 years later at age 11 for children in Michigan stratified by birth weight and home location into eight cases

	Urban community 6-year-olds	Urban community 11-year-olds	Suburban community 6-year-olds	Suburban community 11-year-olds
Low birth weight ($\leq$ 2500 g)	90.1 (15.6)	88.1 (14.7)	107.0 (15.0)	107.8 (14.8)
Normal birth weight (> 2500 g)	99.1 (14.0)	94.1 (13.6)	113.3 (15.4)	112.8 (14.3)

Data source: N. Breslau et al., "Stability and Change in Children's Intelligence Quotient Scores: A Comparison of Two Socioeconomically Disparate Communities," *American Journal of Epidemiology* 154 (2001): 711–717.

Solution Let X be the (approximately) normally distributed IQ score of a randomly chosen six-year-old from the urban community. To find $P(X \leq 110)$, Example 8 tells us it is sufficient to find $P(Z \leq z)$ where $z = (110 - \mu)/\sigma$, μ is the mean of X, and σ^2 is the variance of X. From Table 7.4, we have $\mu = 99.1$ and $\sigma = 14.0$. Therefore,

$$z = \frac{110 - 99.1}{14.0} = 0.779$$

From Table 7.3, we find that $z = 0.78$ corresponds to the probability 0.2823. To this we have to add 0.5 for the area to the left of 0. Thus, the desired probability is

$$P(Z \leq 0.779) \approx P(Z \leq 0.78) = P(Z \leq 0) + P(0 \leq Z \leq 0.78) = 0.50 + 0.2832 \approx 0.78$$

or 78% of the population have an IQ of 110 or less.

To answer this question for the suburban community, Table 7.4 tells us that $\mu = 113.3$ and $\sigma = 15.4$. Therefore, for this community

$$z = \frac{110 - 113.3}{15.4} = -0.214$$

Using Table 7.3 and the symmetry of the standard normal distribution, we get

$$P(Z \leq -0.214) \approx P(Z \leq -0.21) = P(Z \leq 0) - P(0 \leq Z \leq 0.21) = 0.50 - 0.0832 \approx 0.42$$

or 42% of the population have an IQ of 110 or less.

Thus, 78% of normal-birth-weight urban six-year-olds, but only 42% of normal-birth-weight suburban six-year-olds, have an IQ less than 110.

Lognormal distribution

One potential problem with a normally distributed random variable X is that it can be negative; typically, however, biological data only assume positive values (e.g., height or weight). This issue may not be a problem if only the extreme tail of the distribution is associated with a negative value of X, as in the wheat yield example illustrated in Figure 7.24. Sometimes, however, the "normality" of a data set is not apparent until it has been appropriately transformed to take values on $(-\infty, \infty)$. For example, a common transformation for data sets of positive values is taking the natural logarithm of all the data values. Data that exhibit a normal distribution after such a transformation are said to be *lognormally distributed*.

Lognormal Distribution I

A random variable X is **lognormally distributed** if $\ln X$ is normally distributed. In other words, there exist parameters μ and $\sigma > 0$ such that

$$P(\ln X \leq a) = P(X \leq e^a) = \int_{-\infty}^{e^a} \frac{1}{\sqrt{2\pi}\sigma} e^{-\frac{(x-\mu)^2}{2\sigma^2}} \, dx$$

for any real number a.

Example 10   Strep throat incubation periods

The incubation period is the time elapsed between exposure to a pathogen and the time when symptoms and signs are first apparent. For streptococcal sore throat, this period (in hours) has been found to be lognormally distributed; the parameters of the log-transformed distribution are estimated to be $\mu = 4.07$ and $\sigma = 0.42$ (see Figure 7.26). Using these estimates, find the following quantities:

a. The fraction of individuals who start exhibiting symptoms within the first two days

b. The fraction of individuals who start exhibiting symptoms after three days

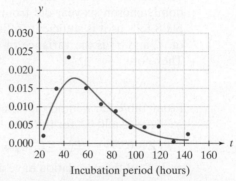

Figure 7.26 Incubation times in hours for a streptococcal sore throat.

Data Source: P. E. Sartwell, "The Distribution of Incubation Periods of Infectious Diseases," *American Journal of Epidemiology* 141 (1995): 386–394.

Solution

a. Let X be the number of hours for a randomly chosen incubation period. For these kind of data, It is usually reasonable to assume (an assumption that can be checked) that $\ln X$ is approximately normal. We want to know $P(X \leq 48)$. Equivalently, $P(\ln X \leq \ln 48)$. The desired z value for this probability is

$$z = \frac{\ln 48 - 4.07}{0.42} \approx -0.47$$

Using z-table (Table 7.3), we get

$$P(Z \leq -0.47) = P(Z \leq 0) - P(0 \leq Z \leq 0.47) = 0.5 - 0.1808 \approx 0.32$$

Hence, 32% of the people exhibited symptoms in the first two days.

b. We want to know $P(X > 72)$. Equivalently, $P(\ln X > \ln 72)$. The desired z value for this probability is

$$z = \frac{\ln 72 - 4.07}{0.42} \approx 0.49$$

Using z-table (Table 7.3), we get

$$P(Z > 0.49) = 0.5 - P(0 \leq Z \leq 0.49) = 0.5 - 0.1879 \approx 0.31$$

Hence, 31% of the people exhibited symptoms after three days.

The following example determines the PDF for the lognormal distribution and explores the effects of the parameters μ and σ on the shape of the distribution.

Example 11   Lognormal PDF

Let X be a random variable such that $\ln X$ has a normal distribution with mean μ and standard deviation σ.

a. Use a change of variables to find the PDF for X.

b. For $\sigma = 1$, plot the PDF of X with $\mu = -1, 0$, and 1. Discuss how μ influences the shape of X's PDF.

c. For $\mu = 1$, plot the PDF of X with $\sigma = 0.5, 1$, and 1.5. Discuss how σ influences the shape of X's PDF.

Solution

a. Since $\ln X$ is normally distributed, the PDF of $\ln X$ is given by

$$\frac{1}{\sqrt{2\pi}\sigma} e^{-\frac{(x-\mu)^2}{2\sigma^2}}$$

To determine the PDF of X, let's begin by finding an expression for $P(X \le a)$ for any positive real number a. Since $X \le a$ if and only if $\ln X \le \ln a$, we obtain

$$P(X \le a) = P(\ln X \le \ln a)$$

$$= \int_{-\infty}^{\ln a} \frac{1}{\sqrt{2\pi}\sigma} e^{-\frac{(x-\mu)^2}{2\sigma^2}} \, dx$$

$$= \int_{0}^{a} \frac{1}{\sqrt{2\pi}\sigma} e^{-\frac{(\ln y-\mu)^2}{2\sigma^2}} \frac{dy}{y} \qquad \text{with the change of variables } y = e^x$$

By the fundamental theorem of PDFs, the PDF of X is given by

$$f(x) = \frac{1}{x\sqrt{2\pi}\sigma} e^{-\frac{(\ln x-\mu)^2}{2\sigma^2}}$$

for $x > 0$.

b. Using technology to plot the PDF of X with $\mu = -1, 0$, and 1 and $\sigma = 1$ yields the following graph:

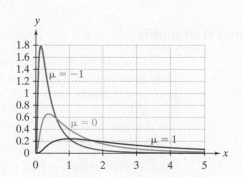

Increasing μ moves the center of the distribution to the right and increases the spread of the distribution about the center.

c. Using technology to plot the PDF of X with $\sigma = 0.5, 1,$ and 1.5 and $\mu = 0$ yields the following graph:

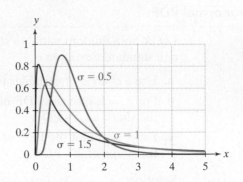

Increasing σ moves the center of the distribution to the left but still increases the spread of the distribution, as represented by the size of the tails (i.e., the area under the curve from $x = 2$ to 3).

In Problem 27 in Problem Set 7.4, you will be asked to show that the mean and variance of the lognormal distribution satisfy the relationships given below.

Lognormal Distribution II

The PDF of the lognormal distribution is defined in terms of two positive parameters $\mu > 0$ and $\sigma > 0$ by the function

$$f(x) = \begin{cases} \dfrac{1}{x\sqrt{2\pi}\sigma} e^{-\frac{(\ln x - \mu)^2}{2\sigma^2}} & for \quad x \geq 0 \\ 0 & otherwise \end{cases}$$

The mean m and variance v of this distribution are given by

$$m = e^{\mu + \sigma^2/2}$$

and

$$v = (e^{\sigma^2} - 1)e^{2\mu + \sigma^2}$$

Example 12   Survival of moths

An entomologist needs adult moths for her wind tunnel studies on how moths navigate their way in flight using pheromones in an odor plume. In a pilot study, she reared the moths from eggs until they eclosed from their pupal stage; then she selected 194 of the healthiest looking individuals for her flight studies. In Table 7.5, the number of moths dying each week is given until the last moth died in the 28th week. Now do the following:

a. Calculate the proportion of moths dying each week and variance of the resulting distribution.

b. Calculate the mean and variance of age at death.

c. Calculate the parameters for the lognormal distribution using the estimates from part **b** and plot the normal and lognormal distributions based on these estimates against the data.

Table 7.5 Number of moths dying each week (values rounded to three decimal places for presentation purpose)*

Week (i)	Number that die	Proportion that dies (p_i)
1	0	0.000
2	3	0.015
3	12	0.062
4	15	0.077
5	26	0.134
6	21	0.108
7	13	0.067
8	20	0.103
9	15	0.077
10	10	0.052
11	7	0.036
12	16	0.082
13	3	0.015
14	3	0.015
15	10	0.052
16	5	0.026
17	2	0.010
18	5	0.026
19	0	0.000
20	0	0.000
21	3	0.015
22	1	0.005
23	1	0.005
24	0	0.000
25	0	0.000
26	0	0.000
27	2	0.010
28	1	0.005
Sum	194	1
Mean		8.42
Variance		25.06

*The sum of the p_is might not be exactly one, depending on rounding errors.

Solution

a. Since the total number of moths at the beginning of the first week is 194, the proportion dying in week i ($i = 1, \ldots, 28$) is the number dying in that week divided by 194. See column 3 of Table 7.5.

b. The mean age at death m is obtained from the calculation

$$m = \sum_{i=1}^{28} (i - 0.5) p_i$$

Note that we selected the midpoint of each week to represent the point at which all individuals die during the week. This, of course, is an approximation, but some approximation must be used because of the discrete nature of the problem. The answer obtained using the given formula is 8.42. The variance v associated with age at death is obtained from the calculation

$$v = \sum_{i=1}^{28} (i - 0.5)^2 p_i - m^2$$

The answer obtained is 25.06.

c. If the observed mean and variance are $m = 8.46$ and $v = 25.15$, then we need to use the relationships $m = e^{\mu+\sigma^2/2}$ and $v = (e^{\sigma^2} - 1)e^{2\mu+\sigma^2}$ to find parameters μ and σ for a lognormal distribution with the observed mean m and variance v. In Problem 38 of Problem Set 7.4, you are asked to show that the resulting equations are

$$\mu = 2 \ln m - \frac{1}{2} \ln(m^2 + v)$$

and

$$\sigma^2 = -2 \ln m + \ln(m^2 + v)$$

Solving these yields $\mu = 1.98$ and $\sigma^2 = 0.30$. The lognormal distribution generated by these parameters is plotted in red in Figure 7.27. In contrast, the normal distribution with mean 8.46 and variance 25.15 is plotted in black Figure 7.27. Clearly, the lognormal distribution provides a much better fit to the data.

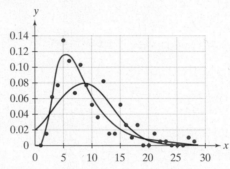

Figure 7.27 Fraction of moths dying each week is plotted over the 28-week period for the actual data (closed circles), as well as the lognormal (red curve) and normal (black curve) distributions that have the same mean and variance as the data.

PROBLEM SET 7.4

Level 1 DRILL PROBLEMS

Assume that a data set is normally distributed with a mean of 0 and a standard deviation of 1. A value X is randomly selected. Find the probability requested in Problems 1 to 4.

1. $P(0 \leq X < 0.85)$ **2.** $P(X \leq 0)$

3. $P(X \geq 0.55)$ **4.** $P(-1.00 < X < 0.75)$

5. In the PDF of the standard normal distribution, find the area under the PDF bounded by the lines $z = 1.20$ and $z = 1.90$ and compare this with the value of $z = 1.90 - 1.20 = 0.70$ in Table 7.3.

6. For X normally distributed with mean $\mu = -1$ and standard deviation $\sigma = 1$, calculate $P(X \geq 0)$.

7. For X normally distributed with mean $\mu = 1$ and standard deviation $\sigma = 2$, calculate $P(X > 0)$.

8. For X normally distributed with mean $\mu = -2$ and standard deviation $\sigma = 2$, calculate $P(-3.00 < X < -1.00)$.

9. For X lognormally distributed with log mean $\mu = -2$ and log standard deviation $\sigma = 2$, calculate $P(e^{-3} < X < e^{-1})$.

10. For X lognormally distributed with log mean $\mu = 0$, calculate $P(0 < X < 1)$.

11. For X lognormally distributed with log mean $\mu = 0$ and log standard deviation $\sigma = 1$, calculate $P(0 < X < 0.5)$.

12. For X lognormally distributed with log mean $\mu = 1$ and log standard deviation $\sigma = 2$, calculate $P(1 < X < 4)$.

13. Example 4 discussed the spatial spread of the large tunicate *Pyura praeputialis* on the Chilean coast. In this example, the fraction of habitat occupied by this tunicate species at time t equals

$$y(t) = \frac{1}{1 + e^{4-1.7t}}$$

where t is measured in hundreds of days.

a. What fraction of habitat was occupied on day $t = 0$?

b. What fraction of habitat was occupied at 100 days?

c. At what point in time will 95% of the habitat be covered?

14. Suppose the fraction of habitat occupied by a tunicate species at time t equals

$$y(t) = \frac{1}{1 + e^{5-2t}}$$

where t is measured in hundreds of days. Let T be the time a randomly chosen location becomes occupied. Find the mean, median, and variance of T.

15. Consider Example 1 with $r = 0.1$ (units per month) and $y(0) = 0.5$ (a relatively slow-spreading disease).

a. Solve the differential equation for $y(t)$.

b. Verify that $y(t)$ is a CDF.

c. Find the probability that a randomly chosen individual is infected with the disease in the next two months.

16. Consider Example 1 with $r = 3$ (units per month) and $y(0) = 0.5$ (a relatively fast-spreading disease).

a. Solve the differential equation for $y(t)$.

b. Verify that $y(t)$ is a CDF.

c. Find the probability that a randomly chosen individual is infected with the disease in fifteen days (i.e., 0.5 months).

17. Consider Example 1 with $r = 1$ (units per month) and $y(0) = 0.1$ (i.e., 10% of the population have the disease).

a. Solve the differential equation for $y(t)$.

b. Verify that $y(t)$ is a CDF.

c. Find the probability that a randomly chosen individual is infected with the disease in one month.

18. Consider Example 1 with $r = 0.5$ (units per month) and $y(0) = 0.3$.

a. Solve the differential equation for $y(t)$.

b. Verify that $y(t)$ is a CDF.

c. Find the probability that a randomly chosen individual is infected with the disease in the 1.5 months.

Use logistic regression to find the best-fitting functions $p(t)$ to the data in the sets $D = \{(t_1, p_1), \ldots, (t_n, p_n)\}$ given in Problems 19 to 22.

19. $D = \{(1, 0.10), (2, 0.15), (3, 0.30), (4, 0.49),$
$(5, 0.58), (6, 0.76), (7, 0.87), (8, 0.95),$
$(9, 0.93), (10, 0.98)\}$

20. $D = \{(1, 0.03), (2, 0.02), (3, 0.08), (4, 0.09),$
$(5, 0.21), (6, 0.30), (7, 0.52), (8, 0.61),$
$(9, 0.84), (10, 0.88)\}$

21. $D = \{(1, 0.01), (3, 0.01), (5, 0.03), (7, 0.03),$
$(9, 0.10), (11, 0.18), (13, 0.29), (15, 0.48),$
$(17, 0.73), (19, 0.85), (21, 0.87)\}$

22. $D = \{(1, 0.16), (3, 0.17), (5, 0.27), (7, 0.34),$
$(9, 0.44), (11, 0.58), (13, 0.63), (15, 0.77),$
$(17, 0.78), (19, 0.85), (21, 0.92)\}$

Level 2 APPLIED AND THEORY PROBLEMS

23. In a large study, human birth weights were found to be approximately normally distributed with mean of 120 ounces and standard deviation of 18 ounces (1 pound = 16 ounces; 1 ounce = 28.35 grams).

a. Find the probability that a randomly chosen baby has a birth weight of 8 pounds or less.

b. Find the probability that a randomly chosen baby weighs between 6 and 8 pounds at birth.

c. Find the probability that a randomly chosen baby weighs more than 9 pounds at birth.

24. A patient is said to be hyperkalemic (high levels of potassium in the blood) if the measured level of potassium is 5.0 milliequivalents per liter (meq/L) or more. In a population of students at Ozark University, the distribution of potassium levels is normally distributed with mean 4.5 meq/L and standard deviation 0.4 meq/L. Estimate the proportion of students who are hyperkalemic.

25. The gestation period of a pregnant woman is normally distributed with mean of 279 days and standard deviation of 16 days.

a. Find the probability that the gestation period is between 263 and 295 days.

b. Find the probability that the gestation period is greater than 303 days.

26. Answer the following questions for the data in Example 9.

a. What is the IQ value that corresponds to the 95th percentile for each of the two six-year-old low-birth-weight groups in Table 7.4?

b. In the normal-birth-weight urban and suburban communities, what is the change from age 6 to age 11 in the estimated proportion of individuals who have an IQ of 140 and above?

c. In the two eleven-year-old low-birth-weight communities, the 50th percentile of the suburban community corresponds to which percentile in the urban community?

In Problems 27 to 30, we emphasize that we are dealing with the lognormal distribution and note that the value e^{μ} is not the mean but the median (see Problem 35) and that *the dispersion parameter σ is not the square root of the variance v of the distribution.*

27. The latent period of disease is the time from a person initially getting infected to the moment the person exhibits first symptoms. In a paper by Sartwell, as referenced in Figure 7.26, Sartwell reported that the latency period (measured in days) of salmonellosis was approximately lognormally distributed. Taking the natural logs of the latency periods that are measured in days, he estimated that $\mu = \ln(2.4)$ and $\sigma = \ln(1.47)$. Using these estimates, find the following quantities:

    a. The fraction of individuals who start exhibiting symptoms within the first three days

    b. The fraction of individuals who start exhibiting symptoms after four days

    c. The fraction of individuals who start exhibiting symptoms between the start of the second day and end of the third day

28. The latent period of disease is the time from a person initially getting infected to the moment the person exhibits first symptoms. Sartwell found that the latency period (measured in days) of poliomyelitis was approximately lognormally distributed. Taking the natural logs of the latency periods that are measured in days, he estimated that $\mu = \ln(12.6)$ and $\sigma = \ln(1.5)$. Using these estimates, find the following quantities:

    a. The fraction of individuals who start exhibiting symptoms within the first two weeks

    b. The fraction of individuals who start exhibiting symptoms after ten days

    c. The fraction of individuals who start exhibiting symptoms between the start of the twelfth day and the end of the fifteenth day.

29. The survival time after cancer diagnosis is the number of days a patient lives after being diagnosed with cancer. In an article titled "Variation in the Duration of Survival of Patients with the Chronic Leukemias," *Blood.* 1960 Mar; 15: 332–349, M. Feinleib and B. McMahon reported that the survival time for female patients diagnosed with lymphatic leukemia (measured in months) was approximately lognormally distributed. Taking the natural logs of the survival times, they estimated that $\mu = \ln(17.2)$ and $\sigma = \ln(3.21)$. Using these estimates, find the following quantities:

    a. The fraction of individuals who survived less than one year

    b. The fraction of individuals who survived at least two years

    c. The fraction of individuals who survived between 1 and 1.5 years

30. The survival time after cancer diagnosis is the number of days a patient lives after being diagnosed with cancer. Feinleib and McMahon, in a study cited in the previous problem, found that the survival time for female patients diagnosed with myelocytic leukemia (measured in months) was approximately lognormally distributed. Taking the natural logs of the survival times, they estimated that $\mu = \ln(15.9)$ and that $\sigma = \ln(2.80)$. Using these estimates, find the following quantities:

    a. The fraction of individuals who survived less than one year

    b. The fraction of individuals who survived at least two years

    c. The fraction of individuals who survived between the start of thirteen months and end of eighteen months (i.e., between 1 and 1.5 years)

31. In looking over her data, the entomologist mentioned in Example 12 found that she had transposed the number of individuals dying in weeks 7 and 8. After fixing this mistake, redo all the calculations covered in Example 12. Note differences to the estimates of the mean and variance associated with the actual data and the log mean μ and log variance σ^2 for the associated log normal PDF.

32. Linguist G. Herdan (see "The Relation Between the Dictionary Distribution and the Occurrence Distribution of Word Length and Its Importance for the Study of Quantitative Linguistics," *Biometrika* 45 (A58): 222–228) found that length of spoken words in $n = 738$ phone conversations was lognormally distributed, with mean $m = 5.05$ letters and variance $v = 2.16$ letters. Find the probability that a randomly chosen word had six or more letters. Hint: See Problem 38.

33. The Gompertz growth equation normalized so that the variable y has the interpretation of a proportion (i.e., $y = 1$ is an equilibrium and upper bound) is given by

$$\frac{dy}{dt} = -r\,y\ln(y)$$

This equation can be used to model a variety of population processes, including tumor growth (y is proportion of maximum size), population growth (y is proportion of environmental carrying capacity), and acquisition of new technologies, as illustrated in the following example.

   The Gompertz equation has been used to model mobile phone uptake, where $y(t)$ is the fraction of individuals who have a mobile phone by time t (say, in years) and r is a parameter that can be fitted to the actual data. Using this model, we can derive a probability density function that represents the time at which an individual acquired her first mobile phone.

To illustrate this idea, assume that $y(0) = 1/e$ (i.e., currently 36.79% of people have mobile phones) and $r = 1$.

a. Solve the differential equation for $r = 1$ and $y(0) = 1/e$.

b. Verify that $F(t) = 1 - y(t)$, where $y(t)$ is the solution found in part **a**, is a CDF.

c. Find the PDF for your CDF.

d. Compute the probability that a randomly chosen individual acquires a mobile phone two years from now.

34. Consider Example 1 with $r > 0$ and $y(0) = y_0 \in (0, 1)$.

a. Solve the differential equation for $y(t)$.

b. Verify that $F(t) = 1 - y(t)$ is a CDF.

35. Show that the PDF of a normal curve has its maximum (i.e., median) at $x = \mu$ and points of inflection at $x = \mu + \sigma$ and $x = \mu - \sigma$.

36. Consider Example 1 with $r > 0$ and $y(0) = y_0 \in (0, 1)$.

a. Verify that $y(t)$ can be written as $y(t) = \dfrac{1}{1 + e^{a-rt}}$ where $a = \ln(1/y_0 - 1)$.

b. Find the PDF for this CDF.

37. For the lognormal distribution defined by

$$f(x) = \begin{cases} \dfrac{1}{x\sqrt{2\pi}\sigma} e^{-\frac{(\ln x - \mu)^2}{2\sigma^2}} & for \quad x \geq 0 \\ 0 & otherwise \end{cases}$$

show that the mean m and variance v are given by

$$m = e^{\mu + \sigma^2/2}$$

and

$$v = (e^{\sigma^2} - 1)e^{2\mu + \sigma^2}$$

38. If $\ln X$ is a normally distributed random variable with mean μ and variance σ^2, and X is a lognormally distributed random variable with mean m and variance v, then show that

$$\mu = 2\ln m - \frac{1}{2}\ln(m^2 + v)$$

and

$$\sigma^2 = -2\ln m + \ln(m^2 + v)$$

7.5   Life Tables

In Section 6.1 we introduced the simplest differential equation model of population growth:

$$\frac{dN}{dt} = rN$$

This model implicitly assumes that all individuals, whether young or old, have the same mortality and fecundity rates. Although this assumption is a useful first approximation, mortality and fecundity are often age dependent. For instance, many animals become sexually mature only after they have reached a particular age. Additionally, the risk of mortality is often higher at younger and older ages. In this section, we consider models that account for age-specific mortality and reproduction.

Survivorship functions

Biology professor Gregory Erickson and colleagues studied fossils of four North American tyrannosaurs—*Albertosaurus*, *Tyrannosaurus*, *Gorgosaurus*, and *Daspletosaurus*. Using the femur bones of these fossils, the scientists estimated that the life spans of the dinosaurs ranged from birth to 28 years. Based on these estimates, the scientists created a *life table* for each of the dinosaurs. These life tables keep track of what fraction $l(t)$ of individuals survived to age t. For example, the life table for *Albertosaurus sarcophagus* (see Figure 7.28) is reported in Table 7.6.

Survivorship Function	A function $l : [0, \infty) \rightarrow [0, 1]$ is a **survivorship function** if

- $l(0) = 1$; that is, all individuals survive to age 0
- $l(t)$ is nonincreasing; that is, if an individual survived to age t, then it survived to all earlier ages
- $\lim\limits_{t \to \infty} l(t) = 0$; that is, all individuals eventually die

Figure 7.28 *Albertosaurus sarcophagus*

Table 7.6 Life table for *A. sarcophagus*

age t (in years)	$l(t)$	age t (in years)	$l(t)$
0	1.00	15	0.32
2	0.60	16	0.27
4	0.56	17	0.24
6	0.54	18	0.19
8	0.51	19	0.16
9	0.48	20	0.10
11	0.46	21	0.08
12	0.43	23	0.05
13	0.40	28	0.02
14	0.38		

Data source: G. M. Erikson et al., "Tyrannosaur Life Tables: An Example of Nonavian Dinosaur Population Biology," *Science* 313 (2006): 213–216.

Example 1   Aging dinosaurs

Use Table 7.6 to do the following:

a. Determine what fraction of dinosaurs died between ages 4 and 6.

b. Determine what fraction of dinosaurs died between ages 11 and 14.

c. Plot $l(t)$ and discuss its shape.

Solution

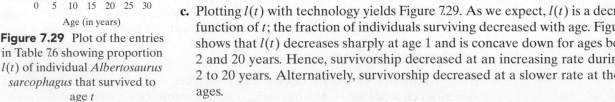

Figure 7.29 Plot of the entries in Table 7.6 showing proportion $l(t)$ of individual *Albertosaurus sarcophagus* that survived to age t

a. Since $l(4) = 0.56$, 56% of the dinosaurs survived to age 4. Similarly, $l(6) = 54\%$ of the dinosaurs survived to age 6. Since $l(4) - l(6) = 2\%$, it follows that 2% of dinosaurs died between ages 4 and 6.

b. Since $l(11) = 46\%$ of the dinosaurs survived to age 11 and $l(14) = 38\%$ of the dinosaurs survived to age 14, $l(11) - l(14) = 8\%$ of those alive at age 2 died between ages 11 and 14.

c. Plotting $l(t)$ with technology yields Figure 7.29. As we expect, $l(t)$ is a decreasing function of t; the fraction of individuals surviving decreased with age. Figure 7.29 shows that $l(t)$ decreases sharply at age 1 and is concave down for ages between 2 and 20 years. Hence, survivorship decreased at an increasing rate during ages 2 to 20 years. Alternatively, survivorship decreased at a slower rate at the older ages.

Survivorship functions have a natural relationship to CDFs of an appropriate random variable, as the following example shows.

Example 2   From survivorship to CDFs and PDFs

Let $l(t)$ be the survivorship function for *A. sarcophagus* and let X be the age at which a randomly chosen *A. sarcophagus* dies. If F is the CDF for X, then determine the relationship between F and l. If X is a continuous random variable, what is the PDF for X?

Solution   Since $l(t)$ is the fraction of individuals that die after age t, $l(t) = P(X > t)$. Alternatively, $F(t) = P(X \leq t)$. Since $P(X > t) = 1 - P(X \leq t) = 1 - F(t)$, we have that $l(t) = 1 - F(t)$ and $F(t) = 1 - l(t)$. If X is a continuous random variable, then by the fundamental theorem of PDFs in Section 7.2, the PDF of X is given by $F'(t) = -l'(t)$. ∎

Using Table 7.6, we can determine how the mortality rates of *A. sarcophagus* vary with age. In particular, imagine (as did a famous movie!) that on a remote island scientists were able to create 100 *A. sarcophagus* babies. The life table implies that of these 100, 60 survive to age 2 and 56 survive to age 4. Hence, 4 of 60 individuals die from age 2 to 4. Thus, the mortality rate over this two-year period is $4/60 = 6.7\%$ and the annual mortality rate is approximately 3.3% per year. Equivalently, we estimate the mortality rate as

$$\frac{1}{4-2}\left(\frac{l(2) - l(4)}{l(2)}\right) = \frac{1}{2}\frac{0.6 - 0.56}{0.6} \approx 0.033$$

In the following example, we compute and interpret the mortality rates for the remaining age classes.

Example 3 Dinosaur mortality rates

Refer to the life table for *A. sarcophagus*.

a. Determine age-specific mortality rates.

b. Discuss which ages were most susceptible and least susceptible to mortality.

Solution

a. For the mortality rate from age 0 to age 2, we have

$$\frac{1}{2-0}\left(\frac{l(0) - l(2)}{l(0)}\right) = \frac{1}{2}\frac{0.4}{1} = 0.2 \text{ per year}$$

We already found that the mortality rate at age 2 is 0.033. To determine the mortality rate at age 4, we can compute

$$\frac{1}{6-4}\left(\frac{l(4) - l(6)}{l(4)}\right) \approx 0.018 \text{ per year}$$

Computing the remaining mortality $m(t)$ rates yields this table:

age t	mortality rate	age t	mortality rate
0	0.200	14	0.158
2	0.033	15	0.156
4	0.018	16	0.111
6	0.028	17	0.208
8	0.059	18	0.158
9	0.021	19	0.375
11	0.065	20	0.200
12	0.070	21	0.188
13	0.050	23	0.120

b. This table suggests the mortality rate in the first year is greatest. For individuals surviving after the first year, mortality risk tends to increase with age and then decrease in the last few years. ∎

In Example 3, we computed mortality rates using the relationship

$$m(t) = \frac{1}{\Delta t}\left(\frac{l(t) - l(t + \Delta t)}{l(t)}\right) \text{ per year}$$

where Δt is the step size between measurements. Multiplying both sides of this equation by $-l(t)$ yields

$$-l(t)m(t) = \frac{l(t + \Delta t) - l(t)}{\Delta t}$$

Taking the limit as Δt approaches 0 provides the following result:

Survivorship-Mortality Equation

If $l(t)$ is the fraction of individuals that survive to age t and $m(t)$ is the mortality rate at age t, then $l(t)$ and $m(t)$ satisfy the equation

$$l'(t) = -m(t)l(t)$$

Equivalently,

$$m(t) = -\frac{l'(t)}{l(t)}$$

Example 4   Constant mortality rates

For many short-lived mammals and birds, the mortality rate $m(t)$ is approximately constant*. Assuming that $m(t) = m$ for all t, determine $l(t)$ and the CDF associated with this survival function. Do they look familiar?

Solution   If $m(t) = m$ is constant, then $l'(t) = -ml(t)$. The general solution to this equation is $l(t) = l(0)e^{-mt}$. Since all individuals survive to age 0, $l(0) = 1$ and $l(t) = e^{-mt}$. In Example 2, we noted that $1 - l(t) = 1 - e^{-mt}$ for $t \geq 0$ is the CDF for the distribution of ages. This CDF corresponds to the exponential distribution with mean $\frac{1}{m}$. Hence, for individuals with a constant mortality rate m per year, life expectancy is $\frac{1}{m}$ years.

Example 5   Mortality rates for humans in the United States

Life tables are an ancient and important tool for human demography. They are widely used for descriptive and analytic purposes in public health, health insurance, epidemiology, and population geography. In recognition of the importance of these life tables, the Max Planck Institute for Demographic Research, the University of California at Berkeley, and the Institut national d'études démographiques developed The Human Life-Table Database, an online resource for human life tables at http://www.lifetable.de/. Data for male and female survivorship functions are illustrated in Figure 7.30. The survivorship function of males is well approximated by the function

$$l(t) = e^{-e^{-4.94 + 0.034t + 0.00032t^2}}$$

where t is measured in years. Compute and interpret the mortality rate $m(t)$ for $l(t)$.

---

*T. A. Ebert, *Plant and Animal Populations: Methods in Demography* (San Diego: Academic Press, 1999).

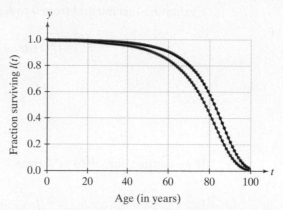

Figure 7.30 Life tables for females (in red) and males (in blue) for the United States in 2005. Black curves correspond to the best-fitting curves to the data.

Data Source: www.lifetable.de

Solution The mortality rate is given by $m(t) = -\dfrac{l'(t)}{l(t)}$. By the chain rule,

$$l'(t) = e^{-e^{-4.94+0.034t+0.00032t^2}}(-e^{-4.94+0.034t+0.00032t^2})(0.034 + 0.00064t)$$

$$= -l(t)e^{-4.94+0.034t+0.00032t^2}(0.034 + 0.00064t)$$

Therefore,

$$m(t) = -\frac{l'(t)}{l(t)} = e^{-4.94+0.034t+0.00032t^2}(0.034 + 0.00064t) \text{ per year}$$

Hence, the instantaneous mortality rate is initially low (approximately 0.02% at $t = 0$) in the first year and increases super-exponentially (approximately 4.5% mortality rate for 75-year-olds) as shown on the left.

Life expectancy

Given a survival function $l(t)$ for a population, we can ask: What is the life expectancy of an individual? To answer this question, let X be the age at which a randomly chosen individual dies. The mean of X is the mean life span of an individual in the population. To compute this mean, recall that the CDF for X is given by $F(t) = 1 - l(t)$ for $t \geq 0$ and 0 otherwise. Hence, the PDF for X (assuming l is differentiable!) is $-l'(t)$ for $t \geq 0$ and 0 otherwise. The mean of X is given by

$$\int_0^\infty -tl'(t)\, dt$$

provided the improper integral is convergent. Let's assume it is convergent. We can simplify the integral by integrating by parts. Define $u = t$ and $dv = -l'(t)dt$, so that $du = dt$ and $v = -l(t)$. This yields

$$\int -tl'(t)\, dt = -tl(t) + \int l(t)\, dt$$

Evaluating this integral from 0 to b and taking the limit as $b \to \infty$ yields

$$\int_0^\infty -tl'(t)\, dt = \lim_{b \to \infty} \int_0^b -tl'(t)\, dt$$

$$= \lim_{b \to \infty} \left(-tl(t) \Big|_0^b + \int_0^b l(t)\, dt \right)$$

$$= \lim_{b \to \infty} -bl(b) + \int_0^\infty l(t)\, dt$$

If we assume $\lim_{b \to \infty} bl(b) = 0$, then

$$\int_0^\infty -tl'(t)\, dt = \int_0^\infty l(t)\, dt$$

Hence, we proved the result shown in Theorem 7.4.

Theorem 7.4   Life expectancy theorem

Let $l(t)$ be a continuously differentiable survivorship function satisfying $\lim_{b \to \infty} bl(b) = 0$. *Let X be the random variable whose CDF is given by $1 - l(t)$ for $t \geq 0$ and 0 otherwise. Then, the mean of X, which is the life expectancy of an individual, equals*

$$\int_0^\infty l(t)\, dt$$

provided that the integral is convergent.

All the mortality functions $l(t)$ we use have the property that $\lim_{b \to \infty} bl(b) = 0$. This limit condition is, of course, much stronger than the requirement that $l(b)$ approach 0 as b increases. This condition, however, is not sufficient to ensure convergence. For example, if $l(t) = \dfrac{1}{t \log t}$ for $t \geq 2$, then $\int_0^\infty l(t)\, dt$ is divergent.

Example 6   Life expectancy of A. *sarcophagus*

Estimate the mean age of *A. sarcophagus* using Table 7.6.

Solution   Using the right endpoint rule and assuming a maximum life span is 30 years, we get

$$\int_0^\infty l(t)\, dt = \int_0^{30} l(t)\, dt$$

$$\approx l(2)2 + l(4)2 + l(6)2 + l(8)2 + l(9)1 + \cdots + l(23)2 + l(28)5$$

$$= 0.6 \cdot 2 + 0.56 \cdot 2 + 0.54 \cdot 2 + 0.51 \cdot 2 + 0.51 \cdot 1 + \cdots + 0.05 \cdot 2 + 0.03 \cdot 5$$

$$= 8.59 \text{ years}$$

Hence, the life expectancy of *A. sarcophagus* is 8.59 years. In the Problem Set 7.5, you are asked to verify that the left endpoint rule provides a more optimistic life expectancy of 10.23 years.

Example 7   Men versus Women

For the data presented in Example 5, the survival function for males is well approximated by

$$l_M(t) = e^{-e^{-4.94 + 0.034t + 0.00032t^2}}$$

and the survival function for females is well approximated by

$$l_W(t) = e^{-e^{-5.057+0.01483t+0.0005t^2}}$$

Use numerical integration to estimate the life expectancy of males and females in the United States.

Solution   Using technology, we get

$$\int_0^\infty l_M(t)\,dt \approx 74.1157 \text{ years}$$

Using technology, we get

$$\int_0^\infty l_W(t)\,dt \approx 79.7978 \text{ years}$$

Hence, females were expected to live five years longer than males.   ■

Example 8   Older is better

Consider a hypothetical population whose mortality rate is

$$m(t) = \frac{3}{1+t} \text{ per year}$$

Determine the life expectancy of this population.

Solution   To determine the life expectancy, we need to find $l(t)$. Since $l(t)$ must satisfy $l'(t) = -m(t)l(t)$ and $l(0) = 1$, we can use separation of variables to solve for $l(t)$:

$$\int \frac{dl}{l} = -\int \frac{3\,dt}{1+t}$$

$$\ln l = -3\ln(1+t) + C = \ln(1+t)^{-3} + C$$

$$l = (1+t)^{-3}e^C = \frac{e^C}{(1+t)^3}$$

Since $l(0) = 1 = e^C$, we get $l(t) = \dfrac{1}{(1+t)^3}$.

To find the life expectancy, we need to compute $\displaystyle\int_0^\infty \frac{dt}{(1+t)^3}$. Using the substitution $u = 1 + t$, we get

$$\int \frac{dt}{(1+t)^3} = \int \frac{du}{u^3}$$

$$= -\frac{1}{2u^2} + C$$

$$= -\frac{1}{2(1+t)^2} + C$$

Therefore,

$$\int_0^\infty \frac{dt}{(1+t)^3} = \lim_{b\to\infty} -\frac{1}{2(1+b)^2} + \frac{1}{2}$$

$$= \frac{1}{2} \text{ years}$$

Reproductive success

So far we have only considered the likelihood of an individual surviving until a certain age. To better understand the dynamics of a population, we also need to know how the reproductive success of individuals depends on their ages. In other words, how many progeny do individuals of a particular age produce on average? In developing the models, we let $b(t)$ denote the average number of progeny produced by an individual of age t. The likelihood $l(t)$ of an individual surviving to age t in conjunction with $b(t)$ provides a considerable amount of information about the demography of a population, as the following example illustrates.

Figure 7.31 The vole
Microtus agrestis

Example 9   Vole life history

Table 7.7 Life table for *Microtus agrestis* where t is measured in weeks, $l(t)$ is the fraction of females surviving to age t, and $b(t)$ is the average number of female offspring produced per week by an individual of age t

t	$l(t)$	$b(t)$
8	0.83	0.08
16	0.73	0.30
24	0.59	0.37
32	0.43	0.31
40	0.29	0.21
48	0.18	0.14
56	0.10	0.08
64	0.05	0.05
72	0.03	0.04

Data source: H. G. Andrewartha and L. C. Birch, *The Distribution and Abundance of Animals*, University of Chicago Press, Chicago, 1954.

In their classic text, *The Distribution and Abundance of Animals*, ecologists H. G. Andrewartha and L. C. Birch created the life table, Table 7.7, for females of the vole species *Microtus agrestis*. Use this table to answer the following question: If you were given 100 female voles of age 0, and you placed them in your backyard, how many female progeny would they produce during their lifetime? Assume that no individuals live beyond 72 weeks and that the entries $b(t)$ in Table 7.7 apply to all females surviving each of the eight-week periods over which the data are discretized.

Solution   Of the 100 females voles, we expect 83% will survive to week 8. Each of these 83 will produce on average 0.08 daughters per week. Hence, in the interval [0, 8], we expect $83 \times 0.08 \times 8 = 53.12$ daughters to be produced. Then, 73% of the female voles survive to week 16. Each of these surviving females will produce on average 0.3 daughters per week from week 8 to week 16. Hence, in the interval [8, 16], we expect $73 \times 0.3 \times 8 = 175.2$ daughters to be produced. Continuing in this manner, we get Table 7.8.

Adding all these daughters yields 588 daughters expected to be produced by 100 female vole. Equivalently, each female vole will produce on average 5.88 daughters.

Table 7.8 Number of daughters (rounded to nearest integer) produced by 100 female voles as they pass through each of the specified age categories and a proportion drop out of each category according to the survival schedule (function) $l(t)$

Age categories	Daughters
[0, 8]	53
[8, 16]	175
[16, 24]	175
[24, 32]	107
[32, 40]	49
[40, 48]	20
[48, 56]	6
[56, 64]	2
[64, 72]	1
Total	588

Example 9 illustrates how to use a life table to determine the average number of daughters produced by a female during her lifetime. To generalize the computations in Example 9 to an arbitrary survival function $l(t)$ and an arbitrary birth function $b(t) \geq 0$, assume that initially there are N females (e.g., $N = 100$ in Example 9) and that Δt is the width of the time intervals for life table (e.g., $\Delta t = 8$ in Example 9). The number of females that survive to age $t_1 = \Delta t$ is $N l(t_1)$. Each of these females produces $b(t_1)\Delta t$ daughters. Hence, by time t_1, there are $N l(t_1)b(t_1)\Delta t$ daughters. The number of females that survive to age $t_2 = 2\Delta t$ is $N l(t_2)$. Each of these females produces approximately $b(t_2)\Delta t$ daughters in the time interval $[t_1, t_2]$. Hence, by time t_2, there are approximately

$$Nl(t_1)b(t_1)\Delta t + Nl(t_2)b(t_2)\Delta t$$

daughters produced. Continuing in this manner, there are approximately

$$Nl(t_1)b(t_1)\Delta t + Nl(t_2)b(t_2)\Delta t + Nl(t_3)b(t_3)\Delta t + Nl(t_4)b(t_4)\Delta t + \cdots$$

daughters produced. Taking the limit as $\Delta t \to 0$ yields the expected number of daughters D to be

$$D = N \int_0^\infty l(t)b(t)\, dt$$

If we now define the *reproductive number* R_0 to be the number of daughters that each individual female is expected to produce in her lifetime—that is $R_0 = D/N$—then we obtain the following relationship:

Reproductive Number

Let $l(t)$ be a survival function and $b(t)$ be a reproduction function. The **reproductive number** for the population, defined to be the average number of daughters produced by a female, is given by

$$R_0 = \int_0^\infty l(t)b(t)\, dt$$

whenever the improper integral is well defined.

Ignoring the role of males, if $R_0 > 1$, then each female more than replaces herself in each generation and the population grows. On the other hand, if $R_0 < 1$, then each female fails to fully replace herself in each generation and the population declines.

Example 10   Reproductive number for painted turtles

Painted turtles are found in Iowa where their favorite pastime is basking in the sun on warm March days. At night, they retire to the bottom of a wetland. Females lay their eggs in late May or June. Using a mark-recapture study, biology professor Henry Wilbur estimated the survival and reproductive functions of painted turtles. He found that $l(t) \approx 0.243e^{-0.273t}$ for $t \geq 1$ and $l(t) \approx e^{-1.69t}$ for $t < 1$. Moreover, he assumed that female turtles are reproductively mature at age 7 and that mature females produce on average 6.6 daughters per year. Using this information, do the following:

a. Estimate the life expectancy of a female painted turtle.

b. Estimate the reproductive number of the painted turtle. Based on this estimate, discuss whether you think the painted turtle population is increasing or decreasing.

Solution
a. To estimate the life expectancy, we need to compute $\int_0^\infty l(t)\, dt$. By the splitting property for integrals, $\int_0^\infty l(t)\, dt = \int_0^1 l(t)\, dt + \int_1^\infty l(t)\, dt$. The first integral equals

$$\int_0^1 e^{-1.69t}\, dt = \frac{1}{-1.69}(e^{-1.69} - 1) \approx 0.4825$$

Since, $\int 0.243e^{-0.273t}\,dt = -\dfrac{0.243}{0.273}e^{-0.273t} + C \approx -0.89e^{-0.273t} + C$, we obtain

$$\int_{1}^{\infty} 0.243e^{-0.273t}\,dt \approx \lim_{b\to\infty} -0.89e^{-0.273b} + 0.89e^{-0.273}$$

$$= 0.89e^{-0.273} \approx 0.6774$$

Therefore, the life expectancy is approximately $0.4825 + 0.6774 \approx 1.16$ years. Hence, the average female turtle is not expected to live to a reproductively mature age, though some do and they reproduce—but enough? The answer lies in the next part.

b. The reproductive number is given by $R_0 = \int_0^\infty l(t)b(t)\,dt$. Since $b(t) = 0$ for $t \le 7$,

$$R_0 = \int_{7}^{\infty} 0.243e^{-0.273t}6.6\,dt$$

Since, ignoring the constant of integration, $\int 0.243e^{-0.273t}6.6\,dt \approx -5.875e^{-0.273t}$, we find

$$R_0 = \lim_{b\to\infty} -5.875e^{-0.273b} + 5.875e^{-0.273\cdot 7}$$

$$\approx 0.8691$$

So a female painted turtle is expected to produce less than one daughter during her lifetime. This suggests that the population of painted turtles is in decline, as individuals are not replacing themselves over their lifetime.

Example 11   Reproductive number in the United States

According to the United Nation's online data website, http://data.un.org/, the reproductive number for women in the United States in the period 2000–2005 is one; on average, a woman produces one daughter during her lifetime. What does this tell us about $b(t)$? Recall, the survival function for women in the United States in 2005 is well approximated by

$$l(t) = e^{-e^{-5.057+0.01483t+0.0005t^2}}$$

Assume that women have a constant birthrate b during their "childbearing years," which in conventional international statistical usage is from age 15 to 49. Estimate b.

Solution   From the definition of R_0 and the reproductive number of a woman in the United States, we get

$$1 = R_0 = \int_{0}^{\infty} l(t)b(t)\,dt = b\int_{15}^{49} l(t)\,dt$$

Using numerical integration, we get

$$\int_{15}^{49} l(t)\,dt \approx 33.33$$

Hence, we get $1 = b \cdot 33.33$ or $b = 1/33.33 \approx 0.03$.

In addition to applications in demography, life tables can be used to understand the spread of disease in a population. As a striking parallel to the demographic process of survivorship and reproduction, consider the following: An individual who contracts a disease will be subject to a maturation process known as a *latent period* and then will become infective, which is akin to reaching sexual maturity. Then, in each period, the infected individual may or may not infect another individual, which is akin to reproduction. Along the way, of course, the infected individual may either recover from the disease or die, which is akin to mortality.

Example 12   Measles epidemics

Table 7.9 Life table for a measles epidemic

Day (t)	Fraction remaining infected $l(t)$	Rate infecting others $b(t)$
1	1.00	0.00
2	1.00	0.00
3	1.00	0.00
4	1.00	0.00
5	1.00	0.00
6	1.00	0.00
7	1.00	0.03
8	1.00	0.06
9	0.94	0.10
10	0.88	0.14
11	0.80	0.18
12	0.74	0.24
13	0.67	0.31
14	0.60	0.38
15	0.53	0.43
16	0.47	0.48
17	0.41	0.48
18	0.35	0.48
19	0.28	0.48
20	0.23	0.48
21	0.17	0.48
22	0.11	0.48
23	0.06	0.48
24	0.00	0.48

Measles is a highly infectious viral disease (genus *Morbillivirus* of the family Paramyxoviridae) that infects, in particular, human infants and adults. An individual infected with measles will become infectious at anywhere from seven to eighteen days and remain infectious for about eight days. Let $l(t)$ be the fraction of individuals infected with measles t days after getting infected (see second column of Table 7.9). The number of new infections that arise from an infected individual (these new infections are equivalent to "births" in the context of the growth of the infected population) depends on many factors, including the rate at which individuals contact other individuals on public transport, at the workplace, and so on. However, in the population of concern, public health officials have determined the number of new cases that infected individuals can be expected to give rise to before they themselves are cured or die; see the third column in Table 7.9.

a. If several infectious individuals are introduced into the population to which these data apply, is an epidemic expected to occur (i.e., is the population of infectious individuals expected to grow)?

b. If the proportion of individuals vaccinated in a population reduces the expected number of individuals infected per infectious individual by this same proportion, then what proportion of the population should be vaccinated to ensure that the disease will not spread?

Solution

a. Since we have cast this problem in terms of life table analysis, whether a measles epidemic will occur depends on the value of R_0 being greater or less than 1. From Table 7.9, it follows that

$$R_0 = \int_0^\infty l(t)b(t)\,dt \approx \sum_{t=7}^{23} l(t)b(t) = 2.29$$

Hence, an infected individual infects, on average, 2.29 other individuals and the population of infected individuals will grow. An epidemic is likely.

b. If a proportion of individuals y are vaccinated, the proportion available to spread the disease is $1 - y$. To control the population, we need to select y to ensure that $R_0 < 1$; that is, we need to solve $R_0 = 2.29(1 - y) < 1$ for y. This implies that $2.29y > 2.29 - 1$ or $y > 1.29/2.29 \approx 0.56$. Hence, at least 56% of the population should be vaccinated to ensure that measles does not spread in the population.

PROBLEM SET 7.5

Level 1 DRILL PROBLEMS

Use Life Table 7.6 for Albertosaurus sarcophagus to compute the quantities in Problems 1 to 4.

1. The fraction of *A. sarcophagus* that died between 14 and 20 years

2. The fraction of *A. sarcophagus* that died between 20 and 28 years

3. The fraction of *A. sarcophagus* that lived at least 6 years

4. The fraction of *A. sarcophagus* that lived at least 8 years

Use Life Table 7.7 for Microtus agrestis to compute the quantities in Problems 5 to 8.

5. The fraction of female voles that lived fewer than 24 weeks

6. The fraction of female voles that lived fewer than 40 weeks

7. The fraction of female voles that lived between 24 and 48 weeks

8. The fraction of female voles that lived between 40 and 64 weeks

Use the survivorship curves for men and women in Example 7 to compute the quantities in Problems 9 to 12.

9. The fraction of women who lived at least 75 years

10. The fraction of men who lived at least 75 years

11. The fraction of women who lived between 25 and 75 years

12. The fraction of men who lived between 25 and 75 years

13. Find the survivorship function $l(t)$ when $m(t) = a + bt$ with $a > 0$ and $b > 0$.

14. Find the survivorship function $l(t)$ when $m(t) = \dfrac{a}{b+t}$ with $a > 0$ and $b > 0$. Discuss how a and b influence the shape of the survivorship function.

15. Use Life Table 7.7 for *Microtus agrestis* to approximate the mortality rates for all age classes of the female vole. Discuss any pattern in the mortality rates that you observe.

16. Use Life Table 7.7 for *Microtus agrestis* to compute the life expectancy of the female vole.

Compute the life expectancy for populations with the (hypothetical) survivorship functions in Problems 17 to 22. Assume t is measured in years. Don't be surprised if one of them turns out to be infinite.

17. $l(t) = e^{-t}$

18. $l(t) = e^{-t/100}$

19. $l(t) = \dfrac{1}{(1+t)^2}$

20. $l(t) = \dfrac{1}{\left(1 + \dfrac{t}{100}\right)^2}$

21. $l(t) = \dfrac{1}{1+t}$

22. $l(t) = \dfrac{1}{\left(1 + \dfrac{t}{20}\right)^3}$

Compute R_0 for populations with the (hypothetical) survivorship and reproduction functions in Problems 23 to 28. Assume t is measured in years.

23. $l(t) = e^{-t}$ and $b(t) = 2$ for $t \geq 1$ and 0 for $0 \leq t < 1$

24. $l(t) = e^{-t/100}$ and $b(t) = t$

25. $l(t) = \dfrac{1}{(1+t)^2}$ and $b(t) = 5$ for $t \geq 5$ and $b(t) = 0$ for $0 \leq t \leq 5$

26. $l(t) = \dfrac{1}{(1+t/100)^2}$ and $b(t) = 0.1$

27. $l(t) = \dfrac{1}{1+t}$ and $b(t) = \dfrac{5}{1+t}$

28. $l(t) = \dfrac{1}{1+\dfrac{t}{2}}$ and $b(t) = \dfrac{3}{1+\dfrac{t}{2}}$

Level 2 APPLIED AND THEORY PROBLEMS

29. Show that $m(t) = -\dfrac{d}{dt}\ln[l(t)]$ provided that $l(t)$ is differentiable.

30. If $\int_0^\infty l(t)\,dt$ is convergent and $b(t) \leq B$ for all t, show that $R_0 = \int_0^\infty l(t)b(t)\,dt$ is convergent.

31. According to the work of Erikson and colleagues reported in Table 7.6, the survivorship function $l(t)$ for the dinosaur species *Albertosaurus* is well approximated by

$$l(t) = 0.6e^{0.039(1 - e^{0.187t})}$$

for $t \geq 2$. This function plotted against the data is shown here:

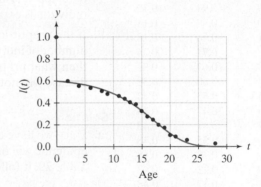

Find and plot the mortality function $m(t)$ for $t \geq 2$.

32. Using $l(t)$ from Example 7, find the mortality function $m(t)$ for females in the United States in 2005.

33. According to the work of Erikson and colleagues, the mortality rate for the dinosaur species *Gorgosaurus* is given by

$$m(t) = 0.0059e^{0.2072t} \text{ per year}$$

Find and plot the survivorship function $l(t)$.

34. According to the work of Erikson and colleagues, the mortality rate for the dinosaur species *Daspletosaurus* is given by

$$m(t) = 0.0018e^{0.2006t} \text{ per year}$$

Find and plot the survivorship function $l(t)$.

35. According to the work of Erikson and colleagues, the survivorship function for the dinosaur species *Tyrannosaurus* is

$$l(t) = e^{0.009 - 0.009e^{0.2214t}}$$

Find and plot the mortality rate $m(t)$.

36. According to a *National Vital Statistics Report* (volume 54, number 14), the life table for people in the United States in 2003 was as follows:

t (years)	$l(t)$
0	1.00
10	0.991
20	0.987
30	0.978
40	0.966
50	0.940
60	0.878
70	0.755
80	0.527
90	0.213
100	0.02

Using right endpoints, estimate the life expectancy of a human.

37. During an outbreak of a SARS-like coronavirus, data were collected that resulted in the construction of the following table:

Day	$l(t)$	$b(t)$
1	1.00	0.20
2	0.98	0.30
3	0.86	0.33
4	0.74	0.38
5	0.62	0.43
6	0.52	0.50
7	0.42	0.60
8	0.40	0.60
9	0.26	0.60
10	0.25	0.60
11	0.12	0.60

Use this table to answer the following questions:

a. If several infectious individuals are introduced into another population, is the epidemic expected to spread?

b. If the proportion of individuals vaccinated in a population reduces the expected number of individuals infected per infectious individual by this same proportion, then what proportion of the population should be vaccinated to ensure that the disease will not spread?

38. Communicable diseases often have at least two stages: a latent stage in which the individual is infected but not infectious, and an infectious stage in which the individual can infect others. For a deadly disease where the time to death is exponentially distributed with mean $1/q$ days, the fraction of individuals surviving t days with the disease is $l(t) = e^{-qt}$. Using differential equations to model the infection with two stages, latent and infectious, the infectiousness of an average infected individual (i.e., the number of people infected per day) is given by

$$b(t) = k\frac{a}{a-c}(e^{-ct} - e^{-at}) \text{ infected per day}$$

where $1/a$ is the mean duration of the latent period, $1/c$ is the mean duration of the infectious period, and k is the rate an infectious individual infects others. For this model find R_0.

39. The parameters of the HIV epidemic vary considerably from country to country. The following table shows survival (including both death and drop-out rates) for treated and untreated segments of the sexually promiscuous population. The numbers reflect the fact that we expect all individuals to die within ten years if they are infected, unless they are treated. In the latter case, we assume that the individuals leave or drop out of the sexually promiscuous population after being part of it for twenty years. Also their infectivity is less for some of the infectivity period because the levels of virus in their body fluids are reduced by treatment. Infectivity comes back later as the efficacy of treatment is reduced over time.

a. Compare the R_0 for the treated and untreated segments of the population. What do you conclude?

b. What levels of condom use in the two subpopulations are needed to control the epidemic, assuming that condom use reduces the probability of transmission by 95%?

Year	Untreated $l(t)$	Untreated $b(t)$	Treated $l(t)$	Treated $b(t)$
1	1	0.5	1	0.5
2	1	0.2	1	0.2
3	1	0.2	1	0.2
4	1	0.2	1	0.2
5	0.95	0.2	0.98	0.1
6	0.9	0.2	0.96	0.05
7	0.8	0.2	0.94	0.05
8	0.65	0.1	0.92	0.05
9	0.45	0.1	0.9	0.05
10	0.2	0.1	0.87	0.05
11	0	0	0.84	0.05
12	0	0	0.81	0.075
13	0	0	0.78	0.1
14	0	0	0.75	0.1
15	0	0	0.72	0.1
16	0	0	0.69	0.1
17	0	0	0.64	0.1
18	0	0	0.54	0.1
19	0	0	0.44	0.1
20	0	0	0.34	0.1
21	Assume all individuals have now left the population of interest			

40. Botswana is a midsized country in central Africa. With a population of just over two million people, it is one of the most sparsely populated countries in the world. Using 2006 data from Human Life-Table database (see Example 5), one can approximate the survival functions of males and females as

$$l_M(t) = e^{-e^{-2.7325+0.05t+0.000012t^2}} \qquad l_W(t) = e^{-e^{-3.307+0.0507t}}$$

These approximations work fairly well as shown below, with males in blue and females in red.

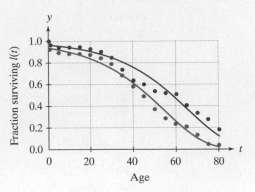

Use numerical integration to estimate the life expectancy for males and females in Botswana.

CHAPTER 7 REVIEW QUESTIONS

1. Consider the following data set corresponding to scores on a test. Let X denote a randomly chosen test score.

Score	Frequency
50–59	6
60–69	14
70–79	26
80–89	10
90–99	4

   a. Construct a histogram.
   b. Find $P(0 \le X \le 89)$.
   c. Find $P(X > 79)$.

2. Let $f(x) = ax^3, 0 \le x \le 4$.
   a. Find a constant a so that it is a PDF.
   b. Find the mean of this PDF.

3. Consider the hyperbolic function

$$F(x) = \begin{cases} \dfrac{x}{k+x} & \text{if } x \ge 0 \\ 0 & \text{elsewhere} \end{cases}$$

   for any $k > 0$.
   a. Show that $F(x)$ is a CDF.
   b. Let X be a random variable with a CDF $F(x)$. Find $P(1 \le X \le 2)$.

4. Use the comparison test to prove that the integral

$$\int_0^\infty t\frac{re^{a-rt}}{(1+e^{a-rt})^2}\,dt$$

   is convergent for any $r > 0$.

5. Determine for which $p > 0$ values the integral

$$\int_2^\infty \frac{dx}{x(\ln x)^p}$$

   is convergent.

6. Use the convergence test to determine whether the given integrals converge or diverge.

   a. $\displaystyle\int_3^\infty \frac{dx}{\sqrt[3]{2x-1}}$ b. $\displaystyle\int_{-\infty}^0 \frac{\sin^2 x\,dx}{1+x^2}$

7. Show that $f(x) = \dfrac{2}{x^2}$ is a PDF on $[1, 2]$ and find its CDF.

8. Compute the mean, variance, and standard deviation for a pair of dice; that is, this data set:

   2, 3, 3, 4, 4, 4, 5, 5, 5, 5, 6, 6, 6, 6, 6, 7, 7, 7, 7, 7, 7,
   8, 8, 8, 8, 8, 9, 9, 9, 9, 10, 10, 10, 11, 11, 12

9. Compute the mean and variance of the random variable with PDF $f(x) = \dfrac{4}{x^5}$ for $x \ge 1$ and $f(x) = 0$ elsewhere.

10. According to Thomson et al. (see Problem 36 in Problem Set 7.1), the elimination constant for lidocaine for patients with congestive heart failure is 0.31 per hour. Hence, for a patient who has received an initial dosage of y_0 mg, the lidocaine level $y(t)$ in the body can be modeled by the differential equation

$$\frac{dy}{dt} = -0.31y \qquad y(0) = y_0$$

   a. Solve for $y(t)$.
   b. Write an expression, call it $F(t)$, that represents the fraction of drug that has left the body by time $t \ge 0$.
   c. If $F(t) = 0$ for $t \le 0$, verify that $F(t)$ is a CDF.
   d. What is the probability that a randomly chosen molecule of drug leaves the body in the first two hours?

e. What is the probability that a randomly chosen molecule of drug leaves the body between the start of the second and start of the fourth hour?

11. The 1999 American Academy of Physician Assistants (AAPA) Physician Assistant Census Survey found that the mean income for a clinically practicing physician's assistant (PA) working full-time was $68,164, with a standard deviation $17,408. Using Chebyshev's inequality, determine a lower bound for the fraction of PAs with an income between $42,052 and $94,276.

12. The time for a mosquito to mature from larva to pupa is approximately exponentially distributed with a mean of fourteen days. Find the probability that a mosquito has matured from larva to pupa in ten days or less. Find the probability a mosquito has taken at least fourteen days to mature from larva to pupa.

13. According to Alexei A. Sharov, Department of Entomology at Virginia Tech, mortality depends on many factors such as temperature, population, and density. Sharov also said that when life tables are built, the effect of these factors is averaged and only age is considered as a factor that determines mortality. Sharov developed a life table for a sheep population in which females are counted once a year, immediately after breeding season:

t (years)	$l(t)$	t (years)	$l(t)$
0	1.00	6	0.626
1	0.845	7	0.532
2	0.824	8	0.418
3	0.795	9	0.289
4	0.755	10	0.162
5	0.699	11	0.060

Data source: A. A. Sharov, "Age Dependent Life Tables," http://home.comcast.net/ ~sharov/PopEcol/lec6/agedep.html

Use this table to compute the life expectancy of a female sheep.

14. Consider a random variable X with the Pareto distribution with parameter $p = 5$.

a. According to Chebyshev's inequality, what is a lower bound for the probability of X being within two standard deviations of its mean?

b. Find the probability that X is within two standard deviations of its mean.

15. Using 2006 data from Human Life-Table Database (see Example 5 of Section 7.5) one can approximate the survival functions of females in Botswana as

$$l_W(t) = e^{-e^{-3.307+0.0507t}}$$

a. Find the fraction of women who live at least forty years.

b. Find the mortality rate at age 40.

16. For loggerhead turtles, females become reproductively active around age 21 years. The fraction of individuals that live to age 21 is 0.0023, and the mortality rate for individuals older than 21 is approximately 0.2 per year. Reproductively active females produce, on average, 160 eggs per year of which half are daughters. Estimate the reproductive number R_0 for this population. What is the fate of the population? How much does one need to reduce the mortality rate of individuals ≥ 21 to reverse this fate?

17. Given a continuously differentiable survivorship function $l(t)$, let X correspond to the lifetime of a randomly chosen individual. What is the PDF and CDF for X?

18. Find parameters a and r in the logistic PDF

$$f(t) = \frac{re^{a-rt}}{(1 + e^{a-rt})^2}$$

such that the mean of this PDF is 1 and the variance is π^2.

19. You are told that a set of data is normally distributed with mean and variance equal to 21 and 64, respectively. Estimate the proportion of these data that have a value greater than or equal to 29.

20. In an experiment involved in rearing cohorts of the human louse, *Pediculus humanus*, researchers Francis Evans and Fredrick Smith at the University of Michigan obtained data on the proportion of individuals that survive over time. From their data, one can calculate the proportion of adults surviving each week of their experiment as follows*: {0.124, 0.301, 0.273, 0.210, 0.083, 0.009}. Since these six values sum to one, they imply that all individuals are dead by the end of the sixth week. In the following figure, the cumulative proportion of adults dying each week is plotted along with the CDF of best-fitting lognormal distribution.

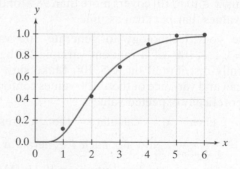

* These values apply to the group of individuals that had already survived the egg and three larval stages, and these data are adapted by us from Francis C. Evans and Frederick E. Smith, "The Intrinsic Rate of Natural Increase for the Human Louse, *Pediculus humanus* L.," *American Naturalist* 86 (No. 830)(1952): 299–310.

If the equation of the best-fitting lognormal PDF is

$$f(x) = \begin{cases} \dfrac{1}{x\sqrt{2\pi}\,(0.523)}e^{-\frac{(\ln x - 0.769)^2}{2(0.523)^2}} & \text{for} \quad x \geq 0 \\ \\ 0 & \textit{otherwise} \end{cases}$$

then what is the mean survival time for these lice? How does this value compare to the value obtained by calculating the mean survival directly from the data, if the six values correspond to survival midway between each week? What accounts for the difference in these two approaches to estimating the mean?

GROUP PROJECTS

Seeing a project through on your own, or working in a small group to complete a project, teaches important skills. The following projects provide opportunities to develop such skills.

Project 7A   Fitting Distributions

Search the Web for a data set consisting of at least several hundred data points. Explore your data as outlined next; provide illustrations to enhance the presentation of your analysis.

1. Draw histograms for several different bin sizes. Select the histogram that results in the smoothest looking probability distribution in terms of being approximated by some curve. Note that if the bin size is too large, the histogram will look like a few big blocks. If the bin size is too small, the histogram will look like a picket fence with lots of missing staves.

2. Calculate the mean and variance from the histogram. Compare these values to the values you get when calculating the mean and variance directly from the data.

3. Calculate the expected fraction of data in each bin of a theoretical histogram obtained from a uniform, logistic, normal, and lognormal distribution that has the same mean and variance as the histogram you constructed from the data.

4. Use a sum-of-squares measure to compare how well the four distributions in step 3 fit the data and discuss your results.

5. Bonus: Search the Web or books for other distributions not dealt with in this chapter and repeat steps 2 and 3 for these distributions.

Project 7B   Play with Logistic Regression

Use an appropriate computer technology to generate a set of data that conforms to the logistic distribution

$$p(x) = \frac{1}{1 + e^{a-rx}}$$

for the case $a = 5$ and $r = 1$ as follows.

1. First verify that $p(0.5) = 0.011$ and $p(10) = 0.993$. Thus, $x \in [0.5, 10]$ covers more than 98% of the range of values that $p(x)$ can assume.

2. Use your technology to generate 100 values x_i, $i = 1, \ldots, 100$, of a random variable X that is uniformly distributed on $[0.5, 10]$. Make sure that the mean and variance of these 100 values conform to the theoretically expected values.

3. For each x_i, calculate the corresponding $p_i = \dfrac{1}{1 + e^{5-x_i}}$. Now for each i generate a value z_i from the uniform distribution on $[0,1]$. (Most technologies refer to this as generating a value at random

between 0 and 1.) If $z_i > p_i$ set $y_i = 0$; otherwise, set $y_i = 1$. Once you have done this for all $i = 1, \ldots, 100$, you will have a data set $D = \{(x_i, y_i) | i = 1, \ldots, 100\}$ with values of x_i between 0.5 and 10 and a value of y_i either 0 or 1.

4. Construct a histogram for these data using six equal bin sizes and the proportion of data points in the bin that have a y_i value equal to 1.

5. Use logistic regression to estimate the parameters $\hat{a}$ and $\hat{r}$ from the best-fitting linear model of the transformed data from the histogram. How close are $\hat{a}$ and $\hat{r}$ to the values 5 and 1, respectively?

6. Now repeat the exercise with 300 points and again with 1000 points. In each case, how close are $\hat{a}$ and $\hat{r}$ to the values 5 and 1, respectively? What do you notice?

7. Write a report that contains your results, and in the concluding section interpret the basic concepts behind this exercise.

Figure 8.1 During a SARS outbreak in China in 2002–2003, children wore face masks to protect themselves. In Section 8.5, we will study models of disease outbreaks, like SARS.

CHAPTER 8

Multivariable Extensions

VINCENT YU/Associated Press

Preview

"For the things of this world cannot be made known without a knowledge of mathematics."

Roger Bacon, Philosopher and Franciscan friar, 1214–1294.
Opus Majus part 4 Distinctia Prima cap 1, 1267.

The exquisite tapestry that is our biological world arises as a result of interacting populations—whether molecules, cells, complex organisms, or groups of organisms. These interactions often create complex patterns of the kind and numbers of individuals that vary over time and space. To gain insight into these patterns that depend on more than one variable requires mathematical concepts that go well beyond what is typically covered in a first course on calculus. The analysis of multivariable models draws upon theories of general linear algebra (also known as matrix theory) and ordinary differential equations, both of which require a term of study to understand these theories in a comprehensive way. It is possible, however, to introduce these topics without invoking many results from advanced theories, by focusing on relatively simple two-dimensional linear and nonlinear processes that have particular relevance to biological and environmental processes.

After studying multivariate functions in the first section of this chapter, we look at linear models (Section 2), consider *eigenvalues* and *eigenvectors* (Section 3), and use these concepts to solve dynamic linear models (Sections 3 and 4). We then explore nonlinear problems (Section 5) using a graphical method called *phase-plane analysis* that provides insights into the structure of solutions without having to explicitly find the solutions. Using these models, we examine outbreaks of diseases (see Figure 8.1), predator-prey interactions, dynamics of gene circuits, accumulation of drugs or toxins in organs and tissues, and electrical activity of neurons.

8.1 Multivariate Modeling

Basic concepts

Throughout this book we have encountered functions of two variables. Recall in Example 5 of Section 1.4, we modeled the height of beer froth in a mug as the function $f(h_0, t) = h_0 0.94^t$ of its initial height h_0 and the time t in seconds since the beer was

591

poured. In Example 2 of Section 4.3, we modeled the growth rate of the Arctic fin whale as the function $f(x, H) = 0.08x(1 - x/500{,}000) - H$ of the number of whales x and the harvesting rate H. In Example 4 of Section 4.4, we modeled the average energy gain of great tits as a function $f(T, t) = 6.3587(1 - e^{-0.0081t})(t + T)$ of the time T to get to a patch of food and the time t spent in the patch. In Section 6.2, we studied differential equations with right-hand sides represented by a general function $h(t, y)$ in the two variables t and y. These examples motivate the following definition.

Function of Two Variables

A **function of two variables** is a rule that assigns one real number z to ordered pairs (x, y) of real numbers. The set of ordered pairs for which the rule is defined is the **domain** of the function and the set of corresponding z values is the **range** of the function. We usually write

$$z = f(x, y)$$

to describe this rule where x and y are called the **independent variables** and z is called the **dependent variable**.

Example 1   Evaluation, domains, and ranges

Consider the function $f(x, y) = \sqrt{x - y}$.

a. Evaluate $f(x, y)$ at the points $(x, y) = (2, 1)$ and $(x, y) = (6, 2)$.

b. Find the domain for f and sketch it in the xy plane.

c. Find the range of f and sketch it on the z axis.

Solution

a. To evaluate $f(x, y)$ at $(x, y) = (2, 1)$, we substitute 2 for x and 1 for y:

$$f(2, 1) = \sqrt{2 - 1} = \sqrt{1} = 1$$

To evaluate $f(x, y)$ at $(x, y) = (6, 2)$, we substitute 6 for x and 2 for y:

$$f(6, 2) = \sqrt{6 - 2} = \sqrt{4} = 2$$

b. Since the square root function is only defined for nonnegative arguments, $f(x, y)$ is only defined when $x - y \geq 0$. Equivalently, $x \geq y$. Hence, the domain is the set of all ordered pairs (x, y) such that $x \geq y$. A plot of the domain in the xy plane is shown in Figure 8.2.

c. The range of the square root function is the nonnegative reals. Hence, the range of f is also the nonnegative reals. A plot of the range on the z axis is shown in Figure 8.2.

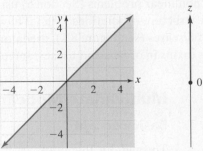

Figure 8.2 Plot of the domain (on left) and range of the function $f(x, y) = \sqrt{x - y}$ examined in Example 1

In ecological systems, species are interconnected through complex interactions. A fundamental interaction between pairs of species is predation, where one species, the predator, benefits by using another species, the prey, as a resource. This involves either killing individual prey (as with predators) or consuming the host from the inside (as with parasites and pathogens). Prey-predator systems, as illustrated in Figure 8.3, can be as diverse as buffalo being preyed upon by lions or T cells succumbing to viruses.

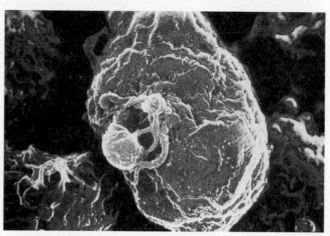

Figure 8.3 Two forms of predation in nature: Lions attacking an African buffalo and a virus attacking a T cell within a human host. Both of these interactions can be modeled by the equation developed in Example 2.

In the next example we examine how the growth rate of a prey population may depend on the density of both prey and predator species. We have seen in previous examples that the density of populations is measured as the number of individuals per unit area or volume. For lions, the density may be the number of individuals per square mile; for viruses, the density may be the number of viral particles per milliliter of blood or other liquid in which the viral particles are found.

Example 2   Predatory impacts on population growth

The growth rate of a population is the rate of change of the population density over time. Hence, the units of growth rate are density per unit time.

a. Assuming that the growth rate of a prey population depends both on its own density x and on the density of a predator species y, find the simplest form for the prey population growth rate model

$$\frac{dx}{dt} = f(x, y)$$

that conforms to the following specifications:

- In the absence of the predator, the prey growth rate is proportional to its density, where we call this proportionality constant r, the *intrinsic rate* of prey growth.
- Each predator attacks and kills prey at a rate proportional to the prey density, where we call this proportionality constant a, the *attack rate* of the predator.

b. Assuming in this simplest model that $r = 1$ and $a = 0.1$, plot the prey population growth rate as a function of x when the predator density is $y = 2$, and plot the prey population growth rate as a function of x when the predator density is $y = 15$.

c. Assuming in this simplest model that $r = 1$ and $a = 0.1$, plot the prey population growth rate as a function of the predator density y when the prey density is $x = 1$.

Solution

a. Since the prey growth is proportional to its density in the absence of the predator, the simplest function with these properties is $f(x, 0) = rx$ where r is the prey's intrinsic rate of growth. The rate at which one predator attacks the prey is proportional to the prey density; each predator decreases the prey growth rate by ax where a is the predator's attack rate. Since the density of predators is y, the net decrease in the prey growth rate due to predation is axy. The simplest model satisfying our assumptions is

$$f(x, y) = rx - axy$$

b. With $r = 1$, $a = 0.1$, and $y = 2$, the prey growth rate is

$$f(x, 2) = x - 0.1x \cdot 2 = 0.8\,x$$

Hence, when $y = 2$, the prey growth rate increases linearly with the prey density as illustrated in Figure 8.4a.

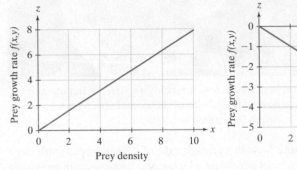

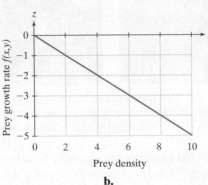

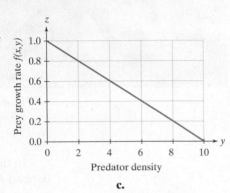

a.   b.   c.

Figure 8.4 Plots of the prey growth rate $z = f(x, y) = x - 0.1xy$ in Example 2 when **a.** the predator density is held constant at $y = 2$, **b.** the predator density is held constant at $y = 15$, and **c.** the prey density is held constant at $x = 1$.

With $r = 1$, $a = 0.1$, and $y = 15$, the prey growth rate is

$$f(x, 15) = x - 0.1x \cdot 15 = -0.5\,x$$

Hence, when $y = 15$, the prey growth rate decreases linearly with the prey density as illustrated in Figure 8.4b.

c. With $r = 1$, $a = 0.1$, and $x = 1$, the prey growth rate is

$$f(1, y) = 1 - 0.1(1)y = 1 - 0.1y$$

Hence, when the prey density is $x = 1$, the prey growth rate decreases linearly with the predator density as illustrated in Figure 8.4c.

Visualizing functions of two variables

In Example 2, we discovered that one way to visualize functions of two variables is plotting "one-dimensional slices," where one of the independent variables is held constant. This approach does not allow us to visualize the function over its entire domain at once. One way to visualize the entire function is to graph it in three-dimensional space, where the xy plane is laid down flat and extended vertically in the "z direction" as shown in Figure 8.5. To plot a function $f(x, y)$ in this three-dimensional space, we treat $f(x, y)$ as the "height" of the function in the z direction above the point (x, y) in the xy plane. By plotting all the points (x, y, z) with $z = f(x, y)$ in this three-dimensional space, we get a *surface*, as illustrated in Figure 8.5, which is the graph of the function $f(x, y)$.

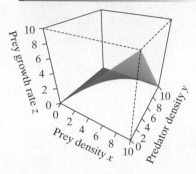

Figure 8.5 Plotted on the left is a point represented by the ordered triplet (x, y, z) in three-dimensional xyz space. Plotted on the right is a surface plot of a function $f(x, y)$ where each point on the surface is given by the ordered triplet (x, y, z) with $z = f(x, y)$.

Graph of a Function of Two Variables	If $f(x, y)$ is a function with domain D, then the **graph of f** is the set of all ordered triplets (x, y, z) such that $z = f(x, y)$. Namely, the graph is given by the set

$$G = \{(x, y, z) : z = f(x, y) \text{ and } (x, y) \in D\}$$

These plots in general are quite challenging to draw by hand. However, we can easily plot them using various forms of technology such as advanced graphing calculators, mathematical software, or online JavaScripts. For example, the graph of the prey growth function from Example 2 is plotted in Figure 8.6.

Figure 8.6 Surface plot of the prey growth function $f(x, y) = x - 0.1xy$ from Example 2.

Example 3   Mix and match

The surface plots of the functions

$$z = 2 - x^2 - y^2, \; z = \sin(x^2/10 + y^2/10), \; z = y - x, \text{ and } z = \sin(x/2)\cos(y/2)$$

are shown out of order in Figure 8.7. Determine which function goes with which surface plot.

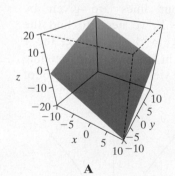
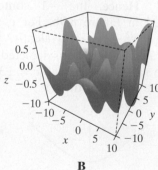
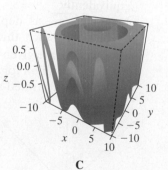
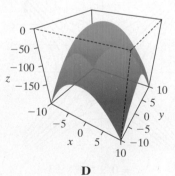

A   **B**   **C**   **D**

Figure 8.7 Mix and match

Solution   The function $z = 2 - x^2 - y^2$ yields downward facing parabolas in the x and y directions. The only surface plot consistent with these features is plot **D**.

The function $z = \sin(x^2/10 + y^2/10)$ is constant along circles centered around the origin and oscillates sinusoidally between the values -1 and 1 as the radii of these circles increase. The only surface plot consistent with these features is plot **C**.

The function $z = y - x$ increases linearly with y and decreases linearly with x. The only surface plot consistent with these features is plot **A**.

There is only one plot left. Hence, by the pigeonhole principle, the function $z = \sin(x/2)\cos(y/2)$ belongs with surface plot **B**.

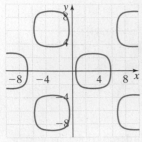

Figure 8.8 Plot of the curves in xy space corresponding to the equation $\sin(x/2)\cos(y/2) = 1/4$.

The function $z = f(x, y)$ defines a surface in three dimensions. If z is given a specific value $z = c$, then the equation $f(x, y) = c$ defines a curve or set of curves in the xy plane. For example, $z = y - x$ which is plot **C** in Example 3, reduces to the equation $y = x + c$ when $z = c$ is given a specific value: a line of slope one with intercept c. However, as we have seen in Section 3.3, equations of the form $f(x, y) = c$ may not simplify to a function of the form $y = f(x)$. For example, $\sin(x/2)\cos(y/2) = c$ represents the intersection of this surface in plot **B** from Example 3 with the plane defined by $z = c$. Using technology to plot $\sin(x/2)\cos(y/2) = c$, Figure 8.8 illustrates that this equation defines several closed curves in the xy plane. Such plots, called contour plots, provide a useful way to visualize functions $z = f(x, y)$.

Contour Lines and Maps

Let $f(x, y)$ be a function of two variables with domain D. A **contour line at $z = c$** of $f(x, y)$ is the set of points $\{(x, y) : f(x, y) = c\}$. A contour line at $z = c$ is also known as a **level curve at $z = c$**. A **contour map for $z = f(x, y)$** is a collection of contour lines for $f(x, y)$ plotted in the xy plane.

Example 4   Plotting contour maps

Sketch contour plots for the following two functions.

a. Consider the function $z = x^2 - y^2$. Plot the contour lines $x^2 - y^2 = c$ for the values $c = 0, \pm 1, \pm 2,$ and ± 3 in the region $D = \{(x, y) | -2 \le x \le 2, -2 \le y \le 2\}$.

b. Consider the function $z = 2 - x^2 - y^2$. Plot the contour lines $c = 2 - x^2 - y^2$ for the values $c = 2, -2, -6, -10,$ and -16 in the region $D = \{(x, y) | -3 \le x \le 3, -3 \le y \le 3\}$.

Solution

a. For $c = 0$, we need to find the set of points (x, y) such that $0 = x^2 - y^2$. Equivalently, $x^2 = y^2$. Hence, the 0 contour lines are given by $y = \pm x$ as shown in Figure 8.9a. For $c = -1$, we need to find the set of points such that $-1 = x^2 - y^2$. Equivalently, $y^2 = x^2 + 1$. Hence, the -1 contour lines are given by $y = \pm\sqrt{1 + x^2}$, as depicted in Figure 8.9b. For $c = 1$, we need to find the set of points such that $1 = x^2 - y^2$. Equivalently, $x^2 = y^2 + 1$. Hence, the 1 contour lines are given by $x = \pm\sqrt{1 + y^2}$, as depicted in Figure 8.9c. Proceeding in a similar fashion for the values $c = \pm 2$ and $c = \pm 3$ yields the contour plot of $z = x^2 - y^2$ illustrated in the margin at the left.

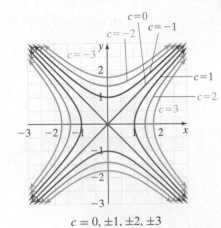

$c = 0, \pm 1, \pm 2, \pm 3$

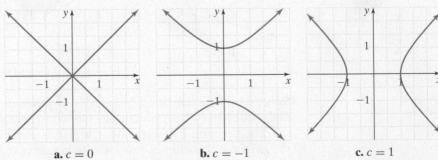

a. $c = 0$ **b.** $c = -1$ **c.** $c = 1$

Figure 8.9 Contour plots for $z = x^2 - y^2$ with contours $z = c$ as labeled.

b. For $c = 2$, we need to find the set of points (x, y) such that $2 = 2 - x^2 - y^2$. Equivalently, $x^2 + y^2 = 0$. Hence, the 0 "contour line" is only a single point at $(0, 0)$. For $c = -2$, we need to find the set of points such that $-2 = 2 - x^2 - y^2$.

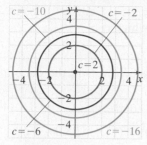

Figure 8.10 Contour plots for $z = 2 - x^2 - y^2$ with contours $z = c$ at $c = 2, -2, -6, -10,$ and -16.

Equivalently, $x^2 + y^2 = 4$. Hence, the -2 contour line is given by circle of radius 2 centered at the origin. For $c = -6$, we need to find the set of points such that $-6 = 2 - x^2 - y^2$. Equivalently, $x^2 + y^2 = 8$. Hence, the -6 contour line is given by circle of radius $2\sqrt{2}$ centered at the origin. Plotting these contour lines and the contours at $c = -10, -16$ yields the contour plot for $z = 2 - x^2 - y^2$ illustrated in Figure 8.10. This contour map corresponds to surface plot **D** in Example 3.

Real world examples

In the next three examples, we apply the concept of contour maps to fitting models to data, understanding the dynamics of genetic circuits, and reading topographical maps. In Section 4.3 we discussed fitting linear models of the form $y = mx + c$ to data, but in Example 6 of that section we restricted our attention to lines through the origin, that is, fitting the one parameter curve $y = mx$ after setting $c = 0$. The general linear regression problem, however, involves two parameters: slope m and intercept c. Specifically, if given a data set $D = \{(x_1, y_1), (x_2, y_2), \ldots, (x_n, y_n)\}$, recall from our definition of the residual sum-of-squares S in Section 4.3 that

$$S = \sum_{i=1}^{n} e_i^2$$

where for linear models $f(x) = mx + c$, we have

$$e_i = y_i - mx_i - c$$

Thus, given data D, the function S is a function of m and c:

$$S(m, c) = \sum_{i=1}^{n} (y_i - mx_i - c)^2$$

To find the best-fitting line to the data, we want to find m and c that minimize the value of $S(m, c)$ as illustrated in the following example.

Example 5   Chirping crickets and linear regression

Peter Cristofono

Figure 8.11 A female cricket of the species *Nemobius fasciatus*

In 1948, George Pierce published a book called *The Songs of Insects*, in which he reported measuring the frequency y of chirping (chirps per second) of a particular cricket (Figure 8.11) at different ambient temperatures x (°F). These are the data he obtained:

$$D = \{(71.6, 15.7), (96.8, 18.4), (85.0, 17.4), (89.6, 19.2), (79.7, 15.2)\}$$

a. Use technology to find the regression parameter values (m^*, c^*) that produce the best-fitting curve through the data and plot the data and this regression line on the same graph.

b. Derive a sum-of-squares function $S(m, c)$ for the data in terms of slope and intercept parameters m and c.

c. Use technology to generate a three-dimensional plot of $S(m, c)$ over the domain $\{0 \le m \le 0.3, 0 \le c \le 10\}$.

d. Use technology to generate a contour plot of $S(m, c)$ over this same domain.

e. Is the location of the regression line point (m^*, c^*), which corresponds to the minimum point $S(m^*, c^*) < S(m, c)$ for all points $(m, c) \ne (m^*, c^*)$, easily identified on either of the plots produced in parts **a** and **b**? If not, why not?

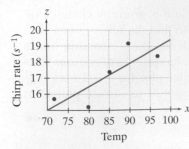

Figure 8.12 Plot of the cricket chirp rate versus temperature data along with the besting-fitting line

Solution

a. Technology can be used to find the best-fitting curve, which is

$$z = 0.147x + 4.7$$

which implies that $(m^*, c^*) = (0.147, 4.7)$. This regression line and the data are plotted in Figure 8.12.

b. From the definition of $S(m, c)$ it follows for the data D that

$$S(m, c) = (15.7 - 71.6m - c)^2 + (18.4 - 96.8m - c)^2 + (17.4 - 85.0m - c)^2$$
$$+ (19.2 - 89.6m - c)^2 + (15.2 - 79.7m - c)^2$$
$$\approx 1487.5 - 14632m - 171.8c + 36102m^2 + 845.4mc + 5c^2$$

c. The following is a technology-produced three-dimensional plot of the surface $z = S(m, c)$.

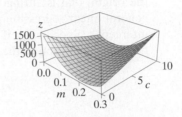

d. The following is a technology-produced contour plot of the surface $z = S(m, c)$ over the same domain as in part **c.** In this plot, lighter blues correspond to larger z values.

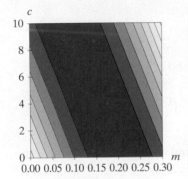

e. Although it is clear from both graphs in parts **c** and **d** that the point $(m^*, c^*) = (0.147, 4.7)$ lies within the area where $S(m, c)$ has its lowest values, the actual point is hard to locate because the plots of $S(m, c)$ are more "pipe-like" than "basin-like." Thus, a flat region exists over which it is hard to find the local minimum at the resolution of the plots produced by technology. Only by using explicit formulas for the best-fitting line was technology able to precisely locate the actual point on the surface of $S(m, c)$ where the minimum $S(m^*, c^*)$ occurs. ■

As we shall see in Section 8.5, contour plots are useful in studying the long-term behavior of nonlinear differential equation models. In the following example, we introduce one of the models examined in Section 8.5 and use a contour plot to determine when the concentration of a protein produced by a genetic circuit is increasing or decreasing.

Example 6   A genetic circuit

Recall that a gene is a sequence of DNA that encodes either protein or RNA molecules that may inhibit or excite the expression of other genes. Collections of genes interacting in this way form "genetic circuits." A simple, yet extremely important type of genetic circuit is one that involves two genes, call them A and B, that inhibit each other's expression. Let x and y denote the concentrations of two proteins encoded by the genes A and B, respectively. Let $f(x, y)$ be the rate of change of the concentration of protein A.

a. Write an expression for $f(x, y)$ based on these assumptions:

- $f(x, y)$ equals the difference between the rate at which gene A produces protein A and the rate at which protein A degrades.

- In the absence of other processes, the concentration of protein A degrades at a rate equal to its concentration.

- Gene A produces protein A at a rate inversely proportional to $1 + y$, where the inverse relationship implies that protein B at concentration y actually inhibits the expression of gene A.
- Gene A produces protein A at a constant rate 3 when protein B is absent.

b. Use a computer or graphing calculator to draw contours of $f(x, y)$ for $z = -1, 0$, and 1 in the nonnegative quadrant of the xy plane, that is, where $x \geq 0$ and $y \geq 0$. Discuss what these contours imply about the rate of change of protein A.

Solution

a. The first assumption implies

$$f(x, y) = \text{rate of production} - \text{rate of degradation}$$

The second assumption implies that the rate of degradation equals x. The third assumption implies that rate of production is proportional to $\dfrac{1}{1 + y}$. In other words, the rate of production is given by $\dfrac{a}{1 + y}$ for some value of $a > 0$. Since the final assumption asserts that the rate of production equals 3 when $y = 0$, we obtain $\dfrac{a}{1 + 0} = a = 3$. Putting together the pieces, we obtain

$$f(x, y) = \frac{3}{1 + y} - x$$

b. Using a computer to generate contours yields the contour plot depicted in Figure 8.13. Intuitively, this plot shows that as the concentration y of the inhibiting protein B increases, the rate of change of protein A's concentration decreases. In particular, when y is sufficiently large, the rate of change is negative and the concentration of protein A would be decreasing. Alternatively, when the concentration x of protein A is too large, the rate of degradation always exceeds the rate of production and the concentration x of protein A would be decreasing; that is, we have a negative rate of change.

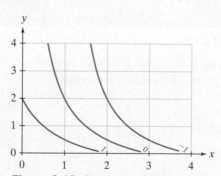

Figure 8.13 Gene circuit contour plots

Contours are used extensively for topographic maps, where the contours correspond to curves of equal elevation.

Example 7   Climbing volcanos

Mount Eden—one of about fifty volcanos in the Auckland, New Zealand, volcanic field—is the highest natural point in the whole of Auckland. The contour plot depicted in Figure 8.14 provides topographic information for Mount Eden with the x, y, and elevation measured in meters.

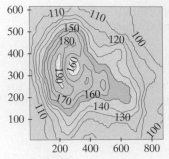

Figure 8.14 Contour plot of Mount Eden with horizontal distances in the east-west and north-south directions measured in meters.

Source: Data for plot were digitized from a topographic map by Ross Ihaka. These data should not be considered as accurate.

a. Suppose you traverse the volcano from west to east at $y = 400$ meters. Describe and plot your elevation as a function of x.

b. Suppose you traverse the volcano from south to north at $x = 400$ meters. Describe and plot your elevation as a function of y.

Solution

a. Going from west to east at $y = 400$, the elevation plotted in the left panel of Figure 8.15 goes rapidly from 110 meters to 190 meters around $x = 200$ meters. Then there is a decline to 130 meters by $x = 450$ meters followed by a slight incline prior to the rapid decline to 90 meters at $x = 800$ meters.

b. Going from south to north at $x = 400$ meters, the elevation plotted in the right panel of Figure 8.15 climbs steadily from 110 meters to a peak of about 180 meters at $y = 300$ meters. After passing the peak along this transect, the elevation decreases steadily from about 180 meters to a minimal elevation of 110 meters at $y = 600$ meters.

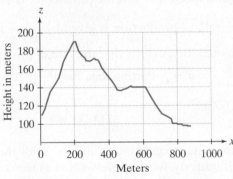

West-east transect at $y = 400$ meters

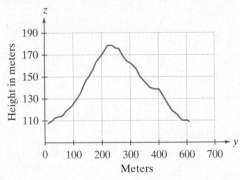

South-north transect at $x = 400$ meters

Figure 8.15 Elevation along transects of Mount Eden, from the topographic map depicted in Figure 8.14

PROBLEM SET 8.1

Level 1 DRILL PROBLEMS

In Problems 1 to 8, evaluate the function $z = f(x, y)$ at the designated point (x, y), and determine the domain and range in each case.

1. $z = 2x - 3y$ at $(1/2, 1/3)$

2. $z = \dfrac{1}{2x - 3y}$ at $(1, 1)$

3. $z = \sqrt{x^2 + 2y^2}$ at $(1, -1)$

4. $z = \sqrt{x + 2y}$ at $(1, 4)$

5. $z = \dfrac{\sqrt{x + 2y}}{x}$ at $(1, -1/2)$

6. $z = \dfrac{\sqrt{x - y}}{x - y}$ at $(2, 1)$

7. $z = \ln y - x^2$ at $(1/2, 1)$

8. $z = \sqrt{16 - x^2 - y^2}$ at $(-3, 2)$

In Problems 9 to 12, consider the function $z = f(x, y) = rx - axy$ from Example 2. For the given values of a and r, plot $z = f(x, y)$ in the xz plane if y is given and plot $z = f(x, y)$ in the yz plane if x is given. Discuss what each of these plots tells you about the effect of one of the species densities on the growth rate of the prey.

9. $a = 0.15, r = 1, y = 2$ **10.** $a = 0.2, r = 0.1, y = 1$

11. $a = 0.5, r = 1.2, x = 2$ **12.** $a = 0.9, r = 1.1, x = 1$

13. Consider the function $z = f(x, y) = \dfrac{3}{1 + y} - x$ from Example 6. For $x = 0$, plot $z = f(x, y)$ in the yz plane for $y \geq 0$. Describe what this graph tells you about the effect of concentration of protein B on the rate of change of protein A.

14. Consider the function $z = f(x, y) = \dfrac{3}{1 + y} - x$ from Example 6. For $y = 0$, plot $z = f(x, y)$ in the xz plane. Describe what this graph tells you about the effect of concentration of protein A on the rate of change of protein A.

In Problems 15 to 20, sketch the requested contour curves for the given function $z = f(x, y)$.

15. $z = y^2$ for $z = 1$

16. $z = x^2 + y^2$ for $z = 4$

17. $z = e^{-0.5(x^2+y^2)}$ for $z = \dfrac{1}{2}$

18. $z = \dfrac{-6y^2}{x^2 + y^2 + 1}$ for $z = 1$ and $z = -1$ when
$D = \{(x, y)| -1 \le x \le 1, -1 \le y \le 1\}$

19. $z = \sin(x + y)$ for $z = 2$ and $z = 1/2$ when
$D = \{(x, y)| -2 \le x \le 2, -2 \le y \le 2\}$

20. $z = 2|\sin x| + y - 1$ for $z = -1$ when
$D = \{(x, y)| -3 \le x \le 3, -3 \le y \le 3\}$

In Problems 21 to 28, use technology to plot the specified functions over the indicated domains and identify one dominant feature that helps one understand the properties of the surface plot in question.

21. $f(x, y) = (x - 3)^2 + (y - 4)^2$,
$D = \{(x, y)|0 \le x \le 6, 0 \le y \le 8\}$

22. $f(x, y) = \sin(x + y^2)$,
$D = \{(x, y)| -3 \le x \le 3, -2 \le y \le 2\}$

23. $f(x, y) = e^{(1-x^2-y^2)}$,
$D = \{(x, y)| -2 \le x \le 2, -2 \le y \le 2\}$

24. $f(x, y) = \dfrac{x - y}{x + y}$,
$D = \{(x, y)| -2 \le x \le 2, -2 \le y \le 2, x \ne y\}$

25. $f(x, y) = \sin x + \sin y$,
$D = \{(x, y)\, -3\pi \le x \le 3\pi, -3\pi \le y \le 3\pi\}$

26. $f(x, y) = \ln(x^2 + y^2)$, $D = \{(x, y)| -5 \le x \le 5,$
$-5 \le y \le 5, x \ne 0$ or $y \ne 0\}$ for the range
$z \in [1, -10]$

27. $f(x, y) = x^2 - y^2$, $D = \{(x, y)| -5 \le x \le 5,$
$-5 \le y \le 5\}$ for the range $z \in [-1, 1]$

28. $f(x, y) = \sin\sqrt{x^2 + y^2}$,
$D = \{(x, y)| -3\pi \le x \le 3\pi, -3\pi \le y \le 3\pi\}$

*For a function of two variables $z = f(x, y)$ with domain D, a point (x_0, y_0) is a **local maximum** if $f(x_0, y_0) \ge f(x, y)$ for all $(x, y) \in D$ near (x_0, y_0). A point (x_0, y_0) is a **local minimum** if $f(x_0, y_0) \le f(x, y)$ for all $(x, y) \in D$ near (x_0, y_0). In Problems 29 to 32, use technology to produce the contour plots over the indicated domains and identify the local maxima and minima that lie within this domain.*

29. $f(x, y) = e^{(1-x^2-y^2)}$,
$D = \{(x, y)| -2 \le x \le 2, -2 \le y \le 2\}$

30. $f(x, y) = \sin x + \sin y$,
$D = \{(x, y)| -3\pi \le x \le 3\pi, -3\pi \le y \le 3\pi\}$

31. $f(x, y) = \sin\sqrt{x^2 + y^2}$,
$D = \{(x, y)| -3\pi \le x \le 3\pi, -3\pi \le y \le 3\pi\}$

32. $f(x, y) = x^2 - y^2$,
$D = \{(x, y)| -5 \le x \le 5, -5 \le y \le 5\}$

Level 2 APPLIED AND THEORY PROBLEMS

33. In Section 8.5 of this chapter we will study the Lotka-Volterra model of competition between two species x and y. Our analysis will involve understanding the growth rate of species 1, $f_1(x, y) = x(1 - x - \alpha y)$, and the growth rate of species 2, $f_2(x, y) = y(1 - y - \beta x)$. Here, x and y are the densities of species 1 and 2, respectively. Use technology to plot these two surfaces on the same axes over the domain $D = \{(x, y)|0 \le x \le 0.7, 0 \le y \le 0.7\}$ for the values $\alpha = 1$ and $\beta = 1/4$. Orient the axes in a way that allows you to see how both surfaces overlap with each other and discuss what you see.

34. In Section 8.5 of this chapter we will study the Lotka-Volterra prey-predator model with the density of prey represented by the variable x and that of predator by the density y. Our analysis will involve understanding the prey growth rate $f_1(x, y) = rx - axy$ and the predator growth rate $f_2(x, y) = abxy - cy$. Use technology to plot these two surfaces on the same axes over the domain $D = \{(x, y)|0 \le x \le 20, 0 \le y \le 10\}$ for the values $r = 1, a = 0.1, b = 0.5,$ and $c = 1$. Orient the axes in a way that allows you to see how both surfaces overlap with each other and discuss what you see.

35. Write the simplest function $z = f(x, y)$ that has the following properties:

   (i) The rate at which a predator at density y kills prey at density x is proportional to x and inversely proportional to the sum $k + ax + by$

   (ii) $\displaystyle\lim_{x\to\infty} f(x, y) = \dfrac{c}{a}$ for any $y \ge 0$

36. The function $z = f(R, N) = \dfrac{aR}{b(1 + (N/c)^d) + R}$ was used by Professor Getz in 2011 (*Ecology Letters* 14:113–124) to model the per capita rate at which a population at density N extracts resources from a resource pool at density R, where $a > 0$ is a maximum extraction rate parameter, b is an extraction efficiency parameter, c is the point at which population density reduces extraction efficiency by half that at the lowest density, and $d \ge 1$ is the abruptness with which density dependence sets in around the point $N = c$. For the parameter values $a = 1, b = 1, c = 1$ and $d = 4$, explore the behavior of the function by

   a. plotting this function over the domain
$D = \{(R, N)|0 \le R \le 10, 0 \le N \le 3\}$

   b. describing the effect of increasing the population density on the per capita rate of resource extraction

   c. describing the effect of increasing the resource density on the per capita rate of resource extraction. How does this effect depend on the population density? Hint: Compare $N = 0$ and $N = 3$.

37. Consider the function

$$z = f(R, N) = \frac{aR}{b(1 + (N/c)^d) + R}$$

presented in Problem 36. If we now set $R = 1, a = 1$, $b = 1$ and $c = 1$, we obtain $z = f(1, N) = \frac{1}{2 + N^d}$, which itself can be considered a function of variables N and d. Plot the function $z = \frac{1}{2 + N^d}$ over the domain $D = \{(N, d) | 0 \le N \le 3, 1 \le d \le 10\}$. What effect does increasing N have on the per capita extraction rate? How does this effect depend on d?

38. Consider the function

$$z = f(R, N) = \frac{aR}{b(1 + (N/c)^d) + R}$$

presented in Problem 36. If we now set $N = 1, a = 1$, and $c = 1$, we obtain $z = f(R, 1) = \frac{R}{2b + R}$, which itself can be considered a function of variables R and b (since it does not depend on d). Plot this function over the domain $D = \{(R, b) | 0 < R \le 10, 0 \le b \le 10\}$. What effect does increasing R have on the per capita extraction rate? How does this effect depend on b?

39. Consider the function

$$z = f(R, N) = \frac{aR}{b(1 + (N/c)^d) + R}$$

presented in Problem 36. If we now set $a = 1, b = 1$, and $d = 4$, we obtain $z = \frac{R}{1 + (N/c)^4 + R}$. By comparing plots of z for the two cases $c = 0.5$ and $c = 1$ over the domain $D = \{(R, N) | 0 \le R \le 10, 0 \le N \le 8\}$, deduce the role that c has in determining the shape of z.

40. In Section 8.5 of this chapter we will study the dynamics of excitable systems that model the electrical activity of neurons. In this model, the variables x and y are used to represent substrates such that high values of x correspond to high firing rates of neurons and high values of y suppress rates of firing. The rate of change of x in this process is represented by the surface $z = f(x, y) = x - x^3/3 - y - I$ where I represents an external current injected into the neuron. Sketch the surface z for the case $I = 5$ over the domain $D = \{(x, y) | -3 \le x \le 3, -3 \le y \le 3\}$.

41. In Example 6 of this section and in Section 8.5 of this chapter we study the dynamics of a switch that turns gene expression on and off. These switches depend on the concentration of two proteins x and y, called repressors.

  a. Let $z = f(x, y)$ be the rate at which protein x changes in time as a function of the two protein

concentrations. Write an expression for $f(x, y)$ that is consistent with the following assumptions:
  - The rate of change is given by the difference in the rate of increase due to gene expression and the rate of decrease due to degradation.
  - The rate of increase due to gene expression is inversely proportional to $1 + y^c$ where c is a positive integer known as the cooperativity constant.
  - The rate of increase due to gene expression when $y = 0$ is a.
  - The rate of degradation is proportional to x, where we assume the proportionality constant equals one.

  b. Plot z obtained in part **a** for the case $a = 10$ and $c = 4$ over the domain $D = \{(x, y) | 0 \le x \le 3, 0 \le y \le 3\}$.

42. Example 5 reported cricket chirp rates measured at different ambient temperatures. Here are data obtained from other crickets:

$$D = \{(84.3, 18.4), (80.6, 17.1), (75.2, 15.5),$$
$$(69.7, 14.7), (82.0, 17.1)\}$$

  a. Use technology to find the regression line solution (m^*, c^*) through these points and plot the regression line along with the data on the same graph.

  b. Derive $S(m, c)$ for these data.

  c. Use technology to produce a contour plot of $S(m, c)$ over the domain $\{0 \le m \le 0.5, -5 \le c \le 0\}$ and locate the regression parameter point (m^*, c^*) on this plot.

43. In Table 1.5 of Problem Set 1.2, we presented data on the weight x_i of individual fiddler crabs and the weight y_i of their large claw. Below is a subset of five data points from this table:

$$D = \{(156.1, 25.1), (238.3, 52.5), (355.2, 104.5),$$
$$(535.7, 195.6), (680.6, 271.6)\}$$

  a. Use technology to find the regression line solution (m^*, c^*) through these points and plot the regression line along with the data on the same graph.

  b. Derive $S(m, c)$ for these data.

  c. Use technology to produce a contour plot of $S(m, c)$ over the domain $\{0.4750 \le m \le 0.4775, -58 \le c \le -57\}$ and locate the regression parameter point (m^*, c^*) on this plot.

44. Below is a topographic (i.e., contour) map of Barro Colorado Island, which is situated in the man-made Gatun Lake in the middle of the Panama Canal. This island is a nature reserve and one of the most intensively studied ecological areas in the world. The contours on the map are 10 meters apart (starting at

20 meters above water level). The highest point on the island is at point B. Plot the height along this transect indicated by the dotted line from point A to point B.

45. Take a closer look at the topographic map of Barro Colorado Island in the previous problem and consider the arc represented by the dotted curve that sweeps from point C to point D. By estimating the distance traversed along this arc from one contour line to the next, draw a distance-elevation profile you would experience walking along this arc.

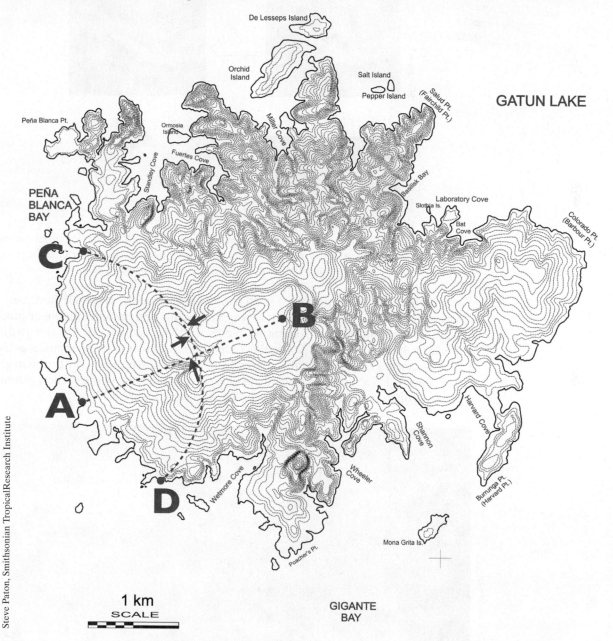

8.2   Matrices and Vectors

The simplest multivariate models are linear: they involve only linear relationships among several different variables. For instance, the number of loggerhead turtles may depend linearly on the number of adult turtles and juvenile turtles in the previous year. Specifically, as depicted in Figure 8.16, the number of adult loggerheads z this year is given by the past year's number of adult loggerheads x multiplied by the fraction p that survived over the year plus the past year's number of juvenile loggerheads y multiplied by the proportion q that matured to adults in the past year. We examine this model later in this section, and explore the implications for saving this endangered species in the next section of this chapter. Another example of a linear multivariate model, which we discuss in detail in Example 1, is the amount of grain produced by a farm depending on the area of its two fields.

Last year   This year

$$z = px + qy$$

Figure 8.16 Linear model of the abundance of loggerhead sea turtles. The total number of adults this year equals the fraction of adults surviving from the past year plus the fraction of juveniles that matured into adults over the past year.

To develop a framework for studying linear models, we need matrices and vectors, which provide a convenient way to describe and manipulate linear models. Matrices first arose from solving problems involving systems of linear equations. Such problems go back to the earliest recorded instances of mathematical activity. For instance, a Babylonian tablet dated from around 300 BC poses the problem given in Example 1. Such problems were possibly solved using one of the earliest examples of a calculating device, the Salamis Tablet, shown in Figure 8.17.

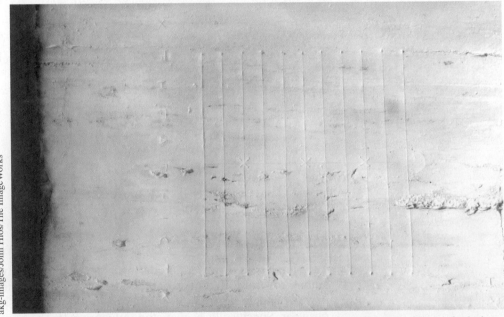

Figure 8.17 Salamis Tablet, 300 BC. This tablet is one of the earliest surviving calculating devices. It worked like an abacus but with gravity holding pebbles in grooves. When pebbles were moved between columns of grooves on the marble surface, the tablet functioned like a four-digit calculator. A tablet such as this might have been used to solve linear equations like the ones in Example 1.

Example 1   Back to Babylon

There are two fields with a total area of 1,800 square yards. One field produces grain at the rate of 2/3 a bushel per square yard while the other field produces grain at the rate of 1/2 a bushel per square yard. If the total yield is 1,100 bushels, what is the size of each field?

Solution   Let x be the size of one field and y the size of the other field in square yards. Since the total area of the two fields is 1,800 square yards, we get $x + y = 1,800$. On the other hand, the first field produces $(2/3)x$ bushels of grain and the second field produces $y/2$ bushels of grain. Since the total yield is 1,100 bushels, we have $2x/3 + y/2 = 1,100$. Hence, we get this system of linear equations:

$$\begin{cases} x + y = 1,800 \\ 2x/3 + y/2 = 1,100 \end{cases}$$

We can solve this linear system of equations in a variety of ways, the simplest being so-called *back substitution*. Namely, we solve for one of the variables in terms of the other in one equation. Then we substitute this solution into the other equation. For example, solving for y in the first equation yields

$$y = 1,800 - x$$

Substituting this expression for y into the second equation yields

$$2x/3 + (1,800 - x)/2 = 1,100$$

Solving for x yields

$$2x/3 - x/2 = 1,100 - 900 = 200$$
$$x/6 = 200$$
$$x = 1,200$$

Hence, the first field is 1,200 square yards and the second field is $1,800 - 1,200 = 600$ square yards.

To develop a systematic approach to solving systems of linear equations, we introduce the concepts of *vectors*, *matrices*, and basic vector and matrix algebra.

Defining vectors and matrices

In Chapters 1–7 we represented points in the xy plane as ordered pairs (x, y) (i.e., ordered by the convention that the first variable is always plotted on the horizontal axis and the second on the vertical axis). Another way of referring to a point in the plane is to represent it as a two-dimensional **vector v** written as a rectangular array with two rows and one column. In the first row, we place x, the horizontal axis variable. In the second row, we place y, the vertical axis variable. Thus, we get

$$\mathbf{v} = \begin{pmatrix} x \\ y \end{pmatrix}$$

The variables x and y are written in italics, as we have done throughout this book for variables that have a single value. The variable **v**, however, is written in boldface type to remind us that it has more than one value—in this case two values, that of its **first element** x and of its **second element** y.

In the xy plane, we can view $\mathbf{v}$ as an arrow whose tail is at $(0, 0)$ and tip is at (x, y), as illustrated in Figure 8.18. Since the vector $\mathbf{v}$ is characterized by its *direction* and its *magnitude*, we can draw $\mathbf{v}$ more generally as an arrow with its tail at any point (a, b) and its tip at $(a + x, b + y)$, as illustrated in Figure 8.19. We do this when we are adding the vector $\mathbf{v}$ to a vector $\mathbf{u} = \begin{pmatrix} a \\ b \end{pmatrix}$. All of these arrows have the same direction and magnitude as the arrow we drew in Figure 8.18. Just like real numbers, we can add two vectors and multiply a vector by a real number. Both of these operations have geometrical interpretations, with addition illustrated in Figure 8.19.

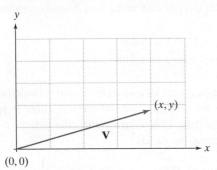

Figure 8.18 Vector $\mathbf{v}$ with first and second row elements x and y

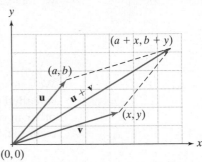

Figure 8.19 Geometrical interpretation of $\mathbf{u} + \mathbf{v}$

Scalar multiplication by a constant $c > 1$ is just stretching ($c > 1$) or shrinking ($0 < c < 1$) the vector by a factor c while maintaining the same direction; multiplying by -1 is just rotating the vector by 180 degrees (i.e., pointing it in the opposite direction without changing its magnitude). We do not focus on these geometrical interpretations in this book. Instead, we focus on the definitions from an algebraic point of view.

Vector Equality, Addition, and Scalar Multiplication

If $\mathbf{u} = \begin{pmatrix} a \\ b \end{pmatrix}$ and $\mathbf{v} = \begin{pmatrix} c \\ d \end{pmatrix}$ are two vectors, we can form these definitions:

Vector equality $\begin{pmatrix} a \\ b \end{pmatrix} = \begin{pmatrix} c \\ d \end{pmatrix}$ if and only if both $a = c$ and $b = d$

Vector addition $\mathbf{u} + \mathbf{v} = \begin{pmatrix} a \\ b \end{pmatrix} + \begin{pmatrix} c \\ d \end{pmatrix} = \begin{pmatrix} a + c \\ b + d \end{pmatrix}$

Scalar multiplication If k is a real number, then

$$k\mathbf{u} = k \begin{pmatrix} a \\ b \end{pmatrix} = \begin{pmatrix} ka \\ kb \end{pmatrix}$$

With this vector notion, we are nearly ready to represent systems of linear equations in a convenient, concise form. We begin by using our definition of vector equality to write the system of linear equations from Example 1 in this form:

$$\begin{pmatrix} x + y \\ \dfrac{2x}{3} + \dfrac{y}{2} \end{pmatrix} = \begin{pmatrix} 1,800 \\ 1,100 \end{pmatrix}$$

These equations have three distinct components: the vector of constants

$$\begin{pmatrix} 1,800 \\ 1,100 \end{pmatrix}$$

on the right-hand side of the equation, the vector of unknowns

$$\begin{pmatrix} x \\ y \end{pmatrix}$$

on the left-hand side of the equation, and the coefficients in front of these unknowns. We can store these coefficients in an array with two rows and two columns:

$$\begin{pmatrix} 1 & 1 \\ \dfrac{2}{3} & \dfrac{1}{2} \end{pmatrix}$$

Such an array is called a 2-by-2 *matrix*. If we define

$$\begin{pmatrix} 1 & 1 \\ \dfrac{2}{3} & \dfrac{1}{2} \end{pmatrix} \begin{pmatrix} x \\ y \end{pmatrix} = \begin{pmatrix} x + y \\ \dfrac{2}{3}x + \dfrac{1}{2}y \end{pmatrix}$$

then our *matrix-vector equation*

$$\begin{pmatrix} 1 & 1 \\ \dfrac{2}{3} & \dfrac{1}{2} \end{pmatrix} \begin{pmatrix} x \\ y \end{pmatrix} = \begin{pmatrix} 1,800 \\ 1,100 \end{pmatrix}$$

is equivalent to our original system of linear equations. We formalize these ideas in the following definitions. We also provide definitions of matrix addition and scalar multiplication, which will be useful later.

Matrices and Matrix Operations

Matrices A 2-by-2 matrix A is an array of the form

$$A = \begin{pmatrix} a & b \\ c & d \end{pmatrix}$$

where $a, b, c,$ and d are real numbers.

Matrix-vector multiplication Multiplication of a vector $\mathbf{v} = \begin{pmatrix} x \\ y \end{pmatrix}$ by the above matrix A (i.e., A times $\mathbf{v}$) is defined to be

$$A\mathbf{v} = \begin{pmatrix} a & b \\ c & d \end{pmatrix} \begin{pmatrix} x \\ y \end{pmatrix} = \begin{pmatrix} ax + by \\ cx + dy \end{pmatrix}$$

Scalar multiplication If α is a real number, then

$$\alpha A = \alpha \begin{pmatrix} a & b \\ c & d \end{pmatrix} = \begin{pmatrix} \alpha a & \alpha b \\ \alpha c & \alpha d \end{pmatrix}$$

Matrix addition If $B = \begin{pmatrix} \alpha & \beta \\ \gamma & \delta \end{pmatrix}$ is another 2-by-2 matrix, then

$$A + B = \begin{pmatrix} a & b \\ c & d \end{pmatrix} + \begin{pmatrix} \alpha & \beta \\ \gamma & \delta \end{pmatrix} = \begin{pmatrix} \alpha + a & \beta + b \\ \gamma + c & \delta + d \end{pmatrix}$$

The trickiest of these operations is matrix-vector multiplication, which corresponds to taking each row of A, multiplying it componentwise against the elements of $\mathbf{v}$, and adding up these terms. Let's get some practice with this operation.

Example 2   Matrix-vector multiplication

Calculate the following matrix-vector products.

a. $\begin{pmatrix} 1 & 0 \\ -1 & 2 \end{pmatrix} \begin{pmatrix} 3 \\ 4 \end{pmatrix}$

b. $\begin{pmatrix} 1 & 1 \\ 2/3 & 1/2 \end{pmatrix} \begin{pmatrix} 1{,}200 \\ 600 \end{pmatrix}$

Solution

a. $\begin{pmatrix} 1 & 0 \\ -1 & 2 \end{pmatrix} \begin{pmatrix} 3 \\ 4 \end{pmatrix} = \begin{pmatrix} 1 \cdot 3 + 0 \cdot 4 \\ -1 \cdot 3 + 2 \cdot 4 \end{pmatrix} = \begin{pmatrix} 3 \\ 5 \end{pmatrix}$

b. $\begin{pmatrix} 1 & 1 \\ 2/3 & 1/2 \end{pmatrix} \begin{pmatrix} 1{,}200 \\ 600 \end{pmatrix} = \begin{pmatrix} 1 \cdot 1{,}200 + 1 \cdot 600 \\ 2/3 \cdot 1{,}200 + 1/2 \cdot 600 \end{pmatrix} = \begin{pmatrix} 1{,}800 \\ 1{,}100 \end{pmatrix}$

An exciting application of matrix models is to conservation biology. Matrix models have been developed for many rare or endangered species as a tool to understand where best to focus conservation efforts. The following example is based on a study that played a key role in determining policy to protect loggerhead turtles. The loggerhead sea turtle (*Caretta caretta*) is a charismatic oceanic turtle species distributed throughout the world (see Figure 8.16). Loggerheads are long-lived species with typical life spans of forty-seven to sixty-seven years. They don't reach sexual maturity until ages 17 to 33, and adults can weigh as much as 1,000 pounds. In the next example, we introduce a simple model to project future population sizes of loggerheads. In the next section, we explore the conservation implications of this model.

Example 3   Projecting future loggerhead abundances

In collaboration with ecological modeler Hal Caswell, Deborah Crouse and Larry Crowder* developed a matrix model for a population of Atlantic loggerheads. Here we consider a simplified version of this model. As in the more complex model, we only keep track of female loggerheads.

Let x and y be the current abundances of juvenile and adult female loggerheads this year, respectively. Approximately, 50% of juveniles survive to the next year and do not mature, and approximately 0.1% of juveniles survive and mature to adults. Adults produce, on average, 80 immature loggerheads each year; 80% of adults survive to the next year.

a. Write an expression involving matrix-vector multiplication to project juvenile and adult abundances for next year.

b. Determine next year's abundances of juveniles and adults if there are currently 10,000 juveniles and 50 adults.

c. Determine the abundance of juveniles and adults two years from now. What trend is the population exhibiting?

Solution

a. Since the number of juveniles next year equals the proportion surviving and not maturing (0.5) multiplied by the current juvenile abundance (x) plus the number

*"A Stage-Based Population Model for Loggerhead Sea Turtles and Implications for Conservation," *Ecology* 68(1987): 1412–1423.

juveniles produced per adult (80) multiplied by the current abundance of adults (y), we get

$$\text{abundance of juveniles next year} = 0.5x + 80y$$

Since the number of adults next year equals the portion of maturing juveniles (0.001) multiplied by the current juvenile abundance (x) plus the proportion of adults surviving (0.8) multiplied by the current abundance of adults (y), we get

$$\text{abundance of adults next year} = 0.001x + 0.8y$$

In matrix-vector notation, we get

$$\begin{pmatrix} 0.5 & 80 \\ 0.001 & 0.8 \end{pmatrix} \begin{pmatrix} x \\ y \end{pmatrix}$$

b. Substituting $x = 10,000$ and $y = 50$ into our matrix-vector equation, we get

$$\begin{pmatrix} 0.5 & 80 \\ 0.001 & 0.8 \end{pmatrix} \begin{pmatrix} 10,000 \\ 50 \end{pmatrix} = \begin{pmatrix} 9,000 \\ 50 \end{pmatrix}$$

Next year, there will be 9,000 juveniles and 50 adults.

c. Substituting the values $x = 9,000$ and $y = 50$ for next year's projected abundances into the matrix-vector equation from part **a**, we get the projected abundances for two years from now:

$$\begin{pmatrix} 0.5 & 80 \\ 0.001 & 0.8 \end{pmatrix} \begin{pmatrix} 9,000 \\ 50 \end{pmatrix} = \begin{pmatrix} 8,500 \\ 49 \end{pmatrix}$$

These projections imply that the total population size starts at $10,000 + 50 = 10,050$ individuals, decreases to $9,000 + 50 = 9,050$ individuals next year, and further decreases to $8,500 + 49 = 8,549$ individuals two years from now. In Example 7 of the next section, we will examine this rate of decline more closely and consider management options for rescuing loggerheads. ◾

Solving matrix equations

For many applied problems, we want to solve matrix equations of the form

$$\begin{pmatrix} a & b \\ c & d \end{pmatrix} \begin{pmatrix} x \\ y \end{pmatrix} = \begin{pmatrix} e \\ f \end{pmatrix}$$

where $a, b, c, d, e,$ and f are known constants and x and y are unknown. One approach to solving these matrix-vector equations is to rewrite them as systems of linear equations and use back substitution as done in Example 1. After exploring a few examples using this approach, we derive a more systematic approach to solving these matrix equations.

Example 4   Feeding fish in a pond

The groundskeeper of a park buys two kinds of food to feed two species of fish found in this park. Call these species red fish and blue fish, of which there are respectively x and y individuals. Each month, each red fish consumes 4 units of the first kind of food and 3 units of the second kind of food, while each blue fish consumes 1 unit of the first kind of food and 2 units of the second kind. Do the following:

a. Write a matrix model that describes how many units of each food the fish need each month for any values of x and y.

b. How many units of each kind of food must the groundskeeper buy each month if $x = 40$ and $y = 25$?

c. Suppose the groundskeeper does not know how many fish there are, but after a month he finds that 100 units of the first kind of food and 120 units of the second kind of food are consumed. How many fish does he have of each species?

Solution

a. If a is the amount of the first kind of food needed each month, then $4x + y = a$. If b is the amount of the second kind of food needed each month, then $3x + 2y = b$. Hence, the matrix model is $A\mathbf{u} = \mathbf{v}$ where $A = \begin{pmatrix} 4 & 1 \\ 3 & 2 \end{pmatrix}$, $\mathbf{u} = \begin{pmatrix} x \\ y \end{pmatrix}$, and $\mathbf{v} = \begin{pmatrix} a \\ b \end{pmatrix}$.

b. The answer to this question is the value of $\mathbf{v}$ when $\mathbf{u} = \begin{pmatrix} 40 \\ 25 \end{pmatrix}$. Using our matrix equation, we get

$$\begin{pmatrix} a \\ b \end{pmatrix} = \begin{pmatrix} 4 & 1 \\ 3 & 2 \end{pmatrix}\begin{pmatrix} 40 \\ 25 \end{pmatrix} = \begin{pmatrix} 4 \cdot 40 + 1 \cdot 25 \\ 3 \cdot 40 + 2 \cdot 25 \end{pmatrix} = \begin{pmatrix} 185 \\ 170 \end{pmatrix}$$

Hence, the answer is 185 units of the first kind of food and 170 units of the second kind of food.

c. Here we are being asked to solve $A\mathbf{u} = \mathbf{v}$ for $\mathbf{u} = \begin{pmatrix} x \\ y \end{pmatrix}$ where $\mathbf{v} = \begin{pmatrix} 100 \\ 120 \end{pmatrix}$. In other words, solve the system of linear equations

$$\begin{cases} 4x + y = 100 \\ 3x + 2y = 120 \end{cases}$$

Solving for y in the first equation gives $y = 100 - 4x$. Substituting this expression for y into the second equation yields

$$3x + 2(100 - 4x) = 120$$
$$-5x = -80$$

Therefore, $x = 16$ and $y = 100 - 4 \cdot 16 = 36$. ■

Example 4, part **c**, shows that sometimes there is a unique solution for $\mathbf{u}$ in the equation $A\mathbf{u} = \mathbf{v}$. However, sometimes there are no solutions or many solutions. We illustrate both possibilities in the following example.

Example 5   No solutions versus infinitely many solutions

Let

$$A = \begin{pmatrix} 1 & 2 \\ 2 & 4 \end{pmatrix}$$

Find all solutions $\mathbf{u}$ to the equation $A\mathbf{u} = \mathbf{v}$ for the given $\mathbf{v}$.

a. $\mathbf{v} = \begin{pmatrix} 1 \\ 3 \end{pmatrix}$ b. $\mathbf{v} = \begin{pmatrix} 0 \\ 0 \end{pmatrix}$

Solution

a. We need to solve this system of linear equations:

$$\begin{cases} x + 2y = 1 \\ 2x + 4y = 3 \end{cases}$$

Solving for x in terms of y gives $x = 1 - 2y$. Substituting this expression into the second equation gives

$$2(1 - 2y) + 4y = 3$$
$$0 = 1$$

which clearly is never true. Hence, there is no solution. We could have also come to this conclusion by noticing that multiplying the first equation $x + 2y = 1$ by 2 gives $2x + 4y = 2$, but the second equation requires $2x + 4y = 3$.

b. We need to solve this system of linear equations:

$$\begin{cases} x + 2y = 0 \\ 2x + 4y = 0 \end{cases}$$

Since the second equation is twice the first equation, it suffices to solve $x + 2y = 0$. We get that $y = \dfrac{x}{2}$ is a solution for any choice of x. Hence, there are an infinite number of solutions of the form $\mathbf{u} = \begin{pmatrix} x \\ x/2 \end{pmatrix}$.

Examples 4c and 5, show that sometimes there may be no, one, or many solutions $\mathbf{u}$ to $A\mathbf{u} = \mathbf{v}$. It would be useful to understand whether some property of A determines which of these outcomes occurs. Consider the general case

$$\begin{pmatrix} a & b \\ c & d \end{pmatrix} \begin{pmatrix} x \\ y \end{pmatrix} = \begin{pmatrix} e \\ f \end{pmatrix}$$

which is equivalent to this system of linear equations:

$$\begin{cases} ax + by = e \\ cx + dy = f \end{cases}$$

Let's assume that $b \neq 0$. Solving for y in terms of x in the first equation gives

$$y = \frac{e}{b} - \frac{ax}{b}$$

Substituting this expression for y into the second equation gives

$$cx + d\left(\frac{e}{b} - \frac{ax}{b}\right) = f$$

Bringing the x terms to the left and the constant terms to the right yields

$$\left(c - \frac{ad}{b}\right)x = f - \frac{de}{b}$$

and multiplying both sides by b gives

$$(ad - bc)x = de - fb$$

This equation has a unique solution for x if $ad - bc \neq 0$, has no solution for x if $ad - bc = 0$ and $de - fb \neq 0$, and holds true for all x if $ad - bc = 0 = de - fb$. When there is a unique solution, it is

$$\begin{pmatrix} x \\ y \end{pmatrix} = \frac{1}{ad - bc}\begin{pmatrix} de - fb \\ af - ce \end{pmatrix}$$

Equivalently, we express this solution as a matrix-vector product:

$$\begin{pmatrix} x \\ y \end{pmatrix} = \frac{1}{ad - bc}\begin{pmatrix} d & -b \\ -c & a \end{pmatrix}\begin{pmatrix} e \\ f \end{pmatrix}$$

Since this solution holds even when $b = 0$ and $ad - bc \neq 0$ (see problem set at the end of this section), we have shown the following result.

Theorem 8.1   Solving matrix equations

Let

$$A = \begin{pmatrix} a & b \\ c & d \end{pmatrix} \text{ and } \mathbf{v} = \begin{pmatrix} e \\ f \end{pmatrix}$$

be given. If $ad - bc \neq 0$, then there is a unique solution to $A\mathbf{u} = \mathbf{v}$ given by

$$\mathbf{u} = A^{-1}\mathbf{v} \text{ where } A^{-1} = \frac{1}{ad - bc}\begin{pmatrix} d & -b \\ -c & a \end{pmatrix}$$

Alternatively, if $ad - bc = 0$, then either there are an infinite number of solutions or no solutions.

To highlight the importance of $ad - bc$ and A^{-1}, a formal definition for each follows.

Determinants and Inverse Matrices

The **determinant of a 2-by-2 matrix** $A = \begin{pmatrix} a & b \\ c & d \end{pmatrix}$ is

$$\det(A) = ad - bc$$

If $\det(A) \neq 0$, then the **inverse** *of A* is

$$A^{-1} = \frac{1}{ad - bc}\begin{pmatrix} d & -b \\ -c & a \end{pmatrix}$$

Now let's solve some systems of linear equations.

Example 6   Using inverses and determinants

For each A and $\mathbf{v}$ below, use Theorem 8.1 to determine whether there is a unique solution to the matrix equation $A\mathbf{u} = \mathbf{v}$. If there is a unique solution, then find this solution.

a. $A = \begin{pmatrix} 3 & 2 \\ 1.5 & 1 \end{pmatrix}$ and $\mathbf{v} = \begin{pmatrix} 1 \\ 2 \end{pmatrix}$ **b.** $A = \begin{pmatrix} 3 & 4 \\ 1.5 & 1 \end{pmatrix}$ and $\mathbf{v} = \begin{pmatrix} 1 \\ 2 \end{pmatrix}$

c. $A = \begin{pmatrix} 4 & 3 \\ 2 & 1 \\ 3 & 2 \end{pmatrix}$ and $\mathbf{v} = \begin{pmatrix} 6{,}600 \\ 1{,}100 \end{pmatrix}$

Solution

a. Since $\det(A) = 3(1) - 2(1.5) = 0$, the matrix A has no inverse and there is no unique solution.

b. Since $\det(A) = 3(1) - 4(1.5) = -3 \neq 0$, the matrix A has an inverse given by

$$A^{-1} = \frac{1}{-3}\begin{pmatrix} 1 & -4 \\ -1.5 & 3 \end{pmatrix} = \begin{pmatrix} -\dfrac{1}{3} & \dfrac{4}{3} \\ \dfrac{1}{2} & -1 \end{pmatrix}$$

The unique solution is given by

$$\mathbf{u} = A^{-1}\mathbf{v} = \begin{pmatrix} -\dfrac{1}{3} & \dfrac{4}{3} \\ \dfrac{1}{2} & -1 \end{pmatrix}\begin{pmatrix} 1 \\ 2 \end{pmatrix} = \begin{pmatrix} -\dfrac{1}{3} + 2\dfrac{4}{3} \\ \dfrac{1}{2} - 2 \end{pmatrix} = \begin{pmatrix} \dfrac{7}{3} \\ -\dfrac{3}{2} \end{pmatrix}$$

To make sure we got the right answer, we verify that $A\mathbf{u} = \mathbf{v}$:

$$A\mathbf{u} = \begin{pmatrix} 3 & 4 \\ 1.5 & 1 \end{pmatrix} \begin{pmatrix} \dfrac{7}{3} \\ -\dfrac{3}{2} \end{pmatrix} = \begin{pmatrix} 7 - 6 \\ \dfrac{7}{2} - \dfrac{3}{2} \end{pmatrix} = \begin{pmatrix} 1 \\ 2 \end{pmatrix} = \mathbf{v}$$

c. Since $\det(A) = 2 - 2 = 0$, A has no inverse and there is no unique solution. ∎

Example 7   Graphical interpretation of solutions

For each of the three problems $A\mathbf{u} = \mathbf{v}$ posed in Example 6, plot the lines representing the two equations on the same graph. Give a graphical interpretation for the possible solutions to the matrix equation.

Solution

a. The matrix equation for the first example is $A = \begin{pmatrix} 3 & 2 \\ 1.5 & 1 \end{pmatrix} \begin{pmatrix} x \\ y \end{pmatrix} = \begin{pmatrix} 1 \\ 2 \end{pmatrix}$, which

yields these two equations:

$$3x + 2y = 1$$
$$1.5x + y = 2$$

Plotting these equations, we obtain the graph shown in Figure 8.20a. Since the lines are parallel and not the same, the lines do not intersect and there is no solution to the matrix equation.

b. The matrix equation for the second example is $A = \begin{pmatrix} 3 & 4 \\ 1.5 & 1 \end{pmatrix} \begin{pmatrix} x \\ y \end{pmatrix} = \begin{pmatrix} 1 \\ 2 \end{pmatrix}$, which

yields these two equations:

$$3x + 4y = 1$$
$$1.5x + y = 2$$

Plotting these equations, we obtain the graph shown in Figure 8.20b. These lines intersect at the point $(7/3, -3/2)$, which is the unique solution to the system of linear equations.

c. The matrix equation for the third example is $A = \begin{pmatrix} 4 & 3 \\ \dfrac{2}{3} & \dfrac{1}{2} \end{pmatrix} \begin{pmatrix} x \\ y \end{pmatrix} = \begin{pmatrix} 6{,}600 \\ 1{,}100 \end{pmatrix}$,

which yields these two equations:

$$4x + 3y = 6{,}600$$
$$\frac{2x}{3} + \frac{y}{2} = 1{,}100$$

Plotting these equations, we obtain the graph shown in Figure 8.20c. Since the lines coincide, there are an infinite number of solutions.

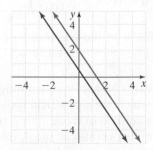
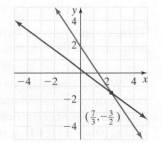
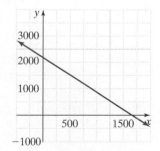

a. Equations yield parallel lines   **b.** Equations yield one solution   **c.** Equations coincide

Figure 8.20 Graphical representation of systems of equations for Example 7

Example 8   How many children are at school?

Ignoring weekends and holidays, 90% of all children at school on any given day are at school the next day. The remaining 10% are at home feeling sick that next day. Of those at home feeling sick on any given day, the next day 80% of them come back to school, 5% of them go to the hospital, and 15% remain at home still feeling sick. Finally, of all the children in the hospital on a given day, 50% of them are in hospital the following day, while the rest of them return home the next day to recover.

Let x and y represent the fraction of children at school and the fraction in the hospital. Assuming that the fraction of children at school, home, and hospital do not change from day to day, write a matrix equation of the form

$$A\mathbf{u} = \mathbf{v}$$

where $\mathbf{u} = \begin{pmatrix} x \\ y \end{pmatrix}$. The entries in A will represent the transitions from one day to the next, Solve for x and y.

Solution   The proportion in school is x and the proportion in the hospital is y. Then the proportion at home is $1 - x - y$. From the description given, 90% of the children in school stay in school and 80% of those children at home come back to school. Therefore, the fraction in school tomorrow is $0.9x + 0.8(1 - x - y)$ $= 0.8 + 0.1x - 0.8y$. Since we have assumed the fraction x does not change from day to day, we get $x = 0.8 + 0.1x - 0.8y$. Equivalently,

$$0.9x + 0.8y = 0.8$$

On any particular day, we have 50% of the children in the hospital staying in the hospital on the next day and 5% of them at home going to the hospital on the next day. Therefore, the fraction in the hospital the next day is $0.5y + 0.05(1 - x - y)$ $= 0.05 - 0.05x + 0.45y$. Since we have assumed that this fraction doesn't change from day to day, we get $y = 0.05 - 0.05x + 0.45y$. Equivalently,

$$0.05x + 0.55y = 0.05$$

Putting the two displayed equations together yields $A\mathbf{u} = \mathbf{v}$ where

$$\mathbf{u} = \begin{pmatrix} x \\ y \end{pmatrix}, \quad A = \begin{pmatrix} 0.9 & 0.8 \\ 0.05 & 0.55 \end{pmatrix}, \quad \mathbf{v} = \begin{pmatrix} 0.8 \\ 0.05 \end{pmatrix}$$

Since $\det(A) = 0.9 \cdot 0.55 - 0.05 \cdot 0.8 = 0.455 \neq 0$, the inverse of A (check this yourself) equals

$$A^{-1} = \frac{1}{0.455} \begin{pmatrix} 0.55 & -0.8 \\ -0.05 & 0.9 \end{pmatrix}$$

Solving for $\mathbf{u}$, we get

$$\mathbf{u} = A^{-1}\mathbf{v} \approx \begin{pmatrix} 0.879 \\ 0.011 \end{pmatrix}$$

Thus, on any day we expect 87.9% of the children to be at school, 1.1% to be in hospital, and the remaining $100\% - 87.9\% - 1.1\% = 11\%$ to be at home. ∎

Wassily Leontief (1905–1999), a Nobel Prize-winning economist, developed the theory of *input-output models*, which are a powerful tool for understanding the economic output of countries in the context of trade among nations. The theory is based on the simple idea that if a level of resources (coal, oil, mining products, agricultural products, electricity, etc.) is represented by a vector $\mathbf{u}$ and $A\mathbf{u}$ represents the amount of resources needed to produce $\mathbf{u}$, then what is left over after production is $\mathbf{d} = \mathbf{u} - A\mathbf{u}$. This left over $\mathbf{d}$ can be used for trade.

Leontief Input-Output Theory

Let **u** be a vector representing resource levels (amounts). Let A be a 2-by-2 matrix such that $A\mathbf{u}$ are the amounts of resources needed to produce **u**. Then the resources **d** available for trade after accounting for production are given by the equation

$$\mathbf{d} = \mathbf{u} - A\mathbf{u}$$

Example 9 Input-output models and the net production of electricity

Coal is used to heat water to drive steam turbines to make electricity. However, some electricity is consumed in the process to run the generating facility. Electricity is also needed to help run coal mines, and some of the coal is used to create heat needed in the buildings that are part of the mine. Suppose, on average, that it takes 0.1 units of electricity and 0.3 units of coal to produce 1 unit of electricity and that it takes 0.4 units of electricity and 0.1 units of coal to produce 1 unit of coal. How many units of coal must be mined for a net production of 1,000 units of electricity so that all the coal is consumed in the process?

Solution   In this problem, let the elements x and y in the vector $\mathbf{u} = \begin{pmatrix} x \\ y \end{pmatrix}$ represent amounts of electricity and coal, respectively. The production matrix A for this problem is then $\begin{pmatrix} 0.1 & 0.3 \\ 0.4 & 0.1 \end{pmatrix}$. Hence, according to the Leontif input-output theory, the number of units of electricity and coal left after the production process, and represented by the vector $\mathbf{d} = \begin{pmatrix} a \\ b \end{pmatrix}$, is given by the model

$$\mathbf{d} = \mathbf{u} - A\mathbf{u} \Rightarrow \begin{pmatrix} a \\ b \end{pmatrix} = \begin{pmatrix} x \\ y \end{pmatrix} - \begin{pmatrix} 0.1 & 0.3 \\ 0.4 & 0.1 \end{pmatrix} \begin{pmatrix} x \\ y \end{pmatrix}$$

where a is the number of units of electricity and b is the number of units of coal (see Figure 8.21).

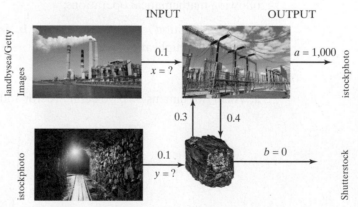

Figure 8.21 A Leontief input-output model can be used to find how much electricity and coal need to be produced to obtain 1000 units of electricity with no coal left over.

This results in the two equations

$$\begin{cases} a = x - 0.1x - 0.3y = 0.9x - 0.3y \\ b = y - 0.4x - 0.1y = -0.4x + 0.9y \end{cases}$$

which can be rewritten in matrix-vector form

$$\begin{pmatrix} a \\ b \end{pmatrix} = \begin{pmatrix} 0.9 & -0.3 \\ -0.4 & 0.9 \end{pmatrix} \begin{pmatrix} x \\ y \end{pmatrix}$$

We want to find x and y such that $a = 1{,}000$ units of electricity available and $b = 0$ coal is available.

Solving this matrix equation using Theorem 8.1, we find

$$\begin{pmatrix} x \\ y \end{pmatrix} = \frac{1}{0.81 - 0.12} \begin{pmatrix} 0.9 & 0.3 \\ 0.4 & 0.9 \end{pmatrix} \begin{pmatrix} 1{,}000 \\ 0 \end{pmatrix} = \frac{1}{0.69} \begin{pmatrix} 900 \\ 400 \end{pmatrix} = \begin{pmatrix} 1{,}304.3 \\ 276 \end{pmatrix}$$

Hence, 276 units of coal must be mined to produce 1,000 units of electricity so that no coal is left over. ∎

Even though we have dealt exclusively in this section with matrix models in two variables, all of the ideas presented extend to any number of variables. To describe this generalization, we need to change our notation, which in two dimensions would be to use the variables x_1 and x_2 rather the x and y. This extends directly to n dimensions (n is any natural number) where by analogy the variables are now $x_1, x_2, \ldots, x_n$.

Example 10   Extending vector and matrix addition and multiplication to n dimensions

If the rectangular arrays with n rows and 1 column $\mathbf{u} = \begin{pmatrix} x_1 \\ x_2 \\ \vdots \\ x_n \end{pmatrix}$, and $\mathbf{v} = \begin{pmatrix} y_1 \\ y_2 \\ \vdots \\ y_n \end{pmatrix}$ are

n-dimensional vectors, and the array

$$A = \begin{pmatrix} a_{11} & a_{12} & \cdots & a_{1n} \\ a_{21} & a_{22} & \cdots & a_{2n} \\ \vdots & \vdots & & \vdots \\ a_{n1} & a_{n2} & \cdots & a_{nn} \end{pmatrix}$$

is a n-by-n matrix, then extend to n-dimensions the definitions in this section of the following mathematical operations:

a. Vector addition   **b.** Scalar multiplication of a vector
c. Matrix multiplication by a vector

Solution
a. As in two dimensions, we define vector addition by adding elements component-wise:

$$\begin{pmatrix} x_1 \\ x_2 \\ \vdots \\ x_n \end{pmatrix} + \begin{pmatrix} y_1 \\ y_2 \\ \vdots \\ y_n \end{pmatrix} = \begin{pmatrix} x_1 + y_1 \\ x_2 + y_2 \\ \vdots \\ x_n + y_n \end{pmatrix}$$

b. As in two dimensions, scalar multiplication of a vector is simply component-wise multiplication:

$$a \begin{pmatrix} x_1 \\ x_2 \\ \vdots \\ x_n \end{pmatrix} = \begin{pmatrix} ax_1 \\ ax_2 \\ \vdots \\ ax_n \end{pmatrix}$$

c. As in two dimensions, matrix-vector multiplication requires that we take each row of the matrix A, which is of dimension n, and the vector $\mathbf{u}$, which is also of dimension n, and multiply each component in each row with the corresponding component in the vector. Adding all of the pairwise multiplications for a particular row of A produces the corresponding component in the new vector, namely:

$$A\mathbf{u} = \begin{pmatrix} a_{11} & a_{12} & \cdots & a_{1n} \\ a_{21} & a_{22} & \cdots & a_{2n} \\ \vdots & \vdots & & \vdots \\ a_{n1} & a_{n2} & \cdots & a_{nn} \end{pmatrix} \begin{pmatrix} x_1 \\ x_2 \\ \vdots \\ x_n \end{pmatrix} = \begin{pmatrix} a_{11}x_1 + a_{12}x_2 + \ldots a_{1n}x_n \\ a_{21}x_1 + a_{22}x_2 + \ldots a_{2n}x_n \\ \vdots \\ a_{n1}x_1 + a_{n2}x_2 + \cdots + a_{nn}x_n \end{pmatrix}$$

PROBLEM SET 8.2

Level 1 DRILL PROBLEMS

Solve each system of equations in Problems 1 to 6 by graphing.

1. $\begin{cases} 2x - 3y = -8 \\ x + y = 6 \end{cases}$

2. $\begin{cases} 2x - 3y = -8 \\ 4x - 6y = 0 \end{cases}$

3. $\begin{cases} 2x - 3y = -8 \\ y = \dfrac{2}{3}x + \dfrac{8}{3} \end{cases}$

4. $\begin{cases} 6x - 4y = 10 \\ 2y = 3x - 5 \end{cases}$

5. $\begin{cases} 3x - y = 1 \\ x - 2y = 8 \end{cases}$

6. $\begin{cases} y = \dfrac{3}{5}x + 2 \\ 3x - 5y = -15 \end{cases}$

In Problems 7 to 12, write the two equations in matrix form and solve for x and y.

7. $x + y = 2$ and $x - y = 0$

8. $3x - y = 4$ and $2x + 2y = 0$

9. $4x + 2y = 6$ and $x - 2y = -1$

10. $-3x + y = -2$ and $2x - 2y = -2$

11. $2x + 3y = 4$ and $3x - 4y = -4$

12. $-3x + 5y = -10$ and $2x + 4y = 2$

Find the determinant and the inverse of the matrices whenever well-defined in Problems 13 to 18.

13. $\begin{pmatrix} 1 & 3 \\ 2 & 6 \end{pmatrix}$

14. $\begin{pmatrix} 2 & -1 \\ -3 & \dfrac{3}{2} \end{pmatrix}$

15. $\begin{pmatrix} 2 & 1 \\ 1 & 2 \end{pmatrix}$

16. $\begin{pmatrix} 2 & -1 \\ 1 & -2 \end{pmatrix}$

17. $\begin{pmatrix} 2 & -2 \\ 1 & 3 \end{pmatrix}$

18. $\begin{pmatrix} \dfrac{1}{2} & -1 \\ 1 & 1 \end{pmatrix}$

Use Theorem 8.1 to solve the matrix equation $\mathbf{v} = A\mathbf{u}$ for the vector $\mathbf{u}$ in Problems 19 to 30.

19. $A = \begin{pmatrix} 4 & -7 \\ -1 & 2 \end{pmatrix}$ and $\mathbf{v} = \begin{pmatrix} -2 \\ 1 \end{pmatrix}$

20. $A = \begin{pmatrix} 4 & -7 \\ -1 & 2 \end{pmatrix}$ and $\mathbf{v} = \begin{pmatrix} 48 \\ -13 \end{pmatrix}$

21. $A = \begin{pmatrix} 8 & 6 \\ -2 & 4 \end{pmatrix}$ and $\mathbf{v} = \begin{pmatrix} 12 \\ -14 \end{pmatrix}$

22. $A = \begin{pmatrix} 8 & 6 \\ -2 & 4 \end{pmatrix}$ and $\mathbf{v} = \begin{pmatrix} -6 \\ -26 \end{pmatrix}$

23. $A = \begin{pmatrix} 2 & 3 \\ 1 & -6 \end{pmatrix}$ and $\mathbf{v} = \begin{pmatrix} 9 \\ -3 \end{pmatrix}$

24. $A = \begin{pmatrix} 2 & 3 \\ 1 & -6 \end{pmatrix}$ and $\mathbf{v} = \begin{pmatrix} 2 \\ -14 \end{pmatrix}$

25. $A = \begin{pmatrix} 1 & 2 \\ \dfrac{1}{2} & -1 \end{pmatrix}$ and $\mathbf{v} = \begin{pmatrix} 2 \\ -1 \end{pmatrix}$

26. $A = \begin{pmatrix} 4 & 2 \\ 1 & -2 \end{pmatrix}$ and $\mathbf{v} = \begin{pmatrix} 4 \\ 1 \end{pmatrix}$

27. $A = \begin{pmatrix} \dfrac{1}{2} & \dfrac{1}{2} \\ \dfrac{1}{3} & \dfrac{2}{3} \end{pmatrix}$ and $\mathbf{v} = \begin{pmatrix} 2 \\ 2 \end{pmatrix}$

28. $A = \begin{pmatrix} \dfrac{1}{2} & 2 \\ 1 & -1 \end{pmatrix}$ and $\mathbf{v} = \begin{pmatrix} -1 \\ 2 \end{pmatrix}$

29. $A = \begin{pmatrix} 3 & -3 \\ 1 & -\dfrac{1}{2} \end{pmatrix}$ and $\mathbf{v} = \begin{pmatrix} \dfrac{5}{2} \\ 1 \end{pmatrix}$

30. $A = \begin{pmatrix} 1 & -\dfrac{1}{2} \\ 2 & -\dfrac{1}{2} \end{pmatrix}$ and $\mathbf{v} = \begin{pmatrix} 3 \\ -2 \end{pmatrix}$

Use the definitions of vector addition and matrix multiplication in n dimensions to compute the quantities requested in Problems 31 to 34.

31. $\mathbf{u} + \mathbf{v}$ where $\mathbf{u} = \begin{pmatrix} 1 \\ 2 \\ 3 \end{pmatrix}$ and $\mathbf{v} = \begin{pmatrix} 0 \\ -1 \\ -2 \end{pmatrix}$

32. $\mathbf{u} + 4\mathbf{v}$ where $\mathbf{u} = \begin{pmatrix} 1 \\ 2 \\ 3 \end{pmatrix}$ and $\mathbf{v} = \begin{pmatrix} 0 \\ -1 \\ -2 \end{pmatrix}$

33. $A\mathbf{u}$ where $\mathbf{u} = \begin{pmatrix} 1 \\ 2 \\ 3 \end{pmatrix}$ and $A = \begin{pmatrix} 0 & 1 & 0 \\ 1 & 2 & 0 \\ 1 & 0 & 0 \end{pmatrix}$

34. $A\mathbf{u}$ where $\mathbf{u} = \begin{pmatrix} -1 \\ 0 \\ 2 \end{pmatrix}$ and $A = \begin{pmatrix} 1 & 7 & 1 \\ 0 & 1 & 1 \\ 2 & 0 & 1 \end{pmatrix}$

Level 2 APPLIED AND THEORY PROBLEMS

35. At the wholesale flower market in Amsterdam, a dealer has two customers only: one from England and one from Germany. The customer from England always orders a dozen roses for every 6 tulips purchased, while the customer from Germany orders a dozen roses for every 9 tulips ordered. On a particular day, the flower wholesaler sells 500 dozen roses and 325 dozen tulips. How many roses and tulips were bought by each of the two customers?

36. A butcher in a particular village makes two kinds of sausage: regular and spicy. In the regular sausage he uses 4 units of fatty meat and 2 units of lean meat, and in the spicy sausage he uses 3 units fatty to 3 units of lean meat. If the butcher uses 6,400 units of fatty meat and 4,400 units of lean meat, how many sausages has he made of each type?

37. In a study of the distribution of frogs in the Amazon jungle, the proportion of frogs of the species *Adenomera andreae* and toads of the species *Bufo dapsilis* in habitats A and B is 15 and 9 individuals (respectively) per unit area in habitat type A and 12 and 17 individuals per unit area in habitat type B. A group of summer school students conducted a survey and collected all the frogs in an area containing only type A and B habitats. If they collected 90 *A. andreae* and 121 *B. dapsilis*, estimate the number of type A and B units of habitat covered.

38. In a study of the distribution of birds in the ponderosa pine forests of the southern Sierra mountains, an ornithologist found that the nest density of Anna's hummingbird (*Calypte anna*) in habitats A and B was, respectively, 37 and 12 individuals per unit area. She also found that the nest density of the Western wood-pewee (*Contopus sordidulus*) in habitats A and B was, respectively, 29 and 11 individuals per unit area. A second ornithologist taking a census in the same region across an area of unknown size counted 418 nests of Anna's hummingbird and 334 nests of the Western wood-pewee. Assuming the data are accurate and the region contains only type A and B habitats, use the counts to estimate the total area covered by the second ornithologist and the relative proportion of type A and B habitats.

39. A pharmaceutical company makes up a vitamin capsule using different recipes for men and women. In each capsule, the recipe calls for 80 units of vitamin B_{12} and 50 units of vitamin C in the capsules for men, and 70 units of vitamin B_{12} and 40 units of vitamin C in the capsules for women. If the manufacturer uses 13 units of vitamin C for every 22 units of vitamin B_{12}, then what is the gender ratio of the customer base?

40. In the animal care facility of a major university, two kinds of rodents are reared for lab experiments: a species of mouse and a species of rat. An investigative journalist who would like to estimate how many of each of these two species are being reared at the university and how many are sacrificed as part of ongoing experiments during a particular period of time obtains the following information: Each mouse is fed 3 units of type A feed and 4 units of type B feed each day. Each rat is fed 5 units of type A feed and 3 units of type B feed each day. If on a particular day the animal care facility uses 11,100 units of type A feed and 9,300 units of type B feed, then how many mice and rats are alive that day? If a week later the amount of feed used drops to 7,500 units of type A and 6,700 units of type B, then how many mice and how many rats were sacrificed during the week in question?

41. To produce 1 unit of electricity requires using 0.2 units of electricity and burning 0.3 units of coal. To mine 1 unit of coal requires using 0.4 units of electricity, but requires no use of coal. How much coal must be mined for a net production of 10,000 units of electricity so that all the coal is consumed in the process?

42. To extract 1 barrel of oil from the ground requires using 0.1 barrels of oil and 0.1 kg of steel. To manufacture 1 kg of steel requires 0.05 barrels of oil and 0.01 kg of steel. How many barrels of oil must be extracted from the ground for a net production of 1 metric ton (1,000 kg) of steel with no oil left over at the end?

43. A sewage treatment plant requires 0.05 units of clean water and 0.2 units of electricity to produce 1 unit of clean water. A power plant requires 0.1 units of clean water and 0.1 units of electricity to produce 1 unit of electricity. What is the minimum amount of clean water that a sewerage plant must produce to supply a neighboring power plant with sufficient clean water to have a million units of electricity available for export?

44. A particular organism requires 0.2 units of active biomass tissue (e.g., biomass belonging to organs and muscle but not fat) and 0.1 unit of energy (from fat) to produce 1 unit of active biomass tissue from the food it eats. It also needs 0.4 units of active biomass tissue and 0.05 units of energy to produce 1 unit of energy (stored as fat). How much active biomass tissue must the organism produce to have 10 units of stored energy ready to expend during hibernation so that it is at the same active biomass tissue state after hibernation as it was before it started to accumulate the excess energy stored in the form of fat?

45. (After an example presented by M. H. Preston in *Elementary Matrices for Economics*) Suppose it takes

- 0.13 units of petroleum, 0.33 units of chemicals, and 0.25 units of transportation to produce one unit of petroleum,
- 0.50 units of petroleum, 0.17 units of chemicals, and 0.25 units of transportation to produce one unit of chemicals, and
- 0.25 units of petroleum, 0.17 units of chemicals, and 0.25 units of transportation to produce one unit of transportation.

a. Write an input-output matrix for this system.

b. Use a process of substitution and elimination to solve the equations to determine the quantities of petroleum, chemicals, and transportation to produce 10 units of petroleum, 28 units of chemicals, and 14 units of transportation.

46. Suppose it takes

- 0.1 units of natural gas, 0.4 units of electricity, and 0.3 units of transportation to produce one unit of natural gas,
- 0.5 units of natural gas, 0.2 units of electricity and 0.3 units of transportation to produce one unit of electricity, and
- 0.3 units of natural gas, 0.1 units of electricity and 0.2 units of transportation to produce one unit of transportation.

a. Write an input-output matrix for this system.

b. Use a process of substitution and elimination to solve the equations to determine the quantities of natural gas, electricity, and transportation to produce 10 units of natural gas, 15 units of electricity, and 5 units of transportation.

47. On a tropical island in the Pacific ocean, days can be cloudy, rainy, or sunny, and the weather is unchanging over the year—that is, there are no seasons. If it is cloudy today, then the probability that it will be cloudy, rainy, or sunny tomorrow is 0.5, 0.2, and 0.3, respectively. If it is rainy today, then the probability that it will be cloudy, rainy, or sunny tomorrow is 0.3, 0.2, and 0.5, respectively. If it is sunny today, then the probability that it will be cloudy, rainy, or sunny tomorrow is 0.3, 0, and 0.7, respectively (i.e., sunny days are never followed by rainy days). Write a matrix model that can be used to solve for the proportion of cloudy, rainy, and sunny days in the year. Solve the three resulting equations through a process of substitution and elimination of variables. Hint: To construct the matrix model you have to follow the logic: If today is cloudy, then there is 0.5 chance that tomorrow is cloudy; if today is rainy, then there is a 0.2 chance that tomorrow is rainy; if today is sunny, then there is a 0.3 chance that tomorrow is cloudy, and so on.

48. In a small tropical seaside town where the weather patterns are the same over the whole year, days can either be cloudy, rainy, or sunny. If it is cloudy today, then the probability that it will be cloudy, rainy, or sunny tomorrow is 0.4, 0.3, and 0.3, respectively. If it is rainy today, then the probability that it will be cloudy, rainy, or sunny tomorrow is 0.3, 0.3, and 0.4, respectively. If it is sunny today, then the probability that it will be cloudy, rainy, or sunny tomorrow is 0.3, 0.2, and 0.5, respectively. Write a matrix model that can be used to solve for the proportion of cloudy, rainy, and sunny days in the year. Solve the three resulting equations through a process of substitution and elimination of variables. (See hint to previous problem.)

49. Prove that Theorem 8.1 is true when $b = 0$.

8.3   Eigenvalues and Eigenvectors

Matrix models are used extensively to describe the dynamics of biological processes. These models are multivariate extensions of the linear difference equations described in Sections 1.7, 2.5, and 4.5. If $\mathbf{u}_n$ describes the state of the biological system at time n, then these models take the form

$$\mathbf{u}_{n+1} = A\mathbf{u}_n$$

In this section, we describe how these models can be used to simulate the growth of "structured populations," such as the African impala in Example 1 and the endangered loggerhead sea turtle introduced in the previous section and studied further in Example 7. We also show how matrix models can be used to understand the distribution of lead in tissues, bones, and the bloodstream of an individual with lead poisoning. Finding solutions to these difference equations reduces to an algebraic problem of finding vectors $\mathbf{v}$ and numbers λ such that $A\mathbf{v} = \lambda\mathbf{v}$. These vectors and numbers are called eigenvectors and eigenvalues.

Main concepts

The next example constructed around a hypothetical impala ranch motivates the main concepts developed in this section.

Example 1   Impala population growth

Impala are the most ubiquitous of the African antelopes in Kruger National Park in South Africa. Piet Meyer, a game rancher, plans to establish a population of impalas on his ranch by introducing x_0 immature females, y_0 mature females, and several mature males. Other ranchers have told Piet that he can expect each female to produce on average 1.2 female fawns every two years (on average males will be produced at a similar rate but we are only focusing on females). Piet can also expect that only 40% of these females mature to age 2, at which age they start to reproduce. He can also expect that during each two-year period 20% of the adult females die.

Figure 8.22 Female impala and a fawn in Kruger National Park.
Source: Photo courtesy of W. M. Getz.

a. Write a model of this population where x_n and y_n are the number of immature and mature females, respectively, in year $2n$.

b. If Piet initially has $x_0 = 100$ immature females in the population, determine the number of immature and mature females that Piet can expect on his ranch for the next 12 years (i.e., $n = 6$ two-year periods) and plot these values in the xy plane. Discuss what you notice about the behavior of the population trajectory.

c. If Piet initially has $x_0 = 50$ immature females and $y_0 = 50$ mature females, determine the number of juveniles and adults Piet can expect on his ranch over the next 6 years ($n = 3$). Discuss how this situation differs from the expected outcome from scenario **b** and how it can be used to calculate the expected number for any arbitrary value of n.

Solution

a. Let x_n and y_n be the number of immature females and mature females, respectively, in year $2n$. Given x_n and y_n, we want their numbers x_{n+1} and y_{n+1} at the end of the next two-year period. Since each mature female produces 1.2 daughters, it follows that $x_{n+1} = 1.2y_n$. Since 40% of the immature females survive to become adults, and 80% of the adults survive over the next two-year period, it follows that $y_{n+1} = 0.4x_n + 0.8y_n$. Hence, the model is given by the equations

$$x_{n+1} = 1.2y_n$$
$$y_{n+1} = 0.4x_n + 0.8y_n$$

In vector-matrix notation, we get

$$\underbrace{\begin{pmatrix} x_{n+1} \\ y_{n+1} \end{pmatrix}}_{\mathbf{u}_{n+1}} = \underbrace{\begin{pmatrix} 0 & 1.2 \\ 0.4 & 0.8 \end{pmatrix}}_{A} \underbrace{\begin{pmatrix} x_n \\ y_n \end{pmatrix}}_{\mathbf{u}_n}$$

b. Let $x_0 = 100$ and $y_0 = 0$. The abundances in the next six two-year periods are given as follows:

$$\mathbf{u}_1 = \begin{pmatrix} x_1 \\ y_1 \end{pmatrix} = \begin{pmatrix} 0 & 1.2 \\ 0.4 & 0.8 \end{pmatrix} \begin{pmatrix} 100 \\ 0 \end{pmatrix} = \begin{pmatrix} 0 \\ 40 \end{pmatrix}$$

$$\mathbf{u}_2 = \begin{pmatrix} x_2 \\ y_2 \end{pmatrix} = \begin{pmatrix} 0 & 1.2 \\ 0.4 & 0.8 \end{pmatrix} \begin{pmatrix} 0 \\ 40 \end{pmatrix} = \begin{pmatrix} 48 \\ 32 \end{pmatrix}$$

$$\mathbf{u}_3 = \begin{pmatrix} x_3 \\ y_3 \end{pmatrix} = \begin{pmatrix} 0 & 1.2 \\ 0.4 & 0.8 \end{pmatrix} \begin{pmatrix} 48 \\ 32 \end{pmatrix} = \begin{pmatrix} 38.4 \\ 44.8 \end{pmatrix}$$

$$\mathbf{u}_4 = \begin{pmatrix} x_4 \\ y_4 \end{pmatrix} = \begin{pmatrix} 0 & 1.2 \\ 0.4 & 0.8 \end{pmatrix} \begin{pmatrix} 38.4 \\ 44.8 \end{pmatrix} = \begin{pmatrix} 53.76 \\ 51.2 \end{pmatrix}$$

$$\mathbf{u}_5 = \begin{pmatrix} x_5 \\ y_5 \end{pmatrix} = \begin{pmatrix} 0 & 1.2 \\ 0.4 & 0.8 \end{pmatrix} \begin{pmatrix} 53.76 \\ 51.2 \end{pmatrix} = \begin{pmatrix} 61.44 \\ 62.464 \end{pmatrix}$$

$$\mathbf{u}_6 = \begin{pmatrix} x_6 \\ y_6 \end{pmatrix} = \begin{pmatrix} 0 & 1.2 \\ 0.4 & 0.8 \end{pmatrix} \begin{pmatrix} 61.44 \\ 62.464 \end{pmatrix} = \begin{pmatrix} 74.9568 \\ 74.5472 \end{pmatrix}$$

Consecutive points $\mathbf{u}_0$, $\mathbf{u}_1$, $\mathbf{u}_2$, $\mathbf{u}_3$, $\mathbf{u}_4$, $\mathbf{u}_5$, and $\mathbf{u}_6$ in the xy plane joined by lines are shown in Figure 8.23.

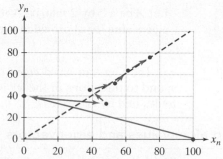

Figure 8.23 Points (x_1, y_1), (x_2, y_2), …, (x_6, y_6) for the impala model plotted in the xy plane

This figure illustrates that population numbers appear to be approaching the line $x = y$ where there would be an equal number of juveniles and adults.

c. If $x_0 = y_0 = 50$, then

$$\begin{pmatrix} x_1 \\ y_1 \end{pmatrix} = \begin{pmatrix} 0 & 1.2 \\ 0.4 & 0.8 \end{pmatrix} \begin{pmatrix} 50 \\ 50 \end{pmatrix} = \begin{pmatrix} 60 \\ 60 \end{pmatrix}$$

$$\begin{pmatrix} x_2 \\ y_2 \end{pmatrix} = \begin{pmatrix} 0 & 1.2 \\ 0.4 & 0.8 \end{pmatrix} \begin{pmatrix} 60 \\ 60 \end{pmatrix} = \begin{pmatrix} 72 \\ 72 \end{pmatrix}$$

$$\begin{pmatrix} x_3 \\ y_3 \end{pmatrix} = \begin{pmatrix} 0 & 1.2 \\ 0.4 & 0.8 \end{pmatrix} \begin{pmatrix} 72 \\ 72 \end{pmatrix} = \begin{pmatrix} 86.4 \\ 86.4 \end{pmatrix}$$

Hence, when starting with equal numbers of immature and mature females, the population continues to have equal numbers of immature and mature females, and the total number of individuals is increasing by a factor of 1.2 in each two-year period, that is, 20% per two-year period. This is equivalent to saying that if we start with $x_0 = y_0 = 50$ individuals then

$$\mathbf{u}_1 = A \begin{pmatrix} 50 \\ 50 \end{pmatrix} = 1.2 \begin{pmatrix} 50 \\ 50 \end{pmatrix}$$

in the first generation. In the second generation we then have

$$\mathbf{u}_2 = A \begin{pmatrix} 1.2 \times 50 \\ 1.2 \times 50 \end{pmatrix} = 1.2 \begin{pmatrix} 1.2 \times 50 \\ 1.2 \times 50 \end{pmatrix} = 1.2^2 \begin{pmatrix} 50 \\ 50 \end{pmatrix}$$

In the third generation we then have

$$\mathbf{u}_3 = A \begin{pmatrix} 1.2^2 \times 50 \\ 1.2^2 \times 50 \end{pmatrix} = 1.2 \begin{pmatrix} 1.2^2 \times 50 \\ 1.2^2 \times 50 \end{pmatrix} = 1.2^3 \begin{pmatrix} 50 \\ 50 \end{pmatrix}$$

and so on up the nth generation where we obtain

$$\mathbf{u}_n = A \begin{pmatrix} 1.2^{n-1} \times 50 \\ 1.2^{n-1} \times 5 \end{pmatrix} = 1.2 \begin{pmatrix} 1.2^{n-1} \times 50 \\ 1.2^{n-1} \times 50 \end{pmatrix} = 1.2^n \begin{pmatrix} 50 \\ 50 \end{pmatrix}$$

Thus, we have a formula that can be applied directly for arbitrary n instead of the process of continued matrix multiplication used in part **b**, where the initial conditions do not satisfy $x_0 = y_0$. ∎

In Example 1, we found that there exists a vector $\mathbf{v} = \begin{pmatrix} 50 \\ 50 \end{pmatrix}$ and a scalar $\lambda = 1.2$, such that $A\mathbf{v} = \lambda\mathbf{v}$ and that the entities $\mathbf{v}$ and λ appear to describe the long-term behavior of the impala population. The fact that $\mathbf{v}$ in the example has equal first and second elements is just fortuitous; in general, the first and second elements of $\mathbf{v}$ need not be equal. The discovery of this vector $\mathbf{v}$ motivates the following definition.

Eigenvalues and Eigenvectors	Let A be a 2-by-2 matrix. A nonzero vector $\mathbf{v}$ and scalar λ are an **eigenvector** and **eigenvalue** for A provided the equation $A\mathbf{v} = \lambda\mathbf{v}$ holds.

The German word *eigen* means *self*. Geometrically, eigenvectors for A correspond to directions in the xy plane that get "mapped onto" themselves after being multiplied by A.

Example 2 Finding eigenvalues and eigenvectors

The following two examples illustrate that if you know either the eigenvalue or the eigenvector of a matrix A, but not both, then finding the other is easy.

a. The matrix

$$A = \begin{pmatrix} 4 & -\dfrac{2}{3} \\ 3 & 1 \end{pmatrix}$$

has eigenvectors $\begin{pmatrix} 1 \\ 3 \end{pmatrix}$ and $\begin{pmatrix} 2 \\ 3 \end{pmatrix}$. Find the eigenvalues associated with these eigen-vectors.

b. The matrix

$$A = \begin{pmatrix} 7 & 0 \\ -8 & -1 \end{pmatrix}$$

has eigenvalues 7 and −1. Find eigenvectors associated with these eigenvalues.

Solution

a. Multiplying A and the first eigenvector yields

$$\begin{pmatrix} 4 & -\dfrac{2}{3} \\ 3 & 1 \end{pmatrix} \begin{pmatrix} 1 \\ 3 \end{pmatrix} = \begin{pmatrix} 2 \\ 6 \end{pmatrix} = 2 \begin{pmatrix} 1 \\ 3 \end{pmatrix}$$

Therefore, the eigenvalue associated with $\begin{pmatrix} 1 \\ 3 \end{pmatrix}$ is $\lambda = 2$.

Multiplying A and the second eigenvector yields

$$\begin{pmatrix} 4 & -\dfrac{2}{3} \\ 3 & 1 \end{pmatrix} \begin{pmatrix} 2 \\ 3 \end{pmatrix} = \begin{pmatrix} 6 \\ 9 \end{pmatrix} = 3 \begin{pmatrix} 2 \\ 3 \end{pmatrix}$$

Therefore, the eigenvalue associated with $\begin{pmatrix} 2 \\ 3 \end{pmatrix}$ is $\lambda = 3$.

b. To find an eigenvector $\mathbf{v} = \begin{pmatrix} x \\ y \end{pmatrix}$ associated with $\lambda = 7$, we need to solve the matrix equation

$$A = \begin{pmatrix} 7 & 0 \\ -8 & -1 \end{pmatrix} \begin{pmatrix} x \\ y \end{pmatrix} = 7 \begin{pmatrix} x \\ y \end{pmatrix}$$

Equivalently, we need to solve the system of linear equations:

$$\begin{cases} 7x + 0y = 7x \\ -8x - y = 7y \end{cases}$$

The first equation holds for any choice of x. Let's choose $x = 1$, in which case, the second equation becomes $-8 - y = 7y$. This implies that $y = -1$. To verify that we found an eigenvector, we multiply A against our candidate for the eigenvector:

$$A = \begin{pmatrix} 7 & 0 \\ -8 & -1 \end{pmatrix} \begin{pmatrix} 1 \\ -1 \end{pmatrix} = \begin{pmatrix} 7 \\ -7 \end{pmatrix} = 7 \begin{pmatrix} 1 \\ -1 \end{pmatrix}$$

If we now choose a different value for x, say $x = c$ where c is some arbitrary constant, then the second equation yields $-8c - y = 7y$, which implies that $y = -c$. In this case, the eigenvector is $\begin{pmatrix} c \\ -c \end{pmatrix}$; again, we can verify that it is an eigenvector for $c \neq 0$ since

$$A = \begin{pmatrix} 7 & 0 \\ -8 & -1 \end{pmatrix} \begin{pmatrix} c \\ -c \end{pmatrix} = \begin{pmatrix} 7c \\ -7c \end{pmatrix} = 7 \begin{pmatrix} c \\ -c \end{pmatrix}$$

Thus, eigenvalues are unique up to multiplication by an arbitrary, nonzero constant. This is the same as saying that the direction of the eigenvector is unique, but its magnitude is arbitrary.

To find an eigenvector $\mathbf{v} = \begin{pmatrix} x \\ y \end{pmatrix}$ associated with $\lambda = -1$, we need to solve this matrix equation:

$$A = \begin{pmatrix} 7 & 0 \\ -8 & -1 \end{pmatrix} \begin{pmatrix} x \\ y \end{pmatrix} = -\begin{pmatrix} x \\ y \end{pmatrix}$$

Equivalently, we need to solve the system of linear equations:

$$\begin{cases} 7x + 0y = -x \\ -8x - y = -y \end{cases}$$

The first equation can only be satisfied if $x = 0$; the second equation is satisfied for all y when $x = 0$. Hence, we choose $x = 0$ and $y = 1$. Recall that we need either x or y to be nonzero. To verify that we found an eigenvector, we can multiple A against what we found:

$$A = \begin{pmatrix} 7 & 0 \\ -8 & -1 \end{pmatrix} \begin{pmatrix} 0 \\ 1 \end{pmatrix} = \begin{pmatrix} 0 \\ -1 \end{pmatrix} = -\begin{pmatrix} 0 \\ 1 \end{pmatrix}$$

In this case, note again that if $\begin{pmatrix} 0 \\ 1 \end{pmatrix}$ is an eigenvector then so is $c \begin{pmatrix} 0 \\ 1 \end{pmatrix} = \begin{pmatrix} 0 \\ c \end{pmatrix}$ for all nonzero values of c. ∎

Example 2 shows that it is straightforward to find eigenvectors if you know the eigenvalues, and vice versa. It does not, however, provide a method for simultaneously finding both. To uncover a method for finding both, we rewrite the equation $A\mathbf{v} = \lambda\mathbf{v}$ as follows:

$$A\mathbf{v} = \lambda\mathbf{v}$$

implies $A\mathbf{v} = \lambda I \mathbf{v}$ where $I = \begin{pmatrix} 1 & 0 \\ 0 & 1 \end{pmatrix}$ is called the **identity matrix**

implies $(A - \lambda I)\mathbf{v} = \begin{pmatrix} 0 \\ 0 \end{pmatrix}$

Thus, solving the equation $A\mathbf{v} = \lambda\mathbf{v}$ is equivalent to solving

$$(A - \lambda I)\mathbf{v} = \mathbf{0} \quad \text{where } \mathbf{0} = \begin{pmatrix} 0 \\ 0 \end{pmatrix}$$

Since we need to find a nonzero $\mathbf{v}$, there is a solution to $(A - \lambda I)\mathbf{v} = \mathbf{0}$ with $\mathbf{v} \neq \mathbf{0}$ if and only if $\det(A - \lambda I) = 0$.

Since

$$\det(A - \lambda I) = \det\left(\begin{pmatrix} a & b \\ c & d \end{pmatrix} - \lambda \begin{pmatrix} 1 & 0 \\ 0 & 1 \end{pmatrix} \right) \qquad \textit{definition of I}$$

$$= \det\left(\begin{pmatrix} a & b \\ c & d \end{pmatrix} + \begin{pmatrix} -\lambda & 0 \\ 0 & -\lambda \end{pmatrix} \right) \qquad \textit{scalar multiplication of a matrix}$$

$$= \det\left(\begin{pmatrix} a - \lambda & b \\ c & d - \lambda \end{pmatrix} \right) \qquad \textit{definition of matrix addition}$$

$$= (a - \lambda)(d - \lambda) - bd \qquad \textit{definition of a determinant}$$

we have shown that solving the equation $\det(A - \lambda I) = 0$ is equivalent to solving the quadratic equation

$$\lambda^2 - (a + d)\lambda + (ad - bc) = 0$$

This gives us the following algorithm for finding the eigenvalues and eigenvectors of a matrix A.

Solving for Eigenvalues and Eigenvectors

The eigenvalues and eigenvectors for a 2-by-2 matrix

$$A = \begin{pmatrix} a & b \\ c & d \end{pmatrix}$$

can be found by implementing the following steps:

Step 1: Find the eigenvalues by solving

$$\det(A - \lambda I) = \lambda^2 - (a + d)\lambda + ad - bc = 0$$

Step 2: For each eigenvalue λ found in step 1, solve the matrix equation $A\mathbf{v} = \lambda\mathbf{v}$ for a nonzero vector $\mathbf{v} = \begin{pmatrix} x \\ y \end{pmatrix}$.

Note that in Step 2, there are an infinite number of solutions, because $A(c\mathbf{v}) = \lambda(c\mathbf{v})$ for any scalar c whenever $A\mathbf{v} = \lambda\mathbf{v}$.

Example 3 Finding eigenvalues and eigenvectors using the two-step method

Find the eigenvalues and associated eigenvectors for the matrix

$$A = \begin{pmatrix} 7 & -2 \\ 15 & -4 \end{pmatrix}$$

Solution First, we find the eigenvalues by solving

$$0 = \det(A - \lambda I) = \lambda^2 - 3\lambda + 2 = (\lambda - 2)(\lambda - 1)$$

Hence, $\lambda = 2$ and $\lambda = 1$ are the eigenvalues.

To find an eigenvector associated with $\lambda = 2$, we proceed to Step 2:

$$A\mathbf{v} = \lambda\mathbf{v}$$

$$\begin{pmatrix} 7 & -2 \\ 15 & -14 \end{pmatrix} \begin{pmatrix} x \\ y \end{pmatrix} = 2\begin{pmatrix} x \\ y \end{pmatrix}$$

$$\begin{pmatrix} 7x - 2y \\ 15x - 4y \end{pmatrix} = \begin{pmatrix} 2x \\ 2y \end{pmatrix}$$

$$\begin{cases} 5x - 2y = 0 \\ 15x - 6y = 0 \end{cases}$$

These equations are multiples of one another, so we only need to solve one. Both equations imply that $y = 5x/2$. Therefore, if we choose $x = 1$, then $y = 5/2$ and a choice of the eigenvector is $\mathbf{v} = \begin{pmatrix} 1 \\ 5/2 \end{pmatrix}$.

To find an eigenvector associated with $\lambda = 1$, we need to solve this system of linear equations:

$$\begin{cases} 6x - 2y = 0 \\ 15x - 5y = 0 \end{cases}$$

These equations are multiples of one another, so we only need to solve one. Both equations imply that $y = 3x$. Therefore, if we choose $x = 1$, then $y = 3$ and a choice of the eigenvector is $\mathbf{v} = \begin{pmatrix} 1 \\ 3 \end{pmatrix}$. ∎

Since the eigenvalues are found by solving a quadratic equation, we can get complex numbers as solutions for the eigenvalue. When this occurs, we enter a realm beyond the scope of this book. For the sufficiently adventuresome, however, we explore this issue further in Problem 56 in Problem Set 8.3.

Our motivation for considering the eigenvalue and eigenvector problem is to analyze the behavior of matrix models of the form

$$\mathbf{u}_{n+1} = A\mathbf{u}_n$$

To see how eigenvalues and eigenvectors help us, suppose A has two distinct real eigenvalues λ and μ and associated eigenvectors $\mathbf{v}$ and $\mathbf{w}$. Given two real numbers a and b, define

$$\mathbf{u}_0 = a\mathbf{v} + b\mathbf{w}$$

Then

$$\mathbf{u}_1 = A\mathbf{u}_0 = a\,A\mathbf{v} + b\,A\mathbf{w} = a\lambda\mathbf{v} + b\mu\mathbf{w}$$
$$\mathbf{u}_2 = A\mathbf{u}_1 = a\lambda\,A\mathbf{v} + b\mu\,A\mathbf{w} = a\lambda^2\mathbf{v} + b\mu^2\mathbf{w}$$
$$\vdots$$
$$\mathbf{u}_n = A\mathbf{u}_{n-1} = a\lambda^n\mathbf{v} + b\mu^n\mathbf{w}$$

Hence, provided that we can express $\mathbf{u}_0$ as a linear combination of the eigenvectors $\mathbf{v}$ and $\mathbf{w}$, we can find a solution to the matrix difference equation. The following theorem clarifies when all solutions are of this form.

Theorem 8.2   Solutions of matrix difference equations theorem

Consider the matrix difference equation

$$\mathbf{u}_{n+1} = A\mathbf{u}_n$$

If $\mathbf{v}$ and $\mathbf{w}$ are eigenvectors corresponding to the eigenvalues λ and μ, then

$$\mathbf{u}_n = a\lambda^n\mathbf{v} + b\mu^n\mathbf{w}$$

is a solution to the matrix difference equation. Moreover, if $\mathbf{v}$ and $\mathbf{w}$ are not collinear (i.e., $\mathbf{v} \neq c\mathbf{w}$ for any c), then every solution of the matrix difference equation can be written in the form $a\lambda^n\mathbf{v} + b\mu^n\mathbf{w}$.

Example 4   Revisiting the impala ranch

In Example 1 we described the number x_n of immature female impalas and the number y_n of mature females impalas on a ranch by the matrix model

$$\mathbf{u}_{n+1} = \begin{pmatrix} 0 & 1.2 \\ 0.4 & 0.8 \end{pmatrix} \mathbf{u}_n$$

where $\mathbf{u}_n = \begin{pmatrix} x_n \\ y_n \end{pmatrix}$ and $2n$ is the number of years that have elapsed.

a. Find the eigenvalues and associated eigenvectors of the matrix $A = \begin{pmatrix} 0 & 1.2 \\ 0.4 & 0.8 \end{pmatrix}$.

b. Find the solution $\mathbf{u}_n$ given $x_0 = 100$, $y_0 = 0$ by expressing $\mathbf{u}_0$ as a linear combination of the eigenvectors that you found in part **a**. Verify that at $n = 4$, this solution agrees with what we found in Example 1, part **b**.

c. Show that $\displaystyle\lim_{n\to\infty} \frac{x_n}{x_n + y_n} = \frac{1}{2}$.

Solution
a. We begin by solving for the eigenvalues that are solutions to

$$0 = \det(A - \lambda I) = \lambda^2 - 0.8\lambda - 0.48$$

Solving for λ yields $\lambda = 1.2$ and $\lambda = -0.4$.

To find an eigenvector $\mathbf{v}$ for $\lambda = 1.2$, we solve the linear equations corresponding to $(A - 1.2I)\mathbf{v} = 0$:

$$\begin{cases} (0 - 1.2)x + 1.2y = 0 \\ 0.4x + (0.8 - 1.2)y = 0 \end{cases}$$

Equivalently, we solve $-1.2x + 1.2y = 0$, which requires that $x = y$. Choosing $x = 1$ yields $\mathbf{v} = \begin{pmatrix} 1 \\ 1 \end{pmatrix}$.

To find an eigenvector $\mathbf{w}$ for $\lambda = -0.4$, we solve the linear equations corresponding to $(A + 0.4I)\mathbf{w} = 0$:

$$\begin{cases} (0 + 0.4)x + 1.2y = 0 \\ 0.4x + (0.8 + 0.4)y = 0 \end{cases}$$

Equivalently, we solve $0.4x + 1.2y = 0$, which requires that $x = -3y$. Choosing $y = 1$ yields $\mathbf{w} = \begin{pmatrix} -3 \\ 1 \end{pmatrix}$.

b. Theorem 8.2 implies that

$$\mathbf{u}_n = a(1.2)^n \begin{pmatrix} 1 \\ 1 \end{pmatrix} + b(-0.4)^n \begin{pmatrix} -3 \\ 1 \end{pmatrix}$$

is a solution to the matrix difference equation for any scalars a and b. We need to find a and b such that

$$\mathbf{u}_0 = a \begin{pmatrix} 1 \\ 1 \end{pmatrix} + b \begin{pmatrix} -3 \\ 1 \end{pmatrix}$$

Since $\mathbf{u}_0 = \begin{pmatrix} 100 \\ 0 \end{pmatrix}$, we need to solve

$$\begin{cases} a - 3b = 100 \\ a + b = 0 \end{cases}$$

The second equation implies that $a = -b$. Substituting $a = -b$ into the first equation yields $-b - 3b = 100$. In other words, $b = -25$. It follows that $a = 25$, and our solution is

$$\mathbf{u}_n = 25(1.2)^n \begin{pmatrix} 1 \\ 1 \end{pmatrix} - 25(-0.4)^n \begin{pmatrix} -3 \\ 1 \end{pmatrix} = 25 \begin{pmatrix} (1.2)^n + 3(-0.4)^n \\ (1.2)^n - (-0.4)^n \end{pmatrix}$$

At $n = 4$, we get

$$\mathbf{u}_4 = 25 \begin{pmatrix} (1.2)^4 + 3(-0.4)^4 \\ (1.2)^4 - (-0.4)^4 \end{pmatrix} = \begin{pmatrix} 53.76 \\ 51.2 \end{pmatrix}$$

This answer for $\mathbf{u}_4$ agrees with what we found in Example 1, part **b**.

c. In part **b**, we found that

$$x_n = 25(1.2)^n + 75(-0.4)^n \quad \text{and} \quad y_n = 25(1.2)^n - 25(-0.4)^n$$

Therefore,

$$\lim_{n \to \infty} \frac{x_n}{x_n + y_n} = \lim_{n \to \infty} \frac{25(1.2)^n + 75(-0.4)^n}{50(1.2)^n + 50(-0.4)^n}$$

$$= \lim_{n \to \infty} \frac{25(1.2)^n + 75(-0.4)^n}{50(1.2)^n + 50(-0.4)^n} \frac{1/(1.2)^n}{1/(1.2)^n}$$

$$= \lim_{n \to \infty} \frac{25 + 75(-5/6)^n}{50 + 50(-5/6)^n} = \frac{25}{50} = \frac{1}{2}$$

In words, the fraction of immature females in the population in the long term is 50%.

In Example 4, part **c**, we found that the fraction of immature females and mature females approached ratios predicted by the eigenvector $\mathbf{v} = \begin{pmatrix} 1 \\ 1 \end{pmatrix}$ associated with the larger eigenvalue $\lambda = 1.2$. It turns out this behavior is quite general. To see why, suppose the matrix A has two real eigenvalues λ, μ and associated eigenvectors $\mathbf{v} = \begin{pmatrix} v_1 \\ v_2 \end{pmatrix}$, $\mathbf{w} = \begin{pmatrix} w_1 \\ w_2 \end{pmatrix}$. Solutions of $\mathbf{u}_{n+1} = A\mathbf{u}_n$ are of the form

$$\begin{pmatrix} x_n \\ y_n \end{pmatrix} = a\lambda^n \begin{pmatrix} v_1 \\ v_2 \end{pmatrix} + b\mu^n \begin{pmatrix} w_1 \\ w_2 \end{pmatrix}$$

Provided the following ratio is well defined,

$$\frac{x_n}{x_n + y_n} = \frac{a\lambda^n v_1 + b\mu^n w_1}{a\lambda^n (v_1 + v_2) + b\mu^n (w_1 + w_2)}$$

If $|\lambda| > |\mu|$ and $a \neq 0$, then dividing the numerator and denominator by λ^n and taking the limit as $n \to \infty$ gives

$$\lim_{n \to \infty} \frac{x_n}{x_n + y_n} = \lim_{n \to \infty} \frac{a\lambda^n v_1 + b\mu^n w_1}{a\lambda^n (v_1 + v_2) + b\mu^n (w_1 + w_2)} \frac{1/\lambda^n}{1/\lambda^n}$$

$$= \lim_{n \to \infty} \frac{a v_1 + b(\mu/\lambda)^n w_1}{a(v_1 + v_2) + b(\mu/\lambda)^n (w_1 + w_2)}$$

$$= \frac{a v_1}{a(v_1 + v_2)} \qquad \text{as } \lim_{n \to \infty} (\mu/\lambda)^n = 0$$

$$= \frac{v_1}{v_1 + v_2}$$

Thus, we have shown the following result.

Theorem 8.3   Stable stage distribution theorem

- -

Assume the matrix A has two real eigenvalues λ and μ and associated eigenvectors $\mathbf{v} = \begin{pmatrix} v_1 \\ v_2 \end{pmatrix}$ and $\mathbf{w} = \begin{pmatrix} w_1 \\ w_2 \end{pmatrix}$. Let $\mathbf{u}_n$ be a solution to $\mathbf{u}_{n+1} = A\mathbf{u}_n$ where $\mathbf{u}_0 = a\mathbf{v} + b\mathbf{w}$ with $a \neq 0$. If $|\lambda| > |\mu|$, then

$$\lim_{n \to \infty} \frac{x_n}{x_n + y_n} = \frac{v_1}{v_1 + v_2} \quad \text{and} \quad \lim_{n \to \infty} \frac{y_n}{x_n + y_n} = \frac{v_2}{v_1 + v_2}$$

*The fractions $\dfrac{v_1}{v_1 + v_2}, \dfrac{v_2}{v_1 + v_2}$ are called the **stable stage distribution**, as they describe asymptotically what fraction of the "population" is in the first and second "stages."*

Let's examine Theorem 8.3 in the context of Fibonacci's model of rabbit population growth that was introduced in Problem 38 of Problem Set 1.7.

Example 5   Fibonacci's rabbits

- -

Fibonacci considered a population of rabbits consisting of x pairs of preadults and y pairs of adults. Each month, each of the y pairs of adults gives birth to one pair of juveniles and each pair of juveniles matures to become a pair of adults. All adults survive to reproduce again one month later.

a. Write a matrix model $\mathbf{u}_{n+1} = A\mathbf{u}_n$ for this rabbit population.

b. Find the long-term growth rate of this population and the long-term fraction of individuals that are juveniles.

c. Use technology to plot $\dfrac{x_n}{x_n + y_n}$ for $0 \le n \le 25$, given that initially there was only one pair of juveniles and one pair of adults.

Solution

a. Let x_n be the number of pairs of juveniles in month n and y_n the number of pairs of adults. Since $x_{n+1} = y_n$ and $y_{n+1} = x_n + y_n$, we get the the matrix model $\mathbf{u}_{n+1} = A\mathbf{u}_n$ with

$$\mathbf{u}_n = \begin{pmatrix} x_n \\ y_n \end{pmatrix} \quad \text{and} \quad A = \begin{pmatrix} 0 & 1 \\ 1 & 1 \end{pmatrix}$$

b. Solving for the eigenvalues, we get that

$$\det(A - \lambda I) = \lambda^2 - \lambda - 1 = 0$$

has solutions $\lambda = \dfrac{1 \pm \sqrt{5}}{2}$. The larger eigenvalue is $\dfrac{1 + \sqrt{5}}{2} \approx 1.618$. Hence, the population in the long term is increasing by approximately 61.8% per month.

Solving for the eigenvector associated with $\lambda = \dfrac{1 + \sqrt{5}}{2}$ requires solving this system of linear equations:

$$\begin{cases} -\dfrac{1 + \sqrt{5}}{2} x + y = 0 \\ x + \left(1 - \dfrac{1 + \sqrt{5}}{2} \right) y = 0 \end{cases}$$

Since these equations are multiples of one another, it suffices to solve the first equation, which gives $y = \dfrac{1 + \sqrt{5}}{2} x$. Setting $x = 1$, we get the eigenvector

$$\mathbf{v} = \begin{pmatrix} 1 \\ \dfrac{1 + \sqrt{5}}{2} \end{pmatrix}$$

Theorem 8.3 implies

$$\lim_{n \to \infty} \frac{x_n}{x_n + y_n} = \frac{1}{1 + \dfrac{1 + \sqrt{5}}{2}} \approx 0.382$$

for any solution with $x_0 > 0$ and $y_n \ge 0$. Hence, in the long term, the fraction of juveniles in the population is approximately 38.2%.

c. Iterating the matrix difference equation $\mathbf{u}_{n+1} = A\mathbf{u}_n$ for twenty-five iterates and plotting the ratio $\dfrac{x_n}{x_n + y_n}$ gives the graph shown in Figure 8.24. As predicted, the ratio converges (rather quickly!) to approximately 0.382.

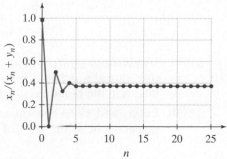

Figure 8.24 Fraction of juveniles in the population as a function of months

Applications

Matrix models are used extensively to model the flow of toxins, chemicals, water, and so on, between compartments of a biological system. One such model considers the accumulation of lead in various parts of a human body. Lead is an environmental pollutant most often found in facilities processing lead-acid batteries, producing lead wires or pipes, or recycling metal. Lead exposure can occur from contact with lead in air, dust, soil, water, and commercial products. Chronic exposure to lead ultimately produces headaches, abdominal pains, memory loss, kidney failure, and male reproductive problems.

Lead enters the human body by inhaling or ingesting lead particulates. From the lungs and gut, lead enters the bloodstream from which it quickly diffuses into tissues such as the liver and kidneys. It also slowly diffuses from the bloodstream into bones. Lead leaves the body through the urinary system, hair, nails, and sweat.

To understand the kinetics of lead within the body, Rabinowitz and colleagues* measured levels of lead in healthy men living in Los Angeles. Using these measurements, they developed and parameterized a model of lead "flow" among the blood, tissue, and bone compartments of the body. In the next example, we consider a discrete time counterpart of their model and, for simplicity, examine separately the blood-tissue subsystem and the blood-bones subsystem, as illustrated in Figure 8.25. We examine all three subsystems simultaneously in Example 8.

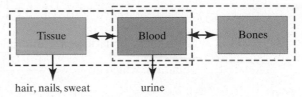

Figure 8.25 Compartmental model of lead in the body. Lead diffuses between the bloodstream, bones, and soft tissues. It is lost through the urinary system, hair, nails, and sweat. In Example 8, the two-dimensional subsystems of blood-bones (blue dashed box) and blood-tissues (red dashed box) are considered separately.

Example 6   Lead in the body

Consider the compartment model depicted in Figure 8.25. In part **a** (below) we model the interaction of the tissue-blood system (dashed red rectangle in Figure 8.25), and in part **b** (below) we model the interaction of the blood-bones system (dashed blue rectangle in Figure 8.25).

a. Let x_n and y_n be the amount of lead (in micrograms) in the blood and tissue, respectively, at the start of month n. The study by Rabinowitz and colleagues provides the following model for calculating the amount of lead a month later:

$$\begin{pmatrix} x_{n+1} \\ y_{n+1} \end{pmatrix} = \begin{pmatrix} 0.39 & 0.04 \\ 0.04 & 0.02 \end{pmatrix} \begin{pmatrix} x_n \\ y_n \end{pmatrix}$$

Use technology to find the eigenvalues and eigenvectors. Determine the asymptotic rate at which lead is leaving this subsystem; also, determine what fraction of lead is in the blood versus the tissues in the long term.

b. Let x_n and y_n be the amount of lead (in micrograms) in the blood and bones, respectively, at the start of month n. The study by Rabinowitz and colleagues

*Rabinowitz, M., G. Wetherill, and J. Kopple. 1973. Lead metabolism in the normal man: Stable isotope studies. *Science* (Wash. D.C.) 182: 725–727.

provides the following model for calculating the amount of lead a month later:

$$\begin{pmatrix} x_{n+1} \\ y_{n+1} \end{pmatrix} = \begin{pmatrix} 0.47 & 0.001 \\ 0.08 & 0.999 \end{pmatrix} \begin{pmatrix} x_n \\ y_n \end{pmatrix}$$

Use technology to find the eigenvalues and eigenvectors. Determine the asymptotic rate that lead is leaving this subsystem; also, determine what fraction of the lead in the blood versus in the bones in the long term. Compare these answers to what you found for part **a**.

Solution

a. First we find the eigenvalues by solving

$$0 = \det(A - \lambda I) = \lambda^2 - 0.41\lambda - 0.0062$$

Using the quadratic formula (or technology), we obtain the eigenvalues 0.394 and 0.016 (approximately). Hence, each month the lead decreases by approximately $(100 - 39.4)\% = 61.6\%$. Solving for the eigenvectors, we get approximately

$$\begin{pmatrix} 99.43 \\ 10.63 \end{pmatrix} \quad \text{and} \quad \begin{pmatrix} 10.63 \\ -99.43 \end{pmatrix}$$

Using technology you might get a different answer for the eigenvectors. Your answer, however, should only differ from ours by a scalar multiple. Since the first eigenvector is associated with the larger eigenvalue, Theorem 8.3 implies that the long term fraction of lead in the tissues is $\dfrac{10.63}{10.63 + 99.43} \approx 0.097$. Hence, most of the lead in this subsystem is in the blood.

b. First we need to find the eigenvalues by solving

$$0 = \det(A - \lambda I) = \lambda^2 - 1.469\lambda + 0.46946$$

Using the quadratic formula (or technology), we obtain the eigenvalues 0.999 and 0.47 (approximately). Hence, each month the lead decreases by approximately $(100 - 99.9)\% = 0.1\%$. This is a much slower rate of decay than in the soft tissue subsystem! This slower rate of decay is illustrated in Figure 8.26, where y_n is plotted for the matrix models considered in parts **a** and **b** with $y_0 = 1000$ and $x_0 = 0$. Solving for the eigenvectors, we get approximately

$$\begin{pmatrix} 0.2 \\ 99.99 \end{pmatrix} \quad \text{and} \quad \begin{pmatrix} -98.88 \\ 14.95 \end{pmatrix}$$

Since the first eigenvector is associated with the larger eigenvalue, Theorem 8.3 implies that the fraction of lead in the bones is $\dfrac{99.99}{99.99 + 0.2} \approx 0.998$ in the long term. Hence, most of the lead in this subsystem is in the bones.

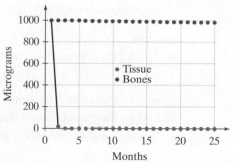

Figure 8.26 Amount of lead in the tissue and the bones plotted as a function of n

One of the more exciting applications of matrix models is to conservation biology. Matrix models have been developed for many rare or endangered species as a tool to understand where best to focus conservation efforts. In Example 3 of Section 8.2, we examined a model of loggerhead turtles. Two important sources of mortality for this species occur at very different life stages. At the earliest ages, hatchlings emerging from eggs laid on beaches are highly susceptible to predation by birds and mammals. Adults may suffocate if they are trapped in fishing nets.

In the 1980s, conservation biologists were faced with the question of whether it would be more effective to focus on protecting nests on beaches or advocating for turtle excluder devices (TEDs) in fishing nets. TEDs reduce mortality in fishing nets by providing turtles with an escape route (Figure 8.27). An important contribution to this debate was a modeling study by Crouse and colleagues referred to in Example 3 of Section 8.2. Using a simplified version of this model, in the next example we evaluate whether TEDs would have a larger impact on saving loggerheads than measures implemented to protect eggs and hatchlings on beaches.

Figure 8.27 Loggerhead escaping a fishing net through a turtle excluder device (TED)

Example 7   Saving the loggerheads

Consider the loggerhead model in Example 3 of Section 8.2. In this model, x_n and y_n are the abundances of juvenile and adult loggerheads at the start of year $n + 1$ and

$$\begin{pmatrix} x_{n+1} \\ y_{n+1} \end{pmatrix} = \begin{pmatrix} 0.5 & 80 \\ 0.001 & 0.8 \end{pmatrix} \begin{pmatrix} x_n \\ y_n \end{pmatrix}$$

Use this model to answer the following questions:

a. What is the long-term growth rate of the population?

b. **Save the old:** If introducing TEDs increases adult survivorship by 10%, what effect does this have on the long-term population growth rate?

c. **Save the young:** If protecting nests increases survivorship of juveniles by 10% (i.e., 0.5 and 0.001 increase by 10%), what effect does this have on the long-term population growth rate?

Solution

a. To find the eigenvalues, we solve

$$0 = \det(A - \lambda I) = \lambda^2 - 1.3\lambda + 0.32$$

which yields $\lambda \approx 0.97$ and 0.33. Hence, the largest eigenvalue in absolute value is 0.97, so the population is decreasing at a rate of approximately 3% per year.

b. Increasing survivorship of mature adults by 10% leads to this new matrix:

$$A = \begin{pmatrix} 0.5 & 80 \\ 0.001 & 0.88 \end{pmatrix}$$

Solving for the eigenvalues of A yields $\lambda \approx 1.03$ and 0.35. Hence, in this case, the largest eigenvalue in absolute value is 1.03, implying that the population would increase at a rate of approximately 3% per year.

c. Increasing survivorship of juveniles by 10% leads to this new matrix:

$$A = \begin{pmatrix} 0.55 & 80 \\ 0.0011 & 0.8 \end{pmatrix}$$

Solving for the eigenvalues of A yields $\lambda \approx 0.99$ and 0.35. Hence, in this case, the largest eigenvalue in absolute value is 0.99, implying that the population would still decrease, but now only at a rate of approximately 1% per year. ∎

As discussed in Example 10 in the previous section, all of the ideas presented here for two-dimensional matrix models extend to higher dimensional matrix models. For example, if $A\mathbf{v} = \lambda\mathbf{v}$ for an n-by-n matrix A, a nonzero n-element vector $\mathbf{v}$, and a scalar λ, then λ is an eigenvector with associated eigenvector $\mathbf{v}$. Solving for the eigenvalues and eigenvectors of higher dimensional matrices is challenging but can be done easily with computational software or online tools. The following example illustrates how to interpret eigenvalues and eigenvectors for the three-compartment version of Example 6 in which we modeled the fate of lead in the human body.

Example 8   Three-compartment model of lead in the body

In Example 6 we considered separately models of the blood and tissue compartments and the blood and bone compartments. Here, we consider all three compartments at once.

Let x_n, y_n, and z_n be the amount of lead (in micrograms) in the blood, tissue, and bones, respectively, at the start of month n. The study by Rabinowitz and colleagues, cited in Example 6 above, provides the following model for calculating the amount of lead a month later:

$$\begin{pmatrix} x_{n+1} \\ y_{n+1} \\ z_{n+1} \end{pmatrix} = \begin{pmatrix} 0.36 & 0.14 & 0.0006 \\ 0.13 & 0.45 & 0.0001 \\ 0.073 & 0.011 & 0.998 \end{pmatrix} \begin{pmatrix} x_n \\ y_n \\ z_n \end{pmatrix}$$

Use technology to find the eigenvalues of the matrix model and the eigenvectors. Provide an interpretation of what the largest eigenvalue and largest eigenvector tell about the long-term abundance of lead in the body.

Solution   Using technology, we get the largest eigenvalue in absolute value is 0.999. Hence, in the long term, only 0.1% of the lead leaves the body each month. The eigenvector associated with the largest eigenvalue is

$$\begin{pmatrix} 1.26 \\ 0.44 \\ 1000 \end{pmatrix}$$

Hence, as suggested by a higher dimensional analogue of Theorem 8.3, we expect in the long term

$$x_n \approx \frac{1.26}{1.26 + 0.44 + 1000} \approx 0.001$$

$$y_n \approx \frac{0.44}{1.26 + 0.44 + 1000} \approx 0.0004$$

$$z_n \approx \frac{1000}{1.26 + 0.44 + 1000} \approx 0.998$$

Thus, in the long term, most of the lead is in the bones. ∎

PROBLEM SET 8.3

Level 1 DRILL PROBLEMS

In Problems 1 to 6, find the eigenvalues of the matrices for the specified eigenvectors.

1. $A = \begin{pmatrix} 1 & 2 \\ 2 & 1 \end{pmatrix}$ with eigenvectors $\begin{pmatrix} 1 \\ 1 \end{pmatrix}$ and $\begin{pmatrix} -1 \\ 1 \end{pmatrix}$

2. $A = \begin{pmatrix} 1 & -2 \\ -2 & 1 \end{pmatrix}$ with eigenvectors $\begin{pmatrix} 1 \\ 1 \end{pmatrix}$ and $\begin{pmatrix} -1 \\ 1 \end{pmatrix}$

3. $A = \begin{pmatrix} 1 & 2 \\ 3 & 2 \end{pmatrix}$ with eigenvectors $\begin{pmatrix} 2 \\ 3 \end{pmatrix}$ and $\begin{pmatrix} -1 \\ 1 \end{pmatrix}$

4. $A = \begin{pmatrix} 1 & -2 \\ 1 & 4 \end{pmatrix}$ with eigenvectors $\begin{pmatrix} -1 \\ 1 \end{pmatrix}$ and $\begin{pmatrix} -2 \\ 1 \end{pmatrix}$

5. $A = \begin{pmatrix} 1/2 & 1 \\ 1 & 1/2 \end{pmatrix}$ with eigenvectors $\begin{pmatrix} 1 \\ 1 \end{pmatrix}$ and $\begin{pmatrix} -1 \\ 1 \end{pmatrix}$

6. $A = \begin{pmatrix} 1/2 & 1 \\ 3/2 & 1 \end{pmatrix}$ with eigenvectors $\begin{pmatrix} 2/3 \\ 1 \end{pmatrix}$ and $\begin{pmatrix} -1 \\ 1 \end{pmatrix}$

In Problems 7 to 12, find the eigenvectors of the matrices for the specified eigenvalues. (Remember that eigenvectors are unique up to multiplication by a constant.)

7. $A = \begin{pmatrix} 2 & 4 \\ 4 & 2 \end{pmatrix}$ with eigenvalues 6 and -2

8. $A = \begin{pmatrix} 5 & 3 \\ 1 & 3 \end{pmatrix}$ with eigenvalues 6 and 2

9. $A = \begin{pmatrix} 1/3 & 2/3 \\ 1 & 2/3 \end{pmatrix}$ with eigenvalues $4/3$ and $-1/3$

10. $A = \begin{pmatrix} 7/2 & -5 \\ 5/2 & -4 \end{pmatrix}$ with eigenvalues $-3/2$ and 1

11. $A = \begin{pmatrix} -1 & -1 \\ 2 & -4 \end{pmatrix}$ with eigenvalues -3 and -2

12. $A = \begin{pmatrix} 5 & -2 \\ 1 & -1 \end{pmatrix}$ with eigenvalues $2 + \sqrt{7}$ and $2 - \sqrt{7}$

In Problems 13 to 24, find the eigenvalues and eigenvectors of the matrices. (Remember that eigenvectors are unique up to multiplication by a constant.)

13. $A = \begin{pmatrix} 1 & 0 \\ 0 & 2 \end{pmatrix}$

14. $A = \begin{pmatrix} 1 & 0 \\ 0 & -1 \end{pmatrix}$

15. $A = \begin{pmatrix} 1 & 2 \\ 0 & 2 \end{pmatrix}$

16. $A = \begin{pmatrix} 2 & 0 \\ -1 & -3 \end{pmatrix}$

17. $A = \begin{pmatrix} 2 & -10 \\ 0 & -10 \end{pmatrix}$

18. $A = \begin{pmatrix} -6 & 1 \\ 0 & -9 \end{pmatrix}$

19. $A = \begin{pmatrix} 2 & 2 \\ 1 & 3 \end{pmatrix}$

20. $A = \begin{pmatrix} 1 & 2 \\ 3 & 2 \end{pmatrix}$

21. $A = \begin{pmatrix} 2 & 6 \\ 2 & -2 \end{pmatrix}$

22. $A = \begin{pmatrix} 3 & 2 \\ 2 & 3 \end{pmatrix}$

23. $A = \begin{pmatrix} -10 & 1 \\ -5 & 8 \end{pmatrix}$

24. $A = \begin{pmatrix} 1 & -7 \\ -8 & 6 \end{pmatrix}$

Iterate the dynamical solution $\mathbf{u}_n = \begin{pmatrix} x_n \\ y_n \end{pmatrix}$ for $n = 0, 1, \ldots, 6$ of the equation $\mathbf{u}_{n+1} = A\mathbf{u}_n$ using the initial vector $\mathbf{u}_0$ and matrix A in Problems 25 to 30.

25. $A = \begin{pmatrix} 0 & 1.2 \\ 0.4 & 0.8 \end{pmatrix}$ and $\mathbf{u}_0 = \begin{pmatrix} 80 \\ 20 \end{pmatrix}$

26. $A = \begin{pmatrix} 0 & 1.2 \\ 0.4 & 0.8 \end{pmatrix}$ and $\mathbf{u}_0 = \begin{pmatrix} 20 \\ 80 \end{pmatrix}$

27. $A = \begin{pmatrix} 0 & 2 \\ 0.25 & 0.6 \end{pmatrix}$ and $\mathbf{u}_0 = \begin{pmatrix} 100 \\ 0 \end{pmatrix}$

28. $A = \begin{pmatrix} 0 & 2 \\ 0.25 & 0.6 \end{pmatrix}$ and $\mathbf{u}_0 = \begin{pmatrix} 0 \\ 100 \end{pmatrix}$

29. $A = \begin{pmatrix} 0 & 2 \\ 0.1 & 0.8 \end{pmatrix}$ and $\mathbf{u}_0 = \begin{pmatrix} 50 \\ 50 \end{pmatrix}$

30. $A = \begin{pmatrix} 0 & 2 \\ 0.1 & 0.8 \end{pmatrix}$ and $\mathbf{u}_0 = \begin{pmatrix} 0 \\ 100 \end{pmatrix}$

In the impala model presented in Example 1, assume these changes: the reproductive rate is b rather than the specific value 1.2; the survival of young to age 2 is s rather than the specific value 40%; and the percentage of mature females that die in every two-year period is m rather than 20%. Construct models for populations with the specified values of b, s, and m in Problems 31 to 36. Iterate the first three years of population numbers $\mathbf{u}_1$, $\mathbf{u}_2$, and $\mathbf{u}_3$ if the initial number of immature and mature females is

$$\mathbf{u}_0 = \begin{pmatrix} 50 \\ 50 \end{pmatrix}.$$

31. $b = 2, s = 50, m = 50$
32. $b = 2, s = 20, m = 40$
33. $b = 1.6, s = 40, m = 50$
34. $b = 1.6, s = 25, m = 40$
35. $b = 1, s = 30, m = 40$
36. $b = 1, s = 30, m = 60$

In Problems 37 to 42, construct a population model for a rodent-like population that breeds in its first and second years but with all individuals dying at the end of the second year. By iterating the solution from the initial

condition $\mathbf{u}_0 = \begin{pmatrix} 70 \\ 30 \end{pmatrix}$ *(e.g., for about six to eight years,*
depending on the parameters of the model), estimate the
proportions p and $(1 - p)$ *of females that are ultimately*
in the first and second age classes, respectively, and the
rate of increase r in the population from one year to the
next.

37. The average female has 2 pups in her first year and 10 in her second year. The percent of female pups that survive their first year to breed is 20. The percent of females that survive to breed in their second year is 50.

38. The average female has 4 pups in her first year and 6 in her second year. The percent of female pups that survive their first year to breed is 15. The percent of females that survive to breed in their second year is 60.

39. The average female has 4 pups in her first year and 5 in her second year. The percent of female pups that survive their first year to breed is 20. The percent of females that survive to breed in their second year is 50.

40. The average female has 4 pups in her first year and 6 in her second year. The percent of female pups that survive their first year to breed is 15. The percent of females that survive to breed in their second year is 40.

41. The average female has 2 pups in her first year and 4 in her second year. The percent of female pups that survive their first year to breed is 20. The percent of females that survive to breed in their second yearb is 50.

42. The average female has 2 pups in her first year and, on average, 8/3 pups in her second year. The percent of female pups that survive their first year to breed is 30. The percent of females that survive to breed in their second year is 50.

In Problems 43 to 46, write the general solution to the
equation $\mathbf{u}_{n+1} = A\mathbf{u}_n$ *for the specified matrix A*

43. $A = \begin{pmatrix} 1/3 & 2/3 \\ 1 & 2/3 \end{pmatrix}$ 44. $A = \begin{pmatrix} 1/5 & 4/5 \\ 1/2 & 1/2 \end{pmatrix}$

45. $A = \begin{pmatrix} 3.25 & -0.75 \\ -2.25 & 1.75 \end{pmatrix}$ 46. $A = \begin{pmatrix} 2 & 2 \\ 2 & -1 \end{pmatrix}$

In Problems 47 to 50, write the solution to the equation
$\mathbf{u}_{n+1} = A\mathbf{u}_n$ *for A and* $\mathbf{u}_0$ *as specified, and calculate the*
stable distribution that the solutions approach in the long
run.

47. $A = \begin{pmatrix} 1/3 & 2/3 \\ 1 & 2/3 \end{pmatrix}$ and $\mathbf{u}_0 = \begin{pmatrix} 30 \\ 20 \end{pmatrix}$

48. $A = \begin{pmatrix} 1/5 & 4/5 \\ 1/2 & 1/2 \end{pmatrix}$ and $\mathbf{u}_0 = \begin{pmatrix} 50 \\ 24 \end{pmatrix}$

49. $A = \begin{pmatrix} 3.25 & -0.75 \\ -2.25 & 1.75 \end{pmatrix}$ and $\mathbf{u}_0 = \begin{pmatrix} 3 \\ 5 \end{pmatrix}$

50. $A = \begin{pmatrix} 2 & 2 \\ 2 & -1 \end{pmatrix}$ and $\mathbf{u}_0 = \begin{pmatrix} 5 \\ -1 \end{pmatrix}$

Level 2 APPLIED AND THEORY PROBLEMS

51. Data collected from a population of rodents reveal that half of all newborn pups survive their first three months to reproduce as adults. The reproduction rate for adults averages 2.4 pups per adult per three-month period. Each adult has a 40% chance of surviving each three-month period to reproduce once again. Write two equations: one for the number of pups and one for the number of adults at time period n. If initially at $n = 0$, there are thirty pups and fifty adults in the population, calculate the number of pups and adults at $n = 1, 2,$ and 3. Use eigenvalue analysis to find out what the fraction of pups in the population will be ultimately and how rapidly the population will increase if the population starts out at this ratio.

52. A species of cat on the pampas in South America has on average five kittens per litter every year. The proportion of kittens that survive to reproductive maturity at age 1 is 1/4. Each year 1/3 of the adults survive to reproduce the following year. Write two equations: one for the number of kittens and one for the number of adults at time period n. If initially at $n = 0$, there are sixty kittens and forty adults in the population, calculate the number of kittens and adults at $n = 1, 2,$ and 3. Use eigenvalue analysis to find out what the fraction of kittens in the population will be ultimately and how rapidly the population will increase if the population starts out at this ratio.

53. The bilby is an endangered Australian marsupial bandicoot that looks a little like a kangaroo rat. Mature bilbies have on average 1.0 young per adult per year. The proportion of young that survive to reproduce at the end of their first year is 0.6. The survival rate of adults depends on seasonal conditions and can be represented by a parameter s whose value lies some where between 0 and 1. Write two equations: one for the number of young and one for the number of adults at time period n. If $s = 1$, use eigenvalue analysis to determine how rapidly the population can ultimately be expected to grow. Of course, s can never be 1, so this growth rate is an upper bound. What will the growth rates be for the two cases $s = 2/3$ and $s = 1/3$? What value of s is needed to ensure that the population will not decline in the long run?

54. Suppose in the model presented in Example 8 in this section that the blood-tissue subsystem model is given by

$$\begin{pmatrix} x_{n+1} \\ y_{n+1} \end{pmatrix} = \begin{pmatrix} 0.5 & 0.04 \\ 0.04 & 0.02 \end{pmatrix} \begin{pmatrix} x_n \\ y_n \end{pmatrix}$$

and that the blood-bone subsystem model is given by

$$\begin{pmatrix} x_{n+1} \\ y_{n+1} \end{pmatrix} = \begin{pmatrix} 0.5 & 0.001 \\ 0.08 & 0.999 \end{pmatrix} \begin{pmatrix} x_n \\ y_n \end{pmatrix}$$

Use technology to find the eigenvalues and eigenvectors of these two models and interpret the results.

55. Demonstrate that the eigenvalues of the matrices $A = \begin{pmatrix} a & 0 \\ b & c \end{pmatrix}$ and $B = \begin{pmatrix} a & b \\ 0 & c \end{pmatrix}$ are a and c and thus independent of the value of b. Matrices whose elements are 0 above the main diagonal are called lower triangular, while those whose elements are 0 below the main diagonal are called upper triangular. Of course, for a 2-by-2 matrix there is only one element above and one below the diagonal, but this idea generalizes to matrices in n dimensions where $n(n - 1)/2$ elements lie above the diagonal and the same number lie below the diagonal.

56. Recall that to find the eigenvalues of $A = \begin{pmatrix} a & b \\ c & d \end{pmatrix}$ we need to solve the quadratic equation $\lambda^2 - T\lambda + D$, where $T = a + b$ is the sum of the diagonal elements of A (also called the *trace* of A) and $D = ad - bc$ is the determinant of A. Show that if $K = T^2 - 4D < 0$, then the eigenvalues are the complex conjugate pair $\lambda_1 = \dfrac{T}{2} + i\dfrac{\sqrt{K}}{2}$ and $\lambda_1 = \dfrac{T}{2} - i\dfrac{\sqrt{K}}{2}$. Also show that the corresponding eigenvectors can be written as $\begin{pmatrix} 1 \\ \dfrac{b}{\lambda_1 - a} \end{pmatrix}$ and $\begin{pmatrix} 1 \\ \dfrac{c}{\lambda_2 - d} \end{pmatrix}$.

In Problems 57 and 58, three-dimensional models are examined.

57. Attempts are being made to halt the spread of a noxious invasive weed species in Texas. Scientists studying this perennial species have classified plant individuals into three size classes: small, medium, and large. For these different size classes, three quantities were estimated: mean fecundity of each class (mean number of seeds produced per plant), annual survivorship (probability an individual of a particular size class survives to the next

year), and graduation rate (fraction of surviving individuals that graduate to the next size class):

	small	medium	large
fecundity	0	30	50
survivorship	0.2	0.5	0.5
graduation rate	0.8	0.6	–

In addition, the scientists estimated the probability that a seed survives to the next year and germinates as a small plant is 0.1. Assume all seeds that don't germinate are lost forever.

a. Write a matrix model for this population.

b. If the population started with ten thousand seeds of which 10% germinated to small plants, estimate the number of plants of each size class two years later.

c. Let A be the matrix for the noxious weed model. For this model, the largest eigenvalue is given by 1.0719 and the associated eigenvector is given by

$$\begin{pmatrix} 0.9792 \\ 0.1797 \\ 0.0943 \end{pmatrix}$$

Assuming that this model is a reasonable representation of the population dynamics, in the long run how rapidly will the population grow? What percentage of the population will ultimately be in the smallest size class?

58. Kevin Crooks* and colleagues (1998) used matrix models to gain insights about cheetah conservation. Here, you will consider a simplified version of their model consisting of four age classes: cubs (first year of life), adolescents (second year of life), young adults (third year of life), and adults (fourth year of life and older). Assume that 6% of cubs survive to adolescence, 70% of adolescents survive to young adulthood, 80% of young adults survive to adulthood, and 80% of adults survive from year to year. Young adult females produce on average two female cubs per year, while adult females produce on average three female cubs per year.

a. Write a matrix model that represents the (female) cheetah population dynamics.

b. For a population with initially 100 female adults, what are the numbers of cubs, adolescents, young adults, and adults two years later?

c. Use technology to find the largest eigenvalue and associated eigenvector for this model. What do they tell you about the long-term behavior of the system?

---

*Crooks, K. R, M. A. Sanjayan, D. F. Doak, 1998. New Insights on Cheetah Conservation through Demographic Modeling. *Conservation Biology*, 12: 889–895.

8.4   Systems of Linear Differential Equations

As we have seen in previous chapters, many biological objects change continuously over time. The volume of a tumor grows continuously over time due to its many cells dividing asynchronously, or it shrinks over time as cells die during chemotherapy. The concentration of a drug in a patient changes continuously as it diffuses between the bloodstream and various organs. In Chapter 6, we explored how to model the processes underlying these changes using differential equations with a single dependent variable. In this section and the next section, we examine differential equations with multiple variables. We begin with the simplest of these equations, systems of linear differential equations. We will see that solving these systems of linear differential equations involves exponential functions, eigenvalues, and eigenvectors. As in the other sections of this chapter, we focus on two-dimensional linear systems and conclude by briefly discussing how to generalize results to n-dimensional linear systems.

We begin our study of linear differential equations by recasting a model of tumor growth from Chapter 4 as a system of linear differential equations.

Example 1   Tumor growth modeled by a linear differential equation

In an experimental study performed at Dartmouth College, two groups of mice with tumors were treated with the chemotherapeutic drug Cisplatin. Before therapy, tumors of proliferating cells (also known as clonogenic cells) increased at a per capita rate of approximately 0.53/day. Upon reaching a volume of approximately 0.5 cm^3, each of the mice was administered a dose of Cisplatin corresponding to 10 mg/kg of body mass. After treatment, 99% of the proliferating cells became quiescent cells (also known as nonproliferating or resting cells). These quiescent cells do not divide and the tumor starts to decrease at a per capita growth rate of approximately 0.87/day. Assume x denotes the volume of proliferating cells in cubic meters and y denotes the volume of quiescent cells in cubic meters; now do the following:

a. Write a pair of differential equations that describe the dynamics of x and y. Indicate the initial values for both variables. Solve the differential equation and plot the total volume of the tumor. Compare your answer to what we found in Example 4 of Section 4.3.

b. Express the model in part **a** in matrix-vector form and the solution in part **a** in vector form.

c. Solve for the eigenvalues and eigenvectors of the matrix in part **b**. Discuss how these relate to the vector form of the solution in part **b**.

Solution

a. Since the per capita growth rate of proliferating cells is 0.53, it follows that

$$\frac{dx}{dt} = 0.53\,x$$

Since the per capita growth rate of quiescent cell is -0.87, it follows that

$$\frac{dy}{dt} = -0.87\,y$$

On the day of treatment, the total volume is 0.5 and 99% of cells are quiescent. Hence, the "initial" conditions for the model are $x(0) = 0.01 \times 0.5 = 0.005$ and $y(0) = 0.99 \times 0.5 = 0.495$. Solving these uncoupled initial value problems yields

$$\begin{cases} x(t) = 0.005e^{0.53t} \\ y(t) = 0.495e^{-0.87t} \end{cases}$$

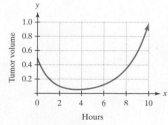

Figure 8.28 Volume of a tumor over time

Plotting the total volume $x(t) + y(t)$ against time yields the graph shown in Figure 8.28, which is equivalent to what we found in Example 4 of Section 4.3.

b. Since our system of differential equations is of the form

$$\begin{cases} \dfrac{dx}{dt} = 0.53x + 0y \\[2mm] \dfrac{dy}{dt} = 0x - 0.87y \end{cases}$$

we write the right-hand side as $A\mathbf{u}$ by defining

$$\mathbf{u} = \begin{pmatrix} x \\ y \end{pmatrix} \quad \text{and} \quad A = \begin{pmatrix} 0.53 & 0 \\ 0 & -0.87 \end{pmatrix}$$

To deal with the left-hand side, we define

$$\frac{d\mathbf{u}}{dt} = \begin{pmatrix} \dfrac{dx}{dt} \\[2mm] \dfrac{dy}{dt} \end{pmatrix}$$

to be componentwise differentiation of $\mathbf{u}$. We can write the solution from part **a** as

$$\mathbf{u}(t) = \begin{pmatrix} 0.005e^{0.53t} \\ 0.495e^{-0.87t} \end{pmatrix}$$

c. Since $\det(A - \lambda I) = (\lambda - 0.53)(\lambda + 0.87)$, the eigenvalues are 0.53 and -0.87. To find the eigenvector $\mathbf{v}$ associated with 0.53 we need to solve this system of linear equations:

$$\begin{cases} 0x + 0y = 0 \\ 0x + (-0.87 - 0.53)y = 0 \end{cases}$$

Hence, we need $y = 0$ and can choose $x = 1$. In other words, $\mathbf{v} = \begin{pmatrix} 1 \\ 0 \end{pmatrix}$. Similarly, one can find that $\mathbf{w} = \begin{pmatrix} 0 \\ 1 \end{pmatrix}$ is an eigenvector associated with the eigenvalue -0.87.

From part **b**, we have

$$\begin{aligned} \mathbf{u}(t) &= \begin{pmatrix} 0.005e^{0.53t} \\ 0.495e^{-0.87t} \end{pmatrix} \\[2mm] &= 0.05e^{0.53t} \begin{pmatrix} 1 \\ 0 \end{pmatrix} + 0.495e^{-0.87t} \begin{pmatrix} 0 \\ 1 \end{pmatrix} \end{aligned}$$

Hence, the solution that we found in part **a** can be expressed in terms of the eigenvalues and eigenvectors of the matrix A. ∎

Example 1 illustrates two key points for this chapter. First, biological processes described by a pair of linear differential equations

$$\begin{cases} \dfrac{dx}{dt} = ax + by \\[2mm] \dfrac{dy}{dt} = cx + dy \end{cases}$$

can be reexpressed in the concise matrix-vector form

$$\frac{d\mathbf{u}}{dt} = A\mathbf{u}$$

where

$$A = \begin{pmatrix} a & b \\ c & d \end{pmatrix}, \mathbf{u} = \begin{pmatrix} x \\ y \end{pmatrix}, \quad \text{and} \quad \frac{d\mathbf{u}}{dt} = \begin{pmatrix} \frac{dx}{dt} \\ \frac{dy}{dt} \end{pmatrix}$$

Notice that $\frac{d\mathbf{u}}{dt}$ corresponds to componentwise differentiation with respect to t of the **u** vector.

Second, Example 1 also suggests that eigenvectors and eigenvalues can be used to find solutions of these linear differential equations. Indeed, if λ and μ are eigenvalues with associated eigenvectors **v** and **w** for A, then Example 1 suggests solutions of the form

$$\mathbf{u}(t) = \alpha e^{\lambda t}\mathbf{v} + \beta e^{\mu t}\mathbf{w}$$

where α and β are constants. If this solution is valid, it will satisfy the vector equation $\frac{d\mathbf{u}}{dt} = A\mathbf{u}$. Indeed,

$$\frac{d\mathbf{u}}{dt} = \alpha e^{\lambda t}\lambda\mathbf{v} + \beta e^{\mu t}\mu\mathbf{w} \qquad \textit{differentiating with respect to t}$$

$$= \alpha e^{\lambda t} A\mathbf{v} + \beta e^{\mu t} A\mathbf{w} \qquad \textit{using the definition of eigenvalue/vector}$$

$$= A(\alpha e^{\lambda t}\mathbf{v} + \beta e^{\mu t}\mathbf{w}) \qquad \textit{using matrix-vector algebra}$$

$$= A\mathbf{u}(t) \qquad \textit{using the definition of } \mathbf{u}(t)$$

Hence, we have proven a key component of the following important theorem.

Theorem 8.4 Solutions of two-dimensional systems of linear differential equations theorem

Consider the system of linear differential equations

$$\frac{d\mathbf{u}}{dt} = A\mathbf{u}$$

*where A is a 2-by-2 matrix. If **v** and **w** are eigenvectors for the matrix A corresponding to the eigenvalues λ and μ, then*

$$\mathbf{u}(t) = ae^{\lambda t}\mathbf{v} + be^{\mu t}\mathbf{w}$$

*is a solution to this system of linear differential equations. Moreover, if **v** and **w** are not collinear (i.e., $\mathbf{v} \neq c\mathbf{w}$ for any c), then every solution of this system of linear differential equations can be written in the form $ae^{\lambda t}\mathbf{v} + be^{\mu t}\mathbf{w}$. The values of the constants a and b can be determined when an initial condition $\mathbf{u}(0)$ is specified by solving for a and b in the equation*

$$a\mathbf{v} + b\mathbf{w} = \mathbf{u}(0)$$

Example 2 Solving linear differential equations

Solve the following linear differential equations $\frac{d\mathbf{u}}{dt} = A\mathbf{u}$ with the given A and given initial conditions.

a. $A = \begin{pmatrix} 1.5 & -0.75 \\ -3 & 1.5 \end{pmatrix}$ and $\mathbf{u}(0) = \begin{pmatrix} 0 \\ 4 \end{pmatrix}$

b. $A = \begin{pmatrix} -1 & 8 \\ -2 & 9 \end{pmatrix}$ and $\mathbf{u}(0) = \begin{pmatrix} 6 \\ 3 \end{pmatrix}$

Solution

a. To solve for the eigenvalues, we solve the quadratic equation

$$\det(A - \lambda I) = \lambda^2 - 3\lambda = \lambda(\lambda - 3)$$

which has solutions $\lambda = 0$ and, naming the second solution, $\mu = 3$. To solve for an eigenvector associated with the eigenvalue $\lambda = 0$, we need to find a solution to $A\mathbf{v} = 0$. Equivalently, solve this system of equations:

$$\begin{cases} 1.5x - 0.75y = 0 \\ -3x + 1.5y = 0 \end{cases}$$

Since these equations (as always!) are scalar multiples of one another, it suffices to find a solution to the first equation, which requires $0.75y = 1.5x$. Equivalently, $y = 2x$. Setting $x = 1$ and $y = 2$ yields the eigenvector

$$\mathbf{v} = \begin{pmatrix} 1 \\ 2 \end{pmatrix}$$

To solve for an eigenvector associated with the eigenvalue $\mu = 3$, we need to find a solution to $A\mathbf{w} - 3\mathbf{w} = 0$. Equivalently, solve this system of equations:

$$\begin{cases} -1.5x - 0.75y = 0 \\ -3x - 1.5y = 0 \end{cases}$$

Since these equations are scalar multiples of one another, it suffices to find a solution to the first equation, which requires $0.75y = -1.5x$. Equivalently, $y = -2x$. Setting $x = 1$ and $y = -2$ yields the eigenvector

$$\mathbf{w} = \begin{pmatrix} 1 \\ -2 \end{pmatrix}$$

According to Theorem 8.4, solutions of $\dfrac{d\mathbf{u}}{dt} = A\mathbf{u}$ are of the form

$$\mathbf{u}(t) = ae^{0t}\begin{pmatrix} 1 \\ 2 \end{pmatrix} + be^{3t}\begin{pmatrix} 1 \\ -2 \end{pmatrix}$$

To find the coefficients a and b, we need to solve $\mathbf{u}(0) = a\mathbf{v} + b\mathbf{w}$; that is, find solutions to

$$\begin{cases} 0 = x(0) = a + b \\ 4 = y(0) = 2a - 2b \end{cases}$$

The first equation implies that $b = -a$. Substituting this expression into the second equation gives $4 = 4a$. Hence, $a = 1$ and $b = -1$. Therefore, the solution to the initial value problem is

$$\mathbf{u}(t) = e^{0t}\begin{pmatrix} 1 \\ 2 \end{pmatrix} - e^{3t}\begin{pmatrix} 1 \\ -2 \end{pmatrix}$$
$$= \begin{pmatrix} 1 - e^{3t} \\ 2 + 2e^{3t} \end{pmatrix}$$

b. To find the eigenvalues, we solve the quadratic equation

$$\det(A - \lambda I) = \lambda^2 - 8\lambda + 7 = (\lambda - 7)(\lambda - 1)$$

which has solutions $\lambda = 7$ and $\mu = 1$ (again we give the second value a different name to distinguish it from the first). To find an eigenvector associated with the eigenvalue $\lambda = 7$, we need to solve the equation $A\mathbf{v} - 7\mathbf{v} = 0$. Equivalently, we need to solve this system of equations:

$$\begin{cases} -8x + 8y = 0 \\ -2x + 2y = 0 \end{cases}$$

It suffices to find a solution to the first equation, which implies $y = x$. Setting $x = 1$ and $y = 1$ yields the eigenvector

$$\mathbf{v} = \begin{pmatrix} 1 \\ 1 \end{pmatrix}$$

To find the eigenvector associated with the eigenvalue $\mu = 1$, we need to find a solution to $A\mathbf{w} - \mathbf{w} = 0$. Equivalently, we need to solve this system of equations:

$$\begin{cases} -2x + 8y = 0 \\ -2x + 8y = 0 \end{cases}$$

The solution is $x = 4y$. Therefore, setting $y = 1$ and $x = 4$ yields the eigenvector

$$\mathbf{w} = \begin{pmatrix} 4 \\ 1 \end{pmatrix}$$

By Theorem 8.4, solutions of $\dfrac{d\mathbf{u}}{dt} = A\mathbf{u}$ are of the form

$$\mathbf{u}(t) = ae^{7t} \begin{pmatrix} 1 \\ 1 \end{pmatrix} + be^{t} \begin{pmatrix} 4 \\ 1 \end{pmatrix}$$

To find the coefficients a and b, we solve $\mathbf{u}(0) = a\mathbf{v} + b\mathbf{w}$. Equivalently,

$$\begin{cases} 6 = x(0) = a + 4b \\ 3 = y(0) = a + b \end{cases}$$

The second equation implies that $b = 3 - a$. Substituting this expression into the first equation gives $6 = a + 12 - 4a = 12 - 3a$. Hence, $a = 2$ and $b = 3 - 2 = 1$. Therefore, the solution to the initial value problem is

$$\mathbf{u}(t) = 2e^{7t} \begin{pmatrix} 1 \\ 1 \end{pmatrix} + e^{t} \begin{pmatrix} 4 \\ 1 \end{pmatrix}$$

$$= \begin{pmatrix} 2e^{7t} + 4e^{t} \\ 2e^{7t} + e^{t} \end{pmatrix}$$

With our newly developed ability to solve systems of linear differential equations, we can analyze more advanced models of tumor growth.

Example 3 Tumor regrowth revisited

In our simple model of tumor regrowth (Example 1), we assumed that all daughter cells of proliferating cells are proliferating cells. This assumption resulted in the linear differential equations being *uncoupled*; the equation for $\dfrac{dx}{dt}$ does not depend on y and the equation for $\dfrac{dy}{dt}$ does not depend on x. Here, we relax this assumption. Let α and β be the per capita rates at which proliferating cells produce proliferating daughter cells and produce quiescent daughter cells, respectively. Let γ be the per capita removal rate of quiescent cells. If x is the volume of proliferating cells and y is the volume of quiescent cells, then we get the following system of differential equations:

$$\begin{cases} \dfrac{dx}{dt} = \alpha x \\ \dfrac{dy}{dt} = \beta x - \gamma y \end{cases}$$

Find the general solution to this system of linear differential equations and write an expression for the total volume of the tumor.

Solution   The matrix for our matrix model $\dfrac{d\mathbf{u}}{dt} = A\mathbf{u}$ is $A = \begin{pmatrix} \alpha & 0 \\ \beta & -\gamma \end{pmatrix}$. Since $\det(A - \lambda I) = (\alpha - \lambda)(-\gamma - \lambda)$, the eigenvalues of A are $\lambda = \alpha$ and $\mu = -\gamma$.

To solve for the eigenvector $\mathbf{v}$ associated with this eigenvalue $\lambda = \alpha$, we solve this system of linear equations:

$$\begin{cases} 0x + 0y = 0 \\ \beta x - (\alpha + \gamma)y = 0 \end{cases}$$

Thus, we can choose $y = 1$ and $x = (\alpha + \gamma)/\beta$ to get $\mathbf{v} = \begin{pmatrix} (\alpha + \gamma)/\beta \\ 1 \end{pmatrix}$.

To solve for the eigenvector $\mathbf{w}$ associated with this eigenvalue $\mu = -\gamma$, we solve this system of linear equations:

$$\begin{cases} (\alpha + \gamma)\,x + 0\,y = 0 \\ \beta x = 0 \end{cases}$$

Thus, we can choose $y = 1$ and $x = 0$ to get $\mathbf{w} = \begin{pmatrix} 0 \\ 1 \end{pmatrix}$.

Theorem 8.4 implies that the general solution to our system of linear equations is

$$\mathbf{u}(t) = ae^{\alpha t}\begin{pmatrix} (\alpha + \gamma)/\beta \\ 1 \end{pmatrix} + be^{-\gamma t}\begin{pmatrix} 0 \\ 1 \end{pmatrix}$$

The total volume of the tumor is given by adding the two components of the solution together:

$$x(t) + y(t) = ae^{\alpha t}\left(\frac{\alpha + \gamma}{\beta} + 1\right) + be^{-\gamma t}$$

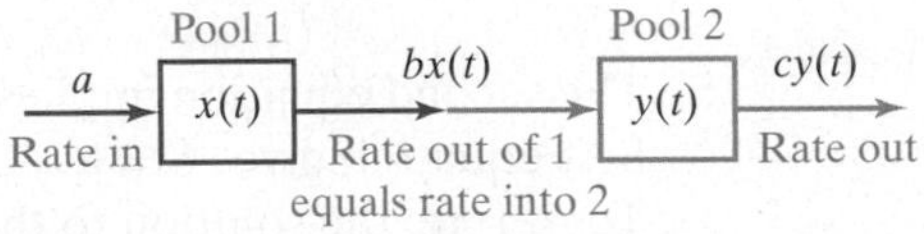

Figure 8.29 Flows in and out of two pools in a sequential two-compartment mixing model

Compartment models

Many interesting problems, as illustrated in Figure 8.29, can be cast in the context of flows from one pool to the next, where the contents of each pool are assumed to be well mixed: namely, in the case of concentrations of molecules in a liquid, the concentration is assumed to be the same everywhere in the pool. For example, an environmental pollutant flows into a dam containing water. Then the water is pumped into a large holding pond where it is tapped for filtering and purification before being sent on to storage tanks for human use. A drug that acts on the central nervous system is intravenously fed into the bloodstream—the first reservoir. It diffuses across the blood-brain barrier, as depicted in Figure 8.30, into the cerebrospinal fluid—the second reservoir—before it affects the target neurons.

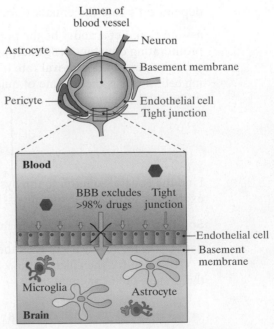

Figure 8.30 Flow of molecules across the blood-brain barrier

The next two examples illustrate the use of compartmental models in pharmacology. The first example illustrates how to model unidirectional flow of morphine across the blood-brain barrier. The second example is based on a clinical study in which a two-compartment model describes the dynamics of morphine-6-glucuronide, an active metabolite produced by morphine.

Example 4   Morphine across the blood-brain barrier

In clinical medicine, morphine is one of the most effective analgesics used to relieve severe or agonizing pain. Intravenous injection is the most common method of administration. Consider a patient receiving an initial injection of 6 mg of morphine into the bloodstream. Let $x(t)$ be the amount of morphine in the blood t hours after the injection and $y(t)$ be the amount of morphine in the patient's cerebrospinal fluid (i.e., primarily the brain) t hours after the injection. Assume the rate morphine moves from the bloodstream across the blood-brain barrier is $x(t)/2$, and the per capita removal rate of morphine from the cerebrospinal fluid is $1/10$.

a. Write the system of linear differential equations for $x(t)$ and $y(t)$.

b. Solve for $x(t)$ and $y(t)$ using the answer to Example 3, and plot the solutions.

Solution

a. Since there is no influx of morphine into the bloodstream and morphine crosses the blood-brain barrier at rate $\dfrac{x}{2}$, we have $\dfrac{dx}{dt} = -\dfrac{x}{2}$. Since morphine enters the brain at rate $\dfrac{x}{2}$ and leaves at a rate $\dfrac{y}{10}$, we have $\dfrac{dy}{dt} = \dfrac{x}{2} - \dfrac{y}{10}$. The model can be written in the matrix form $\dfrac{d\mathbf{u}}{dt} = A\mathbf{u}$ with

$$A = \begin{pmatrix} -1/2 & 0 \\ 1/2 & -1/10 \end{pmatrix} \quad \text{and} \quad \mathbf{u}(0) = \begin{pmatrix} 6 \\ 0 \end{pmatrix}$$

b. Our model from part **a** is equivalent to the tumor regrowth model in Example 3, with $\alpha = -\dfrac{1}{2}, \beta = \dfrac{1}{2}$, and $\gamma = \dfrac{1}{10}$. In this example, the solutions to this system of linear differential equations are of the form

$$\mathbf{u}(t) = ae^{\alpha t}\begin{pmatrix} \dfrac{(\alpha+\gamma)}{\beta} \\ 1 \end{pmatrix} + be^{-\gamma t}\begin{pmatrix} 0 \\ 1 \end{pmatrix}$$

$$= ae^{-t/2}\begin{pmatrix} -0.8 \\ 1 \end{pmatrix} + be^{-t/10}\begin{pmatrix} 0 \\ 1 \end{pmatrix}$$

To solve for the coefficients a, b, we solve $\mathbf{u}(0) = a\mathbf{v} + b\mathbf{w}$ for a and b, which reduces to this system of linear equations:

$$\begin{cases} 6 = -0.8a + b(0) \\ 0 = a + b \end{cases}$$

The first equation implies $a = -7.5$, which together with the second equation implies $b = 7.5$. Hence, the solution is

$$\mathbf{u}(t) = -7.5\,e^{-t/2}\begin{pmatrix} -0.8 \\ 1 \end{pmatrix} + 7.5\,e^{-t/10}\begin{pmatrix} 0 \\ 1 \end{pmatrix} = \begin{pmatrix} 6e^{-t/2} \\ 7.5\,e^{-t/10} - 7.5e^{-t/2} \end{pmatrix}$$

The solution is shown in Figure 8.31, which illustrates there is approximately a five-hour delay between the maximal amount of morphine in the blood (i.e., at the time of injection) and the maximal level of morphine in the brain.

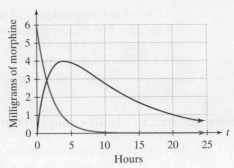

Figure 8.31 Morphine levels in the blood and brain after t hours

Example 5   Morphine-6-glucuronide dynamics from clinical studies

An active metabolite produced by morphine is morphine-6-glucuronide (M6G). This metabolite gets distributed to other tissues, including the brain, through the bloodstream. In a 2002 study published in *Clinical Pharmacology and Therapeutics* (72: 151–162), medical researcher Jörn Lötsch and colleagues reported the development of a two-compartment model for the amount x of M6G in the central bloodstream and the amount y of M6G in the "periphery" (e.g., tissues in the body). Their model is given by

$$\begin{cases} \dfrac{dx}{dt} = -1.65x + 0.82y \\[2mm] \dfrac{dy}{dt} = 0.61x - 0.82y \end{cases}$$

where time t is measured in hours. Concentrations x and y were measured in nanomoles per liter (nmol/l). Since there is no direct loss of M6G from the periphery compartment, the rate at which M6G leaves the periphery compartment equals the rate at which it enters the blood compartment. Conversely, since there is direct loss of M6G from the blood compartment (through the urinary tract), the rate at which M6G leaves the blood compartment is greater than the rate it enters the periphery compartment.

a. Write the model by Lötsch and colleagues in the matrix form $\dfrac{d\mathbf{u}}{dt} = A\mathbf{u}$. Use technology to find the eigenvectors and eigenvalues of A and write down the general solution to this model.

b. In the study by Lötsch and colleagues, eight healthy patients were given an injection of M6G into their blood system that resulted in an initial concentration $x(0) = 250$ and $y(0) = 0$. Find the solution corresponding to this initial condition and plot $x(t)$ for the first nine hours.

c. Plot the fraction $\dfrac{x(t)}{x(t) + y(t)}$ of M6G in the central blood systems over a two-day period. Discuss how this plot relates to eigenvectors of A.

Solution

a. The matrix A for this model is given by

$$A = \begin{pmatrix} -1.65 & 0.82 \\ 0.61 & -0.82 \end{pmatrix}$$

Using technology, we find that the eigenvalues for A are given by $\lambda \approx -2.06$ and $\mu \approx -0.41$, with associated eigenvectors

$$\mathbf{v} \approx \begin{pmatrix} 1 \\ -0.49 \end{pmatrix} \quad \text{and} \quad \mathbf{w} \approx \begin{pmatrix} 1 \\ 1.51 \end{pmatrix}$$

Thus, by Theorem 8.4, the general solution of the system of linear differential equations is

$$\mathbf{u}(t) = ae^{-2.06t}\mathbf{v} + be^{-0.41t}\mathbf{w}$$

b. To solve the initial value problem, we need to find a and b such that $\mathbf{u}(0) = a\mathbf{v} + b\mathbf{w}$. Therefore, we need to solve this system of linear equations:

$$\begin{cases} 250 = a + b \\ 0 = -0.49a + 1.51b \end{cases}$$

Our solution yields $a = 188.75$ and $b = 61.25$. Hence, the solution of the initial value problem is given by

$$\mathbf{u}(t) = 188.75e^{-2.06t}\begin{pmatrix} 1 \\ -0.49 \end{pmatrix} + 61.25e^{-0.41t}\begin{pmatrix} 1 \\ 1.51 \end{pmatrix}$$

Plotting this solution with the data from the study by Lötsch and colleagues (Figure 8.32) produces a very good correspondence between the model and the clinical experiments.

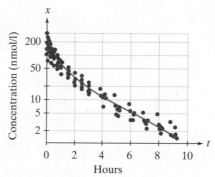

Figure 8.32 Observed concentrations of M6G in the blood plotted against time (hours) and the solution to the differential equation from Example 5

c. The plot $\dfrac{x(t)}{x(t) + y(t)}$ from part **b** is shown in Figure 8.33. Hence, in the long term approximately 40% of M6G in the body is in the blood. This percentage corresponds to the ratio $\dfrac{1}{1 + 1.51} \approx 0.4$ of the entries of the eigenvector $\mathbf{w}$.

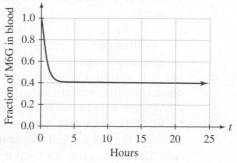

Figure 8.33 Fraction of M6G in the body that is found in the blood plotted as function of time

Stable state distribution and long term behavior

Part **c** of Example 5 suggests that, as in the case with discrete time models, the long term value of $\dfrac{x(t)}{x(t) + y(t)}$ is related to an eigenvector of A. It turns out this behavior is quite general. To see why, suppose the matrix A has two real eigenvalues λ, μ and associated eigenvectors $\mathbf{v} = \begin{pmatrix} v_1 \\ v_2 \end{pmatrix}$ and $\mathbf{w} = \begin{pmatrix} w_1 \\ w_2 \end{pmatrix}$. Solutions of $\dfrac{d\mathbf{u}}{dt} = A\mathbf{u}$ are of the form

$$\begin{pmatrix} x(t) \\ y(t) \end{pmatrix} = ae^{\lambda t} \begin{pmatrix} v_1 \\ v_2 \end{pmatrix} + be^{\mu t} \begin{pmatrix} w_1 \\ w_2 \end{pmatrix}$$

Provided the ratio $\dfrac{x(t)}{x(t) + y(t)}$ is well defined,

$$\frac{x(t)}{x(t) + y(t)} = \frac{ae^{\lambda t} v_1 + be^{\mu t} w_1}{ae^{\lambda t}(v_1 + v_2) + be^{\mu t}(w_1 + w_2)}$$

If $\lambda > \mu$ and $a \neq 0$, then

$$\lim_{t \to \infty} \frac{x(t)}{x(t) + y(t)} = \lim_{t \to \infty} \frac{ae^{\lambda t} v_1 + be^{\mu t} w_1}{ae^{\lambda t}(v_1 + v_2) + be^{\mu t}(w_1 + w_2)} \frac{e^{-\lambda t}}{e^{-\lambda t}} \qquad \textit{multiplying by } 1 = \frac{e^{-\lambda t}}{e^{-\lambda t}}$$

$$= \lim_{t \to \infty} \frac{av_1 + be^{(\mu - \lambda)t} w_1}{a(v_1 + v_2) + be^{(\mu - \lambda)t}(w_1 + w_2)}$$

$$= \frac{av_1}{a(v_1 + v_2)} \qquad \textit{as } \lim_{t \to \infty} e^{(\mu - \lambda)t} = 0$$

$$= \frac{v_1}{v_1 + v_2}$$

Thus, we have shown the following result.

Theorem 8.5   Stable state distribution theorem

Assume the matrix A has two real eigenvalues λ and μ and associated eigenvectors $\mathbf{v} = \begin{pmatrix} v_1 \\ v_2 \end{pmatrix}$ *and* $\mathbf{w} = \begin{pmatrix} w_1 \\ w_2 \end{pmatrix}$. *If $\lambda > \mu$ and $\mathbf{u}(0) = a\mathbf{v} + b\mathbf{w}$ with $a \neq 0$, then*

$$\lim_{t \to \infty} \frac{x(t)}{x(t) + y(t)} = \frac{v_1}{v_1 + v_2} \qquad \text{and} \qquad \lim_{t \to \infty} \frac{y(t)}{x(t) + y(t)} = \frac{v_2}{v_1 + v_2}$$

The fractions $\dfrac{v_1}{v_1 + v_2}, \dfrac{v_2}{v_1 + v_2}$ *are called the* stable state distribution, *as they describe asymptotically what fractions of the "population" are in the first and second "states."*

To illustrate the utility of Theorem 8.5, consider a population living over a large geographical region where environmental conditions vary substantially. For some habitats within a region, environmental conditions may favor population growth; consequently, they are called "source" habitats. For other habitats, environmental conditions may be harsh and retard population growth, and they are called "sink" habitats. For example, ecologist Kerri Vierling* found tallgrass prairies are source habitats for red-winged blackbirds (see Figure 8.34), while hay fields are sink habitats. The next example examines this scenario.

---

*Kerri T. Vierling 2000. Source and sink habitats of red-winged blackbirds in a rural/suburban landscape. *Ecological Applications* 10: 1211–1218.

Figure 8.34 Red-winged blackbirds live in a diversity of habitats, including tallgrass prairies and hay fields. In the grasslands, the blackbirds exhibit positive population growth, whereas in the hay fields, they exhibit negative population growth.

Example 6   Source-sink dynamics

Let x and y denote the density of blackbirds in the source and sink habitats, respectively. The per capita growth rate of blackbirds in prairies is approximately 0.2 per year. The per capita growth rate in hay fields is approximately -0.9 per year. If m is the per capita rate at which individuals move between these habitats, then the source-sink dynamics are given by

$$\begin{cases} \dfrac{dx}{dt} = (0.2 - m)x + my \\ \dfrac{dy}{dt} = mx - (0.9 + m)y \end{cases}$$

We can use this model to understand the effects of movement rates m on the long term fate and spatial distribution of the population.

a. Assume the blackbirds are moving at a low rate $m = 0.2$ between the two habitats. Determine whether in the long-term the population is growing or declining. If it is growing, what fraction of the population lives in the sink habitat in the long term?

b. Assume the blackbirds are moving at a high rate $m = 1$ between the two habitats. Determine whether in the long term the population is growing or declining. Discuss the implications of what you found.

Solution

a. When $m = 0.2$, we use technology to find that the eigenvalues of

$$A = \begin{pmatrix} 0 & 0.2 \\ 0.2 & -1.1 \end{pmatrix}$$

are 0.035 and -1.14. The corresponding eigenvectors are

$$\mathbf{v} = \begin{pmatrix} 1 \\ 0.18 \end{pmatrix} \quad \text{and} \quad \mathbf{w} = \begin{pmatrix} 1 \\ -5.68 \end{pmatrix}$$

Since the larger of these eigenvalues is positive, the population will increase at approximately 3.5% per year in the long term. Theorem 8.5 implies that the fraction of the population living in the sink habitats (the hay fields) is given by the ratio $\dfrac{0.18}{1 + 0.18} \approx 0.15$; namely, approximately 15% of the population is living in the hay fields in the long term.

b. When $m = 1$, we can use technology to find that the eigenvalues of

$$A = \begin{pmatrix} -0.8 & 1 \\ 1 & -1.9 \end{pmatrix}$$

are given by -2.49 and -0.21. Since both eigenvalues are negative, all solutions of $\dfrac{d\mathbf{u}}{dt} = A\mathbf{u}$ decay exponentially to the zero vector. Hence, if the population is moving too quickly between the two habitats, the population will tend to extinction because too many individuals are spending too much time in the sink habitat; in the long-term, 37% are in the sink habitat with $m = 1$ as opposed to 15% when $m = 0.2$. ∎

Linear systems with constant input

When a patient is given a drug, such as morphine, intravenously in a hospital, it may be administered continuously using an infusion pump (also known as a drip), as illustrated in Figure 8.35. Linear differential equations of the form $\dfrac{d\mathbf{u}}{dt} = A\mathbf{u}$ do not account for this input. Indeed, if $\mathbf{u}(0) = \begin{pmatrix} 0 \\ 0 \end{pmatrix}$ for a linear differential equation, then $\mathbf{u}(t) = \begin{pmatrix} 0 \\ 0 \end{pmatrix}$ for all time t. To account for a constant influx into a biological system, we consider models of the form

$$\frac{d\mathbf{u}}{dt} = A\mathbf{u} + \mathbf{b}$$

where $\mathbf{b} = \begin{pmatrix} k \\ \ell \end{pmatrix}$ is an input vector.

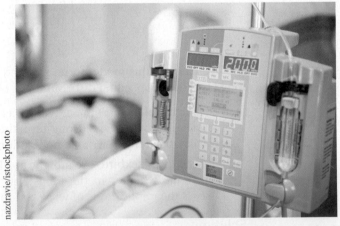

Figure 8.35 Infusion pumps are used to deliver medication or nutrients to patients. Infusion pumps can administer fluids continuously, which would be impractical and unreliable for medical personnel to implement.

Homogeneous and Inhomogeneous Linear Differential Equations

The linear systems of differential equations

$$\frac{d\mathbf{u}}{dt} = A\mathbf{u} + \mathbf{b}$$

is said to be **homogeneous** whenever $\mathbf{b} = \mathbf{0}$ and **inhomogeneous** whenever one or more of the elements $\mathbf{b}$ is nonzero.

To become familiar with these types of models, let us revisit Example 4 with the addition of a constant infusion rate.

Example 7   Continuous infusion of morphine

In Example 4 we considered a patient who received a single injection of morphine into the bloodstream. Now let's consider the same patient subject to a constant infusion of 1 mg/hour into the bloodstream.

a. Write a linear differential equation of the form $\dfrac{d\mathbf{u}}{dt} = A\mathbf{u} + \mathbf{b}$ for the constant infusion of morphine.

b. Find a solution of the form $\mathbf{u}(t) = \mathbf{u}^*$, where the elements of $\mathbf{u}^*$ are constant for all t, to the model in part **a**. Such constant solutions are called *equilibrium solutions*.

Solution

a. In the absence of the infusion of morphine into the bloodstream, we know from Example 4 that

$$\begin{cases} \dfrac{dx}{dt} = -\dfrac{x}{2} \\ \dfrac{dy}{dt} = \dfrac{x}{2} - \dfrac{y}{10} \end{cases}$$

where x is the amount of morphine in the blood, y is the amount of morphine in the cerebrospinal fluid, and time t is measured in hours. To account for a constant infusion of 1 mg/hour into the bloodstream, we need to add this influx into the x compartment of the model:

$$\begin{cases} \dfrac{dx}{dt} = -\dfrac{x}{2} + 1 \\ \dfrac{dy}{dt} = \dfrac{x}{2} - \dfrac{y}{10} \end{cases}$$

In matrix-vector notation, we can write these equations as $\dfrac{d\mathbf{u}}{dt} = A\mathbf{u} + \mathbf{b}$ where

$$A = \begin{pmatrix} -1/2 & 0 \\ 1/2 & -1/10 \end{pmatrix} \quad \text{and} \quad b = \begin{pmatrix} 1 \\ 0 \end{pmatrix}$$

b. If there exists a vector $\mathbf{u}^*$ such that $\mathbf{u}(t) = \mathbf{u}^*$ is a solution for all time t, then

$$\frac{d\mathbf{u}^*}{dt} = \begin{pmatrix} 0 \\ 0 \end{pmatrix}$$

because it is constant and

$$\frac{d\mathbf{u}^*}{dt} = A\mathbf{u}^* + \mathbf{b}$$

because it is a solution to the equation. Equating the right-hand sides of these two equations, we get

$$A\mathbf{u}^* + \mathbf{b} = \begin{pmatrix} 0 \\ 0 \end{pmatrix}$$

Therefore, we want a vector $\mathbf{u}^*$ such that

$$A\mathbf{u}^* = -\mathbf{b}$$

Thus, we need to solve this system of linear equations:

$$\begin{cases} \dfrac{-x^*}{2} = -1 \\ \dfrac{x^*}{2} - \dfrac{y^*}{10} = 0 \end{cases}$$

The solution is $x^* = 2$ and $y^* = 10$.

Part **b** of Example 7 implies that if the patient initially had 2 mg of morphine in her bloodstream and 10 mg in her cerebrospinal fluid, then she would have these amounts of morphine in her bloodstream for as long as she received the continuous infusion of 1 mg/hour of morphine into her bloodstream. However, patients typically have no morphine in their system at the start of a treatment. Therefore, it is important to understand how morphine levels change over time—particularly what levels the morphine may approach during a longer treatment. To resolve such questions, we need to know how to find solutions to the inhomogeneous linear systems model

$$\frac{d\mathbf{u}}{dt} = A\mathbf{u} + \mathbf{b}$$

for a specified matrix A.

Theorem 8.6 Solutions to inhomogeneous linear equations theorem

Consider the inhomogeneous system of linear differential equations

$$\frac{d\mathbf{u}}{dt} = A\mathbf{u} + \mathbf{b}$$

If $\mathbf{z}(t)$ is the general solution to the corresponding homogeneous system, that is

$$\frac{d\mathbf{z}}{dt} = A\mathbf{z}$$

and $\mathbf{u}^*$ is an equilibrium solution to the inhomogeneous system, that is

$$A\mathbf{u}^* = -\mathbf{b}$$

then the general solution to the inhomogeneous system is given by

$$\mathbf{u}(t) = \mathbf{z}(t) + \mathbf{u}^*$$

Proof. See Problem 29 in Problem Set 8.4. ∎

Theorems 8.4 and 8.6, taken together, form the following procedure for solving inhomogeneous systems of linear differential equations.

Solving Inhomogeneous Linear Differential Equations: 2-by-2 Case

The system of linear inhomogeneous differential equations

$$\frac{d\mathbf{u}}{dt} = A\mathbf{u} + \mathbf{b}$$

with specified initial value $\mathbf{u}(0)$ is solved as follows:

Step 1. Find the equilibrium solution $\mathbf{u}^*$ by solving the equation

$$A\mathbf{u}^* = -\mathbf{b}$$

Step 2. Find the general solution $\mathbf{z}(t)$ to the associated homogeneous equation

$$\frac{d\mathbf{z}}{dt} = A\mathbf{z}$$

using Theorem 8.4, which for a 2-by-2 matrix A involves solving for the eigenvalues λ and μ and corresponding eigenvectors $\mathbf{v}$ and $\mathbf{w}$ (providing they are not collinear) to obtain

$$\mathbf{z}(t) = ae^{\lambda t}\mathbf{v} + be^{\mu t}\mathbf{w}$$

Step 3. Solve for the constants a and b in step 2, in terms of the initial condition $\mathbf{u}(0)$, using the equation

$$a\mathbf{v} + b\mathbf{w} + \mathbf{u}^* = \mathbf{u}(0)$$

Step 4. By Theorem 8.6, the solution is

$$\mathbf{u}(t) = ae^{\lambda t}\mathbf{v} + be^{\mu t}\mathbf{w} + \mathbf{u}^*$$

With this procedure in place, we can solve the problem posed in Example 7.

Example 8   Solving the morphine infusion problem

Find the morphine levels in the blood and brain (i.e., cerebrospinal fluid) compartments of the patient described in Example 7 if the patient initially had no morphine in her system. Plot this solution over a twelve-hour period and relate this plot to what we found in part **b** of Example 7.

Solution   Recall from Example 7 that we want to solve $\dfrac{d\mathbf{u}}{dt} = A\mathbf{u} + \mathbf{b}$ where

$$A = \begin{pmatrix} -1/2 & 0 \\ 1/2 & -1/10 \end{pmatrix}, b = \begin{pmatrix} 1 \\ 0 \end{pmatrix}, \quad \text{and} \quad \mathbf{u}(0) = \begin{pmatrix} 0 \\ 0 \end{pmatrix}$$

Step 1 of our procedure requires that we find $\mathbf{u}^*$ such that $A\mathbf{u}^* = -\mathbf{b}$. In part **b** of Example 7, we found that $\mathbf{u}^* = \begin{pmatrix} 2 \\ 10 \end{pmatrix}$.

Step 2 of the procedure requires that we find the eigenvalues and eigenvectors of A. In Example 4, we found $-1/2$ and $-1/10$ are eigenvalues of A with associated eigenvectors

$$\mathbf{v} = \begin{pmatrix} 1 \\ -5/4 \end{pmatrix} \quad \text{and} \quad \mathbf{w} = \begin{pmatrix} 0 \\ 1 \end{pmatrix}$$

Step 3 of the procedure requires that we use the equation $a\mathbf{v} + b\mathbf{w} + \mathbf{u}^* = \mathbf{u}(0) = \mathbf{0}$ to solve for a and b. In other words, we need to solve the system of linear equations

$$a \begin{pmatrix} 1 \\ -5/4 \end{pmatrix} + b \begin{pmatrix} 0 \\ 1 \end{pmatrix} + \begin{pmatrix} 2 \\ 10 \end{pmatrix} = \begin{pmatrix} 0 \\ 0 \end{pmatrix}$$

which reduces to the two equations

$$\begin{cases} a + 2 = 0 \\ -\dfrac{5a}{4} + b + 10 = 0 \end{cases}$$

The first equation implies that $a = -2$. The second equation implies $b = -10 - 2.5 = -12.5$.

Step 4 of the procedure implies that the solution is

$$\mathbf{u}(t) = -2e^{-t/2} \begin{pmatrix} 1 \\ -5/4 \end{pmatrix} - 12.5e^{-t/10} \begin{pmatrix} 0 \\ 1 \end{pmatrix} + \begin{pmatrix} 2 \\ 10 \end{pmatrix}$$

$$= \begin{pmatrix} 2 - 2e^{-t/2} \\ 10 - 12.5e^{-t/10} + 2.5e^{-t/2} \end{pmatrix}$$

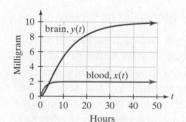

Figure 8.36 Morphine levels (milligrams) in the blood $(x(t))$ and brain $(y(t))$ compartments of a patient after t hours

Plotting $x(t)$ and $y(t)$ against time t yields the graph shown in Figure 8.36. From this figure we see that the morphine levels $x(t)$ and $y(t)$ are approaching the equilibrium values of 2 mg and 10 mg if the infusion is applied for two days.

So far we have considered solutions to systems of two linear differential equations. By analogy, Theorems 8.4 and 8.6 can be easily combined and extended to n dimensions, using the notation introduced in Example 9 of Section 8.2.

Theorem 8.7 Solutions of *n*-dimensional linear differential equations theorem

If the n eigenvalues λ_i with corresponding eigenvectors $\mathbf{v}_i$, $i = 1, \ldots, n$, of the n-by-n matrix A are all unique, and if $\mathbf{u}^*$ satisfies $A\mathbf{u}^* = -\mathbf{b}$ for a given n-dimensional vector $\mathbf{b}$, then for any set of constants $a_1, a_2, \ldots, a_n$,

$$\mathbf{u}(t) = \sum_{i=1}^{n} a_i e^{\lambda_i t} \mathbf{v}_i + \mathbf{u}^*$$

is a solution to the differential equation

$$\frac{d\mathbf{u}}{dt} = A\mathbf{u} + \mathbf{b}$$

If an initial condition $\mathbf{u}(0)$ is specified, then the constants $a_1, a_2, \ldots, a_n$ need to satisfy

$$\sum_{i=1}^{n} a_i \mathbf{v}_i + \mathbf{u}^* = \mathbf{u}(0)$$

In Example 4, we considered a patient receiving a single injection of morphine into the bloodstream and in Example 7 we considered the same patient receiving morphine using an infusion pump or drip system. A third option is for the morphine to be injected intramuscularly. In this case, the morphine enters the patient's blood compartment from the muscle where the injection was given and is then on its way to the brain.

Example 9 Morphine introduced intramuscularly

Suppose the patient considered in Example 4 receives her injection intramuscularly rather than intravenously. The morphine drains at a per capita removal rate of 1/8 per hour from the ventrogluteal muscle where the injection was given into the bloodstream.

a. Write an equation that describes the amount of morphine in the ventrogluteal muscle, blood, and brain compartments of the patient.

b. Using Theorem 8.7, find the solution for the concentration of morphine in the patient's brain compartment. Compare what you find to the injection solution depicted in Figure 8.31 and the infusion solution depicted in Figure 8.36.

Solution

a. We let the first, second, and third elements of the vector $\mathbf{u}$ in the matrix model $\frac{d\mathbf{u}}{dt} = A\mathbf{u}$ represent the amount of morphine in the ventrogluteal muscle, blood, and brain compartments of the patient. Since the per capita flow rate from the muscle to blood compartment is 1/8, the first and second rows of the first column are −1/8 (flow out of muscle to blood) and 1/8 (flow into blood from muscle), respectively. Since the blood and brain compartments now are the second and third compartments in the model, the matrix in Example 4 provides the terms for the second and third rows and the second and third columns of the new model. Finally, the remaining elements in the first row and column are zero, because there is no flow directly between the muscle and brain (i.e., all flows from the muscle to the brain must go through the blood). Thus, the matrix for our three-compartment model is

$$A = \begin{pmatrix} -1/8 & 0 & 0 \\ 1/8 & -1/2 & 0 \\ 0 & 1/2 & -1/10 \end{pmatrix}$$

In terms of initial conditions, the first element of $\mathbf{u}(0)$ is nonzero to reflect the fact that initially the 6 mg of morphine is injected into the muscle. The second and third terms $\mathbf{u}(0)$ are zero to reflect the fact that the blood and brain are initially free of morphine; that is,

$$\mathbf{u}(0) = \begin{pmatrix} 6 \\ 0 \\ 0 \end{pmatrix}$$

b. The eigenvalues and eigenvectors of A, which are found by solving the equation

$$(A - \lambda I)\mathbf{v} = 0$$

can either be solved by hand (the fact that A is lower triangular (see paragraph after this example) means that the eigenvalues are equal to the diagonal terms of A) or by using technology to obtain the solutions $\lambda_1 = -\dfrac{1}{2}$, $\mathbf{v}_1 = \begin{pmatrix} 0 \\ -4/5 \\ 1 \end{pmatrix}$, $\lambda_2 = -\dfrac{1}{8}$, $\mathbf{v}_2 = \begin{pmatrix} -3/20 \\ -1/20 \\ 1 \end{pmatrix}$, and $\lambda_3 = -\dfrac{1}{10}$, $\mathbf{v}_3 = \begin{pmatrix} 0 \\ 0 \\ 1 \end{pmatrix}$. By Theorem 8.7, noting in this case that the system is homogeneous (i.e., $\mathbf{b} = \mathbf{0}$), solutions are of the form $\mathbf{u}(t) = a_1 e^{-t/2}\mathbf{v}_1 + a_2 e^{-t/8}\mathbf{v}_2 + a_3 e^{-t/10}\mathbf{v}_3$. To find the coefficients a_1, a_2, and a_3, we solve $a_1\mathbf{v} + a_2\mathbf{v}_2 + a_3\mathbf{v}_3 = \mathbf{u}(0)$, which reduces to this system of linear equations:

$$\begin{cases} -3a_2/20 = 6 \\ -4a_1/5 - a_2/20 = 0 \\ a_1 + a_2 + a_3 = 0 \end{cases}$$

The first equation implies $a_2 = -40$. Hence, the second equation implies $4a_1/5 = 2$ or $a_1 = 5/2$. Substituting these two values in the third equation yields $a_3 = 40 - 5/2 = 75/2$. Thus, by Theorem 8.7, the solution is

$$\mathbf{u}(t) = \left(\frac{5}{2}\right)e^{-t/2}\begin{pmatrix} 0 \\ -4/5 \\ 1 \end{pmatrix} - 40e^{-t/8}\begin{pmatrix} -3/20 \\ -1/20 \\ 1 \end{pmatrix} + \left(\frac{75}{2}\right)e^{-t/10}\begin{pmatrix} 0 \\ 0 \\ 1 \end{pmatrix}$$

$$= \begin{pmatrix} 6e^{-t/8} \\ 2e^{-t/8} - 2e^{-t/2} \\ \left(\frac{5}{2}\right)e^{-t/2} + \left(\frac{75}{2}\right)e^{-t/10} - 40e^{-t/8} \end{pmatrix}$$

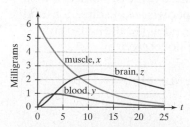

Figure 8.37 Morphine levels (milligrams) in the muscle, blood, and brain compartments after t hours

The solution is shown in Figure 8.37. In comparing the brain time course in this figure, with that of injection directly into the bloodstream, as shown in Figure 8.31, note that time course is less peaked and provides a steadier but lower dose for a longer period of time. However, unlike the drip solution depicted in Figure 8.36, the amount of morphine in the blood ultimately decays out exponentially to zero.

From the general formula for finding the determinant of a n-by-n matrix (see a matrix or linear algebra textbook), one can show for any triangular matrix—that is, any matrix in which all the elements above the main diagonal are zero (lower triangular matrix) or all the elements below the main diagonal are zero (upper triangular matrix)—that the eigenvalues are the diagonal elements (also see Problem 55 in Problem Set 8.3). With this information, it is immediately evident from the lower triangular structure of A that its eigenvalues are $-1/8$, $-1/2$, and $-1/10$. Knowing this it is then relatively easy to find the corresponding eigenvectors.

PROBLEM SET 8.4

Level 1 DRILL PROBLEMS

Write the matrix equation form $\dfrac{d\mathbf{u}}{dt} = A\mathbf{u}$, *where*

$\mathbf{u}(t) = \begin{pmatrix} x(t) \\ y(t) \end{pmatrix}$, *for the differential equations in Problems 1 to 8; then use Theorem 8.4 to write the general solution in terms of two arbitrary constants a and b.*

1. $\begin{cases} \dfrac{dx}{dt} = 2x \\ \dfrac{dy}{dt} = 3y \end{cases}$

2. $\begin{cases} \dfrac{dx}{dy} = -\dfrac{x}{2} \\ \dfrac{dy}{dx} = y \end{cases}$

3. $\begin{cases} \dfrac{dx}{dy} = -x \\ \dfrac{dy}{dx} = 3x - 2y \end{cases}$

4. $\begin{cases} \dfrac{dx}{dy} = \dfrac{x}{3} \\ \dfrac{dy}{dx} = -x + y \end{cases}$

5. $\begin{cases} \dfrac{dx}{dy} = x + 2y \\ \dfrac{dy}{dx} = 2x + y \end{cases}$

6. $\begin{cases} \dfrac{dx}{dy} = x - 2y \\ \dfrac{dy}{dx} = -2x + y \end{cases}$

7. $\begin{cases} \dfrac{dx}{dy} = 2x - 4y \\ \dfrac{dy}{dx} = -x - y \end{cases}$

8. $\begin{cases} \dfrac{dx}{dy} = x + 3y \\ \dfrac{dy}{dx} = x + y \end{cases}$

In Problems 9 to 12, find the particular solutions to the system of linear differential equations $\dfrac{d\mathbf{u}}{dt} = A\mathbf{u}$ *with A and* $\mathbf{u}(0)$ *as specified.*

9. $A = \begin{pmatrix} 1 & 0 \\ 1 & -2 \end{pmatrix}$ and $\mathbf{u}(0) = \begin{pmatrix} 1 \\ 1 \end{pmatrix}$

10. $A = \begin{pmatrix} 2 & -1/2 \\ 0 & -1 \end{pmatrix}$ and $\mathbf{u}(0) = \begin{pmatrix} -1 \\ 1 \end{pmatrix}$

11. $A = \begin{pmatrix} 1 & 3 \\ 1 & 1 \end{pmatrix}$ and $\mathbf{u}(0) = \begin{pmatrix} 0 \\ 1 \end{pmatrix}$

12. $A = \begin{pmatrix} 1 & 2 \\ -1 & -3 \end{pmatrix}$ and $\mathbf{u}(0) = \begin{pmatrix} -1 \\ 1 \end{pmatrix}$

In Problems 13 to 18, find the particular solutions to the system of linear differential equations $\dfrac{d\mathbf{u}}{dt} = A\mathbf{u} + \mathbf{b}$ *with A and* $\mathbf{u}(0)$ *as specified.*

13. $A = \begin{pmatrix} 1 & 0 \\ 1 & -2 \end{pmatrix}$, $\mathbf{b} = \begin{pmatrix} -1 \\ 1 \end{pmatrix}$, and $\mathbf{u}(0) = \begin{pmatrix} 0 \\ 0 \end{pmatrix}$

14. $A = \begin{pmatrix} 2 & 2 \\ 1 & 3 \end{pmatrix}$, $\mathbf{b} = \begin{pmatrix} -1 \\ 2 \end{pmatrix}$, and $\mathbf{u}(0) = \begin{pmatrix} 0 \\ 0 \end{pmatrix}$

15. $A = \begin{pmatrix} 1 & -2 \\ -2 & 1 \end{pmatrix}$, $\mathbf{b} = \begin{pmatrix} -3 \\ 1 \end{pmatrix}$, and $\mathbf{u}(0) = \begin{pmatrix} 1 \\ 0 \end{pmatrix}$

16. $A = \begin{pmatrix} 2 & -4 \\ -1 & -1 \end{pmatrix}$, $\mathbf{b} = \begin{pmatrix} 2 \\ -1 \end{pmatrix}$, and $\mathbf{u}(0) = \begin{pmatrix} 1 \\ 0 \end{pmatrix}$

17. $A = \begin{pmatrix} 1 & 2 \\ 3 & 2 \end{pmatrix}$, $\mathbf{b} = \begin{pmatrix} 0 \\ 0 \end{pmatrix}$, and $\mathbf{u}(0) = \begin{pmatrix} 0 \\ -4 \end{pmatrix}$

18. $A = \begin{pmatrix} -5 & 1 \\ 4 & -2 \end{pmatrix}$, $\mathbf{b} = \begin{pmatrix} 0 \\ 0 \end{pmatrix}$, and $\mathbf{u}(0) = \begin{pmatrix} 1 \\ 2 \end{pmatrix}$

Level 2 APPLIED AND THEORY PROBLEMS

19. A patient initially has 100 mg and 50 mg of morphine, respectively, in her blood and cerebrospinal fluid compartments. The per capita rates at which the drug is transferred across the blood-brain barrier from blood to cerebrospinal fluid and vice versa are $a = 1/2$ and $b = 1/4$. The rates of metabolism of the drug within these pools are $c = 1/5$ and $d = 1/10$, respectively. Find the equilibrium solution and the particular solution to this problem. Use the particular solution to calculate the concentrations of morphine in the blood and the cerebrospinal fluid after two hours.

20. Suppose the patient in Problem 19 initially has no morphine in her system, but morphine is infused into her blood at a rate of 10 mg per hour. Write a model of this process, calculate the equilibrium solution, and calculate the particular solution for the initial condition of no morphine in the body.

21. Recall the sequential pool model considered in this section. In the context of population growth, this model can be generalized to the case where the exchange between the pools is in both directions and the populations in the pools are allowed to grow. Consider a two-pool model given by

$$\frac{dx}{dt} = (a - g)x + hy + c$$

$$\frac{dy}{dt} = gx + (b - h)y + k$$

where x and y are the abundances of individuals in pools 1 and 2, respectively. Provide an interpretation for the parameters $a, b, c \geq 0, g \geq 0, h \geq 0$, and $k \geq 0$ in this model.

22. In a region off the coast of Maine, two neighboring bays have populations of lobsters. In bay 1, the lobster population grows at an annual per capita rate of 0.4, but in bay 2 the lobster population declines at an annual per capita rate of 0.1. Suppose the per capita rate at which lobsters in bay 1 migrate to bay 2 is 0.1; there is no migration in the reverse direction. Also suppose that in a particular year the lobster populations in bay 1 and bay 2 are 2,500 and

1,500, respectively. Write a model for this system under the assumption that a proportion p of lobsters is harvested by fishermen in both bays each year. By solving this system of equations, identify the proportion p so that the combined population of both bays neither grows without bound, nor declines to zero. If this rate is applied from the outset, then how many lobsters can be caught in the long run from both bays?

23. A species of frog inhabits two lakes. The number of frogs in lakes 1 and 2 at time t is $x(t)$ and $y(t)$, respectively. The per capita growth rates of the populations in lakes 1 and 2 are $a = -0.15$ and $b = -0.1$, respectively (i.e., both lakes are population sinks). The per capita rates at which individuals in lake 1 hop across to lake 2 and vice versa are $g = 0.1$ and $h = 0.1$, respectively. The total (i.e., not per capita) rates at which frogs migrate into lakes 1 and 2 from other lakes further upstream and downstream are $c = 10$ and $k = 20$. Write a model for the number of frogs in the population at time t. (Hint: This is an extension of the model in Problem 21 to include the constant inputs). Find the equilibrium for this population. If the population is initially $x(0) = 30$ and $y(0) = 40$, then find the particular solution as a function of time. Calculate approximately how far the population at $t = 10$ is from the equilibrium level in pools 1 and 2.

24. In Example 9, solve for the time course of the morphine in the patient's muscle, blood, and brain over a 24-hour period when 2 mg is injected intravenously and only 4 mg intramuscularly. Plot the time course of morphine in the brain on the same graph as that found for morphine in the brain in Example 9 and comment on the differences between the two.

25. In Example 9, solve for the time course of morphine in the patient's brain over a 24-hour period when only 3 mg is injected intramuscularly, while at the same time the patient is placed on an infusion that delivers 1/8th mg per hour over the course of a 24-hour period. Plot this time course on the same graph as that found for morphine in the brain in Example 9 and comment on the differences between the two.

26. Find the solution to the problem posed in Example 9, except this time assume that the per capita rate at which morphine drains from the muscle to the blood compartment is 1/12 per hour. Plot the time course of morphine in the muscle, blood, and brain compartments over the first 24 hours.

27. Consider a sequence of compartments in which a substance flows from compartment 1 to compartment 2 at rate r_1, from compartment 2 to compartment 3 at rate r_2, from compartment 3 to

compartment 4 at rate r_3, and out of compartment 4 into the environment at rate r_4. Show that a model of this process is given by $\dfrac{d\mathbf{u}}{dt} = A\mathbf{u}$ where

$$A = \begin{pmatrix} -r_1 & 0 & 0 & 0 \\ r_1 & -r_2 & 0 & 0 \\ 0 & r_2 & -r_3 & 0 \\ 0 & 0 & r_3 & -r_4 \end{pmatrix}$$

where the vector

$$\mathbf{u}(t) = \begin{pmatrix} u_1(t) \\ u_2(t) \\ u_3(t) \\ u_4(t) \end{pmatrix}$$

represents the time course of the concentration of substance in each compartment, given some initial concentrations $\mathbf{u}(0) = \mathbf{b}$. Using the fact that the eigenvalues of a lower diagonal matrix are just the diagonal elements themselves, show that the solution is

$$\mathbf{u}(t) = \begin{pmatrix} \dfrac{-(r_1 - r_2)(r_1 - r_3)(r_1 - r_4)}{r_1 r_2 r_3} a_1 e^{-r_1 t} \\[2mm] \dfrac{(r_1 - r_3)(r_1 - r_4)}{r_2 r_3} a_1 e^{-r_1 t} + \dfrac{(r_2 - r_3)(r_2 - r_4)}{r_2 r_3} a_2 e^{-r_2 t} \\[2mm] \dfrac{-(r_1 - r_4)}{r_3} a_1 e^{-r_1 t} - \dfrac{(r_2 - r_4)}{r_3} a_2 e^{-r_2 t} - \dfrac{(r_3 - r_4)}{r_3} a_3 e^{-r_3 t} \\[2mm] a_1 e^{-r_1 t} + a_2 e^{-r_2 t} + a_3 e^{-r_3 t} + a_4 e^{-r_2 t} \end{pmatrix}$$

where a_1, a_2, a_3, and a_4 are solutions to

$$\begin{pmatrix} \dfrac{-(r_1 - r_2)(r_1 - r_3)(r_1 - r_4)}{r_1 r_2 r_3} a_1 \\[2mm] \dfrac{(r_1 - r_3)(r_1 - r_4)}{r_2 r_3} a_1 + \dfrac{(r_2 - r_3)(r_2 - r_4)}{r_2 r_3} a_2 \\[2mm] \dfrac{-(r_1 - r_4)}{r_3} a_1 - \dfrac{(r_2 - r_4)}{r_3} a_2 - \dfrac{(r_3 - r_4)}{r_3} a_3 \\[2mm] a_1 + a_2 + a_3 + a_4 \end{pmatrix} = \begin{pmatrix} b_1 \\ b_2 \\ b_3 \\ b_4 \end{pmatrix}$$

28. In Example 1 of Section 6.3, we examined a simple one-compartment model of the human immunodeficiency virus-type 1 (HIV-1). While at the Los Alamos National Laboratory in New Mexico, scientists Alan Perelson and Patrick Nelson developed a two-compartment extension of this model that simultaneously accounts for the concentration I of infected target cells in the bloodstream and the concentration V of viral particles within the bloodstream. Infected cells burst at a rate $b > 0$ and produce N viral particles upon bursting. Viral particles are cleared from the body at a per capita rate of c. Viral particles infect uninfected cells at a rate a proportional to the density of these particles multiplied by the density of uninfected cells U.

a. Assume the concentration U of uninfected cells is constant and write a linear differential equation model for V and I.

b. Solve for the largest eigenvalue of the model in terms of the model parameters a, b, c, N and U, and determine under what conditions the virus increases exponentially or decays exponentially.

c. Provide a biological interpretation for the expression that you found in part **b**. Hint: Recall that $1/c$ is the mean time a viral particle remains in the bloodstream.

29. Prove Theorem 8.6.

8.5   Nonlinear Systems

In the preceding section, we developed tools to analyze the dynamical behavior of biological systems described by linear differential equations. Many biological systems, however, are best modeled by equations that include nonlinear terms. For example, we saw in Section 6.1 that populations may exhibit exponential growth at low densities but experience nonlinear feedbacks that slow down population growth at higher densities. In Example 2 of Section 8.1, we modeled the rate at which a population of predators consume prey as the product of prey and predator densities. The firing rate of a neuron, as discussed in Example 5 of Section 6.6, depends in a nonlinear way on the level of excitatory and inhibitory signals that the neuron receives. These nonlinearities can generate rich dynamical behaviors that contribute to the complexity and mystery of biological systems. In this section, we provide a brief introduction to this dynamical complexity using a combination of simple graphical methods and numerical computations. Throughout this section, we consider models of the form

$$\begin{cases} \dfrac{dx}{dt} = f(x, y) \\ \dfrac{dy}{dt} = g(x, y) \end{cases}$$

where f and g are functions that describe the rate of change of x and y, respectively.

We begin by examining how prey-predator interactions generate oscillatory population dynamics, a phenomenon observed in many natural prey-predator systems. For example, populations of Canadian lynx (the predator) and snowshoe hare (the prey) exhibited oscillations for nearly a century in North America (Figure 8.38). To understand the dynamics of such interacting populations, we extend Example 2 of Section 8.1 to a two-species model.

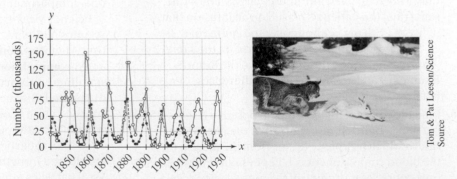

Figure 8.38 Hudson Bay fur-trapping records show that lynx and snowshoe hare populations fluctuate in an eight-to-eleven-year cycle.

Data Source: Taken from Krebs, C. J. R. Boonstra, S. Boutin and A. R. E. Sinclair, "What Drives the 10-year Cycle of Snowshoe Hares?" *BioScience* 51(1)(2001): 25–35. First published by Elton, C., and M. Nicholson, "The Ten-Year Cycle in Numbers of the Lynx in Canada," *Journal of Animal Ecology* 11 (1942): 215–244.

Example 1   Prey-predator dynamics

In Example 2 of Section 8.1, we considered the growth rates of a prey population at density x and a predator population at density y. Under the assumption that the per capita growth rate of the prey population decreases with predator density and the per capita growth rate of the predator population increases with prey density, we arrive at the following model

$$\begin{cases} \dfrac{dx}{dt} = rx - axy \\ \dfrac{dy}{dt} = baxy - cy \end{cases}$$

where $t \geq 0$ is measured in years, $r > 0$ is the intrinsic per capita rate of growth of the prey, $a > 0$ is the predator attack rate, and $c > 0$ is the per capita death rate of the predator. The parameter $b > 0$ is the efficiency at which the predator converts eaten prey into offspring.

Explore the behavior of solutions to these equations with $r = 1$, $a = 1/10$, $b = 1/2$ and $c = 1$ by doing the following:

a. Draw lines in the first quadrant of the xy plane where the prey growth rate $\dfrac{dx}{dt}$ equals zero. For each of the regions bounded by these lines and the xy axes, identify whether the prey population is increasing or decreasing in density.

b. Draw lines (use a different color so as not to confuse the lines with what you found in part **a**) in the first quadrant of the xy plane where the predator growth rate $\dfrac{dy}{dt}$ equals zero. For each of the regions bounded by these lines and the xy axes, identify whether the predator population is increasing or decreasing in density.

c. In each of the regions bounded by the lines you drew and the axes in parts **a** and **b**, the prey is either increasing or decreasing, and the predator is either increasing or decreasing. Sketch an arrow in each of these regions corresponding to how densities of both populations are changing.

d. Discuss what your work in parts **a** through **c** suggests about how the predator and prey numbers change if initially $x(0) = 30$ and $y(0) = 5$.

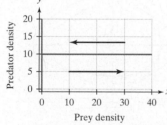

Figure 8.39 Prey nullclines at which $\dfrac{dx}{dt} = 0$ and arrows indicating where $\dfrac{dx}{dt}$ is positive and negative

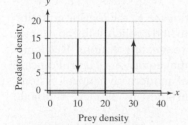

Figure 8.40 Predator nullclines at which $\dfrac{dy}{dt} = 0$ and arrows indicating where $\dfrac{dy}{dt}$ is positive and negative

Solution

a. The growth rate of the prey population is zero when

$$0 = \frac{dx}{dt} = x - \frac{xy}{10} = x\left(1 - \frac{y}{10}\right)$$

Hence, it is zero when $x = 0$ or $y = 10$. When $x > 0$, the prey growth rate is positive for $y < 10$ and negative for $y > 10$. Plotting the zero growth rate lines and arrows where the prey has positive and negative growth rates yields Figure 8.39. Intuitively when there are too many predators, the prey numbers decline. When there are few predators, the prey numbers increase.

b. The growth rate of the predator population is zero when

$$0 = \frac{dy}{dt} = \frac{xy}{20} - y = y\left(\frac{x}{20} - 1\right)$$

Hence, it is zero when $y = 0$ or $x = 20$. When $y > 0$, the predator growth rate is positive for $x > 20$ and negative for $x < 20$. Plotting the zero growth rate lines and arrows where the predator has positive and negative growth rates yields Figure 8.40. Intuitively when there are too few prey, the predator numbers decline. When there are many prey, the predator numbers increase.

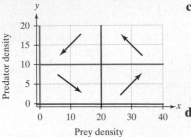

Figure 8.41 x and y isoclines

c. Putting together the zero growth rate lines for the predator and prey populations, we obtain the composite figure in Figure 8.41.

  The direction of the arrows was determined by combining the information from parts **a** and **b**. For example, the arrow in the lower right corner points to the right as the prey growth rate is positive at low predator numbers and points upward as the predator growth rate is positive at high prey numbers.

d. At time $t = 0$, when $x(0) = 30$ and $y(0) = 5$, we obtain $\dfrac{dx}{dt} = x\left(1 - \dfrac{y}{10}\right)$
$= 30\left(1 - \dfrac{5}{10}\right) = 15 > 0$ and $\dfrac{dy}{dt} = y\left(\dfrac{x}{20} - 1\right) = 5\left(\dfrac{30}{20} - 1\right) = 2.5 > 0$. Thus, both the prey and predator have positive growth rates and would increase in density. After some time, however, the predator density would grow to exceed 10 (i.e., enter the upper right corner of Figure 8.41) at which point the prey densities would being to decline. Some time later, the prey densities would fall below 20 (i.e., enter the upper left corner of Figure 8.41) and predator and prey densities would begin to decrease. Again, after some time, the predator densities would fall below 10 (i.e., enter the lower left corner of Figure 8.41) and prey would start to increase again.

Using technology, we can see that our predictions from part **d** of Example 1 are indeed correct. Plotting the solutions for $x(t)$ and $y(t)$ with $x(0) = 30$ and $y(0) = 5$ as a function of t, and plotting the points $(x(t), y(t))$ in the xy plane for increasing values of t, produces the graphs shown, respectively, in the left and right panels of Figure 8.42. These plots exhibit a qualitatively similar behavior to the lynx-hare data in Figure 8.38.

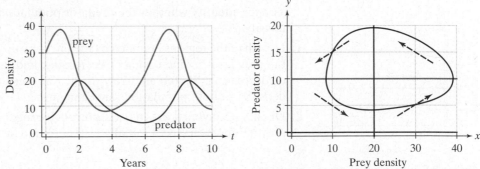

Figure 8.42 Solution to the predator-prey model in Example 1 with $x(0) = 30$ and $y(0) = 5$. The left panel plots $x(t)$ and $y(t)$ against time t. The right panel plots $(x(t), y(t))$ in the xy plane. The dashed arrows and solid lines correspond to the work done in part **c** of Example 1 (see Figure 8.41.)

Using the phase-plane method

Although it is possible to solve certain nonlinear equations (e.g., the logistic equation or the doomsday equation), in general it is impossible to find explicit solutions; therefore, numerical methods must be used to obtain approximate solutions. Recognizing difficulty of finding solutions to nonlinear differential equations, Henri Poincaré (1854–1912), a French mathematician and founder of modern dynamical systems theory, wrote (in *Science and méthode*, english translation by Francis Maitland, London; Thomas Nelson and Sons, 1914):

"Formerly an equation was not considered to have been solved until the solution had been expressed by means of a finite number of known functions. But this is impossible in about ninety-nine cases out of a hundred. What we can always do, or rather what we should always try to do, is to solve the problem qualitatively."

Poincaré's work was the genesis of the qualitative theory of differential equations that examines whether solutions tend to increase or decrease, oscillate, or exhibit more complicated long-term behavior.

Example 1 illustrated a general approach to obtain qualitative understanding. That approach relies on plotting the curves corresponding to setting the right-hand side of differential equations to zero. These curves are called *nullclines* and the method itself is called the *phase-plane method*, as described below.

Phase-Plane Method

The **phase-plane method** for coupled differential equations of the form

$$\begin{cases} \dfrac{dx}{dt} = f(x, y) \\ \dfrac{dy}{dt} = g(x, y) \end{cases}$$

involves the following four steps:

Step 1. Solve the equation $f(x, y) = 0$ to obtain curves in the plane called x **nullclines**. Solve the equation $g(x, y) = 0$ to obtain curves in the xy plane called y **nullclines**.

Step 2. Plot the x nullclines and y nullclines obtained in step 1 in the xy plane in different colors. Identify all points where an x nullcline intersects with a y nullcline. These points are **equilibria**.

Step 3. Identify the different regions bounded by the x nullclines and y nullclines and determine the signs of $\dfrac{dx}{dt}$ and $\dfrac{dx}{dt}$ in each of these regions. Draw arrows indicating the direction of solutions in each of these regions.

Step 4. Sketch some solutions.

A phase-plane analysis can be done without the aid of technology, but a more detailed understanding of the solutions is obtained by using an appropriate technology to plot solutions in the xy plane.* In addition to plotting nullclines, equilibria, and solutions, these technologies often plot a *vector field* for the system of differential equations. These vector fields consist of arrows that the solutions must be tangent to. Whenever $\dfrac{dx}{dt} \neq 0$, the slopes of these arrows are given by $\dfrac{dy}{dt} \Big/ \dfrac{dx}{dt}$.

Example 2   A step-by-step phase-plane analysis

Consider these coupled differential equations:

$$\begin{cases} \dfrac{dx}{dt} = y - x^2 \\ \dfrac{dy}{dt} = 1 - y \end{cases}$$

Perform a phase-plane analysis in the xy plane.

---

*To compute these solutions, you can use the program *pplane*, which at the time of writing this book was available online as a JavaScript applet at http://math.rice.edu/~dfield/dfpp.html.

Solution

Step 1. First we solve for the x nullcline, which is given by the solution to $0 = \dfrac{dx}{dt} = y - x^2$. Hence the x nullcline is $y = x^2$. Next, we solve for the y nullcline, which is given by the solution to $0 = \dfrac{dy}{dt} = 1 - y$. Hence, the y nullcline is $y = 1$.

Step 2. Plotting the x nullcline in blue and the y-nullcline in red yields this graph:

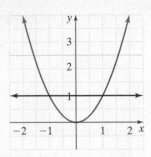

There are two points where the x and y nullclines intersect: $(1, 1)$ and $(-1, 1)$. These two points correspond to equilibria.

Step 3. There are five distinct regions bounded by the nullclines. In the region to the left of the parabola and above $y = 1$, we have $\dfrac{dx}{dt} < 0$ and $\dfrac{dy}{dt} < 0$. Hence, we draw an arrow pointing to the left and downward in this region. In the region inside the parabola and above $y = 1$, we have $\dfrac{dx}{dt} > 0$ and $\dfrac{dy}{dt} < 0$. Hence, we draw an arrow pointing to the right and downward in this region. In the region to the right of the parabola and above $y = 1$, we have $\dfrac{dx}{dt} < 0$ and $\dfrac{dy}{dt} < 0$. Hence, we draw an arrow pointing to the left and downward in this region. In the region above the parabola and below $y = 1$, we have $\dfrac{dx}{dt} > 0$ and $\dfrac{dy}{dt} > 0$. Hence, we draw an arrow pointing to the right and upward in this region. Finally, in the region below the parabola and below $y = 1$, we have $\dfrac{dx}{dt} < 0$ and $\dfrac{dy}{dt} > 0$. Hence, we draw an arrow pointing to the left and upward in this region. Drawing all of these arrows yields this graph:

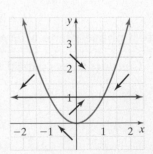

Step 4. Solutions to this system of differential equations will follow roughly the direction of the arrows from step 3. If we start a solution at $(x, y) = (1, -1)$, this solution initially moves to the left and upward, eventually crossing the parabola. Upon entering the parabola (below $y = 1$), the solution switches to moving right and upward towards the green point at $(x, y) = (1, 1)$, which is an equilibrium.

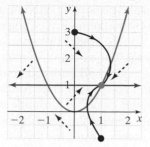

Figure 8.43 Nullclines (blue and red curves), directional arrows (broken), and solutions (black curves) starting at $(3, 0)$ and $(1, -1)$ (black dots), for Example 2, end at the equilibrium $(1, 1)$ (green dot).

Alternatively, if we start a solution at $(x, y) = (0, 3)$, this solution initially decreases and moves to the right, potentially leaving the parabola. If it leaves the parabola (above $y = 1$), then the solution starts moving downward and to the left also towards the green point at $(x, y) = (1, 1)$, which is an equilibrium. Drawing these solutions yields the phase-plane diagram depicted in Figure 8.43 ◼

A detailed, technology-derived version of Figure 8.43 is plotted in Figure 8.44, which provides a more complete picture of the phase-plane diagram for Example 2. This phase-plane diagram illustrates the behavior near the two equilibria $(-1, 1)$ and $(1, 1)$. Solutions starting near $(1, 1)$ approach this equilibrium in forward time. For this reason, $(1, 1)$ is an example of a *stable* equilibrium. In contrast, some solutions starting near $(-1, 1)$ approach $(-1, 1)$ but most do not. For this reason, $(-1, 1)$ is an example of a *saddle* equilibrium. In general, we can classify the behavior of solutions near equilibria into six types, as defined next. Two of these types—the saddle (left side of figure) and stable node (right side of figure)—are evident in Figure 8.44.

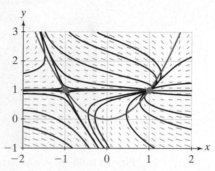

Figure 8.44 Technology-produced, phase-plane diagram for Example 2

Classification of Equilibria in the Phase-Plane

Six qualitatively different types of equilibria can occur in the phase plane:

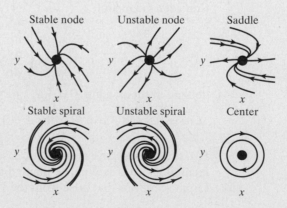

Stable node Solutions approach the equilibrium without winding around it.

Unstable node Solutions move away from the equilibrium without winding around it.

Saddle Solutions approach in one direction and move away in another direction.

Stable spiral Solutions approach the equilibrium while winding around it.

Unstable node Solutions move away from the equilibrium while winding around it.

Center Solutions circle around the equilibrium in closed loops.

To illustrate an application of the phase-plane method and classifying equilibria, let us examine a model of species competition developed by mathematical biologist Alfred Lotka and mathematical physicist Vito Volterra. In this model two species compete for a resource such as food, a place to live, or mates. For example, in rocky intertidal zones on the coast of Washington, barnacles, mussels, and seaweed compete for space on bare rock (Figure 8.45). This competition between species results in each species reducing the growth rate of the other species, which in turn affects the dynamics of the other species. The following example applies the phase-plane method to the classical model by Lotka and Volterra and demonstrates the diversity of dynamical outcomes for competing species.

Figure 8.45 Mussels, barnacles, limpets, and seaweed compete for space in an intertidal pool in Goodman Creek, Washington. The purple starfishes are a major predator of the mussel.

Example 3   Lotka-Volterra competition

Consider the model

$$\begin{cases} \dfrac{dx}{dt} = x(1 - x - \alpha y) \\[2mm] \dfrac{dy}{dt} = y(1 - y - \beta x) \end{cases}$$

defined in the domain $D = \{(x, y) | x \geq 0,\ y \geq 0\}$ for these three cases:

a. $\alpha = 0.5$ and $\beta = 0.6$ **b.** $\alpha = 1.2$ and $\beta = 0.8$ **c.** $\alpha = 1.25$ and $\beta = 1.5$

Use the phase-plane method for case **a** and classify the equilibria for this case. Use technology to create phase-plane diagrams for all three cases and compare the different scenarios that arise from the Lotka-Volterra competition equations.

Solution

a. For the case $\alpha = 0.5$ and $\beta = 0.6$, step 1 of the phase-plane method yields:

- Nullclines for x:

$$x(1 - x - 0.5y) \Rightarrow x = 0 \quad \text{and} \quad y = 2 - 2x \text{ (red)}$$

- Nullclines for y:

$$y(1 - y - 0.6x) \Rightarrow y = 0 \quad \text{and} \quad y = 1 - 0.6x \text{ (blue)}$$

Step 2 of the phase-plane method yields Figure 8.46a. From this phase-plane diagram, we see that $(0, 0)$, $(1, 0)$, and $(0, 1)$ are equilibrium solutions. An additional equilibrium lies in the positive quadrant and is given by solving the equations

$$y = 2 - 2x \quad \text{and} \quad y = 1 - 0.6x$$

which gives $(x, y) \approx (0.71, 0.57)$.

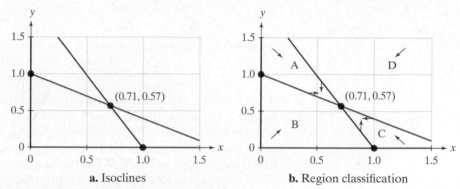

a. Isoclines   **b.** Region classification

Figure 8.46 a. Phase-plane method for the case $\alpha = 0.5$ and $\beta = 0.6$. The two x nullclines are red and the two y nullclines are blue. The black dots are the equilibria where the red and blue nullclines intersect. **b.** The nullclines divide the positive xy quadrant into four regions: A–D. The directions of the solutions are given by the black arrows, including vertical and horizontal arrows where the solutions cross the x and y-nullclines, respectively.

In Step 3 of the phase-plane method, we see that the nullclines divide the positive quadrant into four regions, A–D; their common corner, as illustrated in Figure 8.46b, is the equilibrium $(0.71, 0.57)$. Below the x nullcline $y = 2 - 2x$ we have $\dfrac{dx}{dt} > 0$. This can be verified by selecting a test point $\left(\dfrac{1}{2}, \dfrac{1}{2}\right)$ and evaluating the derivatives to obtain $\dfrac{dx}{dt} = \dfrac{1}{2}\left(1 - \dfrac{1}{2} - \dfrac{1}{2}\dfrac{1}{2}\right) = \dfrac{1}{8}$. Below the y-nullcline $y = 1 - 0.6x$ we have $\dfrac{dy}{dt} > 0$. Again, this can be verified using the same test point $\left(\dfrac{1}{2}, \dfrac{1}{2}\right)$ to obtain $\dfrac{dy}{dt} = \dfrac{1}{2}\left(1 - \dfrac{1}{2} - 0.6\dfrac{1}{2}\right) = 0.1$. Doing these computations for all four regions yields Figure 8.46b.

We classify the equilibria beginning with $(x, y) = (0, 0)$. Solutions move away from this equilibrium in both the vertical and horizontal directions. Therefore, $(0, 0)$ is an unstable node. The equilibria $(\hat{x}, \hat{y}) = (1, 0)$ and $(0, 1)$ are saddles because solutions along the axes approach the equilibrium and other solutions move away from these equilibria (see Problem 32 in Problem Set 8.1). Finally, solutions move toward the interior equilibrium $(0.71, 0.57)$ in both the horizontal and vertical directions. Therefore, this equilibrium is a stable node and corresponds to coexistence of the competing species.

These conclusions can be confirmed using technology to plot solutions starting at different points in the phase plane, as illustrated in Figure 8.47a.

b. Using technology, we can plot phase-plane diagrams for all three cases side-by-side, as shown in Figure 8.47. These three cases represent three qualitatively different phase-plane diagrams produced by the Lotka-Volterra competition model: coexistence (Figure 8.47a), total exclusion by the superior competitor y

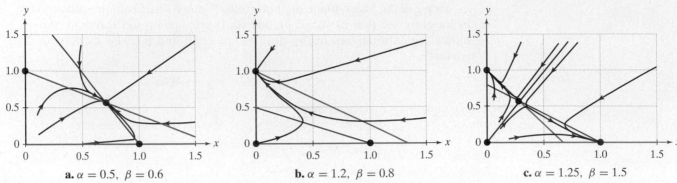

Figure 8.47 Phase-plane diagrams competition model. The x and y nullclines are shown in red and blue, respectively.

(Figure 8.47b), competitive exclusion where the winner depends on the initial values of the competitor (Figure 8.47c). In the case of coexistence, all solutions with both species initially present approach the interior equilibrium. In the case of total exclusion, the nontrivial nullcline of species y lies wholly above the nontrivial nullcline of species x. Whenever species y is present, it excludes species x. Finally, in the case of contingent exclusion, the winner depends on the initial conditions.

The three possible outcomes of competition presented in Example 3 have been observed in biological experiments and in nature. For example, the Russian biologist Georgii Frantsevich Gause demonstrated competitive exclusion in a famous set of experiments involving two different species of paramecia. The data from these experiments are presented in Figure 8.48.

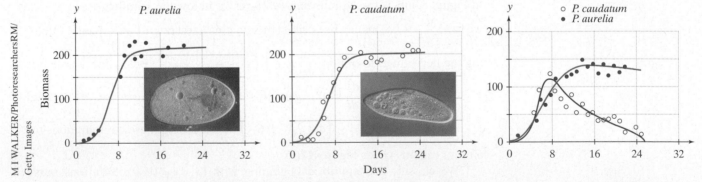

Figure 8.48 Population dynamics of two species of *Paramecium* in isolation (on the left) and in competition (on the right) where *P. aurelia* outcompetes *P. caudatum*.

Data Source: P. H. Leslie 1957. An Analysis of the Data for Some Experiments Carried out by Gause with Populations of the Protozoa, *Paramecium Aurelia* and *Paramecium Caudatum* Biometrika, Vol. 44, pp. 314–327.

Applications in physiology and medicine

Examples 1 and 3 above were drawn from the canons of theoretical ecology. Our next three examples develop applications found in physiology and medicine.

In 1963, Alan Lloyd Hodgkin and Andrew Fielding Huxley received the Nobel Prize in Physiology or Medicine for their work on the initiation and propagation of action potentials in the squid giant axon. A key element of their work was developing and simulating a six-dimensional, nonlinear, differential equation model of the activation and deactivation dynamics of a spiking neuron. The next example presents the two-dimensional FitzHugh-Nagumo model, which is a simplified version of the six-dimensional Hodgkin-Huxley model. It captures several key features, including excitability, of the Hodgkin-Huxley model.

Example 4   Excitable systems

The motivation for the FitzHugh-Nagumo model was to isolate the essential mathematical properties of the electrical activity of a neuron, an example of an excitable system. External input of an electrical current stimulates a neuron, after which it enters an "excited" state. In the FitzHugh-Nagumo model, the variable x represents the level of this excitation. Large positive values of x correspond to high firing rates of the neuron. Negative values of x correspond to the neuron not firing or firing at low rates. After excitation, physiological processes in a neuron will cause it to recover from excitation. The variable y represents this recovery process. Large positive values of y suppress neuronal activity. The domain of the model is the whole plane; that is, $D = \{(x, y)| -\infty < x < \infty, -\infty < y < \infty\}$. A standard form of the model is

$$\begin{cases} \dfrac{dx}{dt} = x - \dfrac{x^3}{3} - y + I \\ \dfrac{dy}{dt} = a(x + b - cy) \end{cases}$$

where I represents the external input of electrical current. Now let's explore the effects of the external input I on the dynamics for the parameters $a = 0.1$ and $b = c = 1$.

a. Perform a phase-plane analysis with no external input; that is, $I = 0$. Use your phase-plane analysis to compare and contrast the behavior of the system for the initial conditions $x(0) = 0$, $y(0) = -0.5$ versus $x(0) = -1$, $y(0) = -0.5$. Use technology to verify your predictions.

b. Perform a phase-plane analysis with an intermediate level of external input of $I = 1$.

c. Perform a phase-plane analysis with a high level of external input of $I = 2$.

Solution

a. With $I = 0$, $a = 0.1$, and $b = c = 1$, the model becomes

$$\begin{cases} \dfrac{dx}{dt} = x - \dfrac{x^3}{3} - y \\ \dfrac{dy}{dt} = 0.1(x + 1 - y) \end{cases}$$

Solving for the x nullcline $0 = \dfrac{dx}{dt} = x - \dfrac{x^3}{3} - y$ yields the cubic curve

$$y = x - \frac{x^3}{3} = x\left(1 - \frac{x^2}{3}\right)$$

which intersects the x axis at $x = 0$ and $x = \pm\sqrt{3}$. For points above this curve (i.e., $y > x - x^3$), $\dfrac{dx}{dt} < 0$. For points below this curve, $\dfrac{dx}{dt} > 0$. Solving for the y nullcline, $0 = \dfrac{dy}{dt} = 0.1(x + 1 - y)$ yields the line $y = x + 1$. For points above this line (i.e., $y > x + 1$), $\dfrac{dy}{dt} < 0$. For points below this line, $\dfrac{dy}{dt} > 0$. Putting this information together yields the phase diagram illustrated in Figure 8.49.

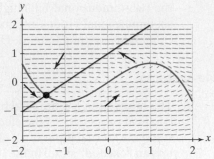

Figure 8.49 Phase-plane diagram for the case $I = 0$

This phase-plane diagram suggests that if the system starts at $x(0) = 0$, $y(0) = -0.5$, then the system gets more excited (i.e., x increases) and recovery variable y increases. When y gets large enough, the system gets less excited (i.e., x decreases), resulting eventually in the recovery variable y also decreasing. Hence, it looks like the system will exhibit at least one large oscillation. Using technology to plot the solution corresponding to this initial condition in the phase plane (left panel in Figure 8.50) and against time (right panel in Figure 8.50) confirms the prediction. The numerical simulation also shows that the system approaches the equilibrium after one oscillation.

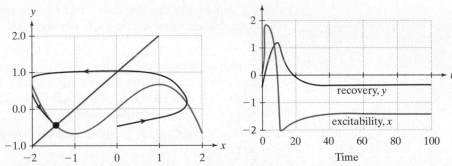

Figure 8.50 Solution plotted in the phase plane (left panel) and functions of time (right panel) for the case $I = 0$ and the initial condition $(x(0), y(0)) = (0, -0.5)$

For the initial condition $x(0) = -1$ and $y(0) = -0.5$, the phase-plane diagram suggests that the system gets less excited (i.e., x increases) and the recovery variable y increases. When x gets small enough, the recovery variable y also decreases. Hence, it looks like the system may not exhibit a large oscillation. Using technology to plot the solution corresponding to this initial condition in the phase plane (left panel in Figure 8.51) and against time (right panel in Figure 8.51) confirms this prediction.

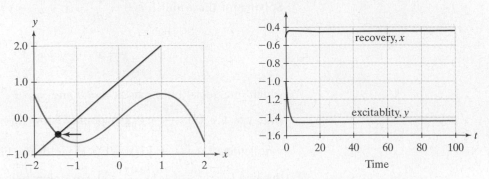

Figure 8.51 Solution plotted in the phase plane (left panel) and as functions of time (right panel) for the case $I = 0$ and the initial condition $(x(0), y(0)) = (-1, -0.5)$

b. With $I = 1$, $a = 0.1$, and $b = c = 1$, the model becomes

$$\begin{cases} \dfrac{dx}{dt} = x - \dfrac{x^3}{3} - y + 1 \\ \dfrac{dy}{dt} = 0.1(x + 1 - y) \end{cases}$$

Solving for the x nullcline $0 = \dfrac{dx}{dt} = x - \dfrac{x^3}{3} - y + 1$ yields the cubic curve $y = x - \dfrac{x^3}{3} + 1$. For points above this curve (i.e., $y > x - x^3$), $\dfrac{dx}{dt} < 0$. For points below this curve, $\dfrac{dx}{dt} > 0$. Solving for the y nullcline, $0 = \dfrac{dy}{dt} = 0.1(x + 1 - y)$ yields the line $y = x + 1$. For points above this line (i.e., $y > x + 1$), $\dfrac{dy}{dt} < 0$. For points below this line, $\dfrac{dy}{dt} > 0$. Putting this information together yields the phase-plane diagram depicted in Figure 8.52.

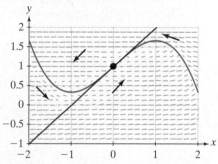

Figure 8.52 Phase-plane diagram for the case $I = 1$

This phase-plane diagram looks quite similar to what we found in part **b** (left panel in Figure 8.53). However, using technology to solve the system for the initial condition $x(0) = 0$, $y(0) = 0.5$ reveals very different long-term behavior; that is, the system undergoes repeated oscillations of excitability and recovery as illustrated in the right panel in Figure 8.53.

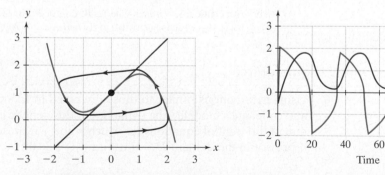

Figure 8.53 Solution plotted in the phase plane (left panel) and plotted as functions of time (right panel) for $I = 1$ and the initial condition $(x(0), y(0)) = (0, -0.5)$

c. The analysis in the case of $I = 2$ is almost identical to that in parts **a** and **b**. However, now the x nullcline has moved vertically one more unit; it is given by $y = x - \dfrac{x^3}{3} + 2$ (left panel in Figure 8.54). Using technology to solve the system for the initial condition $x(0) = 0$, $y(0) = 0.5$ reveals very different long-term behavior; that is, the system remains constantly in an excited state (right panel in Figure 8.54).

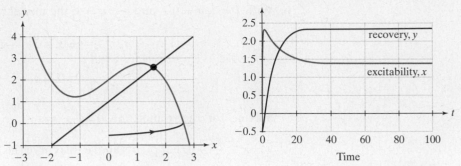

Figure 8.54 Solution plotted in the phase plane (left panel) and as functions of time (right panel) for $I = 2$ and the initial condition $(x(0), y(0)) = (0, -0.5)$

In a study that appeared in *Nature* in the year 2000, Timothy Gardner and colleagues reported construction of a genetic toggle switch in the bacterium *Escherichia coli*. They manipulated the genome of *E. coli* to create a genetic circuit consisting of two repressors and two constitutive promoters (Figure 8.55). Each promoter is inhibited by the repressor that is transcribed by the opposing promoter. Hence, each repressor in this circuit represses the activation of the other gene in this circuit. In Example 6 of Section 8.1, we introduced a model of a genetic circuit. Now we have the tools to analyze the dynamics of such genetic circuits.

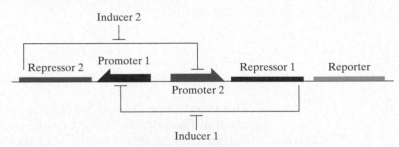

Figure 8.55 Schematic of a genetic toggle switch. In this schematic, repressor 1 inhibits transcription from promoter 1 and is induced by inducer 1. Repressor 2 inhibits transcription from promoter 2 and is induced by inducer 2.

Data Source: Timothy S. Gardner, Charles R. Cantor & James J. Collins (2000) Construction of a genetic toggle switch in *Escherichia coli*. *Nature* 403: 339–342.

Example 5   A genetic toggle switch

Using assumptions similar to those discussed in Example 6 in Section 8.1, Gardner and colleagues modeled the dynamics of this genetic circuit (Figure 8.55) with a system of differential equations for which x and y represent the concentrations of each repressor in the domain $D = \{(x, y) | x \geq 0, y \geq 0\}$:

$$\frac{dx}{dt} = \frac{a}{1 + y^c} - x$$

$$\frac{dy}{dt} = \frac{b}{1 + x^c} - y$$

The parameters $a > 0$ and $b > 0$ represent the rate at which each repressor is synthesized in the absence of the other repressor. The parameter $c \geq 1$ is known as the "cooperativity" of the repressors. We examine this model for $c = 4$, $b = 2$, and $a = 3$ or 10. For each of these values of a, plot the nullclines using technology and sketch the phase-plane diagram.

Solution   For the case $a = 3$, the nullclines are given by $x = \dfrac{3}{1 + y^4}$ and $y = \dfrac{2}{1 + x^4}$. Using technology to plot these nullclines, we get the blue and red curves shown in the left panel in Figure 8.56. We have

$$\frac{dx}{dt} < 0 \quad \text{if} \quad x > \frac{3}{1 + y^4}$$

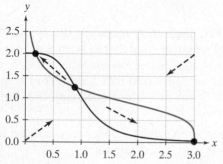 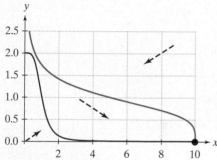

Figure 8.56 Phase-plane diagrams (nullclines and trajectory directions) for the equations of Gardner and colleagues for the cases $a = 3$ (left panel) and $a = 10$ (right panel), with $b = 2$ and $c = 4$ in both cases

and

$$\frac{dx}{dt} > 0 \quad \text{if} \quad x < \frac{3}{1 + y^4}$$

Similarly, we have

$$\frac{dy}{dt} < 0 \quad \text{if} \quad y > \frac{2}{1 + x^4}$$

and

$$\frac{dy}{dt} > 0 \quad \text{if} \quad y < \frac{2}{1 + x^4}$$

Hence, we get the direction arrows shown in the left panel in Figure 8.56. This phase-plane diagram suggests that the interior equilibrium is a saddle and the other two equilibria are stable nodes. Hence, if the initial concentration of either repressor is sufficiently large relative to the concentration of the other repressor, then the system goes to the equilibrium corresponding to a high concentration of the first repressor under consideration.

For the case $a = 10$, the nullclines are given by $x = \dfrac{10}{1 + y^4}$ and $y = \dfrac{2}{1 + x^4}$. Using technology to plot these nullclines, we get the blue and red curves shown in the right panel in Figure 8.56. We have

$$\frac{dx}{dt} < 0 \quad \text{if} \quad x > \frac{10}{1 + y^4}$$

and

$$\frac{dx}{dt} > 0 \quad \text{if} \quad x < \frac{10}{1 + y^4}$$

Similarly, we have

$$\frac{dy}{dt} < 0 \quad \text{if} \quad y > \frac{2}{1 + x^4}$$

and

$$\frac{dy}{dt} > 0 \quad \text{if} \quad y < \frac{2}{1 + x^4}$$

Hence, we get the direction arrows shown in the right panel in Figure 8.56. This phase-plane diagram suggests that the system goes to the equilibrium corresponding to a low concentration of repressor 2 and a high concentration of repressor 1 for all initial conditions.

We conclude this chapter by modeling disease outbreaks. Modeling epidemics using differential equation models has been a growing area of interest over the past 30 years with so-called SEIR models, where S, E, I, and R, represent susceptible, exposed, infected, and removed individuals. Four classic SEIR models are illustrated in Figure 8.57. Some of these models and the disease they best represent are as follows:

- SIR: measles and chickenpox, because individuals who are infected either die or recover with lifelong immunity

- SIS: gonorrhea, because individuals are usually susceptible soon after recovery

- SEIR: diseases with relatively long incubation periods, requiring incorporation of an E phase. Influenza and cold viruses typically have incubations periods of one to three days; measles, one to two weeks; and mononucleosis four to six weeks.

- SEIS: SIS processes that need to be modeled more accurately to account for the incubation period

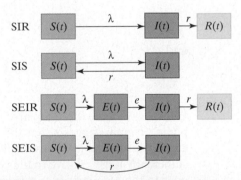

Figure 8.57 Four different idealized types of epidemics involving flows of state transitions among four epidemiological compartments—susceptible (S), exposed (E), infectious (I), and removed (R: dead or immune)—are depicted here in terns of the following rates of progression from one state to another: incident or force of infection rate (λ), progression of infectiousness (e), and recovery (or removal) rate (r).

In the next example, we study the SIR model that has been used to model seasonal cold and influenza epidemics, as well as outbreaks of emergent diseases such as SARS and H1N1.

Example 6   Schools and the common cold

The SIR model of the seasonal spread of the common cold (Figure 8.58) through a population of school children divides the population into three possible disease states: S, I, and R. These letters represent the proportion in the population of

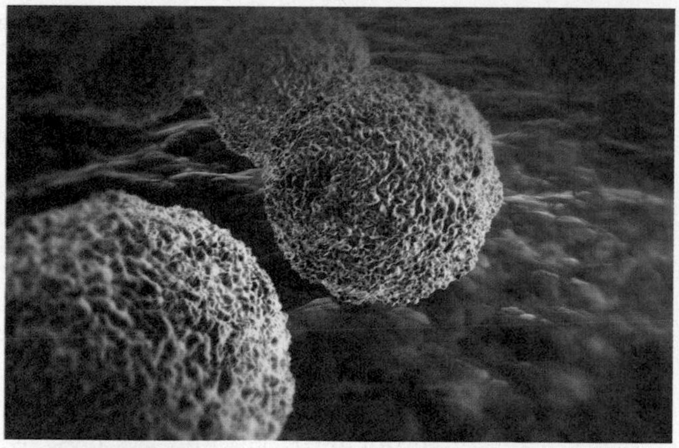

Figure 8.58 Electron micrograph image of the cold virus

susceptible, infectious, and recovered individuals, the latter with immunity for that season. If we assume that the rate of disease transmission per susceptible individual, a quantity known as the **force of infection**, is proportional to the proportion of infected, then the force infection equals βI, where β is a transmission rate parameter. If infected individuals recover at rate r, then disease dynamics are given by:

$$\begin{cases} \dfrac{dS}{dt} = -\beta SI \\[2mm] \dfrac{dI}{dt} = \beta SI - rI \\[2mm] \dfrac{dR}{dt} = rI \end{cases}$$

If $\beta = 1$, use the phase-plane method to infer the behavior of trajectories for the following two cases:

a. $r = \dfrac{1}{2}$ **b.** $r = \dfrac{3}{2}$

Next, answer this question:

c. In a school of 1000 individuals, if one infected individual comes to school on a Monday morning with a cold, what proportion of individuals will ultimately get a cold in case **a** before the epidemic burns itself out?

Solution   Since S, I, and R are proportions of the total school population, individuals can only be in one of the three possible disease states, implying that $S + I + R = 1$. Thus, we can use the relationship $R = 1 - S - I$ to calculate R in terms of S and I. Also, we carry out a phase-plane analysis in the domain $D = \{(S, I) | 0 \le S \le 1, 0 \le I \le 1\}$ of the first two equations since they involve only the variables S and I. To answer parts **a** and **b**, we carry out the first two steps of the phase-plane method for the general model.

Step 1 of the phase-plane method applied to the first two equations yields:

- Nullclines for S:

$$SI = 0 \;\Rightarrow\; S = 0 \;\text{ and }\; I = 0$$

- Nullclines for I:

$$SI - rI = 0 \;\Rightarrow\; I = 0 \;\text{ and }\; S = r$$

The last equation yields a biologically relevant nullcline for the case $r \leq 1$ because S, being a proportion, must satisfy $0 \leq S \leq 1$.

a. For the case $r = \dfrac{1}{2}$, construction of the nullclines yields the left panel in Figure 8.59, which indicates that all points on the S axis are equilibria (i.e., $I = 0$ and $0 \leq S \leq 1$). Furthermore, the nullclines determine two distinct regions separated by the line $S = \dfrac{1}{2}$. In the region to the left of $S = \dfrac{1}{2}$, $\dfrac{dS}{dt} < 0$ and $\dfrac{dI}{dt} < 0$. In the region to the right of $S = \dfrac{1}{2}$, $\dfrac{dS}{dt} < 0$ and $\dfrac{dI}{dt} > 0$. A solution starting out in the right-hand bottom corner (S almost 1 and I slightly above 0) is shown to rise, cross the isocline $S = \dfrac{1}{2}$ horizontally, and then approach the $I = 0$ axis.

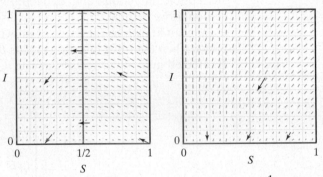

Figure 8.59 Left and right panels correspond to the cases $r = \dfrac{1}{2}$ and $r = \dfrac{3}{2}$ of the SIR model from Example 6. The S nullclines are shown is red and the non-zero I nullcline in blue (left panel only). Since $I = 0$ is also an I nullcline, all points on the S axis are equilibira.

b. For the case $r = \dfrac{3}{2}$, construction of the nullclines yields the right panel in Figure 8.59, which (as in part **a**) indicates that all points on the S axis are equilibria. Unlike part **a**, only one distinct region exists. For all states with $I > 0$ and $S > 0$, $\dfrac{dS}{dt} < 0$ and $\dfrac{dI}{dt} < 0$. Therefore, the proportion of infected individuals in the population always decreases. There are no outbreaks of the cold.

c. To calculate S_∞, assume there exists a function f such that $S(t) = f(R(t))$ for all t. By the chain rule, $\dfrac{dS}{dt} = \dfrac{df}{dR}\dfrac{dR}{dt}$. Therefore, if $\beta = 1$ and $I > 0$, we get

$$\frac{df}{dR} = \frac{\dfrac{dS}{dt}}{\dfrac{dR}{dt}} = -\frac{f(R)I}{rI} = -\frac{f}{r}$$

If we use the initial condition that at $t = 0$, $R = 0$, and $S = S_0$, then we integrate to obtain the solution

$$f(R) = S_0 e^{-R/r}$$

If we make use of the relationship $R = 1 - S - I$, then we obtain

$$S(t) = S_0 e^{-(1-S(t)-I(t))/r}$$

Taking the limit as $t \to \infty$ and using the fact that $\lim\limits_{t \to \infty} I(t) = 0$, we obtain the implicit expression for S_∞:

$$S_\infty = S_0 e^{(S_\infty - 1)/r}$$

For the case $I(0) = 1/1000$ and $R(0) = 0$, we have $S(0) = 0.999$. Thus, for $r = \dfrac{1}{2}$, we can use technology to solve the following equation for S_∞ to obtain

$$S_\infty = 0.999 e^{(S_\infty - 1)/0.5} \quad \Rightarrow \quad S_\infty = 0.203$$

In other words, at the end of this outbreak, $1 - S_\infty = 79.7\%$ of the population caught this strain of the common cold. ■

PROBLEM SET 8.5

Level 1 DRILL PROBLEMS

In Problems 1 to 16, draw phase-plane diagrams for given models and domains D, showing all nullclines and equilibria, and plot the general direction arrows that solutions will follow in regions bounded by the nullclines.

1. $D = \{(x, y) | x \geq 0, y \geq 0\}$
$$\begin{cases} \dfrac{dx}{dt} = 0.2x - 0.5xy \\ \dfrac{dy}{dt} = 0.3xy - 0.05y \end{cases}$$

2. $D = \{(x, y) | x \geq 0, y \geq 0\}$
$$\begin{cases} \dfrac{dx}{dt} = 0.2x - 0.3x^2 - 0.5xy \\ \dfrac{dy}{dt} = 0.3xy - 0.15y \end{cases}$$

3. $D = \{(x, y) | x \geq 0, y \geq 0\}$
$$\begin{cases} \dfrac{dx}{dt} = 0.2x - 0.5x^2 - 0.5xy \\ \dfrac{dy}{dt} = 0.3xy - 0.15y \end{cases}$$

4. $D = \{(x, y) | x \geq 0, y \geq 0\}$
$$\begin{cases} \dfrac{dx}{dt} = 0.2x - 0.3x^2 - \dfrac{0.5xy}{1 + 3x} \\ \dfrac{dy}{dt} = \dfrac{0.3xy}{1 + 3x} - 0.06y \end{cases}$$

5. $D = \{(x, y) | x \geq 0, y \geq 0\}$
$$\begin{cases} \dfrac{dx}{dt} = 0.2x - 0.3x^2 - \dfrac{0.5xy}{1 + 3x} \\ \dfrac{dy}{dt} = \dfrac{0.3xy}{1 + 3x} - 0.07y \end{cases}$$

6. $D = \{(x, y) | x \geq 0, y \geq 0\}$
$$\begin{cases} \dfrac{dx}{dt} = 0.2x - 0.2x^2 - \dfrac{0.5xy}{1 + 4x} \\ \dfrac{dy}{dt} = \dfrac{0.3xy}{1 + 4x} - 0.03y \end{cases}$$

7. $D = \{(x, y) | x \geq 0, y \geq 0\}$
$$\begin{cases} \dfrac{dx}{dt} = -0.25x + 0.05y + 120 \\ \dfrac{dy}{dt} = 0.1x - 0.05y \end{cases}$$

8. $D = \{(x, y) | x \geq 0, y \geq 0\}$
$$\begin{cases} \dfrac{dx}{dt} = 0.2x - 0.3x^2 - 0.2xy \\ \dfrac{dy}{dt} = 0.3y - 0.5y^2 - 0.3xy \end{cases}$$

9. $D = \{(x, y) | x \geq 0, y \geq 0\}$
$$\begin{cases} \dfrac{dx}{dt} = 0.2x - 0.2x^2 - 0.4xy \\ \dfrac{dy}{dt} = 0.3y - 0.3y^2 - 0.5xy \end{cases}$$

10. $D = \{(x, y) | x \geq 0, y \geq 0\}$
$$\begin{cases} \dfrac{dx}{dt} = 0.2x - 0.2x^2 - 0.4xy \\ \dfrac{dy}{dt} = 0.3y - 0.5y^2 - 0.2xy \end{cases}$$

11. $D = \{(x, y) | x \geq 0, y \geq 0\}$
$$\begin{cases} \dfrac{dx}{dt} = 0.2x(1 - 0.6x(1 + y)) \\ \dfrac{dy}{dt} = 0.3y(1 - 0.8y(1 + x)) \end{cases}$$

12. Defined on the whole plane
$$\begin{cases} \dfrac{dx}{dt} = x - \dfrac{x^3}{5} - y \\ \dfrac{dy}{dt} = 0.1(x + 1 - 2y) \end{cases}$$

13. Defined on the whole plane
$$\begin{cases} \dfrac{dx}{dt} = x - \dfrac{x^3}{6} - \dfrac{y}{2} \\ \dfrac{dy}{dt} = 0.3(x + 1 - 2y) \end{cases}$$

14. Defined in the positive quadrant

$$\begin{cases} \dfrac{dx}{dt} = \dfrac{1}{1+y} - x \\[2mm] \dfrac{dy}{dt} = \dfrac{1}{1+x} - y \end{cases}$$

15. Defined in the positive quadrant

$$\begin{cases} \dfrac{dx}{dt} = \dfrac{0.8}{1+y^2} - x \\[2mm] \dfrac{dy}{dt} = \dfrac{0.4}{1+x^2} - y \end{cases}$$

16. Defined in the positive quadrant

$$\begin{cases} \dfrac{dx}{dt} = \dfrac{1.4}{1+3y^2} - x \\[2mm] \dfrac{dy}{dt} = \dfrac{1}{1+3x} - y \end{cases}$$

Level 2 APPLIED AND THEORY PROBLEMS

17. Phage therapy is the use of bacteriophages to treat pathogenic bacteria in a way that is much more specific than antibiotics, because phages are usually species-specific killers. Genetic engineering techniques may be used in the future to design phages able to effectively treat diseases for which current antibiotics are too ineffective to control the growth of bacterial populations—such as *Campylobacter*, *Escherichia* (e.g., *E. coli*), and *Salmonella*—that infect our food supply. For phages to be effective, they need to destroy host bacterial cells at a fast enough rate to drive the bacterial population to zero. Suppose a bacterial species grows logistically on its host at a rate given by the equation $\dfrac{dx}{dt} = 0.3x(1-x)$, where time t is measured in days, the variable $x(t)$ is the density of the population at time t, and $x = 1$ is the population density at which the bacteria stop growing due to overcrowding on the host. Assume a species of phage attacks bacteria at the rate of bx per phage, and each attack produces 100 new phages (also called the burst size, because the host cell bursts when it has this number of phages). Write a model for this bacteria-phage interaction, assuming that free phage die at a per capita rate $c > 0$. Construct a phase-plane diagram for the case $b = 0.5$ and $c = 30$ and discuss the long-term behavior of this system. If b ranges over the values $[0, 1]$, what is the corresponding range of values for c such that c will just be sufficient (i.e., its smallest) to ensure that the phage will drive the bacterial population to zero in the long run?

18. A population of frogs inhabits two lakes. In lake 1, the population grows logistically at a per capita growth rate of 0.3 (time measured in months) at the lowest densities, but with the effects of density dependence causing the population to saturate at the level of 10,000 frogs. In lake 2, the population declines exponentially at a per capita rate of 0.05. Frogs in lake 1 migrate to lake 2 at a per capita rate of 0.1 but frogs in lake 2 never migrate. Write a model of this system and draw the phase-plane diagram for this model showing all nullclines and equilibria. By plotting the general direction that trajectories will follow in regions bounded by the nullclines, deduce the long-term population dynamics.

19. A population of walruses in the Arctic is harvested by an indigenous human population. The walrus population growth equation is given by $\dfrac{dx}{dt} = ax\left(\dfrac{x}{b} - 1\right)\left(1 - \dfrac{x}{K}\right)$, where $a = 0.1$, $b = 500$ tons, and $K = 2,000$ tons. The rate at which the local population is able to harvest walruses is given by $0.0001xy$, where y is the amount of effort expended on the hunt (boats, snowmobiles, etc.). The cost for each unit of effort y is given by a parameter c, and effort will grow or decline at a rate that is equal to the difference between the harvest rate $0.0001xy$ and the total cost rate cy for the harvest. Write a model of this system for the general value c. If the value of $c = 0.04$, then draw the phase-plane diagram for this model showing all nullclines and equilibria. By plotting the general direction that trajectories will follow in regions bounded by the nullclines, deduce the long-term behavior of this population. Repeat the exercise for the cases where $c = 0.1$ and $c = 0.2$. In comparing these three cases, what do you conclude?

20. Suppose a population of wild herbivores (e.g., deer or buffalo), at density N, has a density-independent per capita birthrate a and a density-dependent per capita death rate $bN(1 + cN)$. Write an equation for the growth rate of N and determine the value of the equilibrium in terms of a and b and the conditions under which it is stable. If a disease, such as bovine tuberculosis, is present in the population and the density of infected individuals is I, then the density of susceptible individuals is $N - I$. If the disease transmission rate is proportional to the product of the density of susceptible and infected individuals, then it is given by a term $\beta(N - I)I$, where β is the transmission rate parameter. Assume that the birthrate of infected individuals is the same as that of susceptible individuals, but that the death rate of infected individuals is subject to an additional disease-induced per capita death rate d. Write a model for the dynamics of N and I. For the case $a = 0.3, b = 0.15, c = 1, d = 0.1$, and $\beta = 0.5$ draw the phase-plane diagram for this model showing all nullclines and equilibria. By plotting the general direction that trajectories will follow in regions bounded by the nullclines, show that the long-term behavior of this population is to approach an endemic disease equilibrium (i.e., an equilibrium for which $I > 0$). At this equilibrium, what is the prevalence of the disease and to what extent has the disease suppressed the population below its carrying capacity? (Hint: The equation for N with disease is the same as the one without disease, except that an additional disease-induced mortality term involving I only is included. The equation for I includes transmission, natural mortality, and disease-induced mortality).

21. Consider the following somewhat different version of the FitzHugh-Nagumo model presented in Example 4:

$$\begin{cases} \dfrac{dx}{dt} = -x(x-a)(x-1) - y \\ \dfrac{dy}{dt} = b(x - cy) \end{cases}$$

Draw the isoclines and direction arrows in the phase plane $D = \{(x, y)| -2 \le x \le 2, -2 \le y \le 2\}$ for the case $a = 0.5$ and $c = 1$. Note that the isocline equations do not depend on b. Use technology to generate solutions in the phase plane originating at the four points $(-2, -2), (2, -2), (2, 2)$, and $(-2, 2)$ for the two cases $b = 0.5$ and $b = 0.05$. From this comparison, what do you deduce about the role of b in these equations?

22. Consider the genetic toggle switch model

$$\frac{dx}{dt} = \frac{2}{1 + y^c} - x$$
$$\frac{dy}{dt} = \frac{2}{1 + x^c} - y$$

presented in Example 5 in which the variables x and y represent the concentrations of the two different repressors in the nonnegative quadrant. Use technology to draw the isoclines, direction arrows, and solutions originating at the four points $(0.1, 0), (3, 0), (2.9, 3)$, and $(0, 3)$ in the phase plane $D = \{(x, y)|0 \le x \le 3, 0 \le y \le 3\}$ for the cases $c = 1$ and $c = 4$. When you compare these two cases, what do you deduce about the role of c in the model?

23. *(Endemic disease I)* In Example 6, we modeled the dynamics of the spread of a cold and found that the infection ultimately stops spreading. The reason for this epidemic burnout is that there is no continuous input of susceptible individuals in the population. To address this limitation, we consider the same model augmented to include a constant rate M at which susceptible individuals enter the population and a

per capita rate d at which infected individuals leave the population (e.g., due to natural mortality). In this case, if we set $M = 1$ (i.e., we can interpret the values of S and I relative to the size of M, so generality is not lost so long as we measure S and I in units of M) the equations for S and I become

$$\begin{cases} \dfrac{dS}{dt} = 1 - \beta SI - dS \\ \dfrac{dI}{dt} = \beta SI - rI - dI \end{cases}$$

a. What are the general equations for the nullclines in terms of the model parameters β, b, and r?

b. Solve for the equilibrium solutions and specify the conditions needed for an endemic equilibrium to exist.

c. Produce a phase-plane plot for the case $\beta = 0.1$, $d = 0.02, r = 0.3$ (either by hand or using technology).

24. *(Endemic disease II)* As in the previous problem, we address the limitation of epidemic burnout in Example 6. This time we assume that the rate at which susceptible individuals enter the population is given by a per capita birthrate b, where both susceptible and infected individuals contribute equally to the birth process and, as in the previous problem, both are subject to a per capita natural death rate d. This model has the form

$$\begin{cases} \dfrac{dS}{dt} = b(S + I) - \beta SI - dS \\ \dfrac{dI}{dt} = \beta SI - rI - dI \end{cases}$$

a. What are the general equations for the nullclines in terms of the model parameters β, b, and r?

b. Solve for the equilibrium solutions and specify the conditions needed for an endemic equilibrium to exist.

c. Produce a phase-plane plot for the case $\beta = 0.1$, $b = 0.1, d = 0.02, r = 0.3$ (either by hand or using technology).

25. *(Fugitive species)*. In a classic paper, R. Levins and D. Culver, 1971. Regional Coexistence of Species and Competition between Rare Species *PNAS* 68:1246-1248, modeled two species competing for space. Let x and y denote the fraction of space occupied by species 1 and species 2, respectively. These species exhibit a trade-off in which one species, say species 1, is the dominant competitor and can always dislodge individuals of species 2 by colonizing a site. The other species is the better colonizer, as it produces more offspring per unit time. For $i = 1, 2$, let b_i and d_i be the per capita birth and death rates, respectively, of species i. Under these assumptions,

the model of their growth and interaction dynamics is

$$\begin{cases} \dfrac{dx}{dt} = b_1 x(1 - x) - d_1 x \\ \dfrac{dy}{dt} = b_2 y(1 - x - y) - d_2 y - b_1 xy \end{cases}$$

a. For the case $d_1 = d_2 = 1, b_1 = 2$, and $b_2 = 2.25$, use the phase-plane method to determine the long-term behavior of the model.

b. For the case $d_1 = d_2 = 1, b_1 = 2$, and $b_2 = 5$, use the phase-plane method to determine the long-term behavior of the model.

CHAPTER 8 REVIEW QUESTIONS

1. Suppose $z = f(x, y) = \dfrac{y^2}{9} - \dfrac{x^2}{16}$.

a. Plot the contour for $c = 0$.

b. Plot the contour for $c = 4$.

c. Find $f(2, 3)$, $f(4, -9)$.

d. Sketch the surface.

2. Match each equation with a graph shown in A–D.

a. $z = -\dfrac{10y}{x^2 + y^2 + 1}$

b. $z = (\sin x \sin y)/(xy)$ for $x, y \neq 0$

c. $z = y^2$

d. $z = e^{-x^2 - y^2}$

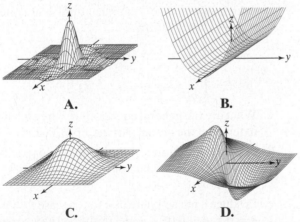

A.   **B.**

C.   **D.**

3. Match each graph with its contour shown in A–D.

a. Contour $z = 0.1$

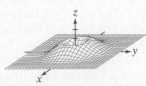

b. Contour $z = 0.5$

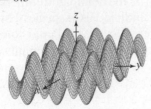

c. Contour $z = 0.6$;

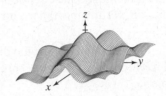

d. Contour $z = 1$

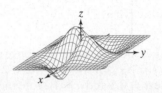

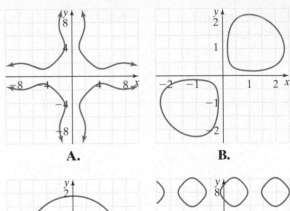

A.   **B.**

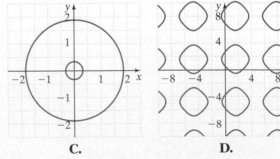

C.   **D.**

4. Draw the requested contour for the given graph.

a. Equation $z = \dfrac{1}{e^{x^2 + y^2}}$; contour $z = 0.1$

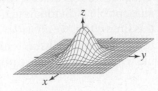

b. Equation $z = -x - y + 1$; contour $z = 1$

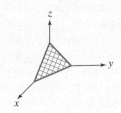

c. Equation $z = \dfrac{x^2}{4} - \dfrac{y^2}{9}$; contour $z = 1$

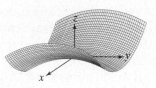

d. Equation $z = x^3 - 3xy^2$; contour $z = 0$

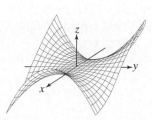

5. Write the two equations in matrix form and solve for x and y.

a. $-\dfrac{2x}{3} + \dfrac{y}{3} = -\dfrac{2}{5}$ and $x + \dfrac{y}{3} = 1$

b. $3x - \dfrac{5y}{4} = -1$ and $\dfrac{2x}{3} + y = -\dfrac{2}{3}$

6. Use Theorem 8.1 to solve the matrix equation
$\mathbf{v} = A\mathbf{u}$ for the vector $\mathbf{u}$ where $A = \begin{pmatrix} -6 & 3 \\ -5/2 & 3/4 \end{pmatrix}$
and $v = \begin{pmatrix} -1 \\ -2 \end{pmatrix}$

7. Consider the economy of a country that depends on two goods, A and B. Production of one unit of good A requires 0.5 units of good A and 0.2 units of good B. Production of one unit of good B requires 0.4 units of good A and 0.1 units of good B. If the total demand for these two goods is 7 units of good A and 4 units of good B, then find the amount of each good that needs to be produced to meet this demand without any surplus.

8. Find the determinant of the matrix $B = \begin{pmatrix} 1/2 & 2/3 \\ 1/4 & 1/3 \end{pmatrix}$.

9. Find the inverse of the matrix $C = \begin{pmatrix} 1/2 & 2/3 \\ 1/4 & -1/3 \end{pmatrix}$.

10. Iterate the dynamical solution $\mathbf{u}_n = \begin{pmatrix} x_n \\ y_n \end{pmatrix}$ for $n = 0, 1, \dots, 6$ of the equation $\mathbf{u}_{n+1} = A\mathbf{u}_n$ using the initial vector $\mathbf{u}_0 = \begin{pmatrix} 0 \\ 1 \end{pmatrix}$ and the matrix
$$A = \begin{pmatrix} 0.9 & 0.5 \\ 0.1 & 0.5 \end{pmatrix}.$$

11. Suppose $A = \begin{pmatrix} 1 & 1 \\ 1 & 3 \end{pmatrix}$ with eigenvectors $\begin{pmatrix} -1 + \sqrt{2} \\ 1 \end{pmatrix}$ and $\begin{pmatrix} 1 + \sqrt{2} \\ -1 \end{pmatrix}$; find the eigenvalues.

12. Find all eigenvalues and eigenvectors for the matrix
$$A = \begin{pmatrix} 1/2 & 1/2 \\ 1/2 & 1 \end{pmatrix}.$$

13. Consider a generalization of the loggerhead model discussed in Example 7 of Section 8.3. Let x_n and y_n be the population abundances of juveniles and adults in year n. A generalized model of the loggerhead population dynamics is given by $\mathbf{u}_{n+1} = A\mathbf{u}_n$, where $A = \begin{pmatrix} 0.5 & 80 \\ 0.001 & s \end{pmatrix}$, $\mathbf{u}_n = \begin{pmatrix} x_n \\ y_n \end{pmatrix}$, and s is the probability that an adult survives to the next year. Find the minimal value of s that guarantees that population will increase over time.

14. Use Theorem 8.4 to find the general solution to these equations in terms of two arbitrary constants a and b:
$$\begin{cases} \dfrac{dx}{dt} = x + 4y \\ \dfrac{dy}{dt} = x - 2y \end{cases}$$

15. Find the particular solution of the system of linear differential equations $\dfrac{d\mathbf{u}}{dt} = A\mathbf{u}$ with $A = \begin{pmatrix} 1 & 1/3 \\ 2/3 & 2/3 \end{pmatrix}$ and $\mathbf{u}(0) = \begin{pmatrix} 2 \\ -1 \end{pmatrix}$.

16. Find the general solution to $\dfrac{d\mathbf{u}}{dt} = A\mathbf{u} + \mathbf{b}$ where $A = \begin{pmatrix} -1 & 0 \\ 4 & -3 \end{pmatrix}$ and $\mathbf{b} = \begin{pmatrix} 5 \\ 0 \end{pmatrix}$.

17. The equations
$$\begin{cases} \dfrac{dS}{dt} = 1 - 0.1SI - 0.05S \\ \dfrac{dI}{dt} = 0.1SI - 0.25I \end{cases}$$
represent an endemic disease process with these characteristics: individuals enter the population at a rate of 1 susceptible individual per unit time; the disease transmission rate is 0.1 susceptible individuals infected per infectious individual per unit time; individuals with the disease die at a rate of 0.05 and

recover to become immune at a rate of 0.2 individuals per infected individual per unit time.

a. What are the equations for the nullclines of this system?

b. Solve for the biologically relevant equilibria of this system.

c. Provide a rough sketch of a phase-plane plot of these equations indicating directions in which trajectories cross the nullclines and move in the phase plane.

18. A patient initially has concentrations 50 mg/L and 70 mg/L of morphine, respectively, in her blood (concentration $x(t)$) and cerebrospinal fluid (concentration $y(t)$). Assume that the per capita rates (units L^{-1} per hour) at which the drug is transferred across the blood-brain barrier from blood to cerebrospinal fluid and vice versa are 0.4 and 0.2, respectively, and that the per capita rates of metabolism of these drugs within blood and the cerebrospinal fluid are 0.1 and 0.2, respectively. Write a model of this process, find the equilibrium solution, and solve for the particular solution for the given initial conditions.

19. A patient initially has no morphine in her body but is receiving a continuous infusion of 1 mg/hour into her bloodstream. The per capita rates at which the drug is transferred across the blood-brain barrier from blood to cerebrospinal fluid and vice versa are $a = 1/3$ and $b = 1/4$. The per capita rates of metabolism of these drugs within these pools are $c = 1/4$ and $d = 1/8$, respectively. Find the equilibrium solution and the particular solution to this problem. Use the particular solution to calculate the amounts of morphine in the blood and the cerebrospinal fluid after one hour.

20. Foxes are introduced to an island that is being overrun by rabbits. Prior to the introduction of the foxes, the rabbits, with density represented by $x(t)$, grew at the lowest densities at an exponential rate of 0.3 per year but leveled off at a carrying capacity of ten rabbits per hectare. The rate at which foxes, with density represented by $y(t)$, preyed on rabbits was 0.2 per rabbit per year per fox, so that the total harvest rate of rabbits was $h(t) = 0.2x(t)y(t)$. Assume foxes gained numbers at a total rate of $0.15h(t)$ per year but died at the per capita rate of 0.15 per year. Write a model for this rabbit-fox interaction. Construct a phase-plane diagram in the domain $D = \{(x, y)|0 \le x \le 10, 0 \le y \le 3\}$ for this interaction and discuss the long-term behavior of this system.

GROUP PROJECTS

Seeing a project through on your own, or working in a small group to complete a project, teaches important skills. The following projects provide opportunities to develop such skills.

Project 8A   Frequency-dependent versus Density-dependent Disease Transmission*

(Refer to Project 6A.)

Consider a population in which N and I represent, respectively, the population density and the density of infected individuals. Let $f(N)$ and $g(N)$ represent the per capita birth and nondisease death rates as a function of population density N, and let $d \ge 0$ and $r \ge 0$ represent, respectively, the disease-induced per capita death rate and the recovery rate, both of which are assumed to be constants. Finally, let $T(N, I)$ represent the pathogen transmission rate. Then, a relatively simple dynamical model of disease caused by the pathogen in question is given by the following systems of equations defined in the domain $D = \{(N, I)|N \ge 0, 0 \le I \le N\}$:

$$\frac{dN}{dt} = Nf(N) - Ng(N) - dI$$

$$\frac{dI}{dt} = T(N, I) - Ig(N) - dI - rI$$

In these simple dynamical models, pathogen transmission resulting from casual contact between infected (I) and susceptible ($S = N - I$) individuals, such as in tuberculosis or influenza, is assumed to be proportional to the density of infected and susceptible individuals, in which case the simplest expression—also known as *mass*

---

*See W. M. Getz and J. Pickering, "Epidemiological Models: Thresholds and Population Regulation," *Am. Nat.* 121(1983): 892–898.

action transmission—is given by $T(N, I) = \beta(N - I)I$. On the other hand, pathogen transmission resulting from behavioral interactions that are independent of density—such as in HIV and other sexually transmitted diseases—is dependent on the proportion of infected individuals that each susceptible individual interacts with at some socially determined rate. In this case the simplest expression—also known as *frequency-dependent*

transmission—is given by $T(N, I) = \beta(N - I)\dfrac{I}{N}$.

Now do the following:

1. Compare the dynamical behavior of solutions to the model for the two types of transmission under the assumption that the per capita birth and death rates are simply the respective constants $f(N) = a > 0$ and $g(N) = b > 0$ that also satisfy $a > b$. Cast your comparison in terms of the qualitatively different situations portrayed in phase-plane diagrams that

arise for different relationships among the various parameters (e.g., see how qualitatively different phase-plane situations are compared for the Lotka-Volterra model discussed in Example 3 of Section 8.5), paying particular attention to the key role that the quantity $\dfrac{d + b + r}{\beta}$ plays in the mass action (or density-dependent) transmission case.

2. How does the behavior portrayed in the phase-plane analysis considered in part **a** change if the death rate has the form $g(N) = b + cN$.

3. How does the behavior portrayed in the phase-plane analysis considered in part **a** change if the transmission rate is generalized to the form $T(N, I) = \dfrac{\beta(N - I)I}{\alpha + (1 - \alpha)N}$ for $0 \leq \alpha \leq 1$. Note that this form includes the mass action ($\alpha = 1$) and frequency-dependent forms ($\alpha = 1$) as special cases.

Project 8B   Population Dynamics of Consumers Subject to Starvation

Consider a consumer population at biomass density $x(t)$, at time t, growing on a resource with the level of resource measured by R (this could be trees extracting photons from sunlight, sheep or grasshoppers eating grass, wolves eating elk, lobsters scavenging detritus on the floor of the ocean, mushrooms growing on a pile of dead leaves, and so on). The growth or decline rate of x depends on the per capita rate at which individuals in the population can ingest resources versus the demand their bodies have for resources due to the physiological processes of respiration, transpiration, excretion, or any other process that leads to loss of biomass. Additionally, biomass is also lost from the population as a result of individuals dying. If the per capita rate of ingestion is denoted by the function $g(R, x)$, the per capita loss rate from physiological processes is simply a constant ℓ, and the per capita death rate is given by m, then the biomass growth rate of the population is $\dfrac{dx}{dt} = (g(R, x) - \ell - m)x$, where R itself may be constant or varying over time. (Note that although we talk about per capita rates, if the biomass is not partitioned into individuals, then the rates are interpreted as per-unit biomass.) If $g(R, x) > \ell$, then individuals are more than meeting their basal metabolic needs and can grow. On the other hand, if $g(R, x) < \ell$, then individuals are not meeting their physiological needs and their condition will likely start to deteriorate as starvation sets in. Suppose we use a variable y to denote the average level of starvation in individuals in the population, then if $g(R, x) - \ell < 0$ we further assume that the level of starvation increases. A simple model would be to

assume that the rate of increase in the starvation level of each individual is proportional to the size of the deficit $g(R, x) - \ell < 0$, where $s > 0$ is the constant of proportionality for this starvation process. Equally, we assume that rate of recovery from starvation of each in individual in the population is proportional to the size of the surplus $g(R, x) - \ell > 0$, where $r > 0$ is the constant of proportionality for the recovery processes. In this case, the variable y representing the average level of starvation of individuals in the population, with $y = 0$ corresponding to the state of no starvation, can be modeled by the equations $\dfrac{dy}{dt} = -s(g - \ell)$ when $g < \ell$ (i.e., y is increasing) and $\dfrac{dy}{dt} = -r(g - \ell)$ when $g > \ell$ and $y > 0$ (i.e., y is decreasing). If we assume that the death rate m depends on y, and in particular $m = \dfrac{d}{1 - y}$ for $d > 0$ so that all individuals must die as $y \to 1$ (since in this case $m(y) \to \infty$), then the model has this form:

$$\frac{dx}{dt} = \left(g(R, x) - \ell - \frac{d}{1 - y} \right) x$$

$$\frac{dy}{dt} = \begin{cases} -r(g(R, x) - \ell) & \text{if } g > \ell \text{ and } y > 0 \\ 0 & \text{if } g > \ell \text{ and } y = 0 \\ -s(g(R, x) - \ell) & \text{if } g < \ell \end{cases}$$

for $D = \{(x, y) | x \geq 0, 0 \leq y < 1\}$

Use phase-plane methods to discuss the behavior of this system for the case $g(R, x) = \dfrac{aR}{b + cx + R}$ in terms of the relative values of the parameters $a > 0, b > 0, c > 0,$ $d > 0, \ell > 0, r > 0,$ and $s > 0$ and the background constant level of the resource $R > 0$ (i.e., relative to the units of x). How would your conclusions be altered if the population were harvested at a constant rate $h > 0$, and also at a per capita rate vx? Can you identify any systems in the literature to which you can apply this model? If so, try and apply the model by selecting values for the parameters that make sense for the system under consideration. In a situation where there are two seasons each year—for example, a wet season and a dry season—discuss what you think would happen if R alternated between a low value in the dry season and a high value in the wet season. (*Low* means that in the long run the population would crash, and *high* means that in the long run the population would be sustained.)

Project 8C   Spread of Venereal and Vector-borne Diseases

The increasing incidence of sexually transmitted diseases—such as gonorrhea, chlamydia, AIDS, and syphilis—is a major health problem in both developed and developing countries. In the United States, for example, more than 2 million people contract gonorrhea annually. Among reportable communicable diseases, its incidence is far greater than the combined totals of syphilis, measles, tuberculosis, hepatitis, plus others. For this reason, we focus on modeling gonorrhea, but many of the principles developed here apply equally well to other sexually transmitted diseases. Gonorrhea takes three to seven days to incubate and can be cured by the use of antibiotics; there is no evidence that a person ever develops immunity.

- In the first phase of this project, consider the effects of gonorrhea in a purely heterosexual population. Let x denote the fraction of promiscuous men who are infected with gonorrhea. Let y be the fraction of promiscuous women who are infected with gonorrhea. Put together an ordinary differential equation (ODE) model that describes how x and y vary over time using the following assumptions:

   Assumption 1: The population density of promiscuous individuals remains fixed over time, say at N individuals per square mile.

   Assumption 2: Only promiscuous members of the population are susceptible to being infected with gonorrhea.

   Assumption 3: A promiscuous man or woman engages in sexual activities with a promiscuous member of the opposite sex at a rate proportional to the density of promiscuous members of that sex.

   Assumption 4: Once infected by gonorrhea, women and men can be cured by antibiotics within (on average) c and d days, respectively.

Be sure to state any other assumptions that you are using to put together your model. Provide a justification or rationale for all of those assumptions.

- Analyze the model. In particular, answer the following questions: What are the equilibrium points of the model you developed? How do the phase-plane diagrams depend on the parameters? For what parameter combinations will there be a continuous epidemic (i.e., an endemic infection)? Does the model ever exhibit cyclic dynamics?

- Interpret and discuss your results. What advice would you give to public health officials who wished to stem a gonorrhea epidemic in an affluent country like the United States or in a relatively affluent city like Hong Kong?

- Develop a less specific model than you already have, say of the form $\dfrac{dx}{dt} = g(x, y)$, $\dfrac{dy}{dt} = h(x, y)$. What assumptions should you place on the form of g and h to have a reasonable model? What can you say about the dynamics of this model (e.g., does it ever exhibit cyclic dynamics)?

- Can you apply any of the previous ideas to diseases that require two hosts? An example is malaria, which is transmitted by mosquitoes.

- Discuss the strengths and weaknesses of your models. Build a more realistic model by incorporating more details into the model. If time permits, try simulating this model.

Project 8D   Glucose-Insulin Dynamics

The concentration of glucose in the blood is regulated by the hormones insulin and glucagon. These hormones are released from clusters of cells in the pancreas known as the islets of Langerhans. In these islets, there are two types of cells, alpha and beta cells. They are sensitive to the glucose concentration in the blood and respond to blood glucose levels by increasing or decreasing the rate of insulin and glucagon production as necessary. As glucose levels increase, the beta cells get excited and respond by increasing the rate of insulin production. Insulin lowers the glucose concentration in the bloodstream in two ways. First, it causes cells in the body to increase the rate at which they utilize glucose. Second, it causes the liver to absorb more glucose and store it as glycogen. On the other hand, as glucose levels fall, the alpha cells respond by increasing glucagon levels. Glucagon causes the liver to transform its stored glycogen into glucose and to release it into the bloodstream.

Normal glucose levels range from 50 to 170 mg per 100 ml. Abnormally high or low glucose levels can cause severe damage to the body and may result in death. The disorder *diabetes mellitus* is characterized by high glucose levels. In type I diabetes, a person's beta cells are destroyed and insulin must be provided to the body via external means. In type II diabetes, a person's cells become less sensitive to insulin and more insulin is required to maintain normal glucose levels. Some diabetes treatments involve daily injections of insulin.

The treatment of diabetes relies on understanding the dynamics of glucose and insulin. In this project you will develop a dynamical model of glucose-insulin concentrations in the body.

- Consider the model developed in 1960 by Bolie. It contains four unspecified functions that depict rates of input and removal of glucose and insulin from the system. The following variables were identified as relevant by Bolie*:

V—extracellular fluid volume

I—rate of insulin entering the system from external sources

G—rate of glucose entering the system from external sources

X—insulin concentration

Y—glucose concentration

$F_1(X)$—rate of insulin degradation

$F_2(Y)$—rate of insulin production

$F_3(X, Y)$—rate liver accumulates glucose

$F_4(X, Y)$—rate tissues utilize glucose

Identify the units for each of these variables and model components. Then write a two-dimensional ODE model of glucose-insulin dynamics.

- In order to analyze the behavior of the model, assumptions about the four unknown functions need to be specified. Determine the appropriate assumptions. For instance, should $F_1(X)$ be a strictly increasing or decreasing function of X? Discuss how these assumptions might be different depending on whether you are modeling a person with type I or type II diabetes. Discuss any additional assumptions being made by the model.

- If you have made reasonable assumptions, then it is possible to study the shape of the nullclines in your model. To understand their shapes, think of the nullclines being given by the graph of Y as a function of X. Use inverse functions and implicit differentiation. In particular, answer the following questions: Are the nullclines increasing or decreasing functions of X? How many points of intersection (i.e., equilibria) can these nullclines have? What effect does increasing I or G in the model have on the nullclines of the model? How do the nullclines of the model differ for a model of type I versus a model of type II diabetes?

- Consider the following questions: If a person's glucose and insulin levels are at a steady state, what happens if the person suddenly eats a large bag of candy? Does her glucose level increase or decrease? What about her insulin level? What happens if the glucose and insulin levels are at a steady state and the person takes a shot of insulin? Takes a shot of insulin and eats a bag of candy? Are the answers to these questions the same if the person has type I diabetes or type II diabetes?

Project 8E   In Vivo HIV dynamics

AIDS, acquired immunodeficiency syndrome, is a disease that kills millions of people every year. HIV, the virus that causes AIDS, is passed from one person to another through sex, sharing needles, or using contaminated blood products. Newborns can have HIV because their mother passed it to them. The virus travels through the bloodstream to many different places in the body. The immune system, which helps the body fight off illness, fights back in three ways: with custom-made antibodies, with macrophages that eat up all foreign

invaders, and with killer T cells that seek out and destroy cells that are already infected with the virus. This defense is coordinated by helper T cells.

But HIV has an ingenious battle strategy: It attacks the T cells themselves, crippling the body's defenses. Here's how it works: HIV has a special shape on its surface that, like a piece of a jigsaw puzzle, fits perfectly into a shape on the T cell. This shape is a protein called CD4. The virus can now enter and infect the cell. The virus's genetic information—called RNΛ—is transcribed into a

---

* V. W. Bolie, 1960. Coefficients of normal blood glucose regulation. *Journal of Applied Physiology* 16: 783–788. Departments of Veterinary Physiology, Veterinary Anatomy, and Electrical Engineering, College of Engineering and Veterinary Medicine, Iowa State University, Ames, Iowa.

form that is identical to the cell's genetic information—called DNA. The virus, now in the form of DNA, hides inside the nucleus of the cell, escaping from the body's defenses. After a while, HIV comes out of hiding and begins to reproduce. The DNA is transcribed into many copies of RNA that produce proteins for the new viruses. The proteins are cut into usable pieces and packaged with the RNA. The new viruses then bud from the cell (see photo).

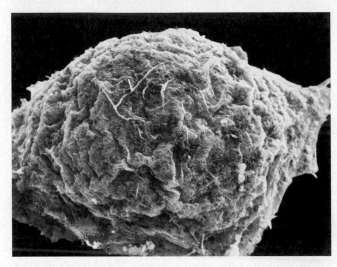

Each new virus may then go on to infect and destroy other T cells, weakening the immune system's defense. Once T cells are reduced to a fraction of those found in healthy individuals (well under a half), the individual can no longer fight off opportunistic infections from otherwise weak pathogens, such as a cytomegalovirus. When this happens, the individual is said to have AIDS. A person with AIDS will likely develop several such opportunistic infections and ultimately die from one of these or from infections of *Mycobacterium tuberculosis*, the bacterium that causes tuberculosis.

In this project, you will develop and analyze a simple three-state variable model of the in vivo dynamics of HIV.

- Develop an ODE model of the HIV dynamics based on the following assumptions:

— There are three stated variables in the body: the concentration of HIV particles (V), the concentration of T cells (T), and the concentration of infected helper T cells (I).
— In the absence of HIV, the T cells exhibit logistic dynamics.
— The rate at which a viral particle encounters T cells is proportional to T.
— Upon encountering a viral particle, a T cell is considered infected.
— An infected T cell has a life expectancy of L days and produces on average N new viral particles.
— Infected T cells do not replicate.
— Viral particles are removed from the body at a rate proportional to their density.

Discuss and try to justify these assumptions as well as any other assumptions you make.

- When initially infected with HIV, the number of T cells in a person's bloodstream drops rapidly in the first two to ten weeks. After this initial drop off, the number of T cells tends to rise and level off for a period of up to ten years. Therefore, when modeling a person during this latency period, it is reasonable to assume that T remains essentially constant and only V and I are dynamic variables. Carry out a phase-plane analysis of this two-dimensional model. Determine how the behavior of the model depends on the parameters.

- Consider the three-dimensional model. Solve for its equilibria and determine whether the virus has a positive per capita growth rate at the virus-free equilibrium. Be sure to answer the following questions: Under what conditions is the virus able or unable to invade the body? If the virus cannot invade the body, what determines the rate at which the entering virus is driven out of the body? If the virus does invade the body, is there an equilibrium that supports all three populations at positive concentrations? Is this equilibrium always stable? What conditions ensure a low viral concentration or a high T cell concentration in the body?

ANSWERS TO SELECTED PROBLEMS

CHAPTER 1

Problem Set 1.1, Page 20

1. a. yes; D: $\{3, 4, 5, 6\}$; R: $\{4, 7, 9\}$ **b.** no **3. a.** no **b.** yes; D: $x \neq 0$; R: $\{-1, 1\}$ **5. a.** no **b.** yes; D: $\mathbb{R}$; R: $y \geq -4$
7. a. yes; D: $\mathbb{R}$; R: $y \geq 0$ **b.** yes; D: $\mathbb{R}$; R: $\mathbb{R}$
9. a. no; D: $-2 \leq x \leq 2$; R: $-4 \leq y \leq 4$ **b.** yes; D: $x \neq \pm 2$; R: $y \neq 1$ **11. a.** no; D: $x \geq -3$; R: $\mathbb{R}$ **b.** yes; D: $\mathbb{R}$; R: $y \geq -80$
13. $D = \mathbb{R}$; $f(0) = 3$; $f(1) = 4$; $f(-2) = -5$
15. $D = \mathbb{R}$, $x \neq 3$; $f(2) = 0$; $f(0) = -2$; $f(-3)$ not defined
17. $D = \mathbb{R}$; $f(3) = 4$; $f(1) = 2$; $f(0) = 4$ **19.** $F(x) = x^2$
21. 8; -7; -16; $M(x) = 3x - 7$ **23.** D: $-6 \leq x \leq 6$;
R: $-6 \leq y \leq 5$; decreasing on $[-6, 0)$; increasing on $(0, 6]$
25. D: $x \neq 2$; R: $y = 5$; constant on $(-\infty, 2)$, $(2, \infty)$
27. D: $x \geq -5$, $x \neq 3$; R: $y \geq -3$, $y \neq 6$; increasing on $[0, 3)$ and on $(3, \infty)$; decreasing on $[-2, 0)$; constant on $[-5, -2]$
29. $D = \mathbb{R}$; $y = 3x - 5$ **31.** D: $\{x \leq 5$; $y = \sqrt{5 - x}$

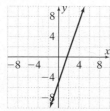   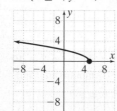

33. $D = \{x \geq 0\}$;   **35.** $A = \frac{1}{16} P^2$
$y = (x + 5)^2 - x^2$
$= 10x + 25$

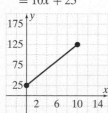

37. $A = 0.1089x + 214.5$ **39.**

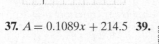

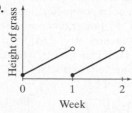

41. a. 25.34 cm/s **d.**
b. 19.01 cm/s
c. $[0, 1.2 \times 10^{-2}]$

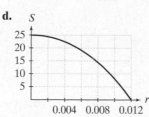

43. a. $n \neq 0$ **c.**
b. positive integers

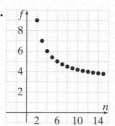

d. It approaches (but does not reach) 3.

45. a. $0 \leq n \leq 12.5$ **b.**

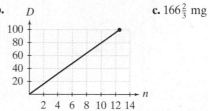

c. $166\frac{2}{3}$ mg

47. a. $(0, 150]$ **b.**

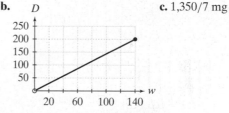

c. $1{,}350/7$ mg

Problem Set 1.2, Page 33

1. $y = \frac{5}{4}x - 2$ **3.** $y = \frac{2}{5}x + 2$

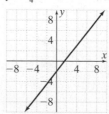   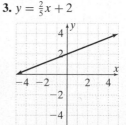

5. $y = -3x + 2$ **7.**

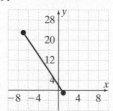   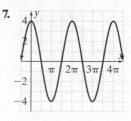

9.

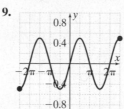

11. $y = 3x$
13. $y = -x + 1$
15. $y = 4$

17. Given $y = mx + b$ with (h, k) a point on the line. That is,

$y = mx + b$	*Given*
$k = mh + b$	*Point (h, k) satisfies the equation.*
$k - mh = b$	*Solve for b.*
$y = mx + (k - mh)$	*Substitute the value for b in the given equation.*
$y - k = mx - mh$	*Subtract k from both sides.*
$y - k = m(x - h)$	*Factor.*

19. $m = 2; y = 2x - 4$ **21.** $p = 2\pi; a = 3; y = 3\cos x$
23. $p = 2; a = 2; y = 2\cos\pi(x - 1)$ **25.** D **27.** A **29.** C

31. $(h, k) = \left(-\dfrac{\pi}{6}, 0\right)$,   **33.** $(h, k) = (0, 0)$,

$a = \dfrac{1}{2}, b = 1, T = 2\pi$ $a = 2, b = 2\pi, T = 1$

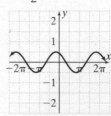   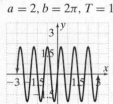

35. $y = \tan 2\left(x - \dfrac{\pi}{4}\right)$;   **37. a.** 48 **b.** 70 **c.** 84 **39.** Best fitting
line is B; 31.65 deaths per 100,000
$(h, k) = \left(\dfrac{\pi}{4}, 0\right)$,

$a = 1, b = 2, T = \dfrac{\pi}{2}$

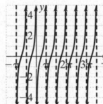

41. a. not quite linear,   **b.** yes; no
but close

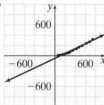

43. a.   **b.** It looks like a quadratic
curve would be better.

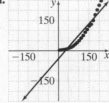

45. a.

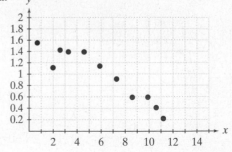

b.

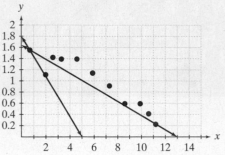

c.   **d.** Answers vary.

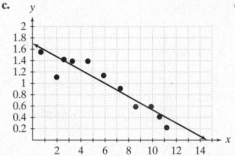

47. a.

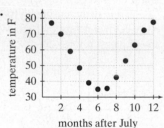

months after July

b. The amplitude is
about 21.25 and the
period is 12 months.

c. $a = 21.25, b = \dfrac{2\pi}{12}$

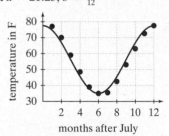

months after July

Problem Set 1.3, Page 46

1. a. power function; $a = \dfrac{1}{3}, b = 1$ **b.** power function; $a = \dfrac{1}{3}$,
$b = -1$ **c.** not a power function **3.** not a power function
5. $y = \dfrac{1}{4}x^{-3/2}$, power function; $a = \dfrac{1}{4}, b = -\dfrac{3}{2}$ **7.** $y = 90^x$, not
a power function **9.** $y = 6x^{-1/2}$, power function; $a = 6$,
$b = -\dfrac{1}{2}$ **11.** It increases by a factor of 10^6. **13.** y is
proportional to the cube of x. **15.** x is proportional to the
sixth power of z.

17.   **19.**

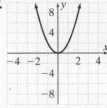   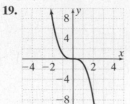

increasing: $(0, \infty)$ decreasing: $(-\infty, \infty)$
decreasing: $(-\infty, 0)$

21.

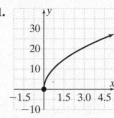

increasing: $(0, \infty)$

23.

$b = 0$
is a power function

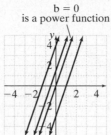

25. $V = \dfrac{1}{3} Sr$; r is doubled **27.** $V = \dfrac{1}{12} \pi h^3$;

$S = \dfrac{\pi}{4}(1 + \sqrt{5})h^2$; S is quadrupled **29.** 33 mg **31.** 91 mg

33. 9 mg **35.** 15,000 lb **37.** $A = 0.003d^{2.99}$

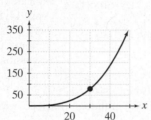

39.

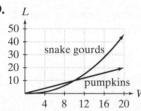

41. $b = 3$; that is, $12^3 = 1,728$
more food than a
Lilliputian would need.

43. a. $b = \dfrac{1}{3}$ **b.**

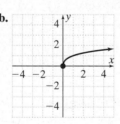

Problem Set 1.4, Page 55

1.

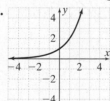

3.

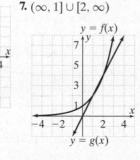

5.

7. $(\infty, 1] \cup [2, \infty)$

$y = f(x)$

$y = g(x)$

9. $[-1.43, 1.89] \cup [6.58, \infty)$ **11.** $x \approx 1.15$
Only the first interval is
illustrated in the
accompanying graph.

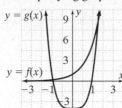

$y = g(x)$

$y = f(x)$

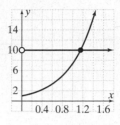

13. $x \approx 0.55$

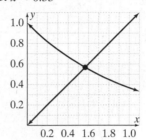

15. $x \approx 1.9$, $x \approx 4.5$

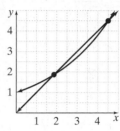

17. $x \approx -1.3$, $x \approx 0.5$

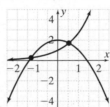

19. \$5,427.43
21. \$1,010.05
23. \$1,054.41
25. $f(x) = 2(5/2)^x$
27. $f(x) = 9(1/9)^x$

29. 168 seconds

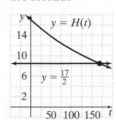

$y = H(t)$

$y = \frac{17}{2}$

31. a. \$120, \$121, \$121.55
b. $100\left(1 + \dfrac{0.2}{n}\right)^n$ **c.** \$122.14

33. a. $(1 + 10)^{365}$, $(1 + 10/2)^{(2 \cdot 365)}$,
$(1 + 10/4)^{(4 \cdot 356)}$ **b.** $(1 + 10/n)^{(n \cdot 365)}$
c. $e^{3,650}$ **35. a.** country #3
b. country #1 (50% growth rate)
c. country #3; decreases by 5%
every year

37. a. 25 mcg **b.** $100\left(\dfrac{1}{2}\right)^{t/7}$ **c.** 23.25 days **39.** $500\left(\dfrac{1}{2}\right)^{t/5,730}$

41. a. $350(1.12)^t$ **b.** 9,362 **43. a.** $1 - (1 - 1/10)^{10} \simeq 0.65$
b. $1 - (1 - 1/10)^{100} \simeq 0.63$ **47. a.**

c. $1 - \left(1 - \dfrac{1}{N}\right)^N$

45. a. $0.5(0.5)^{t/5.7}$ cm^3 **b.** 18.93 days

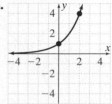

b. Answers vary. **c.** 0.7

49. a.

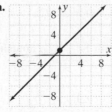

b. Answers vary.
c.

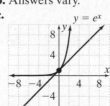

$y = e^x$

Problem Set 1.5, Page 67

1.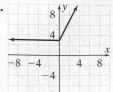

3.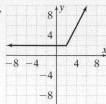

5.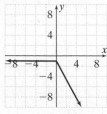

7. $y = \cos x$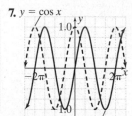

9.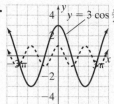

11.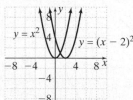

13.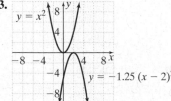

15.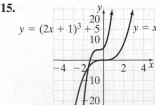

17.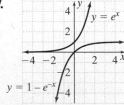

19.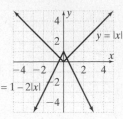

21. a. -3 **b.** not defined **c.** $1{,}190$ **d.** $\dfrac{197}{9{,}700}$ **e.** -5

23. a. period is T; amplitude is A **b.** period is T; amplitude is $4A$ **c.** period is $\dfrac{T}{3}$; amplitude is $2A$ **d.** period is T; amplitude is $2A$

25. $g(x) = 1 - \sin x$; $f(x) = \sqrt{x}$ **27.** $g(x) = e^x$; $f(x) = 1 - x^2$
29. $g(x) = x^2 - 1$; $f(x) = x^3 + \sqrt{x} + 5$

31.

$f + g$; domain	fg; domain	f/g; domain
$x^2 + x - 5$	$2x^3 - 5x^2 - x + 6$	$\dfrac{2x - 3}{(x - 2)(x + 1)}$
D: $x \neq -1$	D: $x \neq -1$	D: $x \neq -1, 2$

33.

$\sqrt{4 - x^2} + \sin(\pi x)$	$\sqrt{4 - x^2}\,\sin(\pi x)$	$\dfrac{\sqrt{4 - x^2}}{\sin(\pi x)}$
D: $-2 \leq x \leq 2$	D: $-2 \leq x \leq 2$	D: $-2 < x < 2$, $x \neq -1, x \neq 0,$ $x \neq 1$

35. $A = 3.1$, $B = \dfrac{\pi}{6}$, $C = -11$, and $D = 2.7$ **37. a.** $z = \dfrac{t + a}{b}$
b. $a = \dfrac{1}{2}, b = \dfrac{1}{2}$ **39. a.** $V = \dfrac{16}{3}\pi$ **b.** $\dfrac{2\pi}{3}t^3$ **c.** $[0, 3]$

41.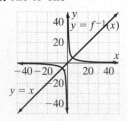

43. May 1, 2011 and August 24, 2011. The sunrise-sunset tables for Los Angeles, indicate the actual dates are April 14, 2011, and August 28, 2011. This offers a good jump-off point for a discussion of mathematical modeling.

45. Between 1844 and 1997, a 153-year period, sunspot intensity went though fourteen cycles. Hence, the period is $\dfrac{153}{14} \approx 10.9$. The minimum number is always around 0 and the black curve indicates an average amplitude of around 70 sunspots. A reasonable fitting sunspot curve is

$y = 70 - 70\cos\left(\dfrac{2\pi x - \pi}{10.9}\right)$, which implies $a = 70$, $b = -70$,

$c = \dfrac{2\pi}{10.9} \approx 0.576$, $d = -\dfrac{\pi}{10.9} \approx 0.288$.

Problem Set 1.6, Page 81

1. not one-to-one **3.** one-to-one;

$f^{-1}(x)$	x
0	11.9
1	17
4	-2
2	4
6	5

5. one-to-one

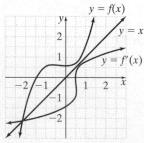

7. one-to-one

9. $y = \dfrac{x}{1 - x}$; D: $x \neq -1$; R: $x \neq 1$ **11.** $y = \sqrt[3]{x + 2} - 1$; D: $\mathbb{R}$; R: $\mathbb{R}$ **13.** $y = -\sqrt{\ln x}$; D: $[-\infty, 0)$, R: $[0, \infty)$ **15. a.** $x = 1$
b. $x = -3$ **17. a.** $x = 3$ **b.** $x = 2$ **19. a.** $x = e^3$ **b.** $x = 10^{4.5}$
21. a. $3x$ **b.** $4x$ **c.** $3x$ **d.** $-5x$ **e.** $-2x$ **23. a.** $e^{x\ln 5}$
b. $e^{-x\ln 2}$ **c.** $e^{\ln 5/x}$ **d.** $e^{x^2\ln 4}$ **e.** $e^{x^e\ln 3}$ **25. a.** $\dfrac{\ln(x + 1)}{\ln 10}$
b. $\dfrac{1 + \ln(x + 1)}{\ln 10}$ **c.** $\dfrac{\ln(x^2 - 2)}{\ln 2}$ **d.** $\dfrac{\ln(2x - 3)}{\ln 7}$ **27.** 543 **29.** 0.5

31.

$10^{-4} \quad 10^{-3} \quad 10^{-2} \quad 10^{-1} \quad 10^{0} \quad 10^{1} \quad 10^{2} \quad 10^{3} \quad 10^{4} \quad 10^{5}$

33.

$10^{-6} \qquad 10^{-4} \qquad 10^{-2} \qquad 10^{0} \qquad 10^{2} \qquad 10^{4} \qquad 10^{6}$

35. a. 0.864 mg/L   **b.** $C(U) = \dfrac{-5{,}000U}{253U - 6{,}039}$ mg/L

37. 6.12 years   **39.** $2.9 \log_2 5$ **41.** $W = 497 L^{3.14}$

43. a.
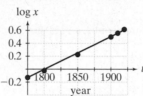
b. $y = 0.74883x + 0.93659$

45. a.
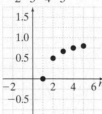
b. $y = 0.0029559x + 1.6313902$

47. $\ln x = 0.005069\, t - 9.127$
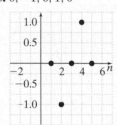

Problem Set 1.7, Page 98

1. $0, \dfrac{1}{2}, \dfrac{2}{3}, \dfrac{3}{4}, \dfrac{4}{5}$
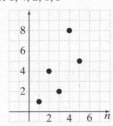

3. $0, -1, 0, 1, 0$
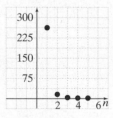

5. $1, 4, 2, 8, 5$
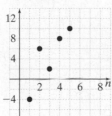

7. $256, 16, 4, 2, \sqrt{2}$
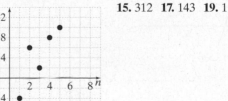

9. $-4, 6, 2, 8, 10$
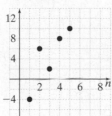

11. 32   **13.** 10
15. 312   **17.** 143   **19.** 1

21. $x = 0, x = \dfrac{1}{2}$
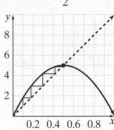

23. $x = 0, x = 2$
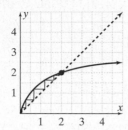

25. $x = 2$
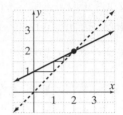

27. $a = 0, a = 5$
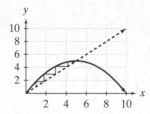

29. $a = -1, a = 0, a = 1$
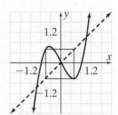

31. $6.25\%; \dfrac{1}{2^n}$

33. a. $a_{n+1} = (1 - c)(A + a_n)$

b. $a = \dfrac{(1 - c)A}{c}$ **c.** Equilibria are greater than A whenever $c < \dfrac{1}{2}$.

35. a. $a \approx 1.096$
b. The fit is very good.
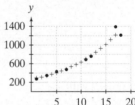

c. 1249, 1219, 1186, 1150, 1110, 1067, 1019, 967, 910, 847, 778, 703, 621, 530, 431, 323, 204, 73, −70, −227; the population becomes extinct in 1995.

37. a. $1, 1 + \dfrac{1}{1} = 2, 1 + \dfrac{1}{1 + \frac{1}{1}} = \dfrac{3}{2}, 1 + \dfrac{1}{1 + \frac{1}{1 + \frac{1}{1}}} = \dfrac{5}{3},$

$1 + \dfrac{1}{1 + \frac{1}{1 + \frac{1}{1 + \frac{1}{1}}}} = \dfrac{8}{5}$ **b.** $\dfrac{1 \pm \sqrt{5}}{2}$ **c.** 1.62   **39. a.** $x_2 = \dfrac{x_1}{1 + x_1},$

$x_3 = \dfrac{x_1}{1 + 2x_1}, x_4 = \dfrac{x_1}{1 + 3x_1}; x_5 = \dfrac{x_1}{1 + 4x_1}$

b. $x_n = \dfrac{x_1}{1 + (n - 1)x_1}$ **c.** Answers vary.

41. a.

$$\text{NUMBER OF ALLELES IN NEXT GENERATION}$$

$$= \dfrac{\text{NUMBER OF A ALLELES IN CURRENT GENERATION}}{\text{TOTAL NUMBER OF ALLELES}}$$

$$x_{n+1} = \dfrac{\dfrac{x_n}{3}(1 - x_n)}{\dfrac{2x_n}{3}(1 - x_n) + (1 - x_n)^2}$$

$$= \dfrac{x_n}{2x_n + 3(1 - x_n)}$$

$$= \dfrac{x_n}{3 - x_n}$$

b. $\dfrac{1}{2}, \dfrac{1}{3}, \dfrac{1}{14}, \dfrac{1}{41}, \dfrac{1}{122}, \dfrac{1}{365}, \dfrac{1}{1,074}, \dfrac{1}{3,281}, \dfrac{1}{9,842}, \dfrac{1}{29,525}$ **c.** $0, 2$

d.

n:	1	2	3	4	5	6	7	8	9	10
$x_{n+1} = \dfrac{x_n}{3 - x_n}$	$\dfrac{1}{2}$	$\dfrac{1}{3}$	$\dfrac{1}{14}$	$\dfrac{1}{41}$	$\dfrac{1}{122}$	$\dfrac{1}{365}$	$\dfrac{1}{1,074}$	$\dfrac{1}{3,281}$	$\dfrac{1}{9,842}$	$\dfrac{1}{29,525}$
$x_{n+1} = \dfrac{x_n}{1 + x_n}$	$\dfrac{1}{2}$	$\dfrac{1}{3}$	$\dfrac{1}{4}$	$\dfrac{1}{5}$	$\dfrac{1}{6}$	$\dfrac{1}{7}$	$\dfrac{1}{8}$	$\dfrac{1}{9}$	$\dfrac{1}{10}$	$\dfrac{1}{11}$

43. Answers vary.

n:	1	2	3	4	5	6	7	8	9	10
$x_{n+1} = \dfrac{x_n}{3 + x_n}$	0.50	0.29	0.18	0.11	0.07	0.05	0.03	0.02	0.01	0.01
all Aa survive	0.50	0.33	0.25	0.20	0.17	0.14	0.13	0.11	0.10	0.09
$\dfrac{2}{3}$ of Aa survive	0.50	0.29	0.18	0.11	0.07	0.05	0.03	0.02	0.01	0.01
$\dfrac{1}{2}$ of Aa survive	0.5	0.25	0.13	0.06	0.03	0.02	0.01	0.00	0.00	0.00
$\dfrac{1}{3}$ of Aa survive	0.5	0.20	0.07	0.02	0.01	0.00	0.00	0.00	0.00	0.00

We conclude that as a diminishing proportion of Aa survive, so the sequences approach zero more rapidly, and if there is no Aa, then the proportion of a immediately goes to zero in the next generation.

Chapter 1 Review Questions, Page 101

1. a. D: $[0.1, \infty)$; R: $[0, \infty)$
b. $f^{-1}(x) = 10^{x^2 - 1}$; D: $[0, \infty)$; R: $[0.1, \infty)$
3. $a = 1.4$

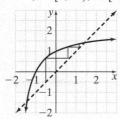

5. The residuals for $y = x$ are $0.7, 0.1$, and 0.2 with the sum of squares 0.54; the residuals for $y = \dfrac{1}{2}x + 1$ are $0.3, 0.4$, and 0.2 with the sum of squares 0.29. Thus, $y = \dfrac{1}{2}x + 1$ is the better-fitting line.

7.

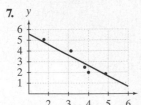

Answers vary; it looks like $m = -1$.

9. $a = 100, b = 0.1$
11. a. $x_{n+1} = \dfrac{9x_n}{10 + 8x_n}$
b. $x_1 = 0.9, x_2 = 0.4709302,$
$x_3 = 0.3078547,$
$x_4 = 0.2223163$

13. a. B   **b.** D   **c.** C   **d.** A   **15. a.** $\dfrac{3}{2}, \dfrac{4}{3}, \dfrac{5}{4}, \dfrac{6}{5}, \dfrac{7}{6}$ **b.** $1, \dfrac{1}{2}, \dfrac{1}{4},$
$\dfrac{1}{8}, \dfrac{1}{16}$ **c.** $2, 3, 5, 7, 11$ **17.** w is linear **a.** $[0, \infty)$ **b.** 20
19. 203.5 kg of wool

CHAPTER 2
Problem Set 2.1, Page 116

1. -3 **3.** 12 **5.** $\dfrac{1}{5}$ **7.** -3 **9.** 6 **11.** $\dfrac{1}{4}$ **13.** -32 feet/second
15. -16 feet/second

17.

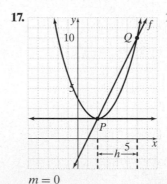

$m = 0$

19.

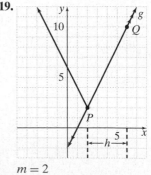

$m = 2$

21.

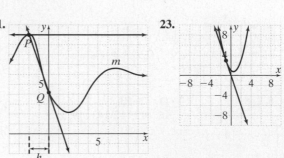

$m = 0$

23.

25.

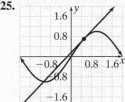

27.

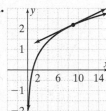

29.

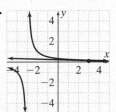

31.

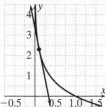

33.

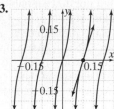

35.

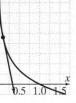

37.

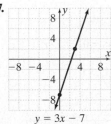

$y = 3x - 7$

39.

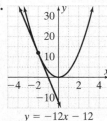

$y = -12x - 12$

41.

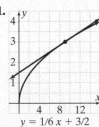

$y = 1/6\,x + 3/2$

43. on $[0, 2]$: increase of 3.19 million people per decade; on $[2, 4]$: increase of 5.64 million people per decade
45. on $[0, 30]$: 0.07 cm of froth is evaporating per second; on $[60, 90]$: 0.0433 cm of froth is evaporating per second

47. at $t = 0$: increase of 2.37 million people per decade; at $t = 2$: increase of 4.18 million people per decade
49. at $x = 0$: estimate 0.06 cm of froth evaporating per second; at $x = 60$: estimate 0.05 cm of froth evaporating per second   **51.** velocities for Ben Johnson:

Distance (in meters)	0	10	20	30	40	50
Velocity (meters per second)	0	5.9	9.6	10.8	11.6	11.9
Distance (in meters)	60	70	80	90	100	
Velocity (meters per second)	12.0	11.9	11.8	11.5	11.1	

53. 1.25 kg/yr

1. a. 0 **b.** 4 **3. a.** 6 **b.** 3 **5. a.** 0 **b.** 1 **c.** does not exist
7. a. 0 **b.** 0 **c.** 0 **9.** $\lim_{x\to1^+} f(x) = 2$ **11.** $\lim_{x\to3^-} h(x) = 2$
13. $\lim_{x\to5^-} f(x) = 13$ **15.** $\lim_{x\to2} h(x) = 10$ **17.** 1 **19.** −1 **21.** $\frac{1}{4}$
23. 0 **25.** 0 **27. a.** 0.1 **b.** 0.01 **c.** 0.001 **29. a.** 0.01
b. 0.0001 **c.** 0.000001

31. a.

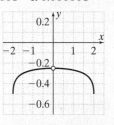

b.

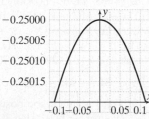

c.

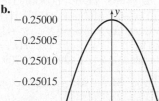

d. The accuracy of the technology has been exceeded.

33. Given $\epsilon > 0$, choose $\delta = \epsilon$. Then,
$$|f(x) - L| = |(x + 1) - 6|$$
$$= |x - 5|$$
$$< \delta = \epsilon$$

35. Given $\epsilon > 0$, choose δ to be the smaller of 1 or $\frac{\epsilon}{2}$. Then,
$$|f(x) - L| = \left|\frac{1}{x} - \frac{1}{2}\right|$$
$$= \left|\frac{2 - x}{2x}\right|$$
$$= \frac{|x - 2|}{2|x|}$$
$$< \frac{\delta}{2} < \epsilon$$

37. Given $\epsilon > 0$, choose δ to be the smaller of 1 or $\frac{\epsilon}{5}$. Then,
$$|f(x) - L| = |(x^2 + 2) - 6|$$
$$= |x^2 - 4|$$
$$= |(x + 2)(x - 2)|$$
$$= |x + 2||x - 2|$$
$$< 5\delta = 5\left(\frac{\epsilon}{5}\right) = \epsilon$$

39. a.

b. Limit exists for all values of a.

41. $f(x) = \begin{cases} \dfrac{x}{75}, & \text{if } 0 \le x \le 3 \\ \dfrac{1}{25} & \text{if } x > 3 \end{cases}$

Problem Set 2.3, Page 142

1. $\lim_{x \to 1} f(x) = -1$ **3.** $\lim_{x \to 0^-} f(x) = -1$; $\lim_{x \to 0^+} f(x) = 1$;
$\lim_{x \to 0} f(x)$ does not exist **5.** $\lim_{x \to 1^-} f(x) = 0$; $\lim_{x \to 1^+} f(x) = 1$;
$\lim_{x \to 1} f(x)$ does not exist **7.** $\dfrac{8}{35}$ **9.** -1 **11.** $\sin 3 + \dfrac{1}{5}$ **13.** $\dfrac{1}{3}$
15. $x = -1$ (jump), $x = 1$ (removable) **17.** $x = 1$ (jump)
19. $f(2) = 3$ **21.** not possible (jump) **23.** Let
$f(x) = -x^7 + x^2 + 4$; $f(1) > 0$, $f(2) < 0$; root for $1 < x < 2$
25. Let $f(x) = \sqrt[3]{x} - x^2 - 2x + 1$; $f(0) = 1 > 0$, $f(1) < 0$;
root for $0 < x < 1$ **27.** Let $f(x) = x2^x - \pi$; $f(1) < 0$, $f(2) > 0$;
root for $1 < x < 2$ **29. a.** -2.709 **b.** -1.562 **c.** -0.194
d. 1.903 **31.** $\dfrac{5}{2}$ **33.** $\dfrac{\pi}{16}$ **39.** Let $a, b,$ and c be chosen.
If x is large enough in the positive direction, we have
$x^3 + ax^2 + bx + c > 0$. Also, if x is large enough in the
negative direction, we have $x^3 + ax^2 + bx + c < 0$. These
follow from the fact that x^3 dominates the quadratic
$ax^2 + bx + c$. Then by the intermediate value theorem,
there must be at least one part of the graph that crosses
the x axis; hence it must have at least one real root.

41. a. $g(N) = \begin{cases} 5 & \text{if } N \geq c \\ \dfrac{100}{1+N} & \text{if } N < c \end{cases}$ **b.** $c = 19$

43. It will happen in 2056. **45.** 2,383; 29,248 **47.** 1,365; 36,603

Problem Set 2.4, Page 153

1. 0 **3.** ∞ **5.** $-\infty$ **7.** ∞ **9.** ∞ **11.** limit does not exist
13. $\dfrac{2}{21}$ **15.** 0 **17.** 0 **19.** $\dfrac{5}{2}$ **21.** 0 **23.** $\dfrac{a}{2}$
25. $2 < x \leq 2 + 10^{(-6)}$ **27.** $1 - 10^{(-6)} \leq x < 1$
29. $\sin^{-1}(10^{-6}) \leq x < 0$ **31.** $x \leq -2\sqrt{5}$ **33.** $x \leq -21$
35. $x > 1{,}000$ **37.** $x > 10^6$ **39. a.** $t > 62$ **b.** $t > 65$
41. a. $y = 25.11$, $x = -19.76$; the horizontal asymptote is the
maximum uptake rate
b.
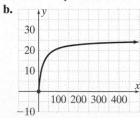
43. $x > 463$ **b.** $x > 4{,}637$
45. the limit is 660, which is
the maximum size the
population will approach as
$t \to \infty$
47. $y = 16$; $x > 1.291$

Problem Set 2.5, Page 169

1. $\dfrac{1}{3}$ **3.** 1 **5.** no limit because of oscillation **7.** no limit
because of oscillation **9. a.** $L = 1$ **b.** $n > 2{,}997$ **11. a.** $L = 0$
b. $n > 1{,}000{,}000$ **13. a.** $L = 0$ **b.** $n > 1{,}000$ **15.** 500,000
17. 20 **19.** ∞ **21.** 1 **23.** 0 **25.** 1 **27.** $\dfrac{2}{3}$ **29.** 1 **31.** no limit;
function not defined after a_2 **33.** $\dfrac{1 + \sqrt{21}}{2}$ **35. b.** 0 **c.** $n \geq 7$
d. $n \geq 997$ **37. a.** frequency approaches 1 **b.** frequency
approaches 0 **39.** 50.0% **41.** 36.6% **43.** 20.7%
45. $r = 0.9$: 0.5, 0.725, 0.904438, 0.982225, 0.997938, 0.99979,
0.999979, 0.999998, 1, 1, $\cdots$; converges to 1
$r = 1.5$: 0.5, 0.875, 1.039063, 0.97818, 1.010196, 0.994746,
1.002586, 0.998697, 1.000649, 0.999675, 1.000162, 0.999919,
1.000041, 0.99998, 1.00001, $\cdots$; converges to 1

$r = 2.1$: 0.5, 1.025, 0.971188, 1.02995, 0.965171, 1.035765,
0.957973, 1.042521, $\cdots$; diverges by oscillation

Problem Set 2.6, Page 180

1. 3 **3.** -2 **5.** $-\dfrac{1}{32}$ **7.** 3 **9.** $\dfrac{1}{6}$ **11.** $3x - y - 2 = 0$
13. $2x + y - 1 = 0$ **15.** $x + 32y + 8 = 0$ **17.** $3x - y + 2 = 0$
19. $x - 6y + 9 = 0$ **21.** not differentiable at $x = 0$. **23.** not
differentiable at $x = \pm 1, x = 0$ **25.** not differentiable
at $x = 2$
27. a.
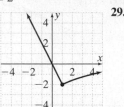
29.
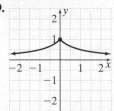

The graph (Problem 29) is continuous but not differentiable
at $x = 0$. **31. a.** 0 ft/s **b.** 4 s **c.** it is falling at 64 ft/s
33. $A'(50) = -0.9$ **35. a.** 9.2 **b.** 14.2 **c.** the prevalence is
increasing more rapidly at higher latitude **37. a.** -28.96
b. the units are particles per milliliter/day. **39. a.** -0.061;
-0.015 **b.** The derivatives are the rate of change of the
amount of glucose consumed per microgram of glucose in the
environment per hour.

Problem Set 2.7, Page 193

1. 0 **3.** $-2x$ **5.** $4x^3$ **7.** $-\dfrac{1}{x^2}$ **9.** 0 **11.** -8 **13.** 32 **15.** -0.01
17. everywhere **19.** $c = 0$

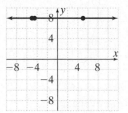
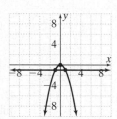

21. $c = \sqrt{2}$
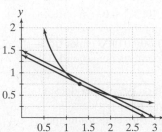
35. $c \approx -2.2, -0.9$
37.
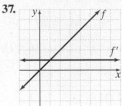
39.
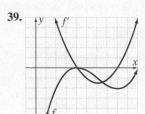

23. E **25.** B **27.** D
29. increasing: $\left(\dfrac{1}{2}, \infty \right)$;
decreasing: $\left(-\infty, \dfrac{1}{2} \right)$
31. increasing: $(-\infty, \infty)$
33. increasing:
$(-3.5, 0) \bigcup (0, \infty)$;
decreasing: $(-\infty, -3.5)$

45. a. increasing for $0 \le t \le 3$; decreasing for $t > 3$

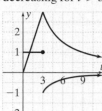

47. a. 38.24
b.

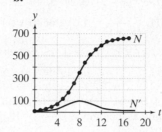

49. From the MVT, there must be some time that the car travels at 72 mph.

Chapter 2 Review Questions, Page 195

1. $x = \pm 1$ average over $(-1, 2)$; tangent at $x = 1$

3. $-\dfrac{2}{x^3}$

5.

7. a. 0 **b.** −2 **c.** $\dfrac{1}{2}$ **d.** $-\dfrac{1}{3}$ **e.** no; yes

9.

11. a. $y = 2$ **b.** Let
$$f(x) = \frac{2x^2 + 1}{x(x-2)}; \text{ vertical}$$
asymptote at $x = 0$:
$$\lim_{x \to 0^-} f(x) = \infty;$$
$$\lim_{x \to 0^+} f(x) = -\infty;$$
vertical asymptote at $x = 2$:
$$\lim_{x \to 2^-} f(x) = -\infty;$$
$$\lim_{x \to 2^-} f(x) = \infty$$

13. a.

t	0	0.1	0.2	0.3	0.4	0.5	0.6	0.7	0.8	0.9	1.0
C	0	0.2	0.4	0.6	0.8	0.9	1.0	0.9	1.0	0.9	0.7
C'	2	2	2	2	1	1	−1	1	1	−1	−2

b.

$C'(t)$ is the rate at which the drug is changing in the patient's blood stream.

15. $L = \dfrac{1}{2}$ and $x > \ln 499{,}999$

17. Since f is continuous on $[0, 1]$, $f(0) = -0.1 < 0$, and $f(1) = \dfrac{1}{e} - 0.1 > 0$, the intermediate value theorem implies there exists a root of f on the interval $(0, 1)$.

19. Since the average population growth rate from 1981 to 1983 is 1.82 million per year, the mean value theorem implies there is a year t_1 between 1981 and 1983 such that $N'(t_1) = 1.82$. Similarly, there is a year t_2 between 1983 and 1985 such that $N'(t_2) = 1.915$. Since $N'(t)$ is continuous on the interval $[t_1, t_2]$, the intermediate value theorem implies there is a time t between t_1 and t_2 such that $N'(t) = 1.85$.

CHAPTER 3

Problem Set 3.1, Page 207

1. a. $7x^6$ **b.** $(\ln 7)7^x$ **3. a.** $15x^4$ **b.** 0 **5. a.** $2x$ **b.** −2 **7.** $5x^4 - 6x$
9. $4e^t - 5$ **11.** $4.78(2.25)^t$ **13.** $2Cx + 5 - 2e^{-2x}$

15. increasing: $(-\infty, 0)$, $\left(\dfrac{2}{3}, \infty\right)$; decreasing: $\left(0, \dfrac{2}{3}\right)$

17. increasing: $(-\infty, -17.9)$, $(-7.8, 5.3)$, $(16.4, \infty)$; decreasing: $(-17.9, -7.8)$, $(5.3, 16.4)$ **19.** increasing: $(-\infty, \ln 2)$; decreasing: $(\ln 2, \infty)$ **21.** $\dfrac{3}{2}x^{1/2}$

23. $-\dfrac{5}{3}x^{-8/3}$, $x \ne 0$ **25.** $\left(\dfrac{3}{2}\right)^t \ln\left(\dfrac{3}{2}\right) - \left(\dfrac{1}{6}\right)^t \ln 6$

27. Write b^x as $f(x) = e^{x(\ln b)}$ so that $f'(x) = (\ln b)e^{x(\ln b)}$ and this can be written as $f'(x) = (\ln b)b^x$. **31. a.** $0.34D - 0.41$; rate of change of weight with respect to dose level **b.** $1.21 \le D \le 8$ **33. a.** 0.8 **b.** 2.43 people/day **35.** $L'(W) = 1.064W^{-0.05}$; $L'(5) = 0.982$ vs. $L'(50) = 0.875$ **37.** 2.88 lb/mass; the additional kilogram, each lifter can be expected to lift with a kilogram increase in the lifter's body weight **39. a.** 41,850 children/year **b.** 2014

Problem Set 3.2, Page 218

1. $30x^4 - 6x^2 + 42x$ **3.** $\dfrac{2(x-2)(2x-3)}{(x^2-3)^2}$ **5.** $2^x(x \ln 2 + 1)$

7. $(x^2 + 3x + 2)e^x$ **9.** $-6L^5 + 4L^3 - 3L^2 + 1$ **11.** $32x + 24$

13. $\dfrac{e^x}{(e^x + 1)^2}$ **15.** $\dfrac{a[(1 + 2^p) - p(\ln 2)2^p]}{(1 + 2^p)^2}$ **17.** $\dfrac{-x^3 - 3x - 4}{3x^3}$

19. $4x + y - 1 = 0$ **21.** $2x + y + 1 = 0$ **23.** $9x - y + 5 = 0$
25. $4x - 5$ **27. a.** 0.177 among all adults who are 63 inches tall; this is the amount that the BMI changes per unit weight increase at a weight of 130 lbs. **b.** −0.54; among all children who weigh 60 lbs, this is the amount that the BMI changes per unit increase in height at a height of 54 inches **29. a.** 98.0%
b. decreasing at a rate of 0.5%/week (at the start of the fifth week of the year) **c.** 97.5%. **31. a.** $\dfrac{(a - b)10^{x-c}}{(1 + 10^{x-c})^2} \ln 10$
b. a is the asymptotic value, b can be used to adjust the intercept (for given values of a and c) and c can be used to make the curve switch more rapidly between the intercept and the asymptotic value. **33.** $f'(0.5) = \dfrac{161}{96} \approx 1.7$;
$f'(2.0) = \dfrac{1.288}{5.043} \approx 0.3$. The values represent the per unit rate of increase in the predation rate of wolves on moose for a unit increase in the moose population abundance of 0.5 and 2 moose per square kilometer, respectively. **35.** increasing on $(0, \infty)$ **37.** increasing on $(0, 17)$ and decreasing on $(17, \infty)$

Problem Set 3.3, Page 229

1. $6(3x - 2)$ **3.** $\dfrac{-8x}{(x^2 - 9)^3}$ **5. a.** $5u^4$ **b.** 3 **c.** $15(3x - 1)^4$

7. a. $15u^{14}$ **b.** $6x + 5$ **c.** $15(3x^2 + 5x - 7)^{14}(6x + 5)$

9. $9(x^4 - x + 5)^8(4x^3 - 1)$ **11.** $-12(1 + x - x^5)^{(-13)}(1 - 5x^4)$

13. $\dfrac{2}{x}$ **15.** $\dfrac{2}{2x + 5}$ **17.** $8x^3(x^4 - 1)^9(2x^4 + 3)^6(17x^4 + 8)$

19. $\dfrac{-x(3x - 2)}{3y^2 - 1}$ **21.** $\dfrac{-y(4x + 3y)}{2x(x + 3y)}$ **23.** $\dfrac{-2}{3}$ **25.** $\dfrac{y - y^2e^{(xy)}}{xye^{(xy)} + 2}$

27. a. $5; m = \dfrac{1}{2}$ **b.** $3; m = \dfrac{3}{2}$ **c.** $m = \dfrac{3}{4}$ **29. a.** $\dfrac{3}{(3x - 1)^2 + 1}$

b. $\dfrac{-1}{x^2 + 1}$ **31.** $5x - 4y - 6 = 0$ **33.** 0.031R cm/yr

35. The predation rate is increasing at 0.477 moose/year.
37. -0.3275 **39.** $y = -0.296x + 0.410,\ y = -1.219x + 1.271$
43. 0.19 inches/month and 0.72 pounds/month

Problem Set 3.4, Page 238

1. $\cos x - \sin x$ **3.** $2\cos 2x$ **5.** $2t - \sin t$
7. $e^{-x}\cos x - e^{-x}\sin x$ **9.** $2\sin\theta\cos\theta$ **11.** $-101x^{100}\sin x^{101}$

13. $(t^2 + 2)\cos t + 2t\sin t$ **15.** $\dfrac{t\cos t - \sin t}{t^2}$

17. $\dfrac{x\cos x - \sin x + 1}{(1 - \sin x)^2}$ **19.** $\dfrac{\cos x - \sin x}{\cos x + \sin x}$

21. $\dfrac{\sin x}{\cos^2 x}$ or $\sec x \tan x$ **23.** $-\csc^2 x$

25. $\dfrac{\sec x(\tan x + \sec x + \csc x + 1)}{\csc x + \cot x}$ **27.** $2\cot x$

29. a. $g(h) = \dfrac{1}{2}(\cos h)(\sin h)$ **b.** $f(h) = \dfrac{1}{2}(1)h$

c. $k(h) = \dfrac{1}{2}(1)\tan h$ **31.** The population is decreasing by
about 18 fish/month.   **33. a.** $r(t) = 1 - \sin t$ **b.** 2π
c. population is increasing on $(0, \infty)$ **35. a.** At noon, the
temperature in the pool is increasing at a rate of $T'(12) = \dfrac{\pi}{3}$
degrees Celsius per hour.   **b.** from 6 A.M. until 6 P.M.

Problem Set 3.5, Page 248

(Note: Answers may vary depending on the point around
which the approximation is made. In some cases, the best point
is obvious in terms of the simplicity of the calculation.)
1. $y = -x + \dfrac{\pi}{2}$; over for $x < \dfrac{\pi}{2}$; under for $x > \dfrac{\pi}{2}$ **3.** $y = 1$;

over   **5.** $y = -\dfrac{4}{25}x + \dfrac{13}{25}$; under   **7.** 5.2 vs. 5.0990195

9. -0.01 vs. -0.10536052 **11.** 0.2 vs. 0.20271   **13.** $\dfrac{1}{6}; 0.001\overline{6}$

15. $\dfrac{1}{2}; -0.1$ **17.** $-1; 0.01$ **19.** $\dfrac{1}{2}; 0.5\%$ **21.** $\dfrac{1}{\ln 2}; 7.2\%$

23. $0; 0\%$ **25. a.** $D(t) \approx 2.02T - 26.38$
b.

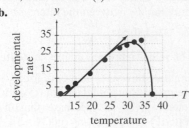

27. 6%   **29.** 0.05 ppm   **31. a.** 1.725   **b.** 8.625%   **33. a.** -1
b. $\pm 2\%$ **35.** $\pm 2\%$ **37. a.** 0.23   **b.** -0.137

39. a. $E = \dfrac{k(a - b)y}{(ky + 1)(b + aky)}$ **b.** $\dfrac{-y}{(y + 1)(y + 2)} \cdot 10\%$

Problem Set 3.6, Page 262

1. $e^{-x}(x - 2)$ **3.** 24   **5.** $-3^{99}\cos 3x$ **7.** $60t^2(7t^2 - 3)$ **9.** 10!

11. $\dfrac{2}{(w + 1)^3}$ **13.** velocity: $s'(t) = 2t - 3$; acceleration:

$s''(t) = 2; s''(t) \neq 0$ **15.** velocity: $s'(t) = \dfrac{1}{3}\sin\dfrac{t}{3}$; acceleration:

$s''(t) = \dfrac{1}{9}\cos\dfrac{t}{3}; s''\left(\dfrac{3\pi}{2}\right) = s''\left(\dfrac{9\pi}{2}\right) = 0$ **17.** velocity:

$s'(t) = (1 - t)e^{-t}$; acceleration: $s''(t) = (t - 2)e^{-t}; s''(2) = 0$
and as the limit as $t \to \infty$ **19.** $y = 1 + x$; underestimates

21. $y = 5 - 4x$: overestimates   **23.** $y = -\dfrac{1}{9}x + \dfrac{5}{9}$;

underestimates   **25.** increasing: $\left(-\infty, -\dfrac{\sqrt{3}}{3}\right) \cup \left(\dfrac{\sqrt{3}}{3}, \infty\right)$;

decreasing: $\left(-\dfrac{\sqrt{3}}{3}, \dfrac{\sqrt{3}}{3}\right)$; concave up: $(0, \infty)$; concave down:

$(-\infty, 0)$; point of inflection: $(0, 1)$ **27.** increasing: $(-\infty, 1)$;
decreasing: $(1, \infty)$; concave up: $(2, \infty)$; concave down:
$(-\infty, 2)$; point of inflection: $(2, 2e^{-2})$ **29.** increasing:
$(-\infty, -1) \cup (-1, \infty)$; concave up: $(-\infty, -1)$; concave down:
$(-1, \infty)$; no point of inflection   **31.** increasing: $\left(-\dfrac{3}{2}, \infty\right)$;

decreasing: $\left(-\infty, -\dfrac{3}{2}\right)$; concave up: $\left(-\infty, -\dfrac{2}{3}\right) \cup (1, \infty)$;

concave down: $\left(-\dfrac{2}{3}, 1\right)$; points of inflection: $\left(-\dfrac{2}{3}, -\dfrac{571}{27}\right)$

and $(1, 2)$ **33.** increasing: $\left(0 + 2n\pi, \dfrac{\pi}{2} + 2n\pi\right) \cup$

$\left(\dfrac{\pi}{2} + 2n\pi,\ \pi + 2n\pi\right)$; decreasing: $\left(-\dfrac{\pi}{2} + 2n\pi,\ 2n\pi\right) \cup$

$\left(\pi + 2n\pi, \dfrac{3\pi}{2} + 2n\pi\right)$; concave up: $\left(-\dfrac{\pi}{2} + n\pi, \dfrac{\pi}{2} + n\pi\right)$;

concave down: $\left(\dfrac{\pi}{2} + n\pi, \dfrac{3\pi}{2} + n\pi\right)$; no points of inflection

35. first: $y = x$ **37.** first: $y = 1 + x$
second: $y = x$ second: $y = 1 + x + x^2$

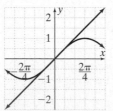   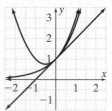

39. first: $y = \dfrac{1}{4}x + 1$; second: $y = 2 + \dfrac{1}{4}(x - 4) - \dfrac{1}{32}(x - 4)^2$

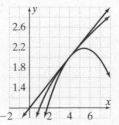

41. f is blue, f' is black, and f'' is red   **43.** f is black, f' is
red, and f'' is blue

45.

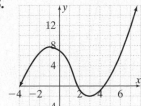

47. If "home improvement" is the function f, then the slogan would say the derivative f' is increasing. **49. a.** concave up: $(1980, 1987)$: concave down: $(1987, 1995)$ **b.** Answers vary; it means that the spread of the epidemic is slowing.

51.

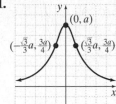

points of inflection: $\left(-\dfrac{\sqrt{3}a}{3}, \dfrac{3a}{4}\right), \left(\dfrac{\sqrt{3}a}{3}, \dfrac{3a}{4}\right)$

concave up: $\left(-\infty, -\dfrac{\sqrt{3}a}{3}\right), \left(\dfrac{\sqrt{3}a}{3}, \infty\right)$

concave down: $\left(-\dfrac{\sqrt{3}}{3}, \dfrac{\sqrt{3}}{3}\right)$

53.

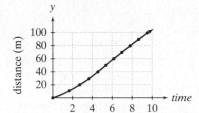

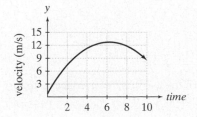

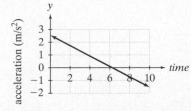

$t_s - 6.27$, $v_{max} = 12.49$; average deceleration is about 0.66

55.

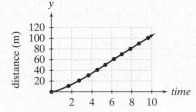

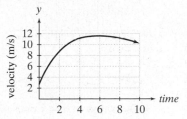

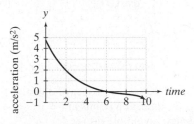

$t_s = 5.81$, $v_{max} = 12.00$; average acceleration is about -0.30 (a negative acceleration is a deceleration)

57. By definition, $f(a) = q(a) = b$ if and only if $q(a) = b$. Since $q'(a) = c$, $q'(a) = f'(a)$ if and only if $f'(a) = c$. Finally, since $q''(a) = 2d$, $q''(a) = f''(a)$ if and only if $d = f''(a)/2$.

Problem Set 3.7, Page 273

1. a. $\lim\limits_{x \to \pi} \dfrac{1 - \cos x}{x}$ is not of the form $\dfrac{0}{0}$ or $\dfrac{\infty}{\infty}$; $\dfrac{2}{\pi}$

b. $\lim\limits_{x \to \pi/2} \dfrac{\sin x}{x}$ is not of the form $\dfrac{0}{0}$ or $\dfrac{\infty}{\infty}$; $\dfrac{2}{\pi}$ **3.** $\dfrac{3}{2}$ **5.** 1 **7.** 0

9. 0 **11.** e^{-6} **13.** 1 **15.** e **17.** $-\dfrac{1}{2}$ **19.** $y = 0$ **21.** $y = 1$

27. a. As $x \to 0$, growth rate is 0 kg/month; as $x \to \infty$, growth rate is 2 kg/month **b.** At three days **c.** growth acceleration: $(0, 0.1)$; growth deceleration: $(0.1, \infty)$

d.

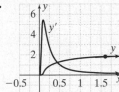

29. x^n grows faster **31.** f and g must be differentiable functions at all points on an open interval containing a (except possibly a itself) where each statement is proceeded by $\lim\limits_{x \to a}$.

Chapter 3 Review Questions, Page 275

1. a. $3x^2 + \dfrac{3}{2}\sqrt{x} + 3\cos 3x$ **b.** $\dfrac{-y}{x + 3y^2}$

c. $\dfrac{2(x^2 - 1)(5x - 1)\ln(x^2 - 1) - 6x^2(x - 2)}{3x^{4/3}(x + 1)(x - 1)(x - 2)^4}$

d. $\dfrac{1}{2}xe^{-\sqrt{x}}(4 - \sqrt{x})$ **e.** $\dfrac{\sqrt{x}}{2}(3\cos 2x - 4x\sin 2x)$

f. $\dfrac{\pi}{4}\sin\left(\dfrac{\pi x}{2}\right)$ **3.** $2(2x - 3)(40x^2 - 48x + 9)$

5. first order: $y = 1$
second order: $y = 1 + x^2$

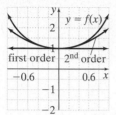

7. graphs vary

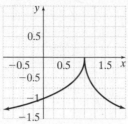

Derivative at $x = 1$ does not exist.

9. increasing: $(-\infty, -25) \cup \left(\dfrac{5}{3}, \infty\right)$; decreasing: $\left(-25, \dfrac{5}{3}\right)$;

concave up: $\left(-\dfrac{35}{3}, \infty\right)$; concave down: $\left(-\infty, -\dfrac{35}{3}\right)$

11. concave up on $(-\sqrt{3}, 0) \cup (\sqrt{3}, 0)$; concave down on $(-\infty, -\sqrt{3}) \cup (0, \sqrt{3})$; inflection points at $t = 0$ $t = -\sqrt{3}$, and $t = \sqrt{3}$, and asymptote $g(t) = 1$

13. a. $H = 100\tan\theta$ **b.** 196 ft   **c.** 27.2% error is the estimate of H.   **15.** $f'(19.98) \approx 0.299$ micrograms/hour

17. $\left(\pm\dfrac{1}{\sqrt{2}}, \dfrac{1}{2}\right), \left(\pm\dfrac{1}{\sqrt{2}}, -\dfrac{1}{2}\right)$ **19.** $y = -200x + 100\pi + 1$

CHAPTER 4

Problem Set 4.1, Page 287

1.

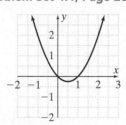

3.

5.

7.

9.

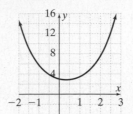

11.

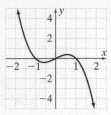

13.

15.

17.

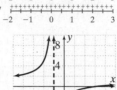

19.

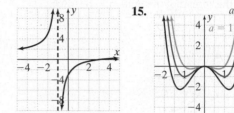

21.

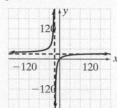

23. As a changes from positive to negative, the graph of the parabola goes from being concave up to being concave down.

25. $y''(x) = \dfrac{(a - b)e^x(1 - e^x)}{(1 + e^x)^3}$
If $b > a$, concave down for $x < 0$; if $b < a$, concave up for $x < 0$; point of inflection at $x = 0$

27.

29. a. $y = 58.7$; $x = -0.76$; **d.**
b. increasing on
$(-\infty, -0.76) \cup (-0.76, \infty)$;
thus, the function is increasing
on its domain
c. up: $(-\infty, -0.76)$; down
$(-0.76, \infty)$

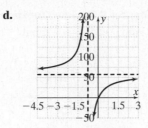

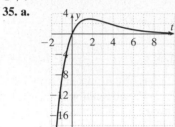

31. $C''(t) = 2.574e^{(-0.3t)} - 28.6e^{(-t)}$; $C''(t) < 0$ for $t < 3.4$ and $C''(t) > 0$ for $t > 3.4$ **33.** Answers vary: one possibility is $\dfrac{2t}{1+t}$; another possibility is $2 - e^{-t}$.

35. a.

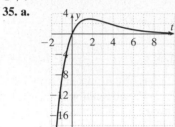

For this problem, the biological domain is $[0, \infty)$.

b. Maximum concentration of 2.92 is achieved at $t = 5 \ln \dfrac{7}{5}$.

Problem Set 4.2, Page 297

1. Local max at $x = -0.6$; global max at $x = 0.8$; local min at $x = -0.1, 1$; global min at $x = -1$ **3.** global min at $x = -1$, local min at $x \approx 0.25$, local max at $x = -2, 1$ **5.** There is one critical point at $x = -\dfrac{3}{8}$ and it is a local minimum. **7.** $t = 0$ is a local minimum, $t = 2$ is a local maximum. **9.** The only critical point is at $x = -1$, at which f is not defined.

11. $x = \dfrac{1}{2}$ is a local minimum, $x = -2$ is a local maximum
13. There is a local maximum at $x = -1$ and a local minimum at $x = 4$. **15.** There is a local maximum at $x = -3$ and local minimum at $x = -1$. **17.** There is a global maximum at $x = 0$ and a global minimum at $x = 2$. **19.** There is a global minimum at $x = 1$ and a global maximum at $x = 0.1, 10$.
21. There is a global minimum at $x = 2$ and there is no global maximum. **23.** There is a global minimum at $x = 1$. **25.** Find all the critical values. Evaluate f at these critical values and at $x = a$. Let M and m be the largest and smallest values, respectively. Find $\lim\limits_{x \to b^-} f(x) = L$, assuming it exists. If $M > L$, then M is the global maximum. If $m \le L$, then m is the global minimum. **27.** Global minimum at $x = 2$; no global maximum. **29.** Global minimum at $x = -1$ and no global maximum **33.** There is a global minimum of 365.1871 at $x \approx 317.8500$ and a global maximum of 371.9310 at $x \approx 324$.
35. The global maximum is 890 at $t = 17$.

37. a. The maximum occurs at $x = r$ because $x = r$ is the only critical value and $P(0) = 0$ and $\lim\limits_{x \to \infty} P(x) = 0$.

b. y

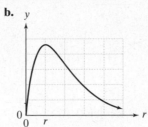

39. $v = \left(\dfrac{w^2}{3A\rho^2 S}\right)^{1/4}$

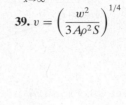

Problem Set 4.3, Page 311

1. $x = 27.6249$ **3.** 32.3667 **5.** 9,600/year **7.** $h = 1.05$
9. $\theta = 1.3720$ **11.** $\theta = 1.2489$ **13.** Elvis should run 12.1 meters along the shore. **15.** Elvis should run $15 - 0.1414d$ meters along the shore.

17. 43,

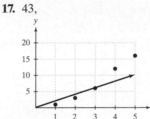

19. 42.46,

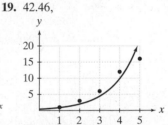

21.

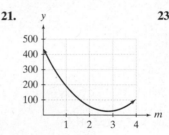

23. Elvis should run $k - 0.1414d$ meters along the shore.

25. a. $V(t) = 0.4388e^{-0.1111t} + 0.0012e^{0.239t}$ **b.** 14.6692 days
c. The model overestimates what the data indicate by about a day. **27.** $r = \sqrt[3]{\dfrac{V}{2\pi}}$ **29.** $N = 124.0326$ **31.** $x = \dfrac{K}{(1+\alpha)^{1/\alpha}}$
33. 65.05 meters **35.** Two crews (1.86 crews) will minimize cost at \$46,000. **37.** If the rate at which energy is accumulated is $E = ar^2 - br^3$, then E is maximized when $r = \dfrac{4a}{b}$. **39.** $P' = \dfrac{-3(2M - 9)}{2M^{1/4}(2M + 3)^2}$ which is zero at $M = 4.5$. This is a maximum since $P' > 0(< 0)$ for M just less (greater) than 4.5. **41. a.** $a^* = 6.43$ **b.**

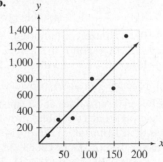

43. $r^* = 0.64$;

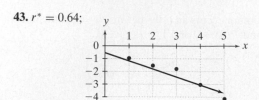

Problem Set 4.4, Page 324

1. 10 seconds **3.** 4.47 seconds **5.** 4.08 seconds

7.

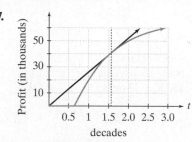

max ≈ 15.7 years

9.

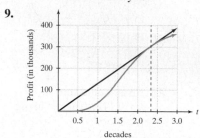

max ≈ 23.4 years

11. Patch 2 **13.** Patch 2 **15.** Patch 1 **17.** 3.11 years
19. 4.93 years **21.** 3.71 years **23.** 2.43 years **25.** 2.38 years
27. $h = 41.6200$ cm **29. a.** 181.66 **b.** 0.54 **31.** 41.48
33. 38.4031

Problem Set 4.5, Page 338

1. Equilibria are 0 (unstable), 0.5 (stable), and 1 (unstable).
3. Equilibria are 0 (unstable) and 0.75 (unstable).
5. Equilibrium is 0 (unstable). **7.** Equilibria are 0 (unstable)
and 1 (stable). **9.** Equilibria are 0 (unstable) and 16 (stable).
11. Equilibria are 0 (stable) and 1 (unstable). **13.** Equilibria
are 0 (unstable) and 16 (stable). **15.** Equilibria are 0
(unstable) and $\frac{3}{4}$ (unstable). **17. a.** Equilibria are $x = 0$ and
$x = 100 - \dfrac{100}{r}$. **b.** 0 is stable provided that $r < 1$ (remember
$r > 0$). **c.** Nonzero equilibrium is positive provided that $r > 1$.
d. $r < 3$ corresponds to stability. **19. a.** Equilibria are 0 and
$r - 1$. **b.** $r < 1$ **c.** $r > 1$ **d.** $r > 1$ **21.** Frequency of allele
approaches $\frac{1}{2}$ (stable), provided that initially the frequency is
between 0 (unstable) and 1 (unstable). **23.** Frequencies of
alleles approach 0 (stable) provided that the initial frequency is
less than 1 (unstable) **25.** 1.4142 **27.** −1.3247 **29.** $x_3 = 0.7391$
31. $x = 0.2592$ **33. b.** 1.3668 **35. a.** 0 (unstable), 82.182
(unstable) **39.** $d \approx 0.4817$ **41.** 21.9251 days **43.** 18.7145 days
45. In the first case, only one real root exists, namely −0.1597.
In the second case, three real roots exist, namely −0.1254,
1.3389, and 1.7865. **47.** $x = 561.1286$ months after April 1974.

Chapter 4 Review Questions, Page 341

1. Max, −2 at $x = -1$; min, −6 at $x = 1$
3.

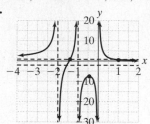

5. a.

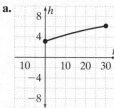

b. Growing most rapidly at
$t = 0$ and least rapidly at
$t = 30$

7. $22,727.15 **9.** 135 days **11. a.** $x_{n+1} = \dfrac{-0.2\,x_n + 1}{-1.2\,x_n + 2}$ **b.** $x = 1$
(unstable) and $x = 0.833$ (stable) **c.** The results imply that as
long as both alleles are present in the population, they will
persist and the frequency of the sickle cell anemia allele will
approach a value of 16.67%. **13.** The bird should spend
$\frac{5}{2} + \frac{1}{2}(\ln 2)$ minutes on the first tree and $\frac{5}{2} - \frac{1}{2}(\ln 2)$ minutes
on the second tree. **15.** $N = 148 \left(\dfrac{1}{1+\theta} \right)^{1/\theta}$ which is an
increasing function of θ **17. a.** $0, \frac{1}{2}(3 \pm \sqrt{5})$
b. $f'(x) = \dfrac{6x}{(x^2 + 1)^2} > 0$ for all $x > 0$ **c.** For initial conditions
$x_0 < \frac{1}{2}(3 - \sqrt{5})$, x_n converges to 0. For initial conditions
$x_0 > \frac{1}{2}(3 - \sqrt{5})$, x_n converges to $\frac{1}{2}(3 + \sqrt{5})$.

19. a.

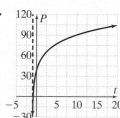

b. $P'(t) = \dfrac{21.364}{0.98t + 1}$

CHAPTER 5

Problem Set 5.1, Page 355

1. $2x + C$ **3.** $x^2 + 3x + C$ **5.** $\frac{6}{5}x^5 + C$ **7.** $\frac{2}{3}x^3 - 5x + C$
9. $2t^4 + \frac{15}{2}t^2 + C$ **11.** $-\frac{5}{x} + C$ **13.** $\sin x + C$
15. $-\frac{3}{2\pi} \cos(2\pi x) + C$ **17.** $3e^x + C$
19. $\frac{2}{5}x^{5/2} + \frac{2}{3}x^{3/2} + \ln|x| + C$ **21.** $3u^2 + 3\sin u + C$
23. $2x + 1$ **25.** $x^2 + 3x$ **27.** $\frac{6}{5}x^5 - \frac{16}{5}$

29. a. $F(x) = x - 2x^2 + 1$
b.

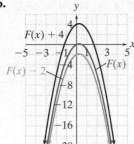

c. $C = -\dfrac{9}{8}$

31.

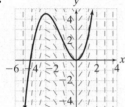

33.

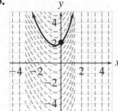

35.

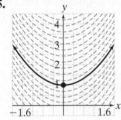

37.

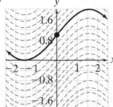

39. $-\dfrac{1}{a}\cos(ax)$ **41. a.** $0.02885x + \dfrac{0.0112}{2\pi}\sin(2\pi x)$
b. 34.7 days   **43. a.** $0.067t - 0.0066\cos(2\pi t) + 0.0066$
b. 14.9 days   **45.** 280 km/h   **47.** 1.24035 s on the moon and 0.5 s
on Earth   **49.** 2,128 people   **51.** $v(R) = \dfrac{a(r^2 - R^2)}{2}$ **53.** yes

Problem Set 5.2, Page 366

1. 2.25   **3.** 3.75   **5.** 27   **7.** 0.859   **9.** 0.635   **11.** 1.183   **13.** 2.2
15. 1.076   **17.** 3.7557   **25.** 63.48 days   **27.** 49.6545   **29.** 8,857;
more accurate   **31.** 14,600 deaths   **33.** 45.4 degree-days
35. 17.5 degree-days   **37.** 299 degree-days   **39.** 37.8 m
41. 950 ft^2

Problem Set 5.3, Page 376

1. $\displaystyle\int_0^1 x\,dx$ **3.** $\displaystyle\int_0^1 (-6 + 9x)\,dx$ **5.** $\displaystyle\int_0^1 (1 - x^2)\,dx$

7. $\displaystyle\lim_{n\to\infty}\sum_{i=1}^n \left(1 + \dfrac{i}{n}\right)^4\dfrac{1}{n}$ **9.** $\displaystyle\lim_{n\to\infty}\sum_{i=1}^n e^{i/n}\dfrac{1}{n}$

11. $\displaystyle\lim_{n\to\infty}\sum_{i=1}^n \left|-1 + \dfrac{2i}{n}\right|\dfrac{2}{n}$

13. 14

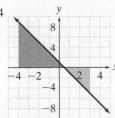

15. 4π

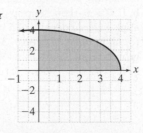

17. $-\dfrac{1}{3}$ **19.** $\dfrac{3}{2}$ **21.** $-\dfrac{1}{2}$ **27.** $\dfrac{2}{3}$ **29. a.** 13   **b.** -29 **31.** 10

33. 1.92; less than   **35.** 1.1885; less than

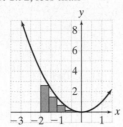   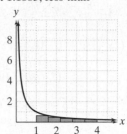

37. $\dfrac{45}{4}$ **39.** true   **41.** false

43. $\displaystyle\int_0^{15} [10 + 10\sin(2\pi x)]\,dx = 150;\ 5\%$

45. $\displaystyle\int_0^{25} 4x\,dx + \int_{25}^{50} (200 - 4x)\,dx;\ 2{,}500$ individuals

47. $\displaystyle\int_{-a}^a f(-x)\,dx = \int_{-a}^0 f(x)\,dx + \int_0^a f(x)\,dx = 0$
as the area under $[-a, 0]$ is equal to the area under $[0, a]$.

51.

n	**Reimann sum**
4	1.896119
8	1.97432
16	1.99357
32	1.998393
2,048	2.000000

Problem Set 5.4, Page 384

1. a. 120   **b.** $8a + 16$ **3. a.** $\dfrac{52}{3}$ **b.** $2 + \dfrac{1}{2}\pi^2$ **5. a.** 18

b. $\dfrac{5}{8} + \pi^2$ **7. a.** $\dfrac{-2^\pi + 2(4^\pi)}{\pi + 1}$ **b.** $-1 + e^2$ **9. a.** $t^4 + t^3 + C$

b. $-2t^4 + \dfrac{5}{2}t^6 + C$ **11. a.** $-3\sin u + C$ **b.** $\dfrac{5}{4}t^4 - \dfrac{2}{3}t^{3/2} + C$

13. a. $\dfrac{2}{15}x^{3/2}(5 + 3x) + C$ **b.** $-\dfrac{1}{2}t^2 + \dfrac{2}{5}t^{5/2} + C$

15. a. $\dfrac{1}{2}x^2 + 2x + C$ **b.** $\dfrac{1}{2}x^2 - x + C$ **17.** $\sqrt{1 + x^2}$

19. $\dfrac{1}{2 + \sin x^2}$ **21.** $-\dfrac{e^x}{x}$ **23.** $f(t) = -2\sin(2x)$ and $a = -1$

25. a. $2\sqrt{x} - 4x + 2$
b.

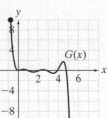

27. a. local maxima:
0.5, 2.5, 4.5; 0.25, 2, 4 local
minima: 0, 1.5, 3.5
b. concave up on $[0, 0.25]$,
$[1, 2], [3, 4]$ **c.** global max at
$x = 4.5$ and global min at
$x = 3.5$.
d.

29. 14.587 cm

31. a. $\displaystyle\int_0^t [18 + 17\sin(2\pi x)]\,dx = 1{,}597$ **b.** 88.4342 days

Problem Set 5.5, Page 393

1. a. 32 **b.** $2\sqrt{3} - 2$ **3. a.** 1 **b.** $\frac{1}{2} \sin \pi^2$ **5. a.** $\frac{128}{5}$

b. $\frac{12}{5} \sqrt[4]{3} - \frac{8}{5} \sqrt[4]{2}$ **7. a.** $\frac{2}{9} x^{9/2} \sqrt{2} + C$ **b.** $\frac{2}{3}(2x^3 - 5)^{3/2} + C$

9. $\frac{1}{10}(2x + 3)^5 + C$ **11.** $\frac{1}{3}(x^2 + 4)^{3/2} + C$ **13.** $\ln(\sin x) + C$

15. $\frac{1}{2} \ln^2 x + C$ **17.** 1,872 **19.** $e - e^{1/2}$ **21.** 0.3045 **23.** $\frac{16}{15}$

25. $e^{\sin x} + C$ **27.** −6 **29.** 44,624 **31.** $N(t) = 10e^{-(\ln 2)e^{-t^2/2}}$

33. 2.4659 moose (on average)

Problem Set 5.6, Page 401

1. $-(x + 1)e^{-x} + C$ **3.** $\frac{-x^2}{4} + \frac{x^2 \ln(x)}{2} + C$

5. $\sqrt{x} \ln x - 2\sqrt{x} + C$

7. $-\frac{3}{13} \cos(3x)e^{2x} + \frac{2}{13} \sin(3x)e^{2x} + C$

9. $-\frac{1}{2}x \cos^2 x + \frac{1}{4} \sin x \cos x + \frac{1}{4}x + C$ **11.** $1 - \frac{5}{e^4}$

13. $6e \ln 3 + 3e - 2$ **15.** $\pi - 2$

17. $-\frac{1}{1,000} \ln|-1,000 + N| + \frac{1}{1,000} \ln|N| + C$

19. $\ln(x - 1,000) + C$

21. $\frac{1}{6} \ln(x - 2) - \frac{1}{2} \ln x + \frac{1}{3} \ln(x + 1) + C$

23. $\frac{1}{2}x \cos(\ln x) + \frac{1}{2}x \sin(\ln x) + C$

25. $\sin(\ln x) - \ln x(\cos(\ln x)) + C$

27. $2 \ln(e^x + 2) - \ln(e^x + 1) + C$

29. $\frac{1}{2}x^2 + \frac{1}{2} \ln(x - 1) + \frac{1}{2} \ln(x + 1) + C$ **31.** $x \ln x - x + C$

33. $\frac{1}{2} e^x \cos x + \frac{1}{2}e^x \sin x + C$ **37.** $13,212

39. $\frac{405}{2} \ln(3) - 240 \ln 2 - \frac{85}{8}$ which yields 45,489 individuals

41. 70.054

Problem Set 5.7, Page 414

1. $L_4 = 1.9688$, $R_4 = 2.7188$, $M_4 = 2.3281$, $S_4 = 2.3333$
3. $L_4 = 0.8942$, $R_4 = 0.2681$, $M_4 = 0.4594$, $S_4 = 0.4548$
5. $L_4 = 0.8453$, $R_4 = 0.7203$, $M_4 = 0.7867$, $S_4 = 0.7854$
7. $L_6 = 0.5111$, $R_6 = 0.2337$, $M_6 = 0.4175$, $S_6 = 0.4029$
9. $L_4 = 0.8453$, $R_4 = 0.7692$, $M_4 = 0.8376$, $S_4 = 0.8358$
11. $L_6 = 1.1607$, $R_6 = 1.1607$, $M_6 = 0.7995$, $S_6 = 1.0356$
13. Using $K_1 = 2$, we need $n > 400$ so $L_{801} \approx 3.2465$
15. Using $K_4 = 1$, $n > 2.05$ so $S_1 = 0.8649$
17. Using $K_2 = 2$, $n > 5.68$, $M_6 \approx 0.7854$
19. Using $K_1 < 3/2$, $n > 750$, $R_{751} \approx 1.63275$
21. Using $k_4 = 20$, we need $n > 1.53526$ so $n = 2$ works,
$S_2 = 0.172767$ **23. a.** $n \geq 213$ **b.** $n \geq 30$ **25. a.** $n \geq 82$
b. $n \geq 30$ **27. a.** $n \geq 1,708$ **b.** $n \geq 2$ **29.** $L_6 \approx 1.75$, $R_6 \approx 2.25$,
$S_6 \approx 2$ **31.** 791,824 **33. a.** $\int_0^{30} 890 \operatorname{sech}^2(0.2t - 3.4)\, dt$

b. 8,882 deaths **35. a.** 22.7792 degree-days **b.** 69.668 days
37. 15,467 cases **39.** Approximately 314 cm².

Problem Set 5.8, Page 426

1. 352.7633 **3.** 222.3130 **5.** 336.1833 **7.** 187,446 **9.** 349,823
11. 42,149 **13.** $2,143,900 **15.** $1,474,100 **17.** 270 ft-lb
19. 12,750 ft-lb **21.** 5 mg **23. a.** $521,580 in account, but paid
in only $43,000 **b.** $410,930 in account, but paid in only
$66,000 **25.** 5.2083 L/min **27. a.** 0.1083 F **b.** 542.9363 cc/min
29. 80 ft-lb **31.** 224,000 Cal, which is very similar to required
work **33.** 138.1 Cal, which requires 13.8 servings of pasta

Chapter 5 Review Questions, Page 428

1. $2\sqrt{x} + C$ **3.** $-\frac{1}{2} + \frac{1}{8}\pi^2$ **5.** $2 \ln 2 - \ln 3$

7. $\ln x - \frac{1}{x} - 1$; 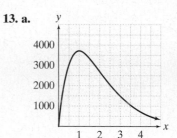 **9.** $2 \sin(4x^2)$
11. 945 patients

13. a.

The graph shows the rate of increase starting low (as only a few are infected), increasing to a maximum, and decreasing to zero (as everyone becomes infected).

b. $10,000\left(1 - \frac{1 + T}{e^T}\right)$

c. $T \approx 1.65$ months

15. $\lim_{n \to \infty} \sum_{i=1}^{n} \tan\left(3 + \frac{4i}{n}\right)\frac{4}{n}$ **17.** $\frac{2}{3}\sqrt{2}$ **19.** $\frac{1}{2}$

CHAPTER 6

Problem Set 6.1, Page 442

1. $\frac{dy}{dt} = ky$ where y is the number of bacteria and k is the
constant of proportionality **3.** $\frac{dT}{dt} = k(A - T)$ where $k > 0$
is the constant of proportionality **5.** If N is the size of the
population, then $\frac{dN}{dt} = ke^{-at}N$ where $k > 0$ and $a > 0$ are
constants. **7.** $\frac{dy}{dt} = ky(P - y)$ where $k > 0$ is the constant
of proportionality **9.** doubles in 1.08 years **11.** halves in
1.69 months **13.** halves in 2,310 years **15.** 6.93 years
17. 921 years **19.** $P = 0, 100$ **21.** $P < 0$ or $P > 100$
23. $P = 0, 1, 100$ **25.** $0 < P < 1$ or $P > 100$ **27.** $p = 0$,
$p = 1$, $p = 2$ **29.** $\lambda = 0.13863$ **31.** 4,103 years **33.** 756 years;
too recent **35. a.** For $P < 5,000$, the population is driven to
extinction; if $P > 5,000$, the population declines to 5,000.
b. If $P > 0$, then the population density monotonically
approaches a density of 5,000. **c.** The population goes extinct
for all initial conditions. **37. a.** In the first two years, the
growth from 0.30% to 1.1% implies $r \approx 0.6496$.
b. $K \approx 72.0\%$ **39. a.** 1.3863 **b.** about 750

Problem Set 6.2, Page 452

9. yes **11.** no **13.** yes **15.** no **17.** $y = \pm\sqrt{\dfrac{1}{C - 2t}}$

19. $y = \sin t + C$ **21.** $y = \ln(t + C)$ **23.** $y = Ce^{(3/2)x^2}$

25. $y = Ce^{2\sqrt{1+x^2}}$ **27.** $y = \left(\dfrac{C + x^{3/2}}{3}\right), C \geq 0$

29. $y = \dfrac{1}{\dfrac{1}{3} - t} - 1$ **31.** $y = \sqrt{\dfrac{11 - 2e^{-t}(t + 1)}{2}}$

33. $\dfrac{1}{2}y^2 + e^y = t + \dfrac{t^2}{2} + \dfrac{1}{2} + e^4$ **35.** $y = \dfrac{1}{1 + e^t}$

37. $g(y) = y + \ln y - 5$

39. a. 2026; an earlier date than predicted in Example 7.

41. a. $y(t) = 100^{1-e^{-t/2}}$

b.

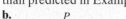
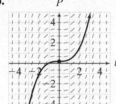

b.

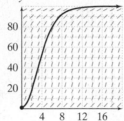

43. $y = a\left(1 - \dfrac{1}{1 + kat}\right)$

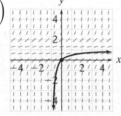

Problem Set 6.3, Page 462

1. $c = 6.9315, P = 1,310,000$ **3.** $c = 4.1589, P = 936,000$
5. $c = 2.7726, P = 873,370$ **7.** 2.385 h **9.** 9.467 h **11.** 20.007 h
13. $21.54(1 - e^{-0.4642t})$ mg **15.** $31.88(1 - e^{-0.3764t})$ mg
17. $12.55(1 - e^{-1.5936t})$ mg **19. a.** yes at $t = 0.5682$; **b.** reduce
inflow to less than 0.46 km³/yr **21. a.** yes at $t = 1.13637$;
b. reduce inflow to less than 0.46 km³/yr **23. a.** yes at
$t = 0.55125$; **b.** reduce inflow to less than 0.36 km³/yr
25. 10,576,000,000; actual, 10,881,909,000 **27.** 2,269,000
29. $\dfrac{a}{b}$; half this value at $t = \ln 2/b$ **31.** 2.87528 mg/h

33. 246.6178 mg/h **35.** 0.882444/h **37.** $y(t) = 600e^{-t/150}$
39. 16.4951 min **41.** females: 0.654196 yr; males: 0.895777 yr

Problem Set 6.4, Page 472

1.

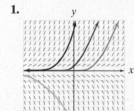

3.

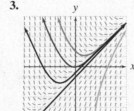

5.

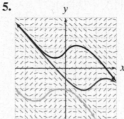

7.

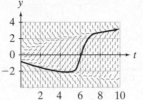

9.

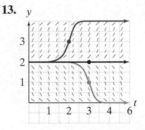

11. a. D **b.** B **c.** C **d.** A

13.

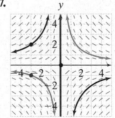

15.

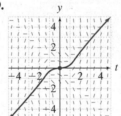

17.

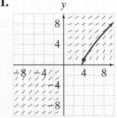

19.

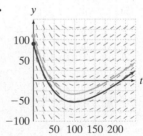

21.

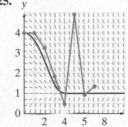

23.

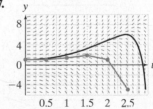

25.

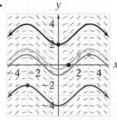

27.

29. $(0, 0.1)$, $(0.5, 4.195)$, $(1.0, 3.59099)$, $(1.5, 4.73438)$, $(2.0, 2.26158)$

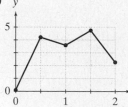

31. $(1, 0)$, $(1.5, 0.0000)$, $(2, -0.5000)$, $(2.5, -0.0000)$, $(3, 0.5000)$, $(3.5, 0.0000)$

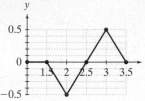

33. b. $(2, 0.8333)$ **35. a.** $\dfrac{dy}{dt} = 100 - 0.3466y$; $y(0) = 0$

c. 288.5

b.

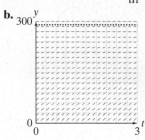

37. a.   **b.**

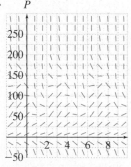 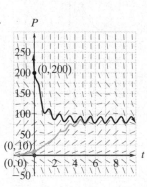

c.

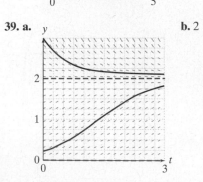

39. a.   **b.** 2

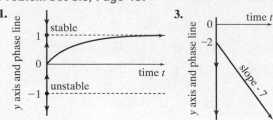

Problem Set 6.5, Page 487

1.   **3.**

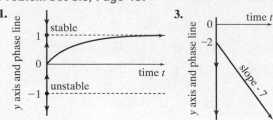

5.

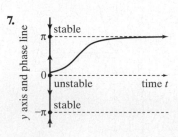

7.

9.

11.

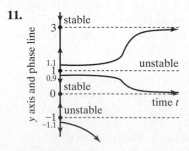

13.

15. $y^* = 2$ is stable   **17.** $y^* = \dfrac{\pi}{4}$ is unstable

19. $y^* = 3$ is stable

21. **23.**

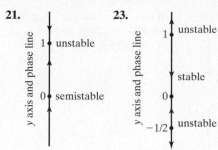

25. **27.**

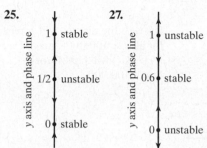

29. a. $\dfrac{dy}{dt} = y(1-y)(1.5y-1)$ **b.**

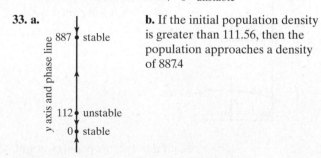

31. a. $\dfrac{dy}{dt} = -Cy(1-y)$ **b.**

33. a. **b.** If the initial population density is greater than 111.56, then the population approaches a density of 887.4

35. It is stable at $V = -65$ and semistable at $V = 40$. It can switch, but the smallest negative current when it is in state $V = 40$ will cause the neuron to switch back to $V = -65$.

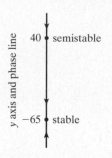

37. Let $x = y - y^*$. Then $\dfrac{dx}{dt} = ax$ and $x(0) = y_0 - y^*$.

Hence, $x(t) = (y_0 - y^*)e^{at}$ and $y(t) = y^* + (y_0 - y^*)e^{at}$

39. By the chain rule,

$$\frac{dy}{dt} = \frac{\dfrac{dN_a}{dt}(N_a + N_A) - N_a\left(\dfrac{dN_a}{dt} + \dfrac{dN_A}{dt}\right)}{(N_a + N_A)^2}$$

$$= r_a y[r_a y + r_A(1 - y)]$$

$$= y(1 - y)(r_a - r_A)$$

41. b. stable if $b > 1$ or $b = 1$ and $c > a/k$

43. a. $a = 1$ **b.** $a = 2$ **c.** $a = 2.5$

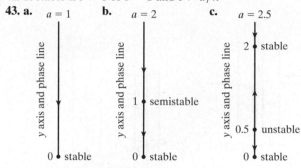

d. Answer vary.

Problem Set 6.6, Page 500

1. $a = 0$ $a = 9$ $a = 25$

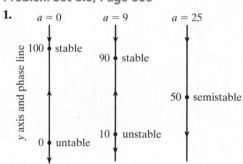

3. $a = -10$ $a = 0$ $a = 10$

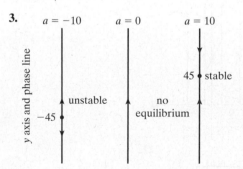

5. $a = 0$ $a = 2$ $a = 4$

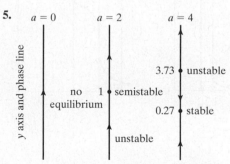

7.

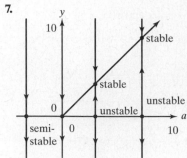

9.

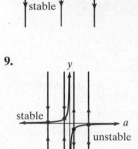

11.

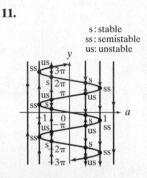

13.

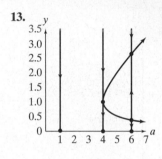

15.

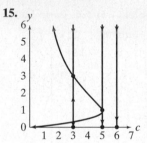

17.

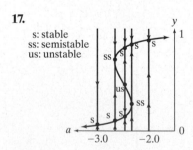

19.

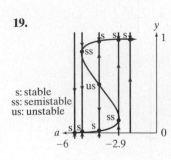

21.

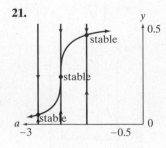

23.

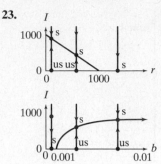

If $b = 1$, the disease will persist for all $r < 1,000$; and if $r = 1$, it will persist for $b > 1/1,000$.

25.

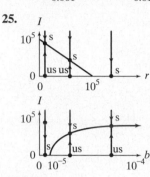

If $b = 1$, the disease will persist for all $r < 100,000$; and if $r = 1$, it will persist for $b > 1/100,000$.

27.

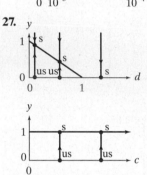

If $c = 1$, the population will persist for all $d < 1$; and if $d = 0$, it will persist for $c > 0$.

29.

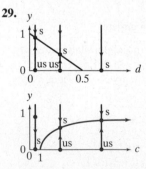

If $c = 1$, the population will persist for all $d < 0.5$; and if $d = 0.5$, it will persist for $c > 1$.

31.

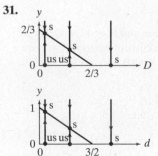

If $d = 1/2$, the population will persist for all $D < 2/3$; and if $D = 0$, it will persist for $d < 3/2$.

33.

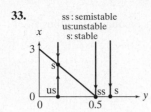

ss : semistable
us:unstable
s: stable

If $y < 0.5$, the population x persists; otherwise, it is driven to extinction for $y \geq 0.5$.

35.

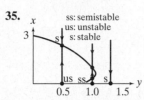

ss: semistable
us: unstable
s: stable

The population is driven to extinction for $y \geq 1$.

37.

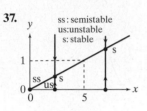

ss : semistable
us:unstable
s: stable

The predator persists at all prey densities $x > 0$ at a level given by the equation $y = 0.2x$.

39. a. $\dfrac{dy}{dt} = -y(1-y)\,((C-B)(n-1)y+C)$

$y = \dfrac{C}{(B-C)(n-1)}$

b.

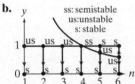

ss: semistable
us:unstable
s: stable

Isoclines are $y = 0$, $y = 1$, and for $B \neq C$ and $n > 1$.

c. If $B = 4$ and $C = 3$, a stable cooperative equilibrium exists only if $n > 4$.

Chapter 6 Review Questions, Page 503

1. $a = 1, b = 2$, and $c = 2$ **3.** $L(t) = 26.1(1 - e^{(-0.25(t+1.64))})$ mm
5. $(2y+1)e^{-2y} + 2(\sin x - \cos x)e^x + 1 = 0$

7.

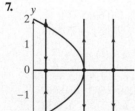

9. 7.182%; the percentage lost is always the same
11. approximately 2,500 individuals   **13.** about ten days
15. stable for $a < 0$ and unstable for $a > 0$

17. a.

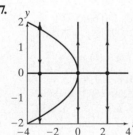

b. As the forest matures, there is a crucial moment when a sudden budworm outbreak occurs.

19. a.

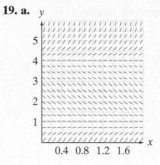

b.

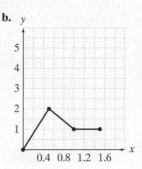

c.

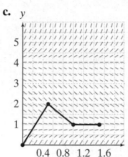

The approximate solution is poor as it crosses the equilibrium $y = 1$, and then lands on $y = 1$. The slope field suggests that the true solution should be increasing from 0 and only asymptotically approach $y = 1$ as $t \to \infty$.

CHAPTER 7

Problem Set 7.1, Page 522

1.

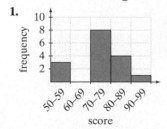

3.

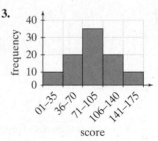

5. a. $\dfrac{3}{16}$ **b.** $\dfrac{3}{16}$ **c.** $\dfrac{3}{4}$ **d.** $\dfrac{1}{16}$ **7. a.** $\dfrac{30}{95}$ **b.** $\dfrac{85}{95}$ **9.** $\dfrac{1}{4}, \sqrt{2}, 1, \sqrt{3}$
11. 3, 0.7937, 0.6300, 0.9086   **13.** 75th percentile for weight and 50th percentile for height   **15.** >99th percentile for weight and greater than the >99th percentile for height

17. $F(x) = \begin{cases} 0 & \text{if } x \leq 0 \\ \dfrac{x}{20} & \text{if } 0 < x \leq 20 \\ 1 & \text{if } x > 20 \end{cases}$

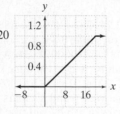

19. a. $y(t) = y_0 e^{-0.25t}$ **b.** $F(t) = 1 - e^{-0.25t}$ for $t \geq 0$
c. $F(1) - F(0) = 1 - e^{-0.25}$ **21. a.** $c = \dfrac{1}{2}$ **b.** $\dfrac{1}{4}$

23. $F(x) = \begin{cases} 0 & \text{if } x < 0 \\ \dfrac{x^2}{4} & \text{if } 0 \leq x < 1 \\ \dfrac{1}{4} + \dfrac{x-1}{2} & \text{if } 1 \leq x < 2 \\ \dfrac{3}{2}x - \dfrac{x^2}{4} - \dfrac{5}{4} & \text{if } 2 \leq x \leq 3 \\ 1 & \text{if } x > 3 \end{cases}$

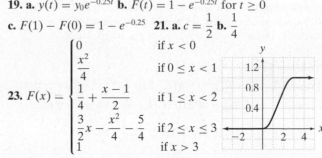

25. a. F is a continuous, nondecreasing, nonnegative function; $\lim_{x \to -\infty} F(x) = 1$; F is a CDF   **b.** $F(1) - F(0) = \dfrac{1}{2}$ and $F(10) - F(2) = 0.24$ **27. a.** 94%   **b.** 0.2%

29. a.

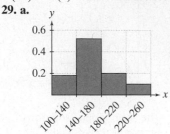

b. 0.82   **c.** 0.90

31. a.

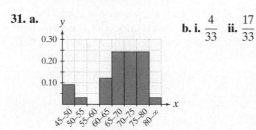

b. i. $\dfrac{4}{33}$ **ii.** $\dfrac{17}{33}$

33. a. $1 - \displaystyle\int_{13}^{14} f(x)\,dx$ **b.** $\displaystyle\int_{12}^{15} f(x)\,dx - \int_{13}^{14} f(x)\,dx$

35. a. f is nonnegative and the area under f is given by $\displaystyle\int_{0}^{2}\left(1 - \dfrac{x}{2}\right)dx = 1$ **b.** $\dfrac{1}{4}$ **c.** $\dfrac{13}{64}$ **37. a.** $y(t) = y_0 e^{-ct}$

b. $F(t) = 1 - e^{-ct}$ **c.** F is a continuous, nondecreasing, nonnegative function; $\lim_{x \to -\infty} F(x) = 0$; F is a CDF

d. $e^{-cr} - e^{-cs}$

Problem Set 7.2, Page 538

1. $\dfrac{1}{4}$ **3.** divergent   **5.** divergent   **7.** 2   **9.** divergent

11. convergent   **13.** convergent   **15.** $F(x) = \dfrac{e^x}{1 + e^x}$

17. $F(x) = \begin{cases} 0 & \text{if } x < 1 \\ 1 - \dfrac{1}{x} & \text{if } x \geq 1 \end{cases}$ **19.** $f(x) = \dfrac{e^{-x}}{(1 + e^{-x})^2}$

21. $f(x) = \begin{cases} e^x & \text{if } x \leq 0 \\ 0 & x > 0 \end{cases}$ **23.** 0.40, 0.96, 1.93 years

25. 0.98, 2.24, 4.47 years   **27.** 2.88, 6.93, 13.8 years
29. Simpson's rule: 0.8862; second integral is approximately $2.8134 \cdot 10^{-8} < 0.000001$ **31.** Apply the definition for a doubly infinite integral and the splitting property to obtain the result   **33. a.** 0.0092   **b.** 0.0092   **c.** 0.9908   **35.** 0.068 of the trips lie between 5 and 20 hours   **37.** Any real paint consists of molecules of some fixed size. Since the horn becomes infinitesimally thin, it would be impossible to fill the horn with paint as the molecules would get stuck on their way down the horn.

Problem Set 7.3, Page 554

1. $\mu = 0.8$, $\sigma^2 = 0.2$, $\sigma = 0.4472$ **3.** $\mu = 1$, $\sigma^2 = 0$, $\sigma = 0$

5. $\mu = 4\dfrac{1}{3}$, $\sigma^2 = 9\dfrac{1}{3}$, $\sigma \approx 3.0551$ **7.** $\mu = 1.94$, $\sigma^2 = 0.684$,

$\sigma = 0.823$ **9.** 1   **11.** 0   **13.** 0   **15.** 2   **17.** $\dfrac{1}{3}$ **19.** 3

21. a. $\mu = 0$, $\sigma = \dfrac{1}{2}$ **b.** Chebyshev's inequality is $\dfrac{3}{4}$, but all of the data points lie in the interval

23. a. $\dfrac{9}{25}$

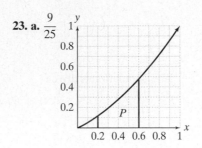

b. $F(x) = \begin{cases} 0 & \text{if } x < 0 \\ x + \dfrac{x^2}{2} & \text{if } 0 \leq x \leq 1 \\ 1 & \text{if } x > 0 \end{cases}$

c. $\dfrac{7}{12}$

25. b. $\mu = \dfrac{4}{3}$, $\sigma^2 = \dfrac{2}{9}$ **27. a.** 157.8   **b.** 215.2   **29.** $\sigma^2 = \dfrac{1}{c^2}$, so $\sigma = \dfrac{1}{c}$, the latter being equal to the mean   **31. a.** $\mu = 14.3$ and 23.2 centuries for mussel/clams and fish, respectively   **b.** 0.067 and 0.042 for mussel/clams and fish, respectively   **33. a.** 30.8   **b.** 0.72   **35. a.** 30.45   **b.** 0.594 compared with 0.6   **c.** 0.10 compared with 0.04   **37.** Let $[c, d]$ be any interval. Then,

$$P(Y \in [c, d]) = P(X \in [a + c, a + d])$$
$$= \int_{a+c}^{a+d} f(x)\,dx = \int_{c}^{d} g(x)\,dx$$

Therefore, the PDF for Y is $g(x)$.

Problem Set 7.4, Page 572

1. 0.3023   **3.** 0.2912   **5.** 0.0864, 0.2580   **7.** 0.6915
9. 0.3830   **11.** 0.244   **13. a.** 0.018   **b.** 0.0911   **c.** 408 days

15. a. $y(t) = \dfrac{1}{1 + e^{-t/10}}$ **c.** 0.0498   **17. a.** $y(t) = \dfrac{1}{1 + e^{2.2-t}}$

c. 0.1315   **19.** $r = 0.671$, $a = 2.89$ **21.** $r = 0.360$, $a = 5.44$
23. a. 0.6716   **b.** 0.5804   **c.** 0.0912   **25. a.** 0.683   **b.** 0.067
27. a. 0.7188   **b.** 0.0924   **c.** 0.7072   **29. a.** 0.3788   **b.** 0.3876
c. 0.1368   **31.** $m = 8.42$, $v = 25.3$, $\mu = 1.98$, and $\sigma^2 = 0.30$
33. a. $e^{-e^{-t}}$ **c.** $e^{-e^{-t}-t}$ **d.** 0.873

Problem Set 7.5, Page 585

1. 0.28   **3.** 0.54   **5.** 0.41   **7.** 0.54   **9.** 73%   **11.** 26%
13. $l(t) = e^{(-at-bt^2)/2}$ **15.** 0.0151, 0.0240, 0.0339, 0.0407, 0.0474, 0.0556, 0.0625, 0.05 for $t = 8, 16, 24, 32, 40, 48, 56, 64$,

respectively   **17.** 1   **19.** 1   **21.** ∞ **23.** 2   **25.** $\dfrac{5}{6}$ **27.** 5

29. $-\dfrac{d}{dt}\ln l(t) = -\dfrac{l'(t)}{l(t)} = m(t)$

31. $m(t) = 0.007293e^{0.187t}, t \geq 2$

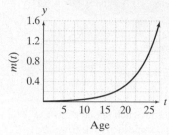

33. $l(t) = e^{0.0285 - 0.0285e^{0.2072t}}$

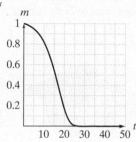

35. $m(t) = 0.0019926e^{0.2214t}$

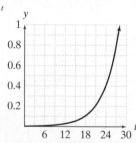

37. a. yes   **b.** $p > 0.59$ **39. a.** untreated $R_0 = 1.76$ and treated $R_0 = 2.02$, so that treatment actually exacerbates the epidemic   **b.** untreated $p = 0.43$ and treated $p = 0.505$

Chapter 7 Review Questions, Page 588

1. a.

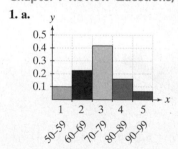

b. 0.93   **c.** 0.23   **3. b.** $F(2) - F(1) = \dfrac{k}{(k+1)(k+2)}$

5. converges for $p > 1$; diverges for $p < 1$

7. CDF: $f(x) = \begin{cases} 0 & \text{if } x < 1 \\ 2\left(1 - \dfrac{1}{x}\right) & \text{if } 1 \leq x \leq 2 \\ 1 & \text{if } x > 2 \end{cases}$

9. mean is $\dfrac{4}{3}$ and variance is $\dfrac{2}{9}$ **11.** $k = \dfrac{3}{2}$, so $P \geq \dfrac{5}{9}$

13. Assuming that all are dead by age 12, the sum is 6.48 (using Simpson's rule).   **15. a.** 76%   **b.** 0.014 per year

17. PDF is $-l'(t)$ and CDF is $1 - l(t)$ **19.** 15.87%

CHAPTER 8

Problem Set 8.1, Page 600

1. $z = 0$, $D = \{(x, y)| -\infty < x < \infty, -\infty < y < \infty\}$, $R = \{-\infty < z < \infty\}$.   **3.** $z = \sqrt{3}$, $D = \{(x, y)| -\infty < x < \infty, -\infty < y < \infty\}$, $R = \{z \geq 0\}$.
5. $z = 0$, $D = \{(x, y)| -\infty < x < \infty, -\infty < y < \infty$ or $y \neq -x/2$ or $y \neq 0\}$, $R = \{-\infty < z < \infty\}$.   **7.** $z = -\dfrac{1}{4}$, $D = \{(x, y)| -\infty < x < \infty, 0 < y < \infty\}$, $R = \{-\infty < z < \infty\}$.

9. $z = 0.7x$

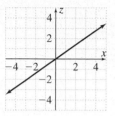

As the prey increase, the population's rate of growth increases.

11. $z = 2.4 - y$

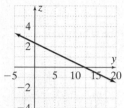

As the predators increase, the population's rate of growth decreases.

13. $z = \dfrac{3}{1+y}$

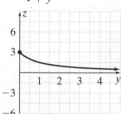

B inhibits production of A, with the rate of change of protein A approaching 0 as B approaches ∞.

15.

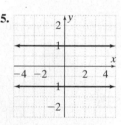

17.

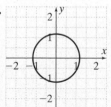

19. No contours exist for $z = 2$.

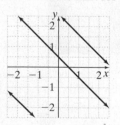

21. Surface has a minimum.

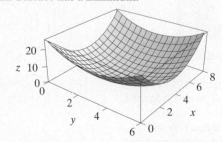

23. Surface has a maximum.

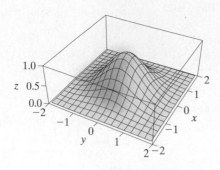

25. Surface is between $z = -2$ and $z = 2$.

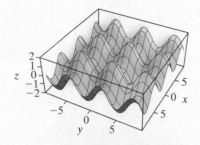

27. Cross sections are hyperbolas.

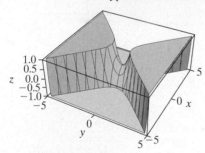

29.

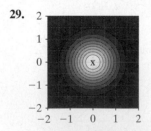

maximum is black X;

31.

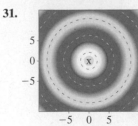

Local minimum at $(0, 0)$ indicated with a black X. Circular ridges of local maxima are shown with dashed black circles. Circular valleys of local minima are shown with dashed white circles.

33.

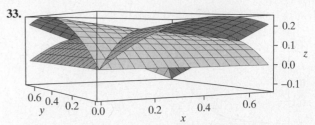

These surface plots indicate that each species has the higher growth rate when its competitor is at low densities, and both species have negative growth rates when densities of both species are high.

35. $z = \dfrac{cx}{k + ax + by}$

37.

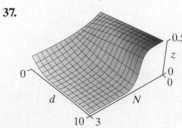

z is an increasing function of N and a decreasing function of d.

39.

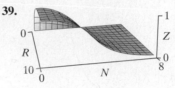

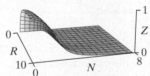

The parameter c acts as a scaling factor for the N axis.

41. a. $f(x, y) = \dfrac{a}{1 + y^c} - x$

b.

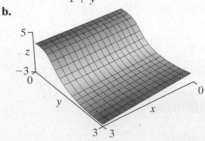

43. a. $(m^*, c^*) = (-57.4, 0.476)$

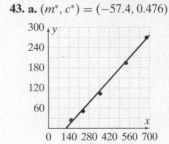

b. $S(m, c) = 126{,}332 - 686{,}362m - 1{,}298.6c + 957{,}512m^2 + 3{,}931.8mc + 5c^2$

c. Black spot on the plot is the location of the point (m^*, c^*).

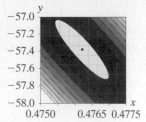

45.

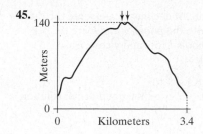

Problem Set 8.2, Page 617

1. $(2, 4)$ **3.** infinitely many solutions **5.** $\left(-\dfrac{6}{5}, -\dfrac{23}{5}\right)$

7. $\begin{pmatrix} 1 & 1 \\ 1 & -1 \end{pmatrix}\begin{pmatrix} x \\ y \end{pmatrix} = \begin{pmatrix} 2 \\ 0 \end{pmatrix}; x = 1, y = 1$

9. $\begin{pmatrix} 4 & 2 \\ 1 & -2 \end{pmatrix}\begin{pmatrix} x \\ y \end{pmatrix} = \begin{pmatrix} 6 \\ -1 \end{pmatrix}; x = 1, y = 1$

11. $\begin{pmatrix} 2 & 3 \\ 3 & -4 \end{pmatrix}\begin{pmatrix} x \\ y \end{pmatrix} = \begin{pmatrix} 4 \\ -4 \end{pmatrix}; x = \dfrac{4}{17}, y = \dfrac{20}{17}$

13. 0; no inverse **15.** 3; $\begin{pmatrix} \dfrac{2}{3} & -\dfrac{1}{3} \\ -\dfrac{1}{3} & \dfrac{2}{3} \end{pmatrix}$ **17.** 8; $\begin{pmatrix} \dfrac{3}{8} & \dfrac{1}{4} \\ -\dfrac{1}{8} & \dfrac{1}{4} \end{pmatrix}$

19. $\mathbf{u} = \begin{pmatrix} 3 \\ 2 \end{pmatrix}$ **21.** $\mathbf{u} = \begin{pmatrix} 3 \\ -2 \end{pmatrix}$ **23.** $\mathbf{u} = \begin{pmatrix} 3 \\ 1 \end{pmatrix}$ **25.** $\mathbf{u} = \begin{pmatrix} 0 \\ 1 \end{pmatrix}$

27. $\mathbf{u} = \begin{pmatrix} 2 \\ 2 \end{pmatrix}$ **29.** $\mathbf{u} = \begin{pmatrix} \dfrac{7}{6} \\ 1 \\ \dfrac{1}{3} \end{pmatrix}$ **31.** $\begin{pmatrix} 1 \\ 1 \\ 1 \end{pmatrix}$ **33.** $\begin{pmatrix} 2 \\ 5 \\ 1 \end{pmatrix}$

35. England: 200 dozen roses, 100 dozen tulips. Germany: 300 dozen roses and 225 dozen tulips **37.** Estimate 0.5 units of type A habitat and 7 units of type B habitat **39.** The pharmacy makes up twice as many tablets for women as it does for men. **41.** 5,882 units of coal **43.** 239,521 units of clean water **45. a.** $\begin{pmatrix} 0.13 & 0.33 & 0.25 \\ 0.50 & 0.17 & 0.25 \\ 0.25 & 0.17 & 0.25 \end{pmatrix}$ **b.** 62.6 units of petroleum, 89.5 units of chemicals, and 59.8 units of transportation **47.** For $\mathbf{u} = \begin{pmatrix} x_1 \\ x_2 \\ x_3 \end{pmatrix}$ with x_1, x_2 and x_3, respectively, representing the proportion of cloudy, rainy, and sunny conditions on any particular day and $\mathbf{v} = \begin{pmatrix} y_1 \\ y_2 \\ y_3 \end{pmatrix}$ with y_1, y_2, and y_3, respectively, representing the proportion of cloudy, rainy, and sunny conditions on the following day: The model is $\mathbf{v} = A\mathbf{u}$ where $A = \begin{pmatrix} 0.5 & 0.3 & 0.3 \\ 0.2 & 0.2 & 0 \\ 0.3 & 0.5 & 0.7 \end{pmatrix}$ and $\mathbf{v} = \mathbf{u}$ because the probabilities are unchanging over time. The solution to the equation $\mathbf{u} = A\mathbf{u}$ is $\mathbf{u} = \begin{pmatrix} 0.38 \\ 0.09 \\ 0.53 \end{pmatrix}$.

Problem Set 8.3, Page 634

1. 3 and -1, respectively **3.** 4 and -1, respectively **5.** $\dfrac{3}{2}$ and $-\dfrac{1}{2}$, respectively **7.** $\begin{pmatrix} 1 \\ 1 \end{pmatrix}$ and $\begin{pmatrix} -1 \\ 1 \end{pmatrix}$, respectively **9.** $\begin{pmatrix} 2 \\ 3 \\ 1 \end{pmatrix}$ and $\begin{pmatrix} -1 \\ 1 \end{pmatrix}$, respectively **11.** $\begin{pmatrix} 1 \\ 2 \end{pmatrix}$ and $\begin{pmatrix} 1 \\ 1 \end{pmatrix}$, respectively

13. The eigenvalues are 1 and 2 and the corresponding eigenvectors are $\begin{pmatrix} 1 \\ 0 \end{pmatrix}$ and $\begin{pmatrix} 0 \\ 1 \end{pmatrix}$, respectively. **15.** The eigenvalues are 1 and 2 and the corresponding eigenvectors are $\begin{pmatrix} 1 \\ 0 \end{pmatrix}$ and $\begin{pmatrix} 2 \\ 1 \end{pmatrix}$, respectively. **17.** The eigenvalues are -10 and 2 and the eigenvectors are $\begin{pmatrix} 1 \\ 1.2 \end{pmatrix}$ and $\begin{pmatrix} 1 \\ 0 \end{pmatrix}$, respectively.

19. The eigenvalues are 4 and 1 and the eigenvectors are $\begin{pmatrix} 1 \\ 1 \end{pmatrix}$ and $\begin{pmatrix} -2 \\ 1 \end{pmatrix}$, respectively. **21.** The eigenvalues are -4 and 4 and the eigenvectors are $\begin{pmatrix} -1 \\ 1 \end{pmatrix}$ and $\begin{pmatrix} 3 \\ 1 \end{pmatrix}$, respectively.

23. The eigenvalues are -9.718 and 7.718 and the eigenvectors are $\begin{pmatrix} 3.544 \\ 1 \end{pmatrix}$ and $\begin{pmatrix} 0.056 \\ 1 \end{pmatrix}$, respectively.

25. $\mathbf{u}_1 = \begin{pmatrix} 24 \\ 48 \end{pmatrix}$, $\mathbf{u}_2 = \begin{pmatrix} 57.6 \\ 48 \end{pmatrix}$, $\mathbf{u}_3 = \begin{pmatrix} 57.6 \\ 61.4 \end{pmatrix}$, $\mathbf{u}_4 = \begin{pmatrix} 73.7 \\ 72.2 \end{pmatrix}$, $\mathbf{u}_5 = \begin{pmatrix} 86.6 \\ 87.2 \end{pmatrix}$, $\mathbf{u}_6 = \begin{pmatrix} 104.7 \\ 104.4 \end{pmatrix}$ **27.** $\mathbf{u}_1 = \begin{pmatrix} 0 \\ 25 \end{pmatrix}$, $\mathbf{u}_2 = \begin{pmatrix} 50 \\ 15 \end{pmatrix}$, $\mathbf{u}_3 = \begin{pmatrix} 30 \\ 21.5 \end{pmatrix}$, $\mathbf{u}_4 = \begin{pmatrix} 43 \\ 20.4 \end{pmatrix}$, $\mathbf{u}_5 = \begin{pmatrix} 40.8 \\ 23 \end{pmatrix}$, $\mathbf{u}_6 = \begin{pmatrix} 46 \\ 24 \end{pmatrix}$

29. $\mathbf{u}_1 = \begin{pmatrix} 100 \\ 45 \end{pmatrix}$, $\mathbf{u}_2 = \begin{pmatrix} 90 \\ 46 \end{pmatrix}$, $\mathbf{u}_3 = \begin{pmatrix} 92 \\ 45.8 \end{pmatrix}$, $\mathbf{u}_4 = \begin{pmatrix} 91.6 \\ 45.8 \end{pmatrix}$, $\mathbf{u}_5 = \begin{pmatrix} 91.7 \\ 45.8 \end{pmatrix}$, $\mathbf{u}_6 = \begin{pmatrix} 91.7 \\ 45.8 \end{pmatrix}$ **31.** $A = \begin{pmatrix} 0 & 2 \\ 0.5 & 0.5 \end{pmatrix}$, $\mathbf{u}_1 = \begin{pmatrix} 100 \\ 50 \end{pmatrix}$, $\mathbf{u}_2 = \begin{pmatrix} 100 \\ 75 \end{pmatrix}$, $\mathbf{u}_3 = \begin{pmatrix} 150 \\ 87.5 \end{pmatrix}$ **33.** $A = \begin{pmatrix} 0 & 1.6 \\ 0.4 & 0.5 \end{pmatrix}$, $\mathbf{u}_1 = \begin{pmatrix} 80 \\ 45 \end{pmatrix}$, $\mathbf{u}_2 = \begin{pmatrix} 72 \\ 54.5 \end{pmatrix}$, $\mathbf{u}_3 = \begin{pmatrix} 87.2 \\ 56.1 \end{pmatrix}$ **35.** $A = \begin{pmatrix} 0 & 1 \\ 0.30 & 0.6 \end{pmatrix}$, $\mathbf{u}_1 = \begin{pmatrix} 50 \\ 45 \end{pmatrix}$, $\mathbf{u}_2 = \begin{pmatrix} 45 \\ 42 \end{pmatrix}$, $\mathbf{u}_3 = \begin{pmatrix} 42 \\ 38.7 \end{pmatrix}$ **37.** $A = \begin{pmatrix} 0.4 & 2 \\ 0.5 & 0.2 \end{pmatrix}$, $p = 0.71$, $r = 0.21$ **39.** $A = \begin{pmatrix} 0.8 & 1 \\ 0.5 & 0 \end{pmatrix}$, $p = 0.71, r = 0.21$

41. $A = \begin{pmatrix} 0.4 & 0.8 \\ 0.5 & 0 \end{pmatrix}$, $p = 0.63, r = -0.14$ (That is, the population is in decline.)

43. $\mathbf{u}_n = a \left(\dfrac{4}{3}\right)^n \begin{pmatrix} 2 \\ 3 \\ 1 \end{pmatrix} + b \left(-\dfrac{1}{3}\right)^n \begin{pmatrix} -1 \\ 1 \end{pmatrix}$

45. $\mathbf{u}_n = a \begin{pmatrix} 1 \\ 3 \end{pmatrix} + b4^n \begin{pmatrix} -1 \\ 1 \end{pmatrix}$

47. $\mathbf{u}_n = 30 \left(\dfrac{4}{3}\right)^n \begin{pmatrix} 2 \\ 3 \\ 1 \end{pmatrix} - 10 \left(-\dfrac{1}{3}\right)^n \begin{pmatrix} -1 \\ 1 \end{pmatrix}$, and the stable distribution is $\begin{pmatrix} 0.4 \\ 0.6 \end{pmatrix}$

49. $u_n = 2 \begin{pmatrix} 1 \\ 3 \end{pmatrix} + 4^n \begin{pmatrix} 1 \\ -1 \end{pmatrix}$, and the stable distribution is $\begin{pmatrix} 0.25 \\ 0.75 \end{pmatrix}$

51. If number of pups and adults at time n is x_n and y_n, respectively, then $\begin{cases} x_{n+1} = 2.4 y_n \\ y_{n+1} = 0.5 x_n + 0.4 y_n \end{cases}$, $A = \begin{pmatrix} 0 & 2.4 \\ 0.5 & 0.4 \end{pmatrix}$,

$\begin{pmatrix} x_0 \\ y_0 \end{pmatrix} = \begin{pmatrix} 30 \\ 50 \end{pmatrix}$, $\begin{pmatrix} x_1 \\ y_1 \end{pmatrix} = \begin{pmatrix} 120 \\ 35 \end{pmatrix}$, $\begin{pmatrix} x_2 \\ y_2 \end{pmatrix} = \begin{pmatrix} 84 \\ 74 \end{pmatrix}$,

$\begin{pmatrix} x_3 \\ y_3 \end{pmatrix} = \begin{pmatrix} 177.6 \\ 71.6 \end{pmatrix}$, $\lambda = 1.32$; the fraction of pups is 0.65

53. If number of young and adults at time n is x_n and y_n, respectively, then $\begin{cases} x_{n+1} = 1 y_n \\ y_{n+1} = 0.6 x_n + s y_n \end{cases}$, $A = \begin{pmatrix} 0 & 1.0 \\ 0.6 & s \end{pmatrix}$.

When $s = 1$, $\lambda = 1.42$, which is a 42% growth rate per year.

When $s = \dfrac{2}{3}$, $\lambda = 1.18$, which is an 18% growth rate per year.

When $s = \dfrac{1}{3}$, $\lambda = 0.96$, which is a 4% decline rate per year.

$\lambda = 1$ implies $s = 0.4$. **55.** Answers vary.

57. a. $A = \begin{pmatrix} 0.0400 & 3.0000 & 5.0000 \\ 0.1600 & 0.2000 & 0.0000 \\ 0.0000 & 0.3000 & 0.5000 \end{pmatrix}$ **b.** $\begin{pmatrix} 481.6000 \\ 38.4000 \\ 48.000 \end{pmatrix}$

c. Growth will be 7.2% with 78.1% in the smallest size class.

Problem Set 8.4, Page 654

1. $A = \begin{pmatrix} 2 & 0 \\ 0 & 3 \end{pmatrix}$; and $u(t) = a e^{2t} \begin{pmatrix} 1 \\ 0 \end{pmatrix} + b e^{3t} \begin{pmatrix} 0 \\ 1 \end{pmatrix}$

3. $A = \begin{pmatrix} -1 & 0 \\ 3 & -2 \end{pmatrix}$; and $u(t) = a e^{-t} \begin{pmatrix} 1 \\ 3 \\ 1 \end{pmatrix} + b e^{-2t} \begin{pmatrix} 0 \\ 1 \end{pmatrix}$

5. $A = \begin{pmatrix} 1 & 2 \\ 2 & 1 \end{pmatrix}$; and $u(t) = a e^{3t} \begin{pmatrix} 1 \\ 1 \end{pmatrix} + b e^{-t} \begin{pmatrix} -1 \\ 1 \end{pmatrix}$

7. $A = \begin{pmatrix} 2 & -4 \\ -1 & -1 \end{pmatrix}$; and $u(t) = a e^{-2t} \begin{pmatrix} 1 \\ 1 \end{pmatrix} + b e^{t} \begin{pmatrix} 1 \\ -\frac{1}{4} \end{pmatrix}$

9. $u(t) = \begin{pmatrix} e^t \\ \dfrac{2e^{-2t}}{3} + \dfrac{e^t}{3} \end{pmatrix}$

11. $u(t) = \begin{pmatrix} \dfrac{-\sqrt{3}}{2} \left[e^{(1-\sqrt{3})t} - e^{(1+\sqrt{3})t} \right] \\ \dfrac{1}{2} \left[e^{(1-\sqrt{3})t} + e^{(1+\sqrt{3})t} \right] \end{pmatrix}$

13. $u(t) = \begin{pmatrix} 1 - e^t \\ -2e^{-2t} + 3 - e^t \\ \hline 3 \end{pmatrix}$ **15.** $u(t) = \begin{pmatrix} \dfrac{9e^{-t} - 2 - e^{3t}}{6} \\ \dfrac{9e^{-t} - 10 + e^{3t}}{6} \end{pmatrix}$

17. $u(t) = \begin{pmatrix} \dfrac{8}{5}e^{-t} - \dfrac{8}{5}e^{4t} \\ -\dfrac{8}{5}e^{-t} - \dfrac{12}{5}e^{4t} \end{pmatrix}$ **19.** Original equilibrium is

$u^* = \begin{pmatrix} 0 \\ 0 \end{pmatrix}$; particular solution is

$\begin{pmatrix} x(t) \\ y(t) \end{pmatrix} = \begin{pmatrix} 43.64e^{-0.131t} + 56.36e^{-0.919t} \\ 99.42e^{-0.131t} - 49e^{-0.919t} \end{pmatrix}$; after two hours,

$\begin{pmatrix} x(2) \\ y(2) \end{pmatrix} = \begin{pmatrix} 42.6 \\ 68.6 \end{pmatrix}$ **21.** a and b are per capita growth rates of the populations in pools 1 and 2, respectively (if either is

negative, it represents a rate of decline); c and k are the rates of flow from the outside into the pools; g and h are the per capita rates of flow from pools 1 to 2 and vice versa, respectively **23.** The model is

$\begin{cases} \dfrac{dx}{dt} = (-0.15 - 0.1)x + 0.1y + 10 \\ \dfrac{dx}{dt} = 0.1x + (-0.1 - 0.1)y + 20 \end{cases}$; the equilibrium solution

is $\begin{pmatrix} x^* \\ y^* \end{pmatrix} = \begin{pmatrix} 100 \\ 150 \end{pmatrix}$; the particular solution is

$\begin{pmatrix} x(t) \\ y(t) \end{pmatrix} = \begin{pmatrix} 9.87e^{-0.33t} - 79.9e^{-0.12t} + 100 \\ -7.71e^{-0.33t} - 102e^{-0.12t} + 150 \end{pmatrix}$;

the solution after $t = 10$ is $\begin{pmatrix} x(10) \\ y(10) \end{pmatrix} = \begin{pmatrix} 76.8 \\ 119.5 \end{pmatrix}$, which for pools 1 and 2 is approximately 24% and 20% under the final equilibrium value

25. $u(t) = \begin{pmatrix} 3e^{-t/8} \\ -\dfrac{5}{4}e^{-t/2} + e^{-t/8} \\ \dfrac{25}{16}e^{-t/2} - 20e^{-t/8} + \dfrac{275}{16}e^{-t/10} \end{pmatrix} + \begin{pmatrix} 0 \\ 1 \\ \dfrac{1}{4} \\ \dfrac{5}{4} \end{pmatrix}$

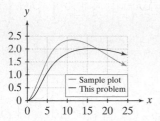

The plot of two solutions shows that now the morphine in the brain rises more slowly, peaks at a lower level, and declines more slowly.

Problem Set 8.5, Page 673

1.

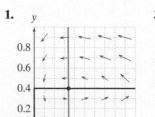

3.

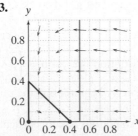

5.

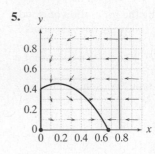

7.

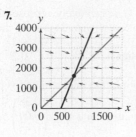

9.

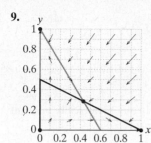

11.

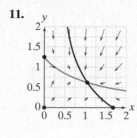

13.

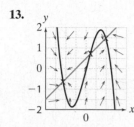

15.

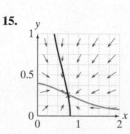

17.
$$\begin{cases} \dfrac{dx}{dy} = 0.3x(1-x) - bxy \\ \dfrac{dy}{dt} = 100bxy - cy \end{cases}$$ in $D = \{(x, y)|x \geq 0, y \geq 0\}$. For $b = 0.5, c = 30$, we have

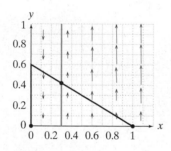

19. The model is $\dfrac{dx}{dt} = ax\left(\dfrac{x}{b} - 1\right)\left(1 - \dfrac{x}{k}\right) - \dfrac{xy}{10000}$, $\dfrac{dy}{dt} = \dfrac{xy}{10000} - cy$.

$c = 0.4, (x, y) \to (0, 0)$ $c = 0.1, (x, y) \to (0, 0)$ if $x(0) < 500$

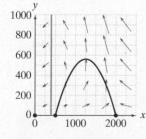   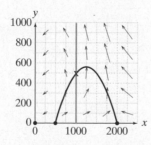

If $x(0) > 500$, (x, y) rotates around $(1,000, 500)$; note this point is not stable. It supports a stable limit cycle, a topic considered in the Group Project 8B.

$c = 0.2, (x, y) \to (0, 0)$
if $x(0) < 500$

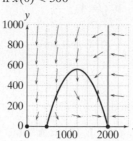

If $x(0) > 500$, $(x, y) \to (2,000, 0)$; the comparison shows that as the cost of hunting rises, the solution goes from hunting the walruses to extinction, through sustainable hunting, to hunting dying out because it is too expensive.

21.

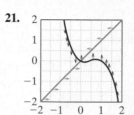

This shows the nullclines along with arrows.

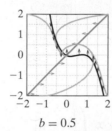   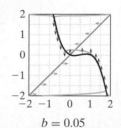

$b = 0.5$ $b = 0.05$

The smaller b allows the x value to hold relatively steady at its upper and lower value for a longer period of time before collapsing to the equilibrium point.

23. a. S—nullclines $I = \dfrac{1}{\beta S} - \dfrac{d}{\beta}$ **b.** If $I^* = 0$, then $\left(\dfrac{1}{d}, 0\right)$ is an equilibrium solution. If $I^* \neq 0$, then $\left(\dfrac{d+r}{\beta}, \dfrac{\beta - d(d+r)}{\beta(d+r)}\right)$ is an endemic equilibrium provided $I^* > 0$, which implies $\beta > \dfrac{d}{d+r}$.   **c.**

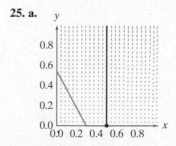

25. a.

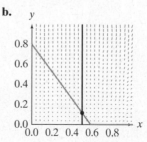

Species 1 displaces species 2.   Species 1 and 2 coexist.

Chapter 8 Review Questions, Page 676

1. a. $c = 0$

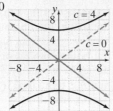

b. $c = 4$

c. $f(2, 3) = \dfrac{3}{4}$;

$f(4, -9) = 8$

d.

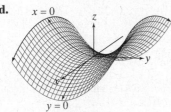

3. a. C   **b.** D   **c.** A   **d.** B

5. a. $\begin{pmatrix} -\dfrac{2}{3} & \dfrac{1}{3} \\ 1 & \dfrac{1}{3} \end{pmatrix} \begin{pmatrix} x \\ y \end{pmatrix} = \begin{pmatrix} -\dfrac{2}{5} \\ 1 \end{pmatrix}$; $x = \dfrac{21}{25}, y = \dfrac{12}{25}$

b. $\begin{pmatrix} 3 & -\dfrac{5}{4} \\ \dfrac{2}{3} & 1 \end{pmatrix} \begin{pmatrix} x \\ y \end{pmatrix} = \begin{pmatrix} -1 \\ -\dfrac{2}{3} \end{pmatrix}$; $x = -\dfrac{11}{23}, y = -\dfrac{8}{23}$

7. 19.19 of good A and 12.97 of good B **9.** $\begin{pmatrix} 1 & 2 \\ 3 & 3 \\ \dfrac{3}{4} & -\dfrac{3}{2} \end{pmatrix}$

11. $2 + \sqrt{2}$ and $2 - \sqrt{2}$, respectively   **13.** $s = \dfrac{21}{25}$

15. $\mathbf{u}(t) = \begin{pmatrix} \dfrac{e^{t/3}}{3} + \dfrac{e^{4t/3}}{3} \\ -2\dfrac{e^{t/3}}{3} + \dfrac{e^{4t/3}}{3} \end{pmatrix}$ **17. a.** S nullcline: $0.1I = \dfrac{1}{S} - 0.05$;

I isocline: $I = 0$ or $S = 2.5$ **b.** Only one biologically relevant equilibrium exists: $(S, I) = (2.5, 3.5)$

c.

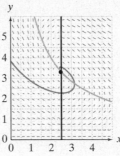

19. Equilibrium is $\mathbf{u}^* = \begin{pmatrix} 2.769 \\ 2.461 \end{pmatrix}$; particular solution is

$\begin{pmatrix} x(t) \\ y(t) \end{pmatrix} = \mathbf{u}^* + (-0.85)e^{-0.786t} \begin{pmatrix} 1 \\ -0.811 \end{pmatrix} + (-1.92)e^{-0.172t} \begin{pmatrix} 1 \\ 1.644 \end{pmatrix}$;

after one hour have $\begin{pmatrix} 0.77 \\ 0.12 \end{pmatrix}$

INDEX

LIMIT FORMULAS

Euler number: $e = \lim\limits_{n \to \infty} \left(1 + \dfrac{1}{n}\right)^n$

Continuously compounded interest: $A = P \lim\limits_{n \to \infty} \left(1 + \dfrac{r}{n}\right)^{nt} = Pe^{rt}$ where A is the future value and P is the initial investment

Trigonometric: $\lim\limits_{x \to 0} \dfrac{\sin x}{x} = 1 \qquad \lim\limits_{x \to 0} \dfrac{\cos x - 1}{x} = 0$

DIFFERENTIATION FORMULAS

PROCEDURAL RULES

Constant multiple $\quad (cf)' = cf'$ for a constant c

Sum rule $\qquad\qquad (f + g)' = f' + g'$

Difference rule $\qquad (f - g)' = f' - g'$

Linearity rule $\qquad (af + bg)' = af' + bg'$ for constants a and b

Product rule $\qquad (fg)' = fg' + f'g$

Quotient rule $\qquad \left(\dfrac{f}{g}\right)' = \dfrac{gf' - fg'}{g^2}$

Chain rule $\qquad\quad \dfrac{dy}{dx} = \dfrac{dy}{du}\dfrac{du}{dx}$

BASIC FORMULAS

Power rule $\qquad\qquad \dfrac{d}{dx}u^n = nu^{n-1}\dfrac{du}{dx}$

Trigonometric rules $\quad \dfrac{d}{dx}\sin u = \cos u \dfrac{du}{dx} \qquad\qquad \dfrac{d}{dx}\cos u = -\sin u \dfrac{du}{dx}$

$$\dfrac{d}{dx}\tan u = \sec^2 u \dfrac{du}{dx} \qquad\qquad \dfrac{d}{dx}\cot u = -\csc^2 u \dfrac{du}{dx}$$

$$\dfrac{d}{dx}\sec u = \sec u \tan u \dfrac{du}{dx} \qquad \dfrac{d}{dx}\csc u = -\csc u \cot u \dfrac{du}{dx}$$

Logarithmic rules $\quad \dfrac{d}{dx}\ln u = \dfrac{1}{u}\dfrac{du}{dx} \qquad\qquad \dfrac{d}{dx}\log_b |u| = \dfrac{\log_b e}{u}\dfrac{du}{dx} = \dfrac{1}{u \ln b}\dfrac{du}{dx}$

Exponential rules $\quad \dfrac{d}{dx}e^u = e^u \dfrac{du}{dx} \qquad\qquad \dfrac{d}{dx}b^u = b^u \ln b \dfrac{du}{dx}$

DIFFERENTIAL EQUATION FORMULAS

The solution to $\dfrac{dy}{dt} = rt$ is $y(t) = y(0)e^{rt}$

The solution to $\dfrac{dy}{dt} = a - by$ is $y(t) = e^{-bt}\left(y(0) - \dfrac{a}{b}\right) + \dfrac{a}{b}$

The solution to $\dfrac{dy}{dt} = ry\left(1 - \dfrac{y}{K}\right)$ is $y(t) = \dfrac{y(0)e^{rt}}{1 + y(0)\frac{(e^{rt}-1)}{K}}$

INTEGRATION FORMULAS

PROCEDURAL RULES

Constant multiple $\displaystyle \int cf(u)\,du = c \int f(u)\,du$ for a constant c

Sum rule $\displaystyle \int [f(u) + g(u)]\,du = \int f(u)\,du + \int g(u)\,du$

Difference rule $\displaystyle \int [f(u) - g(u)]\,du = \int f(u)\,du - \int g(u)\,du$

Linearity rule $\displaystyle \int [af(u) + bg(u)]\,du = a \int f(u)\,du + b \int g(u)\,du$ for constants a and b

BASIC FORMULAS

Constant rule $\displaystyle \int 0\,du = C$

Power rules $\displaystyle \int u^n\,du = \frac{u^{n+1}}{n+1} + C, n \neq -1$

$\displaystyle \int u^{-1}\,du = \ln|u| + C$

Exponential rules $\displaystyle \int e^u\,du = e^u + C$

$\displaystyle \int a^u\,du = \frac{a^u}{\ln a} + C, a > 0, a \neq 1$

Logarithmic rule $\displaystyle \int \ln u\,du = u\ln u - u + C, u > 0$

Trigonometric rules $\displaystyle \int \sin u\,du = -\cos u + C$

$\displaystyle \int \cos u\,du = \sin u + C$

$\displaystyle \int \tan u = -\ln|\cos u| + C$

$\displaystyle \int \cot u\,du = \ln|\sin u| + C$

$\displaystyle \int \sec u\,du = \ln|\sec u + \tan u| + C$

$\displaystyle \int \csc u\,du = -\ln|\csc u + \cot u| + C$

$\displaystyle \int \sec^2 u\,du = \tan u + C$

$\displaystyle \int \csc^2 u\,du = -\cot u + C$

$\displaystyle \int \sec u \tan u\,du = \sec u + C$

$\displaystyle \int \csc u \cot u\,du = -\csc u + C$

Squared formulas $\displaystyle \int \sin^2 u\,du = \frac{1}{2}u - \frac{1}{4}\sin 2u + C$

$\displaystyle \int \cos^2 u\,du = \frac{1}{2}u + \frac{1}{4}\sin 2u + C$